T0188993

Textbooks in Telecommunication Engineering

Telecommunications have evolved to embrace almost all aspects of our everyday life, including education, research, health care, business, banking, entertainment, space, remote sensing, meteorology, defense, homeland security, and social media, among others. With such progress in Telecom, it became evident that specialized telecommunication engineering education programs are necessary to accelerate the pace of advancement in this field. These programs will focus on network science and engineering; have curricula, labs, and textbooks of their own; and should prepare future engineers and researchers for several emerging challenges. The IEEE Communications Society's Telecommunication Engineering Education (TEE) movement, led by Tarek S. El-Bawab, resulted in recognition of this field by the Accreditation Board for Engineering and Technology (ABET), November 1, 2014. The Springer's Series Textbooks in Telecommunication Engineering capitalizes on this milestone, and aims at designing, developing, and promoting high-quality textbooks to fulfill the teaching and research needs of this discipline, and those of related university curricula. The goal is to do so at both the undergraduate and graduate levels, and globally. The new series will supplement today's literature with modern and innovative telecommunication engineering textbooks and will make inroads in areas of network science and engineering where textbooks have been largely missing. The series aims at producing high-quality volumes featuring interactive content; innovative presentation media; classroom materials for students and professors; and dedicated websites. Book proposals are solicited in all topics of telecommunication engineering including, but not limited to: network architecture and protocols; traffic engineering; telecommunication signaling and control; network availability, reliability, protection, and restoration; network management; network security; network design, measurements, and modeling; broadband access; MSO/cable networks; VoIP and IPTV; transmission media and systems; switching and routing (from legacy to next-generation paradigms); telecommunication software; wireless communication systems; wireless, cellular and personal networks; satellite and space communications and networks; optical communications and networks; free-space optical communications; cognitive communications and networks; green communications and networks; heterogeneous networks; dynamic networks; storage networks; ad hoc and sensor networks; social networks; software defined networks; interactive and multimedia communications and networks; network applications and services; e-health; e-business; big data; Internet of things; telecom economics and business; telecom regulation and standardization; and telecommunication labs of all kinds. Proposals of interest should suggest textbooks that can be used to design university courses, either in full or in part. They should focus on recent advances in the field while capturing legacy principles that are necessary for students to understand the bases of the discipline and appreciate its evolution trends. Books in this series will provide high-quality illustrations, examples, problems and case studies. For further information, please contact: Dr. Tarek S. El-Bawab, Series Editor, Professor and Dean, School of Engineering, American University of Nigeria, telbawab@ieee.org; or Mary James, Senior Editor, Springer, mary.james@springer.com.

More information about this series at http://www.springer.com/series/13835

Giovanni Giambene

Queuing Theory and Telecommunications

Networks and Applications

Third Edition

Giovanni Giambene
Department of Information Engineering
and Mathematical Sciences
University of Siena
Siena, Italy

The solutions of all the exercises of this book can be downloaded from https://www.springer.com/us/book/9783030759728.

ISSN 2524-4345 ISSN 2524-4353 (electronic)
Textbooks in Telecommunication Engineering
ISBN 978-3-030-75975-9 ISBN 978-3-030-75973-5 (eBook)
https://doi.org/10.1007/978-3-030-75973-5

This Springer imprint is published by the registered company Springer Nature Switzerland AG
The registered company address is: Gewerbestrasse 11, 6330 Cham, Switzerland

This third edition of the book is dedicated to my son Francesco, my joy.

This book is in loving memory of my father Gianfranco and my uncle Ilvo. A special dedication is to the persons nearest to my heart: my mother Marisa and my wife Michela.

Preface to the Third Edition

Telecommunication today is fundamental, not only for our work but for everyday life. We take this technology for granted, though there is a lot of work behind this system that allows people and machines to talk to each other. Telecommunication is even more crucial now, because of the heavy use of online tools and video conferences to face the present emergency in the form of the COVID-19 pandemic.

The aim of this third edition is on the one hand to update the contents on telecommunication technologies that are more relevant today (for this reason, the past Chap. 2 has been removed and its concepts have been included in Chap. 1) and on the other hand to enrich the analytical approaches to describe more complex methods, such as more details on the PASTA property, the distinction in the analysis between call congestion and time congestion in queues with finite capacity, multi-dimensional Erlang-B, M/G/1 alternative embedding options, deterministic queuing for multi-hop paths, and urn theory, and to describe new networking paradigms (caching and Internet-centric networking, software-defined networking, virtualization, and traffic offloading) and new satellite systems. Finally, many refinements have been included in all the chapters of this book together with new exercises.

Let us take a quick look at the organization of the third edition. The first two chapters deal with network evolution, current systems, and related protocols. Then, the second part of the book addresses more theoretical aspects as follows: Chap. 3 deals with probability theory, Chap. 4 is on Markovian queues of the M/M...type, Chap. 5 addresses M/G/1 queues, Chap. 6 deals with local area networks (both wired and wireless), and, finally, Chap. 7 is devoted to the network of queues.

Siena, Italy

Giovanni Giambene

Preface to the Second Edition

From the invention of the telegraph and the telephone networks, the importance of telecommunication technologies has been evident. Human beings need to interact continuously. The exchange of information of different types is today an absolute necessity. Telecommunications favor the development of countries and the diffusion of knowledge, and they are playing and will play a pivotal role in the society.

Originally, telecommunication systems were conceived as links to transmit information between two points. At present, telecommunication systems are characterized by networks with nodes, where information is processed and correctly addressed to output links, interconnecting nodes. The first telecommunication networks for telegraphy supported the transmission of messages. Then, telephone networks were conceived to establish a physical circuit at call set up to connect source and destination for the whole duration of the conversation. Today's networks are digital and based on the transmission of information organized in blocks, called *packets*, which are either independently routed via the nodes or forwarded through a virtual path from source to destination. Transmission media are typically differentiated on the basis of the network hierarchy; in particular, twisted pairs (copper) or wireless transmissions are needed for the user access, whereas optical fibers are adopted in the core network. Telecommunication systems have reached a worldwide diffusion on the basis of the efforts of international and regional standardization bodies, which have done a significant work, allowing different pieces of hardware to interoperate on the basis of well-defined protocols and formats.

Instead of having a specialized network for each traffic type, the digital representation of information has made it possible to integrate different traffic types efficiently and then services (from voice to video to data traffic, etc.) in the same network.

At present, the network of the networks, that is, the Internet, has a tremendous worldwide-increasing diffusion. The outcome of this impressive process is that the Internet protocol has become the glue, unifying different network technologies, from mobile to fixed and from terrestrial to satellite.

The central issue for modern telecommunication networks is the provision of multimedia services with global-scale connectivity (also including mobile users), guaranteeing several Quality of Service (QoS) requirements, differentiated depending on the application the user is running (i.e., traffic classes). Network resources are precious and costly and must be efficiently utilized. On the other hand, digital information and data traffic worldwide are experiencing an exponential growth that represents a challenge to be addressed by the system designer and the network planners. In this scenario, wireless access will play a major role since wireless connections have surpassed broadband wired ones from 2011.

The design of modern networks requires a deep knowledge of network characteristics, transmission media types, traffic demand statistics, and so on. On the basis of these characteristics, analytical methods can be adopted to determine the appropriate transmission capacity of links, the number of links, the management strategy for sharing resources among traffic classes, and so on.

The main interest of this book is in providing a basic description of important network technologies (in the first part of the book) as well as some analytical methods based on queuing theory to model the behavior of telecommunication systems (in the second part of the book). The aim and ambition is to provide the most important tools of teletraffic analysis for telecommunication networks.

As for Part I of this book, the focus is on network technologies (and related protocols) according to their time evolution. In particular, this part is mainly organized according to a *bottom-up approach*, referring to the ISO/OSI stacked protocol model, since we start from almost-layer 2 technologies (i.e., X.25, ISDN, Frame Relay based, ATM) in Chap. 2 and then we address layer 3 and above technologies in Chap. 3 (i.e., IP routing, MPLS, transport-layer protocols, VoIP, satellite networks). Please note that this chapter numbering refers to the second edition and not the third one. In Part II of this book, queuing systems are studied with a special interest in applying these analytical methods to the study of telecommunication systems. In particular, queuing models are adopted at different levels in telecommunication systems; they can be used to study the waiting time experienced by a given request instanced to a processor or the time spent by a message or a packet waiting to be transmitted on a given link or through a whole network. Note that the behavior of every protocol in every node of a telecommunication network can be modeled by an appropriate queuing process. Our analysis of queuing systems starts from Markov chains, such as the classical M/M/1 queuing model for message-switched networks and the M/M/S/S queue to study the call blocking probability in classical telephone networks. Then, the interest is on more advanced concepts, such as imbedded Markov chains (M/G/1 theory) with related models adopted to study the behavior of ATM switches as well as of IP routers.

This second edition has been enriched and updated for what concerns both new network technologies (Part I) and mathematical tools for queuing theory (Part II). As for Part I, the main improvements are in Chaps. 2 and 3 as follows: (1) better description of policers and shapers for ATM; (2) enriched contents on QoS support in IP networks (e.g., deterministic queuing is introduced to deal with QoS guarantees with IntServ); (3) detailed analysis of TCP congestion control behavior; (4) satellite IP-based networks; (5) VoIP. As for Part II, Chap. 5 on M/G/1 has been substantially improved, detailing more general cases and the relations among different imbedding options. Moreover, Chap. 6 now contains a better explanation of the potential instability of Aloha protocols, updated details on Gigabit Ethernet, and more details on three different approaches for the analysis of random access schemes. Chapter 7 now provides a better description of the conditions for the applicability of the Jackson theorem to real networks. Finally, new exercises have been added to the first part of the book as well as to all the Chapters of the second part of this book. The solution of all the exercises have been removed from the book and provided in a separate *solution manual*, accessible online www.extras.springer.com. Finally, a *collection of slides* has been made available for downloading and represents a valid complementary tool for teaching based on this book www.extras.springer.com.

QoS provision is a key element for both users who are happy of the telecommunication services and network operators. The success of future telecommunication services is heavily dependent on the appropriate modeling of the networks and the application of analytical approaches for QoS support. This is the reason why the analytical teletraffic methods are of crucial importance for the design of telecommunication networks.

Siena, Italy Giovanni Giambene

Preface to the First Edition

From the invention of the first telecommunication systems (i.e., telegraph and telephone networks) the importance of these technologies has been clearly evident. Humans need continuously to interact; the exchange of information of different types at distance is today essential. Telecommunications favor the development of countries and the diffusion of knowledge, and they are playing and will play a pivotal role in the society.

Originally, telecommunications were conceived as links to transmit information between two points. At present, telecommunication systems are characterized by networks with nodes, where information is processed and properly addressed (i.e., switching), and links that interconnect nodes.

The first telecommunication networks due to telegraphy were based on the transmission of messages. Then, telephone networks have been based on the establishment of a physical circuit at call setup in order to connect (for all the duration of the conversation) the source and the destination. Today's networks are digital and based on the transmission of information organized in blocks, called *packets*, that are either independently routed via the nodes or forwarded through a virtual path connecting source and destination. Transmission media are distinguished according to a hierarchy in the network typology; in particular, twisted pairs (copper) or wireless transmissions are used for the user access, whereas optic fibers are employed for core network links. Telecommunication systems have reached a worldwide diffusion based on the efforts of international and regional standardization bodies that have done significant work, allowing different pieces of hardware to interoperate based on well-defined rules.

Instead of having a specialized network for each traffic type, the digital representation of the information has made it possible to integrate efficiently in the same network different traffic types, from voice to video to data traffic, etc.

At present, the network of the networks, that is the Internet, has a tremendous and ever-increasing success. The outcome of this impressive process is that the Internet protocol results as the glue that can unify different network technologies, from mobile to fixed and from terrestrial to satellite.

The crucial point for modern telecommunication networks is the provision of multimedia services with global-scale connectivity (also including mobile users) and guaranteeing several Quality of Service (QoS) requirements, differentiated depending on the application the user is running (i.e., traffic classes). Moreover, network resources are precious and costly and must be efficiently utilized.

The design of modern networks requires a deep knowledge of network characteristics, transmission media types, traffic demand statistics, and so on. On the basis of these data, analytical methods can be adopted to determine the appropriate transmission capacity of links, the number of links, the management strategy for sharing resources among traffic classes, and so on.

The interest of this book is in providing the essential characteristics of current network technologies (i.e., X.25-based, ISDN, Frame Relay-based, ATM-based, IP-based, MPLS, GMPLS, and NGN) as well as some analytical methods based on the queuing theory to be used to study the behavior of telecommunication systems. The aim is to contribute to providing the basis of teletraffic analysis for current telecommunication networks.

Queuing systems are studied in this book with a special interest in applying these analytical methods to the study of telecommunication systems. In particular, queues can be applied at different levels in telecommunication systems; they can be adopted to study the waiting time experienced by a given request instanced to a processor or the time spent by a message or a packet waiting to be transmitted on a given link or through a whole network. In particular, every protocol in every node of a telecommunication network can be modeled through an appropriate queuing process.

Our analysis of queuing systems will start from Markov chains, such as the typical M/M/1 queuing model to be used in message-switched networks and the M/M/S/S queue employed to characterize the call loss behavior of local offices in telephone networks. Then, the interest will be focused on more advanced concepts, such as imbedded Markov chains (M/G/1 theory) with the related models adopted to study the behavior of ATM switches.

QoS provision is a key element both for the users that are happy with the telecommunication service they are adopting and for the network operators. The success of future telecommunication services and networks is heavily dependent on appropriate modeling and analysis to achieve an optimized network design able to guarantee suitable QoS levels for different traffic classes. This is the reason why the analytical methods of teletraffic analysis are of crucial importance for telecommunication networks.

Siena, Italy

Giovanni Giambene

Acknowledgments

The author wishes to thank Prof. Giuliano Benelli of the University of Siena for his support and encouragement.

Contents

Author Biography

Giovanni Giambene (giambene@unisi.it) was born in Florence, Italy, in 1966. He received the Dr. Ing. degree in electronics in 1993 and Ph.D. degree in telecommunication and informatics in 1997, both from the University of Florence, Italy. From 1994 to 1997, he was with the Electronic Engineering Department at the University of Florence, Italy. He was Technical External Secretary of the European Community COST 227 Action ("Integrated Space/Terrestrial Mobile Networks"). From 1997 to 1998, he was with OTE of the Marconi Group (today Selex), Florence, Italy, where he was involved in a GSM development program. In 1999, he joined the Department of Information Engineering and Mathematical Sciences at the University of Siena, Italy. At present, he is an associate professor, teaching the first-level course on fundamentals of telecommunications and the master's-level course on networking at the University of Siena. He has also recently received a national positive evaluation for full professorship.

He has participated in the following international projects:

- COST 290 Action (2004–2008), entitled "Traffic and QoS Management in Wireless Multimedia Networks" (Wi-QoST), vice chair
- The SatNEx I & II network of excellence (EU FP6, 2004–2009), SatNEx III (ESA 2010–2013), and SatNex IV (2015–2019) as work package leader on radio access techniques, cross-layer air interface design, and network coding techniques for satellite systems
- The EU FP7 Coordination Action "Road-mapping technology for enhancing security to protect medical & genetic data" (RADICAL) as work package leader on security and privacy applications for the management of medical data
- The COST Action IC0906 (2010–2014) "Wireless Networking for Moving Objects" (WiNeMO) as national representative; the EU FP7 Coordination Action RESPONSIBILITY.

At present, he is involved in the ESA SatNEX V project (2020-) and the ROMANTICA project financed by ESA ARTES programme (2019–2021), working on resource management for DVB-S2X transmissions.

He is the author of more than 150 papers on internationally recognized journals or conferences (h index = 25 Google scholar, 18 SCOPUS, 14 ISI WoS).

Further details are available on the Web page with the following http://www.dii.unisi.it/~giambene/.

Part I
Telecommunication Networks

Chapter 1
Introduction to Telecommunication Networks

Abstract This chapter starts with a brief look at the history of telecommunications. Then, the ISO/OSI protocol stack model and transmission media are also addressed. The second part of this chapter presents essential details of legacy networks (like, X.25, ISDN, Frame Relay, and ATM) with the aim to provide general concepts on the way to manage network traffic. This chapter is meant to provide a valid background for this book.

Key words: OSI protocol stack, X.25, ISDN, Frame Relay, ATM, Teletraffic engineering

1.1 Milestones in the Evolution of Telecommunications

Before focusing our interest on telecommunication networks, it is essential to take a brief look at the history of telecommunications, referring to the most important steps, which are at the basis of modern telecommunication networks.

After more than 10 years of studies and experimental implementations, Samuel Morse gave on May 24, 1844 a first public demonstration of his telegraph using a wire from the Supreme Court Chamber in the Capitol Building in Washington to Baltimore. Transmissions were of two symbols (i.e., with raised dots and dashes) suitably combined according to a code (called "Morse code"). This simple act can be considered as the start of the telecommunication age. Barely 10 years later, telegraphy was available as a service to the general public. In those days, however, telegraph lines did not cross national borders. Because each country used a different system, messages had to be transcribed, translated, and handed over at frontiers and then retransmitted over the telegraph network of the neighboring country. Since then, therefore, the need emerged to define a system with compatible rules across the national borders, i.e., an *international standard.* The telegraph network was the first worldwide network for data transmissions.

Starting from 1850, many submarine cables were deployed for regional links (telegraph transmissions) around the world. The first successful laying of an Atlantic Ocean submarine cable for telegraph transmissions was completed in 1858 under the direction of Cyrus West Field, who arranged for Queen Victoria to send the first transatlantic message to the US President James Buchanan. Unfortunately, the cable broke after just 3 weeks, and Field did not complete his project until 1866. This was an important achievement for telecommunications over long distances, the first wired connection for telecommunications between America and Europe. Since then, many submarine cables have been laid down everywhere in the world and today are part of the global network backbone, indispensable for the Internet.

In 1876, Alexander Graham Bell demonstrated and patented the telephone for remote transmission of voice. However, the real inventor of the telephone has to be considered Antonio Meucci, who first realized the telephone (he called "teletrophone") that could send voice via wire, although he was too poor to protect his invention with a patent.

Electronic Supplementary Material The online version contains supplementary material available at (https://doi.org/10.1007/978-3-030-75973-5_1).

G. Giambene, *Queuing Theory and Telecommunications*, Textbooks in Telecommunication Engineering,
https://doi.org/10.1007/978-3-030-75973-5_1

Fig. 1.1: Multiple couple of wires supported by poles; one couple for each telephone line

Since 1890s, telephone networks were available with human-operated analog circuit-switching systems (i.e., plugboards). Many wires were needed around cities to reach the switching office, as shown in Fig. 1.1.

In a few years, automatic electromechanical switches became available based on the step-by-step switch patented by Almon Brown Strowger in 1891. Few years were also needed to have a hierarchical organization of the network with local exchanges connected to regional exchanges (to reduce the number of wires circulating a city) and long-distance links between switching offices using the "pupinization" technique, invented by the physician Michael Idvorsky Pupin around 1900. This technique was based on the insertion of inductance coils at regular distances (about 1,800 m) along the transmitting wires to reduce both signal distortion and attenuation.

The progress of the telephone network was significant through the years, reaching all the countries of the world and thus requiring standards for interoperation. Telephone network operations were based on circuit-switching: an end-to-end physical connection has to be established before a conversation may start. This connection has to be released when the phone call ends.

James Clerk Maxwell predicted the existence of electromagnetic waves in 1864 through his very famous equations. In 1888, Heinrich Rudolf Hertz, in Germany, was the first to prove the existence of electromagnetic radiation by building an apparatus to generate radio waves. In 1895, Guglielmo Marconi was successful in sending a radio wave in the famous "hill experiment" in his villa in Italy, during which Marconi transmitted signals at a distance of over 2 km, overcoming the natural obstacle of a hill. From that date, he carried out many other experiments with signals sent even across continents. These experiments represent the birth of wireless telecommunications. The radio transmission of the voice appeared in the early 1900s.

Vladimir Kosma Zworykin, a Russian-born American inventor, working for Westinghouse, and Philo Taylor Farnsworth, a privately backed farm boy from the state of Utah, can be considered as the fathers of television. Farnsworth was the first of the two inventors who successfully demonstrated the transmission of television signals on September 7, 1927, using an electron scanning tube of his design. Farnsworth received a patent for his scanning tube in 1930.

We have to reach an epoch closer to us for considering other important achievements for the transmission of signals over long distances. In particular, in 1945, a RAF electronics officer and a member of the British Interplanetary Society, Arthur Charles Clarke, wrote an article in the Wireless World journal, entitled "Extra-Terrestrial Relays—Can Rocket Stations Give Worldwide Coverage?" describing the use of *manned* satellites having a synchronous motion with respect to the earth in orbits at an altitude of 35,800 km. These characteristics suggested to him the possible use of these GEOstationary (GEO) satellites to broadcast television signals on a vast part of the earth. Clarke's article apparently had small effect. Only in 1955, John R. Pierce of AT&T's Bell Telephone Laboratories described in an article the utility of a communication "mirror" in space, a medium-orbit "repeater" and a 24-h-orbit "repeater." After the launch of Sputnik I in 1957, many persons considered the benefits and the profits associated with satellite communications. However, we had to wait until the years 1962–1964 for the first experimental telephone and TV transmissions via satellites.

In 1948, Claude Elwood Shannon published two seminal papers on Information Theory, containing the basis for data compression (source encoding), error detection, and correction (channel encoding).

Another important medium for the transmission of information at long distances is given by light. In the 1840s, the Swiss physicist Daniel Collodon and the French physicist Jacques Babinet showed that light could be guided by jets of water for fountain displays. The British physicist John Tyndall gave a public demonstration of light guiding capabilities in 1854. In particular, the phenomenon of total internal reflection was exploited to confine the light in a material surrounded by other materials with lower refractive index, such as glass in the air for optical fibers. Since then, different experiments were made to transmit images through optical fibers, but there were many problems related to the use of this medium. The fiber is made with a cover of glass (or plastic) and a transparent cladding of lower refractive index to protect the total reflection surface from contamination. With the invention of the laser in the 1960s, it was recognized the importance of optical transmissions guided by optical fibers. The problem with the first transmission experiments through optical fibers was related to signal losses caused by glass impurities, which drastically limited the transmission range. In 1970, a multimode fiber was reached with losses below 20 dB/km. Moreover, in 1972, a silica-core multimode optical fiber was achieved with 4 dB/km minimum attenuation. At present, multimode fibers can have losses as low as 0.8 dB/km at wavelengths around 1,310 nm, whereas single-mode fibers are available with losses of 0.2 dB/km at wavelengths around 1,550 nm.

The first studies about the Internet started in 1968 with the ancestor ARPANET project. The *number of Internet nodes* grew rapidly as follows:

4 nodes	Year 1969
7 nodes	1970
15 nodes	1971
24 nodes	1972
37 Nodes	1973
More than 100 nodes	1977
More than 200 nodes	1983

In the year 2019, the growth of the Internet has reached more than 92 k autonomous systems, more than 10 M core routers, and more than 1.01 Ghosts.

In 1973, the first local area network, named Ethernet, was invented by Robert Metcalfe (at Xerox), which was capable of a data rate from 1 to 10 Mbit/s. Later it was possible to reach a nominal rate of 100 Mbit/s with the Fast Ethernet technology. Since 1999, the Gigabit Ethernet technologies have permitted to increase the data rate from 1 up to 100 Gbit/s.

Since the initial ARPANET experiments, the Internet was spreading everywhere, starting from a rough suite of protocols and then enriching it with those currently most common, such as Internet Protocol (IP), Transmission Control Protocol (TCP), and HyperText Transfer Protocol (HTTP). More technological details on the historical steps of the Internet are provided at the beginning of Chap. 2. With the widespread diffusion of the Internet, it soon became evident the need for tools to search useful information in it. Starting from 1990, different Web search engines have been designed. It is worth mentioning the definition of the Google search engine in 1997, based on a priority rank, called "PageRank," which assigns a weight to every element of a hyperlinked set of documents, aiming at "measuring" its relative importance within the set [1].

Another important milestone was the definition by the Institute of Electrical and Electronics Engineers (IEEE) in 1999 of the wireless network standard, commonly called WiFi and designated as IEEE 802.11 (with several evolutions/amendments, such as IEEE 802.11 a, b, g, n, ac, ax). The protocols for Mobile *Ad hoc* NETworks (MANETs) were defined by IEEE since the year 2000. MANET is a self-configuring infrastructureless (ad hoc) network of mobile devices connected via wireless links. Similarly, the year 2001 can be related to the first Vehicular *Ad hoc* NETwork (VANET) standards, where moving cars are nodes of a (mobile) network. There are many VANET technologies, such as WiFi IEEE 802.11p, WAVE IEEE 1609, WiMAX IEEE 802.16, Bluetooth, and ZigBee [2].

Today, sensors interconnected to the Internet have strong momentum and they are categorized as the Internet of Things (IoT). Some examples of IoT technologies and related transmission characteristics are detailed below. ZigBee is the most popular industry wireless mesh networking standard for connecting sensors, instrumentation and control systems. ZigBee (based on IEEE 802.15.4) is a low data rate wireless standard (868 MHz, 915 MHz, and 2.4 GHz). The channel data rate ranges from 20 kbit/s to 250 kbit/s. The transmission range depending on the frequency band can be from 200 m to 1 km. LoRa is another wireless technology, specifically designed for long-range, low-power Machine-to-Machine (M2M) and IoT networks. The LoRa range is 2–5 km for urban and 15 km for suburban areas. The adopted frequency band is ISM 868/915 MHz and the standard is IEEE 802.15.4g. The bit-rate ranges from 250 bit/s to 5.47 kbit/s. SIGFOX is another IoT technology that uses ultra narrowband modulation: each message is transferred at 100 or 600 bit/s, depending on the region. Hence, long distances can be achieved while being very robust against the noise.

Since the early 2000s, the term Information and Communication(s) Technology (ICT) has become very popular, highlighting the convergent role and integration of telecommunications (i.e., system protocols), computers as well as necessary applications and storage functionality, which allow users to manipulate information. This evolution brings us to the so-called "Information Society" with the tremendous growth of data stored and the need for broadband communications to efficiently utilize such a knowledge base.

A successful "service" on the Web is Wikipedia founded by Jimbo Wales in 2000; it is a fundamental source of information for everyday life; and it is a free multi-language online encyclopedia, which anyone can edit ("wiki" is a Hawaiian term meaning "fast"). More recently, social networks have acquired great momentum. Among others, we can mention here LinkedIn, the Web site for professional networking launched in 2003 (www.linkedin.com), Facebook, founded by Mark Zuckerberg in 2004 (www.facebook.com), and YouTube (2005), a video-sharing Web site where users can upload, share, and view videos (www.youtube.com).

Finally, very recently, a new ICT approach has acquired increasing importance, i.e., *cloud computing* (the most common examples are of 2007, even if theoretical grounds date back to 1960s). This is concerned with the delivery of computing as a service, whereby shared computing resources, software, and information storage are provided via the Internet, a cloud where functionalities are dispersed. There are different types of cloud computing. In general, end-users access cloud-based applications through a Web browser, using a lightweight desktop or a smartphone, while software and data are stored on Internet servers at remote locations (data centers).

The networks are now based on the concept of virtualization with the separation of the user plane (the part of the network that actually carries the traffic) and the control plane (the part of the network that determines the traffic route and provides the management). Virtualization allows reproducing network functions as logical entities such as logical switches, logical routers, and logical gateways, organized in any topology. The network can be treated as a resource to be assigned dynamically. This approach is essential to differentiate various classes of data traffic (e.g., sensors, city cameras, self-driving cars, real-time communications) and allocate resources based on business priorities and Service Level Agreements (SLAs). Network slicing allows a network operator to define dedicated virtual networks sharing the same physical infrastructure. Each slice entails an independent set of logical network functions optimized to provide the resources for the specific service and traffic. The Software-Defined Networking (SDN) approach entails a programmable network infrastructure. The network intelligence resides in software-based SDN controllers, and network equipment can be configured externally through vendor-independent management software. SDN allows both the centralization of some management functions and the dynamic optimization of the system.

The remainder of this chapter is devoted to some preliminary considerations on telecommunication networks, their taxonomy, a reference model for telecommunications, and the classical, old telephone network. Additional details on networks are discussed in Chap. 2 (geographical networks) and Chap. 6 (local area networks).

1.2 Standardization Bodies in Telecommunications

Many international or regional standardization bodies are involved in the definition of telecommunication networks. They are either government-driven or industry-driven. Among them, we may consider the following:

- International Telecommunication Union (ITU) [3]
- International Organization for Standardization (ISO) [4]
- The Institute for Electrical and Electronics Engineers (IEEE) [5]
- Internet Engineering Task Force (IETF) [6]
- European Telecommunications Standards Institute (ETSI) [7]
- The American National Standards Institute (ANSI) [8]
- The Association of Radio Industries and Businesses (ARIB) in Japan [9]
- Telecommunications Industry Association (TIA) in the USA [10]

- Telecommunications Technology Association (TTA) in South Korea [11]
- The International Electrotechnical Commission (IEC) [12]
- Electronic Industries Association (EIA) in the USA [13]
- The 3rd Generation Partnership Project (3GPP) [14]
- The 3rd Generation Partnership Project 2 (3GPP2) [15].

ITU is the leading organization for the definition of international standards in the field of telecommunications. ITU is an international organization within the United Nations system, based in Geneva, Switzerland [3]. ITU has two main sectors: telecommunications and radio communications. The International Radio Consultative Committee (CCIR) was established at a conference held in Washington in 1927. The International Telephone Consultative Committee (CCIF) was set up in 1924, and the International Telegraph Consultative Committee (CCIT) was set up in 1925. In 1956, CCIT and CCIF were merged to form the International Telephone and Telegraph Consultative Committee (CCITT) to respond more effectively to the needs for the development of telecommunications. In 1992, a plenipotentiary conference held in Geneva significantly remodeled CCITT (renamed ITU) to give it greater flexibility to adapt to the quite complex field of telecommunications. As a result of the reorganization, three sectors were distinguished, corresponding to the three main areas of activity:

- Telecommunication (ITU-T)
- Radio communication (ITU-R)
- Telecommunication Development (ITU-D).

ITU-T Recommendations are organized in series. The most significant ones are listed below:

- *D-series*: General Tariff Principles
- *E-series*: Overall Network Operation, Telephone Service, Service Operation, and Human Factors
- *F-series*: Non-telephone Telecommunications Services
- *G-series*: Transmission Systems and Media, Digital Systems and Networks
- *H-series*: Audiovisual and Multimedia Systems
- *I-series*: Integrated Services Digital Network
- *J-series*: Cable Networks and Transmission of Television, Sound Program, and Other Multimedia Signals
- *K-series*: Protection Against Interference
- *L-series*: Construction, Installation, and Protection of Cables and Other Elements of Outside Plant
- *M-series*: TMN and Network Maintenance: International Transmission Systems, Telephone Circuits, Telegraphy, Facsimile, and Leased Circuits
- *N-series*: Maintenance: International Sound Program and Television Transmission Circuits
- *O-series*: Specifications of Measuring Equipment
- *P-series*: Telephone Transmission Quality, Telephone Installations, and Local Line Networks
- *Q-series*: Switching and Signaling
- *R-series*: Telegraph Transmission
- *S-series*: Telegraph Services Terminal Equipment
- *T-series*: Terminals for Telematic Services
- *U-series*: Telegraph Switching
- *V-series*: Data Communications Over the Telephone Network
- *X-series*: Data Networks and Open System Communication
- *Y-series*: Global Information Infrastructure and Internet Protocol Aspects
- *Z-series*: Languages and General Software Aspects for Telecommunications Systems.

1.3 Telecommunication Networks: General Concepts

Historically, communication systems have started with point-to-point links to directly connect the users needing to communicate utilizing a dedicated circuit. As the number of connected users increased, it became infeasible to provide a circuit to connect every user to every other.[1] Hence, telecommunication networks have been developed with intermediate nodes and interconnections among nodes. A telecommunication network can be defined as a set of equipment elements, transmission media, and procedures in order for two remote user terminals to exchange information (see Fig. 1.2).

[1] In the mesh topology, every node is connected to every node. In the case of n nodes, there is the need of $n(n-1)/2$ bidirectional links [or equivalently, $n(n-1)$ unidirectional links] for a full-mesh topology.

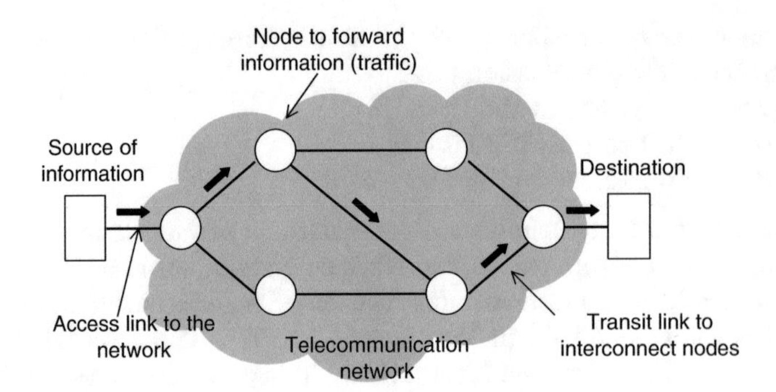

Fig. 1.2: Telecommunication network formed of "intelligent" nodes and links among them for the exchange of information between source and destination pairs

At present, telecommunication networks allow the exchange of "information signals" between whatever point on the earth. These signals can be either the transduction of human voice (analog signal) or data generated by some service, which directly interacts or interfaces with humans. Such a remarkable achievement has been attained through many steps: from the deployment of the classical analog telephone network to the computers and then to the interconnecting network, i.e., the Internet.

Telecommunication networks can be roughly distinguished between *broadcast networks* and *switched networks*. In the first case, all the nodes receive the same information transmitted by a source node. This is the case of radio and television networks. In the second case, the transfer of information (voice, data, etc.) requires routing or switching operations at the different network nodes, which are encountered along the path from the source to the destination. The following study on telecommunication networks mainly refers to switched networks.

Telecommunication networks can also be distinguished based on their extension. In particular, we have Wide Area Networks (WANs) for geographical coverage spanning over countries and continents. Moreover, Metropolitan Area Networks (MANs) are used at the city level. Finally, Local Area Networks (LANs) can provide telecommunication services to a laboratory, a building, a university campus, an industry, etc. Chap. 2 is devoted to the Internet and WANs; instead, Chap. 6 is targeted to LANs and MANs.

The information sent from source to destination along the network can be identified with the generic term of "traffic." Each link along the source-to-destination path in the network conveys traffic that is typically the aggregate contribution of many users. A generic definition of *traffic* should entail the notion of random variables and stochastic processes that will be considered in the second part of this book. Hence, for the sake of simplicity and referring to the transmissions on a link, we can find that a generic traffic is characterized by two quantities:

- The *mean frequency* of information arrival λ (e.g., calls per second in a telephone network or packets per second in a packet data network);
- The *mean duration* of the transmission $E[X]$ of each arrival (e.g., referring to the length of a call or to the transmission time of a packet) on a link.

The product of the mean arrival frequency and the mean transmission time yields the traffic *intensity*, ρ:

$$\rho = \lambda \times E[X] \tag{1.1}$$

ρ is a dimensionless quantity, measured in Erlangs, as detailed in Chap. 4.

In particular, in an old telephone network, the traffic is analog and its intensity is measured as the product of the mean call arrival rate and the mean call duration. The traffic intensity at a local exchange represents the mean number of simultaneously active phone calls. In a data network, the traffic is digital; the traffic intensity at a node can be obtained as the product of the mean packet (or message) arrival rate and the mean packet (or message) transmission time.

When different and independent traffic flows sum at the entrance of a node, the resulting total traffic intensity is equal to the sum of the traffic intensities of the single flows.

Referring to a generic link (i.e., a transmission line), the traffic intensity expresses the percentage of time that the input traffic occupies the link. Hence, the maximum (limit) load condition for a single communication line is represented by the traffic intensity $\rho = 1$ Erlang. Access links in the network are typically characterized by time-varying traffic conditions with low-intensity values (e.g., $\rho < 0.6$ Erlangs). Instead, transit links in the network have more regular traffic with medium–high intensity values (e.g., $\rho \approx 0.8$ Erlangs).

As it is evident from these initial considerations, two nodes not only exchange information generated by traffic sources but also need to exchange *signaling* (i.e., control) messages, which are necessary for the appropriate management of the network. Signaling can be required to establish an end-to-end path in the network for the exchange of information between source and destination. Moreover, signaling may be needed to provide acknowledgments of received data or to request retransmissions.

1.3.1 Transmissions in Telecommunication Networks

Each link in the network is characterized by the transmission of signals, according to the general model shown in Fig. 1.3. In particular, we have a transmitter sending the information through the physical medium of the link and a receiver that can correctly interpret the information. Due to the disturbances and distortions introduced on the signal by the communication channel, a modulator can be used at the transmitter in order "to transpose" the frequency spectrum of the signal in a band suitable to traverse the channel; correspondingly, a demodulator is necessary at the receiver. However, baseband transmissions (i.e., non-modulated) are also possible, for instance, in the case of transmissions on cables.

There are two generic forms of signals evolving in time, which can be transmitted in telecommunication systems (see Fig. 1.4), i.e., *analog* signals and *digital* signals. In the first case, we have a continuously varying signal that represents the electrical transduction of physical data. In the second case, only a few signal levels are possible (e.g., two values corresponding to the representation of bits "0" and "1," but there could also be more than two symbols). Digital signals have the advantage that, since only a few levels are possible, additive noise can be quickly canceled at the receiver using a simple threshold detector (let us refer here to a baseband signal). Finally, digital signals provide a common language, which permits to integrate different media, such as audio, video, and data.

Let us focus on digital transmissions. We refer to the well-known Shannon theorem: in a communication channel, it is possible to transmit up to a maximum bit-rate C (i.e., channel capacity), guaranteeing that, with both suitable coding and digital modulation, the bit error probability can be made as small as needed. In particular, for *a band-limited waveform channel* with additive white Gaussian noise (being N_0 the mono-lateral power spectral density), the *channel capacity* can be expressed as [16]

$$C = W \times \log_2 \left(1 + \frac{P}{N}\right) \left[\frac{\text{bit}}{s}\right] \tag{1.2}$$

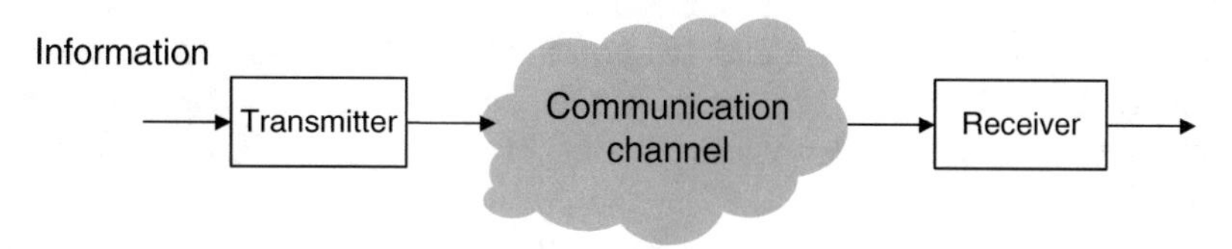

Fig. 1.3: *Transmission* scheme on a link

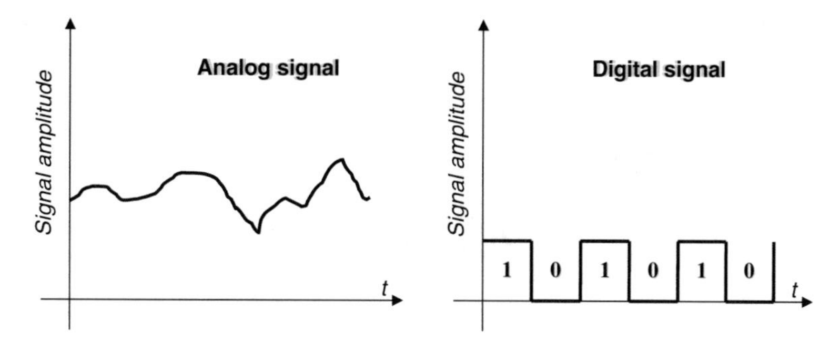

Fig. 1.4: Analog and digital signals

where W is the channel bandwidth, P is the received signal power, and $N = WN_0$ denotes here the noise power received.

From equation (1.2), we can see that generically there is an important relationship between the available bandwidth of the transmission medium, W, and the bit-rate that can be achieved with a certain quality in terms of bit error rate. The capacity formula depends on the channel; for instance, a different capacity expression is obtained for the *classical binary symmetric channel* [16].

The main characteristics of digital transmissions are detailed below:

- Serial or parallel transmissions
- Synchronous or asynchronous transmissions
- Full-duplex or half-duplex transmissions
- Symmetric or asymmetric transmissions
- Constant bit-rate or variable bit-rate (i.e., bursty) transmissions.

Serial transmissions involve sending data bit by bit over a single communication line. In contrast, parallel communications require at least as many lines as the number of bits in a word being transmitted (for an 8-bit word, at least 8 lines are needed). Serial transmissions are beneficial for long-distance communications, whereas parallel transmissions are suitable for short distances (cabling is limited to 5–10 m) or when very high transmission rates are required. The RS-232-C standard (EIA standard EIA-232, ITU V.24) is the classical serial interface for the exchange of information between data terminal equipment and data communications equipment. This standard is characterized by the typical 25-pin D-shaped connectors. It allows transmission speeds from 110 bit/s to 19.2 kbit/s for a distance up to 15 m. The RS-232 standard is an asynchronous interface. Serial ports can be used in personal computers to connect mouse, modem, or special peripherals. Today, RS-232 has been superseded by the Universal Serial Bus (USB) port that is much faster and has connectors that are easier to use. RS-232 ports are still used on programmable boards to upload the operating system on the local memory.

Serial transmissions can be of two different types: synchronous or asynchronous. We refer below to baseband transmissions. Data transmitted between nodes are organized into bits, bytes, and group of bytes, named *packets*. Synchronization involves delimiting and recovering bits, bytes, and packets. The synchronization type depends on the clocks used by the sender and the receiver.

In asynchronous transmissions, transmitter and receiver clocks are independent. Asynchronous transmission is useful for human input/output data (e.g., a keyboard input) with random arrival times and transmission lines characterized by long idle phases. Let us refer to the transmission of a character of one byte (7-bit ASCII code plus a parity bit) at once. Since there is no direct clock information exchanged between the receiver and the transmitter, the receiver must explicitly resynchronize at the first bit of each byte. To achieve such synchronization, additional start and stop bits must be used for sending each byte. Subsequent bits are recovered by estimating bit boundaries. Let us consider the example shown in Fig. 1.5 for the asynchronous transmission of a character (i.e., one byte). The transmission of bit "1" is characterized by a high signal level, whereas the transmission of bit "0" corresponds to a low level. The start bit is a "0" and the end bit is (or bits are) "1" just to be sure that there is at least one transition in the character. Of course, the extra bits to manage the asynchronous transmission reduce the efficiency: 10–11 bits are needed to transmit a character of 8 bits; hence, 27.2% of the link capacity is lost due to the asynchronous protocol.

In synchronous transmissions, there is a global clock or synchronized clocks used in transmission and reception. The transmission unit is a packet of bits, sent together in a stream. The packet contains overhead bits (they are typically concentrated in a header, but some of them could also be in a trailer) and a data payload, as shown in Fig. 1.6. The receiver must resynchronize at each new packet. Suitable bit sequences are at the beginning of a packet so that the receiver can acquire the right synchronism at the packet level (moreover, bits have adequate

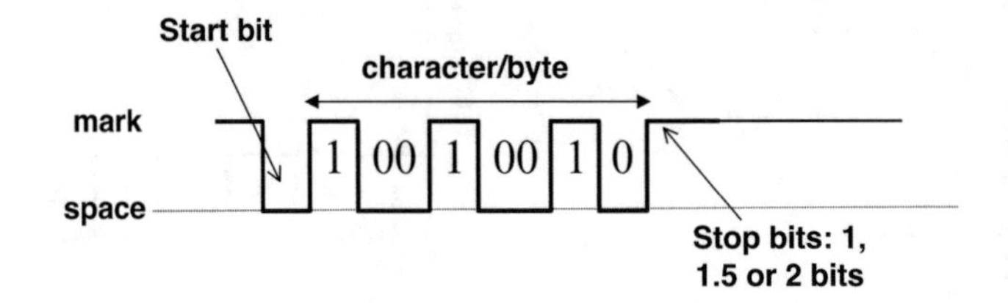

Fig. 1.5: Example of asynchronous transmission (RS 232)

representation to ease the bit synchronization; this is typically accomplished by a suitable line code). Typically, 1–2 bytes are needed for packet synchronization. Since the packet can be sufficiently long, synchronous transmissions allow us to achieve higher efficiency than asynchronous ones. Synchronous communications are well suited to high bit-rate transmissions.

Considering the type of data exchange between source and destination, transmissions can be classified into three different categories:

- Simplex, one way only
- Half duplex, bidirectional, but alternate in time
- Full duplex, bidirectional at the same time and through the same interface.

In bidirectional transmissions, the exchange is symmetric if both parties send a similar traffic load. This is the typical case of phone conversations. Otherwise, we have an asymmetrical situation. A typical example for computer networks is when a client connects to a remote server: the amount of data sent by the client is much lower than that provided by the server (typically, a 1:10 ratio can be considered).

Let us refer to digital traffic sources, characterized by a bit-rate evolving in time, $R(t)$; see Fig. 1.7. $R(t)$ can be modeled as a *stochastic process*, as described in Chap. 4. Digital traffic flows can be roughly distinguished into two broad families: (1) *elastic traffic* (typically referred to data traffic, which can tolerate throughput variations, depending on network conditions) and (2) *inelastic traffic* (typically referred to real-time traffic for which the rate cannot be adjusted depending on network congestion).

Let us refer to real-time traffic and, in particular, to voice or video traffic sources. In both cases, we can consider variable bit-rate traffic sources. In the voice case, we have a constant bit-rate generation during a talkspurt and a negligible traffic generation during a silent pause (ON–OFF voice traffic source). In the video case, bit-rate variations can be obtained since the images to be coded vary so that different compression values are achieved. Very bursty data traffic sources are those related to Internet traffic (background or interactive class), where the bit-rate generated has very low values for long time intervals, but high and sudden peaks are possible. A fixed link capacity assigned to a bursty traffic source based on its peak traffic value can represent a wastage of resources. This is a crucial aspect to take into account when designing a network. If we aggregate the variable bit-rates generated by bursty traffic sources, we obtain a more smoothed traffic (i.e., traffic with lower variations) for which it is easier to predict the required capacity needs. For the network, it is more efficient to aggregate the traffic of bursty sources by exploiting the *multiplexing effect*, as described below.

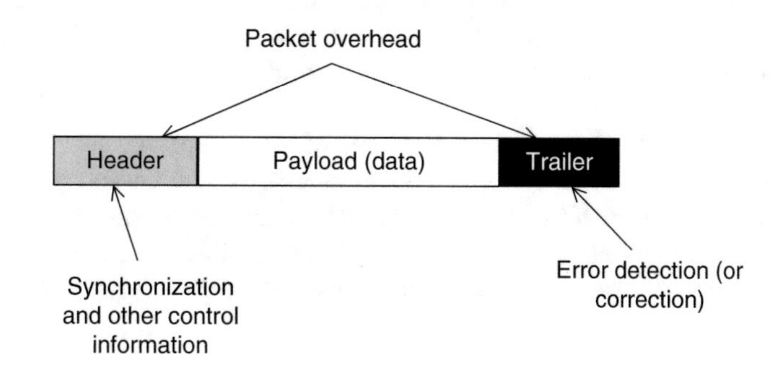

Fig. 1.6: Generic packet format for synchronous transmissions

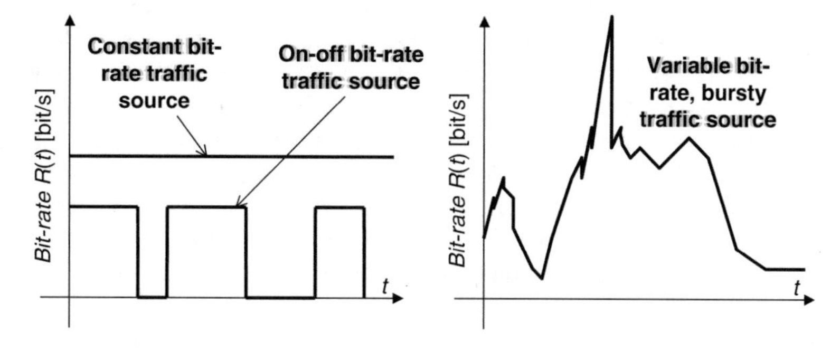

Fig. 1.7: Various examples of digital traffic sources: constant bit-rate, ON–OFF source, and bursty source

Referring to a data traffic source, we can define the *burstiness* β as the ratio between the maximum bit-rate, $R_{\max}$, and the mean bit-rate $E[R]$:

$$\beta = \frac{R_{\max}}{E[R]} \tag{1.3}$$

For an ON–OFF voice traffic source, bit-rate $R(t)$ is equal to $R_{\max}$ in the on phase and equal to 0 in the off phase. Hence, we have $E[R] = R_{\max}\, P_{\text{on}}$, where P_{on} denotes the percentage of the time spent by the source in the on phase (i.e., activity factor). In conclusion, the ON–OFF traffic source has a burstiness degree given as

$$\beta_{\text{on-off}} = \frac{1}{P_{\text{on}}} > 1 \tag{1.4}$$

Assuming that the voice source traffic is transmitted over a digital line of capacity $R_{\max}$ bit/s, the burstiness degree represents the maximum (ideal) number of different ON–OFF voice sources that can be multiplexed onto the digital line. If the various voice sources would be ideally coordinated in their on and off phases, we could have exactly $1/P_{\text{on}}$ voice sources sharing the use of the same line where they transmit alternately.

1.3.2 Switching Techniques in Telecommunication Networks

Historically, three different types of switched networks can be distinguished, depending on the following techniques: *circuit-switching*, *message-switching*, and *packet-switching*. Each of these switching methods is suitable for a specific traffic type, whereas it could not be used (or it could be not efficient to use) for the transfer of other traffic classes. In general, circuit-switching is well suited to traffic, which is regular (almost constant) for a sufficiently long time with respect to the procedures to set up the circuit. In contrast, message- and packet-switching are more appropriate for data traffic and, in particular, for variable bit-rate and bursty traffic.

Circuit-switching is the solution adopted in old telephone networks: when a user makes a phone call towards another user, the network establishes an end-to-end physical (i.e., electrical) connection for all the duration of their conversation. The following subsequent phases characterize a circuit-switched connection and the related service:

- *Circuit setup.* In the case of a phone call, this phase starts when the originating user dials the phone number of the destination and ends when the originating user receives a tone, indicating whether the destination is available or not. In this phase, an end-to-end circuit is built and resources are reserved on the links and at the nodes along the path.
- *Information transfer from a user to the other.* In the case of the telephone service, this phase corresponds to the phone conversation between the two users. During this phase, an end-to-end physical connection is available and no network procedure is involved. Voice is transparently conveyed at the destination by the network.
- *Circuit release.* When the phone call is over (one of the two users closes the connection), the network operates a series of operations to release the resources reserved along the path. These resources can be made available to other users.

Message-switching technology was born in the 1960s. In this case, each message represents an autonomous information unit, typically composed of a variable number of bits. Subsequent messages for the same source–destination pair follow a path decided based on the dynamic state of the network. A network resource (i.e., a link) is used just for the time necessary to transmit a message; soon after it is available to serve other messages. To explain the message-switching technique, let us refer to the example in Fig. 1.8, where terminal A sends a set of messages (i.e., messages M1, M2, and M3) to terminal B. Each message is simply composed of a header and a payload. The header contains the address of source A and the address of destination B. Each message is autonomous, since it contains all the information for routing it to the destination. Each message crosses several nodes and links. When a message reaches a node (i.e., switching element), it is stored in a *buffer* and its header is processed to obtain the destination address. Based on this information, the node determines to which output link (and related node) the message has to be forwarded to reach its destination. Each node is of the "store-and-forward" type.

The telegram network technology was based on message-switching. Message-switching is the right solution for data traffic networks, characterized by bursty traffic. However, this technology has been overtaken by packet-

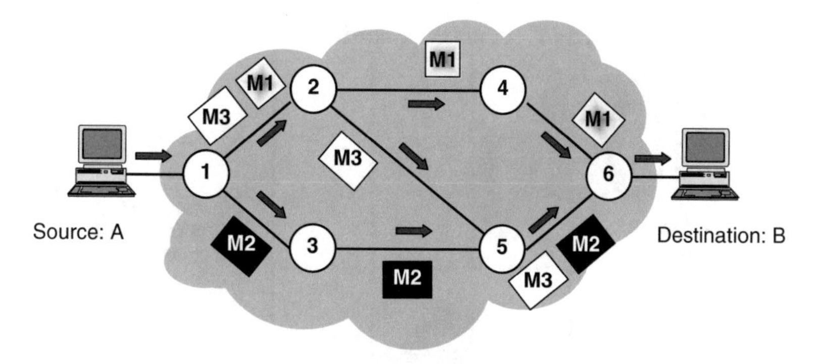

Fig. 1.8: Telecommunication network based on message-switching; messages may have different lengths

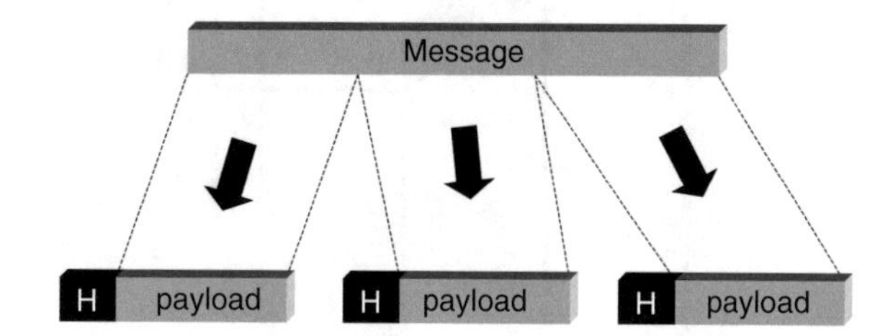

Fig. 1.9: Segmentation in packets

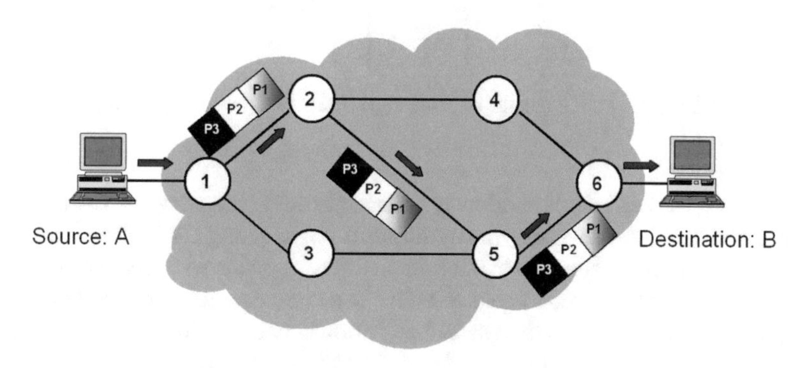

Fig. 1.10: Packet-switching based on virtual circuits

switching, which can achieve better performance in terms of fast switching at nodes and lower transmission delays on links.

Packet-switching was first conceived by Leonard Kleinrock at MIT [17]. Packet-switching can be considered as an evolution of message-switching. In particular, a message is segmented in packets of reduced length, each having a header (control information) and a payload carrying a fragment of the message (see Fig. 1.9).

The header contains many control fields to manage the transmission of data on the links from source to destination. There should also be a counter to determine the number of payload fragments needed to reassemble the original message. Each packet is an autonomous entity.

Packet-switched transmissions may occur according to two different methods: *virtual circuit* and *datagram*. In both cases, buffers are needed at the different network nodes to store the packets to be transmitted on the various output links.

- In the virtual circuit mode, a "logical" path is established in the network from source to destination: there is a setup phase similar to that described for circuit-switched networks. Once the path has been defined in the network, the packet forwarding is very fast from node to node (nodes have not to determine a new route at each new packet, since the flow has a well-defined path). All the packets of a traffic flow have the same route from source to destination (see Fig. 1.10). Therefore, packets are received in the same order of generation; no reordering is needed at the destination. The virtual circuit mode is quite common in telecommunication networks (e.g., ATM networks that will be described later in this chapter or MPLS networks in Chap. 2).

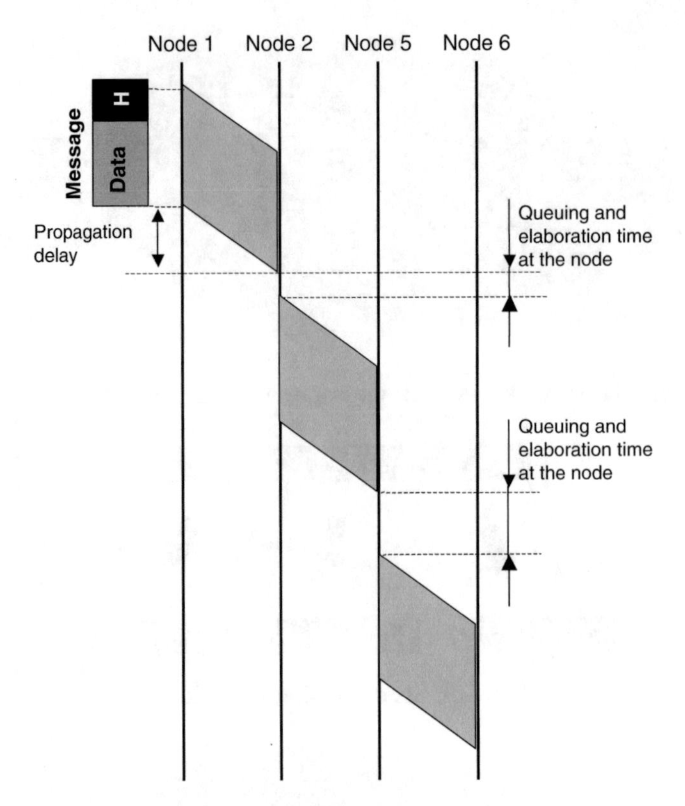

Fig. 1.11: Example of message-switched transmission

- In the datagram mode, each packet is independently routed through the network towards its destination. Hence, packets generated from the same message may have different paths along the network from source to destination. Consequently, packets may arrive at the destination in a different order with respect to that of their generation. The destination node has to reorder the packets using a sequence number contained in the packet header. This transmission mode is similar to message-switching and, hence, we may refer to Fig. 1.8 for a description. The datagram transmission mode is employed in the Internet (see Chap. 2) since it allows some advantages as follows:
  - No circuit must be created before the exchange of data between source and destination.
  - This switching mode is more robust to network faults, malfunctioning, and congestion. The route of packets can be dynamically adapted in response to changing network conditions. On the other hand, in the virtual circuit mode, after a node fault/congestion, all the virtual circuits crossing that node are interrupted/affected.

However, the datagram transmission mode requires that each packet contains the geographical address of the destination that must be processed at each node to find the appropriate output port. In the IPv4 (IPv6) Internet, the address field requires 32 (128) bits. An Internet packet contains the addresses of both source and destination in the header.

Figures 1.11, 1.12, and 1.13 show the time diagrams to compare message-switching and packet-switching techniques in terms of end-to-end delay to deliver the same amount of data from source A to destination B through the network topology shown in Figs. 1.8 and 1.10. In particular, the message is queued and then processed at each node with message-switching. The header is examined to decide where the message has to be forwarded.

With packet-switching, the message is fragmented into many packets, each with header information; in particular, the message originates three packets in Figs. 1.12 and 1.13. These packets are sent in sequence. Each packet is queued and processed independently at each node. The time diagram is different in the case of datagram mode (Fig. 1.12) and in the case of virtual circuit mode (Fig. 1.13). The main difference between these two cases is that in the virtual circuit mode there is an initial setup phase for establishing the end-to-end path, similarly to circuit-switched calls (this phase can be avoided if the flow from A to B occurs on an already-defined path). After this phase, packets are quickly forwarded to the next node without requiring a heavy processing load. The processing

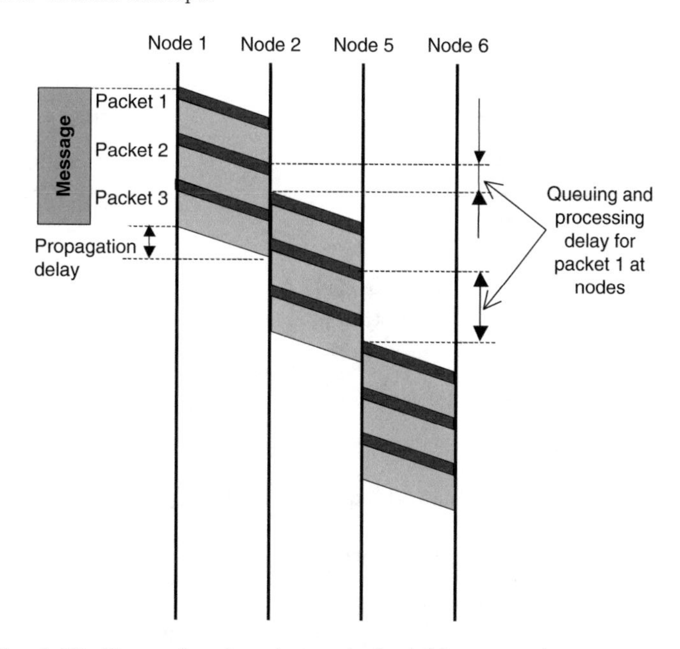

Fig. 1.12: Example of packet-switched (datagram) transmission

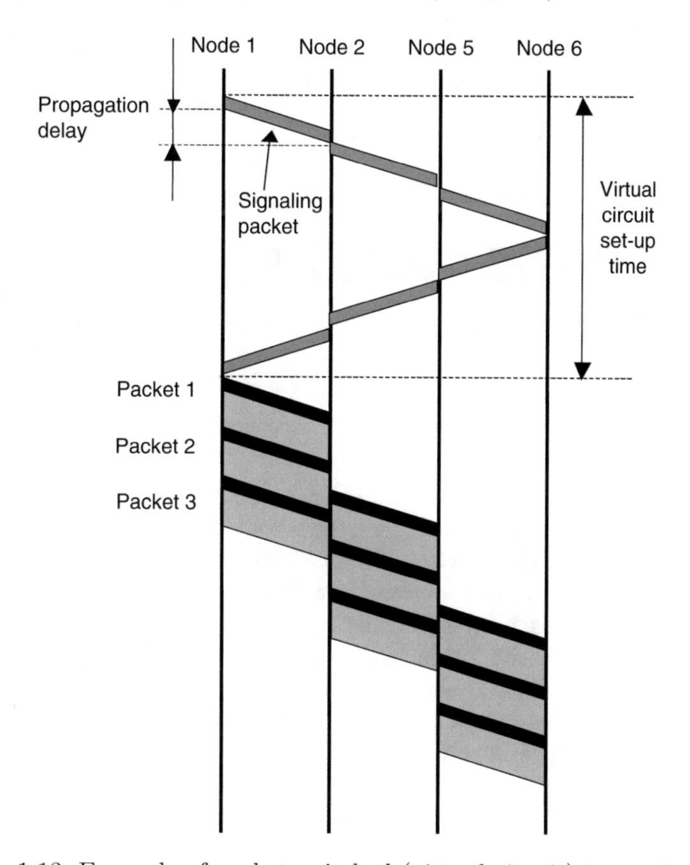

Fig. 1.13: Example of packet-switched (virtual circuit) transmission

of packets in each node is heavier in the datagram mode. Hence, the virtual circuit mode is convenient if a more regular and sufficiently heavy traffic load is sent from node A to node B.

Before ending this section, it is important to summarize the different types of networks through the taxonomy shown in Fig. 1.14.

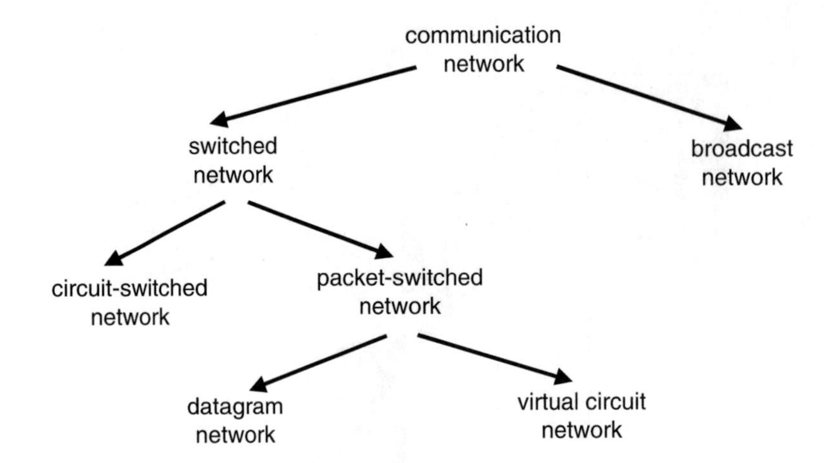

Fig. 1.14: Network taxonomy depending on the type of traffic delivery

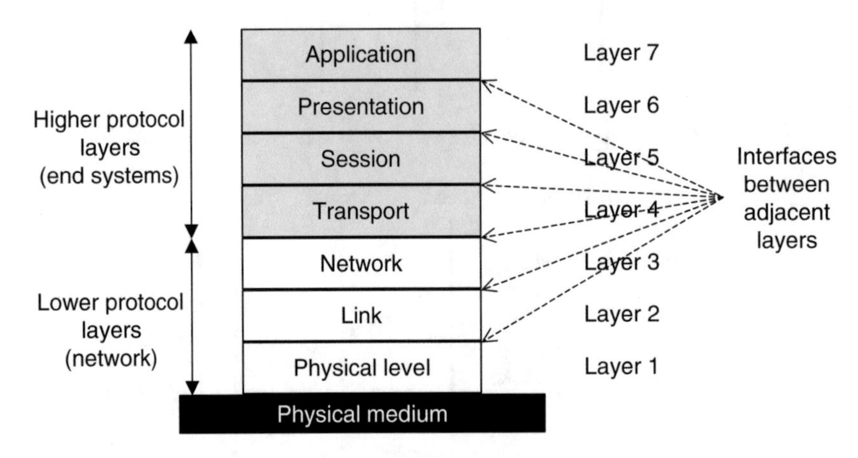

Fig. 1.15: OSI reference model for the protocol stack

1.3.3 The ISO/OSI Reference Model

A suite of protocols must be used to properly exchange data at each interface along a path between two network nodes. These protocols are organized according to a stack. This is the layering approach, namely, dividing a task into smaller pieces and then solving each of them independently. This scheme allows an increasing abstraction level as we move from lower layers to higher ones. Each protocol layer has to perform a suitable function, which permits the above layers to address other aspects. Each layer provides communication services to the layer above. The protocol stack architecture was standardized in the 1970s by the International Organization for Standardization (ISO) [18, 19] with the famous name of OSI (Open System Interconnection) reference model. The target was to define an "open system," meaning that different network elements can interwork independently of the manufacturers. The ISO/OSI protocol stack entails 7 protocol layers, as shown in Fig. 1.15. Lower protocol layers (i.e., physical, link, and network layers) are present in every node of the network, including source and destination, which are called "End Systems." Instead, higher protocol layers (i.e., transport, session, presentation, and application) are present only at the source and destination.

Note that current trends in the design of the protocol stack also envisage interfaces between non-adjacent layers, thus violating the classical ISO/OSI classical structure. This is the *cross-layer design*, recently conceived for wireless networks, where a direct dialogue is also possible between protocols at non-adjacent layers.

Figure 1.16 shows the dialogue between user A and user B; these are the "End Systems," implementing the OSI protocol stack from layer 7 to layer 1. A and B exchange data through a telecommunication network, which is denoted as "Intermediate System." Each network node in the intermediate system supports a reduced protocol stack; typically, only a few layers are implemented (in Fig. 1.16, only layers 1, 2, and 3 are adopted). Starting from source A, data are forwarded progressively from layer 7 to layer 1 and, then, transmitted. Data propagate through

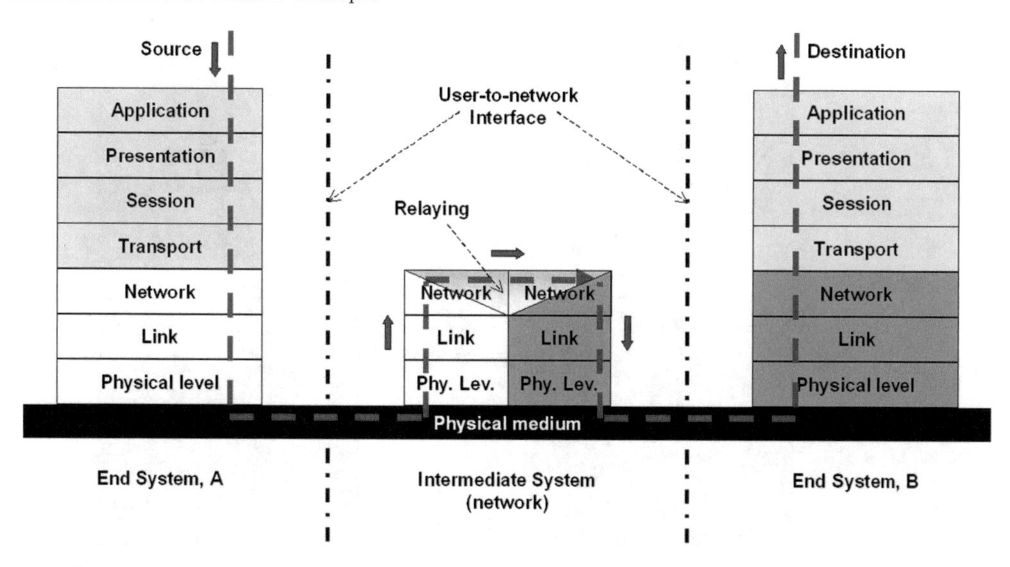

Fig. 1.16: Path followed by the "information" in the communication from A to B through the network and protocol stack at different interfaces (nodes)

the physical medium, thus reaching the next node in the network (i.e., intermediate system). At this node, the information is reprocessed from layer 1 up to layer 3, assuming a network layer switching like in the Internet. When layer 3 is reached, data are not passed to upper layers but are managed at layer 3 to be passed again to the appropriate output link, thus going to layer 2 and physical layer, where transmission is performed. The function performed by layer 3 in the intermediate system in Fig. 1.16 is named "relaying." The protocol stacks on the left and right sides of the node of the intermediate system may be different. Note that the intermediate system can also implement the relaying function at different layers, depending on the network technology. In particular, the relaying function is at layer 1 in circuit-switched networks, at layer 2 in Frame Relay and ATM networks, and at layer 3 in X.25 networks and the Internet.

Let us describe the specific functions of the seven OSI layers:

- Layer 1 is the physical layer, which directly carries out the transmission of bits through the physical medium.
- Layer 2 or data link layer has the primary function to regulate the access to physical layer resources and to recover errors through retransmission techniques (Automatic ReQuest repeat, ARQ, protocols).
- Layer 3 or network layer has the task to route the traffic in the network from source to destination.
- Layer 4 or transport layer performs the end-to-end control of the traffic flow from source to destination. Specific tasks are *flow control* (to avoid overwhelming the destination with too much traffic that it cannot handle) and *congestion control* (to avoid injecting too much traffic in the network, thus causing congestion at an intermediate node, also called "bottleneck").
- Layer 5 or session layer manages the dialogue between the two end application processes.
- Layer 6 or presentation layer is needed to unify the representation of information between source and destination. This protocol interprets and formats data, including compression, encryption, etc.
- Layer 7 or application layer represents the high-level service, having direct interactions with the user.

It is important to remark that the protocol specifications for a layer are independent of the specifications of the protocols at the other layers. In other words, it is possible to change a protocol in a layer with another without having to change anything in the protocols of adjacent layers. Of course, the service provided to the adjacent layers must remain unchanged.

The protocols from the physical layer to the transport one are related to the network infrastructure and deal with telecommunication aspects from the transmission, to error management, to routing, and, finally, to flow and congestion control, whereas protocols of layers 5–7 are mainly related to software elaboration aspects.

Let us refer to a "system" (i.e., a terminal, a host, etc.) implementing the OSI protocol stack. The generic layer $X \in \{1, 2, \ldots, 7\}$ is composed of functional groups, named *entities*. A layer may contain more than one entity. For instance, there will be N-entities at layer $X = 3$. Each entity provides a service to the upper layer through an *interface*. Upper layer entities access to this service through a Service Access Point (SAP); there may be different

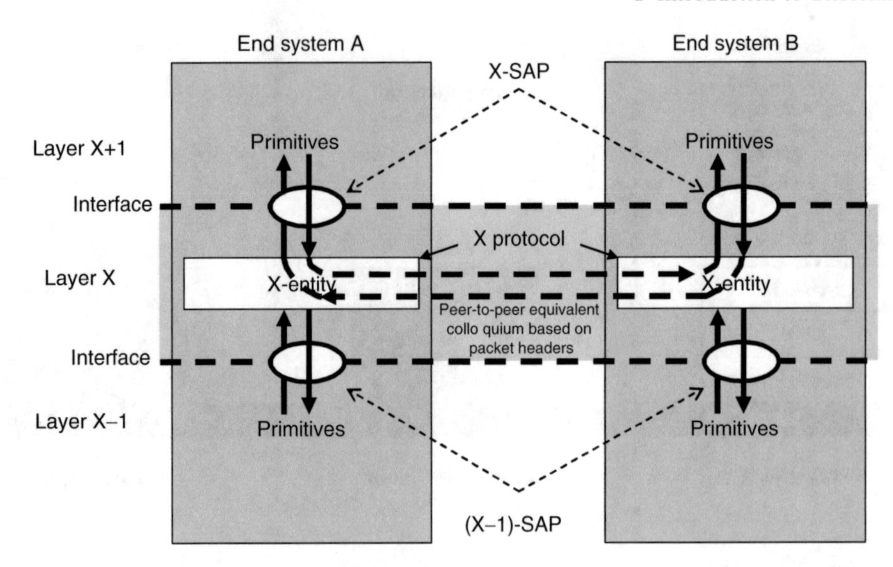

Fig. 1.17: Functional model of a generic OSI layer with the indication of the peer-to-peer colloquium between A and B end systems

SAPs at the interface between two layers. A unique SAP address identifies each SAP. The exchange of messages between two adjacent layers in a stack is made through *primitives*. Each entity also receives services from lower-layer protocols through the lower level SAP. For example, a transport entity (layer $X = 4$) provides a service to upper layers through a T-SAP and receives a service from lower layers through an N-SAP. As for the interaction between "systems," it occurs through the dialogue of entities of the same layer (i.e., peer entities), according to rules, depending on the protocol of the layer considered. The interaction between two systems is depicted in Fig. 1.17. The exchange of commands and instructions between homologous layers on different machines occurs using special standardized messages that form the so-called *signaling*. Each layer communicates logically with its peer, but, in practice, each layer communicates with its adjacent layers in the protocol stack.

A *protocol* is characterized as follows: (1) a set of formats according to which data exchange occurs between peer entities and (2) a set of procedures to exchange data. Standardization bodies define the different protocols, which a system can use to exchange information. The implementation of interfaces is left free to manufactures, provided that they support the primitives characterizing the service (standard). The protocols of a given layer format their messages in transfer units, generically called Protocol Data Units (PDUs). PDUs are exchanged by end systems through the services provided by lower layers.

The PDUs can be very different at various layers, from the user information at layer 7 to the bits transmitted on the physical link at layer 1. Information is exchanged by means of PDUs through SAPs between adjacent layers. For instance, a PDU of layer $X + 1$ is received by the lower layer X through a SAP and is considered as a Service Data Unit (SDU) of layer X. This SDU can in turn be enriched with a header, containing additional control information of layer X (*encapsulation*); we have thus obtained a PDU of layer X. If the SDU received from layer $X + 1$ has a length exceeding the maximum value allowed by layer X, the SDU is fragmented in different segments (the corresponding entity on the receiver side has to reassemble the different segments); conversely, several very short SDUs can be aggregated into a longer one. The process from input PDU to SDU to output PDU repeats at each layer of the OSI protocol stack; see Fig. 1.18. Hence, the PDU of a given layer becomes the SDU of the layer below. For instance, an N-entity receives a T-PDU: layer 3 adds a header to this SDU, thus obtaining an N-PDU. Peer entities have a colloquium as if they were directly exchanging PDUs.

The protocol of a given layer can perform a multiplexing function: the SDUs received from different SAPs can be addressed to the same SAP of the lower layer; otherwise, parallel transmissions can also be employed by using different SAPs towards the lower layer. See Fig. 1.19.

The header added at the generic layer X is needed to manage the protocol of layer X. The process of exchanging information through different layers is detailed in Fig. 1.20. As already explained, data received from the upper layer (in the form of an SDU) are encapsulated with a header (to form a PDU) and passed to the lower layer.

Each protocol layer can provide either a connection-oriented service or a connectionless transfer service with the corresponding peer protocol at the destination. A connection-oriented service is characterized by three phases:

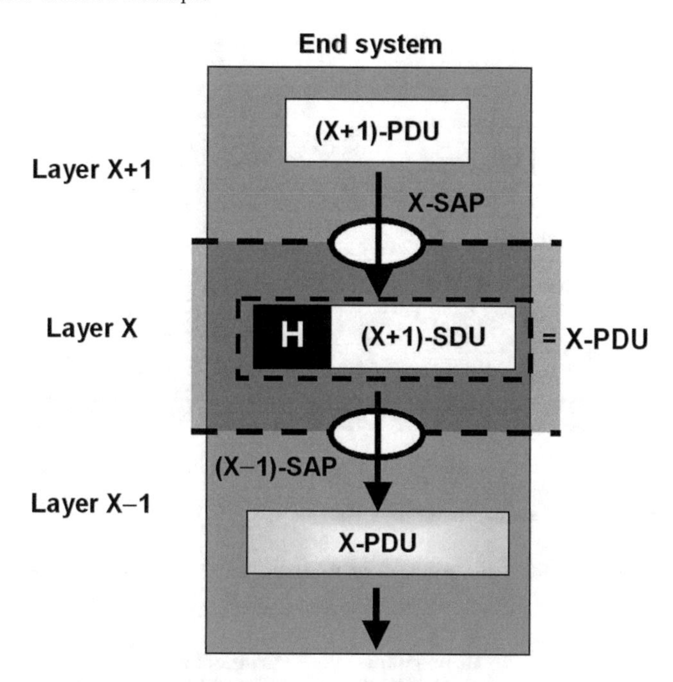

Fig. 1.18: Exchange of data through layer SAPs in the form of PDUs

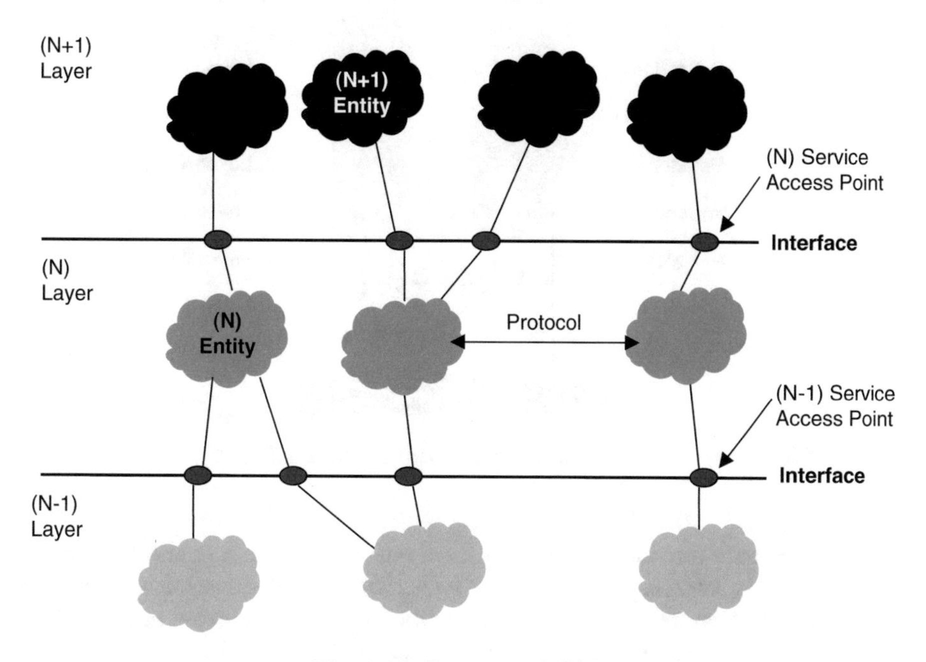

Fig. 1.19: Layers and SAPs

connection establishment, data transfer, and connection release. As soon as the connection is obtained, PDUs are exchanged by specifying the identifier of the connection. Connectionless services are characterized by sending independent PDUs, each typically containing the address of both source and destination. Each PDU has an autonomous route in the network: PDUs of the same service may have different paths to reach the same destination; hence, subsequent PDUs could not be received in order due to different delays. The selection between a connection-oriented service and a connectionless one can be performed at the link, network, and transport layers. In particular, on top of layers 2, 3, and 4, there are two SAPs through which the upper layer can access either connection-oriented or connectionless services. Combining the choices made at the different layers, different typologies of services are possible, as detailed in Fig. 1.21.

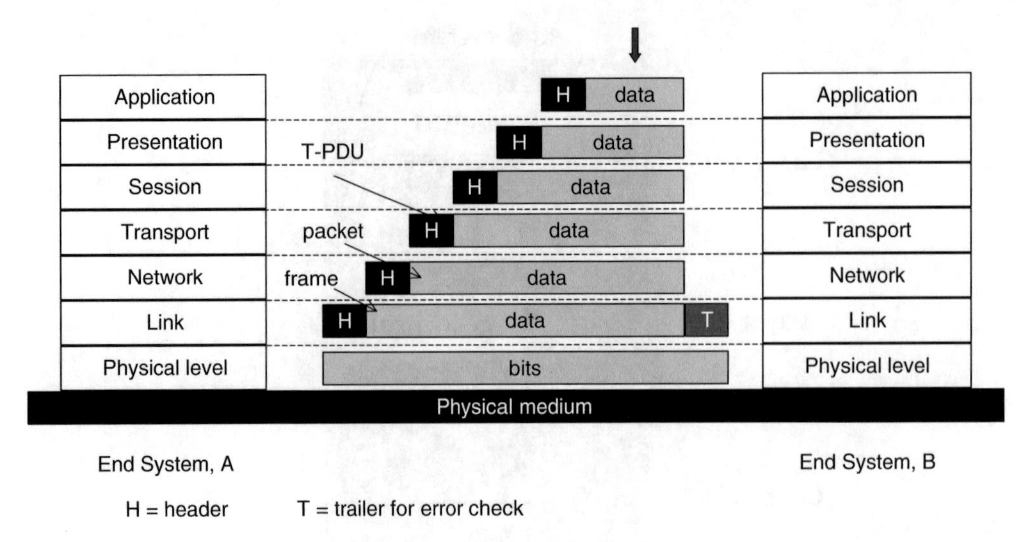

Fig. 1.20: Generation of the PDUs when information goes from layer 7 to layer 1 to be transmitted towards the destination

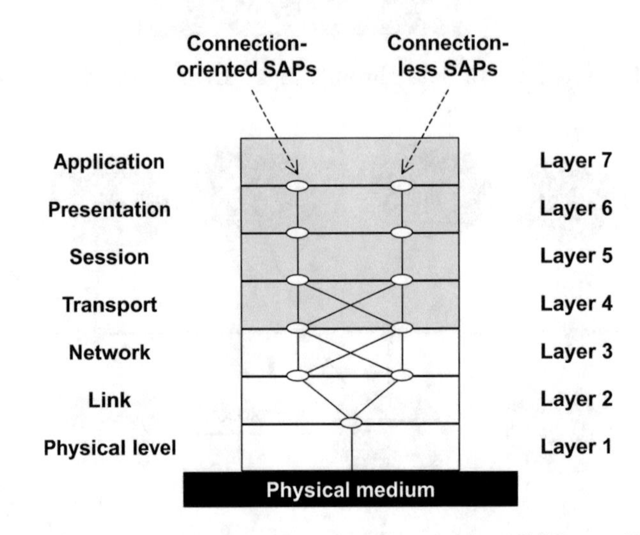

Fig. 1.21: Selection of connection-oriented and connectionless SAPs at different OSI layers

Since the information exchange must occur between two generic terminals connected by the network, an important network functionality is *addressing* that allows identifying the destination to which information has to be delivered. The network level that receives a PDU with the destination address must decide the SAP towards which to forward the information. This is the *routing* functionality. In particular, the layer 3 of each intermediate node has to support two essential functions:

- Routing to select the appropriate output SAP for the PDU, depending on the destination address; this is obtained through a routing table (see the IP routing section in Chap. 2).
- Forwarding to transfer the PDU from the input SAP to the output one.

Table 1.1 provides a classification of the main switched networks (distinguishing between circuit-switched and packet-switched networks) and some protocols, which are identified by the OSI layer of operation. Finally, also the main transport technologies are listed here concerning the different networks. The meaning of the acronyms shown in Table 1.1 will be clarified through the following chapters of the book. Note that, because of the wide variety of network protocols, the list given in this table is largely incomplete, but it is provided here to map the protocols to the appropriate networks and at the appropriate OSI layers.

In many cases, a protocol provides such a strong characterization of a network that it can be practically identified with the network itself. This is the case of the “X.25 network” as well as the case of the “ATM network,”

Table 1.1: Taxonomy of main networks, protocols, and transmission technologies that are described in Chaps. 1 and 2

Networks	*Circuit-switched*		*Packet-switched*
	PSTN, ISDN	ISDN, Digital Network, B-ISDN, Ethernet, LANs, WiFi, Internet, NGN	
Protocols	*Name*	*OSI level(s)*	*Related networks*
	X.25	1, 2 and 3 (user to network interface)	Digital network
	LAP-B	2	X.25-based network
	LAP-D	2	ISDN
	Frame relay	2	Digital network
	Aloha	2	AlohaNET
	IEEE 802.x family	1 and 2	LANs: Ethernet, Token-based, WiFi, Bluetooth, etc.
	ATM	2	B-ISDN
	IP	3	Internet
	ARP	3	Internet
	OSPF	3	Internet
	BGP	3	Internet
	MPLS	2+	Internet
	TCP	4	Internet
	UDP	4	Internet
	RTP	4+	Internet
	FTP	7	Internet
	Telnet	7	Internet
Transmission technologies (layer 1)	*Name*		*Related networks*
	PCM, plesiochronous hierarchy		PSTN, Digital networks
	BRI		ISDN
	PRI		ISDN
	ADSL/ADSL2+/VDSL/VDSL2		PSTN, Internet
	SONET/SDH		B-ISDN, MPLS, Internet
	DWDM		GMPLS, Internet

a synonym of B-ISDN. Finally, we will speak about MPLS-based networks and IEEE 802.x local area networks. The descriptions of these networks are provided in the next sections (X-25, ISDN, ATM), in Chap. 2 (MPLS, etc.), and in Chap. 6 (IEEE 802.x) of this book.

1.3.4 Traffic Engineering: General Concepts

The network needs to be adequately designed to route the traffic properly for each source–destination pair and to allocate suitable capacity on the different links to avoid excessive delays (in packet-switched networks) or blocking phenomena (in circuit-switched networks). Routing should also allow a right balance of traffic load among different possible routes. Link dimensioning is a consequent task of routing. Both network design aspects must be taken into due account to guarantee a certain network performance. Some basic Quality of Service (QoS) metrics for network performance evaluation are as follows:

- End-to-end delay (mean, jitter, and 95th percentile values as described in Chap. 3)
- Packet losses due to buffer congestion and overflow
- Call blocking probability due to the unavailability of resources in circuit-switched networks.

Detailed performance parameters to measure QoS and to define QoS requirements are

- Packet delay (s) at different layers
- Delay jitter (s) at different layers (especially, application); the jitter is measured as the difference in the delay experienced by two subsequent packets of a given flow traveling through the network.
- Throughput (bit/s) at MAC or transport layer
- Packet loss rate (%) at MAC or network layer
- Bit error rate (%) at PHY layer
- Outage probability (% of time) at PHY layer
- Blocking probability (%) at PHY or MAC layer (CAC)
- Fairness (between 0 and 1) at PHY, MAC, or transport layer.

In the field of telephony, QoS was defined by ITU-T Recommendation E.800 (dated back to 1994 and subsequent revisions) [20]. This recommendation defines QoS as a "collective effect of service performance, which determines the degree of satisfaction of a user of the service." According to E.800, QoS depends on the service performance, which is divided into support, operability, "serveability" (the ability of a service to be obtained within specified tolerances and other given conditions), and security. The service performance depends on characteristics such as transmission capacity and availability. In the more recent ITU-T G.1000 Recommendation, new QoS definitions are given. In particular, G.1000 envisages four QoS standpoints: QoS requirements of user/customer, QoS offered/planned by the provider, QoS delivered/achieved by the provider, and QoS perceived by the user/customer.

With the development of the Internet (IP-based traffic), QoS issues have also been addressed by IETF in RFC 2216, according to which QoS refers to "the nature of the packet delivery service provided, as described by parameters such as achieved bandwidth, packet delay, and packet loss rates." QoS in IP-based networks is also addressed by ITU-T Y.1541 Recommendation, where 8 QoS classes are envisaged, also, detailing possible queuing schemes to be adopted at nodes.

The Service Level Agreement (SLA) is a contract between the end-user and the service provider/operator, which defines proper bounds for some of the QoS performance parameters described above. SLA details the responsibilities of an information technology service provider (an Internet Service Provider, a telecommunication operator, etc.), the rights of the users, and the penalties assessed when the service provider violates any element of the SLA. An SLA also defines the service offering itself, network characteristics, security aspects, and evaluation criteria.

The basic approach for QoS support in the classical Internet is over-provisioning: network resources are designed based on the worst-case conditions for the traffic load to provide the SLAs agreed with the users. With the evolution of the Internet, more refined QoS support techniques have been identified. Basic approaches for managing the QoS of different classes (types) are prioritization and resource reservation (e.g., a reserved bit-rate for a real-time traffic flow).

The above network design aspects are covered by *traffic engineering* (or "teletraffic") methods, which encompass measurement, modeling, characterization, and control of multimedia multi-class traffic and the application of analytical approaches to achieve specific network performance objectives [21]. Teletraffic design methods and optimizations (e.g., based on blocking probability, mean throughput, and mean delay) are typically non-linear problems, so that numerical methods are needed to solve them.

QoS is concerned with the consistent treatment of traffic flows at the various nodes in the network. QoS is based on objective numerical metrics. On the other hand, Quality of Experience (QoE) relates to the perceived quality by the user (subjective measure). This applies to voice, multimedia, and data services. ITU-T P.10/G.100 Recommendation defines QoE as "the overall acceptability of an application or service, as perceived subjectively by the end-user." QoE includes complete end-to-end system effects (client, terminal, network, and service infrastructure). The overall acceptability may be influenced by user expectations and the context. QoE is much related to the user experience at the application layer. The Mean Opinion Score (MOS) metric is typically used for QoE assessments, based on subjective estimations made by a pool of users (QoE of telephone voice, video transmissions, etc.). However, many other metrics have also been defined.

1.3.5 Queuing Theory in Telecommunications

In telecommunication networks, queuing theory is used to model a wide range of problems for *teletraffic analysis*. In particular, it is used every time a network resource (a link connecting two nodes, a layer 3 signaling processor, which is in charge of managing incoming data traffic, a network element accessed by hosts, etc.) is shared by competing "requests" (i.e., traffic flows). When service requests arrive temporarily according to a higher rate than the time needed to fulfill each of them, a waiting list is required in each queue, provided that it has enough rooms to store all requests.

Typical problems studied by queuing theory are described below, referring to the OSI protocol layers:

- *OSI Layer 1*: Blocking phenomena of a traffic flow (i.e., a call) due to unavailable resources in at least one link in the path from source to destination.
- *OSI Layer 2*: Queuing is generated by different packets sharing the transmission resources of a link connecting two adjacent nodes (this can also be the case of distributed terminals accessing a shared node).
- *OSI Layer 3*: Queuing is experienced by routing requests at layer 3 signaling processor. Queuing can also occur in the packet forwarding plane.

Different queuing phenomena can be experienced depending on the circuit-switched or packet-switched networks, as detailed below.

The adoption of queuing models is important in circuit-switched networks in which typically no wait is allowed for a free transmission resource. Hence, in case of unavailable transmission resources on a link along the path from source to destination, a call is blocked and cleared. Queuing theory permits to determine the call blocking probability under certain assumptions on the call arrival process.

In packet-switched networks, queuing can be experienced at each node and on each link (OSI layers 3 and 2, respectively). Let us refer to the performance at the packet level (i.e., OSI layer 2): waiting times can be tolerated (within certain limits for real-time traffic flows), but packet losses can still be induced by capacity limitations in buffers. In these cases, queuing theory can be adopted to study the statistics of the number of packets in the queue or of the waiting time experienced by a packet (e.g., distribution of the number of packets in the queue, distribution of the queuing delay, related mean and variance values). Moreover, complex queuing models are needed to study the performance of nodes having to switch input traffic on different output links. The queuing theory will be addressed in the second part of this book.

1.4 Transmission Media

The transmission medium is the physical link between two generic network elements [22]. To achieve the best performance (i.e., high bandwidth and long distance covered), the physical medium has to allow low signal attenuation and low dispersion. Hence, the medium has to achieve low values of input impedance (i.e., low resistance, low inductance, and low capacity) and has to guarantee a high bandwidth for conveying high bit-rate signals. The information is propagated through a transmission medium by an electromagnetic wave. The propagation can be guided or unguided:

- *Guided media*: Waves are guided along a physical path. This is the case of both copper solutions (i.e., twisted pair and coaxial cable) and optical fibers.
- *Unguided media*: There is not a physical path since the electromagnetic wave propagates on air (the atmosphere, the outer space, etc.). This is the case of the so-called "wireless" transmissions (i.e., radio waves or infrared light).

1.4.1 Copper Medium: The Twisted Pair

A typical transmission medium (for low bit-rates and reduced distances) is given by a couple of copper wires; they are manufactured in a number of standardized diameters (the most common diameters are 0.4, 0.5, 0.6, and 0.7 mm). The wires in the cable are twisted together to minimize the electromagnetic induction between different pairs of wires (cross-talk phenomenon). Two pairs or four pairs are typically bundled together. The attenuation per kilometer depends on both the wire diameter and the signal frequency.

For some business locations, a twisted pair is enclosed in a shield, which functions as a ground. This is known as Shielded Twisted Pair (STP). The ordinary wire for the interconnection of the home phone to the local exchange

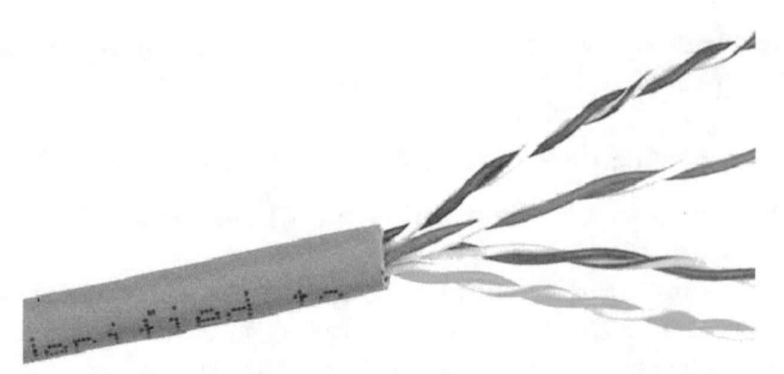

Fig. 1.22: Cable with 4 twisted pairs (UTP category 5)

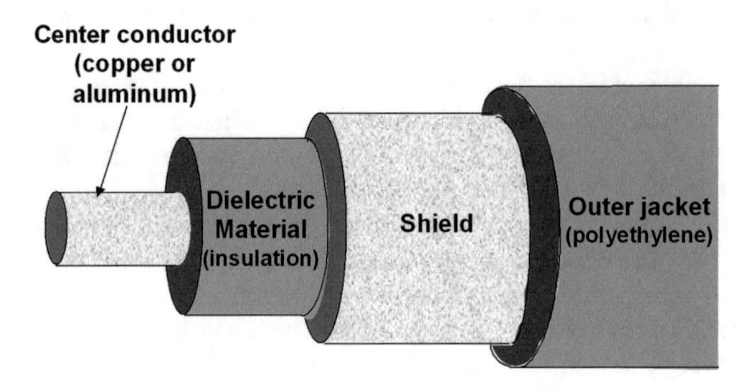

Fig. 1.23: Coaxial cable internal structure

office is the Unshielded Twisted Pair (UTP). UTP is cheap and easy to install but suffers from external electromagnetic interference. UTP cables use the well-known RJ45 connector (e.g., phone line connectors). STP uses a metal braid or sheathing to reduce interference. It is more expensive and harder to handle (thick and heavy).

EIA and TIA have classified and developed standards for several types of UTP cables, distinguished in *categories*. A higher category number requires a tighter twist in the cable so that the cancellation of mutual interference is more effective: the available bandwidth is larger (i.e., the wires have a better transfer function characteristic) as well as the transmission bit-rate. For instance, category 3 is characterized by a twist length from 7.5 to 10 cm and allows a bandwidth up to 16 MHz for use as voice grade in offices. Category 4 permits to achieve a bandwidth of 20 MHz for local area networks. Categories 5 and 5e have a twist length from 0.6 to 0.85 cm and allow up to 100 MHz of bandwidth (see Fig. 1.22). Category 6/6a yields a bandwidth of 200/500 MHz up to 100 m of distance. This cable category is also suitable to support Gigabit Ethernet on shorter distances. Category 7/7a achieves a bandwidth of 600/1,000 MHz up to 100 m of distance for a particular type of STP cables (shielded or foil screened). Category 8 is for an Ethernet cable, which is different from those of previous categories even if the physical appearance is similar. It requires shielded cabling. It can support a speed of up to 40 Gbit/s for a maximum distance of 30 m.

1.4.2 Copper Medium: The Coaxial Cable

A cable consists of one or more coaxial tubes; each of them has an inner conductor surrounded by a tube-shaped outer conductor (see Fig. 1.23), providing a shielding effect concerning adjacent tubes. A photo of (single) coaxial cables is shown in Fig. 1.24. A coaxial cable guarantees bandwidth in the order of hundreds of megahertz (e.g., 400 MHz). The different types of coaxial cables are identified by a code of the type RG-XX (Radio Guide), where XX is a code number. Amplifiers are necessary to reach long distances. Coaxial cables allow a higher traffic capacity than twisted pairs. In the trunk network, coaxial cables are used in pairs, one for each direction of transmission. Today, coaxial cables are no longer installed in the trunk part of the telecommunications network. They have been replaced by optical fiber cables. One of their most common uses today is the distribution of TV signals from antennas.

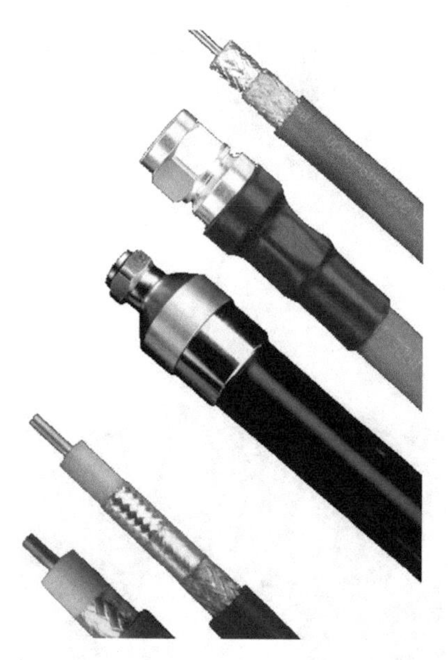

Fig. 1.24: Photo of different types of coaxial cables

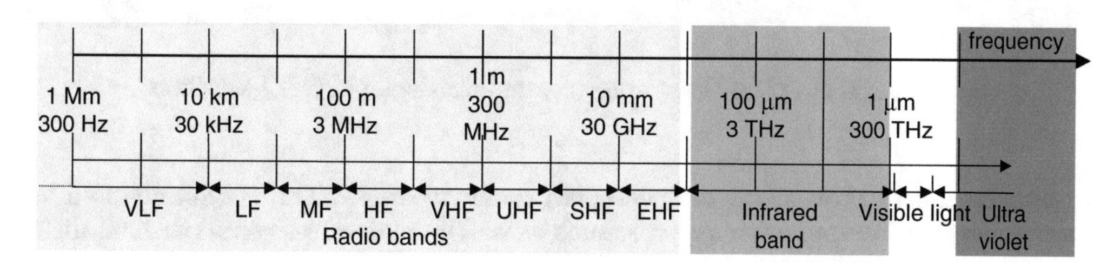

Fig. 1.25: Frequency band representation (frequency axis is in logarithmic scale)

In coaxial cables, the inner conductor consists of a round, solid copper conductor. The outer conductor (i.e., the shield) is made of copper foil or braided wire. The inner conductor must always be centered in the tube; it is kept in position by plastic washers or through compressing the plastic tube slightly at regular distance intervals. To improve shielding performance at low frequencies, a steel tape may be wrapped around the tube.

1.4.3 Wireless Medium

Wireless transmissions concern the radio spectrum and (at higher frequencies) the infrared one and, in some cases, the visible light one (laser links); see Fig. 1.25. These waves propagate at the light speed c (= 300,000 km/s) in air. The relation between radiation wavelength λ and frequency f is

$$\lambda f = c \tag{1.5}$$

Radio transmissions are characterized by wavelength longer than 1 mm. Infrared is electromagnetic radiation having a wavelength in the range from 780 nm to 1 mm. The name is related to the fact that these bands are below (in terms of frequency) the visible red light. Our eyes are only sensitive to a small portion of the electromagnetic spectrum with wavelengths from 400 to 700 nm. Ultraviolet radiation has a wavelength in the range from 10 to 400 nm. X-rays have wavelengths from 0.01 to 10 nm. Finally, gamma radiation has wavelengths lower than 0.01 nm.

Infrared radiation was first discovered around 1800 in an experiment made by the astronomer William Herschel. Then, in 1847, A. H. L. Fizeau and J. B. L. Foucault showed that infrared radiation has the same properties as visible light, being reflected, refracted, and capable of forming an interference pattern. Infrared transmissions are

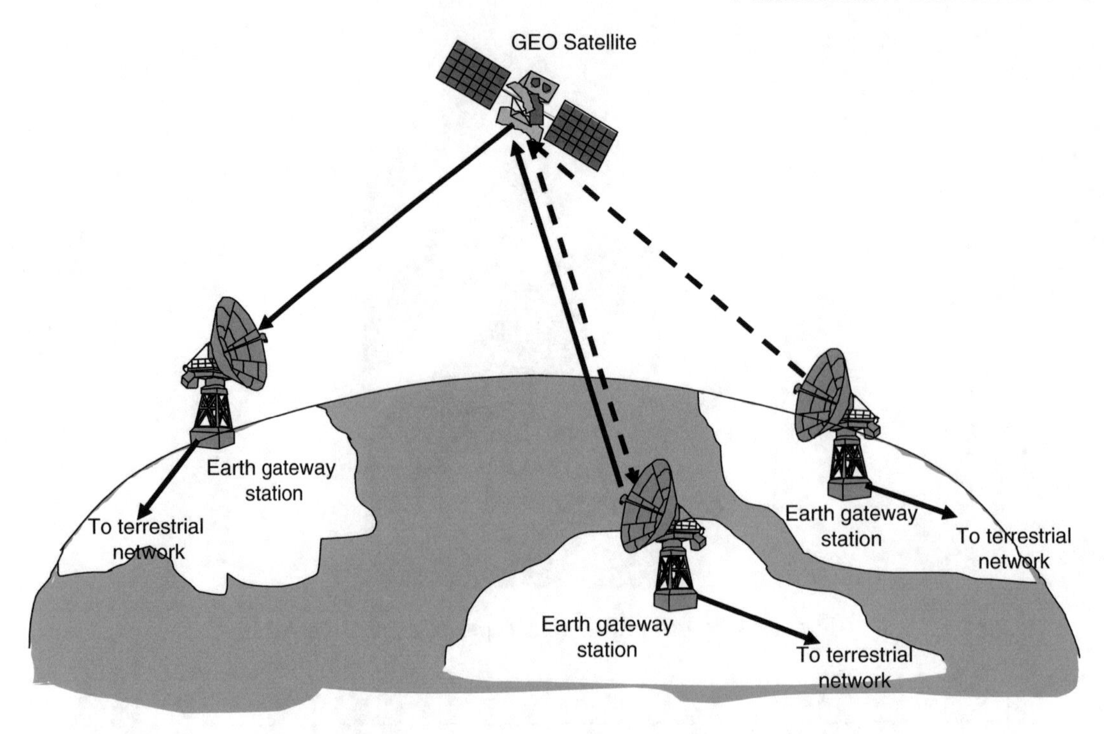

Fig. 1.26: Example of long-range communications via GEO satellites

currently being used for short-distance Line-of-Sight (LoS) communications. This is typically used to interconnect some peripherals to personal computers or laptops, such as mobile phones, printers, personal digital assistants, etc.

The radio spectrum typically goes from 3 kHz to 300 GHz. A complete survey of radio frequency bands, their designation, and their use is provided in Fig. 1.25 and in Table 1.2. Radio transmissions require the use of transmitting antenna and receiving antenna, respectively, for irradiating and capturing the electromagnetic wave. The principal uses of radio wave transmissions are terrestrial microwave links (e.g., interconnecting radio links), cellular systems for mobile phones, broadcast transmissions (i.e., radio and TV diffusion), and satellite communications, as shown in Fig. 1.26. More details on microwave frequency bands (1–30 GHz) are provided in Table 1.3. For instance, satellite communications use frequencies between 2 and 40 GHz (today, the use of higher frequency bands, EHF, is gaining increasing interest even if they are more affected by meteorological events); large bandwidths are available of tens or hundreds of megahertz.

The propagation of a radio wave depends on its frequency. Radio waves with frequencies below 30 MHz are reflected by the different ionized layers of the atmosphere and by the ground. Those radio waves bounce between the atmosphere and the earth so that they can reach long distances; however, capacity is strongly limited to a few hundreds of bit/s. Above 30 MHz, transmissions are not reflected by the atmosphere. This is the case of VHF and UHF frequency transmissions, which are used for TV broadcasting. Transmissions at frequencies above 3 GHz require a LoS path between transmitter and receiver: obstacles of a size comparable with the radiation wavelength severely attenuate the signal (non-LoS conditions).

Let us focus on the attenuation of the radio wave. The propagation of waves in free space is different from guided propagation in cables or optical fibers. These latter transmission media do not lose signal energy as it travels; attenuation is due to absorption or scattering, whereas radio waves propagate in the three-dimensional space, and, as they travel, the surface area they occupy increases as the square of the distance traveled. The power carried by these waves is also spread on a broader surface. Hence, the power of the wave is attenuated according to the square of the distance. The free-space attenuation L_{free} is expressed as

$$L_{\text{free}} = \left(\frac{4\pi D}{\lambda}\right)^2 \tag{1.6}$$

Table 1.2: Radio frequency bands, according to ITU denominations

Band	Name	Frequency	Wavelength
Extremely low frequency is used by the US Navy to communicate with submerged submarines. Signals in the ELF frequency range can penetrate submarine shields. Low transmission rates are allowed by ELF communications.	ELF	Frequencies below 3,000 Hz	10,000–1,000 km
Voice frequency band denotes frequencies within the audio range of the voice.	VF	300–3,000 Hz	1,000–100 km
Very low frequency is used for radio navigation. Many natural radio emissions can be heard in this band. Since a VLF signal can penetrate the water to a depth of 20 m, it is also used to communicate with submarines.	VLF	3–30 kHz	100–10 km
Low frequency is used for AM radio broadcast service. Its main use is for aircraft beacon, navigation, information, and weather systems.	LF	30–300 kHz	10–1 km
Medium frequency is used by regular AM broadcast transmissions.	MF	300–3,000 kHz	1 km–100 m
Since ionosphere often reflects high-frequency radio waves, this range is widely used for medium- and long-range terrestrial radio communications. Many factors influence the propagation: sunlight at the site of transmission and reception, season, solar activity, etc.	HF	3–30 MHz	100–10 m
Very high frequency is commonly used for FM radio broadcast at 88–108 MHz and television broadcast (together with UHF). VHF is also commonly used for terrestrial navigation systems and aircraft communications.	VHF	30–300 MHz	10–1 m
Ultra-high-frequency bands are used to broadcast common television transmissions.	UHF	0.3–3 GHz	1 m–100 mm
Microwaves are electromagnetic waves with a wavelength longer than infrared light, but shorter than radio waves. Microwaves are also known as super high-frequency signals. The boundaries between far infrared light, microwaves, and UHF radio waves are defined differently in various fields.	SHF	3–30 GHz	100–10 mm
Extremely high frequency also called mmWave	EHF	30–300 GHz	10–1 mm

Table 1.3: Details on sub-bands in the microwave and EHF ranges, according to IEEE radar band denominations

Description	Name	Frequency (GHz)
L band is used by some mobile communication satellites	L	1–2
S band is used by weather radar and some mobile communication satellites	S	2–4
C band is primarily used for satellite communications (TV broadcast transmissions). Typical antenna size is in the range of 1.8–3.5 m	C	4–8
X band is primarily used by satellites for telecommunications	X	8–10
Ku band is used by the majority of satellites for digital TV broadcast systems as well as for Internet access systems	Ku	10–18
K band signals are absorbed by water vapor	K	18–26
The 20/30 GHz band is used in satellites for telecommunications	Ka	26–40
This band will be used for satellite digital transmissions	Q	33–50
This band will be used for satellite digital transmissions	V	50–75
This band will be used for satellite digital transmissions	W	75–110

where D is the distance traveled, and λ is the wavelength of the transmitted signal. We can also consider $\lambda \times f = c$, where f is the transmission carrier frequency and c is the light speed.

According to (1.6), the higher the frequency, the bigger the attenuation due to the free-space path loss. Additional attenuation is due to the presence of the atmosphere: attenuation peaks are at 22.3 (within K band) and at 60 GHz (within V band) because of water vapor and molecular oxygen, respectively.

Radio waves propagate at the light speed in air. Hence, not only attenuation but also propagation delay must be taken into due account for long-distance transmissions. This is the case of GEO satellite transmissions, where the propagation delay (from earth to the satellite and then back to the earth) can even reach 280 ms.

1.4.4 Optical Fibers

Optical fibers convey signals in the form of visible light. There are many advantages in using optical fibers for transmitting signals rather than copper cables. In particular, we can consider the following ones:

- Optical fibers entail smaller diameters than copper wires.
- The signal attenuation in optical fibers is much lower than that in copper wires.
- Unlike electrical signals in copper wires, light signals of one fiber do not interfere with those of other fibers in the same cable. This means clearer phone conversations or TV reception.
- Electromagnetic or radio interference or harsh weather conditions do not affect the bit error rate performance of optical fibers. Optical communication systems allow very high data rates and exhibit very low bit error rates. In the Gbit/s range, copper twisted pair communications can achieve a bit error rate of 10^{-5}, while fiber optics typically exhibits a bit error rate in the range from 10^{-11} to 10^{-9}.

An optical fiber is composed of the following parts (see Fig. 1.27):

- *Core*: Thin glass at the center of the fiber, where the light travels.
- *Cladding*: Outer optical material surrounding the core and with lower refractive index, so that the light is reflected back into the core.
- *Buffer coating*: Thermoplastic coating to protect the fiber from damage and moisture.

The glass fiber has a glass core with a surrounding glass cladding. The core consists of doped glass, whereas the cladding is made of pure quartz glass. Typically, the diameter of the cladding is 125 μm. The diameter of the core is different for different types of fibers: 8, 10, 50, or 62.5 μm. Hundreds or thousands of these optical fibers are arranged in bundles in optical cables. These bundles are protected by the outer cable covering, called the jacket or buffer; see Fig. 1.28.

The difference in densities between core and cladding allows us to exploit the principle of total internal reflection. As optical radiation passes through the fiber, it is constantly reflected during the propagation through the center core of the fiber. The resulting energy fields in the fiber can be described as a discrete set of electromagnetic

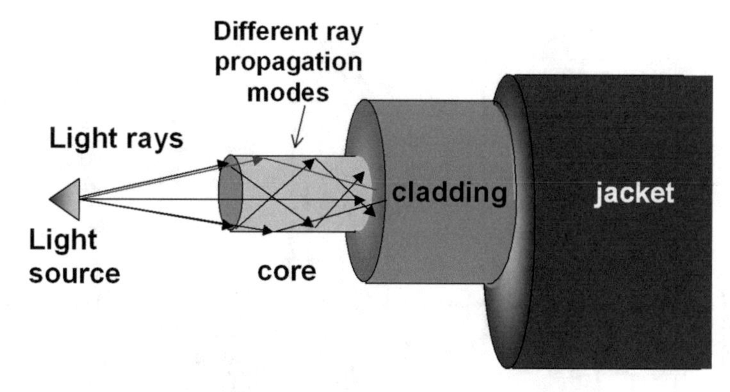

Fig. 1.27: Optical fiber, internal structure

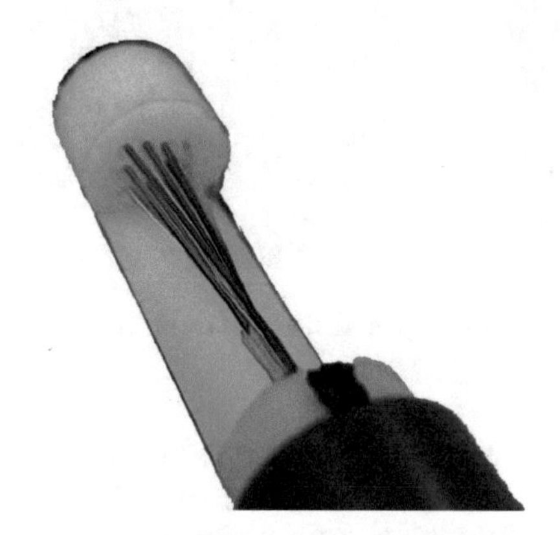

Fig. 1.28: Optical fibers from a bundle. Many bundles are arranged in an optical cable

waves propagating axially through the fiber, called the guided *modes* of the fiber. In single-mode fibers, only one radiation ray propagates in the fiber.

The characteristics of optical fibers evolved in time, as detailed by the attenuation as a function of the light wavelength in Fig. 1.29. Referring to the attenuation curve obtained with technologies of 1980, three low attenuation regions were identified in terms of the wavelength. They were the first, the second, and the third window.

The 850 nm region ("first window") was initially attractive because the technology for light emitters and detectors was already available (i.e., Light-Emitting Diodes, LEDs, and silicon detectors, respectively). As the technology evolved, the first window became less attractive because of its relatively high attenuation of 2 dB/km. The "second window" at 1,310 nm allowed reduced attenuation of about 0.5 dB/km. In late 1977, Nippon Telegraph and Telephone (NTT) developed the "third window" technology at 1,550 nm. It offered the theoretical minimum optical loss for silica-based fibers, about 0.2 dB/km. Even a "fourth window" near 1,625 nm has been identified that, however, has higher optical attenuation than the third window. This fourth window has expanded the spectrum usable for multi-wavelength multiplexing.

Two types of optical fibers are available:

- *Single-mode fibers* have small cores (about 9 μm in diameter) and use lasers transmitting infrared light (wavelength from 1,300 to 1,625 nm).
- *Multimode fibers* have larger cores (about 62.5 μm in diameter) and use LEDs transmitting infrared light (wavelength from 850 to 1,300 nm). Some optical fibers can be made of plastic. These optical fibers are distinguished between "step-index" and "graded-index" (referring to the variation of the refraction index in the fiber from the center to the outer part).

Each wavelength has some advantages. Longer wavelengths offer better performance but always have higher costs. The shortest link lengths can be handled with multimode fibers and wavelengths of 850 nm (the less expensive solution). Single-mode fibers at 1,310 nm are used for medium distances ranging from 2 to 40 km. The longest distances require single-mode fibers at 1,550 nm and optical multiplexing techniques.

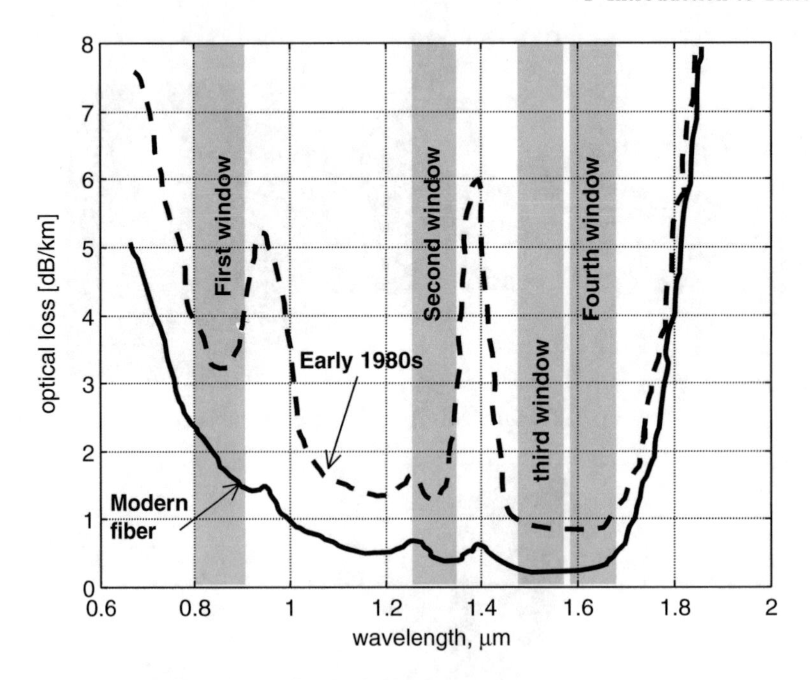

Fig. 1.29: Optical fiber attenuation curve

In 1990, Bell Labs transmitted a 2.5 Gbit/s signal over 7,500 km without regeneration. The system used a soliton laser and an Erbium-Doped Fiber Amplifier (EDFA). In 1998, they were able to send 100 simultaneous optical signals, each with a data rate of 10 Gbit/s, at a distance of about 400 km. This result was at the basis of the Dense Wavelength-Division Multiplexing (DWDM) technology, which has increased the total data rate carried by one fiber today up to 30 Tbit/s per fiber pair (terabits per second, 1 T = 10^{12}) in experimental tests (MAREA transatlantic submarine cable by Acacia Communications operator[2]), combining multiple wavelengths into one optical signal. A common DWDM commercial technology uses 80 wavelengths, each providing a 100 Gbit/s Ethernet link.

In modern glass optical fibers, the maximum distance is not significantly limited by the material absorption, but rather by the spreading of the optical pulses traveling in the fiber (dispersion phenomenon). Dispersion increases with the length of the fiber. It is common to characterize a fiber by the bandwidth–distance product, expressed in MHz × km. This quantity measures the goodness of the fiber since there is a trade-off between bandwidth and distance reached due to the dispersion effect.

The term "dark fiber" denotes unused optical fibers: when fibers are deployed by the operators, the common approach is to install more fibers than needed (concerning the current demand) to support the future increase in traffic.

The jacket color is sometimes used to distinguish multimode cables from single-mode ones. The standard TIA-598C recommends the use of a yellow jacket for a single-mode fiber and orange or aqua for a multimode fiber, depending on the type.

1.5 Multiplexing Hierarchy

Human voice and hearing range from about 20 Hz to about 14 kHz. When the telephone system was designed, it was decided for economic reasons to reduce the bandwidth available to just the necessary one, which permits to a have good quality and to recognize the persons. Hence, the net phone bandwidth ranges from 300 to 3,400 Hz; such restricted bandwidth allows us to capture most of the energy of the voice signal. In the analog telephony,

[2] These cables use couples of fibers to communicate in both directions. In total, 8, 12, 20 couples of fibers are possible with today's technology. Submarine cables are laid down in the deep sea. They have different length; for instance, the MAREA cable length is 6,640 km and the Asia-America Gateway cable is 20,000 km long.

voice is channelized at 4 kHz (net band plus guard-bands = gross band) and conveyed by the voice-grade phone line to the local exchange office. Since the spectrum of the voice signal limited to the maximum frequency $f_{\max} =$ 4 kHz, it is necessary to take one voice sample for every $T_c = 1/(2 \times f_{\max}) = 1/8{,}000$ s = 125 μs on the basis of the Nyquist sampling theorem. Each sample value is expressed by a 13-bit code word. A *companding* (logarithmic) characteristic is used to compress the dynamics of samples [23]. Two companding laws are possible: A-law for Europe and μ-law for the USA and Japan. The obtained value is quantized with 8 bits (7 bits in the USA). Hence, 8 bits every 125 μs correspond to a bit-rate of 64 kbit/s (56 kbit/s in the USA). This is the digital voice representation of the Pulse Code Modulation (PCM) system, which is the basis of any digital voice transmission. PCM is standardized in the ITU-T G.711 Recommendation [23].

Note that 125 μs is the frame duration value for all time division multiplexing systems (both US and ITU-T standards), which allow the transport of many multiplexed voice traffic flows. The frame duration represents the time periodicity of the resource allocation to the different users. The frame duration of 125 μs is used in both PDH (see Sect. 1.5.2) and SDH/SONET digital transmission hierarchies (see Sect. 1.7.5).

A communication channel typically has a sufficiently wide bandwidth to carry many elementary signals (e.g., voice signals) simultaneously. Therefore, it is essential to fully exploit the bandwidth of the physical medium for greater efficiency and cost reduction. The procedure according to which the signals of different users are transmitted through the same physical medium (a cable, an optical fiber, etc.) without generating mutual interference is called *multiplexing*; the corresponding device is called multiplexer.

There are two classical multiplexing schemes: Frequency Division Multiplexing (FDM) and Time Division Multiplexing (TDM), which separate the different transmissions in frequency or time, respectively.

1.5.1 FDM

Each signal occupies its frequency band for the entire duration of the transmission. Frequency bands can be allocated permanently or on demand. FDM techniques are used for radio and TV broadcasts. The user signal spectrum is limited using filtering; however, guard-bands are needed between adjacent transmissions to avoid interference.

Analog trunks of the telephone network use a form of FDM, which is described as follows. The various telephone signals are amplitude modulated (AM) with carriers at different frequencies spaced by 4 kHz (the net voice bandwidth ranges from 300 to 3,400 Hz; a total bandwidth of 4 kHz is considered including some guard-bands for filtering purposes). Instead of a "full" AM, a Single-Sideband Suppressed-Carrier (SSB-SC) signal is used to save bandwidth (see Fig. 1.30). All the SSB-SC signals properly transposed in frequency and spaced of 4 kHz are added and transmitted together to form the FDM signal. The carrier frequency separation should be sufficient to ensure that there is no spectral overlap between adjacent bands. Adequate band-pass filtering must be used to demodulate the signals.

ITU-T has recommended a hierarchy for FDM in telephony, as shown in Table 1.4. In particular, a single voice *channel* occupies 4 kHz. The first FDM multiplexing level is obtained by multiplexing 12 channels to form a *group*. Five groups are multiplexed to form a *supergroup*. Five supergroups are multiplexed to form a *mastergroup*. Finally, three mastergroups are multiplexed to form a *supermastergroup*.

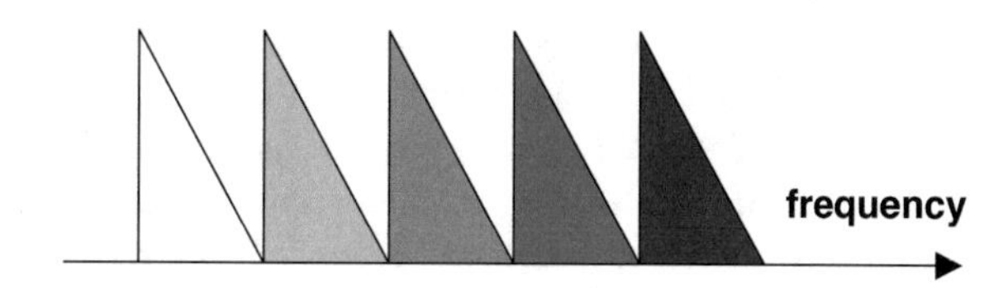

Fig. 1.30: Example of multiplexed signals in frequency

Table 1.4: ITU-T FDM multiplexing hierarchy

Name	Frequency range (kHz)	Number of channels
Channel	0–4	1
Group (12 channels)	60–108	12
Supergroup (5 groups)	312–552	60
Mastergroup (5 supergroups)	812–2,044	300
Supermastergroup (3 mastergroups)	8,516–12,388	900

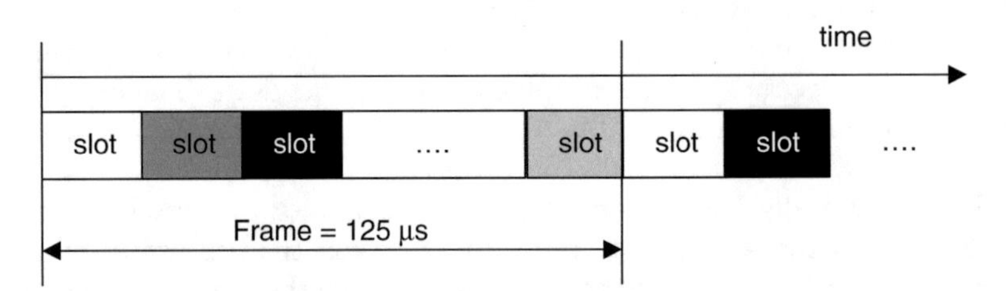

Fig. 1.31: Example of signals multiplexed in the time. We have a frame containing many slots. Each slot permits to convey one byte (i.e., a voice sample in digital telephone systems)

Table 1.5: Digital multiplexing hierarchies in different regions

Level	North America	Japan	International (ITU)
0	64 kbit/s (DS0)[a]	64 kbit/s[a]	64 kbit/s[a]
1	1.544 Mbit/s (T1/DS1)[b]	1.544 Mbit/s (J1)	2.048 Mbit/s (E1)[c]
2	6.312 Mbit/s (DS2)	6.312 Mbit/s (J2)	8.448 Mbit/s (E2)
3	44.736 Mbit/s (T3/DS3)	32.064 Mbit/s (J3)	34.368 Mbit/s (E3)
4	139.264 Mbit/s (DS4)	97.728 Mbit/s (J4)	139.264 Mbit/s (E4)
5	400.352 Mbit/s	565.148 Mbit/s	565.148 Mbit/s

[a] 1 voice circuit (i.e., one digital user channel)
[b] 24 user channels
[c] 30 user channels

1.5.2 TDM

In this case, there is a frame structure of 125 µs. All signals use the same frequencies for the duration of the transmission. The slots can be allocated permanently or on demand. TDM is used in digital telephony and data communications. In the simplest example, we may consider that each slot conveys the digitized version of a voice sample (see Fig. 1.31).

TDM standardization has different characteristics in North America, Europe, and Japan. In particular, T-carrier is the generic designator for any of several digitally multiplexed carriers, initially developed by Bell Labs and used in North America and Japan. The E-carrier system, where "E" stands for Europe, is compatible with the T-carrier and is used almost everywhere else in the world. The comparison of the different legacy TDM hierarchies is shown in Table 1.5. Additional multiplexing hierarchies for higher bit-rates are defined for fiber optic transmissions and related SONET/SDH technologies (see Sect. 1.7.5) or Ethernet technologies (see Chap. 6).

Legacy multiplexing hierarchies shown in Table 1.5 can be interpreted as follows. Referring to the ITU standard, 32 voice channels (practically, 30 voice channels plus two control channels, as detailed below) are multiplexed to obtain an E1 signal; 4 E1 signal are multiplexed to form one E2; 4 E2 are multiplexed to have an E3; 4 E3 are multiplexed to have an E4; and 4 E4 are multiplexed to obtain an E5. As for the North America TDM hierarchy, one T2 signal conveys 4 T1 (6.312 Mbit/s); a T3 signal transports 6 T2 (44.736 Mbit/s).

Let us describe in detail the technique adopted to multiplex E signals according to the Plesiochronous Digital Hierarchy (PDH) [23–26]. The basic data transfer rate is E1 at 2.048 Mbit/s. The exact data rate of the 2.048 Mbit/s E1 data stream is controlled by a clock in the data generating equipment. The exact rate is allowed to vary some percentage (±50 ppm) either side. Hence, different 2.048 Mbit/s E1 data streams can probably run at slightly different rates (they are not perfectly synchronized). To move multiple E1 streams from one place to another, they are multiplexed in groups of 4 to achieve the E2 signal. This is done by taking 1 bit from stream #1, followed by 1 bit from stream #2, then #3, and then #4, and so on, cyclically. Since the four E1 signals may have some discrepancy in the relative synchronization, it may occur that the multiplexer will look for the next bit of an E1 flow, when it has not arrived yet. Hence, to compensate for these absences, the transmitting multiplexer adds additional bits called "justification" or "stuffing" bits. In this case, the multiplexer signals to the receiving multiplexer that a bit is "missing." This allows the receiving multiplexer to correctly reconstruct the original data for each of the 4 E1 streams and at the different plesiochronous rates. The resulting E2 data stream from the above process is at 8.448 Mbit/s. Similar techniques are adopted for the higher levels of the multiplexing hierarchy.

The PDH multiplexing approach entails some problems when a given flow has to be extracted from a higher-level hierarchy; for instance, an E1 flow has to be extracted from an E2 signal. If the multiplexed flows were strictly synchronous, consecutive instances of a given E1 flow would be regularly spaced in time in the multiplexed flow. However, clock deviations and the insertion of justification bits disrupt such a possibility. Hence, it is impossible to demultiplex the E1 flow alone, simply based on synchronous timing. With PDH, the only solution is to demultiplex the whole structure to extract E1, determining where justification bits are inserted. The entire structure must then be multiplexed again and retransmitted. This is quite a complex process.

1.5.3 The E1 Bearer Structure

E1 has a capacity of 2.048 Mbit/s and employs line encoding in order both to eliminate the continuous component from the digital baseband transmission and to help a fast synchronization to the signal. E1 time slots are numbered from 0 to 31. The periodic use of one time slot (i.e., 8 bits) per frame corresponds to a capacity of 64 kbit/s.

The E1 signal can be structured or unstructured. In the unstructured case, a 2.048 Mbit/s capacity is provided. Instead, in the structured case, framing is necessary for allowing that any equipment receiving the E1 signal can synchronize and correctly extract the individual channels. Let us refer to a structured E1 signal, named PCM-30, where there are 30 information channels at 64 kbit/s. In particular, we have the following:

- *Time slot 0* carries a frame alignment signal as well as a remote alarm notification, five national bits, and optional Cyclic Redundancy Check (CRC) bits.
- *Time slot 16* carries out-of-band signaling.

In PCM-30, time slots 1–15 correspond to channels 1–15 and time slots 17–31 correspond to channels 16–30. These time slots are "clear channels": no bits are robbed for signaling purposes.

Finally, a short note on the T1 carrier having 24 slots for 64 kbit/s channels. Three versions are allowed:

- 24 phone signals at 56 kbit/s (7 bits/sample plus one signaling bit)
- 23 data channels at 64 kbit/s (8 bits) plus one signaling channel
- Unstructured flow.

1.6 The Classical Telephone Network

The old telephone network (named Public Switched Telephone Network, PSTN) is the concatenation of the world's public telephone networks, operated by various telephone companies and administrations (Telecom Operators and Public Telephone and Telegraph, PTT, Operators) around the world. PSTN is also known as the Plain Old Telephone System (POTS).

PSTN is based on the circuit-switching technique:[3] an end-to-end path must be established reserving resources for the entire duration of a phone call. Free transmission resources must be reserved on each link from switch to switch (see Fig. 1.32). These resources are dedicated to this conversation for all its duration. If there is no available resource on a link on the path, the call is blocked and refused (assuming that there are no resources available on alternate routes); thus, the originating phone user hears a busy tone.

When telephony began, a simple network architecture was used: only local exchanges with directly connected subscribers. The only possibility was to switch telephone calls between subscribers connected to the same local

[3] Today, the telephone network is IP-based (see Chap. 2), where the IP layer is packet-switched.

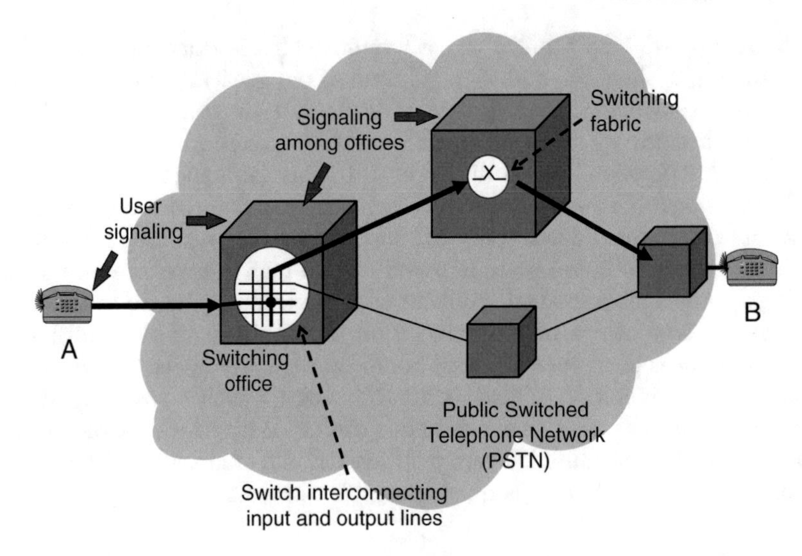

Fig. 1.32: Classical telephone network and resources involved in a phone call from user A to user B

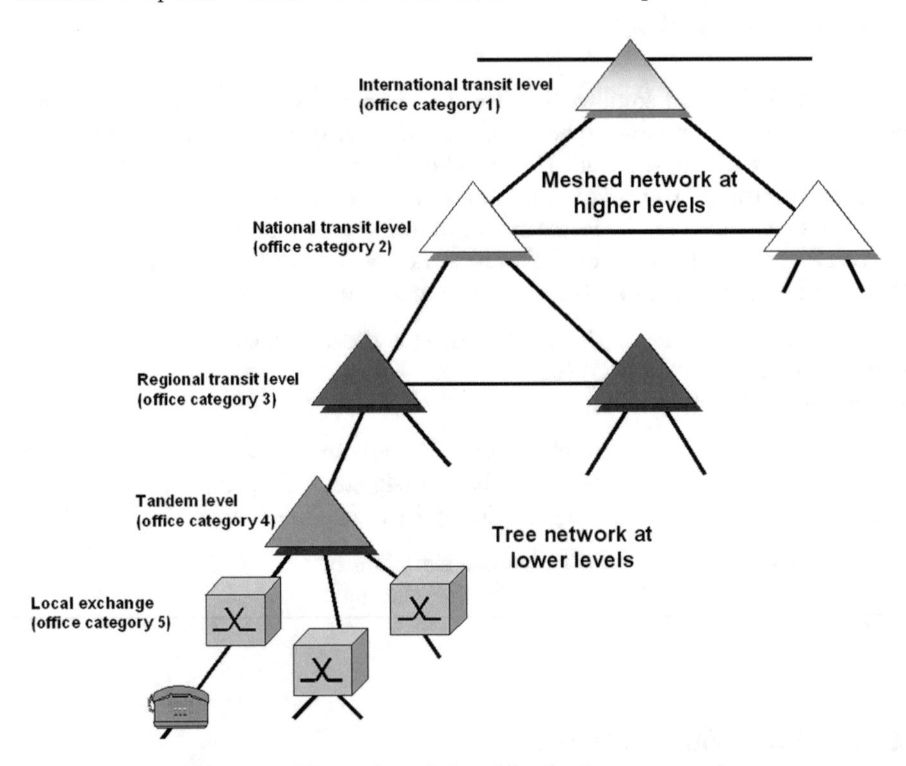

Fig. 1.33: Hierarchy of the old telephone network

exchange (basically, in the same town). It was soon realized that it would be quite complicated to interconnect a local exchange with all other local exchanges according to a mesh topology. The solution to this problem was to introduce a hierarchy in the network. Nodes were conceived at different levels. As a result, not all nodes needed direct connections to all other nodes. A common choice is to have *five* levels (even if the structure of the network and the number of levels may vary from operator to operator), as shown in Fig. 1.33.

Local exchanges are used to connect subscribers (*local loops*), while the task of regional transit exchanges is to transfer traffic upward in the PSTN hierarchy and to switch the traffic between local exchanges. Moreover, a tandem exchange is necessary in most cases in metropolitan areas to transfer traffic between several different local exchanges. A tandem exchange usually does not transfer traffic upward in the PSTN, but only between adjacent local exchanges.

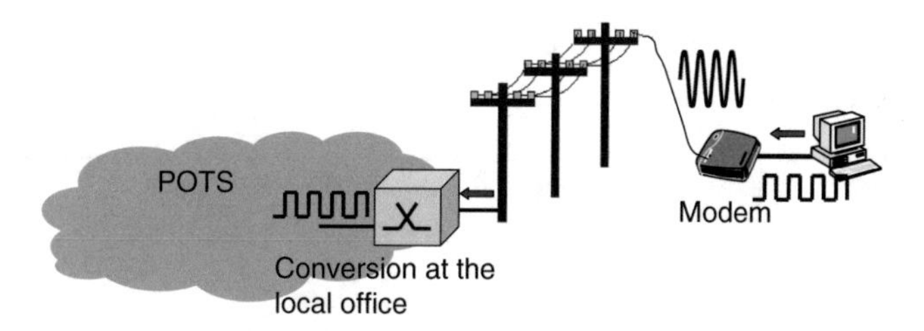

Fig. 1.34: Digital transmissions using modems in POTS

The local loop uses the twisted pair as the transmission medium. Loading coils are added within the subscriber loop to improve the voice transmission (transfer function flattened) within the 4 kHz bandwidth, while increasing the attenuation at higher frequencies.

At the highest levels of the hierarchy, links are called *trunks* and use coaxial cables, optical fibers, or microwave radio links. Trunks transport many signals using suitable multiplexing techniques. The level of the multiplexing hierarchy increases as we move from local to regional, to national, and to international levels.

Finally, there are also private networks within large companies, which are linked to the PSTN through Private Automatic Branch eXchange (PABX) systems.

Initially, the telephone network was based on analog technologies and traffic and on FDM multiplexing. Then, PSTN became fully digital, except for the part from the user to the first local exchange (here, the signal is analog and carried by twisted pairs). The basic voice circuit is at 64 kbit/s in the digital PSTN. Multiplexing is achieved by means of TDM according to the previously described hierarchies.

Between PSTN switches, signaling is digital using the Signaling System No. 7 (SS#7). Signaling is needed to define the end-to-end path of the circuit across different switches and to allow each switch involved to route the call internally in the proper way.

It is possible to estimate the traffic intensity contributed by each user to the PSTN. We can consider, for example, that a user spends on average 45 min a day (= 1,440 min) making or receiving calls. Hence, this user is busy for a percentage of time equal to $45/1{,}440 \approx 0.031$ Erlangs (or, equivalently, 31 mErlangs).

As a concluding note, it is important to remark that POTS is today an obsolete network. The Next Generation Network (NGN) with IP-based transport supports all the services, including telephony. The transition from POTS to NGN has been carried out worldwide.

1.6.1 Digital Transmissions Through POTS

The access line for POTS customers was based on analog technology. Therefore, a modulation was used to carry the digital signal in the voice band up to the first local exchange, where the signal was demodulated to its original digital format and transferred to the digital network. Each user needed a device named *modem*, which can modulate a digital signal, to transmit it and to demodulate the received signal, thus recovering the original digital format (see Fig. 1.34). Classical dial-up modems had to set up a circuit-switched call to an Internet Service Provider through the POTS.

Today's modems still modulate the signal but use a different bandwidth from the voice (voice and Internet traffic signals can coexist simultaneously on the access line). The modulation data signal is used to open a packet data session with a local point-of-presence of an operator from which the user traffic enters the Internet. This is the case of the ADSL access that will be discussed below in more detail. The fiber optic access technology does not need a modem to connect to the Internet.

The modem approach for data transmissions through POTS is not very efficient for two reasons: (1) a circuit must be dedicated to the data traffic even during intervals when no data are exchanged (this may be a significant loss of efficiency in the presence of bursty data traffic) and (2) data traffic undergoes digital-to-analog conversion when entering the network (and vice versa when leaving the network) even if the core network adopts a digital technology. With ISDN (see Sect. 1.7.2), digital (baseband) access is allowed directly from user premises.

Table 1.6: POTS-band ITU-T modem evolution

Year	Speed	Modulation
1960s	Very low rate modems: 300 bit/s (V.21) and 1,200 bit/s (V.22)	FSK and QPSK
1968	2.4 kbit/s (V.26)	QPSK
1972	4.8 kbit/s (V.27)	8-PSK
1976	9.6 kbit/s (V.32)	16-QAM + TCM
1986	14.4 kbit/s (V.32bis)	64-QAM + TCM
1989	19.2 kbit/s (V.33bis)	64-QAM + TCM
1993	28.8 kbit/s (V.34)	DMT
1998	56 kbit/s downstream (V.90)	PAM (downstream)
2000	56 kbit/s (V.92, a V.90 improvement)	PAM (downstream)

FSK, Frequency Shift Keying; *QPSK*, Quadrature Phase Shift Keying; *8-PSK*, 8-Phase Shift Keying; *QAM*, Quadrature Amplitude Modulation; *TCM*, Trellis Coded Modulation; *DMT*, Discrete Multitone Modulation; *PAM*, Pulse Amplitude Modulation

The available phone bandwidth of 4 kHz in POTS poses a significant limitation to the bit-rate of the digital signal, which has to be modulated in the above bandwidth. The evolution of modem technologies (and standards) is described for the classical 4 kHz phone bandwidth in Table 1.6.

The 4 kHz limitation for POTS modems does not depend on the twisted pair medium but on the presence of a filter[4] at the first local exchange, which "selects" the 4 kHz phone bandwidth. Without such a filter, the twisted pair could have a bandwidth of hundreds of megahertz (or even more with today's category 7 and category 8 cablings) that reduces with the distance. The attenuation of the twisted pair is a critical parameter, limiting the covered distance without repeaters. The frequency response of a twisted pair (without any filtering) is determined by the *skin effect*: as the transmission frequency increases, the electric current becomes more confined on the conductor surface, thus reducing the "equivalent section surface" of the conductor and increasing the ohm resistance.

Moreover, the transfer function of a transmission medium is not perfectly constant (in modulus) over all frequencies of the signal. This fact entails that a short impulse sent across the medium is received as enlarged over time (i.e., time dispersion). Consequently, Inter-Symbol Interference (ISI) practically limits the maximum bit-rate achievable by a transmission.

Different digital transmission techniques are available that make better use of the twisted pair capacity, as shown in Table 1.7. In particular, we may refer to the Asynchronous Digital Subscriber Line (ADSL) technique. With ADSL, no loading coils are used in the subscriber loop. The ITU-T G992.1 ADSL standard is based on Discrete MultiTone (DMT) transmissions (see also Sect. 6.4.4). With DMT, the available bandwidth in the twisted pair is divided among 256 carriers (i.e., sub-channels), with a carrier spacing of 4.3215 kHz, so that the total occupied bandwidth is 1.1 MHz. The first six carriers are not used in order to separate adequately the DMT signal of ADSL from the 0−4 kHz phone band. Hence, the ADSL spectrum starts at 26 kHz. Among the remaining 250 carriers, 218 are used for downstream transmissions to the user and 32 are employed for upstream transmissions from users. The frequency occupancy on the phone line is depicted in Fig. 1.35. Each carrier conveys an n-QAM signal, where the number "n" of adopted QAM symbols may vary from 4 to 1,024; the n value increases for the carriers at frequencies experiencing lower attenuation. The binary information to be sent is divided among the sub-channels. ADSL tests the signal-to-noise ratio of each sub-channel to determine its maximum speed; this process is known as DMT. ITU-T G.992.3 Recommendation, also referred to as ADSL2, extends the data rate capability of basic ADSL up to 12 Mbit/s downstream and up to 3.5 Mbit/s upstream. ADSL2 uses the same bandwidth as ADSL but achieves higher throughput using improved modulation techniques. Actual speeds mainly depend on the distance from the DSLAM to the user equipment (see also Sect. 1.7.4.2 for DSLAM definition as an ATM traffic concentrator). ITU-T G.992.5 Recommendation, also referred to as ADSL2+, reaches a maximum theoretical speed of 24 Mbit/s (download)/1.4 Mbit/s (upstream). Still, this value may reduce depending on the distance from the DSLAM (the maximum distance is 2 km). ADSL2+ uses a double downstream bandwidth (i.e., 2.2 MHz) compared to ADSL and ADSL2 (i.e., 1.1 MHz).

VDSL as ADSL uses the DMT approach. The main difference between ADSL and VDSL is the bandwidth used. ADSL and ADSL2 have available a 1.1 MHz band divided into 256 channels (see Fig. 1.35). ADSL2+ uses

[4] For long distance loops, the standard practice of telephone companies was to add loading coils, which extend the distance covered by a line by flattening the frequency response in the 2–3 kHz regions. However, these loading coils significantly attenuate the frequency response above these frequencies. These loading coils have not to be used for today's loops transporting digital signals.

Table 1.7: Non-PSTN-band (xDSL) modems for high bit-rate transmissions on twisted pairs

Technology	Description	Bit-rate	Mode	Applications	Distance
DSL	Digital Subscriber Line	160 kbit/s	Symmetric	ISDN services, voice and data	8–10 km
HDSL (2 pairs)	High data rate Digital Subscriber Line	2.048 Mbit/s	Symmetric	E1 services, WAN, access to LAN	5.5 km
SDSL	Single-line Digital Subscriber Line	2.048 Mbit/s	Symmetric	As HSDL	
ADSL	Asymmetric Digital Subscriber Line	Down: 1.5–9 Mbit/s Up: 16–640 kbit/s	Asymmetric	Access to the Internet. Multimedia and interactive traffic	1–5.5 km
ADSL2+	ADSL version 2+	Down (max): 24 Mbit/s Up (max): 1.4 Mbit/s	Asymmetric	Access to the Internet. Multimedia and interactive traffic	<2 km
VDSL	Very high data rate Digital Subscriber Line	Down (max): 55 Mbit/s Up (max): 3 Mbit/s	Asymmetric	As HSDL. High-definition TV, use with FTTC	200–900 m
VDSL2	VDSL version 2	Down (max): 200 Mbit/s Up (max): 100 Mbit/s	Symmetric and asymmetric	As VDSL	300 m

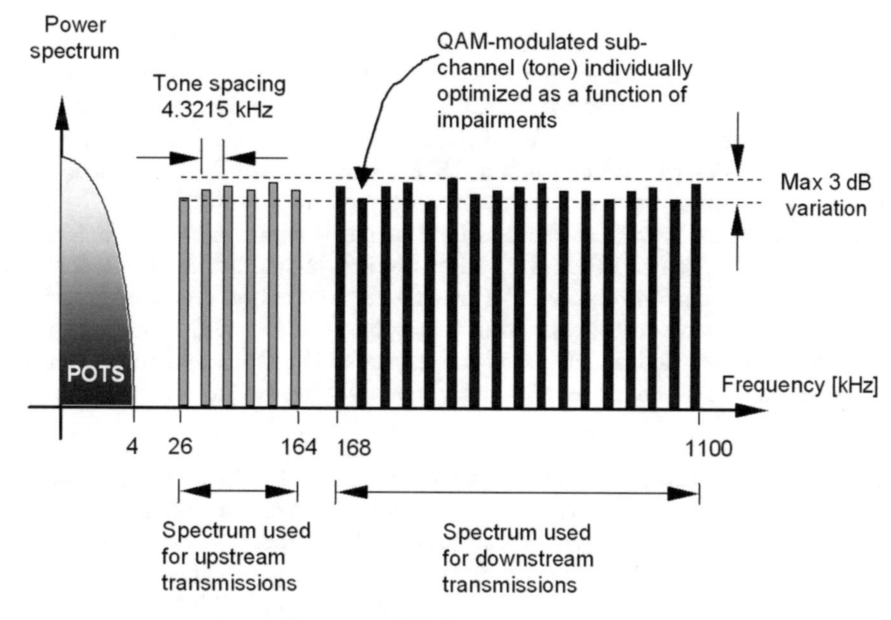

Fig. 1.35: ADSL transmissions

a 2.2 MHz band, divided into 512 channels. VDSL can use a band of either 8 MHz, 12 MHz, 17 MHz, or 30 MHz. The use of these wider bands allows higher bit-rates. ITU-T ratified VDSL2 with the G993.2 Recommendation in 2006. VDSL2 is still based on DMT, which makes VDSL2 spectrally compatible with existing ADSL and ADSL2+. Also, with these most recent technologies, the speed depends on the distance from the DSLAM. Typically, VDSL2 achieves 200 Mbit/s downstream and 100 Mbit/s upstream. These capacity values can further increase with the most recent evolution of the standard, known as VDSL2-Vplus (ITU G.993.5, 2019), reaching up to 300 Mbit/s downstream and 100 Mbit/s upstream.

A splitter filter is required at user premises to separate the voice signal bandwidth from that of the data signal. At the local exchange, the digital transmission is extracted and addressed towards a data network.

Today, the twisted pair of the user can connect to a cabinet with a DSLAM in a close zone (a few hundreds of meters) from which conversion to optical signal and fibers depart to connect to the Internet. This is the classical Fiber-To-The-Curb (FTTC) case, where optical fiber cables run from central office equipment to a communication switch (DSLAM) located within 300 m of a home or enterprise. The transmission on the copper part can be carried out using VDSL2 or the newer G.fast technology; G.fast has been standardized by ITU-T in 2014 (G.9701 Recommendation) and can reach a speed of up to 1 Gbit/s for a short range of 100–150 m. The evolution of this mixed type of access is the Fiber-To-The-Home (FTTH), where the optical fiber arrives directly at the user premises. In this case, the optical technology adopted is Ethernet Passive Optical Network (EPON) or Gigabit-capable Passive Optical Network (GPON).

1.7 Survey of Legacy Digital Networks

This section aims to survey the first digital networks, which were suitable for transporting data (or, in general, multimedia) traffic [27], and to present the main principles of networking and traffic management. In particular, we will consider data networks based on X.25 [28], ISDN, Frame Relay [29–31], and, finally, B-ISDN and ATM networks [32–34]. This section contains some basic concepts (as flow control, traffic shaping and policing, buffer management, etc.) that will be useful to understand the Internet protocols discussed in Chap. 2.

1.7.1 X.25-Based Networks

X.25 is an ITU-T Recommendation defined in 1976 and subsequently refined [27]. This specification defines the protocols for synchronous transmissions between a user terminal (here, named Data Terminal Equipment, DTE) and the first network equipment (here, named Data Circuit-terminating Equipment, DCE). The packet data network connecting all the DCEs is based on Packet-Switching Exchange (PSE) elements. No details are given on the protocols employed in the network interconnecting DCEs. However, the X.75 protocol by ITU-T (specifying the protocols for the communication between two packet-switched data networks) can be used in this network [35]. Even if X.25 defines the protocol stack at the user interface, it is common to use the term "X.25 network" to denote the whole network with DCEs and PSEs. X.25 addresses are defined in the ITU-T X.121 Recommendation. Typical applications of X.25 included automatic teller machine networks and credit card verification networks.

The subdivision of X.25 protocols into layers was at the basis of the OSI model. In particular, X.25 is a connection-oriented protocol, which defines the first three layers of the OSI architecture, that is, physical, data, and network layers, called in this standard as *physical*, *frame*, and *packet* layers, respectively. These layers are described below.

1. *Physical layer*: It is based on the X.21 protocol, which is similar to the serial transmissions of the RS-232 standard (ITU V.24). X.21 is an ITU recommendation for the operation of digital circuits [36]. The X.21 interface uses eight interchange circuits (i.e., signal ground, DTE common return, transmit, receive, control, indication, signal element timing, and byte timing); their functions are defined in Recommendation X.24 [37] and their electrical characteristics are described in Recommendation X.27 [38].
2. *Data link layer*: It employs the Link Access Protocol-Balanced (LAP-B), a subset of the High-Level Data Link Control (HDLC) protocol in its balanced version, meaning that both parts can start a new transmission without needing the authorization of the other part.
3. *Network layer*: The Packet Layer Procedure (PLP) is adopted. The transfer of information between two DTE devices attached to a packet-switched network depends on PLP. The PLP layer communicates between DTE devices using units, called packets.

Note that information and control messages share the same protocol layers in X.25; this is what is called "in-band signaling."

In the X.25 protocol stack, layer 2 provides error control. Moreover, both layers 2 and 3 implement two independently operated flow control techniques. Flow control is needed to avoid overwhelming the receiver with too much data. Error control is adopted to verify whether the data have been received correctly; in the presence of errors, a retransmission is requested. Error and flow controls entail a heavy overhead for X.25. LAPB is a bit-oriented

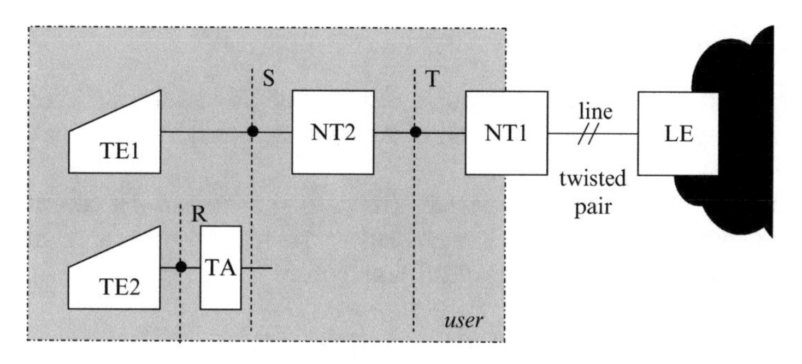

Fig. 1.36: User access architecture to the ISDN network (basic rate access case)

protocol, which ensures that frames are correctly ordered and error-free. LAPB adopts an ARQ scheme to recover the erroneous frames on each link (in the LAPB frame, there are two bytes—Frame Check Sequence field—used for error detection). Both Go-Back-N and Selective Repeat schemes can be adopted to manage retransmissions. A sliding window technique is integrated with the ARQ scheme to operate flow control, assuming a maximum window size of n frames: the sender can send up to n frames before stopping transmissions, waiting for an acknowledgment (which allows sliding the window).

The layer 3 protocol (PLP) supports a flow control task to ensure that a source DTE does not overwhelm the destination DTE and to maintain timely and efficient delivery of packets. Flow control is operated for each virtual circuit, differently from LAPB, which provides flow control independently of virtual circuits (it does not know what a virtual circuit is; LAPB just controls all the traffic on a link). The destination DTE has to send an acknowledgment for each packet received. PLP adopts a sliding window flow control mechanism like that used by LAPB [27]; the PLP max window size is either 8 or 128 packets. The PLP protocol is connection-oriented with two possible services: Switched Virtual Circuit (SVC) and Permanent Virtual Circuit (PVC). In the first case, the exchange of data between source and destination requires the setup of a path, which connects these network endpoints; a release procedure must be performed when the call ends. Each node stores the packets in a buffer before processing and transmitting them on the appropriate output link. This method is referred to as *store-and-forward.*

X.75 is a signaling system used to interconnect packet-switched network elements (such as X.25) on international circuits [35]. On layer 2, X.75 uses LAPB in the same way as X.25. On layer 3, X.75 is almost identical to X.25.

LAPB error control (with retransmissions) on each hop, and hop-by-hop flow control, entails a significant protocol overhead. Putting "intelligence into the network" made sense in the mid-1970s, when very simple terminals were available. Today, the adoption of a quasi-error-free transmission medium (like optical fibers) favors pushing "intelligence to the edges." This is the reason why the X.25 technology quickly disappeared.

1.7.2 ISDN

Numeric access from user premises is provided by Integrated Services Digital Network (ISDN) [29–31], thus allowing a unified system to support voice and different types of data transitions. A computer can thus be connected to the network with a baseband link, without using a modem. This technology was standardized by ITU-T in the 1980s. The ISDN network employs the twisted pair medium (as PSTN) for the access of users; moreover, ISDN substitutes the common channel Signaling System No. 7 (SS#7) with an enriched signaling set. ISDN supports both circuit-switching and packet-switching, an essential characteristic to manage different service types with related digital traffic flows.

The end-user is connected to the ISDN network (i.e., the Local Exchange, LE) using a twisted pair, which arrives at a Network Termination 1 (NT1). Moreover, the Terminal Equipment (TE) uses a Network Termination 2 (NT2) to connect to NT1. A non-ISDN terminal can also be connected using a Terminal Adaptor (TA); see Fig. 1.36. NT1 supports all the functions of a network termination. In particular, it operates at OSI layer 1 (termination of the transmission line, management of the clock, channel multiplexing on the line). NT2 has the functionalities of layers 1, 2, and 3; for instance, NT2 can be an ISDN Private Automatic Branch eXchange (PABX). NT2 functionalities cannot be divided between TE and NT1. TE has all the seven layers of the OSI protocol stack.

There are two different channel types in ISDN:

- Channel B at 64 kbit/s. It transparently transports the flux of bits from one end to another in the network according to circuit-switching. Hence, only the physical layer is needed for B-channels in the switches within the network.
- Channel D at 16 or 64 kbit/s. This channel is packet- (message-) switched. Hence, at each node of the network, all the first three OSI layers (i.e., 1, 2, and 3) are needed to manage the traffic coming from a D-channel. This channel is used to send both signaling messages and user packet data.

There are two basic types of ISDN access structures:

- Basic Rate Interface (BRI) [39], which consists of two 64 kbit/s B-channels and one 16 kbit/s D-channel for a total bit-rate of 144 kbit/s: 2B + D. This basic service is intended to meet the needs of most individual users.
- Primary Rate Interface (PRI) [40] for users requiring a higher capacity. This channel structure has 23 B-channels in the USA and 30 B-channels in Europe plus one 64 kbit/s D-channel (totally, 1,536 kbit/s in the USA and 1,984 kbit/s in Europe): 23B + D and 30B + D, respectively.

There are three different types of services [41]: *bearer services*, *teleservices*, and *supplementary services*. A bearer service has the task to transfer digital information between endpoints (S or T) across the network. Bearer services are described in Recommendations from I.230 to I.233. Bearer services entail protocols for OSI layers 1, 2, and 3. The network acts as a relay system operating at layers 1, 2, or 3. A teleservice entails an end-to-end communication accessed at S or T reference points. Teleservices involve OSI protocols from layer 1 to layer 7. Teleservices rely on bearer services for the transport of information from one end to another end of the network. Typical examples of teleservices are (ITU-T I.240 and I.241 Recommendations) telephony, videotelephony, and facsimile. Supplementary services are provided together with a bearer service or a teleservice to improve it. Many supplementary services are defined to support bearer services of the circuit type (ITU-T Recommendations from I.251 to I.257), such as calling number notification, group calls, etc.

1.7.2.1 ISDN Protocol Stack

ITU-T I.320 Recommendation defines the protocol stack for reference points S and T [42]. The OSI reference model was mainly related to X.25, where signaling was managed by the same protocol stack as the information traffic ("in-band" signaling). Hence, the X.25 approach is incompatible with circuit-switching, where once a circuit is established, information is transparently conveyed by the network that, in this case, acts as a relay system at level 1. To overcome these limitations of the X.25 approach, the ISDN protocol stack has been conceived with two parallel stacks: one for information traffic (also called User Plane) and the other for signaling traffic (also called Control Plane). At each layer, we have two protocols, one for the user plane and the other for the control plane. ISDN adopts an "out-of-band" signaling approach; see Fig. 1.37. In a circuit-switched connection (ITU-T I.320 Recommendation), we have both user and control planes at each node. However, the user plane stack related to channel B is reduced to only the physical layer (physical relay). In contrast, the control plane of channel D has a complete stack, where, practically, only layers 1, 2, and 3 are used (for instance, Q.931 is a layer 3 protocol for channel D, also including higher-layer functions).

1.7.2.2 Layer 1 Protocol

In the definition of the physical layer, there is no distinction between channels D and B. According to Recommendation I.431 [40], PRI uses the same layer 1 of the 2 Mbit/s E1 numeric transmission (ITU-T G.703 and ITU-T G.704 Recommendations, respectively, on the electric interface and frame structure [43, 44]). PRI is characterized by a point-to-point configuration (i.e., a single terminal directly connected to the network). Instead, the physical layer of BRI has required an ad hoc solution, detailed in the ITU-T I.430 Recommendation [39]. The most general BRI access structure is based on a *passive bus* where many TEs can be connected; this is the so-called *multi-point* access architecture.

In general, an NT1 can operate both in point-to-point configuration and in multi-point configuration. In a multi-point configuration, the maximum distance is 200 m (short bus) or 500 m (extended bus); instead, in a point-to-point configuration, the maximum distance is 1,000 m.

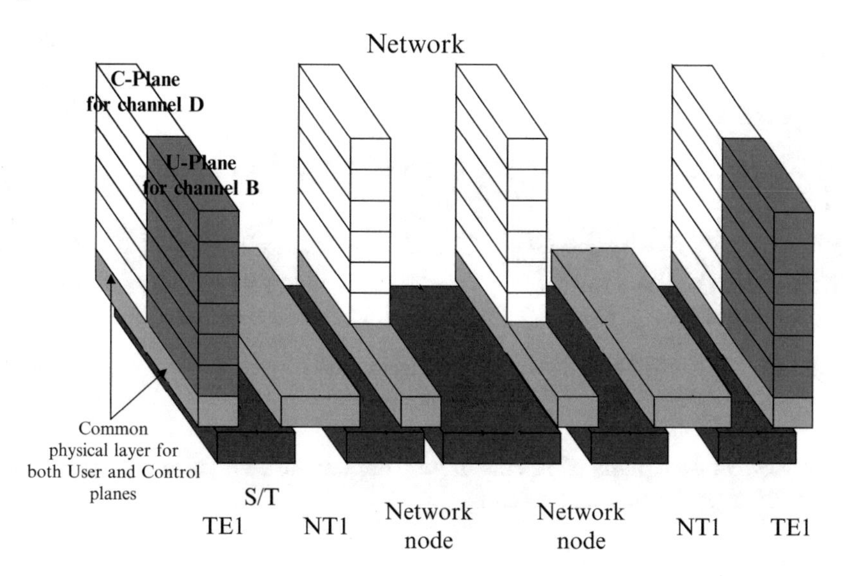

Fig. 1.37: Example of protocol stacks at different interfaces for a circuit-switched ISDN connection

1.7.2.3 Layer 2 Protocol

The ISDN protocols specified by the recommendations for layers 2 and 3 are valid only for D-channels. As for layer 2, ITU-T Q.920 and Q.921 Recommendations are considered [45, 46]. The layer 2 protocol is based on HDLC and its frame structure. In particular, the protocol is named Link Access Procedure on the D-channel (LAPD) and has the specific task of allowing the communication between peer layer 3 entities. A layer 3 entity is identified by a Service Access Point (SAP). There are different types of SAPs, each denoted by a suitable SAP Identifier (SAPI): SAPI = 0 is related to signaling (e.g., Q.931 signaling), SAPI = 16 is used for X.25 packet data traffic, SAPIs from 32 to 62 denote Frame Relay data, SAPIs different from 16 and 32–62 are used for call control messages, and finally, SAPI = 63 is adopted for management messages. To distinguish different TEs in a multi-point connection, a suitable Terminal Endpoint Identifier (TEI) is defined. In the LAPD header, each layer 2 connection is therefore identified by SAPI + TEI, which together form the Data Link Connection Identifier (DLCI), the address field of a LAPD frame. The SAPI field has 6 bits (numbers from 0 to 63) and TEI has 7 bits (numbers from 0 to 127). TEI 127 is used for group broadcast: a frame transmitted by the network with TEI = 127 is received by all the terminals, which are connected to the related network termination.

1.7.2.4 Layer 3 Protocol

Layer 3 is specified in ITU-T Q.930, Q.931, and Q.932 Recommendations for signaling traffic on channel D [47]. These protocols have to manage the exchange of end-to-end signaling for channel B. When a call arrives at user premises using multi-point connections, all the terminals (e.g., different ISDN phones) must be alerted. As soon as the first terminal is activated, the other terminals are released. In the case of data packet traffic on channel D, the X.25 layer 3 protocol (i.e., PLP) is used.

1.7.3 Frame Relay Networks

This network technology is based on a layer 2 protocol, named Frame Relay, which can be considered as a variant of the LAPD protocol used in ISDN. Frame Relay was one of the "fast packet-switching" technologies introduced in the early 90s. Frame Relay is a sort of modification and evolution of ISDN. Frame Relay entails lower overhead and achieves higher performance than previous protocols. Digital networks employing Frame Relay at layer 2 are called Frame Relay networks. The ITU-T Recommendations are published as I.233 [48], Q.922 Annex A [49], and Q.933 [50].

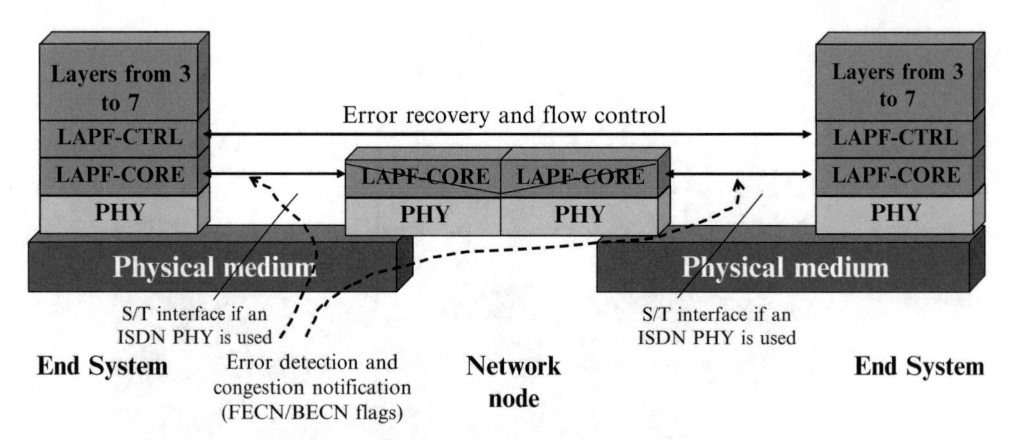

Fig. 1.38: Frame Relay service: user plane protocols in internal network nodes and at end systems. Note that error recovery and flow control are performed end-to-end

In X.25 networks, error control was performed hop by hop because these networks were based on unreliable physical medium with considerable bit error rates from 10^{-5} to 10^{-3}. With the adoption of optical fibers, the error rates are drastically reduced (bit error rates from 10^{-9} to 10^{-6}), thus making it useless to perform error recovery on every link. This is the reason why Frame Relay performs end-to-end error recovery (no hop-by-hop error recovery). Such simplification allows improving the data throughput performance of the network.

Frame Relay is used in point-to-network, point-to-point, star, full-mesh, and partial mesh topologies. The access to a Frame Relay network is allowed both to terminals (hosts) and to network equipment, provided that they support the Frame Relay protocol stack. In this case, a Frame Relay Access Device (FRAD) is interposed between the host and the network, thus having the new interface named FR-UNI.

Frame Relay is a connection-oriented protocol with virtual circuits: an end-to-end connection must be established before data can be transferred. Switching is performed at layer 2, differently from X.25 networks, where switching is performed at layer 3. The protocol stack employs a user plane (data, information flow) and a control plane (signaling). Hence, signaling is out of band as in ISDN and differently from X.25. The Frame Relay protocol stack is described below:

- *Physical layer*: It is common for user and control planes. It is based on ISDN physical resources (one B-channel, one ISDN BRI access according to I.430 [39], one ISDN PRI according to I.431 [40], etc.).
- *Layer 2*: User and control planes typically adopt different protocols both related to ITU-T Q.922 Recommendation. In particular, the control plane employs the LAPF protocol defined in Q.922. In contrast, the user plane adopts LAPF at end nodes and a subset of LAPF, named LAPF-core (i.e., the lower part of the full LAPF protocol, which is defined in Annex A of Recommendation Q.922 [49]) at intermediate nodes. The typical functions of LAPF-core are framing, multiplexing/demultiplexing of virtual circuits, error detection, address, and management of congestion events. Note that the upper part of the LAPF protocol, named LAPF-control, is used to operate end-to-end error recovery (ARQ protocol) and flow control. At intermediate nodes in the network, the user plane only terminates the LAPF-core; this is the classical *frame relay service*, as shown in Fig. 1.38. However, it is also possible that the network adopts both LAPF-core and LAPF-control (i.e., a full LAPF protocol) in the user plane, as in the control plane; in this case, the network provides a *frame switching service.*
- *Layer 3*: On the control plane, the Q.933 protocol [50] is adopted, derived from the Q.931 protocol of ISDN networks. This protocol is responsible for the management of virtual circuits. On the user plane, only end systems have a full layer 3 protocol.

User and control planes convey data organized in layer 2 messages called *frames.* They are "routed" through virtual circuits using the address field, named Data Link Connection Identifier (DLCI). The DLCI field has only a local meaning; it can be changed at each node according to the path defined during the setup phase. The frames on the control plane have the same format as the LAPF frames on the user plane. Some fields of the frame header are described below:

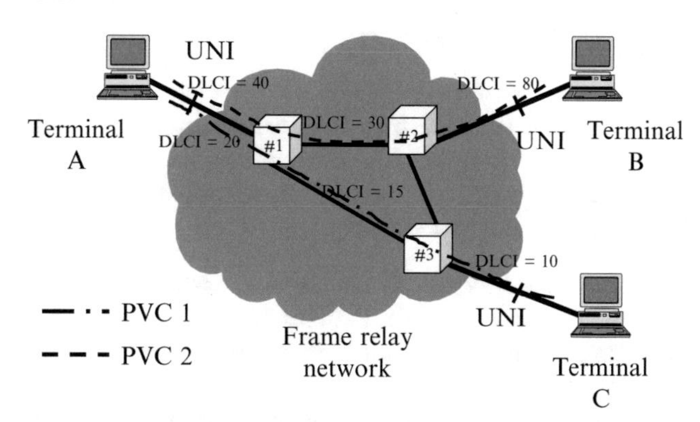

Fig. 1.39: Characterization of virtual channels and use of DLCI in Frame Relay networks

- DLCI of different lengths, depending on the three different formats (10, 16, and 23 bits, respectively).The DLCI field with all bits equal to 0 (i.e., DLCI = 0) is reserved for a channel conveying signaling for all the virtual connections on the same link. The DLCI field with all bits equal to 1 (e.g., DLCI = 1,023 in the 10-bit DLCI case) is used for a channel transporting management information for the link.
- Forward Explicit Congestion Notification (FECN) bit: If it is set to 1 by an internal node of the Frame Relay network, it denotes a congestion situation on the related link on the path towards the destination of the frame.
- Backward Explicit Congestion Notification (BECN) bit: If it is set to 1 by an internal node of the Frame Relay network, it denotes a congestion situation on the link where the frame is sent, but in the opposite direction.
- Discard Eligibility (DE) bit: If it is set to 1 by an access node of the Frame Relay network, it authorizes to discard the related frame with priority (with respect to those with DE = 0) in internal nodes when they are congested. The setting of the DE bit requires a traffic policing function implemented at the entrance nodes of the Frame Relay network. The discard of packets marked with DE = 1 requires a buffer management function at intermediate nodes.

Frames are produced by a source with FECN = 0, BECN = 0, DE = 0. The DE bit can be modified at the first (access) node of the Frame Relay network. FECN and BECN bits can be modified at any internal node of the Frame Relay network.

The frame payload has a variable length with a maximum value of 4,096 bytes. However, the Frame Relay forum developed an implementation agreement setting the maximum payload size at 1,600 bytes for interoperability reasons (this frame size can easily support the largest Ethernet frame for LANs).

Finally, every frame has a 2-byte Frame Check Sequence (FCS) trailer field, which is used to detect errors in the frame received (cyclic redundancy check).

End-users are interconnected using virtual circuits, which can be either PVC or SVC, as already used for X.25. A PVC is a permanent connection between two endpoints that is set up by the operator. This connection always exists, meaning that there is a circuit used for this PVC at each node along the path in the network. Instead, an SVC is a temporary connection between two endpoints, which is set up upon request of one of the parties. This connection can be released when it is not needed, similarly to a phone call. Referring to the Frame Relay network example in Fig. 1.39, we can note that one path (i.e., one end-to-end virtual channel) is characterized by the DLCI values of the links crossed at the different nodes. For instance, PVC 2 connecting terminal A to terminal B is characterized by the following associations at each node along the path:

$$(\text{Terminal A, DLCI} = 40) \cup (\text{Node\#1, DLCI} = 30) \cup (\text{Node\#2, DLCI} = 80).$$

PCVs can be used when there is stable traffic between endpoints (e.g., interconnections of different locations belonging to the same organization); otherwise, SVC connections are more efficient since they are set up on demand, thus allowing better multiplexing of resources among competing traffic flows. SVCs are typically used for public access. The Q.933 layer 3 protocol of the control plane is in charge of supporting the setup of a virtual path, its maintenance/control, and its release when the call ends.

1.7.3.1 Traffic Regulation (Policing)

We are considering here the case where a variable bit-rate traffic source has an access line to the Frame Relay network with a capacity denoted by Access bit-Rate (AR), which is typically much higher than the maximum traffic load generated by the source. During the connection establishment phase, the following flow control parameters are defined to monitor and regulate the input traffic flow:

- *Measurement interval*, T_c, i.e., the time interval on which we measure the source traffic to determine whether it is conformant to specifications. T_c is the time periodicity according to which the input traffic is controlled.
- *Committed burst size*, B_c, denoting the maximum number of bits that the network can accept and convey in a time T_c from a given source.
- *Excess burst size*, B_e, representing the maximum number of excess bits in T_c (with respect to the B_c value) that the network will try to convey to destination without any special guarantee.

Based on the above parameters, the capacity that the Frame Relay network assures to a terminal traffic flow is denoted as Committed Information Rate (CIR) and can be expressed as

$$\mathrm{CIR} = \frac{B_c}{T_c} \quad \left[\frac{\mathrm{bit}}{\mathrm{s}}\right] \tag{1.7}$$

The extra capacity that the network can provide, denoted as Excess Information Rate (EIR), is expressed as

$$\mathrm{EIR} = \frac{B_e}{T_c} \quad \left[\frac{\mathrm{bit}}{\mathrm{s}}\right] \tag{1.8}$$

The frames sent in a T_c interval and requiring the extra capacity (of the B_e bits in T_c) are *marked* with $\mathrm{DE}=1$, so that they can be discarded at an intermediate node if it experiences buffer congestion.

Of course, the access capacity AR must fulfill the condition below:

$$\mathrm{CIR} + \mathrm{EIR} \leq \mathrm{AR} \left[\frac{\mathrm{bit}}{\mathrm{s}}\right] \tag{1.9}$$

Higher values of T_c are preferable for users since they allow sending bursts of data. From the network standpoint, lower T_c values are preferred since they permit both a better control on the traffic injected into the network and a better statistical multiplexing of traffic flows.

To summarize, the frames generated by a source are monitored on a T_c time interval basis according to a traffic policing function exerted at the access to the network. As long as the number of bits generated in T_c is lower than or equal to B_c, frames are accepted in the network with $\mathrm{DE}=0$; if the bits generated in T_c exceed B_c but are lower than or equal to $B_c + B_e$, frames are accepted in the network with $\mathrm{DE}=1$; and if the bits generated in T_c exceed $B_c + B_e$, frames are *discarded*. Then, the measurement process of the bits generated by the source restarts in the next T_c interval, and so on. This situation is depicted in Fig. 1.40, where B_t denotes the maximum number of bits that the access line can convey in T_c (i.e., $B_t = \mathrm{AR} \times T_c$).

1.7.3.2 Congestion Control

In the Frame Relay network, flow control is end-to-end operated to limit the traffic load injected into the network. The traffic generated by a source is controlled at the entrance of the network according to the previously described traffic regulator. Congestion control is a crucial part in telecommunication networks since the occurrence of congestion leads to buffer overflows and the consequent loss of frames (an end-to-end ARQ scheme is needed), unpredictable delays, and the reduction of network throughput. Congestion control is end-to-end operated. The network is in charge of monitoring congestion at transit nodes and reporting it to the end terminals, which have the responsibility to react accordingly. Mainly, two techniques are available to manage buffer congestion [51]:

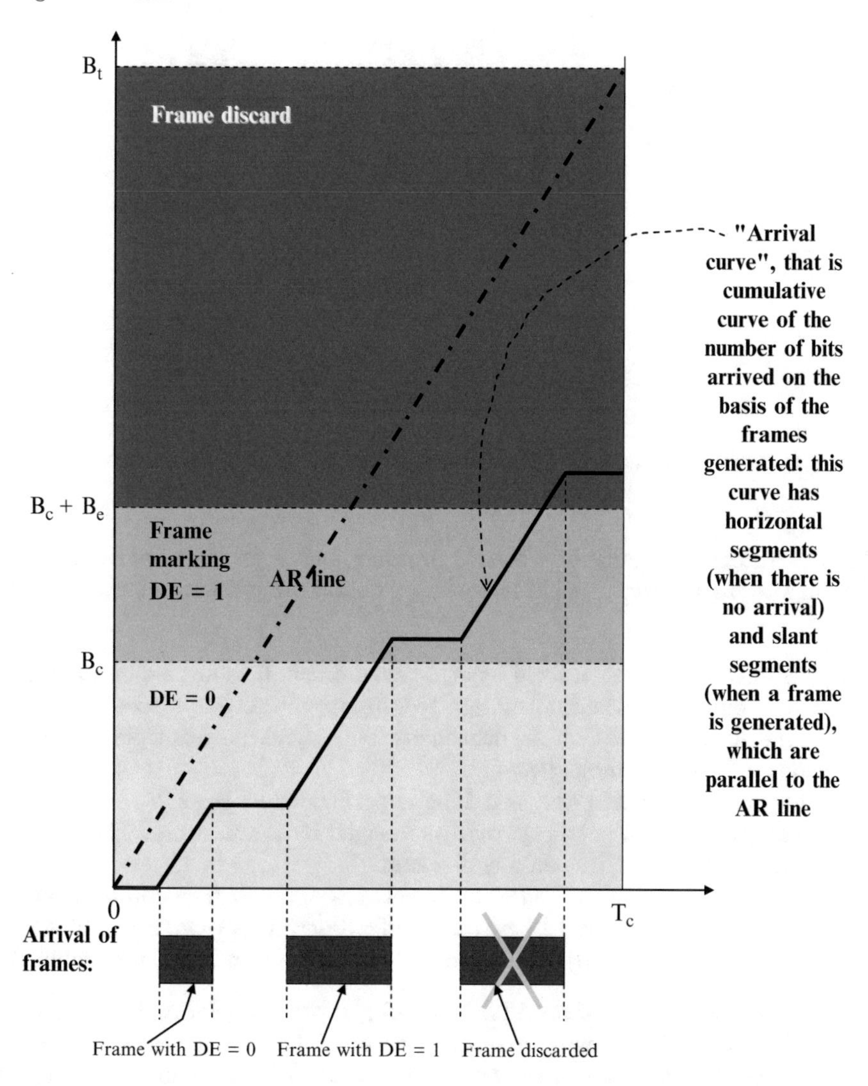

Fig. 1.40: Management of source traffic entering a Frame Relay network

- Each node controls the occupancy of its buffers; when a threshold value is exceeded for the buffer of a given link, a procedure is started to notify congestion to all virtual channels using this link. Hence, FECN is set to 1 for all the frames sent by this node through the bottleneck link; moreover, BECN is set to 1 for all the frames received by this node through the bottleneck link. Let us refer to Fig. 1.41, referring to the virtual circuit from terminal A to terminal B. Let us assume that node #4 reveals congestion on the link towards node #2. Hence, FECN is set to 1 at node #4 for all the frames that are sent from node A to node B; moreover, BECN is set to 1 at node #4 for all the frames that from node B are sent back to node A. BECN notifies the sender that there is congestion in the network and that a bit-rate reduction is needed. FECN can be used by the destination device in the case that its upper layer protocols can control the traffic injected by the source through an end-to-end procedure. This is the typical case of the TCP protocol, as described in Sect. 2.8.1.
- If a link is congested (i.e., the related transmission buffer is full), the related node can discard frames starting from those having DE = 1 for which the network does not guarantee correct delivery.

1.7.4 B-ISDN and ATM Technology

The broadband evolution of ISDN (i.e., Broadband ISDN, B-ISDN) was defined in the ITU-T Recommendation I.150 [52]. Asynchronous Transfer Mode (ATM) denotes a technology for the transmission of multimedia traffic on

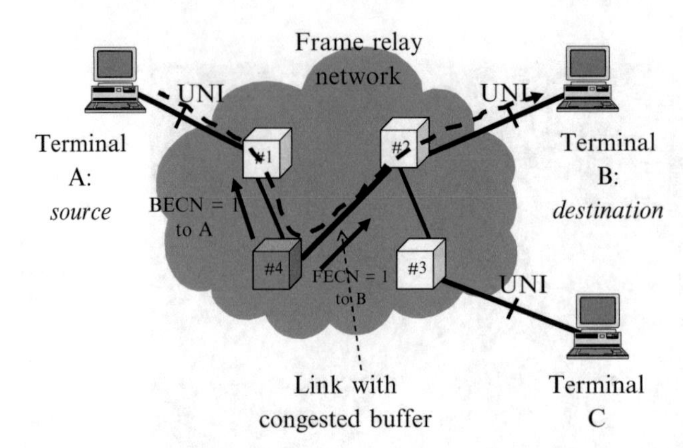

Fig. 1.41: Use of BECN and FECN in the presence of congestion on a bottleneck link

B-ISDN [32–34]. ATM represents the name of a layer 2 protocol, but it provides such a strong characterization of the network that we can also use the term "ATM network." The following list summarizes the main characteristics of an ATM network:

- The basic transmission unit is a packet of fixed length, called a *cell*. It is formed of a payload of 48 bytes and a header of 5 bytes, which contains all the information to support the ATM protocol.
- The transmission on the links is based on asynchronous Time Division Multiplexing, an innovative solution with respect to previous network technologies.[5]
- An ATM network is connection-oriented and switching is performed at layer 2.
- The payload of an ATM packet (cell) is transparently managed by the network. There is no error control[6] and no flow control at intermediate nodes, but only end-to-end.
- Multimedia traffic classes can be managed by the ATM network. They correspond to different applications (i.e., services). Each traffic class is characterized by certain traffic descriptors (e.g., mean bit-rate behavior) and has guaranteed some Quality of Service (QoS) requirements (e.g., maximum delay, maximum delay jitter, etc.).

Due to the connection-oriented nature of an ATM network, before a sender and a receiver can exchange data, an end-to-end path must be established using a setup procedure. During this setup phase, not only a path is established, but it is also verified that resources on the involved links are enough to support the new traffic, guaranteeing for it (and for the already-active connections) the contractual QoS levels. This is a Connection Admission Control (CAC) procedure. If this verification is successful, the new connection is activated; otherwise, it is refused. ATM networks manage both switched virtual paths (formed upon request) and semipermanent virtual paths (i.e., paths configured by the operator and that are active for a long time to provide a fixed end-to-end connectivity). The end-to-end established path is not physically switched but is logically formed and identified by some "labels," denoting the links between the different network elements. This is the reason why paths are "virtual" in ATM networks. This approach is similar to the management of paths with the use of DLCIs with Frame Relay.

An ATM network is typically composed of two different network elements:

- Multiplexers/demultiplexers and
- Switches.

Let us refer to the typical ATM network architecture. A multiplexer receives the packet data traffic from different input TDM lines and queues data to be sent on a single TDM output link according to the asynchronous TDMA scheme (i.e., no rigid assignment of output TDM resources—slots—to input lines). A multiplexer typically allows passing from low utilization input lines to high utilization output lines, i.e., a traffic concentrator, exploiting

[5] In Asynchronous Time Division Multiple Access (A-TDMA), we have different packet data traffic sources sharing the slots of the TDMA frame without a fixed, predetermined allocation (this would be the case of Synchronous-TDMA, S-TDMA). A traffic source can have assigned different slots and a different number of slots from frame to frame to adapt to varying traffic load conditions. A-TDMA improves the utilization of the transmission line resources and entails lower delays than S-TDMA by exploiting the multiplexing effect.

[6] Typically, the transmission medium used (i.e., optical fiber) is quite reliable. Hence, bit error rates are on the order of 10^{-10} (and lower). In these circumstances, it is not efficient to check the correctness of the cell payload at each hop, but only end-to-end.

the statistical multiplexing of (bursty) traffic sources. A demultiplexer performs the opposite operation. We may expect that multiplexers and demultiplexers are close to the end systems just to concentrate or to split the traffic.

A switch connects TDM input lines to TDM output lines. The switch processor must analyze each packet of each input line. A more detailed description of the switches and their internal architectures is provided in Sect. 1.7.4.3.

The cell header contains the description of the virtual circuit, characterized by two fields: Virtual Path Identifier (VPI) and Virtual Channel Identifier (VCI). During the virtual path setup phase (or during the circuit configuration process in the case of permanent paths), each switch is suitably instructed so that it can forward an incoming cell having a certain VPI + VCI to an output link corresponding to a new VPI + VCI couple, which is updated in the cell header. An end-to-end virtual circuit is formed of a VPI and a VCI on each link: the virtual circuit in the cell header is updated at each hop. The resources of a link are shared among some virtual paths (VPIs); moreover, a path "multiplexes" several virtual channels (VCIs).

The physical links used by ATM are typically based on optical fibers. More details on the ATM physical layer and physical medium are provided in Sect. 1.7.5.

The protocol stack of ATM is three-dimensional as follows:

- *User plane*, for the end-to-end transfer of information traffic.
- *Control plane*, supporting signaling traffic for virtual path setup, for CAC of a new connection, for the maintenance of a connection, and, finally, for the release of a connection.
- *Management plane*, for operation and maintenance functions and for the coordination of the different planes.

Both user and control planes are characterized by two (stacked) layer 2 protocols (i.e., ATM Adaptation Layer, AAL, and ATM layer) and the physical layer. End systems have a complete protocol stack from the physical layer to layer 7. Instead, intermediate nodes (i.e., multiplexers and switches) have only the lower layers (i.e., ATM and PHY).

1.7.4.1 Cell Format

In previous data networks (i.e., X.25 and Frame Relay), the switched unit was a packet (or frame) of a variable length. In contrast, in the ATM case, a fixed-length packet, called "cell," has been defined as a result of a complex standardization process that took different aspects into account, such as efficient utilization of transmission resources and delay to cross a node.

The ATM cell is formed of a 5-byte header and a 48-byte payload. The header reduces the transmission efficiency, since header bits do not carry information but are necessary for the management of the ATM protocol. In the ATM case, the percentage of wasted resources due to the header is about equal to 9.43%.

The structure of an ATM cell is represented in Fig. 1.42 by distinguishing the format at the User-to-Network Interface (UNI) and that at the Network-to-Network Interface (NNI). In the first case, we have the interface for the user access to the network; in the second case, we refer to the interface between two internal network elements. The cell structure definition is contained in ITU-T I.361 Recommendation.

Let us describe the different fields of an ATM cell, referring to Fig. 1.42 (starting from the top):

- GFC (Generic Flow Control) is present in the UNI case, but not present in the NNI one. GFC is used to support a flow control scheme for the input traffic of the user towards the network (not in the opposite direction).
- VPI is a field of 8 bits for the UNI cell or of 12 bits for the NNI cell. It identifies a virtual path between two nodes.
- VCI is a field of 16 bits (both UNI and NNI cell format), which is used to identify the virtual channels of a given virtual path.
- Payload Type Identifier (PTI) is a field of 3 bits used to describe the cell type and to transport some control information. PTI permits to describe the content of the cell payload, among the following three cases: information data, Operation, Administration, and Maintenance (OAM), and Resource Management (RM) signaling. The most significant bit discriminates between information data (bit equal to 0) and all the other cases (bit equal to 1). Moreover, in the case of a data cell, the second bit set to 1 in the PTI field is used to notify that the cell crossed a switch with congestion along the path towards destination. This is the Explicit Forward Congestion Indication (EFCI); see also Sect. 1.7.5.1. Finally, the last bit of the PTI field in the case of information data is the AUU bit (ATM-User-to-ATM-User), which is used by the AAL5 protocol to denote the last cell (AUU = 1) of a cell train deriving from the segmentation of the same higher-layer packet. When the destination receives

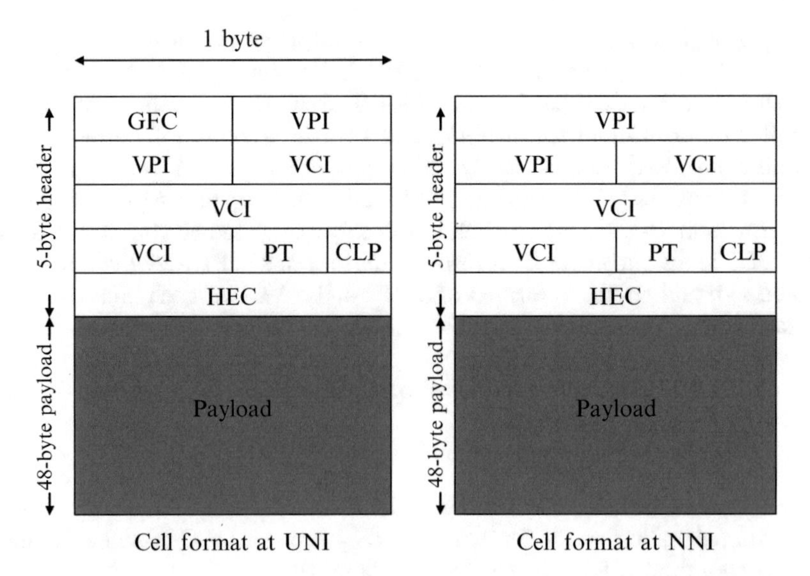

Fig. 1.42: Cell format (each row corresponds to one byte) for both UNI and NNI interfaces

an ATM cell with EFCI = 1, it marks the congestion indication in RM cells (having PTI = 110) sent in the opposite direction to notify the source. This mechanism is exploited only by the Available Bit Rate (ABR) traffic class (see next sub-section) to inform the source to reduce the traffic injection according to a reactive control scheme.

- Cell Loss Priority (CLP) bit to denote whether the cell has low (CLP = 1) or high (CLP = 0) priority. Different priority levels can be assigned to cells, so that only low-priority cells can be dropped in case of congestion in the queues of ATM nodes. Hence, even in the presence of congestion, high-priority cells are delivered to the destination with high probability. The CLP bit can be set either by the sender to differentiate the priority among cells or by the access node if the connection violates its traffic contract with the network.
- Header Error Control (HEC) is a field of one byte for the parity check of the cell header at each hop. This code allows revealing errors and correcting single errors in the header. Due to the high reliability of the transmission medium (typically, optical fiber), it is not convenient to check the integrity of the entire cell (this task will be performed only end-to-end). Instead, only the header is verified: if the cell header is correct (or with a single error that can be corrected), the cell is further forwarded (the network is sure to forward the cell on the intended path); otherwise, the cell is discarded (higher-layer protocols at the end system will be in charge of recovering this loss). VPI and VCI are updated at each node according to the virtual circuit-switching approach. Hence, even the parity check (HEC) field has to be recomputed at each hop.

As for the payload, different formats are possible depending on the AAL protocol.

1.7.4.2 ATM Protocol Stack

The ATM protocol stack (lower layers) is detailed in Fig. 1.43. In particular, we have the following:

- The physical layer is divided into two sublayers: Physical Medium (PM) and Transmission Convergence (TC). PM is in charge of physical layer-related functions such as the electro-optic conversion of bits and bit timing. TC, among other tasks, generates the HEC field of the cell.
- The ATM layer performs the following tasks: it operates flow control at UNI using GFC; it generates the first 4 bytes of the ATM cell header and adds them to the payload (transmission phase at the traffic source) or removes them from the cell (reception phase at the traffic destination); it translates the VPI & VCI fields from input to output of a switch; it performs the multiplexing (and demultiplexing) of the cells of different VPIs and VCIs on the same shared physical resources.
- The AAL layer has the following tasks: end-to-end transfer of messages of various lengths with cells of fixed length, management of erroneous cells and lost cells, flow control and congestion control, timing of the transported flow, and multiplexing of different traffic flows on the same ATM connection. AAL is only end-to-end

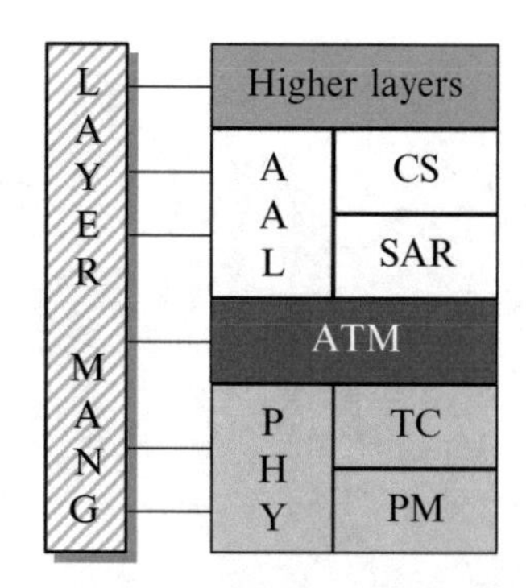

Fig. 1.43: ATM protocol stack

operated, that is, by the end nodes and not at the intermediate ones. The AAL layer is subdivided into two different sublayers: Segmentation And Reassembly (SAR) and Convergence Sublayer (CS). In transmission, SAR divides the PDUs received from the CS sublayer into smaller units (SAR-SDUs) that, with some added control, form the SAR-PDUs fitting with the cell payload length (segmentation); in reception, SAR reobtains the PDU for the CS sublayer.

The traffic classes are differentiated based on time-criticality, bit-rate behavior, and type of connection. AAL1 typical application is the support of services with circuit emulation (the network provides a dedicated end-to-end circuit). AAL1 is used for Constant Bit-Rate (CBR) real-time traffic for audio, video, and, in general, isochronous applications. AAL1 does not allow the multiplexing of different ATM connections. The AAL2 protocol is adopted for real-time Variable Bit-Rate (rt-VBR) connection-oriented traffic. AAL2 can be used for voice and video packet services. AAL2 allows the multiplexing of different AAL2 flows on the same ATM connection with given VPI and VCI fields through suitable flow identifiers. In the AAL2 case, there is not the SAR sublayer, but the CS one is more complex. AAL3 and AAL4 have practically the same characteristics. They can be used for non-real-time Variable Bit-Rate (nrt-VBR) traffic for connection-oriented (e.g., Frame Relay) or connectionless services. The AAL3/AAL4 protocol allows the multiplexing of different flows on the same connection using suitable flow identifiers. Finally, AAL5 is the most straightforward and efficient adaptation protocol. It is either connectionless or connection-oriented. It is well suited to local area network emulation (Available Bit-Rate, ABR, class), nrt-VBR, and IP-ATM interworking (ABR and Unspecified Bit-Rate, UBR, classes). AAL5 adopts a cumulative overhead at the CS PDU level: an 8-byte trailer. The CS PDU payload has a variable length of up to 65,535 bytes. Finally, a CRC field is used for revealing errors in the entire CS PDU. AAL5 does not support the multiplexing of different AAL flows on the same ATM connection.

More details on CBR, rt-VBR, nrt-VBR, ABR, and UBR services are provided in Sect. 1.7.5.1.

The ADSL/ADSL2+/VDSL/VDSL2 (xDSL in general) transmissions described in Sect. 1.6.1 could use the ATM at layer 2 to allow a high-speed access to the Internet. The user from his/her home adopts an appropriate modem to transmit the xDSL signal on the twisted pair to the local office; here, a DSL Access Multiplexer (DSLAM) terminates the xDSL session and sends the traffic in the ATM network using a permanent virtual circuit till the first router to access the Internet (point-of-presence).

1.7.4.3 ATM Switches

The switch is the crucial element of the ATM network architecture [53, 54]. It operates at layer 2 (ATM layer). It realizes the virtual circuit-switching by receiving a cell on an input port with a given VPI + VCI and by switching it (according to routing instructions defined in the path setup phase) to an output port with, in general, a new couple of values for VPI and VCI. However, there can be cases with only changes of VPI (see below) or even cases in which the pair VPI + VCI does not change. Two typical ATM switch architectures are detailed in Figs. 1.44 and 1.45. In the first case, we have an ATM cross-connect switch where a cell only changes its VPI from input to output. Instead, in the second case (the most common situation for ATM switches), a cell can have both VPI and VCI modified from input to output. The ATM cross-connect switch can be considered as a first, simplified

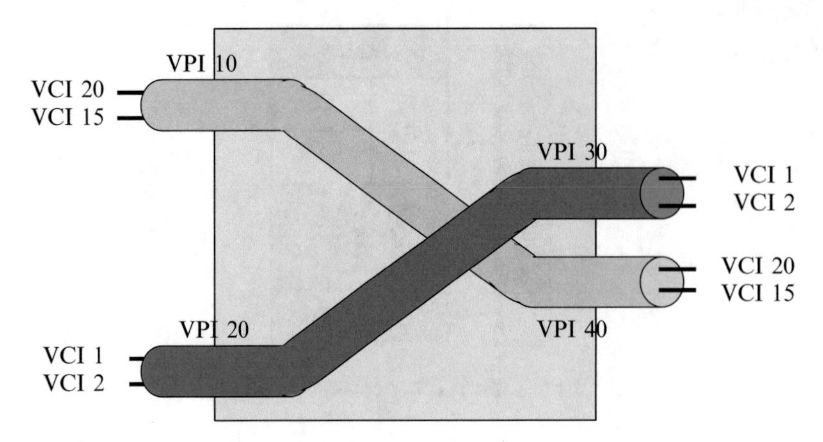

Fig. 1.44: ATM cross-connect switch

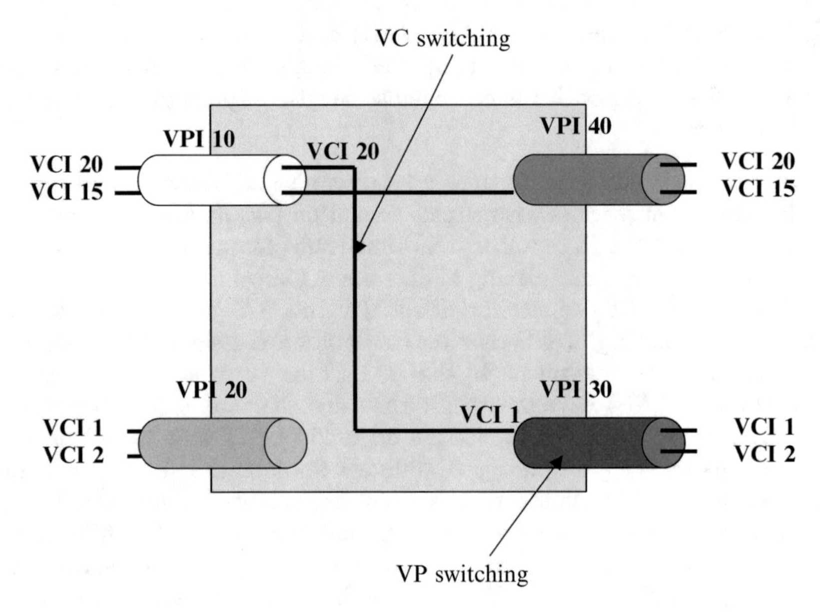

Fig. 1.45: ATM switch. We have highlighted that the input VCI = 20, VPI = 10 is switched to the output VCI = 1, VPI = 30

implementation of an ATM switch and can manage at most 4,096 (= 2^{12}, as the VPI field contains 12 bits) input virtual circuits.

ATM virtual packet-switching is possible because during the setup phase, each switch updates its switching (routing) table with the association between the input port and the input VPI + VCI of a traffic flow and the corresponding output port with the output VPI + VCI. Since the output cell has a new pair VPI + VCI with respect to the input one, the HEC field needs to be recomputed; see Fig. 1.46.

An ATM switch is composed of the following parts: (1) an input port block to interface input lines, (2) the switching fabric, and (3) the output port block to interface output lines. When a cell is received from a line, its header (actually its VPI and VCI) is processed by the input port to determine to which output port is destined based on the routing table. Within the node, it may occur that different cells simultaneously need to be addressed to the same output port, thus contending for output resources (i.e., a conflict in the use of output resources). In these circumstances, only one cell must be selected at a time, while the other cells need to be buffered; this is the classical Head-Of-Line (HOL) problem. Buffers can be placed in either input ports (see Fig. 1.46) or output ones. A third solution adopts buffers at the switching fabric level (central buffering). The last approach uses buffers for both input and output.

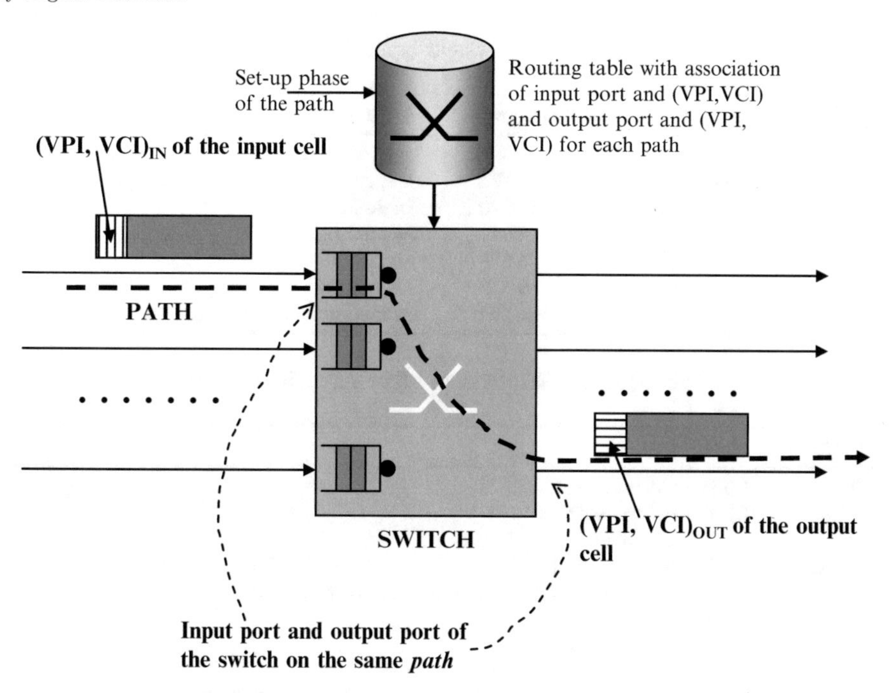

Fig. 1.46: Switching procedures at a node (switch); input and output ports are associated with (VPI, VCI) couples

1.7.5 ATM Physical Layer

Two different modalities are available for the transmission of cells on the physical medium, according to ITU-T I.432 Recommendation [55]. Referring to the User-to-Network Interface (UNI; either public or private), we have the following:

- *Sequence of cells*: The transmission of cells is carried out directly on the physical medium without using a specific frame structure. A continuous stream of cells is sent. Periodical insertion of OAM cells is needed. This solution may be adopted for private UNI.
- *SDH or SONET*: ATM traffic streams are multiplexed in complex transmission structures, where a pointer identifies each stream. More details on these transmission structures are provided below.

ITU-T Recommendations I.432.1, I.432.2, I.432.3, I.432.4, and I.432.5 specify the physical layer characteristics at ATM UNI interfaces, considering the following bit-rates [55]: 155.52 and 622.08 Mbit/s (I.432.2), 1.544 and 2.048 Mbit/s (I.432.3), 51.84 (I.432.4), and 25.6 Mbit/s (I.432.5).

Different physical layers can be used for ATM networks. Correspondingly, various media are available, such as single-mode or multimode optical fiber, shielded or unshielded twisted-pair (STP, UTP), and coaxial cable. Details on the physical layers and media are provided in Tables 1.8 and 1.9 for public and private UNI, respectively. For instance, the access capacity of 155.52 Mbit/s can be achieved by both the cell sequence approach (private UNI) and the SDH/SONET one (public and private UNI). The transmission bit-rates shown in Tables 1.8 and 1.9 are related to maximum distances, depending on the adopted medium.

The SONET standard was published by the American National Standards Institute (ANSI) [56]. The Synchronous Digital Hierarchy (SDH) standard [57] was defined in ITU-T G.707, G.708, and G.709 Recommendations [58–60]. These are standards for the aggregation of data traffic on backbones. SONET and SDH *transport technologies* are not directly related to ATM but can be used to transport ATM cells. SDH is used in Europe, and SONET is used in the USA and Japan. There are slight differences in the frame format between SONET and SDH. Synchronous Transfer Signal (STS) denotes the electrical specifications of the various levels of the SONET hierarchy. Synchronous Transfer Mode (STM) is the analogous term for the SDH hierarchy. In SDH/SONET, data transmission is organized in frames[7] of 125 µs. The base signal for SONET is STS-1, and the base signal for SDH is STM-1.

[7] PDH, SDH, and SONET are all based on 64 kbit/s digital voice channels of the PCM type.

Table 1.8: ATM physical layer for public UNI

Frame format	Bit-rate (Mbit/s)	Media
DS1	1.544	Twisted pair
DS3	44.736	Coaxial pair
STS-3c, STM-1	155.520	Single-mode fiber
E1	2.048	Twisted pair
E3	34.368	Coaxial pair
J2	6.312	Coaxial pair
$N \times T1$	$N \times 1.544$	Twisted pair and coaxial pair

Table 1.9: ATM physical layer for private UNI

Frame format	Bit-rate (Mbit/s)	Media
Cell stream	25.6	UTP-3 (phone wire) or STP
STS-1	51.84	UTP-3 (phone wire)
FDDI	100	Multimode fiber
STS-3c, STM-1	155.52	UTP-5 (data grade UTP)
STS-3c, STM-1	155.52	Single-mode fiber, multimode fiber, and coaxial pair
Cell stream	155.52	Multimode fiber and STP
STS-3c, STM-1	155.52	UTP-3 (phone wire)
STS-12, STM-4	622.08	Single-mode fiber and multimode fiber

SDH and SONET allow direct synchronous multiplexing: several lower bit-rate signals can be directly multiplexed onto a higher speed SDH or SONET signal without intermediate stages of multiplexing. A single multiplexed signal is called *tributary* or *container* for SONET and SDH, respectively.

Before SDH and SONET, the digital transmission hierarchy was based on the PDH technology, as already introduced in Sect. 1.5. When a PDH multiplexer is trying to multiplex different signals onto one data stream, it has to consider that the clocks of all incoming tributaries are not perfectly synchronized; the rise and fall times of pulses are not coincident in the tributaries. A PDH multiplexer reads data from all the incoming streams at the maximum allowed speed according to a cyclic process. It may happen that when the multiplexer services a stream, the bit of this stream has not yet arrived, because this stream has a slower clock; then, the multiplexer stuffs the data stream with "dummy bits" (or "justification bits"). This process is known as "plesiochronous operation" (from the Greek, "almost synchronous"). The multiplexer has a means of notifying the receiving end that stuffing has taken place so that extra bits can be discarded when demultiplexing the flows. The problem with PDH multiplexing is that a lower-level data stream is extracted from a higher-order one only if the demultiplexer performs all the operations made by the multiplexer that created the higher-level flow. As a result, all the flows need to be demultiplexed. This operation is called Add/Drop; it is a complex task, and the related equipment is quite expensive.

Differently from PDH, SDH/SONET transport networks are tightly synchronized: atomic clocks are used to synchronize the clocks of the networks. The reality is that perfect synchronization is practically impossible in large-scale geographical networks: temperature variations and different cable lengths always cause a residual drift in the clocks of tributaries. This is the reason why SDH/SONET adopts a new approach for multiplexing tributary signals in a higher-order one: pointers are used to individuate tributaries in the payload. Hence, it is possible to manage tributaries not running at the same clock rate and/or not aligned with the clock of the multiplexer. In particular, SDH/SONET adopts a pointer, describing the start of a tributary flow in the STS/STM frame payload. Hence, it is not necessary for the multiplexer to get the tributary signals in synchronism or to stuff the frame with bits. If a tributary signal clock slips over time with respect to the multiplexer clock, the SDH/SONET multiplexer simply recalculates the pointer for each new frame. Each byte of the tributary signal, and thus, the tributary signal itself, is visible in the frame. Hence, it is possible to extract a single tributary signal out of the primary signal, not needing to demultiplex all the flows as with PDH.

Table 1.10: Characterization of ATM services

	CBR	rt-VBR	nrt-VBR	ABR	UBR
Bandwidth guarantee	Yes	Equivalent	Equivalent	Minimum	No
Real-time traffic	Yes	Yes	No	No	No
Data bursty traffic	No	Yes	Yes	Yes	Yes
Congestion notification	No	No	No	Yes[a]	No

[a]ABR is the only traffic class, which foresees a congestion notification to invite the traffic source to reduce its traffic injection into the network (i.e., reducing the bit-rate)

1.7.5.1 Management of Traffic

In ATM networks, flow control and error control are not operated at intermediate nodes, but only end-to-end. It is essential to control not only the quality of the traffic but also its quantity in order not to congest some network nodes with the consequent increase in the delays experienced by all the related virtual circuits. Therefore, suitable techniques must be used to prevent congestion conditions.

In circuit-switched networks, congestion control is simply operated during the setup phase of the end-to-end link; in fact, it is necessary to check the availability of resources on all the links along the source-to-destination path (CAC technique). Such an approach is not sufficient in ATM networks, since traffic sources may generate variable bit-rate: their loads are unpredictable. In addition to this, the adoption of packet-switching causes that links are shared by several traffic flows having variable congestion levels. The traffic management problem is complicated by the fact that there can be different types of traffic sources with different characteristics and QoS requirements. Hence, each traffic flow must have guaranteed a given bandwidth (later in this section, we will expand this concept in terms of *equivalent bandwidth* [61–65]) in the different links of the path to fulfill its QoS levels.

In ATM, the traffic can be with or without QoS guarantees. CBR and VBR belong to the first case, and ABR and UBR belong to the second case. Referring to QoS-guaranteed traffic, two different types of techniques can be considered: *preventive control* (e.g., traffic load control) and *reactive control* (i.e., congestion control). Preventive control is used to decide whether a new connection can be admitted in the network (CAC technique), to smooth its traffic and to monitor the input traffic on the link to avoid unacceptable traffic peaks (Usage Parameter Control, UPC). Reactive control entails an action taken when a congestion event has occurred; the problem of this approach is that it implies an end-to-end delay before a repair action can start.

ATM networks can implement one or a combination of the following control functions to meet the QoS objectives of connections.

- Preventive control:
  - CAC
  - Resource reservation into the network
  - Traffic shaping
  - UPC, i.e., traffic policing
  - Traffic scheduling at nodes
- Reactive control:
  - Explicit Forward Congestion Indication (EFCI), together with end-to-end feedback signaling to notify the source to reduce the traffic rate.

Before starting the description of these control techniques, we need both to characterize the traffic sources in terms of traffic descriptors and to define their QoS parameters. The characterization of these services is summarized in Table 1.10. Note that a fixed bandwidth is reserved in the network for CBR sources, whereas an equivalent bandwidth must be available on all the links of the path to accept an rt-VBR or an nrt-VBR traffic source. Minimum end-to-end bandwidth is guaranteed for ABR sources, but even a larger bandwidth can be dynamically assigned to them, if available. Finally, there is no capacity guarantee for UBR traffic sources.

The equivalent bandwidth for a given traffic source, B_{eq}, is a complex parameter to be derived; it represents the bandwidth needed to guarantee some QoS levels for the generated traffic. Different equivalent bandwidth formulas are available, depending on the characteristics of the traffic source and the QoS requirements. There is a rich

Table 1.11: Connection traffic descriptors

	Acronym	Definition
Peak Cell Rate	PCR	Maximum rate according to which cells will be sent in the network
Sustainable Cell Rate	SCR	Mean rate (long term value) according to which cells will be sent in the network
Minimum Cell Rate	MCR	Minimum acceptable rate of cells in the network
Maximum Burst Size	MBS	Maximum number of cells that can be sent together (in a burst) at the line rate (PCR)
Cell Delay Variation Tolerance	CDVT	Maximum acceptable difference in the delay of output cells at a node (related to queuing delays; similar to jitter)

Table 1.12: QoS parameters

	Acronym	Definition
Cell Loss Ratio	CLR	Percentage of lost (or late) cells
Cell Transfer Delay	CTD	End-to-end delay for the transmission of a cell (maximum and mean value)
Cell Delay Variation	CDV	Variance of the end-to-end transmission delay of a cell
Cell Error Ratio	CER	Percentage of erroneous cells
Cell Misinsertion Rate	CMR	Percentage of erroneously delivered cells (routing error) among all the cells of a flow

literature on the equivalent bandwidth. For more details, the interested reader could refer to [61–65]. For instance, the equivalent bandwidth of an rt-VBR traffic source can be determined, assuming that this traffic arrives at a queue having a service rate of B_{eq} bit/s. Due to this service capability, the traffic experiences a delay, which is a random variable. Since rt-VBR is real-time traffic, a QoS requirement is represented by a deadline, i.e., a maximum delay within which each cell has to be transmitted. The B_{eq} value can be determined by imposing a constraint on the probability that the service delay exceeds the deadline, thus causing packet dropping. Hence, we can refer to the following example of B_{eq} characterization:

$$B_{\text{eq}} : \text{Prob}\left\{\text{service delay}\left(B_{\text{eq}}\right) > \text{deadline}\right\} \leq 5\% \tag{1.10}$$

For an rt-VBR source, we can generally consider that $\text{SRC} \leq B_{\text{eq}} \leq \text{PRC}$.

Traffic descriptors detailed in Table 1.11 are used to characterize the traffic generated by a given source. Referring to this table, PCR denotes the maximum bit-rate allowed to the source and SCR corresponds to the mean bit-rate. Hence, the source burstiness factor is $\beta = \text{PCR}/\text{SCR}$; of course, a CBR source has $\beta = 1$. The greater the traffic source burstiness, the higher the multiplexing gain by aggregating many sources of this type on the same link.

In ATM networks, there are many parameters to describe the QoS requested by a traffic source; some of them are defined in Table 1.12. These parameters are measured at the receiver.

For traffic with QoS guarantees, the user and the network stipulate a *traffic contract*, also called Service Level Agreement (SLA). Such traffic contract specifies a traffic conformance algorithm and the expected QoS provided by the network under some traffic characteristics as defined by the descriptors (e.g., PCR, SCR, MBS, MCR, and CDVT). The guaranteed QoS level can be in terms of maxCTD, CDV, CLR, etc. Note that CDV is measured as follows: CDV = maxCTD − minCTD. The network agrees to meet or exceed (for some small percentage of the time) the QoS negotiated as long as the traffic source complies with the contract [66]. UBR traffic does not require a traffic description, since it has no QoS guarantee. ABR traffic has guaranteed just MCR. ABR and UBR traffic classes should have no impact on the QoS provided to guaranteed QoS traffic classes.

ATM layer functions (e.g., cell multiplexing) may alter the traffic characteristics of connections by introducing some Cell Delay Variation (CDV). When cells from two or more connections are multiplexed, the cells of a given connection may be delayed because of the presence of cells of other connections. Similar problems are due to the insertion of OAM cells in a traffic flow. Consequently, some randomness may affect the interarrival time between consecutive cells of a given connection, as monitored at the UNI. The delay tolerance parameter CDVT regulates the upper bound to this delay variation [67]: the CDVT allocated to a particular connection provides a limit to the delay differences among the cells belonging to the same traffic flow; this is similar to the jitter for Internet traffic.

Connection Admission Control

CAC is a control operated by the network at the setup of a new connection to verify whether the QoS requirements can be fulfilled for both the new connection and the connections already in progress [67]. CAC procedures, based on traffic descriptors (see Table 1.11), permit to allocate resources and to derive parameter values for UPC operation. Several CAC techniques can be considered; they are generally categorized into two broad groups: (1) CAC based on bandwidth aspects and (2) CAC based on CLR considerations. In what follows, an example is provided about CAC dependent on bandwidth aspects.

Let us refer to VBR traffic sources (bursty traffic) on a shared access link to the ATM network. It would be highly inefficient to reserve the bandwidth corresponding to the PCR value for each VBR connection; hence, it is important to allocate the equivalent bandwidth for each VBR flow [61–65]. Let C denote the capacity of the link, and let $B_{\mathrm{eq,\,i}}$ be the equivalent bandwidth of the ith VBR connection on the same link. A new VBR traffic source with equivalent bandwidth B_{eq} fulfills the CAC condition and, hence, is admitted if the following condition is fulfilled:

$$\sum_i B_{\mathrm{eq,\,i}} + B_{\mathrm{eq}} \leq C \qquad (1.11)$$

Otherwise, the new connection is rejected.

Usage Parameter Control or Traffic Policing

CAC operated in the setup phase is an important control to guarantee the QoS, but it cannot protect from the risk that an admitted traffic source overloads the network. Therefore, the main purpose of UPC is to protect the network resources from malicious as well as unintentional misbehavior, which can affect the QoS of already-established connections. UPC entails a monitoring action performed for each traffic flow (connection). UPC is based on ITU-T Recommendations I.356 [68] and I.371 [66]. There can be both temporary traffic bursts produced by VBR sources or persistent traffic loads violating the contract stipulated with the network and verified in the CAC phase. To cope with these problems, UPC techniques are used on the network side of UNI. UPC is intended to ensure the conformance of a connection with the negotiated traffic contract. The connection traffic descriptors contain the necessary information for *testing the conformance* of the cells generated. Conformance applies to cells as they pass UNI: cells are tested according to some algorithm so that the network may decide whether a connection is compliant or not. The UPC function is implemented in the *policer* on the network side of UNI [67]. Referring to Fig. 1.47, the policer shall be capable of

- passing a cell that is conformant to connection traffic descriptors and
- discarding a cell if it is not conformant to connection traffic descriptors; alternatively, if the tagging option is allowed for a connection, the policer shall be capable of converting CLP from 0 to 1 for a non-conformant cell, which is accepted into the network.

The action operated by a policer is quite similar to the flow control scheme adopted in Frame Relay networks (see Sect. 1.7.3.1).

ITU-T and ATM Forum have defined the Generic Control Rate Algorithm (GCRA) to be used for the conformance test. GCRA can be considered as a virtual scheduling algorithm [67]. A GCRA test can be suitably defined for each ATM traffic class. GCRA can be used to control the peak cell rate (PCR). Otherwise, GCRA can be used to verify whether the cell rate is within some requested bounds on a given time window (i.e., SCR control). Moreover, different GCRA schemes can also be combined to obtain a more complex conformance test based on multiple parameters (e.g., PCR and SCR). GCRA algorithms (suitable not only for policing but also for traffic shaping) are of the following types:

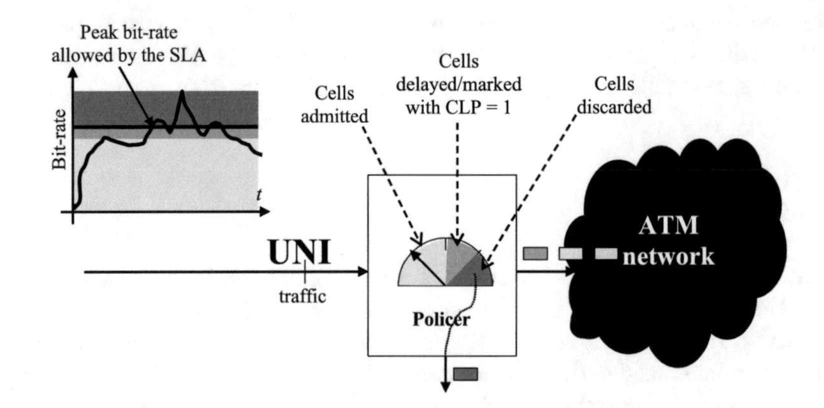

Fig. 1.47: Traffic policer based on a UNI conformance test, which monitors input traffic and compares it with the traffic contract (SLA)

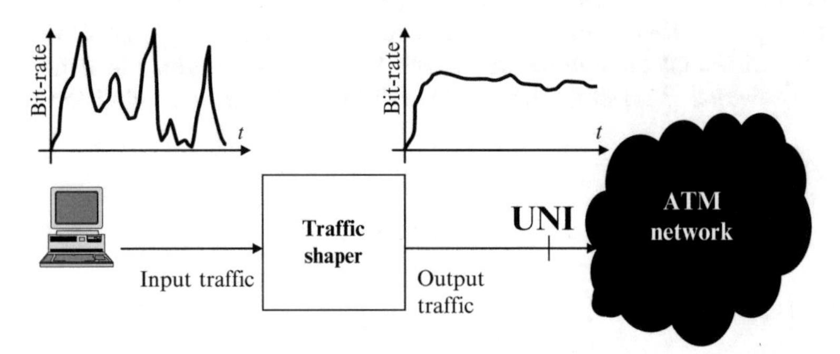

Fig. 1.48: Traffic shaper

1. PCR policing for CBR sources
2. Combined SCR and MBS policing for VBR sources without limits on PCR
3. Combined PCR, SCR, and MBS (or CDVT) policing for VBR sources with PCR, SCR, and burst (or delay, packet loss) limitations.

The effectiveness of CAC schemes depends on the fulfillment of the traffic contracts for the different traffic flows, as monitored by the policers.

Traffic Shaping

Shaping is a traffic source-side mechanism, which alters the traffic characteristics of a stream of cells to match its SLA. Traffic shaping allows us to control outgoing traffic, thereby eliminating bottlenecks because of data rate mismatches. Each connection is subject to traffic shaping in an ATM network. Let us refer to traffic shaping on the terminal side at UNI; it consists in filtering the input traffic of a source to reduce its burstiness. At the output of this regulator, the traffic offered to the network (UNI interface) is more regular and smoothed (almost constant). Avoiding burstiness is an essential need for the networks, since sudden traffic peaks may cause congestion at node buffers and high delays. However, traffic may have a residual burstiness at the output of the shaper. This is important, because a shaper that completely smoothes the traffic may entail unacceptable delays in the delivery of the cells. The traffic shaper action on the input traffic is depicted in Fig. 1.48.

Policers and shapers usually have common structures. They identify traffic descriptors violations in identical ways. They are both based on a conformance test,[8] but differ in the way they respond to violations. A policer monitors input cells (does not regulate them) and drops or marks them if the conformance test fails (i.e., the traffic

[8] It is possible that the same type of test is used in both shaper and policer.

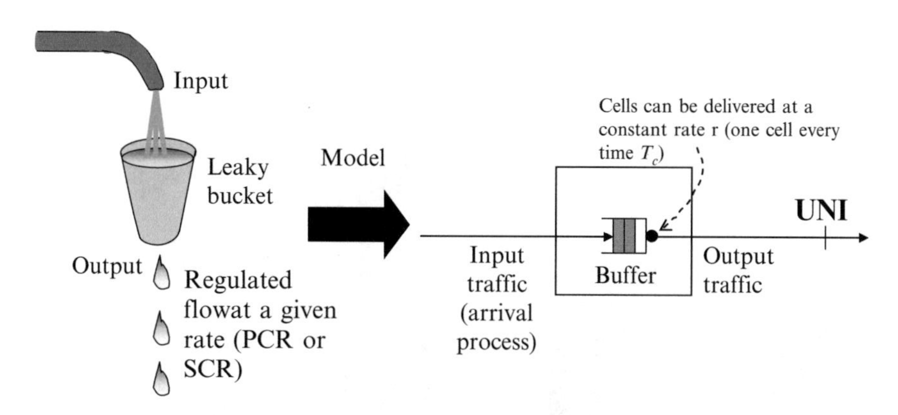

Fig. 1.49: Leaky bucket traffic shaper: conceptual scheme and model based on queuing theory

contract is exceeded). Instead, a shaper includes not only an algorithm for conformance test but also a queue: if cells exceed the traffic contract, they are queued, not dropped.

Traffic shaper and policer should work in tandem. A good traffic shaping scheme should make it easier to detect misbehaving flows at the entrance of the network.

The determination of the parameters for the traffic shaping algorithms is a quite complex task due to the multiplexing of different traffic sources. The shaper can determine the conformance time for the transmission of each cell arriving on a link. However, there can be conflicts when multiple cells, from different connections, become eligible for transmission in the same time slot. As a result, the shaper can have a backlog of conformant cells, particularly when traffic arrives from multiple input links. These collisions can distort the shaped traffic flows and increase delays, even for conformant cells.

In what follows, we will examine two typical traffic shapers: the *leaky bucket* regulator and the *token bucket* regulator; with slight modifications, they can also be used as policers. *Dual leaky bucket* and *dual token bucket* schemes will be shortly discussed as well.

Traffic shaping techniques have gained considerable importance in MPLS and in Integrated or Differentiated Services for QoS support in IP networks, as described in Sect. 2.5.1.

Leaky Bucket Shaper

In this case, the traffic shaper is simply a buffer, which can deliver cells at a predetermined rate. The output cell rate is regulated at a given value at the expense of increased delays experienced by the cells (the greater the input traffic burstiness, the higher the delays). A typical regulation could be based on PCR for type #1 GCRA or on SCR for type #2 GCRA, referring to the list of regulators in the previous sub-section; see Fig. 1.49. The buffer should have limited rooms if we want to constraint the delay caused by the shaper (i.e., the contribution to CTD). This constraint is fulfilled at the cost of some losses for those cells arriving at a full leaky bucket buffer (overflowing cells are discarded).

The analytical model of a leaky bucket regulator is a G/D/1 queue (see Chap. 5), where "G" refers to a general input arrival process of cells, "D" is related to the deterministic time to deliver a cell (i.e., time T_c, according to Fig. 1.49), and "1" means that one cell is delivered to the network at a time. Actually, we should consider a queue of finite length.

Token Bucket Shaper

This traffic shaping scheme adopts both a token bucket and a data queue (it becomes a policer if there is no data queue). Tokens are put into the bucket at a specific rate r. The bucket has a maximum capacity b of tokens (i.e., bucket depth). If the bucket is full, newly arriving tokens are discarded. Each token represents the permission for the source to transmit a certain number of bits in the network. To send a cell, the regulator must remove from the bucket a number of tokens corresponding to the cell size. If the bucket does not contain enough tokens to transmit a cell, the cell either waits until the bucket has enough tokens, or the cell is discarded, or the cell is marked and

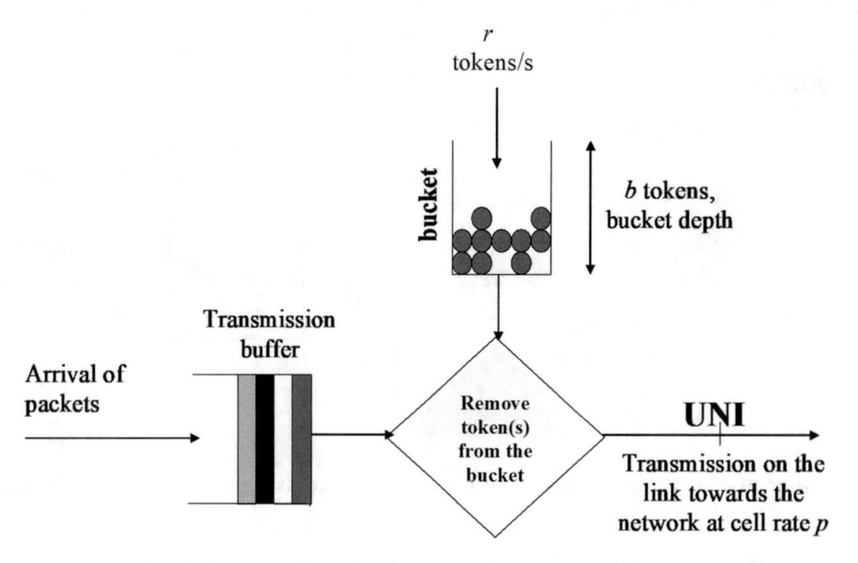

Fig. 1.50: Token bucket shaper according to the $(r,\ b,\ p)$ model

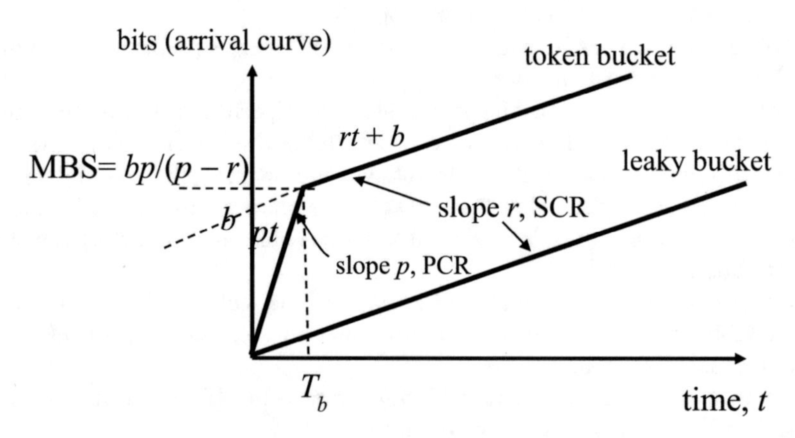

Fig. 1.51: Arrival curve model of the output traffic for a token bucket shaper, characterized by PCR, SCR, and MBS; comparison with the arrival curve of the leaky bucket shaper

transmitted. If the bucket is already full of tokens, incoming tokens overflow and are not available for future cells. The token bucket regulator permits to maintain some traffic burstiness at the output; this is an advantage with respect to the leaky bucket shaper. The largest burst (i.e., MBS, the maximum length of a burst of data, which are transmitted at the maximum speed, PCR) a source can send into the network with the token bucket shaper is proportional to the bucket depth.

The model of the token bucket regulator is characterized by $(r,\ b,\ p)$, where r denotes the rate at which tokens are accumulated, b is the depth of the bucket, and p is the maximum transmission rate (PCR); see Fig. 1.50. For instance, let us simply consider that a token is permission to transmit one cell. The token bucket regulates the output traffic, guaranteeing a regime cell rate of r cells/s, that is, SCR. The bucket depth b allows the transmission of up to b cells at the maximum rate p $(> r)$, thus having some burstiness for the output traffic (MBS). The arrival curve of a traffic source denotes the cumulative number of bits generated as a function of time. The arrival curve of a token bucket-regulated source is shown in Fig. 1.51, and the asymptotic burstiness index of the output flow (upper bound to traffic) is $\beta = p/r$. The token bucket regulator described here implements a GCRA algorithm of type #2, according to the categorization in the previous sub-section. The analysis of the token bucket shaper will be provided in Sect. 2.5.1 about the Guaranteed Service of IntServ.

Another variant of the token bucket regulator is the dual token bucket, which adopts two cascade token buckets.[9] Two alternatives are possible: (1) the *single-rate token bucket* with SCR regulation and where the tokens overflowed

[9] Both classical token bucket and dual token bucket schemes can be described by means of the set of three values $(r,\ b,\ p)$.

from the first bucket are placed in a second bucket to transmit excess traffic peaks (GCRA algorithm of type #2) and (2) the *dual-rate token bucket* with token rates corresponding to both PCR and SCR regulations (GCRA algorithm of type #3).

Traffic Scheduling

Traffic scheduling is a fundamental function for ATM networks to share the physical transmission resources among competing flows with conflicting QoS requirements. Scheduling must guarantee to preserve some form of priority among traffic classes. Typically, different transmission queues are needed to manage the different traffic classes at a multiplexer. Different scheduling and priority schemes can be adopted [69]. For instance, a priority level and Weighted Fair Queuing (WFQ) can be used to define the service order of the queues and the service time for each of them. Note that the WFQ service discipline, also known as Packet-by-packet Generalized Processor Sharing (PGPS), is an approximation of the *ideal* Generalized Processor Sharing (GPS) scheme, where the available capacity is bit by bit divided among the active (fluid) flows of traffic, according to their different weights. Another interesting scheduling method is represented by the Earliest Deadline First (EDF) scheme; it is a form of a dynamic priority scheduler, where the priority of each cell is assigned as it arrives. Specifically, a deadline is assigned to each cell based on the delay guarantee associated with the flow to which the cell belongs. The EDF scheduler selects to service (i.e., to transmit on the link) the cell with the closest deadline.

Congestion Control by Means of Buffer Management

Buffer management is a type of preventive congestion control, which selects the cells to be discarded from a buffer in the attempt to prevent congestion. Overload situations are natural since network operations are asynchronous: cells can compete for the same time slots on a link and therefore need to be stored in a waiting list. Buffering, on the other hand, has to be limited, due to its cost and the impact on latency. A good ATM switch should achieve a trade-off between latency (buffer size) and cell loss rate. CAC and UPC cannot avoid congestion situations in the network. Hence, to manage overload situations, preventive control can be used to discard cells from congested buffers selectively. In particular, we consider the two following buffer management techniques, which protect high-priority cells (CLP = 0) with respect to low-priority ones (CLP = 1):

- In a *push-out mechanism*, all cells are allowed to enter the buffer until there are available rooms. Let us now consider a cell arriving at a full buffer: if this cell has a low priority, it is discarded; instead, if this cell has a high priority, it is discarded only if there is no low-priority cell in the buffer, which can be discarded to make room for the new high-priority cell.
- The *threshold mechanism* allows all cells to enter the buffer as long as the number of waiting cells is lower than a given threshold. When the number of waiting cells exceeds such limit, newly arriving cells with low priority are discarded; instead, high-priority cells are admitted as long as there are rooms available in the buffer.

Both schemes have similar performance. However, the threshold mechanism is preferred because it is simpler than the push-out one. A similar approach is used in the Internet with the Random Early Detection (RED) scheme; see Sect. 2.5.2.

More refined schemes control the cells to be dropped rather than having them dropped at random. In situations where a higher-layer packet (ALL level) is segmented in cells, the drop of a single cell entails the need to resend the entire higher-layer packet. In such circumstances, it is convenient to continue to drop all the cells from the same packet in the presence of congestion. In particular, we can consider the Early Packet Discard (EPD): If a BOM cell arrives and the buffer occupancy is above a certain threshold, all the cells of the same packet are discarded beforehand.

Reactive Schemes for Congestion Control

A reactive scheme can be used to manage congestion events: a congested network node may set the EFCI flag in the cell header so that this indication can be notified to the destination (forward congestion notification). Then, this indication received can trigger a signaling protocol at the end system to inform the source to reduce the traffic injection into the network. This reactive scheme is supported only by the ABR traffic class in ATM. It is also possible that the congestion notification is sent directly to the source by a congested intermediate node; this is a backward congestion notification scheme. Both forward and backward schemes suffer from network-wide delays in reacting to congestion events.

1.8 Main Lessons Learned on Resource Management and Teletraffic Engineering

The study of these legacy network technologies (i.e., X.25, ISDN, Frame Relay, and ATM) is crucial to understand some fundamental concepts on networking that we will find again when dealing with the Internet in the next chapter. We survey here those approaches that have evolved with time starting from some old technologies and that are very useful to understand their use in today's networks.

First of all, we consider the layering of the protocols according to a stack and the separation of roles among distinct layers. This is the OSI standard approach, which is fundamental for networking and facilitates the interoperability of devices of different vendors. This approach is used by all the legacy technologies and related standardizations (i.e., X.25, ISDN, Frame Relay, and ATM). The Internet protocol stack has slightly modified this model (as we will see in the next chapter), but its main concepts still remain. In this sense, especially the characterization of lower layers is valid with PHY layer dealing with transmissions of data, link layer addressing the sharing among users of transmission resources on a link, and network layer dealing with the interconnection of nodes to distribute the information end-to-end.

Then, traffic digitalization is also another fundamental need that allows traffic types of different nature (voice, video, web pages, sensors data, etc.) to share the same transmission resources using multiplexing schemes.

An interesting concept for traffic engineering is the management of data traffic according to traffic classes that specify distinct management for distinct QoS levels (OSI layer 3). This is a basic concept in networking, especially today that data traffic is continuously increasing, and there is the need to manage it in the best way also in the critical conditions of congestion events. The different classes are characterized by different QoS requirements in terms of packet loss rate, mean and maximum packet delay, and mean and maximum delay jitter. QoS support is today essential, and we will see in the next chapter the approaches that are available for the current Internet. Today QoS classes are also very important for the design of mobile networks (5G and beyond 5G). The system can support the QoS for every single flow in the network, guaranteeing for it hard QoS requirements (i.e., maximum delay lower than a threshold), or the system can guarantee the QoS level based on average values (e.g., mean delay lower than a threshold) at the level of a whole traffic class.

Another important basic concept is connection-orientation versus connectionless. In legacy networks, a connection-oriented approach requires a connection setup phase at the beginning of the communication. This approach can provide some advantages because when the traffic flow is set in the network, its forwarding can be faster. Connection-oriented links are based on virtual circuits that use virtual identifiers, actually labels, that are switched hop by hop and allow the fast identification of the end-to-end path. On the other hand, the connection orientation may require extra complexity, and then it cannot be suitable for the whole Internet. The connection setup phase can implement a CAC process to decide whether the new flow can be admitted or not to preserve the QoS of the flows currently admitted into the system.

In the legacy technologies, we have seen interesting traffic management techniques like traffic shaping and traffic policing (OSI layer 2 or 3) that can also contribute to protecting the QoS levels for the different traffic flows in the network. On the one hand, it is important to control the rate with which the users inject data in the network. On the other hand, the network has to protect itself by avoiding that, for any reason, a single traffic source can get more resources than agreed in the service level agreement. Also, in the mobile access systems of today, a shaping/policing scheme is adopted that is volume-based so that a user can inject traffic in or receive traffic from the network for a maximum amount per month. Beyond this limit, the traffic gets dropped in part.

Traffic scheduling at the nodes of the network can help to enforce the QoS; this is particularly needed at edge nodes. The schedulers take account of the different QoS constraints of the traffic flows and decide the next packet to be transmitted on the link. Scheduling techniques can be implemented at both layers 2 and 3 of the OSI model. Scheduling schemes need to account for the relative priority, the deadlines, and the minimum throughput needs for the traffic flows sharing a transmission resource (i.e., a link).

System congestion is a problem of today's networks. It can be very critical because congestion first entails extra delays and, in the end, may lead to packet losses because of buffer overflow. To circumvent congestion issues, possible teletraffic approaches are flow control, buffer management, and alternative/dynamic routing. Here, we can refer to the methods adopted by Frame Relay and ATM as possible starting points on how to deal with network congestion. In particular, these networks can support simple types of congestion control where a congested node or the destination node can feedback a congestion signal (OSI layer 3 and above) to the traffic source to inform it to lower its injection rate because of congestion. This is a quite interesting and general approach that will be used by the transmission control protocol of the Internet protocol suite. In addition to this, we have seen that there can be some strategies to reduce the number of packets in a full or almost full buffer to allow the admission of high-priority traffic. This is the typical task of buffer management. We will find something similar in the next chapter when dealing with the Internet, where the RED scheme can be adopted. In particular, after a certain level of buffer occupancy, IP packets can be dropped with an increasing packet loss rate to allow an anticipated signal of congestion, thus triggering an implicit mechanism that informs the source about the congestion event.

Finally, the evolution of the networks from analog techniques to digital ones highlights the increasing need for network softwarization that is today adopted jointly with other fundamental technologies, like cloud computing and virtualization. This approach also facilitates the differentiation of the QoS among traffic class and the rapid configuration and adaptation of the network and its routing conditions, thus also easing the implementation of traffic engineering schemes for the whole network.

References

1. Brin S, Page L (1998) The anatomy of a large-scale hypertextual web search engine. Comput Network ISDN Syst 30:107–117
2. Hoffmann O (2010) Radiocommunications: from the basics to future developments part 3: advances in wireless LANs, Tutorial at MobiLight 2010, 12 May 2010, Barcelona (slides available at the following address: http://www.ict-omega.eu/)
3. ITU official Web site with http://www.itu.int/home/index.html
4. ISO official Web site with http://www.iso.org/
5. IEEE official Web site with http://www.ieee.org/
6. IETF official Web site with http://www.ietf.org/
7. ETSI official Web site with http://www.etsi.org/
8. ANSI official Web site with http://www.ansi.org/
9. ARIB official Web site with http://www.arib.or.jp/english/
10. TIA official Web site with http://www.tiaonline.org/standards/
11. TTA official Web site with http://www.tta.or.kr/
12. IEC official Web site with http://www.iec.ch/
13. EIA official Web site with http://www.eia.org/
14. 3GPP official Web site with http://www.3gpp.org/
15. 3GPP2 official Web site with http://www.3gpp2.org/
16. Proakis JG (1995) Digital communications. McGraw-Hill, New York, NY
17. Kleinrock L (1964) Ph.D. thesis published by McGraw-Hill. Communication nets
18. ISO/IEC 7498-1 standard, information technology – open systems interconnection – basic reference model: the basic model
19. ITU-T Recommendations, series X.2000 on Open System Interconnection
20. ITU-T (2008) E.800: definitions of terms related to quality of service. Sept 2008 http://www.itu.int/rec/T-REC-E.800-200809-I/en
21. Iversen VB (2010) Teletraffic engineering handbook. ITU-T Study Group 2 http://oldwww.com.dtu.dk/teletraffic/handbook/telenook.pdf
22. Stallings W (2003) Data and computer communications. Prentice Hall, Englewood Cliffs, NJ
23. ITU-T. Pulse code modulation (PCM) of voice frequencies. G.711 Recommendation
24. ITU-T. Physical/electrical characteristics of hierarchical digital interfaces. G.703 Recommendation
25. ITU-T. Synchronous frame structures used at 1544, 6312, 2048, 8488, and 44,736 kbit/s. G 704 Recommendation
26. ITU-T. Frame alignment and cyclic redundancy check (CRC) procedures relating to basic frame structures defined in Recommendation G.704. G.706 Recommendation

27. Stallings W (2003) Data and computer communications. Prentice Hall, Upper Saddle River, NJ
28. ITU-T. Interface between Data Terminal Equipment (DTE) and Data Circuit-terminating Equipment (DCE) for terminals operating in the packet mode and connected to public data networks by dedicated circuit. Recommendation X.25, October 1996
29. Helgert HJ (1991) Integrated services digital networks: architecture, protocols, standards. Addison-Wesley, New York
30. Stallings W (1992) Advances in ISDN and broadband ISDN. IEEE Computer Society Press, Los Alamitos, CA
31. Stallings W (1995) ISDN and broadband ISDN with frame relay and ATM. Prentice-Hall, Upper Saddle River, NJ
32. de Prycker M (1991) Asynchronous transfer mode: solution for broadband ISDN. Ellis Horwood, Chichester
33. Onvural RO (1994) Asynchronous transfer mode networks: performance issues. Artech House, Inc., Norwood, MA
34. Karim MR (1999) ATM technology and services delivery. Prentice-Hall, Upper Saddle River, NJ
35. ITU-T (1996) Packet-switched signalling system between public networks providing data transmission services. Recommendation X.75, October 1996
36. ITU-T (1992) Interface between data terminal equipment and data circuit-terminating equipment for synchronous operation on public data networks. Recommendation X.21, September 1992
37. ITU-T (1988) List of definitions for interchange circuits between Data Terminal Equipment (DTE) and Data Circuit-terminating Equipment (DCE) on public data networks. Recommendation X.24, November 1988
38. ITU-T (1996) Electrical characteristics for balanced double-current interchange circuits operating at data signalling rates up to 10 Mbit/s. Recommendation X.27, October 1996
39. ITU-T (1995) Basic user-network interface – layer 1 specification. Recommendation I.430, November 1995
40. ITU-T (1993) Primary rate user-network interface – layer 1 specification. Recommendation I.431, March 1993
41. ITU-T (1988) Principles of telecommunication services supported by an ISDN and the means to describe them. Recommendation I.210, November 1988
42. ITU-T (1993) ISDN protocol reference model. Recommendation I.320, November 1993
43. ITU-T (2001) Physical/electrical characteristics of hierarchical digital interfaces. Recommendation G.703, November 2001
44. ITU-T (1998) Synchronous frame structures used at primary and secondary hierarchical levels. Recommendation G.704, October 1998
45. ITU-T (1993) Digital subscriber Signalling System No.1 (DSS1) – ISDN user-network interface data link layer – General aspects. Recommendation Q.920, March 1993
46. ITU-T (1997) ISDN user-network interface – data link layer specification. Recommendation Q.921, September 1997
47. ITU-T (1998) Digital subscriber Signalling System No. 1 (DSS 1) – ISDN user-network interface layer 3 specification for basic call control. Recommendation Q.931, May 1998
48. ITU-T (1991) Frame mode bearer services. Recommendation I.233 [It deals with Frame Relay and Frame Switching], October 1991
49. ITU-T (1992) ISDN data link layer specification for frame mode bearer services. Recommendation Q.922 [This recommendation describes the LAPF protocol; Annex A is on frame relay], February 1992
50. ITU-T (1993) ISDN Digital subscriber Signalling System No. 1 (DSS 1) – signalling specification for frame mode basic call control. Recommendation Q.933 [This recommendation deals with user-to-network signaling for Frame Relay services; it is based on Q.931, which deals with circuit/packet switched services], March 1993
51. ITU-T (1991) Congestion management for the ISDN framerelaying bearer service. Recommendation I.370
52. ITU-T (1992) B-ISDN asynchronous transfer mode functional characteristics functional characteristics. Recommendation I.150
53. Pandya AS, Sen E (1998) ATM technology for broadband telecomm networks. CRC Press, Boca Raton, FL
54. Arpaci M, Copeland JA (2000) Buffer management for shared-memory ATM switches. IEEE Commun Surv Tutor 3(1):2–10
55. ITU-T (1993) B-ISDN user-network interface – physical layer specification. Recommendations of the I.432 series (I.432.1, I.432.2, I.432.3, I.432.4, I.432.5)
56. ANSI. SONET – basic description including multiplex structure, rates and formats. T1.105, 2001
57. The ATM Forum (1993/1994) ATM user-network interface specifications 3.0 and 3.1. Prentice-Hall, Upper Saddle River, NJ

58. ITU-T (2007) Network node interface for the synchronous digital hierarchy (SDH). Recommendation G.707, January 2007
59. ITU-T (1991) Sub STM-0 network node interface for the synchronous digital hierarchy (SDH). Recommendation G.708, April 1991
60. ITU-T (2012) Interfaces for the optical transport network (OTN). Recommendation G.709, February 2012
61. Guérin R, Ahmadi H, Naghshineh M (1991) Equivalent capacity and its application to bandwidth allocation in high-speed networks. IEEE J Sel Area Comm 9:968–981
62. Elwalid A, Mitra D (1993) Effective bandwidth of general Markovian traffic sources and admission control of high speed networks. IEEE/ACM T Network 1:329–343
63. Kesidis G (1994) Modeling to obtain the effective bandwidth of a traffic source in an ATM network. MASCOTS, Los Alamitos, CA, pp 318–322
64. Chang C-S, Thomas J (1995) Effective bandwidth in high speed digital networks. IEEE J Sel Area Comm 13:1091–1100
65. Gibbens RJ, Hunt PJ (1991) Effective bandwidths for the multi-type UAS channel. Queueing Syst 9:17–28
66. ITU-T (1994) Traffic control and congestion control in B-ISDN. Recommendation I.371
67. ATM Forum (1999) Traffic management specification. Version 4.1. AF-TM-0121.000. ftp://ftp.atmforum.com/pub/approved-specs/af-tm-0121.000.pdf
68. ITU-T (1993) B-ISDN ATM layer cell transfer performance. Recommendation I.356
69. Guérin R, Peris V (1999) Quality-of-service in packet networks: basic mechanisms and directions. Comput Network 31:169–189

Chapter 2
IP-Based Networks and Future Trends

Abstract The Internet has become essential for today's life. This chapter aims to describe the main Internet protocols for the network layer: IP addressing, Routing algorithms (like the Dijkstra one), Distributed Routing protocols, IntServ (and the related deterministic queuing theory), and Diffserv. The second part of this chapter describes the transport layer and, in particular, the TCP protocol; some essential characteristics of this protocol are provided, and some variants like Tahoe, Reno, NewReno, and Cubic are described. The third part of this chapter deals with recent network advances, like virtualization, network softwarization, Internet-centric networking, network coding, and satellite communication systems.

Key words: IP addressing, Dijkstra algorithm, Routing protocols, IntServ, DiffServ, Transport layer protocols, TCP, Network coding, Satellite systems

2.1 Introduction

A growing number of people are using the Internet, the network of the networks; this is also evident from the different bandwidth-intensive applications supported by the Internet and by considerable number of Internet books, videos, etc., which have become available during these years. The widespread diffusion of social networks (Facebook, YouTube, etc.), peer-to-peer traffic, and cloud applications has further contributed to the impressive Internet use growth. In addition to this, video traffic (video downloads, streaming) represents today more than 58% of the total downstream traffic volume on the Internet.

Consumer Internet traffic amounted to 100 exabytes per month in 2017 (1 exabyte = 10^{18} bytes). In 2022, the global consumer IP traffic is expected to pass 300 exabytes per month. There will be more than 5 billion of Internet users by 2023 [1].

This chapter focuses on the protocols and the network technologies for the Internet.

2.2 The Internet

J. C. R. Licklider of the Massachusetts Institute of Technology (MIT) proposed a global network of computers in 1962 and moved to the Defense Advanced Research Projects Agency (DARPA) to lead a project to interconnect Department of Defense (DoD) sites in the USA. L. Kleinrock of MIT (and, later, University of California, Los Angeles, UCLA) developed the theory of packet-switching, which is at the basis of Internet traffic. In 1965, L. Roberts of MIT connected a Massachusetts computer with a California computer using a dial-up telephone line. He showed the feasibility of wide area networking, but also that the telephone circuit-switching was inadequate

---

Electronic Supplementary Material The online version contains supplementary material available at (https://doi.org/10.1007/978-3-030-75973-5_2).

G. Giambene, *Queuing Theory and Telecommunications*, Textbooks in Telecommunication Engineering,
https://doi.org/10.1007/978-3-030-75973-5_2

for this traffic, thus confirming the importance of the Kleinrock packet-switching theory. These pioneers (as well as other people) are the actual founders of the Internet. The Internet, then known as ARPANET, was brought online in 1969, initially connecting four major sites (computers), under a contract held by the renamed Advanced Research Projects Agency (ARPA).

Once the initial sites were installed, representatives from each site meet together to solve technical problems concerning the interconnection of hosts through protocols. A working group, called Network Working Group (NWG), was in charge of defining the first "rules" (i.e., protocols) of the network. The open approach adopted by the first NWG meeting continued in a more formalized way by using meeting notes, called Request For Comments (RFC). These documents are intended to keep members updated on the status of several things concerning Internet protocol. They were also used to receive responses from researchers.

The Internet was designed to provide a communication network able to work even if some sites are destroyed. The early Internet was used by computer experts, engineers, scientists, and librarians. There were no personal computers and no massive use in those days. Different "initial" applications and protocols were conceived to exploit ARPANET. The e-mail was adopted for ARPANET in 1972. The telnet protocol, to log on a remote computer, was defined in 1972 [2]. The FTP protocol, enabling file transfers between Internet sites, was published as RFC 354 in 1972 [3, 4], and from then further RFCs were made available to update the characteristics of the FTP protocol. Today, RFCs are the method used to standardize every aspect of the Internet; they are freely accessible in the ASCII format through the Internet Engineering Task Force (IETF) Web site [5]. RFCs are approved after a very strong review process. IETF is an open, all-volunteer organization (started its activities in 1983), with no formal membership or membership requirements. It is divided into a large number of working groups, each dealing with a specific Internet issue.

In 1974, a new suite of protocols was proposed and implemented in the ARPANET, based on the Transmission Control Protocol (TCP) for end-to-end communications. In 1978, a new Internet design approach was conceived with the division of tasks between two protocols:

- The new Internet Protocol (IP) for routing packets and device-to-device communications (i.e., host-to-gateway or gateway-to-gateway).
- The TCP protocol for reliable, end-to-end communications.

Since TCP and IP were initially conceived as working in tandem, this protocol suite is commonly denoted as TCP/IP [6, 7].

As long as the number of Internet sites was small, it was easy to keep track of the available resources. But as more and more universities and organizations connected, the Internet became harder to track. There was the need for tools to index the available resources. Starting from 1989, significant efforts were pursued in this direction. In particular, T. Berners-Lee and others at the European Laboratory for Particle Physics (i.e., CERN) laid the basis to share documents using *browsers* in a multi-platform environment. In particular, three new technologies were incorporated into his proposal: (1) the HyperText Markup Language (HTML) used to write documents (also named "pages") for the Internet; (2) the HyperText Transfer Protocol (HTTP), an application layer protocol to transmit documents in HTML format; (3) a browser client software program to receive and interpret HTML documents and to display the results. His proposal was based on *hypertext*, i.e., a system of embedding links, that is addresses of Internet resources (like files), in the text to refer to other Internet documents.

In 1991, the World Wide Web was born because the first really friendly interface to the Internet was developed at the University of Minnesota; it was named "gopher," after the University of Minnesota mascot, the golden gopher. In 1993, the development of the graphical browser, called Mosaic, by M. Andreessen and his team at the National Center For Supercomputing Applications (NCSA), a research institute at the University of Illinois, gave a substantial boost to the Web. Starting from this browser, new ones rapidly spread and made the Web a worldwide success. Further developments to the Web were represented by the Web search engines, as already discussed in Chap. 1 (Sect. 1.1).

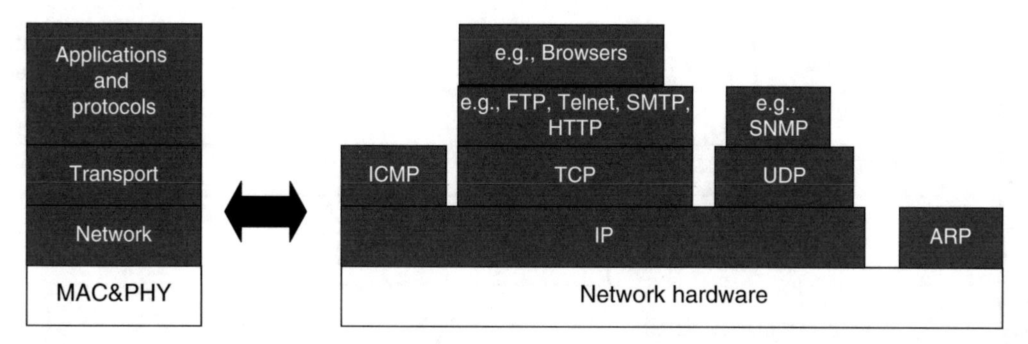

Fig. 2.1: Simplified Internet protocol suite. The acronyms in this figure will be described along with this chapter; this figure will be taken as a reference

2.2.1 Introduction to the Internet Protocol Suite

TCP/IP's goal was to interconnect different physical networks to form what appears to the user as a universal network. Such a set of interconnected networks is called the *Internet* [8–11]. Communication services are provided by Internet protocols, which operate between the link layer and the application one. The architecture of the physical networks is hidden to the users.

To be able to interconnect two networks, we need a "computer" that is attached to both networks and that can forward packets from one network to another and vice versa; this device, called *router*, has two essential characteristics:

- From the network standpoint, a router is a normal host.
- From the user standpoint, routers are invisible to the users who only see a larger network.

Each host has an address assigned, the *IP address*, to identify it in the Internet. When a host has multiple network adapters, each adapter has a separate IP address.

2.2.2 TCP/IP Protocol Architecture

Although there is no universal agreement on how to describe TCP/IP with a layered model, it is generally regarded as composed of fewer layers than the seven layers of the classical OSI model. Most TCP/IP descriptions define three to five functional levels in the protocol architecture [12]; a four-layer TCP/IP model is shown in Fig. 2.1.

As in the OSI model, data are passed down through the stack when they are sent to the network and passed up through the stack when they are received from the network. Each layer treats the information it receives from the layer above as *data* and adds its *header* in front of that information to ensure the proper management of these data. The operation to add the header (containing control information) is called *encapsulation*.

The *network layer* is the lowest layer of the TCP/IP protocol hierarchy. The protocols of this layer provide the means to route data to other network devices. Unlike higher-level protocols, network layer protocols must know the details of the underlying network (its packet structure, addressing, etc.) to correctly format the data being transmitted to comply with local network constraints.

The Internet protocol stack has a layered architecture resembling an *hourglass* (see Fig. 2.2): the reason for this denomination of the Internet protocol model is that there are many PHY and MAC layer protocols and there are many application and transport layer protocols, while on the waist of the hourglass at the network layer there is just the IP protocol. The hourglass model expresses the concept that the IP protocol is the glue, the Internet's fundamental building block. The protocols of the waist are those to which we are referring when talking about the Internet "ossification"; this is seen today mostly as a limit to flexibility and security because all information is forced through a small set of mid-layer protocols.

The Internet Protocol (IP), originally defined in RFC 791 [6] is the heart of the Internet protocol suite and the most important protocol of the network layer. IP provides the basic packet delivery service for the networks. All the higher-layer protocols (and the related data flows) use IP to deliver data. Its functions include:

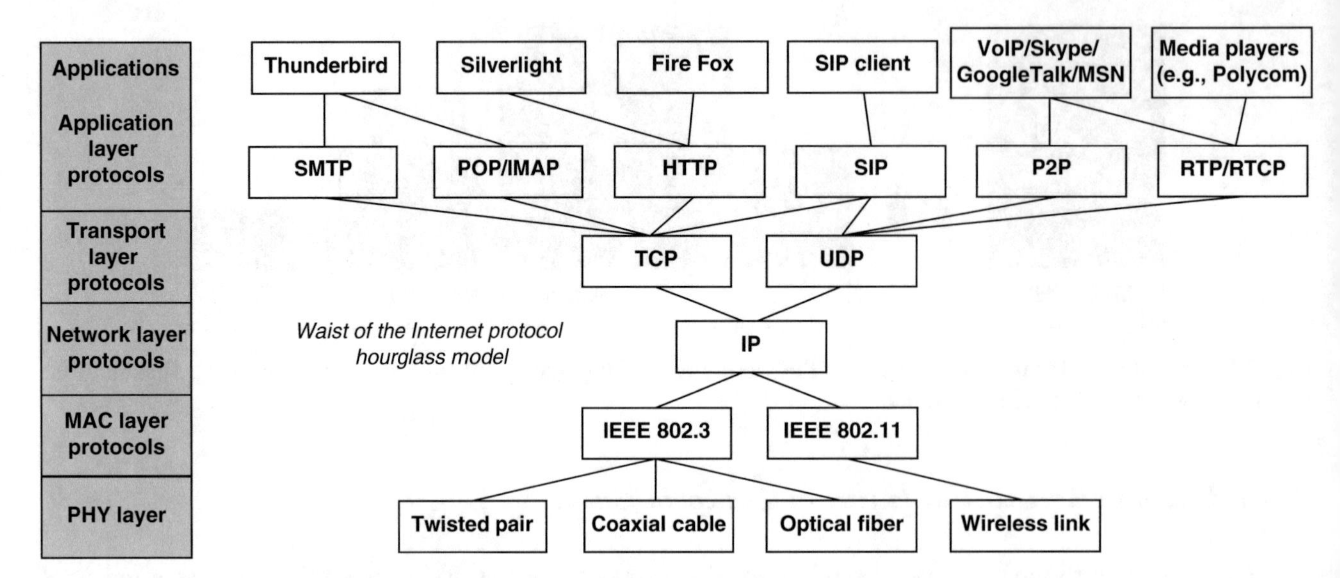

Fig. 2.2: The Internet protocol stack and the hourglass model (note that not all the protocols have been shown at the different layers, but just some of them)

- Defining the IP packet (i.e., a datagram, the basic transmission unit in the Internet).
- Defining the Internet addressing scheme.
- Moving data between network and transport layers.
- Routing datagrams to remote hosts.
- Performing fragmentation and reassembly of datagrams.

IP is an *unreliable protocol* because it does not perform error detection and recovery for transmitted data. This does not mean that we cannot rely on this protocol. IP can be relied upon to deliver data accurately to the destination, but it does not check whether data are received correctly or not. Higher-layer protocols of the Internet protocol stack are in charge of providing error detection and recovery if required.

The protocol layer just above network one is the *host-to-host transport layer*. This name is commonly shortened as the *transport layer*. The two most essential protocols at the transport layer are Transmission Control Protocol (TCP) and User Datagram Protocol (UDP). TCP provides a reliable, connection-oriented, byte-stream data delivery service; error detection and error recovery (through retransmissions) are performed end-to-end. UDP provides a low-overhead, unreliable, connectionless datagram delivery service. Both protocols exchange data between application and network layers. Applications programmers can choose the service that is most appropriate for their specific needs.

UDP gives application programs direct access to a datagram delivery service, like the delivery service provided by IP. This allows applications to exchange messages over the network with a minimum protocol overhead.

Applications requiring the transport protocol to provide reliable data delivery use TCP since it verifies that data are accurately delivered across the network and in the right sequence.

The *application layer* is at the top level of the TCP/IP protocol architecture. This layer includes all processes that use transport protocols to deliver data. There are many application layer protocols. Most of them provide user services; new services are constantly being added at this layer. The most popular and implemented application layer protocols are:

- Telnet: The network terminal protocol, which allows us to log on hosts spread in network.
- FTP: The File Transfer Protocol used for file transfer.
- SMTP: The Simple Mail Transfer Protocol, which delivers electronic mail.
- HTTP: The Hypertext Transfer Protocol, delivering Web pages over the network.
- Domain Name System (DNS): This is a service to map network name addresses (e.g., in the form www.domain.country) to the corresponding numeric IP addresses.

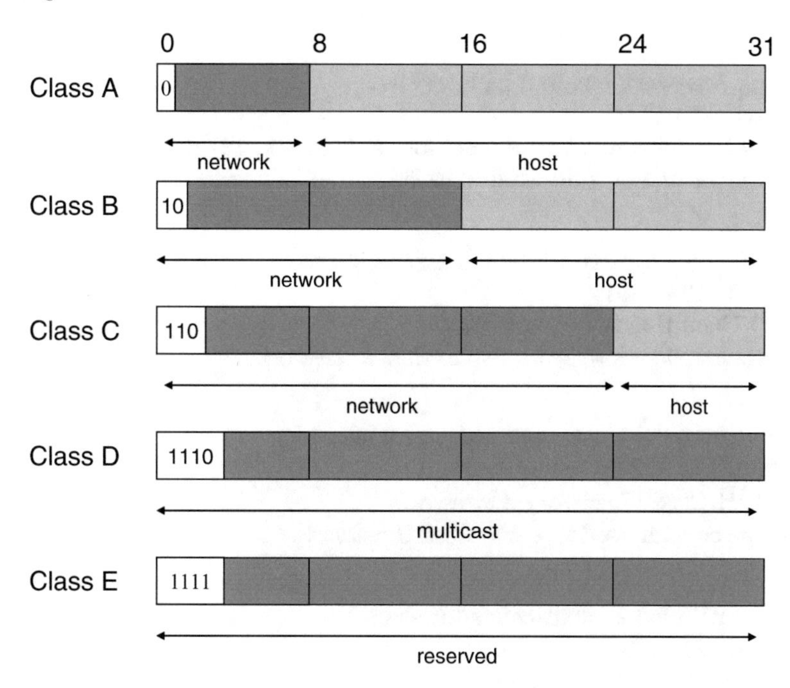

Fig. 2.3: IPv4 address classes

- Network File System (NFS): This protocol permits to share files among various hosts in the network.
- Finally, the Open Shortest Path First (OSPF), which is a layer 3 routing protocol, includes a transfer protocol for the exchange of routing information among routers and as such (even with some debate) can also be considered as an application layer protocol.

2.3 IP (Version 4) Addressing

IP addresses are used to route datagrams in the network and to allow their correct delivery to the destination. An IP version 4 (IPv4) address is formed of 32 bits, written by dividing the bits into groups of 8 and taking the corresponding decimal number. Each of these numbers is written separated by a dot (i.e., dotted-decimal notation) and can range from 0 to 255. For example, 1.160.10.240 could be an IP address. The specification of IP addresses is contained in RFC 1166 [13]. An IP address can be divided into a pair of numbers (the length of these fields depends on the IP address class):

IP address = <network identifier> + <host identifier>.

There are five classes of IP addresses, as described in Fig. 2.3. Classes are introduced to divide the space of IP addresses into groups of a limited number of addresses (i.e., that can support a limited number of hosts). This is carried out for efficient use of IP addresses and takes the name of "classful" IPv4 addressing.

For classes A, B, and C, the address of a network has all the host bits equal to "0," whereas the broadcast address of a network is characterized by all the host bits equal to "1." The number of hosts addressable in a network is therefore related to the number of available combinations for the bits of the host field minus two addresses for network and multicast purposes.

Class A
- First bit set to “0” plus 7 network bits and 24 host bits.
- Initial byte ranging from 0 to 127.
- Totally, 128 (= 2^7) Class A network addresses are available (0 and 127 network addresses are reserved).
- 16,777,214 (= $2^{24} - 2$) hosts can be addressed in each Class A network.

Class B
- First two bits set to “10” plus 14 network bits and 16 host bits.
- Initial byte ranging from 128 to 191.
- Totally, 16,384 (= 2^{14}) Class B network addresses.
- 65,534 (= $2^{16} - 2$) hosts can be addressed in each Class B network.

Class C
- First three bits set to “110” plus 21 network bits and 8 host bits.
- Initial byte ranging from 192 to 223.
- Totally, 2,097,152 (= 2^{21}) Class C network addresses.
- 254 (= $2^8 - 2$) hosts can be addressed in each Class C network.

Class D
- First four bits set to “1110” plus 28 multicast address bits.
- Initial byte ranging from 224 to 247.
- Class D addresses are used for multicast flows.

Class E
- First four bits set to “1111” plus 28 reserved address bits.
- Initial byte ranging from 248 to 255.
- This address class is reserved for experimental use.

Note that the special range of addresses 169.254.0.1-169.254.255.254 within class B is used for the Automatic Private IP Addressing (APIPA) that automatically enables computers to self-configure an IP address when their Dynamic Host Configuration Protocol (DHCP) server is not reachable.

A router receiving an IP packet extracts its IP destination address and classifies it by examining its first bits. Once the IP address class has been determined, the IP address can be broken down into network and host bits. Intermediate routers ignore host bits and only need to match network bits within their routing table to route the IP packet along the correct path in the network. Once a packet reaches its target network, its host field is examined for the final local delivery.

IPv4 addressing space is limited: this is a significant problem because of the continued spread of the Internet. To address this issue, possible approaches are IP subnetting (see Sect. 2.3.2), the use of private IP addresses (see Sect. 2.3.3), and the new IP version 6 (see Sect. 2.3.6).

2.3.1 IPv4 Datagram Format

Data transmitted over the Internet using IP addresses are organized in variable-length packets, called IP datagrams. Let us consider here the IPv4 datagram format, defined in RFC 791 [6]. An IPv4 datagram is divided into two parts: the header and the payload. The header contains addressing and control fields, while the payload carries the actual data to be sent. Even though IP is a relatively simple, connectionless, “unreliable” protocol, the IPv4 header carries some control information that makes it quite long. It is at least 20-byte long and can be even longer with the options. The IP datagram format is shown in Fig. 2.4, where each row corresponds to four bytes (i.e., a word of 32 bits). The meaning of the different header fields is explained below.

- Version (4 bits): Identifies the IP version of the datagram. For IPv4, obviously, this field contains the number 4. The purpose of this field is to ensure compatibility among different devices, which may be running different IP versions. In general, a device running an older IP version will reject datagrams created by newer implementations.
- IHL, Internet Header Length (4 bits): Specifies the IP header’s length in 32-bit words. This length includes any optional field and padding. The normal value of this field when no options are used is 5 (i.e., 5 words of 32 bits, corresponding to 20 bytes).

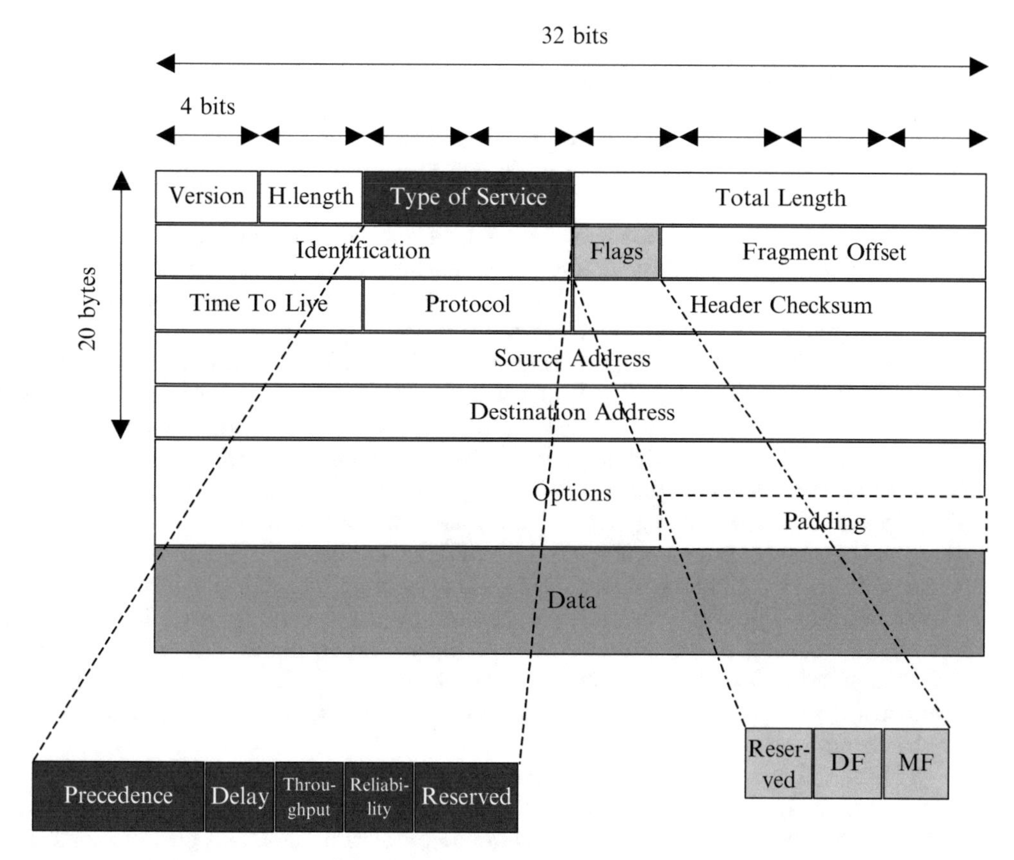

Fig. 2.4: IPv4 datagram format

- ToS, Type of Service (8 bits): A field carrying information to support quality of service features, such as prioritized delivery of IP datagrams. The ToS byte is divided into four subfields, as shown in Fig. 2.4:
  - The first three bits are used for the *precedence* field (value of 0 for a normal priority, up to 7 for control messages).
  - The *delay* bit specifies whether a low delay is required for the datagram transfer ($D=1$) or if the delay is not critical ($D=0$).
  - The *throughput* bit $T=1$ when high throughput is needed, instead of $T=0$ if the throughput is not a critical issue.
  - The *reliability* bit $R=1$ when high reliability is required, instead of $R=0$ if reliability is not needed.
  - The last two bits are unused.

The ToS byte has never been used as originally defined. A great deal of experimental, research, and deployment work has focused on how to use these 8 bits (ToS field), which have been redefined by IETF for use by Differentiated Services (DiffServ) and by Explicit Congestion Notification (ECN); see also the following Sects. 2.5, 2.7.8.2, and 2.7.8.3.

- TL, Total Length (16 bits): This field specifies the total length of the IP datagram in bytes. Since this field is 16 bits wide, the maximum length of an IP datagram is 65,535 bytes (typically, they are much smaller to avoid fragmentation due to MAC layer constraints). The most common IP packet length is 1500 bytes to be compatible with the maximum Ethernet payload size.
- Identification (16 bits): This field contains a 16-bit value, common to each fragment belonging to the same message. It is filled in for originally unfragmented datagrams if they have to be fragmented at an intermediate router along the path. The recipient uses such a field to reassemble messages to avoid accidental mixing of fragments coming from different messages since the IP datagrams can be received out of order.

- Flags (3 bits): It contains three control flags, but only two of them are used: Do not Fragment (DF) flag and More Fragments (MF) flag. If DF = 1, the datagram should not be fragmented. MF = 0 denotes the last fragment of a datagram.
- Fragment Offset (13 bits): When a message is fragmented, this field specifies the position of the current data fragment in the overall message. It is specified in units of 8 bytes (64 bits). The first fragment has an offset of 0.
- TTL, Time To Live (8 bits): This field specifies how long a datagram is allowed to "live" in the network in terms of router hops. Along the path from source to destination, each router decrements the TTL value of 1 before forwarding the related datagram. If TTL becomes zero, the datagram is not forwarded, but discarded, assuming that the datagram has taken a too long (wrong) route (e.g., a loop).
- Protocol (8 bits): This field identifies the higher-layer protocol carried out in the datagram. The values of this field were originally coded in IETF RFC 1700 [14]. For instance, the TCP protocol has a code equal to 6 (see Sect. 2.8.1); the Internet Control Message Protocol (ICMP) has a code equal to 1 (see Sect. 2.4).
- Header Checksum (16 bits): This is not the complex Cyclic Redundancy Check (CRC), typically used by data link layer protocols to protect the whole packet. This field is a checksum computed only on the IP packet header to provide basic protection against routing errors. The checksum is calculated by considering "the 16 bit one's complement of the one's complement sum of all 16-bit words in the header" [6]. In particular, the 16 bit one's complement sum is obtained by dividing the header into blocks of 16 bits; these blocks are summed (note that checksum bits are considered equal to 0 for this calculation), and the carry (if any) is summed to the result. Finally, the bits of the resulting binary number are complemented to obtain the bits of the checksum field. Since the TTL field changes at each hop, the checksum must be recalculated at each hop. The device receiving the datagram performs the checksum verification and, in the presence of a mismatch, discards the datagram as damaged (since the datagram could be misrouted).
- Source Address (32 bits): This is the 32-bit IP address of the originator of the datagram.
- Destination Address (32 bits): This is the 32-bit IP address of the intended recipient of the datagram. Even though routers may be the intermediate destinations of the datagram, this field always refers to the ultimate destination.
- Options (variable length): Several types of options may be included after the standard header of IP datagrams. The options field can be used for traceroute (i.e., determination of the path in the Internet), maximum TCP packet size determination, etc.
- Padding (variable length): If one or more options are adopted and the number of bits used for them is not a multiple of 32, some zero bits are padded to obtain a header length multiple of 32 bits.
- Data (variable length): The data to be transmitted in the datagram, either an entire higher-layer message or a fragment.

A few additional notes are needed on the checksum. IPv4 uses the checksum to verify the correctness of the header (IPv6 does not adopt any checksum control that is left to upper layers). Note that even TCP and UDP protocols use a checksum, but in this case the checksum is used to verify the correctness of both header (including a pseudo-header) and payload. The checksum is computed in a similar way as that described above by organizing data in 16-bit words: it is the one's complement of the one's complement sum of all 16-bit words. The limit of this checksum is that if two errors occur in the same position in two 16-bit words, no error can be revealed. On the contrary, the layer 2 CRC approach represents a powerful mechanism to detect errors. CRC is based on a cyclic code. For instance, a CRC of 4 bytes (called Frame Check Sequence, FCS) is used to protect Ethernet frames, as shown in Chap. 6.

2.3.2 IP Subnetting

Due to the explosive growth of the Internet, the use of IP addresses became inflexible to allow easy changes to local network configurations. These changes might occur when:

- A new physical network is installed in a location.
- The growth of the number of hosts requires splitting the local network into separate subnetworks.

Table 2.1: Network organization in subnets

<network number>	<subnet number>	<host number>
172.16	8	1, . . . , 255
172.16	15	1, . . . , 255

To avoid requesting additional IP network addresses in these cases, the concept of *subnets* was introduced: the main network now consists of a set of subnetworks (or subnets). The host field of the IP address of the main network is further subdivided into a subnetwork number and a host number. The IP address is organized as follows:

<network number> + **<subnet number>** + **<host number>**.

The combination of the subnet number and the host number is often called "local address" or "local part." "Subnetting" is implemented in a transparent way to remote networks. A certain host A within a network that has subnets is aware of subnetting, but a host B in a different network is unaware of them: B still regards the entire local part of the IP address of A as a host number. We consider that the subnetworks of a given network are interconnected via at least one router, which adopts a *subnet mask*, a sort of "filter" to identify these subnetworks. The router uses a mask to identify the subnetwork a given IP address belongs to. The mask is formed of a certain number of higher-order bits equal to "1," whereas the remaining lower-order bits are equal to "0." When a packet arrives, the router performs the AND operation between the IP destination address of the packet and the available mask(s) to determine the subnetwork the packet belongs to. In this case, both IP address and mask(s) are considered in binary format. Such operation permits us to extract the subnetwork address from the IP address (the result of the AND operation in binary format can also be expressed in the dotted-decimal notation for an easier representation). If the AND operation yields a match with one of the subnetworks connected to the router, the router forwards the packet through the appropriate interface towards the subnetwork; here, *direct routing* is used to deliver the packet on a local network by employing its MAC address (the Address Resolution Protocol, ARP, is used to convert the IP address to the MAC address). If the above match fails, the router has to send the packet towards the Internet (this is the classical routing case, also called *indirect routing*, which is based on routing tables); this happens when an IP packet is sent from a local host towards the Internet (i.e., the router has not to forward this packet to one of its connected subnetworks).

When we speak about the *default subnet mask* for a given address class, we consider a mask not modifying the length of the host part of the IPv4 address, i.e., not dividing the network into subnets. The following default subnet masks are defined:

- 255.0.0.0 for Class A
- 255.255.0.0 for Class B
- 255.255.255.0 for Class C

Within a subnetwork, the host address with all the bits equal to "0" is used for the subnetwork address. Instead, the host address with all the bits equal to "1" denotes the subnetwork broadcast address.

Let us consider the following simple example: given the Class B network, 172.16.0.0, we can extend the network part of the address from 16 to 24 bits by setting the last two bytes of the subnet mask equal to 1111111100000000 (i.e., 255.255.255.0 in dotted-decimal notation). Then, we can have, for example, two subnets: 172.16.8.0 and 172.16.15.0, each with up to 254 host addresses, as shown in Table 2.1. An alternative representation of an IP address of a subnet is using the number of bits "1" in the mask appended at the end of the IP address with a slash character. This permits us to identify the mask. Hence, referring to the above example, we could have for instance the following address to indicate a host in the subnet 172.16.8.0: 172.16.8.1/24.

When subnets with non-default masks are used, we speak about "classless" IP addressing. In classless addressing, any number of bits can be used for <network number> + <subnet number> and the slash notation can be used to easily represent this.

We consider now another example. We want to use the Class B address 131.15.0.0 for the network shown in Fig. 2.5, which is divided into two subnetworks interconnected through a router. The first subnetwork includes the hosts H1, H2, H3, and H4; the second one includes the hosts H5, H6, H7, H8, H9, H10, and H11. The bridge has no impact on IP addressing since it is a device operating at layer 2 of the ISO/OSI model. On the contrary, each router port needs an IP address.To efficiently assign IP addresses to the network in Fig. 2.5 from the given Class B network address, we adopt a subnetting approach: we enlarge the default Class B subnet mask by adding another

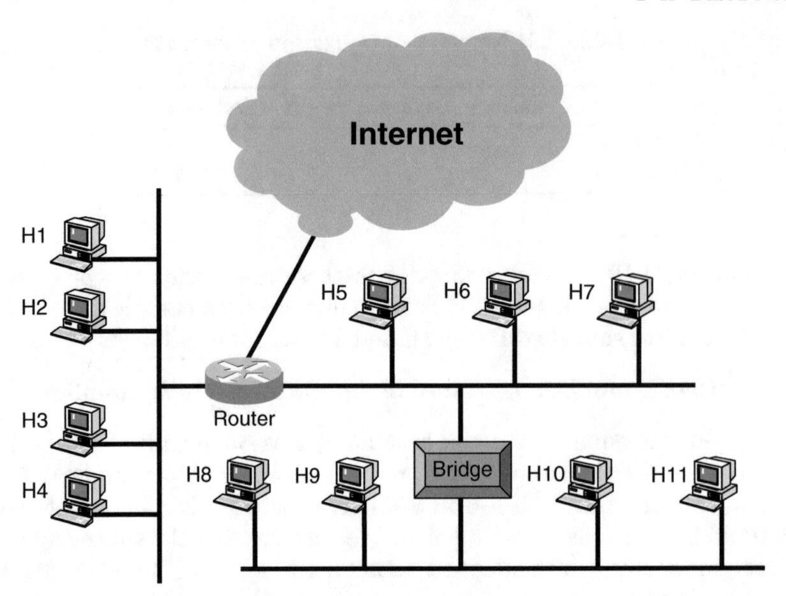

Fig. 2.5: Network divided into two subnetworks

byte to the network number. The default subnet mask is 255.255.0.0 for Class B, i.e., the first two bytes of the IP address indicate the network number. Hence, we adopt a new subnet mask: 255.255.255.0. A possible addressing choice could be the following one:

SUBNET #a: H1, . . . , H4
Subnetwork address: 131.15.2.0
Address of the router port connected to subnet #a: 131.15.2.5
H1 address: 131.15.2.1
H2 address: 131.15.2.2
H3 address: 131.15.2.3
H4 address: 131.15.2.4

SUBNET #b: H5, . . . , H11
Subnetwork address: 131.15.4.0
Address of the router port connected to subnet #b: 131.15.4.1
H5 address: 131.15.4.2
H6 address: 131.15.4.3
H7 address: 131.15.4.4
H8 address: 131.15.4.5
H9 address: 131.15.4.6
H10 address: 131.15.4.7
H11 address: 131.15.4.8

The port of the router connected to the Internet has address: 131.15.0.1.

Let us consider having a host with address 210.20.15.90/30. Based on the slash notation, such host belongs to a subnetwork with a mask equal to 255.255.255.252 (the last byte is equal to 11111100). We have to determine the subnetwork address. The subnetwork belongs to a Class C network with address 210.20.15.0. Since the mask divides 1 byte into two parts, the subnetwork address can only be determined by considering the binary representation of both host address and subnet mask and performing the AND operation. Only the last byte of the address is affected and we focus on it:

Host address	...90	= ...01011010	
Subnet mask	...252	= ...11111100	
Subnet addr.		= ...01011000	→ 88 (decimal representation)

Hence, the subnetwork address in dotted-decimal format is 210.20.15.88. Since only 2 bits are left free by the mask, this subnetwork has 2^2 addresses. Among these addresses, two addresses are special ones: network address and broadcast address; it is possible to have just two hosts in this subnetwork.

Now let us refer to the above Class C network address: 210.20.15.0. Considering still the above mask 255.255.252, we are interested in determining how many subnetworks can be obtained. Since the subnetwork number is 6 bit-long, there are $2^6 = 64$ combinations. Among these subnets, there are two special cases: the subnet Zero (i.e., all the subnet address bits are equal to 0) and the all-ones subnet (i.e., all the subnet address bits are equal to 1). In the past, it was suggested not to use these special subnets: the subnet Zero would have an address coincident with the entire Class C network address; moreover, the all-ones subnet would have a broadcast address coincident with the broadcast address of the whole Class C network. This constraint is now removed, as shown in RFC 1878.

2.3.3 Public and Private IP Addresses

So far, we have considered "public" IPv4 addresses, that is geographically used IP addresses having a general meaning. Public IP addresses are used by those systems that need to be reached by the entire Internet, such as

- Web servers.
- e-mail servers (POP, SMTP protocols).
- Database servers.

A local institution could purchase one or more IP subnetworks to interconnect to the Internet, but this approach is very expensive because of the limited number of available IP public addresses. To implement a local network (Intranet), there is no need for public IP addresses. Private IP addresses (i.e., internal IP addresses to the Intranet) could be adopted locally, using functionality to translate them into a public address at the Gateway (GW) towards the public Internet. Private IP addresses have no global validity (RFC 1918). Private IP addresses belong to the following ranges:

- **10**.0.0.0–**10**.255.255.255 (Class A).
- **172.16**.0.0–**172.32**.255.255 (Class B).
- **192.168**.0.0–**192.168**.255.255 (Class C).

The Network Address Translator (NAT) gateway permits us to connect to local networks using private IP addresses (RFC 1631 and 2663). The advantages of using NAT are:

- The NAT limits the number of public IP addresses to connect a local network to the Internet.
- The private address space is wide, thus allowing some flexibility.

There are two main NAT translation modes:

- Dynamic translation: a large number of internal users share a single external IPv4 address; this is the case of the "one-to-many NAT," below referred to as "IP masquerading." This is the approach typically adopted.
- Static translation: a block of external addresses is translated into a block of the same size of internal addresses.

The "one-to-many NAT" operates as follows. The NAT-gateway must have a public IPv4 address. When a client on the local network sends IP packets to an Internet server, they contain IP source and destination addresses and the port to be used. In particular, the NAT uses the following data in the IP packet:

- Source IP address (e.g., 192.168.10.45).
- TCP or UDP source port (e.g., 2510).

The following modifications are made in the IP packet sent by the NAT-gateway (*IP masquerading*):

- The source IP address is replaced with the external (public) IPv4 address of the gateway, for instance, 70.15.0.5.

- The source IP port is replaced with a new port not used by the gateway; for instance, 28136 (port in the range of dynamic ports; see Sect. 2.8.3).

The gateway-NAT will record the modifications made in the IP packet in its state table so that it can perform the inverse operation for the return packets. Both the local client and the Internet server are unaware of these modifications: for the local host, the NAT is simply the Internet gateway; instead, the Internet server does not know that the local host has actually sent the packet: for the Internet server it is as if the packet was directly sent by the NAT. When the Internet server responds to the IP packet received from the local client, it sends the packets to the IP address of the gateway-NAT (70.15.0.5) and to the modified port (28136). Then, the NAT will search its state table for a match with an already-established connection. Hence, the NAT can recover the local IP address (192.168.10.45) and the actual source port (2510). The original IP address and the source port are restored before delivering the received packet to the local client.

The NAT approach also protects the local network client identities from external attacks.

With IPv6, there is no need for a NAT to reduce the number of public IP addresses. Nevertheless, some NAT-like functions are still needed to protect the identity of local clients from external attacks (security issue). Recent IETF work is considering that this functionality could be supported by IPv6-to-IPv6 network prefix translation (RFC 6296), where a 1:1 mapping is supported between "inside" and "outside" IPv6 prefixes. Finally, some form of IPv6 private addresses is supported by site-local unicast addresses, as specified in Sect. 2.3.6.

2.3.4 Static and Dynamic IP Addresses

The IP address is assigned to a host or at the booting time or permanently with a fixed configuration. In the first case, we speck of dynamic IP address; instead, we speak of static IP address in the second case.

Static IP addresses are manually assigned to a computer by an administrator. The exact procedure depends on the platform. This contrasts with the management of dynamic IP addresses, typically assigned by a server, using the DHCP protocol. In some cases, a network administrator may implement dynamically assigned static IP addresses: a DHCP server is used, but it is specifically configured always to assign the same IP address to a computer. In this way, static IP addresses are managed centrally.

Dynamic IP addresses are most frequently used in Local Area Networks (LANs) and wireless LANs: DHCP frees the network administrator from assigning an IP address to each host manually. Dynamic addresses assigned by DHCP can also be private IP addresses.

The use of dynamic IP addresses allows many devices of a network to share a limited address space in the case where only a few of them are simultaneously active. This applies to Internet Service Providers (ISPs), in which a given IP address can be assigned to different users at different times.

2.3.5 An Example of Local Area Network Architecture

Figure 2.6 depicts an example of IPv4 LAN architecture, including the following main elements:

- Web servers.
- A DNS server (client servers query the DNS to translate alphanumeric addresses into IP addresses).
- A gateway (a router interconnecting to the Internet and typically having even layer 4 protocols).
- A firewall-NAT (to protect the network).
- A DHCP server (for assigning dynamic IP addresses to hosts).

A single layer-2 network can be virtually divided into multiple broadcast domains, called Virtual Local Area Networks (VLANs). A VLAN is a broadcast domain created by one or more switches. IP packets can pass between VLANs only through one or more routers. A VLAN is a group of terminals, servers, and other network resources, which behave as if they were connected to a single network segment. A LAN segment is a *collision domain*: collisions remain within the segment (see also Sect. 6.5.1). The area within which broadcasts and multicasts are confined is called a *broadcast domain*. LAN segments interconnected through bridges or switches are in the same broadcast domain. VLANs allow a network manager to divide a LAN into different broadcast domains logically.

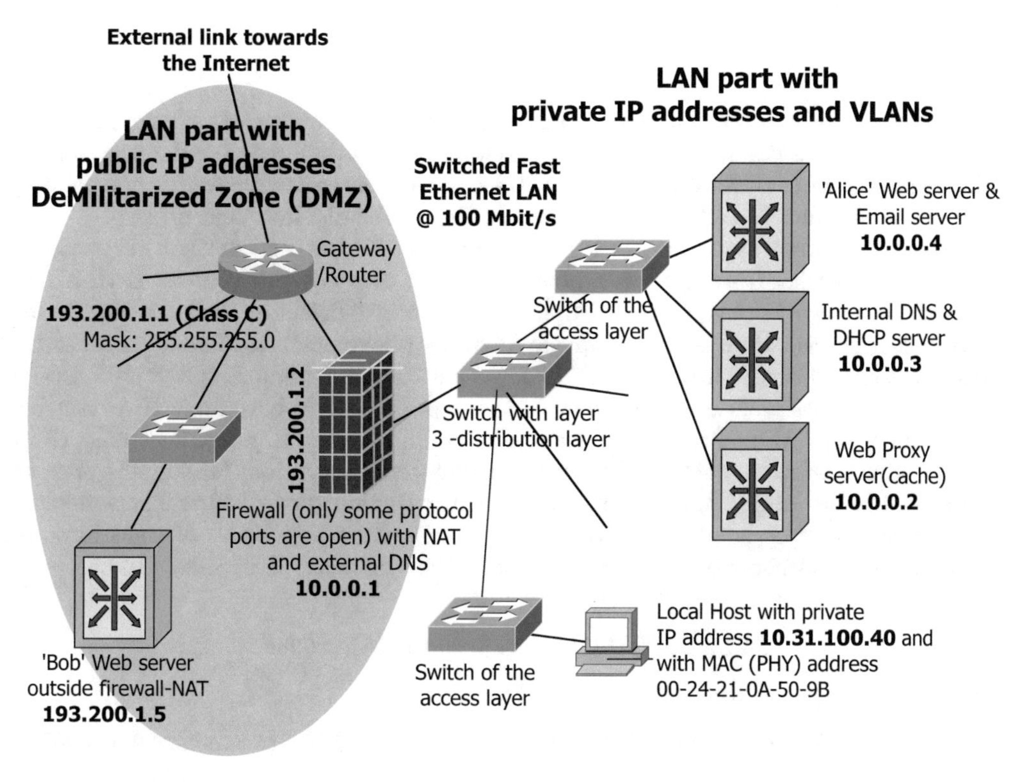

Fig. 2.6: Example of LAN with public and private addresses. This architecture adopts a firewall-NAT and a gateway

One of the most important VLAN protocols is IEEE 802.1Q. A VLAN tag of 4 bytes is inserted in the Ethernet frame header after the source MAC address. Different VLANs can be used within a LAN to support different traffic types, such as voice, data, wireless, etc. Different VLANs (layer 2) correspond to different subnets (layer 3). For instance, the VLAN with subnet 10.4.2.0/24 could be used for voice traffic (VoIP), instead the VLAN with subnet 10.4.0.0/24 could be used to support data traffic. VLANs permit us to improve performance in the presence of multicast and broadcast traffic, mainly destined to a portion of the LAN. Moreover, VLANs help to create workgroups and manage security.

A LAN design is based on the 80/20 rule: 80% of the traffic remains within the LAN and only 20% is routed outside the network. Many organizations have centralized their resources: Internet Web servers, e-mail servers, and other servers are in the same segment of the network, as shown in Fig. 2.6. Hence, most of the LAN traffic has to be routed to this portion of the LAN. Because routing introduces more latency than switching, the 20/80 rule has dictated the need for a faster Layer 3 technology, namely, Layer 3 switching. A Layer 3 switch is a router and switch together suitably designed.

A hierarchical model has been defined to design LANs. In particular, three layers are considered (see Fig. 2.6).

- The Access Layer is where the end-user connects to the network. Access Layer switches generally have a high number of low-cost ports per switch, and VLANs are usually configured at this Layer. In a distributed environment (80/20 rule), servers and other such resources are kept in a suitable portion (VLAN) of the Access Layer. Switching is performed at the access layer.
- The Distribution Layer provides end-users with access to the Core (backbone) Layer. Security and QoS are usually configured at the Distribution Layer. This layer ensures that packets are routed adequately between subnets (VLANs) in the LAN.
- The Core Layer is the backbone of the network. The Core Layer is concerned with switching data quickly, efficiently, and reliably between all other network layers. In a centralized environment (20/80 rule), servers are placed in a portion of the Access Layer so that the Core Layer must switch traffic from all other Access Layers to this part.

2.3.6 IP Version 6

Internet Protocol version 4 (IPv4) was standardized in the 1970s. The number of unassigned Internet addresses is running out, so a new addressing scheme has been developed and is designated as Internet Protocol version 6 (IPv6). Since the beginning of the 1990s, hundreds of RFCs have been written, covering several aspects, such as expanded address space, simplified header format, flow labeling, authentication, and privacy.

IPv6 represents an evolutionary step from IPv4: IPv6 can be installed as a normal software upgrade in Internet devices and is interoperable with IPv4. IPv6 is replacing IPv4 in core network routers. IPv6 is designed to run well on high-performance networks (e.g., Gigabit Ethernet) and to be efficient at the same time in low-bandwidth networks (e.g., wireless systems). IPv6 is defined in the following documents: RFC 2460 (updated by RFC 8200), "Internet Protocol, Version 6 (IPv6)" and RFC 2373, "IP Version 6 Addressing Architecture." Besides increasing the address space, other essential features of IPv6 are: (1) the possibility to have large IP packet payloads (jumbograms) for better efficiency; (2) quality of service marking and flow labels to prioritize traffic; (3) network layer security using IPsec, which includes protocols for authentication and encryption; (4) support of mobility.

An IPv6 address is 128-bit long (instead of 32 bits as in IPv4); this allows 340 trillion trillion trillion (2^{128}) of addresses. For a compact representation, an IPv6 address is written as a series of eight hexadecimal strings separated by colons; each hexadecimal string has four hexadecimal symbols and represents 16 bits. An IPv6 address example is:

2001:0000:0234:C1AB:0000:00A0:AABC:003F.

In IPv6, there are three types of addresses:

- *Unicast*: An address used to identify a single interface. Based on reachability conditions, the following types of unicast addresses are available:
  - Global unicast address. This is an address that can be reached and identified globally. A global unicast address consists of a global routing prefix, a subnet ID, and an interface ID. The current global unicast address allocation uses the range of addresses that start with the binary string 001.
  - Site-local unicast address. This is an address that can be reached and identified only within the customer site (similar to an IPv4 private address). Such addresses have the *prefix* 1111111011, a subnet ID, and an interface ID.
  - Link-local unicast address. This is an address that can be reached and identified only by nodes attached to the same local link. These addresses have the prefix 1111111010 and an interface ID.
- *Anycast*: The anycast address is a global address assigned to a set of interfaces belonging to different nodes. A packet destined to an anycast address is routed to the nearest interface (according to the routing protocol measure of distance), one recipient. An anycast address must not be assigned to an IPv6 host, but it can be assigned to an IPv6 router. According to RFC 2526, a type of anycast addresses (reserved subnet anycast addresses) is composed of a 64-bit subnet prefix, a 57-bit code (all "1s" with one possible "0"), and an anycast ID of 7 bits. Anycast addresses can be used for implementing new services, such as (1) selection of the nearest server for a given service; (2) DNS and HTTP proxies addressing, thus avoiding to know local addresses.
- *Multicast*: As in IPv4, a multicast address is assigned to a set of interfaces belonging to different nodes. A packet destined to a multicast address is routed to all interfaces identified by that address (i.e., many recipients). IPv6 multicast addresses use the prefix 11111111 and have a group ID of 112 bits.

The IPv6 header is shown in Fig. 2.7 according to RFC 2460 [15].The IPv6 header fields in Fig. 2.7 are described below:

- Version (4 bits): Indicates the protocol version so that it contains the number 6.
- Traffic Class (8 bits): This field is used for quality of service support, similar to the Type of Service byte in IPv4. In particular, the six most significant bits are used for Differentiated Services (DiffServ). The remaining two bits are used for Explicit Congestion Notification (ECN).
- Flow label (20 bits): It is used for the management of traffic flows.
- Payload length (16 bits): Indicates the size of the data field.
- Next header (8 bits): Identifies the type of header immediately following the IPv6 header.
- Hop limit (8 bits): Such value is reduced by one to each node that forwards the packet. When the hop limit field reaches zero, the IP packet is discarded.

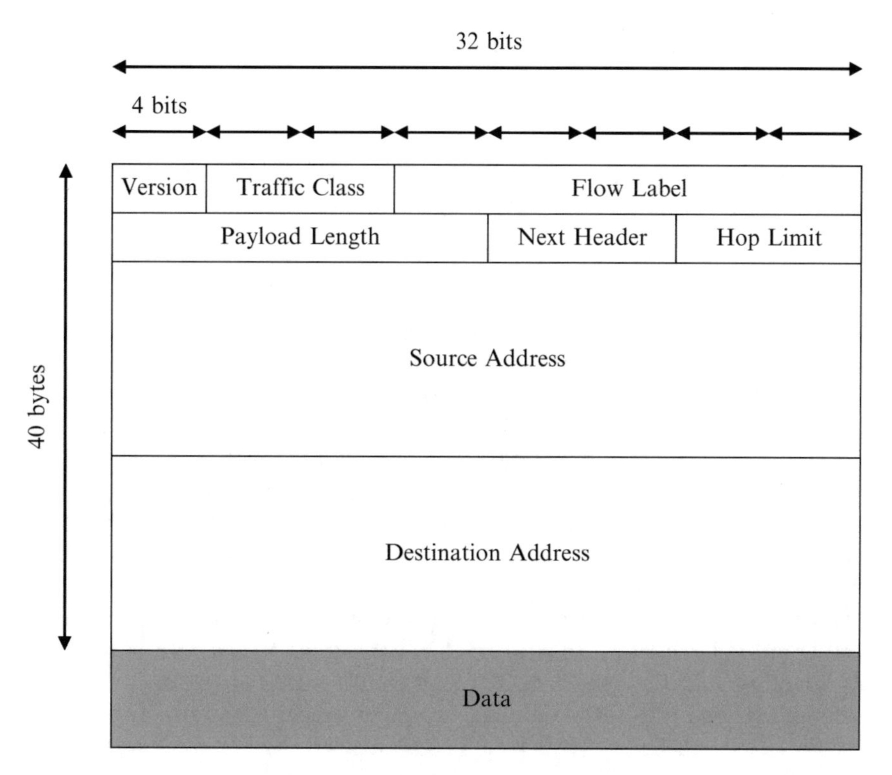

Fig. 2.7: IPv6 header format

- Source address (128 bits): The address of the originator of the packet.
- Destination address (128 bits): The address of the intended recipient of the packet.

Multiple extension headers can be present in the same IPv6 packet header (e.g., Hop-by-Hop header, Destination header, Routing header, Fragmentation header, Authentication header, and Encapsulating Security Payload header). In particular, the fragmentation header is used by the source to indicate that the packet was fragmented to fit within the Maximum Transmission Unit (MTU) size; see also the following Sect. 2.8.1 on TCP. In IPv6, unlike IP4, packet fragmentation and reassembly are performed by the end nodes rather than by routers; this solution further improves the IPv6 efficiency. IPv6 uses ICMP error reports to determine the MTU to be used along a path.

The decision to eliminate the checksum in the header of IPv6 datagrams derives from the fact that error control is typically already performed at layer 2, and this is sufficient because of the low error rate in the current networks. Better performance is thus achieved since routers no longer need to recompute the checksum of each packet.

Let us now focus on the IPv6 deployment strategy. Any successful strategy requires to implement IPv6 to coexist with IPv4 for a certain period. The following strategies have been envisaged for managing the complex and prolonged transition from IPv4 to IPv6.

- Dual-stack backbone: In dual-stack backbone deployment, all routers in the network maintain both IPv4 and IPv6 protocol stacks. Applications may choose between IPv4 and IPv6.
- IPv6 over IPv4 tunneling: In this solution, IPv6 traffic is encapsulated in IPv4 packets to be transmitted over an IPv4 backbone. This solution enables IPv6 end systems and routers to communicate across an existing IPv4 network.

2.4 Domain Structure and IP Routing

Routing is a fundamental function of the IP layer. It provides the mechanisms to ensure that the routers interconnect different physical networks so that the exchange of data is possible from a source host to a destination. The Internet is such a large collection of nodes that one routing protocol cannot handle the update of tables of all routers; this task would be computationally unfeasible. Therefore, the Internet is divided into Autonomous Systems (AS): an AS is a group of networks and routers under the authority of a single administration. An AS is also sometimes referred to as a *routing domain* or simply domain. The routing problem is thus divided into smaller (easier) subproblems inside the AS or between ASs. The administration of an AS appears to other ASs to have a single and consistent interior routing plan and presents a coherent description of which networks are reachable through it. In particular, routing functions can be distinguished as:

- Intra-domain routing protocols (or Interior Gateway Protocol, IGP), i.e., routing within an AS.
- Inter-domain routing protocols (or Exterior Gateway Protocol, EGP), i.e., routing between ASs.

All interior routing protocols have the same basic functions: they determine the best route for each destination within an AS and distribute routing information among the routers of the AS. From the standpoint of exterior routing, an AS can be viewed as a monolithic block. Moving routing information into and out of these monoliths is the task of exterior routing protocols. The routing information passed between ASs is called *reachability information*. It is merely information about which networks can be reached through a specific AS. An important feature of exterior routing protocols is that most routers do not make use of them. Exterior protocols are required only when an AS exchanges routing information with other ASs. Only those gateways that connect an AS to another AS need to run an exterior routing protocol. Unless we have to provide a similar service level, there is probably no need to run exterior routing protocols.

A *domain name* is a label, which identifies a realm of administrative autonomy, authority, or control on the Internet. Domains are organized according to a hierarchical (tree) structure with sub-domains: the first-level of domains are top-level domains, such as .com, .net, .org, and country-code top-level domains. Below these top-level domains, there are second-level and third-level domains, which represent LANs needing to be interconnected to the Internet. The registration of the domain names is usually administered by domain name Registries, which sell their services to the public. Domain names are determined based on suitable rules and procedures. The Internet Assigned Number Authority (IANA) administers the root domains, that is, the domains at the top of the hierarchy. Upon request of the administrators, the Registry associates a name with the long and difficult-to-memorize numerical IP address of the domain/network. This association is stored in an archive (database of assigned names) that all computers connected to the Web must query to reach a domain. This service is called Domain Name System (DNS). The DNS is a large, distributed, hierarchical database, which resides on various computers and contains names and IP addresses of various hosts and domains on the Internet.

Within interior routing protocols, we can distinguish two sub-cases: *direct routing* and *indirect routing*. If the destination host is attached to the same network of the source host, an IP datagram can be transmitted simply by encapsulating it within the physical network frame. This is called direct routing. Instead, indirect routing is used when the destination host is not in the same network as the source host: the only way to reach the destination is via one or more routers. A host can recognize whether a route is direct or indirect by comparing the network and the subnet parts of source and destination addresses. If they match, the route is direct and the source host can identify the destination through the Address Resolution Protocol (ARP).[1]

For indirect routes, routing entails knowing next router's IP address on the path towards the destination network. Each router keeps an *IP routing table* with mappings between destination IP addresses and IP addresses of *next-hop* routers. Three types of mappings can be found in this table:

- Direct routes for locally attached networks.
- Indirect routes for networks reachable via one or more routers.
- A default route, which contains the IP address of a router to be used for all IP addresses not covered by direct and indirect routes.

Routing tables are generated and maintained by routing protocols, running in all routers that are synchronized to one another.

Routers' fundamental function is present in all IP implementations: an incoming IP datagram, specifying a destination IP address other than a local IP address is treated as a normal outgoing IP datagram. This outgoing IP datagram is subject to the IP forwarding algorithm at the router, which selects the next hop for the datagram based on a local table of the router. This next hop can be towards any of the networks physically attached to the router. Then, the result is that the router has to forward the IP datagram from one network to another, as shown in Fig. 2.8.

[1] ARP is an Internet protocol, which dynamically determines the physical hardware (MAC) address corresponding to an IP address in case of direct routing.

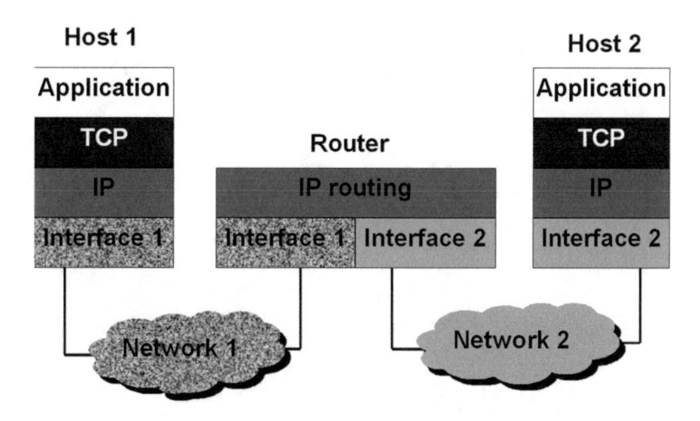

Fig. 2.8: Example of routing of IP datagrams between two different networks

In managing the routing of IP datagrams, some error reporting should be implemented by routers via ICMP messages. They should be able to report the following errors back to the source host:

- Unknown IP destination network using an ICMP Destination Unreachable message.
- Redirection of traffic to a more suitable router through an ICMP Redirect message.
- Congestion problems (i.e., too many incoming datagrams for the available buffer space) by an ICMP Source Quench message.
- If the "Time To Live" (TTL) of an IP datagram has reached zero, this is reported with an ICMP Time Exceeded message.
- Also, the following base ICMP operations and messages should be supported:
  - Parameter problem
  - Address mask
  - Timestamp
  - Information request/reply
  - Echo request/reply

2.4.1 Routing Algorithms

The desirable properties of routing protocols are correctness, simplicity, robustness, stability, fairness, and optimality (e.g., the shortest path concerning some "distance metric"). The network is considered an *oriented graph* with nodes (= routers) and edges (= links between routers), where edges have a direction associated with them. There is a *weight* (a type of "distance") associated with each link connecting nodes in the network. A weight equal to infinity means that there is no link between two nodes. The weight of a path (involving different routers in the networks) is calculated as the sum of the weights of all the edges (links) in the path. A path from x to y is the "shortest" one if there is no other path connecting x and y with a lower total weight. If the weight is 1 for each link, the routing metric is *hop count* (related to the TTL field of the IP packet header).

Chapter 7 will also consider multipath routing as those techniques that can combine the use of multiple paths exploiting the meshed connectivity to deliver the traffic to the destination; this study will address traffic engineering issues and also exploit some network-of-queues modeling, and this is the reason it is in Chap. 7. Here, we consider only single-path routing.

Shortest path routing is based on the Bellman's famous *optimality principle*: if router j is on the optimal path from router i to k, then even the optimal path from j to k is on the same path (see Fig. 2.9). Consequently, the set of optimal paths from all the routers to a tagged router forms a tree, named *sink tree* (or *minimum spanning tree*) for the tagged router. This principle is for instance exploited by the Dijkstra algorithm [16], an example of shortest path routing algorithm.

A classification of routing algorithms can be as follows:

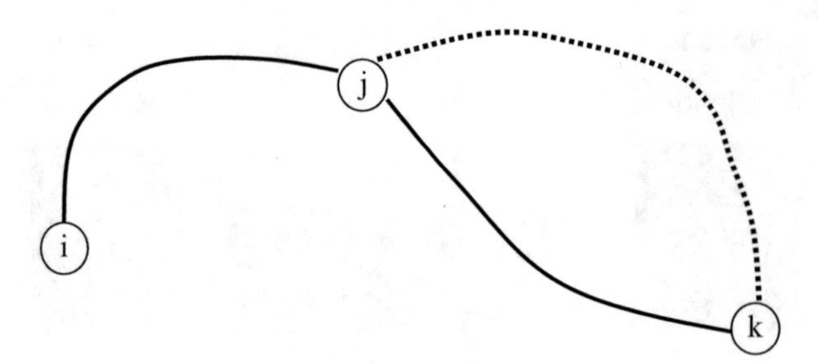

Fig. 2.9: Selection of the shortest path from node i to node k

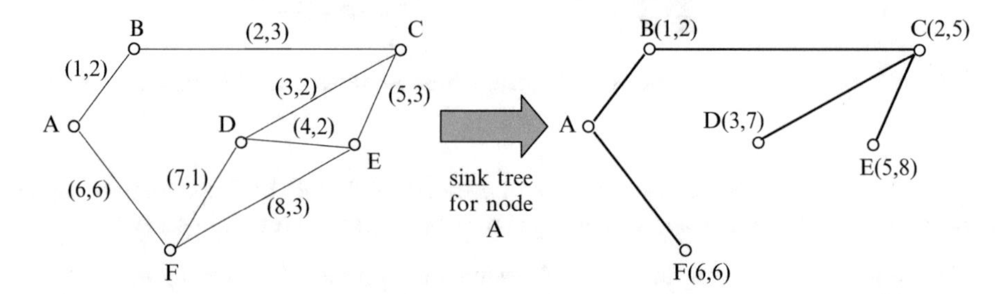

Fig. 2.10: A generic link in the network is labeled by (a, c), where "a" is the link number and "c" is the link cost or weight. The example of the sink tree for node A is provided: a single path is selected to reach each node from A; a generic node is labeled with the number of its last link of the path and with the total cost

- *Centralized routing*: The routing algorithm is performed once for the whole network at the Routing Control Center (RCC), thus generating the different routing tables of the routers.
- *Decentralized routing*: The algorithm is running in parallel at the different routers and converges to the definition of their routing tables. Each router knows the address of its neighbors and knows the cost to reach them. These algorithms require a signaling protocol for the exchange of information among adjacent routers to contribute globally to the creation of routes.

Routing algorithms can also be classified into two broad categories:

- *Static algorithms*, where routes never change after their initial definition.
- *Adaptive algorithms*, which employ dynamic information (e.g., current network topology, load, and delay) to update routes.

There are many shortest path routing algorithms in the literature [e.g., Dijkstra (for link-state routing), Bellman–Ford (for distance-vector routing), A* Search, Prim, Floyd–Warshall, Johnson, and Perturbation theory] [17]. We describe below the Dijkstra algorithm [16], also known as Shortest Path First (SPF), a centralized routing scheme. The *weight* of a path (involving different routers in the networks) is calculated as the sum of the weights of the links in the path; each hop could count for 1 if a simple hop-count metric is used. A path from x to y is the "shortest" one if there is no other path connecting x and y with a lower total weight. Let us recall that a *sink tree* is a tree connecting a given node to all the other network nodes with the shortest paths (see the example in Fig. 2.10).

The Dijkstra algorithm determines the sink tree for each node and operates by extending the paths for increasing distances from a given source node. This process is repeated for all the nodes. The sink tree of a node can be easily converted into the routing table of that node (i.e., a table showing the next hop for each destination to be reached from that node).

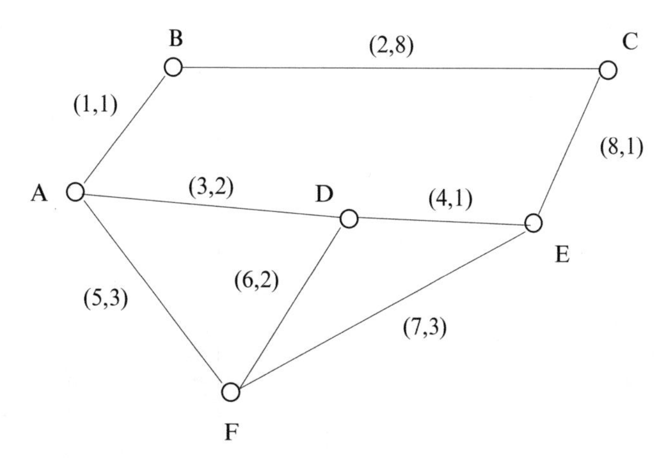

Fig. 2.11: Network with bidirectional links, labeled as (a, c), where "a" is the link number and "c" is the link cost

The Dijkstra algorithm to determine the sink tree of a generic node (called here "A") is based on the steps below and uses a table containing couples (a, c) of arc numbers and costs of all the links in the network; in what follows, we consider that all the *links are bidirectional.*

1. Let us start from the generic node A as source and we label all the other nodes of the graph with infinite costs.
2. We examine the nodes linked directly to node A and we relabel them with link numbers and link costs to A.
3. We create an *extension of the path*: According to the labeled nodes, we add to the tree the node, which has the smallest cost and consider it as part of the shortest path tree for node A. If there are more possibilities with the same cost, we select the link with the lowest arc number. Let C denote the node selected and added to the tree at this step.
4. We *relabel the nodes*: the nodes that can be relabeled are those not yet added to the shortest path tree, but linked to node C. Let B denote a generic node belonging to this set for potential relabeling. The new label of B is formed of the number of the link connecting B–C and a new cost given by the sum of the B–C link cost with the cost in the label of C. The new label will actually substitute the old one if and only if the cost of the new label is lower than that of the old label (it can also happen that there are no label changes at a given step).
5. We go back to step #3, until all nodes of the graph are added to the shortest path tree of A.

The Dijkstra algorithm completes a sink tree in a maximum number of iterations equal to $n - 1$, where n is the number of network nodes. Therefore, the computational complexity of the Dijkstra algorithm for a network of n nodes is $O(n^2)$ {using a suitable data structure, complexity can be reduced to $O[L + n \times \log(n)]$, being L the number of links in the network}. Note that there is a certain degree of redundancy in the sink trees of the different nodes. Hence, there is probably no need to completely recompute the sink trees of all the nodes in the network; some tree parts can be reused based on the optimality principle.

Let us examine the following example of the Dijkstra algorithm application. It is requested to determine the sink tree of node A for the network shown in Fig. 2.11, where each link is labeled by a number and a cost.

We start the Dijkstra algorithm by labeling all nodes with infinite costs. We take node A as a reference and relabel all nodes connected to A, that is, nodes B, D, and F.

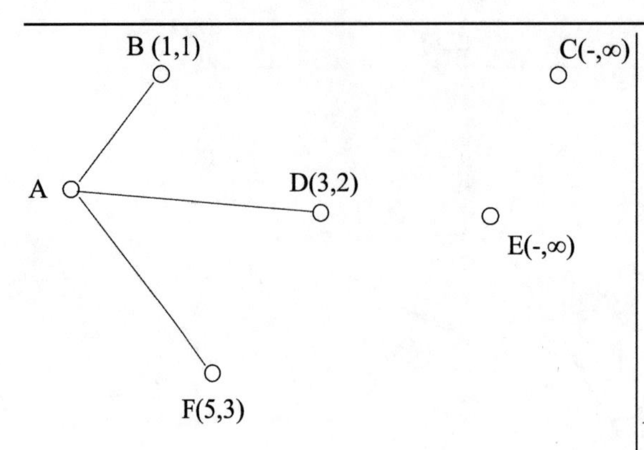

Among nodes B, D, and F, we select to connect node B, which has the lowest cost. Then, we need to relabel the nodes connected to node B (i.e., node C).

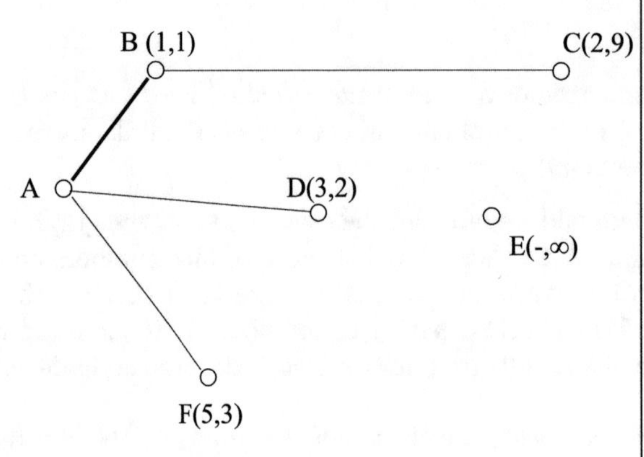

Among nodes C, D, and F, we select to connect node D, which has the lowest cost. Then, we need to relabel the nodes connected to node D (i.e., nodes E and F). However, the label of node F does not change because the cost would increase.

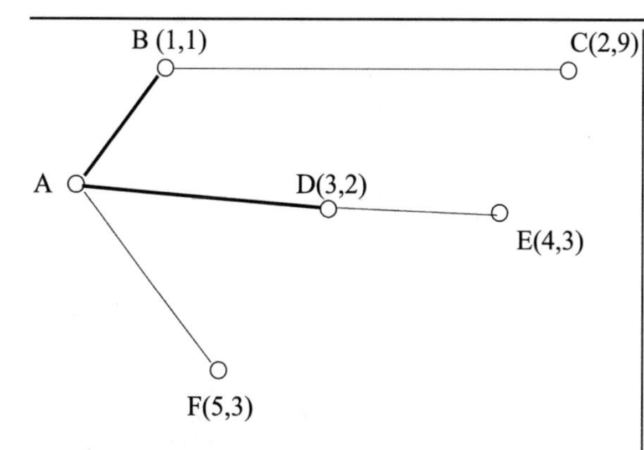

Among nodes C, E, and F, we select to connect node E, which has a lower cost than C and the same cost, but a lower link number than F. Then, we need to relabel the nodes connected to node E. However, the label of node F does not change because the cost would increase. Instead, the label of node C changes because the cost reduces.

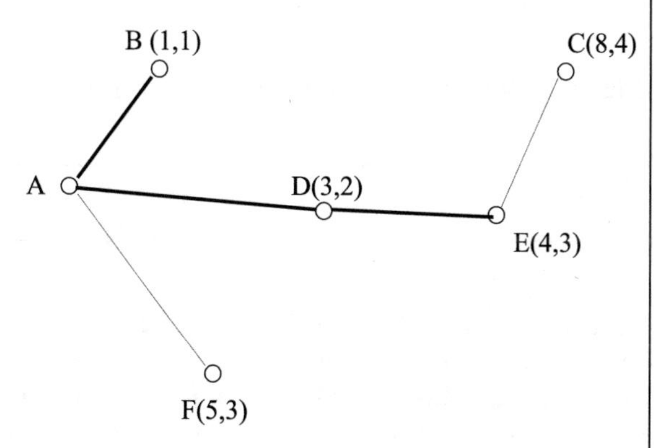

Between nodes C and F, we select to connect node F, which has the lowest cost. Then, we need to relabel the nodes connected to node F. However, the labels of nodes D and E cannot change because these nodes already belong to the sink tree.

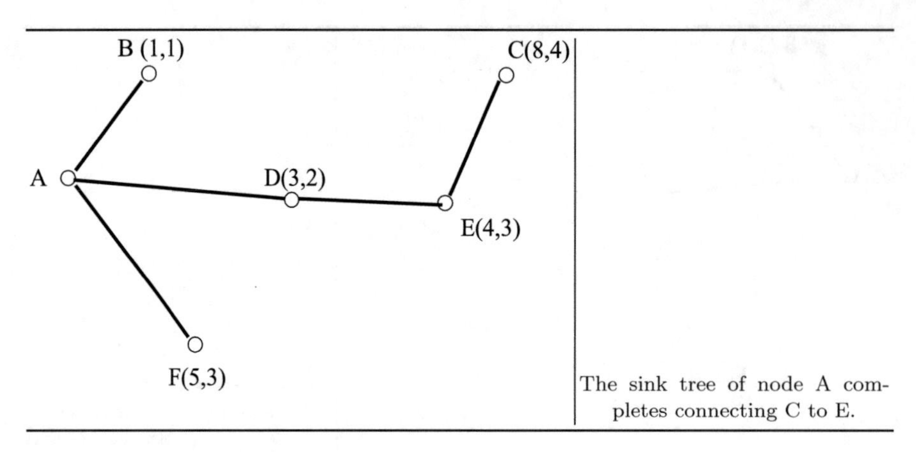

The sink tree of node A completes connecting C to E.

We can also represent the sink tree of node A in a tabular form as follows:

Destination node	B	C	D	E	F
Path from A	AB	ADEC	AD	ADE	AF
Cost	1	4	2	3	3

The routing table for node A can be derived immediately from the sink tree as shown below:

Destination node	B	C	D	E	F
Next hop from A	B	D	D	D	F

Once we have solved the shortest path routing using the Dijkstra algorithm, we can distribute the traffic flows on the links from the source node to the destination node. For instance, referring to the input traffic at node A and the sink tree we have achieved, we can consider the traffic flows to be allocated in the network links. In particular, let λ_{Aj} be the generic input flow at node A destined to node $j \in \{\mathrm{B}, \mathrm{C}, \mathrm{D}, \mathrm{E}, \mathrm{F}\}$. Assuming that each link has a capacity C_{link}, we need to verify if it can support the allocated traffic due to the λ_{Aj} contributions. Let us see Fig. 2.12, where these traffic flows are routed according to the shortest path. For the sake of simplicity, let us assume that all the traffic flows λ_{Aj} are equal. Then, the bottleneck link is the link AD for which we have to verify the condition:

$$\lambda_{\mathrm{AC}} + \lambda_{\mathrm{AD}} + \lambda_{\mathrm{AE}} < C_{\text{link}} \tag{2.1}$$

This example could be completed by allocating the traffic flows originated at other nodes of the network to the different links.

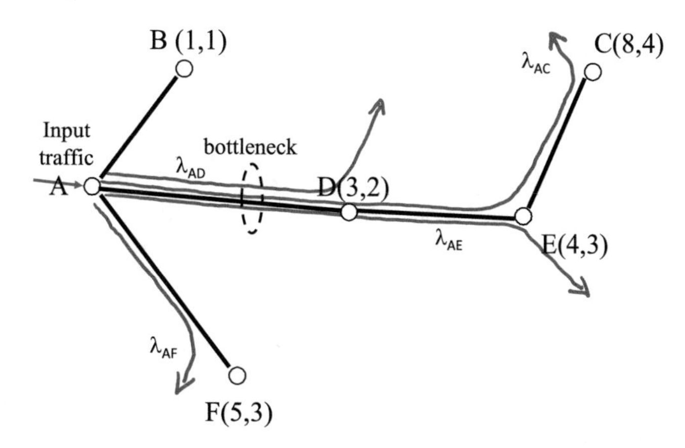

Fig. 2.12: Network with the shortest paths for the traffic flows from node A

Another static routing technique is the *flooding scheme*. Flooding entails that a router sends each arriving packet on every output link except the link from which the packet has arrived. Flooding is a distributed routing scheme. This routing scheme is quite simple to be implemented and requires limited processing capabilities at the routers: practically, flooding does not use routing tables. Flooding can be used as a benchmark scheme for other routing algorithms. Flooding always uses the shortest path because it uses any possible path in parallel; consequently, no other protocol can achieve lower delays. Flooding also has practical applications in ad hoc wireless networks and sensor networks, where all messages sent by a station can be received by all other stations in the transmission range. The problem with flooding is that it makes use of network resources in a redundant way: flooding involves an increasing number of links as long as we move away from the source (initial) node. Flooding can cause congestion: flooding has the drawback of generating a virtually infinite number of packets. There are some techniques to avoid this problem: a counter is used in each packet; source router ID and sequence number are used in each packet; selective flooding is adopted, where packets leaving a router are transmitted only on those output links going in the right directions.

Let us now focus on the main intra-domain routing algorithms. In particular, we refer to two adaptive, distributed routing algorithms:

- *Distance-Vector Routing* (based on the Bellman–Ford algorithm). Each router maintains a table giving the best known distance to every destination and the output port to be used for sending there. These tables are updated iteratively, exchanging distance vectors with neighbor routers; routers are neighbors if they are directly connected. The vector sent by a router contains all the known distances from other routers in the network. The distributed process is as follows. Upon receipt of the distance vector from a certain neighbor router A, router B updates the distances (obtained summing the distances in the vector of A with the distance from B to A) for the other routers in the network considering paths going through A. Then, all the information on routes received from neighbors is merged to create the new routing table of B. In particular, for each destination router, the next hop (i.e., a neighbor node) is selected based on the shortest distance criterion. After some iterations exchanging distance vectors among neighbor nodes, the routing tables at the nodes converge to stable values. This algorithm was used in the ARPANET but exhibited problems in the case of link failures because the process to update the tables may be too long to converge, thus causing possible routing loops.
- *Link-State Routing* (adopting the Dijkstra algorithm). Each router is responsible for contacting its neighbors and learning their names. Each router constructs a packet called the Link-State Packet (LSP), containing the list of neighbor routers with their names and *costs*. There are several options to define the cost of a link (not only or simply the distance). The LSP is transmitted to all other routers utilizing flooding. Each router stores the most recently generated LSP from each other router. Based on this information exchange, each router builds and maintains a database describing the topology and link costs for the whole network. Hence, each router uses the Dijkstra algorithm to determine the shortest paths based on the information found in its database. Link-state routing achieves some performance advantages with respect to distance-vector routing.

More details on distance-vector and link-state routing algorithms are provided below.

2.4.1.1 Distance-Vector Routing

Distance-vector routing is a distributed and iterative protocol, based on the Bellman–Ford algorithm. Each router in the network sends a vector on all its links (interfaces) containing the IP addresses of the destinations it can reach and the related distances. A generic neighboring router stores the distance vector after having summed these distances with its distance from the vector originating node. The routing table is computed through a fusion of the distance vectors obtained from all neighboring nodes according to the following method: for each possible destination, we compare all distance vectors of neighboring nodes and select as next hop the neighboring node with the shortest total distance to destination.

In a network with n nodes and L links, the computational complexity of the Bellman–Ford routing algorithm is $O(n \times L)$. Hence, in a full-mesh network with $n(n-1)/2$ bidirectional links, the complexity of the Bellman–Ford routing algorithm is $O(n^3)$, which is a value greater than that of the Dijkstra algorithm, i.e., $O(n^2)$.

With the distance-vector algorithm, each router starts with a set of routes for those networks or subnetworks to which it is directly connected, and possibly with some additional routes to other networks or hosts if the network topology is such that the routing protocol would be unable to provide the desired routing correctly. These routes are kept in a routing table, where each entry identifies a destination network or host with the "distance" to that network, typically measured in "hops" (*hop-count metric*).

In distance-vector routing, routers update their tables in three circumstances: (1) when routers are initialized; (2) on a periodical basis; (3) when routers have changes in their routing tables. Let us refer now to the first case.

The distance vector sent by the router at the first iteration only contains the distances to adjacent and connected routers (1 hop). At each new iteration, the number of hops (i.e., number of entries) in the distance vector increases by 1 until the network diameter is reached. In order to better understand how the distance-vector algorithm operates at each iteration, let us refer to the example in Fig. 2.13, where router R3 updates its routing table based on the distance vectors received from neighboring routers R1, R2, and R4. The distance vector provides the cost to reach different networks, denoted by their IP addresses. For instance, let us consider the routing towards the network with IP address 12.0.0.0 (this network is not shown in the figure). R3 receives the cost (distance) 4 from R1 and knows that the cost of the link R3–R1 is equal to 5; hence, the total cost to reach network 12.0.0.0 from R3 through R1 is $4+5=9$. Similarly, R3 derives the total cost to reach network 12.0.0.0 through router R2 as $0+15=15$ (R2 can directly reach network 12.0.0.0 so that the hop metric is 0). Finally, R3 computes the total cost to reach network 12.0.0.0 through R4 as $15+5=20$. In conclusion, R3 updates its routing table, selecting the path through R1 to reach network 12.0.0.0 with a cost 9. R3 updates its routing table in the same way for all the destination networks received from adjacent routers. The routing table thus obtained contains the distance vector that R3 will send to all its neighboring routers at the next iteration of the algorithm.

The *convergence time* of the routing algorithm is defined as the time needed so that each router in the network has a consistent and stable routing table.

Distance vectors can be more complex than described so far. In particular, the distance vector could be extended to contain not only a distance for each destination but also a direction in terms of the next-hop router; this could be useful to avoid some routing loops. This is the reason why most distance-vector routing protocols (e.g., Routing Information Protocol, RIP) send their neighbors the entire routing table.

Distance-vector updates are sent to adjacent routers on a periodical basis (= iteration time), ranging from 10 to 90 s. When a vector arrives at router B from router A, B examines the set of destinations it receives and the distances for each of them. B will update its routing table if:

- A knows a shorter way to reach a destination.
- A provides a destination that B has not in its table.
- The distance of A to a destination, already routed by B through A, has changed.

Referring to the network in Fig. 2.13 and assuming that routing tables are fully stabilized, we explain how changes in topology are managed by the distance-vector routing algorithm. Let us consider a first case when network 45.0.0.0 goes down: router R2, in its next scheduled update, marks network 45.0.0.0 as unreachable and passes this information, thus starting a new phase to converge towards new routing tables. Let us refer now to another case, where router R2 fails (instead of network 45.0.0.0). Router R3 still has entries in its routing table where R2 is the next hop for networks 30.0.0.0 and 45.0.0.0, but this information is no longer valid: R3 will continue to forward packets through R2 for some destinations, but they will be unreachable. This problem can be handled by setting *a route invalidation timer* for each entry in a routing table. For example, when router R3 first hears about 45.0.0.0 and enters the information into its routing table, R3 sets a timer for that route. At every regularly

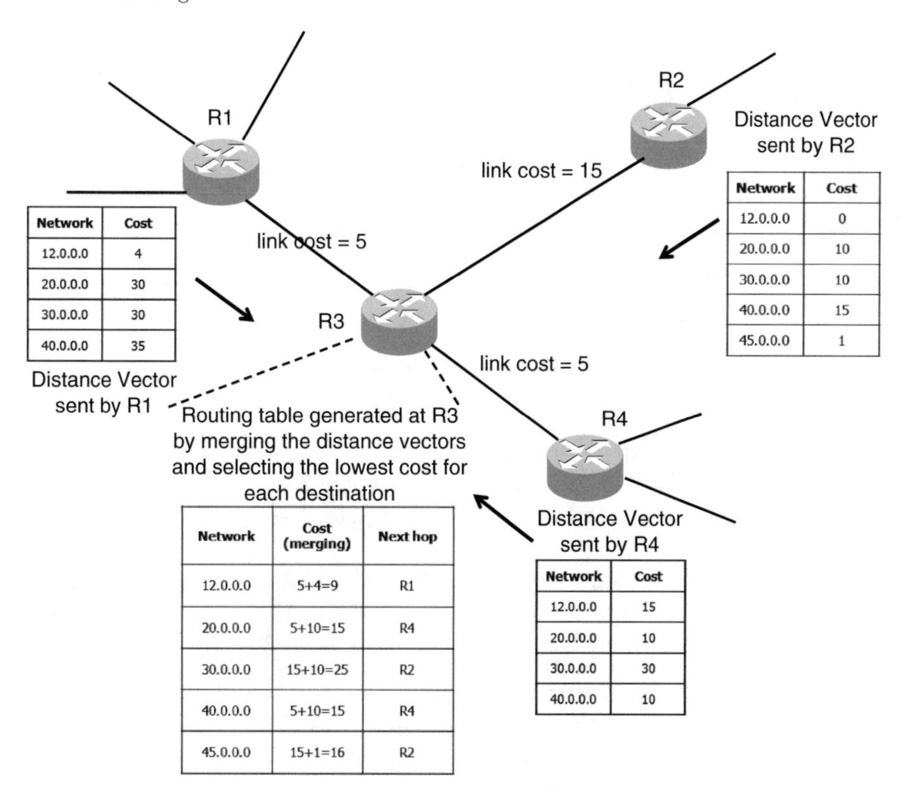

Fig. 2.13: Example of operation of the distance-vector routing algorithm and generation of the routing table of R3 at a certain iteration

scheduled update from router R2, R3 discards the already-known update and resets its timer for that route. If router R2 goes down, R3 will no longer receive updates about 45.0.0.0. Hence, when the timer will expire, R3 will flag the route as unreachable and will pass this information in the next updates. Typical route invalidation timers range from three to six update periods. A router would not want to invalidate a route after missing a single update because this situation could be the result of an update packet corrupted or of network congestion. At the same time, if the timer has an excessive duration, the convergence towards new (correct) routing tables might take too long.

Distance-vector routing is easy to implement, but it has some disadvantages:

- This routing algorithm is slow to converge; hence, there is a scalability issue because large networks require longer times to propagate routing information and then to converge.
- When routes change quickly, that is, a new connection appears or an old one fails, the routing topology may not stabilize to match the changed network topology: routing information propagates slowly from one router to another and, while it is propagating, some routers may have incorrect routing information. This may cause *routing loops* and the related *count-to-infinity problem*, as detailed later with an example.
- Another disadvantage is that each router has to send to every neighbor distance-vector updates (or even the entire routing table) at regular intervals. Of course, we can use longer intervals to reduce the network load, but such an approach causes problems related to how well the network responds to topology changes.
- This routing algorithm can adopt the hop-count metric, which does not take link speed (bandwidth), delay, and reliability into account. For instance, this algorithm uses a path with hop count 2 crossing two slow-speed lines and does not select a path with hop count 3 crossing three other networks, which could be substantially faster.

Let us explain the classical count-to-infinity problem (routing loops, routing instability) referring to a basic distance-vector algorithm: we consider distance vectors containing only costs to reach different networks (i.e., no next-hop indication in the distance vector). We refer to the situation depicted in Fig. 2.14 with three routers and linear topology: A is linked to B and B is linked to C. The hop-count metric is adopted: the cost of each link is "1" (see Table 2.2). B calculates its distance from C equal to 1. A calculates its distance from C equal to 2. We assume that the link between B and C breaks down at a certain instant or that C does not work. The following events occur in sequence to cause the count-to-infinity problem:

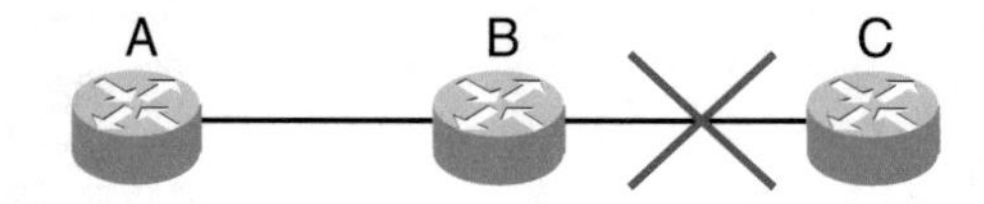

Fig. 2.14: An example to illustrate the count-to-infinity problem with Distance Vector Routing

Table 2.2: Hop metric for the network in Fig. 2.14 before the link breaks down between nodes B and C

Source	A	A	B	B	C	C
Destination	B	C	A	C	B	A
Cost	1	2	1	1	1	2

1. B decides that C is unreachable because B does not receive periodic distance-vector updates from C.
2. B must redetermine its distance from C. B decides that it is now 3 hops away from C (but this is false) based on the distance vector received from A, which contains the (old) distance 2 from C, and also knowing that B is 1 hop away from A.
3. Since B has changed its distance vector, it sends this info to its remaining neighbors (i.e., A).
4. Upon receiving a modified distance vector from B, A recalculates its distance vector and concludes that C is now 4 hops away (i.e., 3 hops away from B that is 1 hop away from A).

A and B continue the process #2–#4 by exchanging messages and computing the distance from C that grows indefinitely. They recognize that the best route to C is through the other node: packets for C get bounced between A and B until they are dropped when TTL = 0. This simple example clarifies the convergence issues with distance-vector routing. The count-to-infinity problem is mainly due to the impossibility to differentiate between "good" and "bad" route cost updates: updates do not contain enough information in this case.

The RIP protocol (defined in RFC 1058 and RFC 2453) is an intra-domain routing scheme based on the distance-vector algorithm; it adopts the hop count as a cost metric. The RIP protocol envisages a maximum number of hops equal to 15 to reach a destination; this permits us to stop routing loops and limits the network size where RIP can be adopted. RIP sends periodic updates every 30 s; however, updates can also be triggered by some changes in the network. The route invalidation timer of RIP is set to 180 s. RIP is the interior routing protocol most-commonly used on UNIX systems.

Different techniques have been proposed to address the count-to-infinity problem. Let us now refer to the extended distance vector, containing the next hop.

A variant of distance-vector routing is represented by *split-horizon routing*, where a router does not advertise the cost of a destination to a neighbor if this neighbor is the next hop towards that destination. This approach allows to solve the count-to-infinity problem in some cases; referring to the example in Fig. 2.14, all the previous steps #2–#4 should be avoided in this case.

Another approach, called *split-horizon routing with poisoned reverse*, is adopted by RIP, where each router includes in its messages towards an adjacent router the paths learned from that router but using a metric equal to 16 (equivalent to infinity). If router X routes traffic to Z via the neighboring router Y, then X sends to Y a distance X–Z equal to 16 (equivalent to infinity). In this way, Y does not route traffic to Z through X. This approach accelerates the convergence of routing tables. Poisoned reverse is performed by a router when it learns about an invalid route (broken link): a routing update is sent with a cost equal to 16 for this route. Explicitly telling a router to ignore a route is better than not telling it. The router also starts a *hold-down timer* to prevent that regular update messages reinstate a route, which was declared as invalid. Hold-down timers instruct routers to ignore any update for a specific period of time. This prevents routing loops, but, on the other hand, there is a significant increase in the convergence time. Poisoned reverse solves the routing loops with only two nodes. Let us explain poisoned reverse referring to the example in Fig. 2.14. When the link between B and C is broken, B sends an update to A that the hop metric to C is now 16. Hence, A knows that C is unreachable and updates its routing table. Then, A will advertise back on the same interface that C is 16 hops away, even though split-horizon does not usually allow the route to be advertised back on the same interface. The goal is to make sure that every possible device knows about the poisoned route.

Another variant of distance-vector routing is *path-vector routing*, where each entry in the distance vector is annotated with the path used to obtain the cost. The count-to-infinity problem is solved by path-vector routing, but the drawback is that signaling is heavy because of the use of large path vectors. The Border Gateway Protocol

(BGP), defined in RFC 1105, is an EGP based on path-vector routing: the distance vector includes the distance to each destination and the path related (this is the so-called *network reachability information*), thus making it possible to take constraints on paths (routing policies) into account.

2.4.1.2 Link-State, Shortest Path First Routing

Distance-vector routing (i.e., RIP) was used in the ARPANET until 1979. According to the drawbacks explained above, the Internet's growth pushed the distance-vector routing protocol to its limits. The alternative is a class of protocols known as Link-State, Shortest Path First. The main features of these routing protocols are described below.

- A set of physical networks is divided into a number of areas.
- All the routers within an area have identical databases.
- Each router database describes the complete topology of an area (i.e., which routers are connected to which networks). The database is called Link State information DataBase (LSDB).
- Each router uses its database to derive the set of optimum paths to all destinations so that it can build its routing table. A shortest path routing algorithm is adopted to determine the optimum paths.

When a link-state router boots, it needs first to discover the routers to which it is directly connected. For this purpose, each router sends a Hello message every N seconds to all its interfaces. This message contains the router address. Hello messages are sent only to neighbors that are connected directly to the router. A router never forwards the Hello messages received. Hello messages can also be used to detect link and router failures. A link is considered to have a failure if no Hello message is received from the corresponding router for a certain period of time.

Once a router has discovered its neighbors, it must reliably distribute its local links to all the routers in the network to allow them to compute their description of the network topology. This is achieved as follows. Each router periodically sends Link-State Packets (LSPs) to all routers employing controlled flooding, where duplicate LSPs are not forwarded. An LSP lists the neighboring routers and their costs (an LSP does not contain the whole routing table). Multiple routing metrics can be used to define the cost of each link. Once all the routers have received all the LSPs, the routers construct a map of the network in a database (LSDB). This database describes both the topology of the router domain (i.e., map of the network) and the routes to networks outside the domain. By means of this network map, each router locally runs a routing algorithm (typically the Dijkstra algorithm) to determine its shortest path to each router and network that can be reached. Then, the routing table is built based on the sink tree. When a network link changes its state (on to off or vice versa), LSPs are flooded through the network. All routers realize the change and recompute their routes accordingly.

In a network with n routers and L links, link-state protocols have a message complexity of $O(n \times L)$; instead, the computational load to build the routing table of a node is $O(n^2)$ with the Dijkstra algorithm.

Compared to distance-vector protocols, link-state protocols send updates when there is news and may send regular updates to ensure neighboring routers that a connection is still active. More importantly, the information exchanged by LSPs is the distance from adjacent routers, but not the whole routing table. This means that link-state algorithms reduce the overall broadcast traffic and can take better routing decisions using an improved routing metric, which can be more sophisticated than the distance or the hop count. In particular, the link cost can be based on link bandwidth, delay, reliability, and load. Link-state algorithms achieve a faster route convergence than distance-vector ones. Link-state routing protocols are robust to router failure events: When a failure occurs, new LSPs are flooded and each router recalculates its routing table. However, link-state algorithms entail a heavier computational load (more memory-intensive and processor-intensive) than distance-vector routing protocols. Finally, link-state algorithms can also suffer from route oscillations.

Open Shortest Path First (OSPF) defined in RFC 2328 and RFC 5340, respectively, for OSPF Version 2 for IPv4 and OSPF Version 3 for IPv6 (and progressively updated by multiple RFCs, like RFC 8042 for RFC 2328 and RFC 8362 for RFC 5340) and Intermediate System to Intermediate System (IS–IS) defined in ISO/IEC 10589:2002 and RFC 1142 (with the update in RFC 7142) are two very common link-state protocol implementations for intra-domain routing. OSPF is widely used in large enterprise networks. Instead, IS–IS is more common in large service provider networks.

2.4.1.3 Exterior Routing Protocols

In a distributed architecture, the ASs require both interior and exterior routing protocols to make intelligent routing choices.

The characteristics of the main exterior routing protocols are provided below.

Exterior Gateway Protocol

We must not confuse an exterior gateway protocol (generic term) with the actual Exterior Gateway Protocol (EGP), a particular (old) exterior routing protocol [18]. A gateway running EGP announces that it can reach networks, which are part of its AS. It does not announce that it can reach networks outside its AS. For example, the gateway of a given AS could even reach the entire Internet through its external connections, but since only one network is contained in its AS, it only announces one network with EGP.

Before sending routing information, EGP first exchanges Hello and I-Heard-You (I-H-U) messages. These messages permit the EGP gateways to establish a dialogue. Gateways communicating via EGP are called "EGP neighbors," and the procedure to exchange Hello and I-H-U messages is called "acquiring a neighbor."

Once a neighbor is acquired, routing information is requested via a poll. The neighbor responds by sending a packet containing reachability information, called "update." The local system includes the routes from the update into its local routing table. If the neighbor fails to respond to three consecutive polls, the local system assumes that the neighbor is broken and removes these neighbor routes from its table.

Unlike interior protocols, EGP does not attempt to choose the best external route. EGP updates contain distance-vector information, but EGP does not evaluate this information. The routing metrics of different ASs are not directly comparable: each AS can use different criteria to determine these values. Therefore, EGP leaves the choice of the best route to someone else. When EGP was designed, the network relied on a group of trusted *core gateways* to process and distribute the routes received from all ASs. These core gateways were expected to have the necessary information to choose the best external routes. EGP reachability information passed into core gateways was combined and passed back to ASs. The adoption of core gateways allows for consistency in the routing decisions taken in different ASs. However, this approach (based on a centrally controlled group of gateways) does not scale well and is therefore inadequate for the rapidly growing Internet. As the number of ASs and networks connected to the Internet grew, it became difficult for the core gateways to keep up with the increasing workload. This is one reason why the Internet moved to a more distributed architecture that leaves the burden of processing routes to each AS. Another reason is that no central authority controls the Internet: the Internet is composed of many equal networks (ASs). Due to these issues, EGP is no longer popular.

Border Gateway Protocol

RFC 1771 defines the BGP [19], the leading exterior routing protocol. BGP is based on the OSI Inter-Domain Routing Protocol (IDRP). BGP adopts a policy-based routing, where routing decisions are taken considering also non-technical reasons (e.g., political, organizational, or security considerations). BGP permits the AS to choose among routes on the basis of *routing policies* without relying on a central routing authority. This is an important feature in the absence of core gateways.

Routing policies are not part of the BGP protocol. Policies are provided externally as configuration information. The National Science Foundation (NSF) provides Routing Arbiters (RAs) at the Network Access Points (NAPs), where large ISPs interconnect.[2] The RAs can be queried for routing policy information. Most ISPs also develop private policies based on bilateral agreements they have with other ISPs. BGP can be used to implement these policies by controlling the routes it announces to others and the routes it accepts from others. The network administrator enforces the routing policy by configuring the router.

BGP is implemented on top of TCP (described in Sect. 2.8.1): BGP routers connect each other by using TCP for reliable delivery of messages. BGP uses the "well-known" TCP port 179 (see also Sect. 2.8.3). BGP neighbors

[2] Peering locations are places where the networks of different ASs interconnect. Public peering locations were known as NAPs, but today are most often called Internet Exchange Points (IXPs). See also Sect. 2.10.1.

are called peers. Once connected, BGP peers exchange OPEN messages to negotiate session parameters, such as the BGP version to be used.

The BGP update message lists the destinations that can be reached through a specific path and the attributes of the path. BGP is a path-vector protocol since it provides the entire end-to-end path of a route in the form of a sequence of AS Numbers (ASNs). ASNs defined initially as 16-bit integers can now be represented with 32 bits (RFC 4893). Multiple update packets may be sent to build a routing table. The knowledge of the complete AS path eliminates the possibility of routing loops and count-to-infinity problems. BGP peers send each other complete updates when the connection is first established. After that, only the changes are notified. If there are no changes, just a small 19-byte keep-alive message is sent to indicate that the peer and the link are still operational. BGP is very efficient in using network bandwidth and system resources.

2.4.2 Routing Implementation Issues

In order to prepare and keep updated the routing tables, a routing protocol sends and receives signaling packets containing routing information to and from other routers. In some cases, routing protocols can themselves run over routed protocols. A routing protocol running over a particular transport mechanism of layer N cannot be considered as a protocol of layer $N + 1$. For instance, the OSPF routing protocol runs directly over IP and has its reliable transmission mechanism. RIP runs over UDP over IP. Instead, BGP runs over TCP over IP.

The routing protocols are often implemented on UNIX-based routers by using a "daemon."[3] Routing daemons initialize and dynamically maintain the kernel routing table by communicating with daemons on other systems to exchange information according to routing protocols. Daemons can be of two types:

- *Routed*: Pronounced "route D." This is the most common routing daemon for interior routing. It adopts the Routing Information Protocol (RIP).
- *Gated*: Pronounced "gate D." This is a more sophisticated daemon on UNIX-based systems for interior and exterior routing. It can employ RIP as well as several additional protocols, such as OSPF, BGP, and others in a single package.

Only one of them (i.e., routed or gated) can run on a router at any given time. The gated software combines interior and exterior routing protocols into one software package. Most sites use UNIX systems only for simple routing tasks for which RIP is usually adequate. Large and complex routing applications requiring advanced protocols are handled by dedicated router software. Many of the advanced routing protocols are available in gated for UNIX systems. Gated also has the following features:

- Gated combines the routing information learned from different protocols and selects the best routes.
- Routes learned via an interior routing protocol can be announced through an exterior routing protocol, which permits the externally announced reachability information to adapt depending on changing interior routes dynamically.
- Routing policies can be implemented to control accepted routes and advertised routes.
- All protocols are configured from a single file (/etc/gated.conf) using a consistent syntax.
- Gated is continuously upgraded to contain the most up-to-date routing software.

2.5 QoS Provision in IP Networks

The introduction of real-time traffic on the Internet (e.g., Voice over IP, VoIP) calls for new solutions to provide QoS: the classical IP best-effort traffic is no longer sufficient. Real-time traffic (as well as other applications) require priority treatment to achieve a good performance. In IP networks, user QoS requirements are specified in the ITU-T Y.1541 Recommendation in terms of different parameters (i.e., packet transfer delay, IP packet delay variation, IP packet loss ratio, and IP packet error ratio) for eight traffic classes; Class 0 has the most stringent

[3] In Unix and other computer operating systems, a *daemon* is a particular class of computer programs running in background, rather than under the direct control of a user. These processes run independently of users, who are logged-in. Usually, daemons have names ending with a "d."

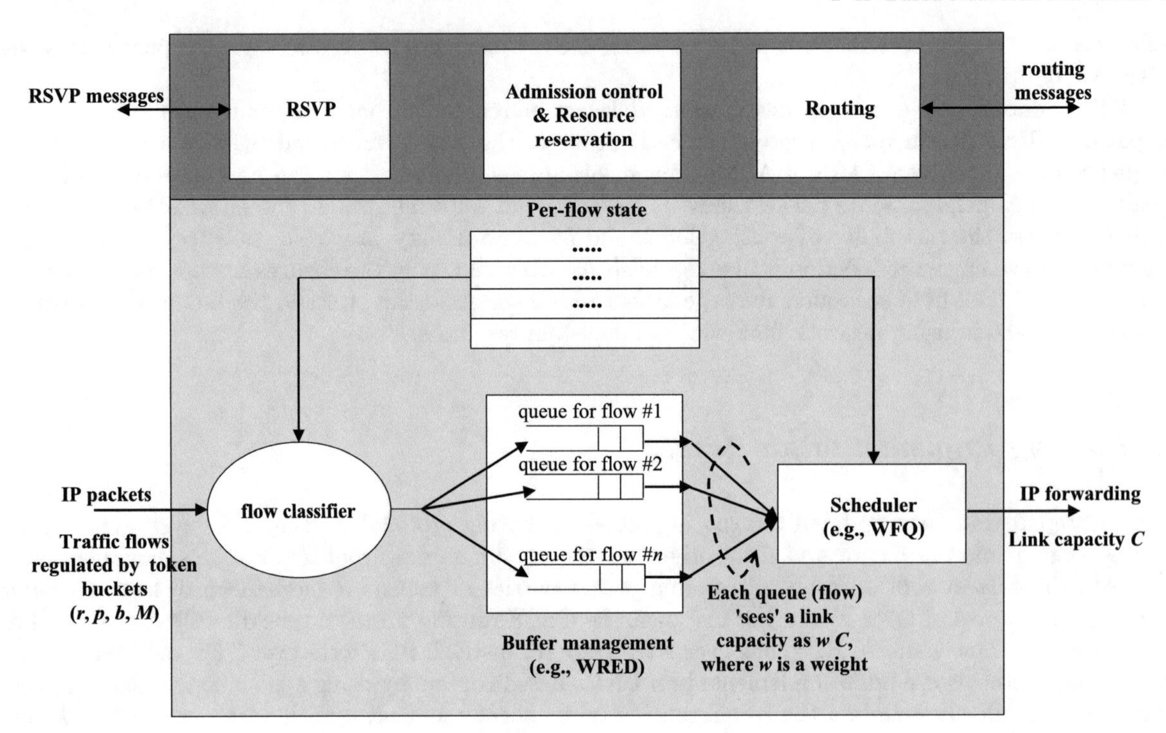

Fig. 2.15: Model of an IntServ router with both forwarding plane (below) and control plane (above)

requirements, while Class 5 has the less stringent requirements (the classical best-effort traffic). For instance, Class 0 is used for real-time highly interactive applications, sensitive to jitter such as VoIP and video. The requirements for Class 0 are: mean packet delay lower than 100 ms, delay variation lower than 50 ms, and packet loss ratio lower than 10^{-3}. Finally, Class 7 is used for applications highly sensitive to losses, such as television, high-capacity TCP transfers, and TDM circuit emulation. The requirements for Class 7 are: mean packet delay lower than or equal to 400 ms, delay variation lower than or equal to 50 ms, packet loss ratio lower than or equal to 10^{-5}. In this recommendation, queuing mechanisms at the nodes and the conditions for routing paths are also specified for each traffic class.

The key mechanisms available today to support QoS in IP-based networks are Integrated Services (IntServ) and Differentiated Services (DiffServ). The corresponding detailed descriptions are provided below.

2.5.1 IntServ

The IntServ main concept is to reserve resources for each flow through the network [20, 21]. IntServ adopts an explicit *setup mechanism* involving the routers in the definition of source-to-destination paths. Each flow can request a specific QoS level. RSVP (Resource reSerVation Protocol) is the most widely used *resource reservation* mechanism for setting up source-to-destination paths (RFC 2205 and RFC 2210). RSVP permits a fine bandwidth control. The main drawback of RSVP is due to the use of per-flow state and per-flow processing at the routers, thus having scalability issues in large networks (heavy processing and signaling load). IntServ adopts separate queues at routers for the different traffic flows (per-flow buffer management). IntServ can provide deterministic QoS guarantees. The IntServ node (router) architecture is described in Fig. 2.15.

IntServ supports two services types:

- Guaranteed Service (GS) specified in RFC 2212
  - Targets *real-time inelastic applications.*
  - Uses per-flow traffic characteristics and service requirements.
  - Requires admission control (CAC) at each router along the path.
  - Can deterministically guarantee a maximum delay and no packet losses.
- Controlled-Load Service (CLS) specified in RFC 2211
  - Targets adaptive real-time applications, which can adapt to network conditions within a certain performance window and that can tolerate a certain degree of loss and delay.
  - Uses a traffic description and an average bandwidth needed for each traffic flow: CAC and policing are performed based on these data. There is no actual bandwidth reservation in this case, but just an implicit reservation resulting from the CAC procedure.
  - Requires admission control (CAC) at each router along the path.
  - CLS does not provide any quantitative guarantee on delay bounds.

RSVP is a transport-level protocol for reserving resources in IP networks. RSVP must be present at the sender, receiver, and intermediate routers. RSVP performs per-flow reservation with soft state maintained at intermediate routers. RSVP uses two types of Flow Specs to notify routers how to set up a path:

- Traffic Specification (T-Spec), which describes the traffic characteristics of the sender according to a token bucket model:
  - Bucket rate and sustainable rate, r (bits/s)
  - Peak rate, p (bits/s)
  - Bucket depth, b (bits)
  - Maximum packet size, M (bits) that can be accepted
  - Minimum policed unit, m (bits): any packet with a size smaller than m will be counted as m bits.
- Request Specification (R-Spec) is used only in the GS case and contains the amount of bandwidth to be reserved, according to the following details:
  - Service rate, R (bits/s): the amount of bandwidth to be reserved for a traffic flow.
  - Slack term, S (μs): the extra amount of delay (tolerance) with respect to the end-to-end delay requirement, which can be tolerated by the source. A network element can utilize this slack term to reduce the bandwidth reservation for a traffic flow.

GS provides quantitative QoS guarantees for traffic flows. In particular, GS provides guaranteed bandwidth (reservation), strict bounds on end-to-end delay, and no packet loss for conformant flows. GS can manage applications with stringent real-time delivery requirements, such as audio and video applications. In order for a new traffic flow to be admitted (CAC), both T-Spec and R-Spec (called together FLOWSPEC) are used by RSVP. A *downstream* procedure (from source to destination) is adopted to determine the path and the bandwidth R to be reserved (R is computed and included in R-Spec at destination) for a specific traffic flow characterized by T-Spec. In particular, the value of R is determined according to the (r, p, b, M) token bucket model specified by T-Spec. Then, an *upstream* procedure (from destination to source) is performed where each router admits the new flow based on its R-Spec and the resources currently allocated to other (active) flows. Each router allocates bandwidth R and a specific buffer space B to each GS flow admitted.

IntServ is supposed to use General Processor Sharing (GPS) to schedule traffic at the routers, where each flow uses a distinct buffer. A GPS scheduler is a theoretical concept, a benchmark, not really implementable because it schedules traffic on a bit basis, coherently with a fluid model. Hence, in real networks, where traffic is organized in packets, the Weighted Fair Queuing (WFQ) scheduler is used to approximate the GPS behavior, as shown in Fig. 2.15. At a given node, GPS serves several flows in parallel, as if there was a certain bit-rate allocated to each of them. During a period of duration t, GPS guarantees that a flow with some backlog in the node receives an amount of service at least equal to $R \times t$, where R is the rate allocated to the flow at the node.

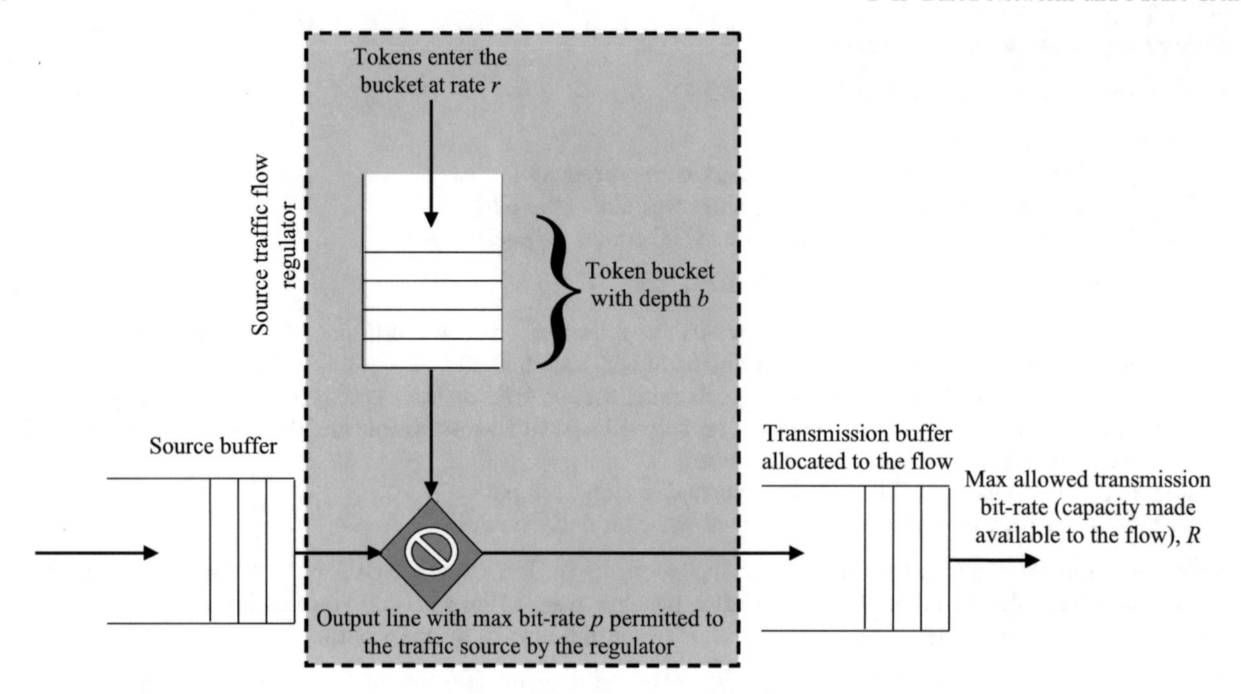

Fig. 2.16: IntServ GS approach; (r, p, b) token bucket shaper applied to a source to control/to model the bit-rate injected

Let us first study how the GS service can guarantee the QoS by adopting the simplified (r, p, b) token bucket model in Fig. 2.16. We make the following assumptions: (1) we consider a *fluid-flow traffic model* so that a source generates traffic according to a variable bit-rate continuous-time process $\rho(t)$ [in this model, traffic arrives bit by bit; we do not consider packets arriving according to a certain point process];[4] (2) 1 token enables the transmission of 1 bit; (3) we start with an empty buffer and a full bucket with b tokens; (4) $p > R > r$, where R denotes the service rate to be allocated to the traffic flow (reservation made by RSVP); (5) we neglect propagation delays and do not consider packet sizes in this model so that parameters m and M are not used (for numerical formulations it is as we had: $m = M = 0$). We do not use the slack term, considering that the delay requirement has to be fulfilled in the strict sense.

The Maximum Burst Size (MBS) transmitted by the token bucket shaper at the maximum rate p is determined as follows, referring to the burst time interval T_b:

$$\mathrm{MBS} = T_\mathrm{b} p = b + rT_\mathrm{b} \tag{2.2}$$

Hence, given the token bucket parameters r and b, we obtain T_b (assuming $p > r$) as

$$T_\mathrm{b} = \frac{b}{p - r} \tag{2.3}$$

The MBS bits sent in T_b result as

$$\mathrm{MBS} = T_\mathrm{b} p = \frac{bp}{p - r} \tag{2.4}$$

After time T_b, the output rate is regulated by r, the arrival rate of new tokens.

Let $\alpha(t)$ denote the *arrival curve*, which is the total number of bits generated up to time t. The arrival curve is a non-decreasing function of time. According to our notation and the fluid-flow traffic model, we have

[4] In real systems, there is always a granularity in the arrival process (at the level of packets) so that the arrival traffic is a discrete-time process with a finite set of bit-rate values.

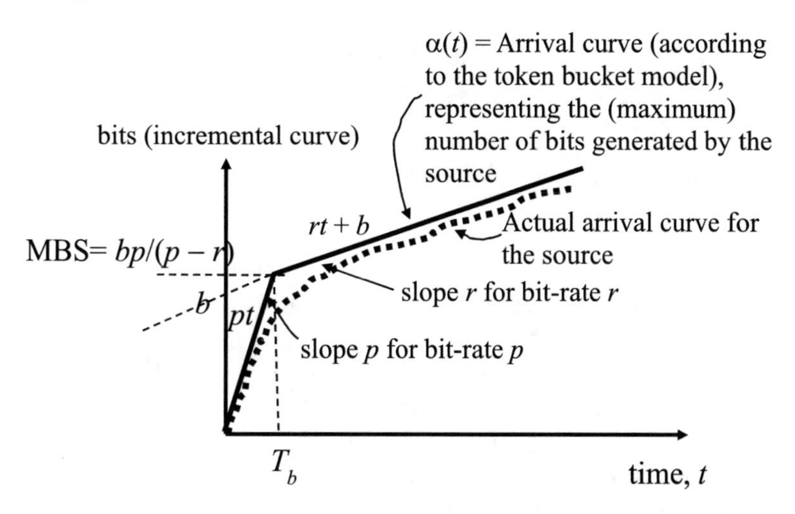

Fig. 2.17: Behavior of the bit-rate generated by the source regulated by the token bucket (r, p, b)

$$\alpha(t) = \int_0^t \rho(t)\mathrm{d}t \tag{2.5}$$

Figure 2.17 shows the behavior of the arrival curve of the source regulated by the token bucket (fluid-flow case): this continuous piecewise-linear function has to be intended as an upper bound to the number of bits actually arrived up to time t from the regulated source, as shown in Fig. 2.17. In reality, $\alpha(t)$ is the upper bound to the actual arrival curve generated by the traffic source in the generic interval of duration t, $[\tau, \tau + t]$.

Based on the situation depicted in Fig. 2.17, the arrival curve $\alpha(t)$ of a source regulated by the token bucket shaper is characterized as

$$\alpha(t) = \min\{pt, rt + b\} \tag{2.6}$$

The regulated source provides bits at the input of a network node, as shown in Fig. 2.18. The *departure curve* $\beta(t)$ denotes the number of bits departing from the node up to time t based on the allowed (maximum) output rate R ($R > r$) characterizing the *service curve* $\sigma(t) = Rt$.

The departure curve $\beta(t)$ is as follows:

$$\beta(t) = \min\{\sigma(t), \alpha(t)\}, \quad t > 0 \tag{2.7}$$

Figure 2.19 describes the relation between curves $\alpha(t)$, $\beta(t)$, and $\sigma(t)$.

The occupancy of the buffer (backlog) at a generic instant t, $B(t)$, can be expressed as

$$B(t) = \alpha(t) - \beta(t) \tag{2.8}$$

Referring to the situation depicted in Fig. 2.19, point X corresponds to the maximum delay to cross the node $D_{\max}$ and the largest buffer occupancy at the node $B_{\max}$, which can be formally expressed as follows in the case $r < R < p$:

$$D_{\max} = t^* - T_{\mathrm{b}} = \frac{b}{R} \times \left(\frac{p-R}{p-r}\right) \le \frac{b}{R} \tag{2.9}$$

$$B_{\max} = pT_{\mathrm{b}} - RT_{\mathrm{b}} = b \times \left(\frac{p-R}{p-r}\right) \le b \tag{2.10}$$

Hence, in a perfect fluid model, a flow conformant to a token bucket of rate r and depth b will have its delay bounded by b/R, provided that $R \ge r$ [22, 23].

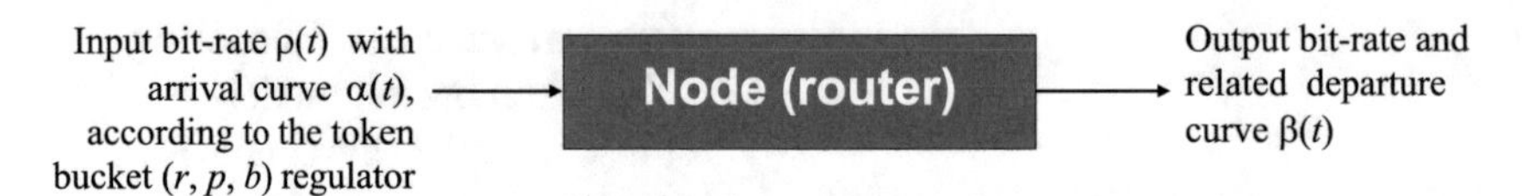

Fig. 2.18: Input and output processes for a node, characterized by cumulative input/output bit-rate behaviors $\alpha(t)$ and $\beta(t)$

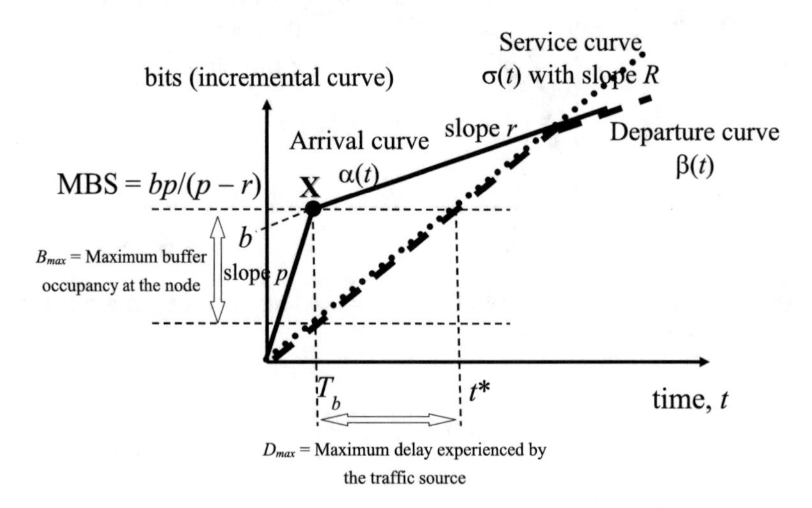

Fig. 2.19: Behavior of the departure curve describing the total number of bits transmitted by the node as a function of time t, $\beta(t)$, and relation with the arrival curve according to the token bucket (r, p, b)

In a real system, we have deviations from this perfect fluid model. In particular, two error terms C and D are used to model the intrinsic delay experienced traffic in crossing the node. In addition to this, the packetization effect of the flow needs to be taken into account using the maximum packet size M. Then, we have to consider the full token bucket model with parameters (r, p, b, M). With respect to what is shown in Fig. 2.19, the arrival curve $\alpha(t)$ now starts from the M value at $t=0$ [i.e., $a(0)=M$] assuming[5] $b > M$ and the service curve $\sigma(t)$ is just shifted on the right of[6] $T_0 = \frac{C}{R} + D$. However, the method to derive $D_{\max}$ and $B_{\max}$ is still the same: $D_{\max}$ is obtained as the maximum "horizontal aperture" between $\alpha(t)$ and $\beta(t)$ curves and $B_{\max}$ is given by the maximum "vertical aperture" between $\alpha(t)$ and $\beta(t)$ curves. Therefore, the generalization of (2.9) and (2.10) to express $D_{\max}$ and $B_{\max}$ are as follows (representing together the two possible cases $r < R < p$ and $r < p < R$ [24]):

$$D_{\max} = \frac{1}{R}\frac{b-M}{p-r}(p-R)^+ + \frac{M}{R} + T_0 \tag{2.11}$$

$$B_{\max} = b + rT_0 + \left(\frac{b-M}{p-r} - T_0\right)^+ \left[(p-R)^+ - p + r\right] \tag{2.12}$$

where we use the following notation $(x)^+ = \max(x, 0)$. Note that these maxima values are obtained in correspondence with angular points of the diagram with the arrival curve and the service curve that however has not been displayed here.

Let us clarify better the meaning of C and D parameters. The term C is a rate-dependent error term. An example of such an error term is the need to account for the time taken serializing a datagram segmented into MAC packets sent at a certain rate. The term D is a rate-independent error term representing the worst-case non-rate-based transit time variation through the node. It is generally set at boot or configuration time. An example of D is a slotted network, in which guaranteed flows are assigned particular slots in a cycle of slots. In this case,

[5] In this refined token bucket model, we assume that at time $t=0$ the regulator allows the transmission of a whole packet of size M at an infinite speed if $M < b$. Combining the token bucket contractual constraint $b + rt$ with the physical limitations $M + pt$, the resulting arrival curve is $\alpha(t) = \min(pt + M,\ rt + b)$.

[6] The service curve is now $\sigma(t) = \max[0,\ R(t - T_0)]$.

D would represent the maximum amount of time a flow's data, once ready to be sent, might have to wait for a slot. It is also possible to include propagation delays in D even if they have a negligible contribution.

The graphical approach described here to study delay and buffer bounds belongs to a discipline called *network calculus* (i.e., *deterministic queuing systems*) and can also be applied to other traffic regulation problems (e.g., leaky bucket shapers) or can be used to study the problem of the dejitter buffer for audio/video streaming (typically, there is the need to store let us say a few seconds of the stream to be sure to bridge gaps in the flow of the same or lower duration) [24].

Let us now study the IntServ GS case, considering a generic traffic flow characterized by the full token bucket parameters (r, p, b, M) of T-Spec. IntServ GS has to allocate a bandwidth R for all the link crosses along the path from source to destination to fulfill a given delay constraint $\Delta_{\max}$ and a certain buffer capacity B to be determined at each node based on the maximum backlog to avoid packet losses.[7]

The reservation of the resources of a flow at each node along the path is performed by RSVP during the setup phase. According to RFC 2212 [25], a typical RSVP procedure can be summarized as follows, involving several nodes along the path and not just a single node, as considered in the previous study related to Figs. 2.18 and 2.19 [25]. In the setup phase, the sender transmits downstream PATH messages towards the destination. Each router along this path updates the PATH message. The PATH message contains T-Spec, which is not altered in transit, and the Advertisement SPECification (ADSPEC), which contains the progressive sums of C_i and D_i error terms that are accumulated at the nodes along the path.

When the destination (hereafter called "receiver") gets the PATH message, it knows T-Spec (r, p, b, M), the number of hops as well as ΣC_i and ΣD_i accumulated by the routers along the path. Then, the receiver can compute the bandwidth requirement R considering $D_{\max}$ in (2.11) where

$$T_0 = \frac{\sum_i C_i}{R} + \sum_i D_i \tag{2.13}$$

to have that $D_{\max}$ is lower than $\Delta_{\max}$ taking the effect of the concatenated nodes into account ("pay bursts only once"), as explained in [24–26].

Then, an upstream resource reservation process is initiated by the receiver, involving all the routers in the path back to the source. In particular, the receiver sends the RESV message to the first router in the upstream direction, containing both T-Spec (r, p, b, M) and R-Spec with the computed value of R and a slack term S that however is not considered in this explanation, $S = 0$ (the slack term could be used to admit a flow also when in some hops in the path it is not possible to reserve the whole bandwidth R specified in R-Spec). This router performs a CAC control to verify whether it is able to reserve both rate R and buffer capacity B, recomputed at each router based on the partial T_0 terms with the sums of C_i and D_i up to them along the path using (2.13). This procedure entails verifying that the sum of the reserved rates is lower than the total available bandwidth and that the non-reserved buffer capacity is greater than or equal to B. If the CAC verification is positive, the router passes the RESV message upstream to the next router along the path, which repeats the reservation process. On the other hand, if the CAC verification fails, the router discards the reservation and informs the source.

Inside the network, the regulated flows can be altered because of queuing effects along the path. A traffic flow entering the network as conformant at a certain node may be no longer conformant at some downstream node. Therefore, inside the network, reshaping points can be needed, delaying datagrams until they conform to T-Spec.

The CLS model is defined in RFC 2211. CLS provides a traffic flow with a QoS approximating the QoS that the same flow would receive from an unloaded best-effort network, assuming that the flow is compliant with its traffic contract (SLA). CLS operates as follows. A description of the traffic flow characteristics (mainly T-Spec and an estimation of the mean bandwidth requested; R-Spec is not used in this case) must be submitted to a router along the source–destination path to request the CLS service. The router has a CAC module to estimate whether the mean bandwidth requested is available for the traffic flow. In the positive case, the new flow is accepted, and the related resources are *implicitly reserved.* With the CLS service, there could be some packet losses for the flows admitted and no delay bound guarantees. The CLS service model provides only statistical guarantees:

- A very high percentage of packets is successfully delivered.
- Data packets experience small average queuing delays.

[7] Referring to the idealized model where $C = D = 0$ for all the nodes along the path, it would be enough to reserve a buffer capacity $B_{\max}$ (upper bounded by b) to the ingress node because the other nodes would have no need to queue the arriving data that are served with the same rate R at each node along the path from source to destination.

The important difference from the traditional Internet best-effort service is that the CLS flow does not deteriorate noticeably as the network load increases. CLS can be supported by RSVP signaling. CLS is not suited to those applications requiring very low latency.

The IntServ approaches can be too heavy to be adopted in core networks but can be suitable for some access networks (with a reduced number of flows), where flow-based traffic management is possible.

2.5.2 DiffServ

IntServ and especially RSVP have some implementation issues, such as:

- Scalability: Maintaining per-flow states at routers is difficult in high-speed networks because of the very large number of flows.
- Only two classes: We should provide more qualitative service classes with "relative" service differentiation (Platinum, Gold, Silver, etc.).
- Heavy protocol: Many applications only need to specify a service qualitatively.

To achieve scalability, the DiffServ architecture envisages aggregate traffic flows rather than single flows as IntServ. Most of the complexity is outside of the core network in the edge devices, which process lower traffic volumes and lower numbers of flows. DiffServ is based on a simple model, where packets entering the network are classified at *edge routers* according to a small number of aggregate flows or classes. These classes are characterized by the DiffServ Code Point (DSCP) field contained in the Type of Service (ToS) byte of the IPv4 header (see Fig. 2.4) or in the Traffic Class byte of the IPv6 header (see Fig. 2.7). DSCP is of 6 bits; the first 3 bits of DSCP correspond to the IP precedence field. Class-based queuing is performed at routers. Each DiffServ router within the core network forwards the packets according to a Per-Hop Behavior (PHB), which corresponds to the DSCP. No per-flow state has to be maintained at core routers, thus improving scalability. DiffServ traffic management is as follows:

- At the edge routers of a DiffServ region: Each flow is analyzed operating classification, DSCP marking, policing, and shaping.
- At the core routers within a DiffServ region: Buffering, scheduling, and forwarding are differentiated based on PHBs.

DiffServ tags the traffic directly with *in-band QoS markings*. Instead, IntServ adopts an *out-band approach for QoS support* based on the RSVP protocol. DiffServ provides probabilistic QoS guarantees to aggregate traffic flows and uses a CAC algorithm based on SLAs between subscribers and service providers or between two service providers. Three different PBHs can be considered with the related DSCPs [27, 28]:

- Expedited Forwarding (EF), defined in RFC 3246, offers some quantitative QoS guarantees to aggregate flows. The EF traffic is for guaranteed bandwidth, low jitter, low delay, and low packet losses. EF traffic is managed by a specific queue at the routers. EF implies traffic isolation: the EF traffic is not influenced by the other traffic classes (AF and BE). Non-conformant EF traffic is dropped or shaped. EF traffic is often strictly controlled by CAC (admission based on peak rate), policing, and other mechanisms. The recommended DSCP for EF is 101110.
- Assured Forwarding (AF) is defined in RFC 2597 and RFC 3260. AF is not a single traffic class, but four subclasses: AF1, AF2, AF3, and AF4. Hence, we can expect to have four AF queues at the routers. The service priority for these queues at the routers is: AF1 > AF2 > AF3 > AF4. Within each subclass (i.e., within each queue), there are three drop precedence values from a low drop level 1 up to a high drop level 3 (with related DSCP coding) to determine which packets will be dropped first in each AF queue if congested: the drop precedence order for the generic queue AFx, $x \in \{1, 2, 3, 4\}$, is AFx3 before AFx2 before AFx1. The packets of a generic AFx class queue are sent in FIFO order. AF is used to implement services that differ from each other (e.g., gold, silver). Non-conformant traffic is remarked but not dropped. AF is suitable for services that require a minimum guaranteed bandwidth (additional bandwidth can only be used if available) with possible packet dropping above the agreed data rate in case of resource shortage.
- Best Efforts (BE).

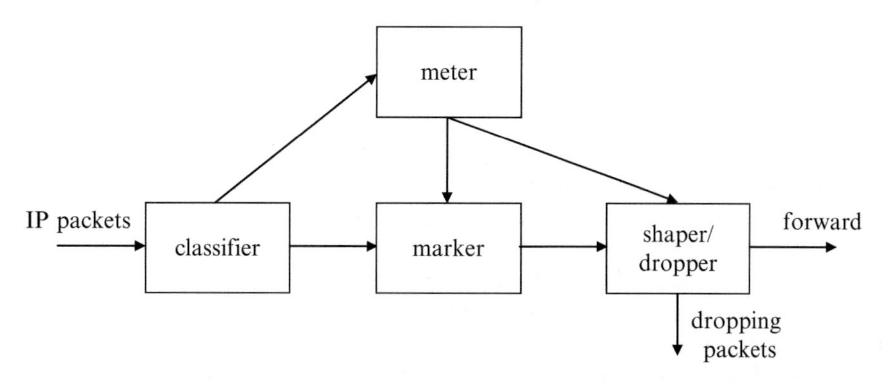

Fig. 2.20: DiffServ edge router functionalities (forwarding plane)

In the routers, we can consider that there are a total of six queues (i.e., EF, AF1, AF2, AF3, AF4, and BE), which are serviced considering their relative priorities.

A DiffServ edge router supports the following functions (see Fig. 2.20 above):

- *Classification*: It selects the packets according to different aspects, such as protocol type, IP precedence or DSCP if available, packet length, etc.
- *Metering*: It checks whether the traffic is conformant with the negotiated profile (SLA) according to a token bucket approach.
- *Marking*: It writes/rewrites the DSCP value in the packet header; in the AF case, it is also possible to increase the drop precedence for non-conformant traffic.
- *Conditioning* (*shaping*): It delays some packets and then forwards or discards exceeding packets.

Core routers perform only forwarding functions based on the PHB assigned to the IP packet and corresponding to the DSCP in the header. With DiffServ, no per-flow state information has to be maintained at routers, and this is a significant advantage with respect to IntServ.

DiffServ routers adopt suitable scheduling and buffer management solutions, as described below.

- *Scheduling*: Rather than using queues with strict priorities, more balanced scheduling algorithms such as fair queuing or weighted fair queuing are likely to be used.
- *Buffer management*: To prevent problems associated with drop-tail events (i.e., arriving packets get dropped when the queue is full regardless of the flow type or importance), Random Early Detection (RED) or Weighted RED (WRED) active queue management algorithms [29] are often used. In the RED case, a single FIFO queue is considered; when the average queue length exceeds a minimum threshold, packets are dropped randomly according to a probability depending on the average queue length. If a maximum threshold is exceeded, all newly arriving packets are dropped. The RED algorithm reduces the synchronization of TCP flows sharing a buffer utilizing the randomness of packet losses. WRED drops packets selectively based on the drop precedence. If a congestion event occurs, the highest class traffic (i.e., AF1) has priority, while the packets with the highest drop precedence are discarded first.

DiffServ is now the most common approach for QoS support in IP networks, also including the case of DiffServ for MPLS networks (see Sect. 2.7).

Finally, it is interesting to note that one possible approach to support end-to-end QoS in IP networks (access network ↔ core network ↔ access network) is to use connection-oriented resource reservation (i.e., IntServ) in the access part and service differentiation (i.e., DiffServ) in the core part of the network. In this case, border routers between IntServ and DiffServ domains must implement an appropriate QoS mapping [30, 31].

The Common Open Policy Service (COPS) protocol has been defined by IETF in RFC 2748 as a way to support *policy control* for a QoS environment in IP networks. COPS is a simple query–response protocol that allows Policy Decision Points (PDPs) to communicate network policy information decisions (i.e., the allocation of network traffic resources according to desired service priorities) to Policy Enforcement Points. The COPS protocol has been developed to complement the resource-related CAC of IntServ with a policy-related CAC. However, the concept of policy control applies to both IntServ and DiffServ networks, even if different signaling and CAC models are needed. Referring to the IntServ RSVP case, the network nodes running RSVP are the Policy Enforcement Points, while a centralized element acts as a PDP (i.e., a *resource broker*).

2.6 IP Traffic Over ATM Networks

When IP networks were defined, the need for suitable lower-layer technologies to efficiently transport IP traffic was soon evident. This section deals with the transport of IP-based traffic on ATM networks. The next section will consider a new technology suitably developed.

The concept of *adjacency* can be applied to each OSI level. In particular, there is a layer 3 adjacency for two IP routers interconnected; there is a layer 2 adjacency for two ATM nodes connected by virtual circuits; there is a layer 1 adjacency for interfaces connected to the same physical transmission medium. Finally, we speak about *interoperability* when two nodes work together but at different OSI levels.

The first standard document for IP traffic over ATM was RFC 1483 [32], which dealt only with the problem of the inefficient mapping of IP datagrams in ATM (short) cells. In particular, the use of AAL5 was proposed. However, AAL5 does not support the multiplexing of different higher-layer traffic flows (protocols) into the same virtual connection. Then, two methods have been defined for multiplexing IP traffic over AAL5 [32].

- The first method to multiplex multiple protocols on a single ATM virtual circuit adopts the Logical Link Control (LLC) encapsulation: the protocol related to an AAL5 PDU is identified by prefixing the AAL5 PDU with an IEEE 802.2 LLC header. In this method, called "LLC Encapsulation," we have IP on top of LLC, on top of AAL5, on top of ATM: IP/LLC/AAL5/ATM. This solution requires a reduced number of ATM Virtual Circuits (VCs) with respect to the following method.
- The second approach implicitly performs the multiplexing of higher-layer protocols by using different VCs: a single VC (i.e., a VPI/VCI pair) is used for each protocol. This method, called "VC-Based Multiplexing," entails minimal bandwidth and processing overheads (there is no need to include explicit multiplexing information in the AAL5 PDU payload). The drawback is that there are many VCs to be managed in the network, which may cause high costs and complexity.

Another problem was related to the support of IP routing in ATM networks. The two following approaches were proposed:

- "Overlay Model": There is a rigid separation between IP network and ATM network. IP routing and ATM switching operate independently. The ATM network is only used to interconnect IP routers and typically adopts Permanent Virtual Circuits (PVCs). There are two basic approaches for the overlay model, that is the "Classical IP over ATM" {CIP, in RFC 1577 and in RFC 2225 [33, 34]} defined by IETF and the "LAN Emulation" (LANE) defined by the ATM Forum (the LLC encapsulation method is adopted in this case).
- "Integrated Model" or "Peer Model": IP + AAL5-ATM/PHY. This is an evolution of the first approach with the aim of reducing the functional redundancies between IP and ATM layers for what concerns routing and switching (as in the previous overlay model). In this case, ATM and IP are at the same layer. The integrated model had some problems due to the complexity of designing ATM switches with IP routing functionality. Another problem was due to the development of non-standard solutions.

Let us now concentrate on the CIP approach, that is IP/ALL5/ATM/PHY. Suppose the ATM network is not entirely meshed. In that case, the IP datagram is reassembled and segmented again in each intermediate node encountered (i.e., a router over ATM) using the ATM SAR sublayer. This may result in a waste of processing resources at each router that could also cause congestion if the router is not designed correctly. This is the reason why the *networks adopting the CIP approach tend to have a full-mesh topology*: the routers in the network are directly connected via ATM links (layer 2 adjacency) of the permanent or switched type (PVC or SVC). Figure 2.21 shows an IP network (with address 175.20.0.0), where routers are interconnected according to a full-mesh ATM topology. The routers and the hosts connected to this network form a Logical IP Subnet (LIS). The communication between hosts belonging to different LIS requires crossing one or more routers. Using the PVCs configured in the LIS, the mapping between IP addresses of the LIS and PVC addresses (VPI/VCI) can be preset in the routers. Instead, in the case of SVCs, the correspondence between IP addresses of the LIS and ATM addresses has to be determined through an ATM Address Resolution Protocol (ATM-ARP) server.

In the example in Fig. 2.21, a host connected to LAN 195.31.230.0 has to communicate with a host connected to LAN 138.31.0.0. In the case of a full-mesh LIS configured to use only PVCs with n routers, $L = n(n-1)$ monodirectional virtual circuits need to be pre-configured at layer 2. The ATM level complexity is $O(n^2)$. There is a limit to the number of layer 2 adjacencies that can be managed by an ATM switch. Moreover, the IP routing protocol of the OSPF type needs to exchange $O(n \times L) = O(n^3)$ signaling LSP messages to configure the routing tables. Thus, the complexity of using a full-mesh topology of PVCs increases significantly with n. Moreover, when an ATM

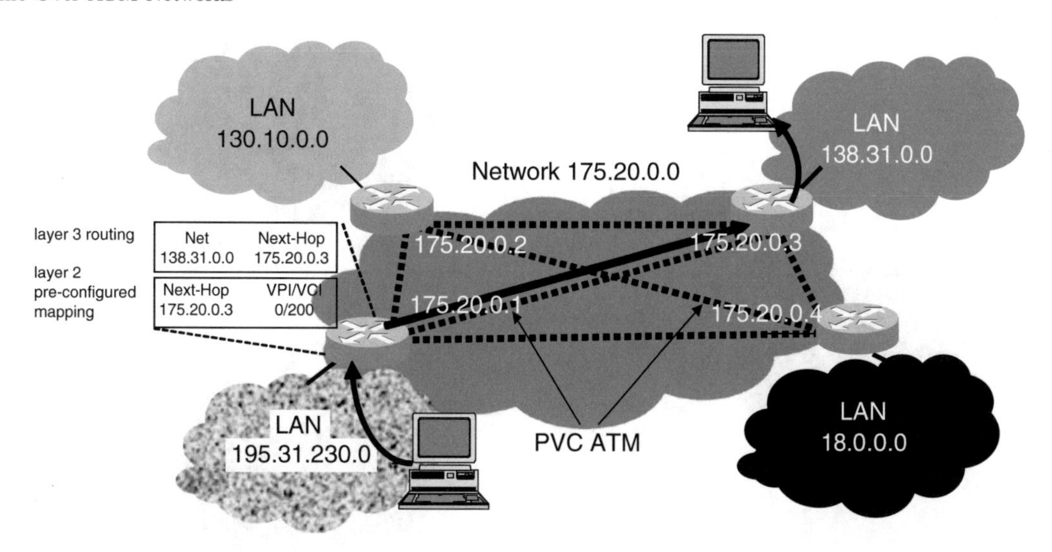

Fig. 2.21: Representation of an IP network (175.20.0.0) over ATM with a full-mesh topology of PVCs interconnecting routers

physical link fails, all PVCs using that link fail and many routers have to update their routes simultaneously. This entails from $O(n^3)$ to $O(n^4)$ routing messages to be exchanged among routers to reconfigure the paths. This is what is called "routing storm," which may cause network routing to become unstable after a single link failure event. Experiments have shown that the IP routing protocol has severe convergence problems and huge processing loads in full-mesh networks with more than 20 nodes.

Before concluding this section, it is important to highlight two advantages of the overlay approach:

- Flexibility in allocating capacity: it is possible to configure ATM PVCs of different capacities even in an asymmetrical way.
- Exploitation of ATM potentialities to support differentiated QoS levels for distinct IP traffic flows.

CIP is an IETF standard [33, 34]. Based on RFCs 1577 and 2225, the two methods shown in the next sections have been proposed to reduce the number of ATM layer adjacencies (i.e., ATM virtual circuits) used to manage the IP traffic according to the CIP approach.

2.6.1 The LIS Method

A LIS is an IP subnetwork (such as a department or a workgroup) consisting of hosts. A LIS is characterized as follows (see Fig. 2.22):

- All the members of a LIS have the same IP subnetwork address.
- All the members of a LIS are connected directly to each other at the ATM level (layer 2 adjacency): it is not necessary to perform IP routing within a LIS since internal routers identify themselves to each other using ATM PVCs or SVCs.
- Hosts or routers, external to the LIS, are accessible through a border router (IP routing).
- Inter-LIS communications are performed through one or more routers.

There can be several LISs in the same ATM network, but routers are needed to interconnect them. Each LIS includes an ATM-ARP server that resolves IP-to-ATM addresses. When a host is turned on, it connects with the ATM-ARP server that requests the host IP and ATM addresses to be stored in the ATM-ARP lookup table for future use. Hosts and routers have to contact the ATM-ARP to resolve IP addresses into ATM addresses (ATM SVC cases).

There are some inefficiencies in the LIS approach. In particular, if two LISs are in the same ATM network, a host in one LIS must go through a router to communicate with a host in the other LIS, even if the underlying ATM network can set a virtual circuit, which directly connects both hosts.

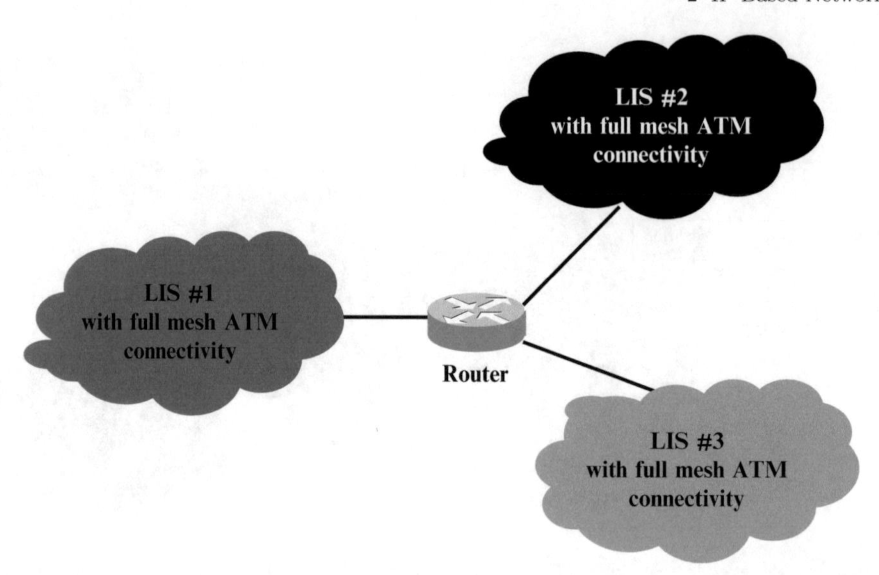

Fig. 2.22: LIS approach for IP traffic over ATM networks: there is a border router to connect (at layer 3) the different LIS domains

Subsequently, IETF defined a CIP improvement under the name Next Hop Routing Protocol (NHRP), where direct ATM virtual circuit connections can be set up between hosts in different LISs that, in this case, are named Local Address Groups (LAGs). More details are provided below.

2.6.2 The Next Hop Routing Protocol

An IP system at the edge of an ATM network needs to find the ATM address of the optimal next hop for a destination IP address. A first solution to this problem is provided by the ATM-ARP server of the LIS method in RFC 1577, which, however, operates within one IP subnet. This technique does not scale well to large multi-organization networks. NHRP proposes a solution to deal with subnet-based routing within the same ATM network. NHRP has been conceived from RFC 1577 by substituting the LIS concept with the LAG one. Instead of using an ATM-ARP server, NHRP adopts a server, called NHRP Server (NHS). Hosts are configured with the address of the NHS; the hosts have to register to the related NHS. Each NHS maintains a table with the IP-ATM mapping of all the hosts belonging to the LAG or reachable through a router connected to the LAG and managed by the NHS. Typically, border routers are coincident with NHSs.

When a host (sender) has to transmit an IP packet (and hence there is the necessity to resolve an IP destination address), it sends a request to the related NHS (i.e., the NHS serving the same LAG of our host). If the NHS can directly resolve the destination IP address (i.e., the destination address is in the same LAG of the sender), the NHS provides the IP-to-ATM mapping to the sender, whereas if the destination IP address is not in the same LAG, the NHS sends a mapping request towards the next NHS along the IP path towards the destination (this is the reason of the name "Next Hop" given to this protocol). This procedure is repeated until an NHS is reached that re-solves the address in its LAG and finds the corresponding ATM address. Hence, such NHS returns the mapping to the origination NHS via the same path used to propagate the request. In this way, the NHSs encountered can update their tables with this new IP-to-ATM mapping (so that it can be made available for future use, if needed). Hence, NHRP allows an effective mapping mechanism when different subnetworks (i.e., LAGs) are involved. Once the sender knows the ATM address of the receiver, it can establish an end-to-end connection with the destination, called "shortcut," to transfer IP datagrams between them. Before a shortcut is established, data will be forwarded through routers as in the classical CIP.

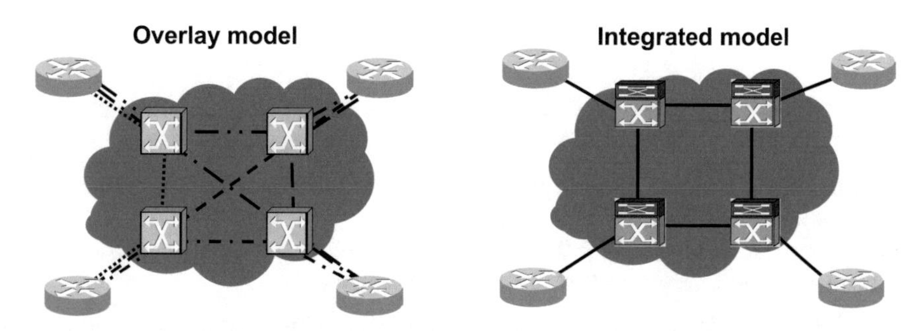

Fig. 2.23: The overlay model (on the *left*) is based on layer 2 adjacencies so that a full-mesh ATM topology is needed to achieve the best efficiency. Instead, the integrated model (on the *right*) routes traffic at the IP level and uses the ATM network only for the fast packet-switching

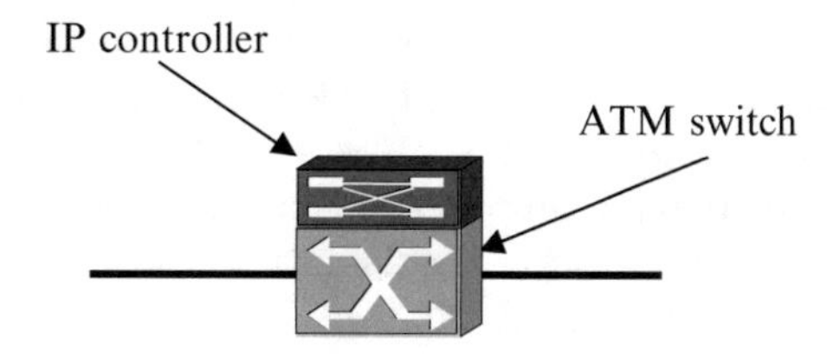

Fig. 2.24: IP + ATM switch conceptual structure: an IP layer (*control component*) on top of a layer 2 fast packet switch (*forwarding component*); a direct integration of these two functions is achieved

2.6.3 The Integrated Approach for IP Over ATM

The idea behind the integrated approach is the elimination of ATM switching in managing IP traffic. Hence, we have a simple IP network where IP datagrams are conveyed by ATM cells on virtual circuits, determined according to the IP routing protocol (e.g., RIP, OSPF, IS–IS). The comparison between overlay and integrated models is depicted in Fig. 2.23.

The practical realization of the integrated approach implies two fundamental requirements:

- To insert IP routing intelligence in ATM switches. Hence, hybrid machines need to be built that from now on will be called "IP + ATM switches," with both IP routing and ATM switching (see Fig. 2.24).
- To define a new protocol to bind the labels (i.e., VPI/VCI fields) of ATM cells to the path determined according to IP routing. The IP routing table also contains the ATM addresses of the next hop for each IP destination address.

All major manufacturers have produced hardware and software able to fulfill both of the above requirements, but they have proposed proprietary solutions. For instance, we can mention the following manufacturers with proprietary solutions for the integrated approach:

- Toshiba: Cell Switch Router (CSR)
- Ipsilon: IP Switching
- Cisco: Tag Switching
- IBM: Aggregate Route-Based IP Switching (ARIS)
- Telecom Finland: Switching IP Through ATM (SITA)
- Cascade: IP Navigator
- NEC: IP Switching Over Fast ATM Cell TranspOrt (IPSOFACTO)

In these cases, we cannot properly speak of ATM networks since only the fast packet-switching of ATM is used. Even if these integrated systems have many common aspects, they are not interoperable and this is a significant limit. All these solutions use the control software of an IP router and integrate it into the hardware of an ATM switch. As for the *control component*, each IP + ATM switch adopts an IP routing protocol (e.g., RIP, OSPF, IS–IS). Finally, as for the *forwarding component*, the IP + ATM switches use conventional ATM hardware and label (i.e., VPI/VCI) switching for sending cells into the network.

2.6.3.1 IP Switching

The solution achieving the best integration of IP and ATM is the "IP switching," developed by Ipsilon Networks, Inc. The IP switch communicates the information needed for the management of traffic using two protocols:

- General Switch Management Protocol (GSMP): This protocol is in charge of mapping a given IP input traffic flow to a particular output port of the ATM switch.
- Ipsilon Flow Management Protocol (IFMP): This protocol is used to exchange control information of IP traffic flows among switches (e.g., the QoS requested by a given flow).

At system start-up, virtual circuits are established among adjacent IP switches employing predefined VPI/VCI (i.e., ATM PVCs). An IP flow is characterized using the IP header fields (e.g., IP source, IP destination, but also requested QoS). As soon as the first datagram of an IP flow reaches the IP switch, it is classified based on local routing procedures (the switch also operates at layer 3). Then, the controller instructs the switch so that (from now on) it can switch the packets of the given IP flow to a suitable output VPI/VCI.

There is a scalability problem in using such an approach in geographical areas: if the number of flows to be switched increases, the IP switch must become faster and faster and with high memory capabilities.

Many telecommunication operators adopt the IP + ATM technology in their network backbones. Since the proprietary solutions made by different manufacturers were not interoperable, a standardization effort was made to unify the different solutions. The final outcome was the definition of the Multi-Protocol Label Switching (MPLS) technology, where label switching is used based on IP routing. The following section provides a survey on MPLS.

2.7 Multi-Protocol Label Switching Technology

At the beginning of 1997, a Working Group was established in IETF to define a label switching standard to simplify and speed up the forwarding of IP packets using labels managed at layer 2. It was decided to call such technology "Multi-Protocol Label Switching" (MPLS). Starting from the integrated approach, the target was to develop a standard for:

- An efficient integration of network layer traffic with different layer 2 technologies, not only ATM.
- Increasing the speed in forwarding IP traffic at the nodes.
- Enriching the IP routing with new functionalities (e.g., traffic engineering aspects).
- A greater scalability in IP networks, enabling them to convey huge traffic loads and providing services like Virtual Private Networks (VPN) to user groups.
- Introducing mechanisms for QoS support in IP networks, which typically provide best-effort services.

MPLS is a connection-oriented protocol to route IP traffic over different layer 2 technologies such as ATM, Frame Relay, and Ethernet. The fundamental elements of an MPLS network domain are described below (Fig. 2.25):

- Label Edge Routers (LERs): These are high-speed routers placed at the boundary of the MPLS network (domain) and used to manage the associations between destinations and labels (i.e., the label-switched path).
- Label Switch Routers (LSRs): These high-speed routers are used within the core network to switch IP packets based on the labels they convey.

An MPLS domain is physically inserted inside an IP network, receiving IP traffic from it and delivering IP traffic to it after having routed it along a path. This forwarding scheme is novel and based on fixed-length labels, which are prefixed to any IP datagram. The label has only a local meaning to properly forward a datagram at a node. A label summarizes several routing information concerning the datagram, such as

- Destination
- Precedence
- Belonging to a VPN
- QoS
- Traffic engineering information

At present, the evolution of MPLS is represented by Generalized Multi-protocol Label Switching (GMPLS), which permits label switching for IP traffic to be used on different lower-layer technologies, including optical transport networks.

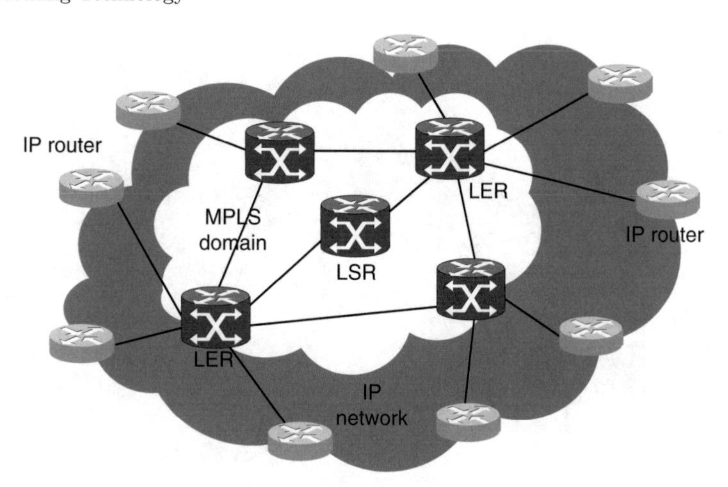

Fig. 2.25: An MPLS network is physically inserted into an IP network, but its operation is ideally distinguished. LERs (at the borders) receive IP datagrams; LERs label these datagrams based on their destinations and forward them through the MPLS domain along label-switched paths (like tunnels)

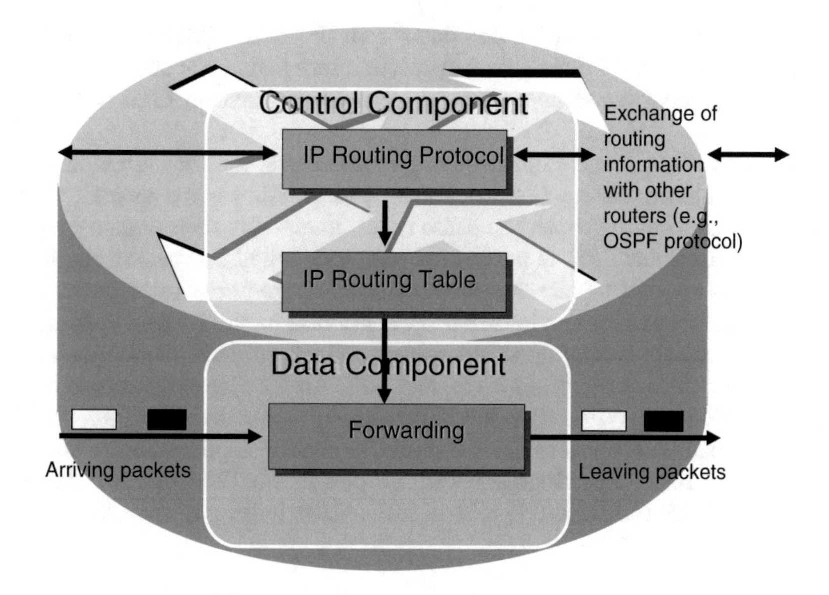

Fig. 2.26: Functional elements of an IP router

2.7.1 Comparison Between IP Routing and Label Switching

IP routing can be functionally divided between *data component* and *control component* (see Fig. 2.26).

- The data component is in charge of the actual forwarding of IP packets from input to output across a switch or router. The forwarding table maintained by a router and the information carried by the IP packet header are used to forward the packets to the next hop. The router uses the packet header information to select a particular entry in its forwarding table. In particular, an exhaustive search is performed on all the IP addresses contained in this table before deciding on the next hop.
- The control component is responsible for building and maintaining the forwarding table. It consists of one or more routing protocols with the related procedures to update the forwarding tables.

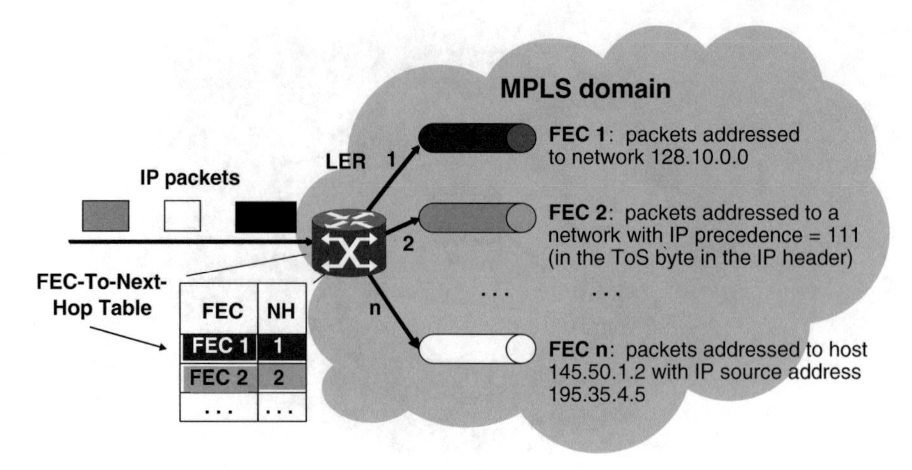

Fig. 2.27: The definition of an FEC depends on several aspects, such as the IP address of the destination, the IP precedence level, the existence of a source–destination reserved path, and traffic engineering considerations

Each router makes an independent forwarding decision for an IP packet. Each router analyzes the packet header (i.e., destination IP address) and, based on a routing algorithm, independently chooses the next hop for the packet [35]. Note that the IP packet header contains considerably more information than needed to determine the next hop.

In MPLS, choosing the next hop is the composition of two functions: the first function partitions all the packets into a set of Forwarding Equivalence Classes (FECs); the second function maps each FEC to a next hop. An FEC corresponds to a group of packets sharing both the same requirements for their transport and the same path. The assignment of a packet to a particular FEC is performed just once when the packet reaches the ingress LER. An FEC corresponds to a path in the MPLS domain with suitable characteristics in terms of available bandwidth, priority, etc. An FEC is defined based on the information contained in the IP packet header (e.g., the IP address of the destination). Different packets belonging to the same FEC are subject to common forwarding decisions. An example of packet classification at an LER is shown in Fig. 2.27. An FEC corresponds to a local label for each hop along the path in the MPLS domain. When a packet is forwarded to its next hop (i.e., an LSR), the local label is sent along with it. When the packet reaches an LSR internal to the MPLS domain, its label is examined to decide the output interface to forward this packet together with its new label. Therefore, inside the MPLS domain, it is not necessary to scan the whole IP routing table and the IP packet header is not used; the forwarding procedure is immediate.

2.7.2 Operations on Labels

The MPLS network realizes a tunnel for the IP packets belonging to a given FEC. The *label binding* performed in the path setup phase is the association of a local label to an FEC at each LSR along the path. The set of label bindings for the different hops from input to output of an MPLS network forms the Label-Switched Path (LSP). The LSP is related to an FEC and depends on: (1) QoS requirements; (2) the dynamical condition of the network (i.e., the congestion of the network when the LSP is established); (3) traffic engineering aspects. Once defined, the LSP is not modified. LSPs are unidirectional.

An LER operates at the edge of the MPLS network and typically supports multiple ports connected to different networks (e.g., Frame Relay, ATM, and Ethernet). It is the responsibility of the LER to recognize the FEC corresponding to a given input IP packet and then to assure that the packet is forwarded in the corresponding LSP by imposing the appropriate label on top of the packet. The label-to-FEC correspondence has to be unique.

Each LSR builds a table to specify how a packet must be forwarded, as described later in Sect. 2.7.5. Packet forwarding schemes have been defined for all types of layer-2 technologies with a different label encoding in each case. MPLS handles labels just like all other virtual circuit identifiers in other virtual circuit-switching technologies. For instance, layer 2 circuit identifiers of Frame Relay or ATM (i.e., DLCIs for Frame Relay networks or VPIs/VCIs for ATM networks) can be directly used as labels.

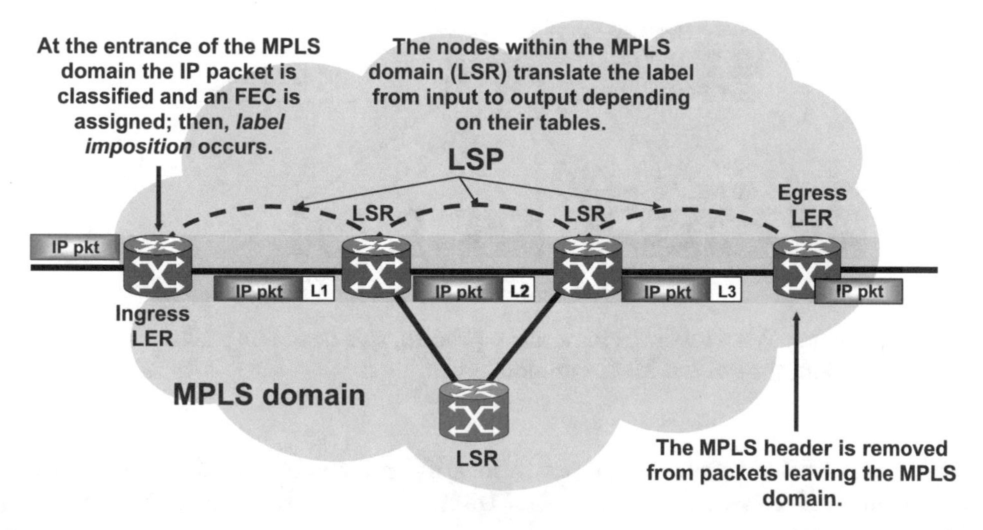

Fig. 2.28: MPLS switching is based on label swapping at the LRSs according to their tables. At each LSR internal to the MPLS domain, a new label is recomputed and replaces the previous one. These labels allow the packet to be routed along the right path, depending on the IP packet FEC

Let us consider an IP packet entering an MPLS domain, as in Fig. 2.28. When a packet arrives at the first router of the MPLS domain, *ingress LER*, the IP packet header is analyzed. Let us assume that data in the IP packet header match an already-defined FEC with related LSP in an LER table. Then, the ingress LER inserts (i.e., *pushes*) an MPLS header in the packet: this is the *imposition* of label L1 on top of the IP packet in Fig. 2.28. Subsequent LSRs in the MPLS domain update the MPLS header by *swapping* the label (L1 against L2, L2 against L3). Finally, the last router of the LSP, called *egress LER*, removes (i.e., *pops*) the MPLS header (i.e., L3) so that subsequent MPLS-unaware IP routers can handle the packet. Referring to the example in Fig. 2.28, we say that the MPLS domain allows a tunnel of "layer 1," meaning that we have a *single* MPLS domain, whose labels are swapped at the LSRs.

When different MPLS networks interoperate, the OSI layer 3 is used to allow the exchange of packets (i.e., they are interconnected by border routers).

2.7.3 MPLS Header

MPLS can be considered as a new "shim" protocol between data link and network layers. This is because the MPLS header (containing the label) is between the MAC header and the network layer packet. Hence, referring to the OSI model, MPLS can be considered as a protocol of a hypothetical 2.5 layer. MPLS just provides *encapsulation* for network layer packets. As defined in RFC 3031 (updated in RFC 6178 and RFC 6790), MPLS is intended to be "multi-protocol" for both the protocol layers above and those below it [35].

The shim MPLS header is processed at each LSR with a very low computational load. The MPLS header (see Fig. 2.29) is composed of four fields of fixed length: label field (20 bits), EXP field (3 bits), S flag (1 bit), and TTL field (8 bits). More details on these fields are provided below.

- Label: This 20-bit field carries the actual value of the local label. The characterization of this field depends on the protocol used to assign and distribute the labels among LSRs.

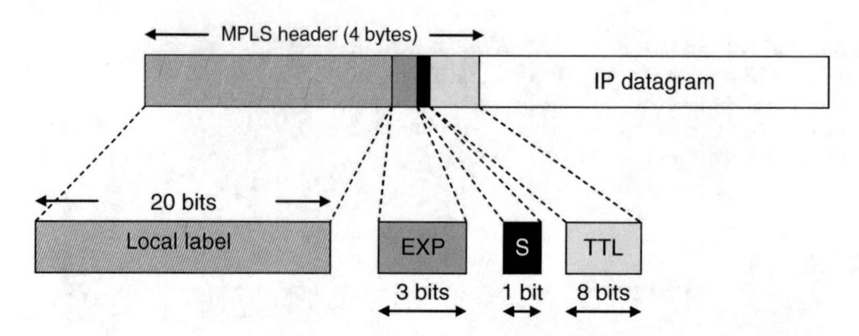

Fig. 2.29: 32-bit MPLS header. With the label imposition process, this header is prefixed to the IP datagram when it reaches the LER at the entrance of the MPLS domain

- EXP: These three bits have an experimental use to identify traffic classes or network congestion. If all these three bits are used for traffic class specification, eight traffic classes can be defined; this is important if DiffServ is adopted for QoS support.
- S: This bit is used for label stack functions, that is, when multiple MPLS headers are stacked ($S = 1$ denotes the last node in the domain: at the next hop, the IP datagram leaves the current MPLS domain).
- TTL: It is a counter decremented at each hop. This field reproduces at the MPLS level the same hop count used for IP datagrams.

The MPLS forwarding procedure is based on a 20 bit label, which is shorter than an IP address so that MPLS can speed up the delivery of IP datagrams. Moreover, the same protocol can be adopted for both unicast and multicast traffic, thus improving the scalability: a label can be used for a single route, for a group of routes, or for a multicast tree.

2.7.3.1 Management of the Time-to-Live Field

Each IP datagram has a header with a TTL field; each time this datagram crosses a router, its TTL value is reduced by 1. If TTL becomes equal to 0 before the datagram reaches its destination, the packet is discarded. Such functionality has been consistently extended to MPLS through the 8-bit TTL field in the MPLS header.

At the ingress LER in the MPLS domain, the TTL value in the IP datagram header is copied in the corresponding field of the MPLS header. When an IP datagram travels across an MPLS domain, crossing the different LSRs of its LSP, it must leave the MPLS tunnel with the same TTL value that would have been crossing the same number of IP routers. Hence, the TTL field in the MPLS header is reduced by 1 at each LSR. When the datagram leaves the MPLS domain, the value of the TTL field of the MPLS header is copied in the corresponding field of the IP header.

The TTL field can be used for two different purposes in MPLS:

- Avoiding loops.
- Limiting the maximum number of LSRs crossed by an IP packet in an MPLS domain.

2.7.4 MPLS Nested Domains

When an IP datagram crosses different nested MPLS domains (i.e., areas managed by different ISPs, with one MPLS domain included into another), it has a stack of MPLS (multiple) headers, each of them requiring 32 bits, and organized in a Last Input First Output (LIFO) way. Each level of the stack is used for a domain. The label stack allows that the MPLS domains are organized in a hierarchical way. MPLS can be used for routing at both low level (e.g., between individual routers of an ISP) and at the higher domain-by-domain level. Figure 2.30 below depicts a stack of MPLS headers.

The operation to be performed on the label stack before forwarding the packet to the next LSR/router may be to swap the top label stack entry with another, or to pop an entry off of the label stack, or to push one additional

Fig. 2.30: MPLS header with label stack

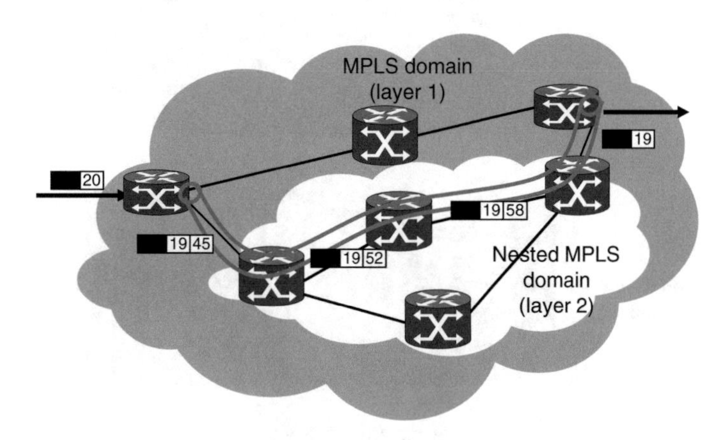

Fig. 2.31: Use of the label stack in a nested MPLS domain scenario

entry at the top of the label stack [36]. Figure 2.31 describes the typical situation of an MPLS hierarchical domain: when a packet enters an MPLS domain, which is contained in another MPLS domain, a new MPLS header is appended to the packet. In particular, we have the arrival of a packet with label 20 (label of layer 1) at the first LSR of the MPLS domain. Such LSR firstly swaps the label from 20 to 19 and inserts the packet into the layer 2 tunnel by pushing the label 45 of the nested MPLS domain (label of layer 2). The intermediate LSR in the nested MPLS domain swaps the label of layer 2 (without considering the layer 1 label). Finally, the output LSR clears (i.e., pops) the layer 2 label and sends an output packet with label 19 of layer 1.

2.7.5 MPLS Forwarding Tables

MPLS forwarding tables contain information on the next hop organized according to the Next Hop Label Forwarding Entry (NHLFE), which provides instructions on how to forward a packet for which a label has already been assigned. NHLFE contains the following information: the packet next-hop address, the output interface, the output label, and the operation to be performed on the label (e.g., swapping, popping). The MPLS forwarding table, also called Forwarding Information Base (FIB), is specific for each MPLS router (i.e., LER or LSR). A FIB can be of two different types:

- FIB with FEC-to-NHLFE (FTN) mappings, that is, the correspondences between incoming packet FECs and NHLFE entries.
- FIB with Incoming Label Map (ILM) containing the mappings between labels carried by incoming packets and NHLFE entries.

LERs use FIB with FTN. Instead, LSRs (internal to the MPLS domain) use FIB with ILM. The use of FIBs is explained as follows (see Fig. 2.32).

- Suppose a packet with no label arrives at an edge MPLS router (LER case). The MPLS router first determines the FEC of the packet, then looks up in the FIB for the FTN entry, matching the packet FEC. This FTN contains a label and an NHLFE, specifying the next hop for the packet. The MPLS router pushes an MPLS header, which contains the label read in the FTN and forwards the packet according to the NHLFE.
- Now consider that a labeled packet arrives at an internal MPLS router (LSR case). The MPLS router searches in the FIB for an ILM entry matching the input packet label and reads the corresponding NHLFE. The NHLFE can either indicate that the MPLS header must be swapped against a new label, or popped. After swapping or popping operations are completed, the MPLS router forwards the packet according to the NHLFE.

Let us consider a given LSR #A (see Fig. 2.33) with routing table in Table 2.3 and a possible organization of the corresponding "extended" FIB (with both FTN and ILM parts) in Table 2.4.

The corresponding FIB of ILM type is shown in Table 2.5.

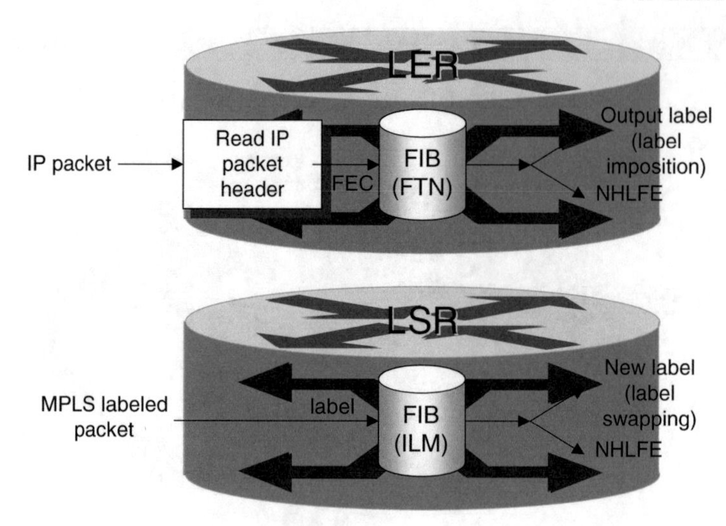

Fig. 2.32: Label management at an LER and at an LSR

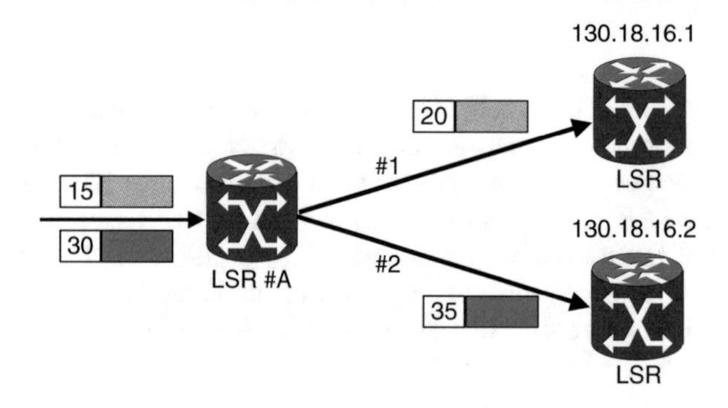

Fig. 2.33: Example of forwarding operations performed by a generic LSR #A

Table 2.3: Routing table example for LSR #A

Destination network address	Next hop
193.20.16.0	Output interface: #1 Next hop address: 130.18.16.1
194.80.18.0	Output interface: #2 Next hop address: 130:18.16.2
....	....

Table 2.4: FIB for LSR #A

FIB of ILM type (Input label, Output label, Output interface)

FEC	Input label	Output label	Output interface	IP of the Next Hop
193.20.16.0	15	20	#1	130.18.16.1
194.80.18.0	30	35	#2	130:18.16.2
...	...	...	...	...

NHLFE entries (Output label, Output interface, IP of the Next Hop)

FIB of FTN type (FEC, Output label, Output interface)

Table 2.5: ILM for LSR #A

Input label	NHLFE
15	Output label: 20 Output interface: #1 Next hop: 130.18.16.1 Type of operation on the label: swap
30	Output label: 35 Output interface: #2 Next hop: 130.18.16.2 Type of operation on the label: swap
...	...

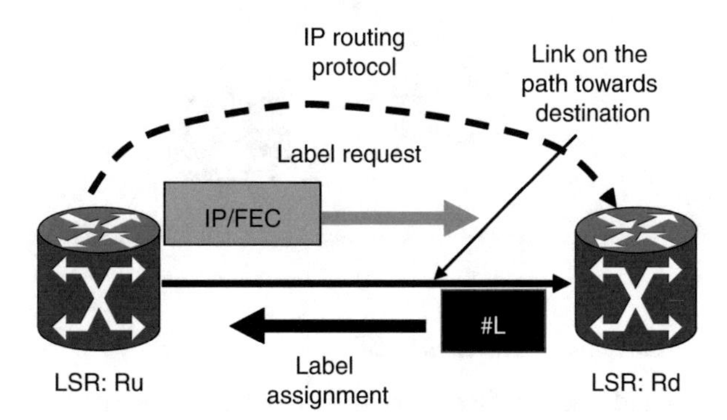

Fig. 2.34: Label assignment and distribution

2.7.5.1 NHLFE Details

Let us provide more details on the NHFLE entry. NHLFE contains the following data:

1. The next hop for the packet (i.e., the output interface).
2. The operations to be performed on the packet label stack:

- Replace the top label with a specified new label (swapping).
- Pop the label stack (if present/needed).
- Replace the top label with a specified new label, and then push one or more specified new labels onto the label stack.

3. The data link encapsulation to be used to transmit the packet.
4. The way to encode the label stack for transmitting the packet.
5. Any other information needed to manage the packet properly.

2.7.6 Protocols for the Creation of an LSP

When there is no already-defined LSP for a traffic flow, a protocol has to be used to create the LSP when an IP packet of this flow arrives at an LER. In particular, this protocol is in charge of FEC-to-label bindings and building the FIB at each MPLS router. The LSP must be defined before transmitting the traffic flow in the network. The MPLS standard does not impose any specific protocol; the only requirement is that labels are "downstream-assigned" and that bindings are distributed in the "downstream-to-upstream" direction. In MPLS, upstream LSR and downstream LSR are relative terms, which always refer to a prefix (i.e., an FEC). Let us refer to the situation depicted in Fig. 2.34: the decision to bind a label to a particular FEC to be used for the traffic from LSR Ru to LSR Rd is made by LSR Rd, which is downstream for that traffic; then Rd informs Ru of the label binding. Ru and Rd are *label distribution peers.* This process has to be extended if more LSRs are involved in the source-to-destination path, thus requiring that a label binding is made at each hop in the upstream direction.

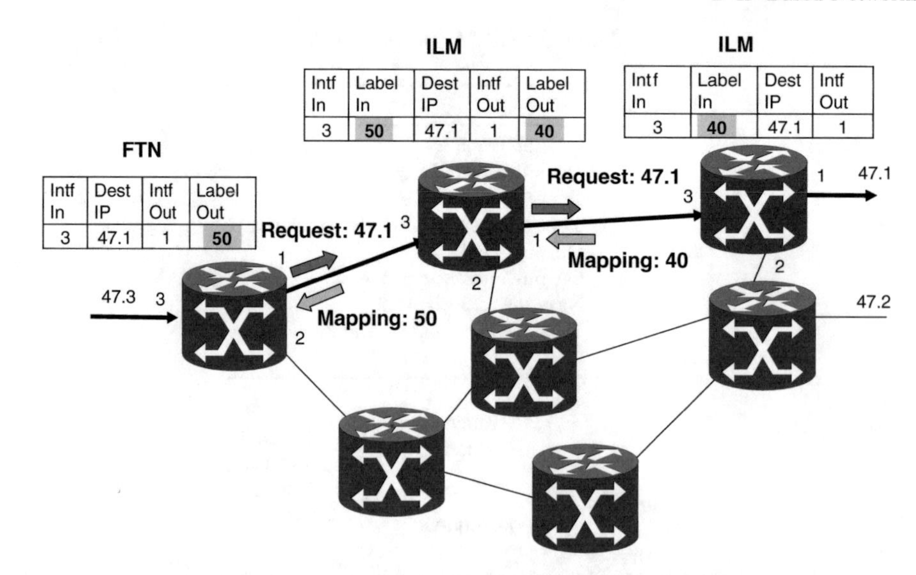

Fig. 2.35: LSP route selection: example of hop-by-hop routing based on the IP address of the destination, "47.1.0.0"

Two options are available to select the route of an LSP:

- *Hop-by-hop routing*: Each LSR independently selects the next hop for a given FEC according to an IP routing protocol (e.g., OSPF). In this case, we assume that the IP routing protocol has already determined the routing tables at the LSRs. Based on this procedure, MPLS builds its forwarding tables (i.e., FIB) by operating label bindings along the path in the upstream direction from the destination back to the source. See Fig. 2.35.
- *Explicit Routing* (ER): The ingress LSR specifies the list of nodes traversed by the LSP according to a source routing approach. Along the path, the resources may be reserved to ensure QoS for the data traffic. ER can define different paths with respect to those of conventional IP routing to account for traffic engineering aspects; the paths could also be non-optimal for some aspects.

Existing IP routing protocols, such as BGP, have been enhanced to piggyback the label information within their messages. The RSVP protocol has also been extended to support the exchange of labels.

RFC 3036 (then substituted by RFC 5036) defines a protocol, called Label Distribution Protocol (LDP), to bind labels to hops for a given FEC [37]. Two LSRs using the LPD protocol for label bindings are said to be "LDP peers"; among them, there is an "LDP session" for the exchange of label associations. LDP needs a negotiation phase among LSRs before exchanging labels.

A summary of the different protocols to manage labels is provided below:

- LDP maps unicast IP destinations into labels.
- RSVP-Traffic Engineering (RSVP-TE) is the classical RSVP protocol with increased capabilities to advertise LSPs.
- Protocol-Independent Multicast (PIM) supports label mapping for multicast traffic.
- BGP can be used to distribute labels in MPLS VPNs.

Finally, the main characteristics of MPLS routing can be summarized as follows:

- MPLS combined with a standard routing protocol of the Interior Gateway Protocol (IGP) type, such as OSPF or IS–IS, provides a reliable and scalable IP routing in networks of whatever type (e.g., ATM, Frame Relay).
- MPLS combined with Border Gateway Protocol (BGP) provides VPN scalable services (each site can host thousands of VPNs).
- MPLS routing can support QoS for IP traffic flows, independently of the underlying technology.
- MPLS routing can take traffic engineering concepts into account for efficient use of network resources. Routing may depend on traffic type and node congestion. For instance, MPLS can reroute the traffic towards underutilized nodes.

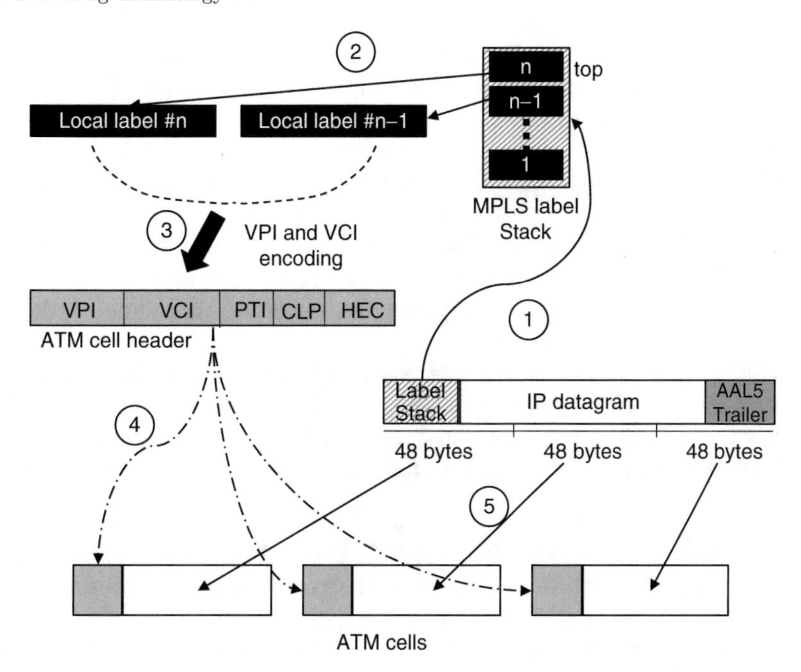

Fig. 2.36: From the MPLS header of an IP datagram to the VPI/VCI encoding of ATM (ALL5) cells, as performed by an LER or an LSR in an ATM network within the MPLS domain. The *numbers encircled* in the figure highlight the order of the different steps

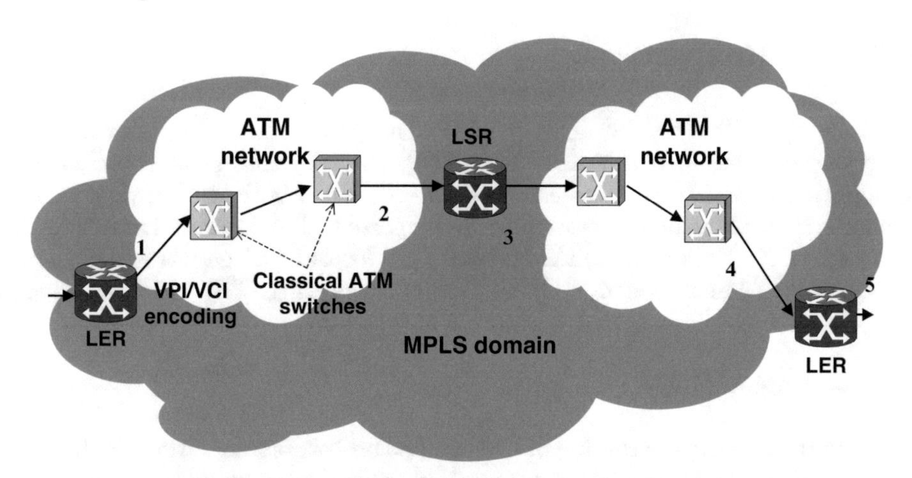

Fig. 2.37: Example of use of ATM networks in an MPLS domain: LSPs need to have MPLS routers (i.e., numbers 1, 3, 5) at their borders; these routers can operate on the MPLS header and, in particular, to perform label swapping, to update the TTL field, and to encode the VPI and VCI fields. All the intermediate nodes in the LSP (within the ATM networks numbered as 2 and 4) are normal ATM switches. Hence, from the ingress LER to the output one, TTL is reduced by 3

2.7.7 IP/MPLS Over ATM

It is possible to use ATM switches as LSRs, as described in RFC 3031 [35]. As a matter of fact, MPLS forwarding is similar to ATM switching: ATM switches use the input port and the incoming VPI/VCI value as indexes into a "cross-connect" table from which they obtain the output port and the outgoing VPI/VCI value. Hence, if the labels can be directly encoded into the VPI/VCI fields managed by ATM switches, then these switches become LSRs by using some software upgrades. We will refer to these ATM switches as "ATM-LSRs." There are three ways to encode labels into the VPI/VCI fields of the ATM cell header, assuming to use AAL5 and an MPLS label stack (see Figs. 2.36 and 2.37):

- Switched Virtual Circuit (SVC) encoding: the VPI/VCI fields are used to encode the label at the top of the label stack. With this encoding technique, each LSP is obtained as an ATM SVC, and the label distribution protocol becomes the ATM signaling protocol. In this case, the ATM-LSRs cannot perform "push" or "pop" operations on the MPLS header.
- Switched Virtual Path (SVP) encoding: The VPI field is used to encode the label at the top of the label stack and the VCI field is used to encode the second label of the stack (if present). This technique adopts ATM Virtual Path (VP) switching: LSPs are realized as ATM SVPs, with the ATM signaling protocol used as the label distribution protocol. However, this technique cannot always be used (e.g., if the network includes an ATM VP through a non-MPLS ATM network). When this encoding technique is adopted, the ATM-LSR at the exit of the VP performs a "pop" operation.
- SVP Multi-point encoding: The VPI field encodes the label at the top of the label stack, part of the VCI field encodes the second label of the stack (if present), and the remainder of the VCI field is used to identify the ingress LSP. With this technique, conventional ATM VP-switching capabilities can be used to provide multipoint-to-point VPs (merging of VPs): packets from different sources are distinguished by using different VCIs within the VP.

The time spent encoding the ATM header (VPI and VCI fields) from the MPLS header at the MPLS domain entrance is widely regained using the fast-forwarding procedures within the MPLS domain. Note that the MPLS TTL field is not to be modified when the IP datagram is fragmented in ATM cells to traverse the MPLS-ATM network. Consequently, an LSP supported by ATM is considered as a single hop (in terms of crossed nodes) at the IP/MPLS level: the value of the TTL field is, therefore, scaled-down of 1; more details are provided in the example in Fig. 2.37.

2.7.8 MPLS Traffic Management

IP does not provide any guarantee on the delivery of IP datagrams; routers can discard packets without any notification. It is up to the higher-level protocols (e.g., TCP) to verify that packets are correctly received. This classical TCP/IP approach to traffic management does not guarantee both the delivery times and the capacity provided to incoming traffic. Only best-effort traffic is supported. The interest here is in describing MPLS's potentialities to manage traffic classes with differentiated QoS levels in IP networks. In particular, we focus on two important aspects of traffic management, such as Traffic Engineering (TE) and QoS provision.

2.7.8.1 MPLS Traffic Engineering

The TE concept deals with adapting the routing of traffic based on network conditions with the joint goals of good user performance, efficient usage of network resources, and service guarantees (i.e., providing some redundant backup paths for high service availability) [38]. Classical routing protocols do not have enough information to achieve these purposes; path computation simply based on IGP metric (e.g., shortest path metric) is not enough because it may cause that some parts of the network are overloaded and that other parts are underutilized. Hence, TE approaches can be adopted to define routes avoiding these problems. In general, TE techniques can be used for the following purposes:

- To maximize the utilization of links and nodes in the network.
- To engineer the links to achieve the required QoS in terms of delay.
- To spread network traffic across different network links; this strategy is crucial if we want to minimize the impact of a single failure.
- To have some spare link capacity for rerouting the traffic in case of failure.
- To implement the policy requirements requested by the network operator.

Traditional TE approaches modify routing metrics. MPLS supports TE with the MPLS-TE technique that allows greater potentialities: it can define paths by considering the dynamic conditions of network congestion; paths can be redefined to bypass congested nodes. More details on MPLS-TE are provided in the following section in relation to QoS support.

2.7.8.2 QoS Approaches in MPLS

IP supports a "native" mechanism to pass QoS information to all the routers in the network: QoS information can be inserted in the ToS field of the IPv4 header using the 3-bit priority level (IP precedence). The problem with the use of ToS is the analysis time at nodes.

The first method for providing QoS in MPLS networks is based on FECs. As already stated, the LSP of a given FEC is determined by taking the following aspects into account: QoS, dynamic network conditions, and traffic engineering requirements. The MPLS capacity to guarantee QoS to different traffic classes is related to the ability of LERs to analyze the input IP traffic so that it can be assigned from the beginning to the proper FEC class with QoS guarantee. In particular, the IP packet is analyzed at the ingress LER to determine whether it belongs to a simple data flow or rather to a real-time traffic flow with a suitable precedence level. Then, the destination IP address is examined. Based on all these data, the packet is associated with one of the following FECs:

- If the packet belongs to real-time traffic, then it is associated with a guaranteed-bandwidth FEC, exclusively used for real-time traffic.
- If the packet comes from a privileged user, the packet is associated with an FEC reserved to that user. The corresponding LSP is defined so that the output LER from the MPLS domain is the closest one to the final destination of the IP flow.
- If the packet is generated by an unprivileged user, it is routed along an LSP shared with other users of the same type; this packet is not associated with a special FEC.

The second approach for QoS provision in MPLS networks is to adopt DiffServ [27, 39], described in the previous Sect. 2.5.2. Accordingly, network resources are managed and QoS is provided on a per-flow basis. DiffServ DSCP uses the six most significant bits of the Type of Service (ToS) byte of the IPv4 header or the six most significant bits of the Traffic Class byte of the IPv6 header. The two least significant bits in these bytes can be used for Explicit Congestion Notification (ECN). The DSCP of DiffServ can be coded in the MPLS header in two different ways:

- If it is not necessary to map more than eight PHB levels, the EXP field can be used (EXP has 3 bits, whereas DSCP has 6 bits). MPLS packets with a given EXP setting are treated as IP packets with a given IP precedence.
- If there are more than eight PHBs, it becomes necessary to use FECs with adequate QoS support, as discussed at the beginning of this sub-section.

MPLS can establish LSPs with QoS support through LDP [37], RSVP-TE [40], or Constraint-based Routing—LDP (CR-LDP) [41]. When using LDP, LSPs have no associated bandwidth. However, when using RSVP-TE or CR-LDP, a bandwidth can be assigned to each LSP that can be defined, taking traffic engineering issues into account. Moreover, we can consider MPLS Traffic Engineering (MPLS-TE), which combines extensions to OSPF or IS–IS, to distribute link resource constraints, with the label distribution protocols RSVP-TE or CR-LDP. In MPLS-TE, a traffic flow has some requirements in terms of bandwidth, media type, a priority versus other flows, etc. MPLS determines the routes for traffic flows based on: (1) the resources the traffic flow requires and the resources available in the network (CAC); (2) the shortest path that meets the requirements of the traffic flow.

Let us consider MPLS-TE implemented together with DiffServ. MPLS-TE provides CAC in addition to the PHB offered by DiffServ. MPLS-TE avoids sending more traffic on a particular path than the available bandwidth and queues higher-priority traffic ahead of lower-priority one. The problem with MPLS-TE is that it does not perform CAC on a per-QoS class basis. This issue is solved by MPLS-TE with DiffServ-Aware Traffic Engineering (DS-TE): this technique makes MPLS-TE aware of DiffServ so that one can establish separate LSPs for different traffic classes, taking the available bandwidth for each class into account.

2.7.8.3 Congestion Notification in MPLS

ECN is supported by the two least significant bits of the ToS byte in the IPv4 header or of the Traffic Class byte in the IPv6 header. This is obtained by setting the ECN bits in the IP datagram header as follows:

- ECT (*ECN capable transport*) is a flag that, if set to 1, enables the use of the following bit dealing with network congestion.
- CE (*Congestion Experienced*) is a flag to notify the occurrence of congestion on the LSP if set to 1.

Table 2.6: Management of CE and ECT flags of the IP header through an MPLS domain experiencing congestion or no congestion

ECT flag of an input IP datagram	MPLS ECN bit value	ECT flag of the output IP datagram
NOT ECN capable (ECT = 0)	1	ECT = 0, CE = 0
ECN capable (ECT = 1)	0	ECT = 1, CE = 0
ECN capable (ECT = 1)	1	ECT = 1, CE = 1

The MPLS header should use the EXP field to convey these 2 bits. However, if we carefully examine the different combinations for ECT and CE bits, we note that not all the four possibilities are used:

1. ECT = 0, CE = 0: There is no congestion notification and hence, no congestion can be notified.
2. ECT = 1, CE = 0: There is congestion notification, but there is no congestion.
3. ECT = 1, CE = 1: There is congestion notification, and there is congestion.
4. ECT = 0, CE = 1: This case is not used because it is meaningless.

MPLS can encode these three different cases employing a single EXP bit so that the two remaining EXP bits can be used, for instance for PHB encoding of DiffServ (this is only possible if four PHBs are used; otherwise, FECs with QoS provision should be adopted). Hence, MPLS uses a single ECN bit in the EXP field to notify congestion as follows [42]:

- ECN bit = 1 in case of NOT ECN capable OR CE on the LSP.
- ECN bit = 0 in case of ECN capable AND no CE on the LSP.

This mapping, further described in Table 2.6, must be applied when an IP datagram enters the MPLS domain and when an IP datagram leaves the MPLS domain.

Alternatively to the use of an EXP bit for congestion notification, LSPs can be constructed so as to avoid congestion. FECs are assigned to LSPs also based on the dynamic congestion conditions of the nodes encountered in the MPLS domain. FECs will not be assigned to congested LSPs. Moreover, if during the LSP construction phase or in a subsequent forwarding phase on this path, network congestion is revealed, MPLS signaling is sent from the congested LSRs to the label distribution peers of the LSP to start the procedures to modify the LSP.

2.7.9 GMPLS Technology

GMPLS represents an extension of MPLS to support the technologies of optical transport networks. GMPLS no longer needs that the label is carried as a logical identifier of the data flow, but can be *implicit.* For example, time slots (in SDH/SONET) and wavelengths (in DWDM) can be labels. In these cases, the label switching operations become like "switch this incoming wavelength onto this outgoing wavelength." Therefore, GMPLS is the ideal solution for optical transport networks; many extensions to the protocol have been specified.

As optical cross-connects have become cost-effective, ISPs have started to carry IP traffic directly on the optical transport medium, bypassing any intermediate layer, such as SDH/SONET and ATM; this is possible by utilizing GMPLS. The protocol stack evolution from MPLS to GMPLS is depicted in Fig. 2.38.

As for the management of labels, GMPLS is practically equivalent to MPLS. GMPLS extends some basic functions of MPLS and, in some cases, also adds new ones. In particular, the most significant innovations have been made to the procedure to request, assign, and notify the labels (e.g., LDP protocol), to create bidirectional and symmetrical LSPs, and the propagation of error signaling. A bidirectional and symmetrical LSP has the same LSRs, the same traffic engineering requirements, error detection and correction methods, and resource management in both directions. The bidirectional paths must be set up independently using LDP.

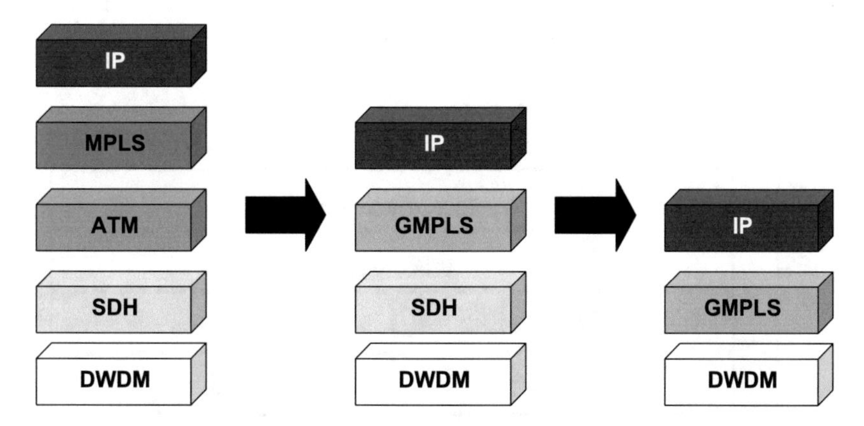

Fig. 2.38: Protocol stack evolution for optical transport networks carrying IP traffic

2.8 Transport Layer

In the previous part of this chapter, we have focused on layer 3 topics, such as addressing, routing, traffic management, and QoS support. At this point, we need to consider an intermediate level between IP layer and computer applications: this is the transport layer of the OSI model (layer 4), containing the protocols that allow client–server applications.

What happens when a user launches an FTP application and types in an IP address so that the directory of a remote server appears? The user only specifies the IP address of a remote server from which to download data. In what follows, we examine a set of protocols that support exactly this task, operating at the transport layer.

The transport layer turns the unreliable and very basic service provided by IP into one worthy of the term "communication." The services listed below can optionally be provided at the transport level. However, not all applications want all these services.

- *Connection-orientation.* Even if the network layer provides a connectionless service, the transport layer often provides a connection-oriented service.
- *Same order delivery.* The network layer generally does not guarantee that packets are received at the destination in the same order in which they were transmitted. Nevertheless, the transport layer ensures delivery in sequence to higher levels.
- *Error-free data.* The underlying network may be noisy and data may be received corrupted. The transport layer can deal with this problem: Typically, it makes a checksum of the received data to detect errors. The transport layer may also manage the retransmission of packets that have been lost.
- *Flow control.* The amount of memory on a computer is limited, and without flow control a "powerful" computer might flood another computer with so much data that it cannot handle them. Nowadays, this is not a big issue, as memory is cheap while bandwidth is expensive, but at the time of initial networks, this was a more critical issue. The flow control operated by the transport layer allows the receiver to stop the transmission before it is overwhelmed (see the following Sect. 2.8.1.1).
- *Byte orientation.* Rather than dealing with packets, the transport layer views communication as a stream of bytes. The transfer of data is characterized by a sequence number, a counter of the bytes.
- *Ports.* Ports are essentially ways to address multiple applications in the same host. More than one network-based application can be running at the same time on a host. Each application listens to its ports for the data to be exchanged. More details on the use of ports and the assigned port numbers will be provided in Sect. 2.8.3.

Here we focus on two transport layer protocols: the Transmission Control Protocol (TCP) and the User Datagram Protocol (UDP). TCP is a complex protocol that provides connection-oriented, reliable data transfer to the application layer. TCP operates end-to-end flow control and congestion control on the data delivered to the destination. TCP uses its built-in messaging mechanisms to ensure such control. UDP behaves differently from TCP. UDP provides a connectionless, unreliable transfer to the application layer. TCP is suitable to support data applications, which do not tolerate losses but tolerate delays (elastic traffic); instead, UDP is more suitable for real-time applications (inelastic traffic), which can tolerate some packet losses.

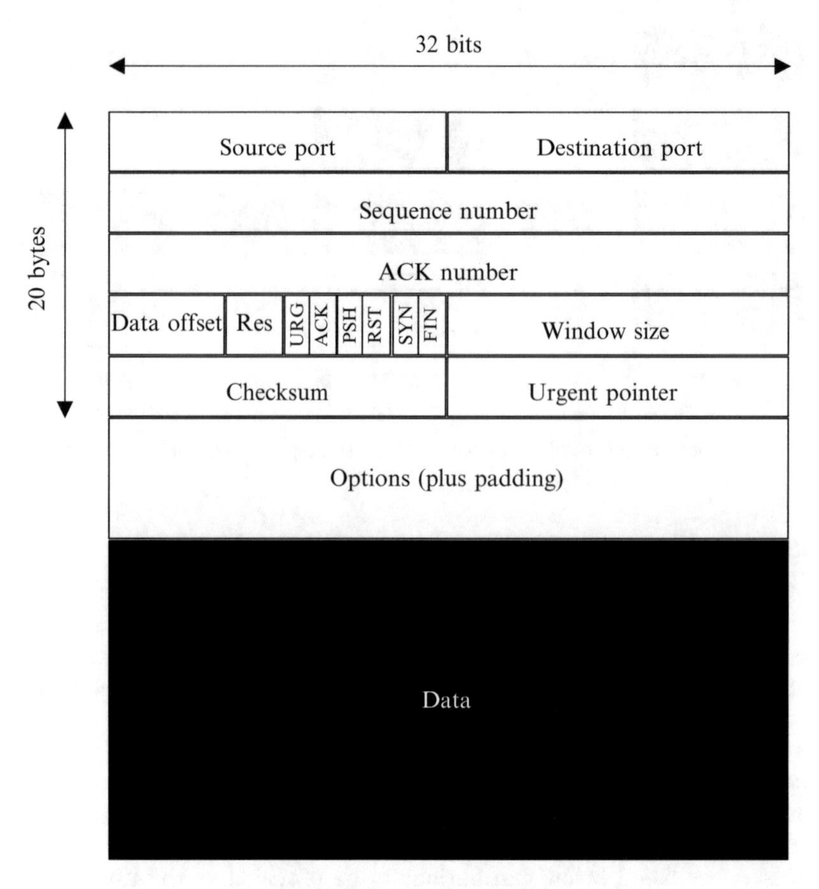

Fig. 2.39: TCP header and data

2.8.1 TCP

TCP originally defined in RFC 793 [7] adds many functions to IP networks, as detailed below.

- *Byte-streams.* TCP data are organized as a stream of bytes, similarly to a file. The datagram nature of the network is transparent to TCP.
- *Reliable delivery.* Sequence numbers are used to determine which data have been transmitted and received. TCP manages the retransmissions if it determines that some data have been lost.
- *Congestion control.* TCP dynamically learns the end-to-end delay conditions of a network and adjusts its operation to maximize the throughput without causing congestion (i.e., buffer overflows) within the network.
- *Flow control.* TCP manages the traffic injection at the sender to avoid buffer overflows at the destination host. Fast senders will be periodically stopped to keep up with slower receivers.
- *Full-duplex operation.* A TCP session can be considered as two independent byte streams, traveling in opposite directions between the two endpoints. During connection start and close sequences, TCP can exhibit asymmetric behaviors (i.e., data transfer in the forward direction, but not in the reverse one, or vice versa).

TCP is the transport layer protocol used by those applications requiring a reliable data transfer. TCP is an end-to-end protocol that presents to upper layers a buffer (this buffer is part of a *socket*; see also Sect. 2.8.3 for more details) where applications can write data (sender-side) or from which applications can read data (receiver-side), thus masking the complexities of lower-layer communications. For instance, TCP is adopted by FTP (file transfer) and HTTP (Web browsing) that have different characteristics in terms of traffic persistency (i.e., elephant versus mice TCP connections).

The combination of TCP header and TCP data in one packet is called TCP *segment.* Figure 2.39 describes the format of all valid TCP segments, organized in words of 32 bits. The size of the TCP header without options is 20 bytes. A TCP segment header contains the following fields:

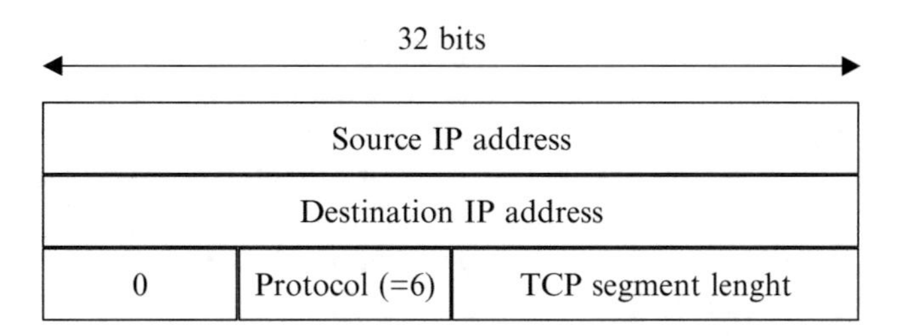

Fig. 2.40: 96-bit TCP pseudo-header (TCP has a protocol code equal to 6)

- *Source and destination ports* (16 bits each): TCP port numbers of both the sender and the receiver. TCP and UDP ports are assigned separately.
- *Sequence number* (32 bits): The sequence number of the first byte in the data part of the segment.
- *Acknowledgment number* (32 bits): If the ACK control bit is set, this field contains the value of the next sequence number (in bytes) the destination of the TCP flow is expecting to receive. This field (also referred to as "ACK") is used to inform the sender of the last segment received in sequence. In particular, *the ACK scheme is cumulative*: if the ACK contains the number $N+1$ it means that all the bytes up to sequence number N have been received correctly. See also Sect. 2.8.1.2.
- *Data offset* (4 bits): Since the TCP header has a variable length (due to the "options" field), the data offset denotes the total number of 32-bit words in the TCP header. This field is used to indicate where data start. Due to the use of 4 bits, the maximum header length is constrained to 15 words (i.e., 60 bytes), thus leaving 40 bytes for the "options" field.
- *Reserved* (6 bits): These are bits reserved for future use (they must be equal to 0).
- *Flags* (6 bits): The flag field consists of six 1-bit flags:
  - Urgent pointer (URG): If this flag is set, the urgent pointer field (see below) contains a valid pointer. If the urgent pointer flag is 0, the value of the urgent pointer field is ignored.
  - Acknowledgment valid (ACK bit): This flag is set when the acknowledgment number field is used; only during the three-way handshake procedure, the ACK flag is not set.
  - Reset (RST): The reset flag is used to abort a connection quickly.
  - Push (PSH): When dealing with some applications (e.g., real-time or highly interactive applications), it is more convenient to immediately deliver a short message, not waiting to fill in a large segment. The application can set the push option in writing data in the sender socket so that the TCP segment is sent promptly with the PSH flag set in the header. Upon receiving this packet with the PSH flag set, the receiver knows to forward the segment up to the application immediately.
  - Synchronization (SYN): This flag is used during the three-way handshake procedure to initiate a new TCP connection.
  - Finish (FIN): This flag is used to close a TCP connection.
- *Window* (16 bits): The number of bytes beginning with the one indicated in the acknowledgment field that the TCP receiver is able to accept. This field, set by the TCP receiver, is sent back to the sender in a TCP segment. If a receiver cannot accept more data, it notifies a window equal to zero. This field is used to implement a flow control mechanism.
- *Checksum* (16 bits): It is a parity check for the whole TCP segment, including the pseudo-header shown in Fig. 2.40, conceptually prefixed to the TCP header. The pseudo-header contains the Source Address, the Destination Address, the code of the protocol generating the segment (i.e., the value 6, the code of the TCP protocol), and the TCP segment length.
- *Urgent pointer* (16 bits): This field indicates the position of urgent data (if present), which should be processed immediately. A typical example is when a telnet user types CTRL-C to abort the current process.
- *Options* (variable length): This field can be used to detail several options. Optional header fields are identified by an option kind number (from 0 to 254). Each option can use up to three fields with related lengths: option kind (1 byte), option length (1 byte), and option data (variable). The maximum length of the "options" field is 40 bytes. An example of option is the maximum segment size option, which permits the sender and receiver to agree on larger segment size.
- *Padding* (variable length): This is a padding field to have that the TCP header length is a multiple of 32 bits.

Layer 2 packets have a maximum payload size, named Maximum Transmission Unit (MTU) that is available to convey IP datagrams. All link layer protocols have an MTU value; for instance, MTU is 1500 bytes for Ethernet (1500 bytes is the maximum payload capacity of an Ethernet frame) and 9180 bytes for AAL5 of ATM [33, 34]. The Maximum Segment Size (MSS) is the maximum length of the TCP segment payload (i.e., the TCP segment without the TCP header). A too-large TCP segment with respect to the layer 2 packet payload would require fragmentation in many layer 2 packets and a reassembly procedure at the destination. Fragmentation entails a loss of efficiency. Hence, to avoid fragmentation, we need that MSS = MTU − 40 bytes, considering that 20 bytes are used for the IP header and additional 20 bytes are used for the TCP header. Note that a TCP segment with MSS = 2000 bytes and the DF flag set in the header of the related IP datagram cannot pass an interface on a router with MTU = 1500 bytes: the router will discard the datagram by returning an ICMP Destination Unreachable message with a code meaning "fragmentation needed and DF set." Let Path MTU (PMTU) denote the minimum of the MTU values of the hops on the source-to-destination path: a TCP segment with length PMTU will not be fragmented on that path.

A TCP connection is established through the so-called three-way handshake procedure, as shown in Fig. 2.41 (where a comparison is also made with the simpler transfer approach allowed by the UDP protocol). This procedure is needed to synchronize both ends of the communication; this is obtained by exchanging the *initial sequence numbers* each end host wants to use for its data transmissions, as well as other parameters to control how the connection operates. All messages transmitted during the three-way handshake phase are just TCP headers (no data) with the proper flags set. The three-way handshake procedure starts with a TCP client sending an SYN message with an initial client-side sequence number X (typically a value randomly selected by the sender; this is useful to differentiate connections) to a remote server. Moreover, the TCP header can also contain information on the client-side MSS; this is achieved by adopting a suitable MSS option in the header. When the remote server receives the connection request, it responds with an SYN message having ACK = $X + 1$ and another initial server-side sequence number Y (typically a value randomly selected by the receiver); the reply can also contain the MSS value of the server.[8] Finally, the client responds using an SYN message containing ACK = $Y + 1$.

Even with the above procedure, a TCP segment transmitted by the client or by the server can be segmented if it has to traverse an intermediate network with a lower MTU value than those of the networks of client and server. If we want to avoid such segmentation to maximize the efficiency of the data transfer, we can use the Path MTU Discovery (PMTUD) protocol defined in RFC 1191 [43]. This algorithm attempts to discover the MTU value, which may be used without fragmentation along an IP path. The basic idea of PMTUD is that a source host initially assumes that the PMTU of a path is the known MTU of its first hop and sends PMTU segments on that path with the DF bit set to force the non-fragmentation of the datagram. If the datagrams are too large to be forwarded without fragmentation by a certain router along the path that router will discard them and return ICMP Destination Unreachable messages. If this router complies with RFC 1191, its MTU value is contained in the ICMP error message. Otherwise, upon receiving this ICMP message, the source host reduces its assumed PMTU for the path and the procedure repeats until the correct PMTU value is discovered.

In TCP/IP networks, the distribution of the IP packet size (i.e., MTU) is typically tri-modal: 40-byte packets (the minimum only-header TCP packet), 576-byte packets (for TCP implementations not using the PMTUD protocol), and 1500-byte packets (the classical maximum Ethernet payload size). Of course, other packet sizes are also possible because of partly filled packets.

Finally, a "four-way" handshake procedure is used to tear down a connection. Four messages are exchanged because each host has to send a FIN message request and receive an ACK to close the communication in one direction.

TCP flows can be classified as follows considering the duration and the amount of data exchanged:

- *Long-lived flows* due for instance to FTP file transfer ("elephant TCP flows" or persistent flows).
- *Short-lived flows* due for instance to HTTP page transfer ("mice TCP flows" or non-persistent flows).

[8] The MSS values to be used for a TCP connection (by both sides) can be defined during the TCP three-way handshake procedure by the two end systems. Each end system *notifies* an MSS value (typically a host bases its MSS value on its outgoing interface MTU size) using the MSS option in the initial SYN message sent; the other end system makes use of the notified MSS value when it sends TCP segments. If one end does not receive an MSS option from the other end, a default MSS value of 536 bytes (i.e., MTU of 576 bytes) is assumed. However, RFC 1122 states that the use of the MSS option is mandatory in the connection set up phase.

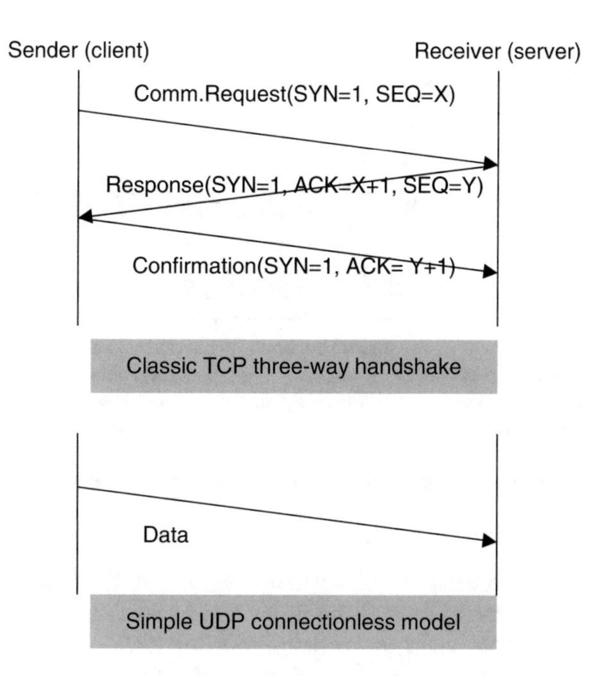

Fig. 2.41: Connection-oriented (i.e., TCP) and connectionless (i.e., UDP) data transfer procedures

2.8.1.1 TCP Flow Control Based on a Sliding Window Approach

Let us recall that TCP adopts a buffer (i.e., a part of the socket operating in the system kernel) between application and network layers at both sender and receiver. Data received from the network are stored in this buffer, from whence the application can read at its own pace. As the application processes the data, the transport-layer buffer space at the receiver is freed up to have room to store new data coming from the network. The window field in the TCP header specifies the space available at the receiver, i.e., the receiver buffer space minus the amount of data currently stored in it. Hence, the window size value (set by the receiver in the TCP segments it sends back to provide ACKs) permits the TCP receiver to inform the TCP sender about the space currently available in its buffer. This is the reason why this window is also referred to as *receiver window* or *advertised window*, rwnd.

Since 16 bits are used for the window field in the TCP header, the maximum window size is $2^{16} = 65{,}536$ bytes; this is an upper bound to the quantity of data that can be transmitted all together, even without receiving ACKs. Referring to the classical Ethernet IP packets of 1500 bytes, the maximum window size entails an upper bound equal to 44 pkts to the amount of in-flight data. This limit could cause the underutilization of the network capacity in the presence of high propagation delays; in these circumstances, the window scale option can be used to multiply the window value in the TCP header, as specified in RFC 1323. If this option is used, it has to be agreed upon during the initial three-way handshake phase.

TCP implies a bidirectional traffic flow from sender to receiver: Data (i.e., TCP packets) are going downstream from sender to receiver; instead, ACKs go back, upstream from receiver to sender. Here, ACK is a particular TCP segment sent by the receiver back to the sender and containing the current rwnd value in the window field (this ACK packet is an IP packet of at least 40 bytes to convey the TCP header).

The window field is used by TCP to implement a flow control algorithm. The window value actually specifies a *sliding window* to control the transmission of new data. The current window value represents the maximum amount of data, which can be sent without waiting for new ACKs. The operation of the sliding window algorithm is described below (see Fig. 2.42):

1. Transmit all the new segments allowed by the current value of the window.
2. Wait for an ACK to arrive; several packets can be acknowledged with the same ACK due to the cumulative ACK scheme of TCP.
3. When an ACK arrives at the sender, slide the window depending on the amount of data acknowledged and set the window size to the value advertised by the ACK; the transmission resumes from the packet following the last packet transmitted.

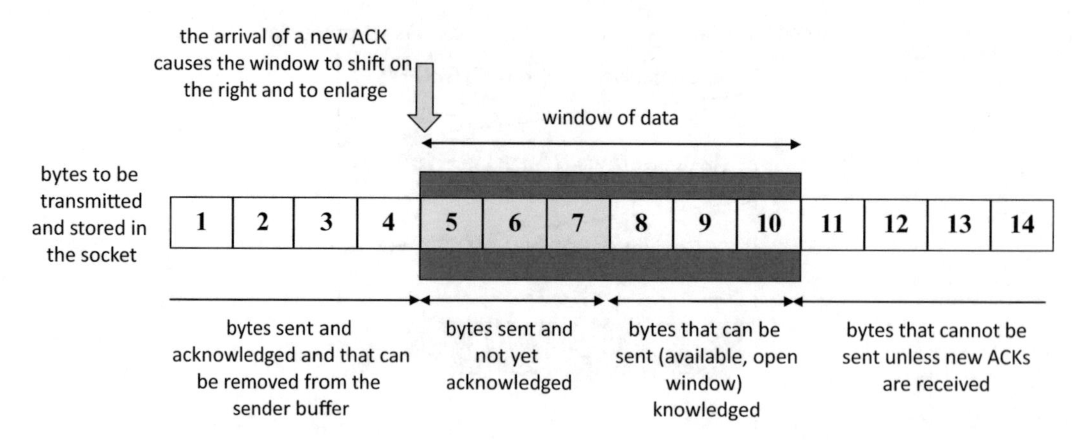

Fig. 2.42: Sliding window concept. The window limits the number of in-flight packets

In this algorithm, the window value used is not actually rwnd, but the minimum between rwnd and another window (called congestion window, cwnd) that will be described in Sect. 2.8.1.3.

According to the sliding window approach, *the transmission of TCP packets is clocked by the receipt of new ACKs.* TCP is based on the principle of "conservation of packets": if a connection is fully exploiting the available network capacity, a new packet cannot be transmitted into the network until a (previous) packet is received at destination. TCP implements this principle using ACKs to clock outgoing packets: the receipt of an ACK implies that a packet has been correctly received at the destination so that the window can slide and new data can be sent. We can say that TCP is "self-clocking": each arriving ACK can trigger the transmission of a new segment. Finally, let us notice that, in transferring a window of data, ACKs arrive at the sender at least separated by the packet transmission time.

The Round-Trip Time (RTT) is the time required for a packet to travel from a specific source to a specific destination and back again in the form of an ACK. RTT includes the packet transmission time, the queuing delay at the nodes crossed, and the physical end-to-end propagation delay. The Round-Trip propagation Delay (RTD) is the minimum possible RTT, counting only the physical end-to-end propagation delay. Sometimes RTT and RTD are confused. The ping command of ICMP provides measurements of RTT. The prevailing medium for long-distance transmissions is represented by optical fibers, where the propagation speed is about two-thirds of the light speed, i.e., 200,000 km/s. Hence, for instance, a distance of 5000 km entails an RTD value of 50 ms. More details are provided in Sect. 2.8.1.9.

A Retransmission TimeOut (RTO) timer is started for each packet sent: the sender waits for the ACK of a given packet for a maximum time equal to RTO. If the ACK does not arrive within RTO, the sender retransmits all the packets starting from the one for which RTO has expired (*go-back-N approach*).

2.8.1.2 Cumulative ACKs and the Impact of Packet Losses

TCP adopts a cumulative acknowledgment scheme: an ACK with sequence number $N+1$ confirms the correct receipt of packets containing bytes up to sequence number N, which represents the maximum byte number received in order. An ACK can also confirm the correct delivery of more TCP packets in sequence. When a packet is transmitted, the sender triggers timer RTO. If the packet ACK is not received within RTO, a packet retransmission is performed; thus, we can state that the TCP protocol is reliable.

Figure 2.43 clarifies the use of cumulative ACKs, referring to a case with packets carrying 1000 bytes, RTD = 2 packet transmission units, a congestion window cwnd (see next Sect. 2.8.1.3) corresponding to six packets, and a receiver window rwnd > cwnd. Note that the number associated with each packet in Fig. 2.43 represents the highest-order byte transported by the packet (actually, this is not the sequence number as defined in Sect. 2.8.1). After the correct delivery of the first packet, ACK 1001 is sent. After the receipt of the second packet, ACK 2001 is sent. Then, let us assume that packet 3000 is lost because of a buffer overflow at an intermediate router. Hence, when packet 4000 is received, an ACK is sent, which acknowledges the highest byte number received in order, i.e., again 2001: this is a Duplicate ACK (DUPACK). DUPACKs do not permit to slide the window of

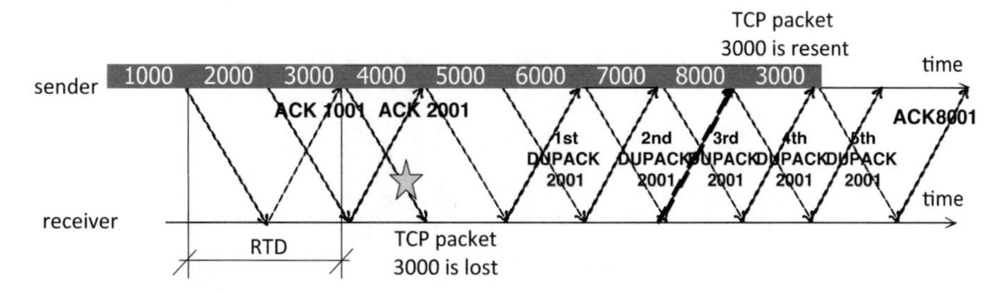

Fig. 2.43: The cumulative ACK scheme and the effect of packet losses

data transmission (cwnd). However, since cwnd corresponds to six packets, the transmission can continue after packet 3000 with the packets 4000, 5000, 6000, 7000, and 8000; with this cwnd value, no further packets can be sent without receiving new ACKs. When these packets are received correctly, further DUPACKs 2001 are sent. If the packet loss is not recovered quickly, it may happen that an RTO expires. In most recent TCP versions, the sender can use a new mechanism to recognize a packet loss if no ACK is received for this packet for a sufficiently long time. This is done to anticipate the RTO scheme. In particular, it is decided that a packet loss has occurred when the third DUPACK is received. This is the case considered in Fig. 2.43: after packet 8000 is sent, the third DUPACK is received, packet 3000 is resent and correctly received; then, ACK 8001 is sent, being 8000 the highest sequence number received in order. The receipt of ACK 8001 permits the transmission window to slide.

TCP "sees" an end-to-end erasure channel, where packet losses are basically due to two causes:

1. A TCP segment is dropped as erroneous when it is received with some bit *errors* so that checksum fails.
2. A TCP segment is lost if it *overflows* from a congested buffer of an intermediate router.

TCP does not distinguish between these two cases and assumes (in the congestion control algorithm presented in the next sub-section) that all the packet losses are due to congestion, thus reducing the traffic injection rate at the sender.

2.8.1.3 TCP Congestion Control

In October 1986, the Internet had its first congestion collapse event. Congestion entails: (1) Packet losses due to buffer overflows; (2) Retransmissions to recover packet losses; (3) Drastic reduction of the traffic carried at destination (i.e., *throughput*). TCP needs a protocol allowing a host (i.e., the *sender*) to inject data into the network towards a destination (i.e., the *receiver*) without any coordination with other hosts, but only based on its perception of the network congestion. In 1988, Van Jacobson proposed the TCP congestion control, subsequently incorporated into the TCP Tahoe version. The TCP congestion control treats the network as a *black box* and uses two algorithms to probe network resources and to gradually increase the amount of data injected based on the ACKs sent by the receiver [44]; these algorithms are called "slow-start" and "congestion avoidance."

In TCP, congestion control and flow control are integrated into the same mechanism based on a sliding window approach. The congestion window (cwnd) is managed by the sender and represents its perception of network congestion; instead, the receiver window (rwnd) represents the amount of buffer space available at the receiver. Both cwnd and rwnd are dynamically updated through ACKs. Rwnd is signaled through the ACKs generated at the TCP layer of the receiver. Cwnd is updated when the sender receives a new ACK according to either slow-start or congestion avoidance, depending on the comparison between cwnd and the slow-start threshold (ssthresh) value, as described below.

Note that cwnd, rwnd, and ssthresh have values expressed in bytes, but their values are considered to be converted and updated in MSS units for the following two algorithms. When a TCP connection starts, the "slow-start" algorithm begins with cwnd = 1 [MSS unit], the initial window value. Correspondingly, the initial ssthresh value is typically (default) set equal to the initial rwnd value (i.e., 65,535 bytes). However, for the following considerations (see also Fig. 2.44), a lower initial ssthresh value is used to avoid a too bursty initial injection of traffic. The sender can transmit a quantity of data that is the minimum between the current cwnd and rwnd and then has to stop transmissions, waiting for new ACKs. In what follows, for the sake of simplicity, we assume rwnd > cwnd so that the injection of traffic in the network depends only on cwnd.

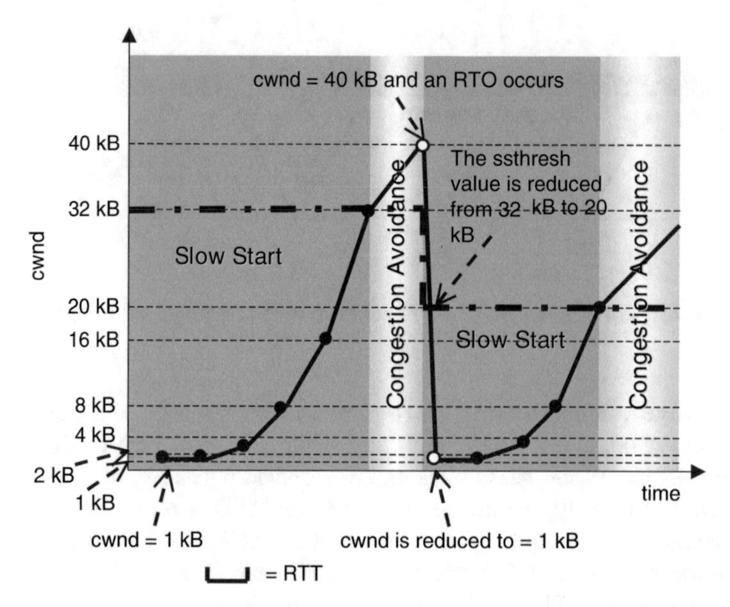

Fig. 2.44: Cwnd behavior of a basic TCP congestion control algorithm in the presence of a packet loss (the initial ssthresh value is equal to 32 kB; the MSS is 1 kB)

- If cwnd < ssthresh, the "slow-start" algorithm is adopted: when a new ACK is received, the following cwnd update is performed: cwnd = cwnd + 1 [MSS unit]. Correspondingly, cwnd doubles (i.e., *exponential increase*) on an RTT basis. Despite its name, the "slow-start" algorithm tries to enlarge cwnd in a sufficiently fast, but controlled way. TCP originally had no congestion control mechanism: a source just started by sending a full window of data. With respect to such a choice, a cwnd increase according to an exponential law represents a slow-start.
- As soon as cwnd goes beyond ssthresh, the "congestion avoidance" algorithm is invoked: when a new ACK is received, the following cwnd update is performed: cwnd = cwnd + 1/cwnd [MSS unit]. Hence, cwnd increases by 1 for each block of cwnd segments sent in an RTT time. In conclusion, cwnd increases by one segment (i.e., *linear increase*) on an RTT basis. This solution permits to probe the bandwidth still available in the network after the slow-start phase.

In the slow-start phase, cwnd increases by Δ packets in a time equal to $\log_2(\Delta)$ in RTT units. In the congestion avoidance phase, cwnd increases by Δ packets in a time equal to Δ in RTT units.

If the ACK of a transmitted segment is not received with RTO, it is assumed that the segment has been lost because of network congestion. RTO is continuously updated and represents a filtered version of RTT with some margins, proportional to the RTT standard deviation. When timer RTO of a given segment expires, ssthresh is set equal to half of the current minimum value between cwnd and rwnd and cwnd is reset to its initial value (i.e., 1 MSS) to force the "slow-start" algorithm. Then, the sender retransmits all packets in sequence, starting from the packet for which timer RTO has expired, according to a go-back-N approach. When a segment loss is detected using an RTO expiration, the above mechanism drastically reduces the TCP traffic injection, assuming network congestion.

Even if the traffic injected in the network on an RTT basis depends on the minimum between cwnd and rwnd (also called *in-flight size*), we consider here that rwnd is typically so large that it has no impact on limiting the traffic injection that therefore depends only on cwnd (if rwnd would be lower than cwnd, even if cwnd increases according to the above algorithms, the traffic injection does not depend on cwnd but on rwnd). Hence, cwnd represents the number of packets sent as a function of time (RTT basis); this is proportional to the bit-rate injected by the senders as a function of time. The integral of cwnd as a function of time yields the arrival curve of the TCP-based traffic flow.

An example of cwnd behavior as a function of time (in RTT units) is depicted in Fig. 2.44, referring to a basic TCP version. Cwnd starts from the initial value of 1 kB and experiences a "slow-start" phase, where cwnd has an exponential increase ($y = 2^x$, where x is expressed in RTT units). As soon as cwnd reaches the ssthresh value (= 32 kB), cwnd increases linearly ($y = x$) according to the "congestion avoidance" algorithm. Let us assume that,

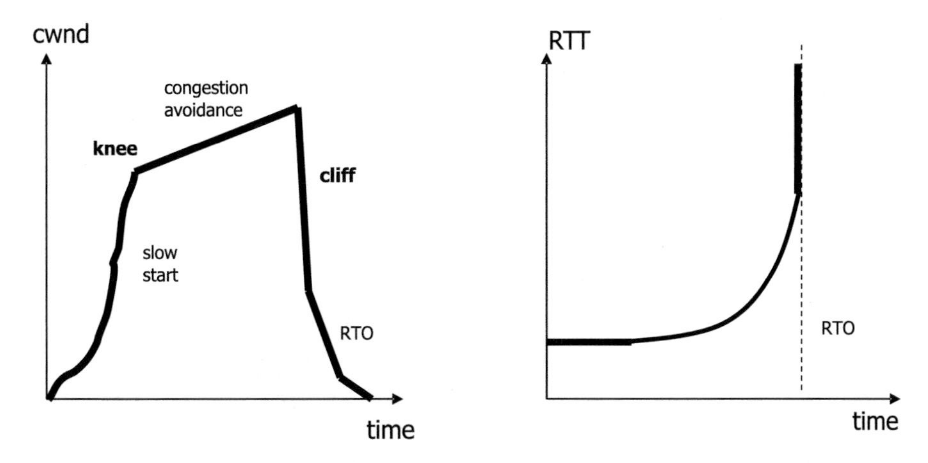

Fig. 2.45: Generic examples of cwnd and RTT behavior for classical TCP versions

when cwnd = 40 kB, timer RTO of a given TCP segment expires because of network congestion. Hence, TCP sets ssthresh to 20 kB, resets cwnd = 1 kB, and triggers a new "slow-start" phase.

Figure 2.45 compares the classical behaviors of cwnd and RTT as a function of time. As cwnd increases, RTT increases as well when there are queuing phenomena at the intermediate buffers. RTT can have a sudden peak when cwnd approaches its maximum value. A congestion event (i.e., basically a packet loss) happens so that cwnd and RTT are restarted.

Note that, according to RFC 3390, it is allowed (option) to increase the initial window according to the following formula: min (4×MSS, max (2×MSS, 4380 bytes)). Moreover, the window scale option of RFC 1323 can be used to set the maximum possible rwnd so that it can be larger than 65,535 bytes.

The different operating systems use distinct settings for both the initial cwnd and the maximum rwnd, as shown in the following examples:

- Microsoft Windows XP: Initial cwnd of 1460 bytes and maximum possible (initial) rwnd of 65,535 bytes.
- Microsoft Windows 7: Initial cwnd of 2920 bytes (i.e., more than one segment) and maximum possible rwnd of $65{,}535 \times 2^2$ bytes.
- Ubuntu 9.04: Initial cwnd of 1460 bytes and maximum possible rwnd of $65{,}535 \times 2^5$ bytes.
- MAC OS X Leopard 10.5.8: Initial cwnd of 1460 bytes and maximum possible rwnd of $65{,}535 \times 2^3$ bytes.

Different variants of the TCP congestion control algorithm have been proposed in the literature; see Sect. 2.8.1.8.

2.8.1.4 TCP RTO Algorithm

TCP performs a reliable delivery of data by retransmitting segments, which are not received correctly. When a packet is transmitted, timer RTO is triggered and assigned to it. If RTO expires for a given packet without receiving an ACK, retransmissions are restarted from this packet according to a go-back-N approach. Correspondingly, ssthresh is set equal to half of the current minimum between cwnd and rwnd and cwnd is reset to its initial value (i.e., 1 MSS) to force the "slow-start" algorithm.

RTO should be larger than RTT (the time interval between the transmission of a segment and the receipt of its ACK), but not much bigger than RTT in order not to waste time before reacting to a loss. Hence, an accurate dynamic determination of the RTO value is essential to the TCP performance. RTO is computed by estimating the mean RTT value and a sort of standard deviation of RTT. The latest RFC dealing with the RTO algorithm is RFC 6298 [45]. To compute the current RTO value, the TCP sender maintains two state variables: Smoothed RTT (SRTT), an average RTT value, and RTT VARiation (RTTVAR), a sort of standard deviation of RTT. When packets are sent over a TCP connection, the sender measures how long it takes for them to be ACKed, producing a sequence of RTT measures: $R(0)$, $R(1)$, $R(2)$, etc. With each new measure $R(i)$, SRTT is updated as follows (low-pass filter):

$$\text{SRTT}\,(i+1) = (1-\alpha) \times \text{SRTT}(i) + \alpha \times R(i) \tag{2.14}$$

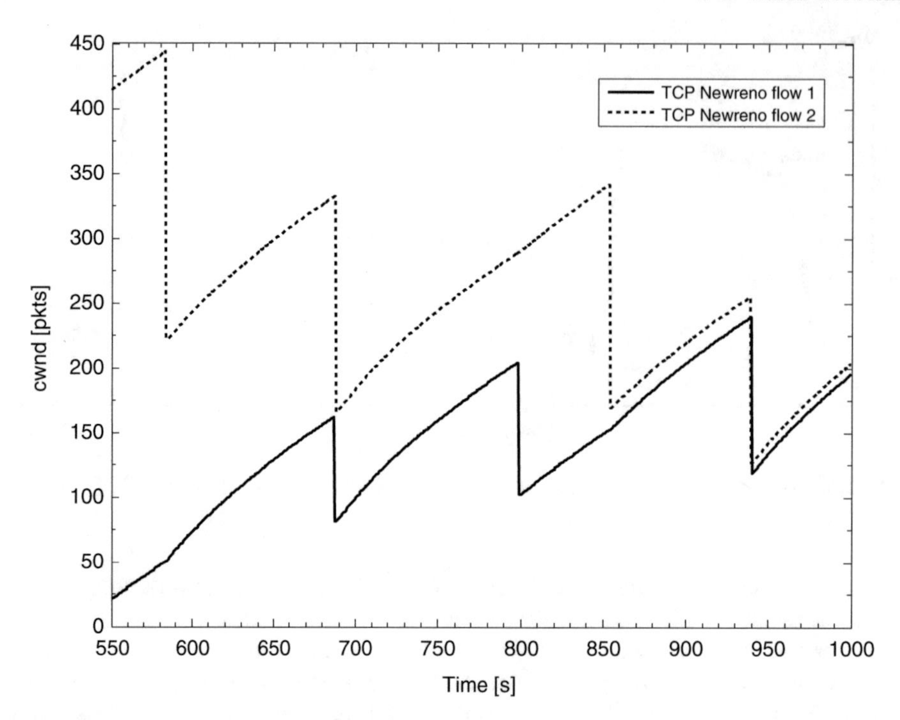

Fig. 2.46: Example of two TCP (NewReno) flows sharing a bottleneck link and experiencing synchronized losses due to the drop-tail discipline

where α is a constant between 0 and 1; note that SRTT(1) is made equal to $R(0)$.

Another formula is used to update RTTVAR [i.e., a filtered version of the difference $\mathrm{SRTT}(i) - R(i)$], using coefficient β from 0 to 1 ($\beta \neq \alpha$) as

$$\mathrm{RTTVAR}\,(i+1) = (1-\beta) \times \mathrm{RTTVAR}\,(i) + \beta\,|\mathrm{SRTT}\,(i) - \mathrm{RTT}\,(i)| \tag{2.15}$$

Recommended values are $\alpha = 1/8$ and $\beta = 1/4$. After the ith RTT measure, RTTVAR and SRTTT are both updated. Subsequently, the $i+1$-th value of RTO is obtained as follows:

$$\mathrm{RTO}\,(i+1) = \max\left\{\mathrm{SRTT}\,(i+1) + \max\left[G, K \times \mathrm{RTTVAR}\,(i+1)\right], 1\text{ s}\right\} \tag{2.16}$$

where $K = 4$ is a constant and G represents the clock granularity (i.e., a "tick," which is typically equal to 500 ms, as explained below). According with RFC 6298, it has been made explicit in (2.16) the fact that RTO has to be greater than or equal to 1 s.

At the beginning (when the first window of data is sent), RTO is made equal to 1 s, which is also the minimum possible RTO value, according to RFC 6298. Then, as soon as the first RTT measurement, $R(0)$, is available, it is used to compute RTO(1) as follows: $\mathrm{RTO}(1) = \max\{R(0) + \max(G, K \times R(0)/2), 1\text{ s}\}$. The next RTT values are then used to update RTTVAR and SRTT and then to compute RTO according to (2.16).

Whenever an RTO expiration occurs, RTO is increased by some factor before retransmitting data. This is a backoff scheme. When the overload condition disappears, TCP reduces its RTO value to its normal SRTT-based value. Typically, RTO is doubled at each expiration according to an exponential backoff. As soon as the ACK of new data is received so that a new RTT measurement is available, the previous formula (2.16) is reused to compute the RTO value, which is therefore significantly reduced.

At each RTO, ssthresh is halved so that if multiple RTOs occur in sequence (this could be the case of massive losses or channel disruption for a sufficiently long time in the order of seconds), we can easily reach a situation where ssthresh is 1 packet so that cwnd restarts from 1 packet with the congestion avoidance algorithm.

RTT is measured as a discrete variable in multiples of a tick (coarse grain), where 1 tick is equal to 500 ms in many implementations. According to RFC 6298, the minimum RTO should be 1 s (i.e., 2 ticks). Moreover, according to RFC 1122, there is the possibility to restrict the maximum RTO at a value not lower than 60 s.

The use of the TCP timestamp option allows a better TCP RTT estimation and then a more accurate definition of the RTO value.

2.8.1.5 Buffer Management Techniques and TCP Behavior

Different TCP flows sharing the same bottleneck link receive loss indications at roughly the same instants: if the buffer of the bottleneck link adopts the *drop-tail scheme*, these flows will most likely experience simultaneous packet losses due to buffer congestion; these are the so-called synchronized losses. This phenomenon causes all the TCP flows sharing the same buffer (on a certain bottleneck link) to decrease their cwnds simultaneously. Hence, there are intervals in which the bottleneck link bandwidth is significantly underutilized. Figure 2.46 shows an example of synchronized losses for two TCP (NewReno version; see Sect. 2.8.1.8) flows, sharing a common bottleneck link with buffer drop-tail scheme; see Sect. 2.8.1.9.

To avoid synchronized losses, *active queue management schemes* can be adopted at the buffers. For instance, we can consider the RED policy (see Sect. 2.5.2): when the buffer approaches congestion, RED can introduce random packet losses, thus removing the synchronization of the cwnds of the different flows.

2.8.1.6 TCP Deadlock Events

Deadlocks are complex events, which cause a block in the data transmission; these events may happen under special circumstances in the TCP case, where sender and receiver are both waiting for the other to finish so that none of them can send new data. Some TCP deadlock events are due to implementation (known) problems. Interesting examples are summarized below.

The following case refers to a slow receiver. If the receiver buffer is full of data, then it sends an ACK to the sender containing a window size rwnd = 0. This stops sender transmissions. The receiver sends a window update segment (with rwnd > 0) when it has space available in its buffer. If this window update segment is lost, then a deadlock occurs. This problem can be overcome adopting a sender-side persistence timer. This timer is started when a segment is received with rwnd = 0. When the persistence timer expires, a probe segment is sent to the receiver. The receiver responds with another window size either equal to 0 or non-zero so that the transmission can resume in the second case.

Another deadlock problem could be caused by a circular-wait condition between sender and receiver due to the adoption of the Nagle algorithm[9] (RFC 896) jointly with the adoption of the delayed acknowledgment scheme (RFC 813). In particular, the Nagle algorithm considers a small segment as having a length lower than the connection MSS, and usually limits the number of outstanding small segments to one to avoid inefficiency. Moreover, the delayed acknowledgment strategy prevents a receiver from acknowledging small segments by delaying ACKs until they can be piggybacked onto either a data segment or a window update packet. When the sender and the receiver socket buffer size fall in a certain region, the sender will not send small segments due to the Nagle algorithm, and the receiver will not send ACKs because of the delayed ACK algorithm. This deadlock can be solved using a delayed ACK timer [46].

2.8.1.7 A Model for the Study of the TCP Behavior

Let us refer to the network model shown in Fig. 2.47, where there is the TCP sender, the TCP receiver, and the path between sender and receiver is characterized by a *bottleneck link* (at an intermediate router) with a capacity of B packets and bit-rate denoted as Information Bit-Rate (IBR). A single TCP flow case is considered here: capacity B and bit-rate IBR are fully available for this flow. Note that IBR refers here to IP packets, having a size equal to the TCP Maximum Segment Size (MSS) plus TCP/IP headers of at least 40 bytes (i.e., 40 bytes plus additional bytes of TCP or IP options if present in the respective headers). In this model, we neglect the impact of sockets (meaning that these have a capacity much greater than B; for instance, rwnd $= \infty$) so that the traffic injection in the network depends only on cwnd.

RTT is the time needed from the packet transmission to the receipt of its ACK. RTT depends not only on the physical round-trip propagation delay from source to destination but also on the buffer congestion of the bottleneck link, which in turn depends on the cwnd value used by the sender. Let us recall that RTT is not

[9] The Nagle algorithm is used to avoid sending small segments in the network. The generation of small packets can be due to either some applications or a slow receiver asking to continuously reduce the window (*silly window syndrome*) up to the point that the data transmitted is smaller than the packet header, making data transmissions extremely inefficient. On slow links, many small packets can potentially lead to congestion. The Nagle algorithm works by combining a number of small packets and sending them all together.

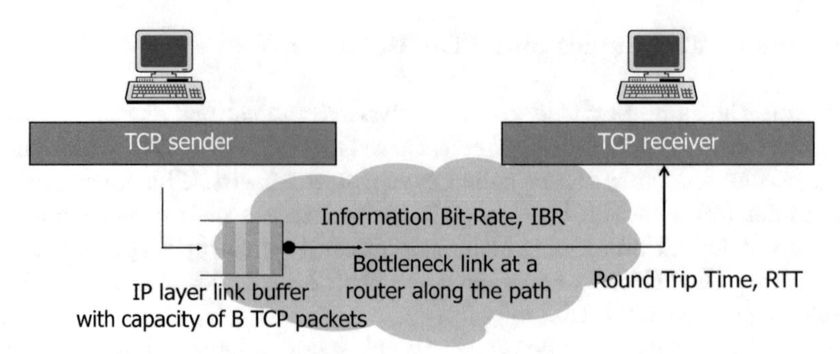

Fig. 2.47: Model of the system for the reliable delivery of data ("elephant" FTP case)

constant but increases with cwnd. $\mathrm{RTT_m}$ denotes the minimum RTT value due to physical conditions (i.e., the physical round-trip propagation delay, RTD).

TCP performance (throughput) does not depend directly on the transfer bit-rate IBR of the link, but rather on the product of IBR and RTT, as specified by the Bandwidth-Delay Product (BDP). This value represents the maximum amount of in-flight data in the system. BDP can be expressed in TCP segments, according to the following formula:

$$\mathrm{BDP} = \frac{\mathrm{RTT} \times \mathrm{IBR}}{\mathrm{MTU}}\ \text{[pkts]} \tag{2.17}$$

where MTU denotes the IP-level maximum packet size, obtained as the MSS summed to the TCP/IP headers (e.g., 40 bytes). MTU is considered here expressed in bits.

BDP depends on RTT, which has no constant value since it varies with cwnd [47]: $\mathrm{BDP} = \mathrm{BDP(cwnd)}$. $\mathrm{BDP_m}$ is the minimum BDP value when RTT is equal to $\mathrm{RTT_m}$. In the following study, when we will speak about BDP, we will refer implicitly to $\mathrm{BDP_m}$ with $\mathrm{RTT_m} = \mathrm{RTD}$. Based on this approach, the derivation of BDP according to (2.17) is much simpler because RTT becomes constant and equal to $\mathrm{RTT_m}$, which depends only on the physical characteristics of the source–destination path and not on the variable congestion conditions.

The initial TCP cwnd behavior depends on the initial ssthresh value, whose maximum value (default value) corresponds to $2^{16} - 1 = 65{,}536$ bytes. The initial ssthresh should be determined to achieve a trade-off between two contrasting issues: a low value of the initial ssthresh, which is below or around the $B + \mathrm{BDP}$ value, forces TCP to prematurely switch into the Congestion Avoidance phase, thus causing the sender to probe gently for available network bandwidth instead of rapidly increasing the traffic injection. Instead, suppose the initial ssthresh value is greater than $B + \mathrm{BDP}$ (for instance, this could be the case of Linux operating systems). In that case, the first slow-start phase usually ends with multiple packet losses, and an RTO expiration may also happen. RFC 5681 recommends that the initial ssthresh value should be set arbitrarily high (i.e., equal to the 65,536 value) so that the network conditions determine the sending rate, rather than some arbitrary host limits. However, if the sender could have a certain knowledge of the network path, it would be convenient to set the initial ssthresh value about equal to BDP in order not to create initial congestion in the path bottleneck and to maximize the overall performance.

The maximum cwnd is reached when the pipe is full, intending that the packets on the fly are filling both the propagation delay and the bottleneck link buffer: $\mathrm{cwnd_{max}} = B + \mathrm{BDP}$, where this BDP is here computed using RTD instead of RTT in (2.17). If cwnd becomes larger than $\mathrm{cwnd_{max}}$, there are packet losses due to buffer overflow at the bottleneck link. Then, cwnd is soon reduced by the TCP mechanism to recover the losses. See also the microanalysis study carried out in Sect. 2.8.1.9.

The *throughput* Γ represents the utilized bandwidth, a sender-side measurement of the bit-rate generated by a given TCP traffic source. The *goodput* γ is a receiver-side measurement of the bit-rate received for a certain TCP traffic source. The goodput is lower than the throughput since the throughput also considers retransmissions. Both throughput and goodput depend on the buffer capacity B and information bit-rate IBR of the bottleneck link and the RTT value. In the following sub-sections, throughput formulas are derived.

Telecommunication networks with large BDP values (e.g., $\mathrm{BDP} > 50$ pkts) are, for instance: satellite-based communication systems (large BDP because of large RTD) and broadband optical fiber communication systems (large BDP because of large IBR). Networks with high BDP values are also called "Long, Fat pipe Networks"

(LFN). These networks are critical from the TCP throughput standpoint since TCP cannot fully exploit the BDP since the cwnd value is practically limited by the 16-bit representation of the window field in the TCP header. In addition to this, when a packet loss occurs, the classical TCP version requires a long time to recover the cwnd value reached before the loss. During this time interval, network resources are underutilized. For instance, assuming to use TCP NewReno (see the next Sect. 2.8.1.8) and to operate in congestion avoidance with a linear increase of cwnd, if an isolated packet loss happens when cwnd $= 2 \times$ BDP (here $B =$ BDP), a time equal to BDP in RTT units is needed to recover the original cwnd value. Hence, this recovery time can be significantly long.

2.8.1.8 Different Versions of TCP Congestion Control

Different TCP congestion control versions have been defined. The versions proposed are more than those supported by RFCs because the definition of a new TCP version with an RFC requires a careful approval process: the Internet community approves only those versions, which are stable, reliable, scalable, and fair. We can consider the following main historical steps for the definition of different TCP versions and related RFCs:

- 1981: The basic RFC 793 for TCP. In this version, there is not cwnd, but only rwnd. When there is a packet loss, we have to wait for an RTO expiration to recover the packet loss based on a go-back-N scheme.
- 1986: Slow Start and Congestion Avoidance algorithms defined by Van Jacobson and supported by TCP Berkeley [48].
- 1988: Slow Start, Congestion Avoidance, and Fast Retransmit (three DUPACKs) supported by TCP Tahoe [48]. Van Jacobson first implemented TCP Tahoe in the 1988 BSD release (BSD stands for Berkeley Software Distribution, a computing library).
- 1990: Slow Start, Congestion Avoidance, Fast Retransmit, and Fast Recovery supported by TCP Reno according to RFC 2001. In 1990, Van Jacobson implemented the 4.3BSD Reno release [44].
- 1996: Use of the SACK option for the selective recovery of packet losses according to RFC 2018 [49], followed by RFC 2883 [50].
- 1999: RFC 2582 describing the original version of TCP NewReno. RFC 2582 also includes the *Slow-but-Steady* and *Impatient* variants of TCP NewReno with differentiated timer RTO management when there are multiple packet losses in a window of data.
- 2004: RFC 3782 describing an improved version of TCP NewReno [51]: the *careful* variant of NewReno with Fast Retransmit and Fast Recovery algorithms and better management of retransmissions after an RTO expiration.
- 2012: RFC 6582 dealing with the latest version of TCP NewReno specification.

Almost all previous versions are differentiated based on the law according to which cwnd is managed when an ACK is received or when a packet loss is recognized. Let us describe below in detail the essential characteristics of the main TCP versions.

TCP Tahoe

Tahoe refers to the TCP congestion control algorithm proposed by Van Jacobson in his paper in [48]. TCP Tahoe adopts slow-start and congestion avoidance. If the sender receives three DUPACKs, Tahoe assumes that there was a packet loss and reacts as if an RTO expiration had occurred: Tahoe performs a "fast retransmit" phase, halving the slow-start threshold with respect to the current window value, and reducing the congestion window to 1 MSS to restart from a slow-start phase, forgetting everything on the segments sent after the first non-ACKed one (go-back-N approach). Note that a packet loss is not decided at the first DUPACK but at the third one in order not to react too fast, mainly because the IP network is connectionless and out-of-sequence packets (generating DUPACKs at the receiver) could be misinterpreted as packet losses. This is the reason why this protocol waits till the third DUPACK before starting the packet retransmission phase.

TCP Vegas

TCP Vegas is a TCP congestion avoidance algorithm, which exploits packet delay, rather than packet loss, as a signal of congestion. It was developed in 1994 by Lawrence Brakmo and Larry L. Peterson at the University of Arizona.

Vegas continuously monitors the status of the network and increments or decrements cwnd to prevent packet losses due to congestion. Vegas reveals a congestion (before it occurs) by considering quantity $\Delta =$ (expected throughput − actual throughput) × RTTbase, where expected throughput = cwnd/RTTbase, actual throughput = cwnd/RTT, RTTbase is the minimum of RTT, RTT_m, and RTT is the currently measured RTT. Hence, we have

$$\Delta = \text{cwnd} \times (1 - \text{RTTbase}/\text{RTT})$$

The purpose of the Vegas algorithm is to control the amount of data in the buffer of the bottleneck link according to threshold parameters α (too few data in the buffer) and β (too much data in the buffer). The default values of α and β are 1 and 3, respectively. Vegas adopts a slow start phase where cwnd doubles only every 2 RTTs. Then, the following algorithm is used in the congestion avoidance phase; this algorithm is sensitive to the current measured RTT value (where RTTbase $\leq$ RTT) as representative of the congestion.

As far as we have negligible congestion, RTT is small and close to RTTbase so that $\Delta < \alpha$. In these conditions, Vegas increases cwnd linearly (one packet during the next RTT). When RTT increases so that $\alpha < \Delta < \beta$, Vegas leaves the cwnd unchanged. Finally, if there is congestion so that RTT increases too much, we have that $\Delta > \beta$ and then Vegas decreases cwnd linearly (1 packet during the next RTT). According to this mechanism, TCP Vegas detects congestion before it happens based on an increase in the RTT values, while TCP Reno and NewReno detect congestion only after it has actually occurred (with consequent packet losses). The Vegas algorithm depends heavily on the accurate estimation of the RTTbase value.

TCP Vegas does not work well in the presence of other TCP flows based on Tahoe or (New)Reno: Vegas reduces cwnd to avoid congestion, while Tahoe and (New)Reno make use of the additional bandwidth. This is a classical inter-protocol unfairness issue, as discussed later in this section.

TCP Reno

TCP Reno (Jacobson 1990, RFC 2001) adopts Slow Start, Congestion Avoidance, Fast Retransmit, and Fast Recovery: when three DUPACKs are received (i.e., four identical ACKs are received in sequence), a segment loss is assumed and a Fast Retransmit/Fast Recovery (FR/FR) phase starts, which can be summarized as follows:

- ssthresh is set equal to flightsize/2. According to RFC 2581, the flightsize is the amount of data that has been sent but not yet acknowledged. We can roughly consider that flightsize = min{cwnd, rwnd}, and neglecting the effect of rwnd, practically, flightsize $\approx$ cwnd.
- The last unacknowledged segment is soon retransmitted (fast retransmit algorithm).
- cwnd = ssthresh + ndup, where initially ndup = 3 due to three DUPACKs to start the FR/FR phase. This inflates the congestion window by the number of segments that have left the network and that are cached at the receiver.
- Each time another DUPACK arrives, increment cwnd by the segment size (cwnd = cwnd + 1). This inflates the congestion window due to the additional segment, which has left the network. Then, transmit a packet, if allowed by the new cwnd value.
- When the first non-DUPACK (i.e., a "full ACK," acknowledging all packets sent or a "partial ACK," acknowledging some progress in the sequence number in case of multiple packet losses in a window of data) is received, cwnd is set to ssthresh to deflate the window and the FR/FR phase ends.
- Then, a new congestion avoidance phase starts.

This approach allows us to recognize a packet loss in advance with respect to an RTO expiration, which would cause a drastic reduction in throughput. TCP Reno performs well in the presence of sporadic packet losses, but when there are multiple losses in the same window of data an RTO expiration may occur as with TCP Tahoe; this problem has been solved by the TCP NewReno version described below.

TCP NewReno

TCP NewReno, initially defined in RFC 2582 (year 1999), RFC 3782 (year 2004), and RFC 6582 (2012) is based on an improved FR/FR algorithm. In the presence of multiple packet losses in a window of data, NewReno avoids unnecessary multiple FR/FR phases and manages all these losses using a single FR/FR phase.

The basic algorithm defined in RFC 2582 did not attempt to avoid unnecessary multiple FR/FR phases when an RTO occurs during an FR/FR phase (i.e., FR/FR then RTO expiration and then an unnecessary FR/FR phase triggered by DUPACKs). This is especially the case of a spurious (not-needed) RTO, triggered not because of a real packet loss, but just because RTT has experienced a sudden increase that RTO cannot cope with: spurious RTOs cause many DUPACKs and three DUPACKs trigger an FR/FR phase (where cwnd is halved).[10]

With the "(more) careful" version of TCP NewReno as specified in RFC 2582, FR/FR is disabled (i.e., DUPACKs are ignored) after an RTO until all previously transmitted packets are acknowledged. A "less careful" version of this restriction allows FR/FR when DUPACKs arrive. The "(more) careful" version avoids unnecessary FR/FR phases after a spurious RTO expiration, while the "less careful" variant may still have unnecessary FR/FR phases. RFC 3782 specifies the "careful" variant of the FR/FR algorithm as the basic version of TCP NewReno. It is based on the new variable "recover." It is initially set to the initial sequence number. At each invocation of the FR/FR algorithm or at each RTO expiration, "recover" is made equal to the maximum order of the segment sent. The use of the "recover" variable is fundamental to have a single FR/FR phase with multiple packet losses and to avoid unnecessary FR/FR phases after an RTO ("careful" version).

A partial ACK acknowledges some but not all the packets that were outstanding at the start of the FR/FR phase (i.e., the sequence number of a partial ACK is lower than "recover"). With TCP Reno, the first partial ACK causes TCP to leave the FR/FR phase by deflating cwnd back to ssthresh. Instead, partial ACKs do not take TCP out of the FR/FR phase with TCP NewReno: the partial ACKs received during the FR/FR phase are treated as an indication that the packet immediately following the acknowledged packet has been lost and needs to be retransmitted.

The main characteristics of the FR/FR algorithm of NewReno are summarized below:

- When three DUPACKs are received, FR/FR is started if the sender is not already in the FR/FR procedure.
- A segment retransmission is soon performed starting from the first unacknowledged segment.
- The management of ssthresh and cwnd is similar to Reno.
- When an ACK is received that does not acknowledge all outstanding data, as indicated in the "recover" variable, this ACK is considered a partial ACK. The FR/FR phase continues retransmitting one lost segment per RTT until all lost segments have been retransmitted.
- The FR/FR phase ends when a full ACK is received, acknowledging all the packets, which were outstanding when the FR/FR phase began.
- A new congestion avoidance phase is performed starting with cwnd = ssthresh, where ssthresh is equal to half of the cwnd value just before starting the FR/FR phase.
- After an RTO expiration, record the highest sequence number transmitted in the "recover" variable and exit the FR/FR procedure.

Moreover, there are two more variants of the TPC NewReno algorithm, denoted as "Slow-but-Steady" and "Impatient." They differ in the way they manage timer RTO during the FR/FR phase.

- The Slow-but-Steady variant resets timer RTO after receiving each partial ACK and continues to make small adjustments to the cwnd value. The TCP sender remains in the FR/FR mode until it receives a full ACK. Typically no RTO occurs.
- The Impatient variant resets timer RTO only after receiving the first partial ACK. Hence, in the presence of multiple packet losses, the Impatient variant attempts to avoid long FR/FR phases by allowing timer RTO to expire. All the lost segments are recovered according to a go-back-N scheme in a slow-start phase.

RFC 3782 recommends the Impatient variant over the Slow-but-Steady one. For instance, let us consider $\text{RTO} \approx 2 \times \text{RTTs}$; assuming RTO is equal to 1 s (minimum RTO value), this condition relating RTO and RTT is valid in a GEO satellite scenario where $\text{RTD} = 0.5$ s. Then, let us consider having multiple losses in the same window of data: an FR/FR phase is started in about 1 RTT when three DUPACKs are received. After 1 RTT, we consider that the first partial ACK is received, which resets RTO: we now have 2 further RTTs before RTO

[10] In general, DUPACKs can be generated due to many reasons, such as a segment loss, segments received out-of-sequence, retransmission of packets already received at the destination.

expires. The FR/FR procedure has 3 RTTs before RTO expires and can recover 3 packets. Hence, if the total number of lost packets in the same window of data is greater than 3, there is an RTO expiration with the Impatient version [52]. Instead, in the Slow-but-Steady case, the FR/FR phase continues slowly recovering 1 packet per RTT without RTO expirations; when the FR/FR phase ends, a congestion avoidance phase starts.

SACK Option in TCP

TCP Reno and NewReno retransmit at most 1 lost packet per RTT during the FR/FR phase so that the pipe can be used inefficiently in the presence of multiple losses. To recover more quickly the lost packets, the Selective ACK (SACK) option can be used, as detailed in RFCs 2018 and 2883 [49, 50]. With SACK, the receiver can inform the sender of all segments that have arrived successfully so that the sender retransmits only those segments that have been lost. The support for SACK is negotiated between sender and receiver at the beginning of a TCP connection. In particular, the SACK-permit option is used in the three-way handshake phase: both sender and receiver need to agree on the use of SACK. Note that SACK uses an optional field in TCP headers. SACK does not change the meaning of the ACK field in TCP headers. A *block* is a contiguous group of correctly received bytes: the bytes just before the block and just after the block have not been received. The receiver uses the SACK option to inform the sender about non-contiguous blocks of data that have been received and queued. If the SACK is enabled, the SACK option should be used in all ACKs not acknowledging the highest sequence number in the receiver queue. A block is specified in the SACK option using the first and the last sequence number of the block: the SACK option specifying n blocks has a length of $8 \times n + 2$ bytes. Hence, a maximum of four blocks can be specified using the 40 bytes of TCP options.

We refer below to the implementation of SACK combined with TCP Reno by S. Floyd that requires a new state variable, called "pipe" [53, 54]. This Reno-SACK algorithm can be summarized as follows:

- Whenever the sender enters the FR/FR phase (after three DUPACKs received), it initializes a variable "pipe," which estimates how many packets are outstanding in the network and sets cwnd to half of its current size.
- If pipe > cwnd, no packet can be sent since the number of in-flight packets is larger than the cwnd value.
- Pipe is decremented by 1 packet (or 2 packets) when the sender receives a DUPACK (or a partial ACK if SACK is used in the NewReno case).
- Whenever pipe becomes lower than cwnd, it is possible to send packets, starting from those lost (holes reported by SACK) and then new ones. Thus, more than 1 lost packet can be sent in 1 RTT.
- Pipe is incremented by 1 packet when the sender transmits a new packet or retransmits an old one.
- The FR/FR phase ends when a full ACK is received.

Although the above describes a special implementation case of SACK, this option can also be used with other TCP versions. SACK is enabled by default on most Linux distributions.

Additive Increase Multiplicative Decrease (AIMD) and Multiplicative Increase Multiplicative Decrease (MIMD) algorithms

The AIMD algorithm is a further alternative for managing cwnd in the congestion avoidance phase. AIMD combines a linear growth of cwnd with a multiplicative reduction when a congestion event (i.e., packet loss) occurs. Cwnd is increased by a fixed amount for every RTT, but when congestion is detected, the transmitter decreases cwnd by a multiplicative factor; for example, cwnd is reduced to cwnd/2 after a loss. With AIMD, the law used to update cwnd can be generally expressed as follows:

$$\begin{aligned} \text{cwnd} &= \text{cwnd} + \frac{a}{\text{cwnd}} && \text{upon an ACK arrival} \\ \text{cwnd} &= \text{cwnd} \times (1 - b) && \text{upon a loss detection} \end{aligned} \tag{2.18}$$

TCP Reno and NewReno ($a = 1$ and $b = 1/2$) and High-Speed TCP (HSTCP) are examples of AIMD algorithms. HSTCP is an adaptive AIMD algorithm proposed in RFC 3649 for networks with large BDPs [55]: the increment and the decrement of cwnd in response to the receipt of an ACK or to a packet loss depend on the current cwnd value.

Multiple TCP flows using AIMD and sharing the same bottleneck link converge to use equal amounts of bandwidth (fairness).

However, in high-speed networks (LFN networks), the AIMD algorithm could still be too slow to increase cwnd, thus leading to inefficient use of resources. To overcome this drawback, the MIMD algorithm has been proposed, where the cwnd update law can be generally expressed as follows:

$$\begin{aligned} \text{cwnd} &= \text{cwnd} + \alpha \quad \text{upon an ACK arrival} \\ \text{cwnd} &= \text{cwnd} \times (1 - \beta) \quad \text{upon a loss detection} \end{aligned} \tag{2.19}$$

Scalable TCP (STCP) is an example of MIMD algorithm, where the following α and β values are suggested: $\alpha = 0.01$ and $\beta = 0.125$ [56]. STCP gets its name because the time it takes to recover from a loss occurred when cwnd $= W^*$, i.e., the time cwnd takes to return to W^* does not depend on the W^* value.

One problem with MIMD algorithms is that multiple TCP flows using MIMD and sharing the same bottleneck link may not converge to share the bandwidth equally, thus leading to unfairness.

Other TCP Versions

There are many other variants of the TCP congestion control algorithm. The details of some of them are provided below. Only a few of these TCP variants are supported by RFCs. Different operating systems make use of different TCP versions. TCP Westwood apart, all the TCP variants described below are characterized by a more aggressive behavior of cwnd than TCP NewReno to recover more quickly from loss events. This behavior, however, entails unfairness issues with classical TCP versions. These new TCP variants are well suited to LFN networks or error-prone networks.

In TCP Westwood (and Westwood+), there is still the slow-start phase and the congestion avoidance one. The main innovation is estimating the available bandwidth based on the rate of ACKs received (low-pass filtering). This estimation is used to determine cwnd and ssthresh values after a congestion event (three DUPACKs or RTO). Unlike TCP Reno, which halves cwnd after three DUPACKs, TCP Westwood sets ssthresh to the value corresponding to the estimated capacity, and cwnd is made equal to ssthresh to avoid the slow-start phase. When an RTO expires, ssthresh is set as in the case of three DUPACKs, but cwnd is reset to 1. TCP Westwood is well suited to applications in error-prone links (i.e., wireless scenario).

FAST TCP aims at maintaining a constant number of packets in the queues of the network. The number of packets in the queues is estimated by measuring the difference between the observed RTT and the base RTT (i.e., the minimum observed RTT for the connection, RTD), as in TCP Vegas. If too few packets are queued, the sending rate is increased, while if too many packets are queued, the rate is decreased. The difference between TCP Vegas and FAST TCP depends on how the rate is adjusted when the number of packets stored is too small or large. TCP Vegas adopts fixed-size adjustments, while FAST TCP uses adaptive steps to improve the speed of convergence and the stability: larger steps when the system is far from equilibrium and smaller steps close to equilibrium.

TCP Veno combines the congestion control algorithms of Reno (reactive congestion control) and Vegas (proactive congestion control). In particular, in the presence of a packet loss, Veno can decide if it is likely to be due to congestion or random bit errors (*congestive state* or *non-congestive state*). In the first case, the classical Reno approach is used for cwnd and ssthresh (ssthresh $\leftarrow$ cwnd/2, cwnd $\leftarrow$ ssthresh); instead, in the second case, a different multiplicative reduction of cwnd is adopted to reduce the rate less aggressively (in particular, ssthresh $\leftarrow$ cwnd $\times$ 4/5 and the rest is as in Reno). The Vegas estimation on the network congestion is adopted to determine whether packet losses are likely to be due to network congestion or random errors.

Compound TCP is a Microsoft implementation of TCP, which simultaneously adopts two different congestion window algorithms to achieve a good performance in LFNs and not compromise the fairness. Like FAST TCP and TCP Vegas, Compound TCP uses an estimation of the queuing delay as a measure of congestion. Compound TCP maintains two congestion windows: a regular AIMD window and a delay-based one. The size of the actual cwnd is the sum of these two windows. The AIMD window is increased in the same way as TCP Reno. If the delay is small, the delay-based window increases rapidly to improve the utilization of the network. When queuing occurs, the delay-based window gradually decreases to compensate for the increase in the AIMD window. The aim is to keep their sum almost constant, thus approximating the BDP value.

The cwnd behavior with BIC TCP has different phases with logarithmic (binary search), exponential (max probing), and additive increases. In the presence of a packet loss, a multiplicative decrease is performed. BIC TCP is well suited to LFN networks.

CUBIC TCP is less aggressive than BIC TCP. To determine cwnd, CUBIC adopts a cubic function of time since the last congestion event, with the inflection point corresponding to the window value before that event. The cwnd cubic function has first a concave part and then a convex part. CUBIC TCP has a very slow cwnd increase in the transition between the concave and the convex growth regions. This scheme allows the network to stabilize before CUBIC starts looking for more bandwidth. CUBIC does not rely on the receipt of ACKs to increase cwnd: the cwnd value is dependent only on the time since the last congestion event. CUBIC TCP achieves RTT fairness (see the next Sect. 2.8.1.9) among flows since the window growth is independent of RTT. The goodput of CUBIC TCP is quite robust to random packet losses. CUBIC TCP reacts to a packet loss by reducing cwnd by a multiplicative factor equal to 0.8. CUBIC TCP is the default TCP version in Linux kernels (2.6.19 or above).

Finally, Bottleneck Bandwidth and Round-trip propagation time (BBR) is a TCP congestion control algorithm proposed by Google in 2016. The algorithm uses both the bandwidth as the maximum observed delivery rate and the propagation delay as the minimum observed RTT over a certain interval. These values, estimated separately, are used by BBR to build a network model, like TCP Vegas. Each cumulative or selective acknowledgment for a received packet produces a rate sample, which records the amount of data delivered over the RTT interval. The main objective of BBR is to ensure that the bottleneck remains saturated but not congested. The BBR algorithm has four different phases: Start-up, Drain, Probe Bandwidth, and Probe RTT. In these phases, the injection of data is controlled by a *pacing parameter* that is around 1 (slightly below 1, 1, or slightly above 1) and multiplies the estimated bandwidth. BBR achieves a throughput improvement with respect to other TCP versions.

2.8.1.9 Evaluation of TCP Performance and Comparisons

In normal conditions, when no loss occurs, Tahoe, Reno, and NewReno behave identically. These three classical TCP versions behave differently in their congestion control algorithms when a packet loss occurs (or multiple packet losses occur). All these versions recognize a packet loss when the third DUPACK is received. Subsequently, Tahoe scales cwnd down to 1 segment, uses a slow-start phase, and retransmits all segments after the lost one according to a go-back-N approach. Instead, Reno and NewReno reduce cwnd to half of its value, retransmit the lost packet, and enter fast recovery. Reno and NewReno differ on when they exit fast recovery. Reno exits when the first non-duplicate ACK is received. NewReno instead leaves fast recovery only with the receipt of a new ACK confirming the successful delivery of all segments sent before fast retransmit was triggered: multiple segment losses could be recovered during fast recovery with NewReno. At the end of the fast recovery phase, cwnd is restored to the ssthresh value to perform a congestion avoidance phase.

On the basis of the above, Tahoe has worse performance than Reno and NewReno in the presence of isolated (uncorrelated) losses. However, the situation is reversed in the presence of multiple losses in a window of data (correlated losses) [57]. Tahoe reacts to multiple losses by performing almost soon a fast retransmit and then a slow-start phase with cwnd = 1 (without waiting for an RTO). Instead, Reno can have a phase with flat cwnd, which may also result in an RTO or multiple RTOs (in this case, both cwnd and ssthresh reach 1 so that cwnd has a congestion avoidance phase starting from 1). On the other hand, NewReno can have a flat cwnd behavior without RTO in the Slow-but-Steady variant.

Microanalysis: cwnd Behavior

The microanalysis is the study of the TCP behavior (in terms of cwnd, RTT, RTO, sequence number, and ACK number) with the finest time granularity to verify the reaction to important events, like packet losses and RTO expirations. This study is opposed to the macroanalysis, which deals with the evaluation of the macroscopic TCP behavior based on time averages (e.g., average throughput and average goodput).

Let us refer to the network model adopted in Fig. 2.47 with the bottleneck link characterized by a layer 3 buffer with capacity B and bit-rate IBR. In the microanalysis of the cwnd behavior, we consider, for the sake of simplicity, as if all the packets of a cwnd were "generated" together in a bunch (i.e., meaning "arrived" together at the layer 3 buffer) and that all the related ACKs are received together in an RTT time. According to this simplified approach, we can study the cwnd behavior at regular intervals of length RTT. Note that this model entails some degree of approximations and a certain granularity in the packet generation process. The packets of a cwnd are not "generated" altogether since the generation is controlled by the arrival of ACKs that allow the window to slide

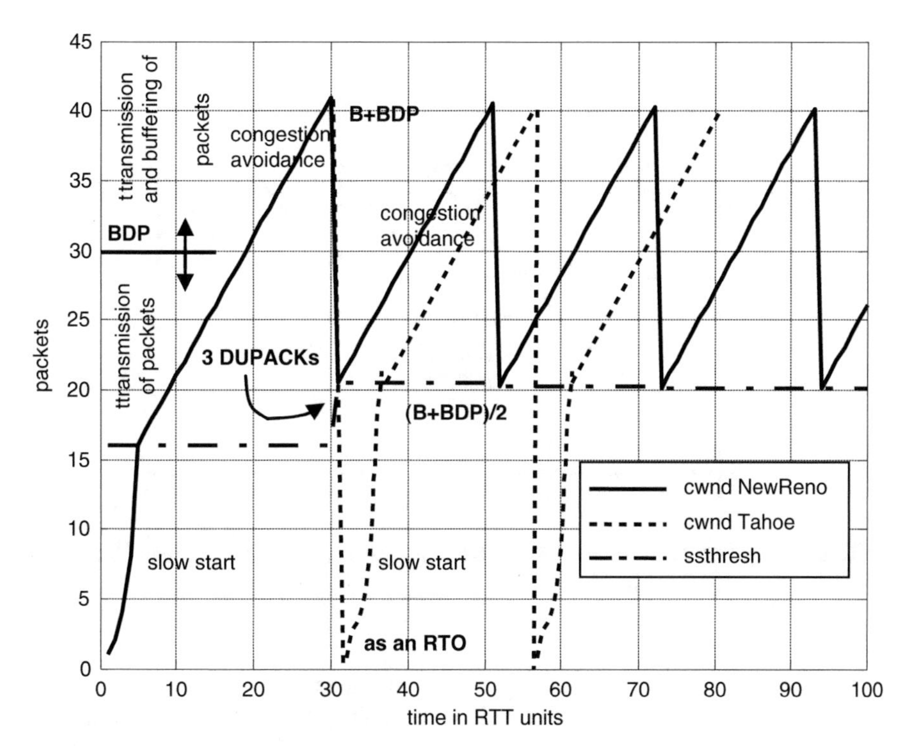

Fig. 2.48: Cwnd and ssthresh behaviors for TCP NewReno and TCP Tahoe (model with periodic losses), considering sockets' buffer size much larger than $B + \text{BDP}$. The TCP NewReno curve would also be valid for TCP Reno

and to transmit new packets. The ACKs of a window of data do not arrive altogether at the sender but spaced at least by the packet transmission time, $t_{\text{pkt}} = \text{MTU/IBR}$. Then, the packets of a cwnd are "generated" spaced in time of t_{pkt}, which is also the time separation of the ACKs received if we neglect the contribution of the ACK packets transmission time. Therefore, as long as cwnd $<$ BDP, no packet is waiting for service in the bottleneck link queue.

We do not consider delayed ACKs in the model: one ACK is sent back for each TCP segment received.[11] In the congestion avoidance phase, cwnd increases linearly on an RTT basis until a congestion event occurs. Hence, we expect that also the RTT value may increase with cwnd because of the gradual occupancy of buffer B. Therefore, if we plot cwnd as a function of time in seconds, the cwnd behavior is concave down, especially if B is not too small and comparable with BDP. This is because RTT increases with respect to RTD (fixed value, RTT_{m}) as buffer B fills up. In the graphs of cwnd as a function of RTT, we do not see the impact of the RTT variation because the abscissa unit of measurement is RTT itself.

An example of cwnd behaviors for TCP NewReno and TCP Tahoe is shown in Fig. 2.48, referring to a case with a single TCP flow and a low initial ssthresh value (i.e., ssthresh $<$ BDP); in particular: BDP $= 30$ pkts, $B = 10$ pkts, rwnd $= \infty$ (it would be enough to have rwnd $> B + \text{BDP}$), initial ssthresh $= 16$ pkts. The cwnd behavior is deterministic in the case of a single flow in the network shown in Fig. 2.47. Let us recall that, according to our model, the cwnd value corresponds to the number of packets injected as in a bunch on an RTT basis. In both Tahoe and NewReno cases, TCP congestion control starts at time 1 RTT with a slow-start phase and a cwnd exponential increase on an RTT basis. After 5 RTTs, cwnd reaches the initial ssthresh value, and a congestion avoidance phase starts with a linear cwnd increase on an RTT basis: the cwnd curve has a slope of 45°. Then, cwnd increases, reaching the maximum value of B packets in the buffer plus BDP in-flight packets: $\text{cwnd}_{\text{max}} = B + \text{BDP}$. In the next RTT, after cwnd has reached its maximum, we have a single packet loss, which is soon recognized based on three DUPACKs received; ssthresh is made equal to half of the current cwnd value [i.e., ssthresh $= (B + \text{BDP})/2$] and cwnd is made equal to ssthresh in the NewReno case (here we neglect the short time spent in the FR/FR phase to recover the isolated loss). In contrast, cwnd is made equal to 1 in the TCP Tahoe case. From this point onwards, TCP Tahoe and TCP NewReno behave differently.

[11] When the receiver implements *delayed ACKs*, the number of ACKs sent by the receiver roughly halves so that the sender opens cwnd more slowly (approximately of 1 segment for every two RTTs in the congestion avoidance phase).

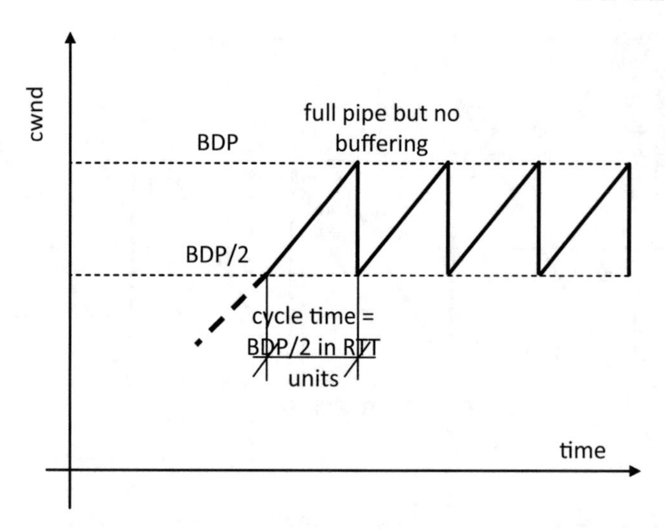

Fig. 2.49: TCP NewReno cwnd behavior in a case without a buffer at the bottleneck link, no delayed ACKs, and no packet losses caused by the physical medium

In the Tahoe case, cwnd has a periodic waveform (due to periodic packet losses when cwnd overcomes $\text{cwnd}_{\max}$), involving a slow-start phase from $\text{cwnd} = 1$ to $\text{cwnd} = (B + \text{BDP})/2$ and a congestion avoidance phase from $\text{cwnd} = (B + \text{BDP})/2$ to $\text{cwnd} = B + \text{BDP}$. The time that the cwnd of Tahoe takes to growth after a packet loss from $\text{cwnd} = 1$ to $\text{cwnd}_{\max}$ (*Tahoe cycle time*), T_{Tahoe}, is obtained as

$$T_{\text{Tahoe}} \approx \log_2\left(\frac{B + \text{BDP}}{2}\right) + \frac{B + \text{BDP}}{2}\,[\text{RTT}] \tag{2.20}$$

In the NewReno case, the cwnd behavior is according to a periodic sawtooth waveform (due to periodic losses) between $(B + \text{BDP})/2$ and $B + \text{BDP}$. The time that the cwnd of NewReno takes to grow after a packet loss from $\text{cwnd} = (B + \text{BDP})/2$ to $\text{cwnd}_{\max}$ (*NewReno cycle time*), T_{NewReno}, is

$$T_{\text{NewReno}} \approx \frac{B + \text{BDP}}{2}\,[\text{RTT}] \tag{2.21}$$

The cycle time represents the recovery time after a packet loss. We can thus note that $T_{\text{Tahoe}} > T_{\text{NewReno}}$. Correspondingly, TCP NewReno has an average throughput higher than TCP Tahoe, computed as the sum of the cwnd_i values up to a certain number $i = n$ of RTT intervals divided by the sum of the corresponding RTT_i values up to $i = n$. An approach to analytically derive the throughput is provided below when dealing with macroanalysis.

In the case of a higher initial ssthresh value, that is ssthresh $> B + \text{BDP}$, the traffic injection in the network during the initial slow-start phase is huge so that we may have a significant initial loss of packets (multiple packet losses in the same window of data), which may cause a phase during which cwnd has a flat behavior or where there is even an RTO expiration. In any case, after a certain time, the periodic regime behavior of cwnd is recovered, as described before.

Therefore, we can conclude that the appropriate selection of the initial ssthresh value can have a significant impact on determining the best behavior of TCP in the start-up phase. For this purpose, the Hoe algorithm can be used to identify an appropriate initial ssthresh value [58].

Macroanalysis: TCP Throughput Without Packet Losses

Let us derive the average throughput of TCP Reno/NewReno in a simple case without a buffer at the bottleneck link (i.e., $B = 0$), considering the sockets size larger than BDP and a constant RTT value equal to the minimum (i.e., RTD). We refer to the regime cwnd behavior shown in Fig. 2.49 for a single-flow case (i.e., no cross-traffic).

The regime cwnd behavior is a triangular waveform (sawtooth pattern) between BDP/2 and BDP. The average throughput can be approximated as

$$\Gamma = \frac{\overline{\text{cwnd}}}{\text{RTT}} \left[\frac{\text{bit}}{\text{s}}\right] \tag{2.22}$$

where $\overline{\text{cwnd}}$ is derived by taking the average of the regime cwnd behavior in Fig. 2.49, where cwnd is considered as a continuous function of time.

$$\overline{\text{cwnd}} = \frac{\int_{\text{BDP}/2}^{\text{BDP}} x\text{d}x}{\text{BDP} - \text{BDP}/2} = \frac{3}{4}\text{BDP} = \frac{3}{4}\frac{\text{RTT} \times \text{IBR}}{\text{MTU}} \,[\text{pkts}] \tag{2.23}$$

where pkts here refers to IP packets and MTU is expressed in bits.

By substituting $\overline{\text{cwnd}}$ of (2.23) in (2.22), we achieve the following TCP throughput measured at layer 3 (we should multiply the formula below by the reduction factor MSS/MTU if we like to consider the throughput at layer 4):

$$\Gamma = \frac{3}{4}\text{IBR} \left[\frac{\text{bit}}{\text{s}}\right] \tag{2.24}$$

Since the maximum bit-rate available for the TCP flow (upper bound) is equal to IBR, we can conclude that the average throughput is 3/4 of the maximum bit-rate: a TCP Reno/NewReno flow without a buffer can achieve maximum utilization of $3 \times 100/4\% = 75\%$. This is a relatively low value (the link is underutilized), which justifies the need for a buffer with capacity $B > 0$ on the bottleneck link.

If $B > 0$, system dynamics are more complex since we have queuing phenomena at the buffer and RTT is not constant [59]. According to [47], RTT is constant and equal to RTD when $\text{cwnd}(t) \leq \text{BDP}$; otherwise, if $\text{cwnd}(t) > \text{BDP}$, the buffer of the bottleneck link starts to accumulate a number $Q(t) = \text{cwnd}(t) - \text{BDP}$ of packets. Hence, $\text{RTT}(t)$ results as

$$\text{RTT}(t) = \begin{cases} \text{RTD} + \dfrac{\text{MTU}}{\text{IBR}} Q(t), & \text{if } \text{cwnd}(t) > \text{BDP} \\ \text{RTD}, & \text{if } \text{cwnd}(t) \leq \text{BDP} \end{cases} \tag{2.25}$$

where MTU/IBR represents the IP packet transmission time t_{pkt} on the bottleneck link.

When $\text{cwnd}(t) > \text{BDP}$, cwnd and $Q(t)$ have corresponding increases. Hence, we can state that if cwnd has a sawtooth behavior in time (Reno/NewReno cases), also RTT has a certain sawtooth behavior.

According to [47], we study the TCP throughput, using a different approach with respect to (2.22), considering the *cycle time* (*renewal theory approach*) between two consecutive packet losses [with cwnd growing from $(B + \text{BDP})/2$ to $B + \text{BDP}$]. In particular, the throughput is derived for TCP Reno/NewReno in cases where the sockets' buffer size does not limit TCP traffic injection, dividing the number of packets in a cycle time by the cycle time itself, computed using the previous RTT formula (2.25), where the effects of the queuing at the bottleneck link are accounted for. The number of packets transmitted in a cycle is obtained by using the continuous approximation as used for (2.23)

$$\text{number of packets in a cycle} = \int_{\frac{B+\text{BDP}}{2}}^{B+\text{BDP}} x\text{d}x = \frac{3}{8}\left(B + \text{BDP}\right)^2 \tag{2.26}$$

Moreover, we also use the continuous approximation to determine the sum of RTTs in the cycle for the case $B \leq \text{BDP}$, where the RTT duration is given by (2.25)

$$\text{cycle duration} = \int_{\frac{B+\text{BDP}}{2}}^{B+\text{BDP}} \text{RTT}\,(t)\,\text{d}t$$

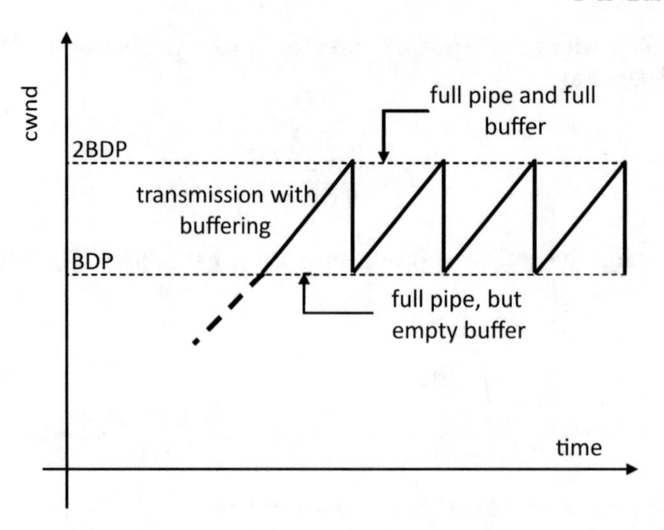

Fig. 2.50: TCP NewReno cwnd behavior in the case with optimal buffer $B = \mathrm{BDP}$

$$
\begin{aligned}
&= \int_{\frac{B+\mathrm{BDP}}{2}}^{\mathrm{BDP}} \mathrm{RTD}\,\mathrm{d}t + \int_{\mathrm{BDP}}^{B+\mathrm{BDP}} \mathrm{RTD}\,\mathrm{d}t \\
&\quad + \frac{\mathrm{MTU}}{\mathrm{IBR}} \int_{\mathrm{BDP}}^{B+\mathrm{BDP}} [x - \mathrm{BDP}]\,\mathrm{d}x \\
&= \mathrm{RTD}\frac{B+\mathrm{BDP}}{2} + \frac{\mathrm{MTU}}{\mathrm{IBR}}\left[\frac{(B+\mathrm{BDP})^2 - (\mathrm{BDP})^2}{2} - \mathrm{BDP}\times B\right] \\
&= \frac{\mathrm{MTU}}{2\,\mathrm{IBR}}\left[(B+\mathrm{BDP})^2 - \mathrm{BDP}\times B\right]
\end{aligned}
\tag{2.27}
$$

Analogously, we can determine the sum of RTTs in the cycle for the case $B \geq \mathrm{BDP}$ considering that now all the cases are with cwnd $> \mathrm{BDP}$ for $\mathrm{RTT}(t)$ in (2.25). We have

$$
\begin{aligned}
\text{cycle duration} &= \int_{\frac{B+\mathrm{BDP}}{2}}^{B+\mathrm{BDP}} \mathrm{RTT}\,(t)\,\mathrm{d}t \\
&= \frac{\mathrm{MTU}}{\mathrm{IBR}} \int_{\frac{B+\mathrm{BDP}}{2}}^{B+\mathrm{BDP}} x\mathrm{d}x \\
&= \frac{3}{8}\frac{\mathrm{MTU}}{\mathrm{IBR}}\,(B+\mathrm{BDP})^2
\end{aligned}
\tag{2.28}
$$

Then, in the case of no cross-traffic, we determine the throughput as the ratio of the number of packets in a cycle and the cycle duration combining the two cases of (2.27) and (2.28) [47]:

$$
\Gamma = \begin{cases} \dfrac{3}{4}\mathrm{IBR} \times \dfrac{(B+\mathrm{BDP})^2}{(B+\mathrm{BDP})^2 - B\times\mathrm{BDP}}, \text{ if } B \leq \mathrm{BDP} \\ \mathrm{IBR}, \text{if } B > \mathrm{BDP} \end{cases} \quad \left[\frac{\mathrm{bit}}{\mathrm{s}}\right]
\tag{2.29}
$$

If the buffer size B is too low, the TCP can only use a fraction of the available bandwidth (i.e., IBR). If $B = 0$, (2.29) gives $\Gamma \equiv 3 \times \mathrm{IBR}/4$, as already obtained in (2.24) with a different approach.

The optimal buffer value is the minimum B value, which allows us to keep the pipe filled continuously (i.e., the pipe never becomes empty) to exploit the link at the maximum rate, IBR. A *rule-of-thumb* is to consider $B = \mathrm{BDP}$ packets based on the network model in Fig. 2.47. Indeed, when $B = \mathrm{BDP}$, formula (2.29) yields $\Gamma \equiv \mathrm{IBR}$.

If $B < \mathrm{BDP}$, the link is said to be *under-buffered*, while, if $B \geq \mathrm{BDP}$, the link is said to be *over-buffered* and the bottleneck link is utilized at 100% (i.e., at IBR). Assuming no cross-traffic, the cwnd behavior with the optimal buffer size is shown in Fig. 2.50.

The socket buffer size (at both sender and receiver) should be equal to $B + \text{BDP}$ packets in order not to limit the traffic injection, according to our model [47]. Let us recall that BDP is used as BDP_m computed with the physical round-trip propagation delay RTD so that $\text{BDP}_\text{m} + B$ corresponds to the actual BDP value of the system that is computed with RTT instead of RTD. The socket buffer size depends on the settings of the operating system. For instance, the default TCP socket size is 64 kB for MAC OS X; this is the maximum possible value with the standard settings of the 16-bit representation of the window size.

Additional buffering (beyond the optimal settings of $B = \text{BDP}$) in the network could be used to increase better system robustness against congestion events. However, it was recognized that too much buffering and filled buffers might cause problems for delay-sensitive applications [60]. This is the so-called *bufferbloat issue* (i.e., the trend towards configuring increasingly large buffers on the Internet that would spend most of their existence mainly being full) that results in significantly reduced responsiveness of applications because of excess buffering of packets within the network [60]. Bufferbloat causes high latency and appreciable jitter. With excessive buffering, a packet loss in reliable links occurs on the complete saturation of the bottleneck buffer instead of the moment when the congestion really occurred. This has an impact on TCP behavior.

Macroanalysis: TCP Throughput with Packet Losses, the Square-Root Formula

In [61], a model has been proposed to study the TCP throughput in the presence of random packet losses. This formula has been derived under the following simplifying assumptions:

- No buffering ($B = 0$)
- Single flow, no cross-traffic
- RTT is constant
- Each correctly received packet is ACKed (no delayed ACK is used)
- *Periodic single losses* with rate p. The loss occurs when cwnd reaches the value of W (pkts)[12]
- No RTO expiration occurs
- TCP Reno/NewReno version is taken as a reference
- This analysis is carried out at regime when cwnd has a periodic sawtooth behavior based on the congestion avoidance algorithm

Even in this case, we study the window evolution based on cycles. A cycle is the time interval between two consecutive packet losses. In particular, we have a *periodic cwnd behavior* according to a sawtooth waveform between $W/2$ and W (congestion avoidance phase). Hence, this behavior is quite similar to that shown in Fig. 2.49, previously assumed to derive the throughput with $B = 0$. In a cycle of duration $W/2$ in RTT units, the number of packets sent is obtained by integrating $\text{cwnd} = t$ over time t:

$$\text{packets sent per cycle} = \int_{W/2}^{W} t \, \mathrm{d}t = \frac{3}{8}W^2 \tag{2.30}$$

Since there is a single packet loss in a cycle time where $3W^2/8$ packets are generated, the loss rate p can be expressed as

$$p = \frac{1}{\frac{3}{8}W^2} \tag{2.31}$$

Using (2.31), we can solve W as a function of p as

$$W = \sqrt{\frac{8}{3p}} \tag{2.32}$$

[12] In absence of losses caused by the PHY layer, $W = \text{BDP}$ if $B = 0$. However, because of packet losses the maximum value of cwnd is below BDP. This is the reason why, symbol W has been used here.

In this case, similarly to (2.29), the average throughput Γ is obtained by dividing the number of packets sent per cycle by the cycle duration $W/2$ in RTT units; then, measuring the throughput at layer 3, we need to multiply by MTU (expressed in bits):

$$\Gamma = \mathrm{MTU} \times \frac{3W^2/8}{[W/2] \times \mathrm{RTT}} = \frac{\mathrm{MTU}}{\mathrm{RTT}} \frac{3}{4} W \quad (2.33)$$

By substituting the W expression as a function of p of (2.32) in (2.33), we achieve the following result, known in the literature as the *square-root formula*:

$$\Gamma = \frac{\mathrm{MTU}}{\mathrm{RTT}} \frac{C}{\sqrt{p}} \left[\frac{\mathrm{bit}}{\mathrm{s}}\right] \quad (2.34)$$

where $C = \sqrt{3/2} \approx 1.22$.

In the presence of random losses with mean rate p, we can reuse the square-root formula provided to consider $C = 1.31$ for TCP Reno/NewReno. Let us make the following considerations to compare the TCP throughput sensitivity to both RTT and p variations:

- If RTT doubles, the throughput value is reduced by a factor 0.5.
- If p doubles, the throughput value is reduced by a factor 0.7.

Hence, we can conclude that TCP throughput is more sensitive to p variations than to RTT variations. In any case, both the increase in RTT and the increase of p have a negative impact on the TCP throughput.

If p is too high (typically $p > 0.1$), the occurrence of RTO expirations in the TCP behavior cannot be neglected, and the square-root formula cannot be applied. TCP throughput analyses are also provided in the literature taking RTO effects into account; these aspects are beyond the scope of this book, but the interested reader may refer to [59] for more details.

The throughput of the square-root formula increases when p reduces. To avoid absurd situations where throughput $\Gamma \to \infty$ when $p \to 0$, we need to impose the physical upper bound to the throughput Γ value, which is determined by the IBR value;[13] the corresponding goodput γ value (see Sect. 2.8.1.7) is determined by multiplying the throughput by a factor $(1 - p)$, denoting the probability that a transmitted packet is correctly received.

$$\Gamma = \min\left\{\frac{\mathrm{MTU}}{\mathrm{RTT}} \frac{C}{\sqrt{p}}, \mathrm{IBR}\right\} \left[\frac{\mathrm{bit}}{\mathrm{s}}\right]$$
$$\gamma = (1 - p) \times \Gamma \left[\frac{\mathrm{bit}}{\mathrm{s}}\right] \quad (2.35)$$

When measuring the throughput at the transport layer, MTU has to be substituted with MSS and IBR has to be substituted with IBR $\times$ MSS/MTU in the previous formula.

Let us determine the minimum value of p for which (2.34) can be applied by considering the following condition where $C = \sqrt{3/2}$:

$$\frac{\mathrm{MTU}}{\mathrm{RTT}} \sqrt{\frac{3/2}{p}} = \mathrm{IBR} \Rightarrow \sqrt{\frac{3/2}{p}} = \mathrm{BDP} \quad (2.36)$$

Solving for p, we achieve the minimum value of p below which we can approximately consider that TCP reaches the maximum throughput/goodput.

$$p = \frac{3/2}{\mathrm{BDP}^2} \quad (2.37)$$

A more refined throughput/goodput formula than the square-root one, considering the effects of RTO expirations, buffer size (and the related maximum cwnd value), and the RTT variations during the cycle, has been provided in [59]. The details of this approach are beyond the scope of this book.

---

[13] More correctly, if $B = 0$, the maximum rate would be $3 \times \mathrm{IBR}/4$ and not IBR.

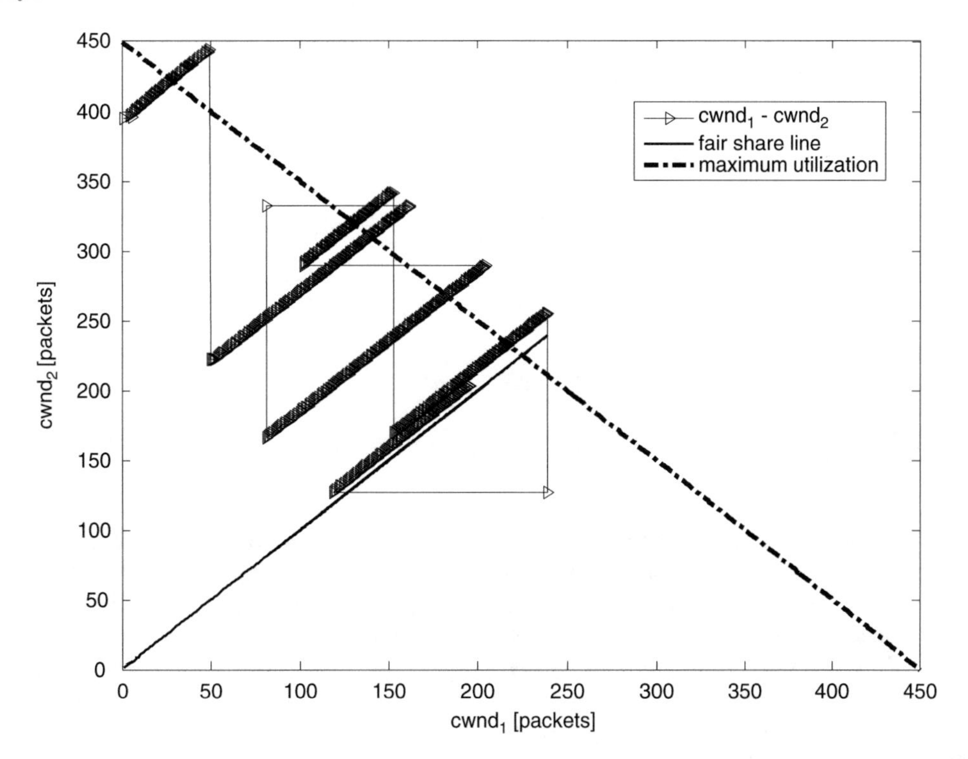

Fig. 2.51: Example of two TCP NewReno (AIMD) flows converging to a fair sharing of resources. When a TCP flow is fully exploiting a common bottleneck link, a second TCP flow is started: the two cwnds oscillate until they almost equally share the available bandwidth

TCP Fairness

Let us refer to a situation where two TCP flows share a common bottleneck link. Typically, a dumbbell network topology is considered for these studies. The congestion window values for these two flows are cwnd_1 and cwnd_2. Figure 2.51 plots the behavior of cwnd_2 versus cwnd_1 in a case where both flows are of the NewReno type (this is a graphical representation equivalent to that in Fig. 2.46). In Fig. 2.51, we have the following curves:

- The cwnd_2 curve versus cwnd_1 for the two flows sharing the bottleneck link.
- The *fairness line* with $\text{cwnd}_1 = \text{cwnd}_2$, representing the line along which a perfect resource sharing is achieved between the two competing flows.
- The *maximum utilization line* with $\text{cwnd}_1 + \text{cwnd}_2 = \text{cwnd}_{\max} = B + \text{BDP}$, representing the line along which the bottleneck link capacity is fully utilized.

In the beginning, flow #2 has a maximum cwnd_2 value and the other flow #1 is switched off ($\text{cwnd}_1 = 0$). Then, flow #1 is inserted and the two cwnds behave, so that flow #1 exploits the packet loss events of flow #2 to increase its cwnd_1 value. Some losses are synchronized because of the adoption of the drop-tail policy at the bottleneck link buffer. These events occur when both cwnd_1 and cwnd_2 have a sudden reduction. After some time, both cwnd values converge close to the fairness line and the maximum utilization line; this is an optimal situation.

Let us consider a case where n different flows share a bottleneck link. We refer here to the case that all these flows need the same bandwidth. Fairness can be measured through the *Jain fairness index*, which considers the average throughput (or the average goodput) of each flow Γ_i from $i = 1$ to n as

$$\Phi = \frac{\left(\sum_{i=1}^{n} \Gamma_i\right)^2}{n \sum_{i=1}^{n} \Gamma_i^2} \tag{2.38}$$

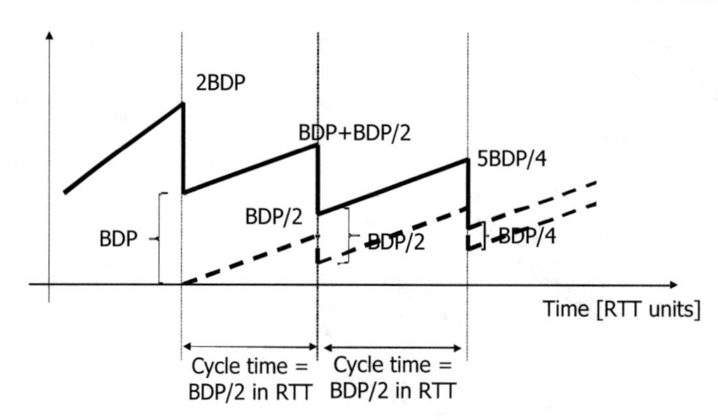

Fig. 2.52: Model for the behavior of two TCP NewReno flows sharing a bottleneck link

If all the n TCP flows sharing a bottleneck link achieve the same throughput (i.e., IBR/n at the IP layer), the Jain fairness index is maximum and equal to 1. Instead, the minimum fairness value $1/n$ is achieved when there is only one flow utilizing the whole bandwidth (IBR), and the other flows do not generate traffic.

Fairness can be considered among TCP flows of the same type or among TCP flows using different congestion control algorithms. In particular, *intra-protocol* (*inter-protocol*) *fairness* evaluates how TCP flows of the same variant (different TCP variants) interact with each other. We expect that different TCP flows of the same type reach a fair sharing of resources (e.g., Φ quite close to 1); this is an essential prerequisite for designing a good TCP variant. It is more challenging to achieve this goal in the presence of different TCP versions, where typically the most aggressive version prevails. For instance, we have inter-protocol fairness issues when the resources of the bottleneck link are shared between TCP Tahoe and TCP Reno/NewReno: TCP Tahoe is penalized. Another unfair case is when TCP CUBIC and TCP NewReno share a bottleneck link: TCP CUBIC takes almost the whole bandwidth.

Let us consider a different type of fairness related to concurrent TCP flows, sharing a bottleneck link, and experiencing different RTT values. The RTT value characterizes the cwnd growth time: the TCP flow with the shortest RTT value takes the shortest time to exploit the available bandwidth. A good TCP protocol should allocate the bandwidth fairly among those connections. The *RTT fairness index* is defined as the ratio of the throughputs of two flows with different RTTs; the optimal ratio would be 1, meaning that both flows equally share the available bandwidth. The MIMD algorithm (e.g., Scalable TCP) is RTT unfair: the TCP flow experiencing the lowest RTT grabs the full capacity. According to [62], the RTT fairness index of two TCP flows of the same type with different RTTs sharing a bottleneck link with synchronized losses can be expressed as $\Gamma_1/\Gamma_2 = (\mathrm{RTT}_2/\mathrm{RTT}_1)^{1/(1-d)}$, where d is a TCP protocol-related constant (e.g., $d = 0.5$ for TCP AIMD versions like TCP NewReno, $d = 0.82$ for HSTCP, and $d = 1$ for STCP[14]). The RTT unfairness increases with d.

Finally, we consider the coexistence of TCP flows and non-TCP flows (i.e., flows managed by other transport protocols than TCP). In this case, we speak of *TCP friendliness*, i.e., the capacity of non-TCP flows not to alter the behavior of TCP flows too much. A TCP-friendly flow means a flow that behaves like a TCP flow under congestion.

The *convergence time* is the time needed, starting from a single (elephant) TCP flow saturating the bottleneck link ($\mathrm{cwnd}_1 = \mathrm{cwnd}_{\max}$, $\mathrm{cwnd}_2 = 0$) to reach a condition where a new started TCP flow reaches a fair sharing of the bottleneck link capacity (i.e., $\mathrm{cwnd}_2 \approx \mathrm{cwnd}_1$). Let us estimate the convergence time in the case of TCP Reno/NewReno. We make the following simplifying hypotheses: (1) the shared bottleneck link has a buffer $B = \mathrm{BDP}$; (2) the second flow starts when the first one has the maximum cwnd value, $\mathrm{cwnd}_1 = \mathrm{cwnd}_{\max}$ (worst-case); (3) losses are synchronized because of the drop-tail policy adopted at the shared bottleneck link buffer; (4) both flows are in the congestion avoidance phase with linear cwnd increases on an RTT basis. Under these conditions, we may refer to Fig. 2.52, which describes a possible model for the behaviors of cwnd_1 and cwnd_2. In any case, this behavior is ideal since losses cannot always be synchronized. The cwnd values behave according to cycles with a duration equal to BDP/2 in RTT units. At the end of each cycle, the difference $\mathrm{cwnd}_1 - \mathrm{cwnd}_2$ is halved compared to the previous cycle. Hence, approximately $\log_2(\mathrm{BDP})$ cycles are needed to achieve convergence. The product of the number of cycles and the cycle duration yields the TCP NewReno convergence time T under

[14] Protocol-dependent constant d is used to generalize the square-root formula in (2.34) as $\Gamma = \frac{\mathrm{MTU}}{\mathrm{RTT}} \frac{C}{p^d}$.

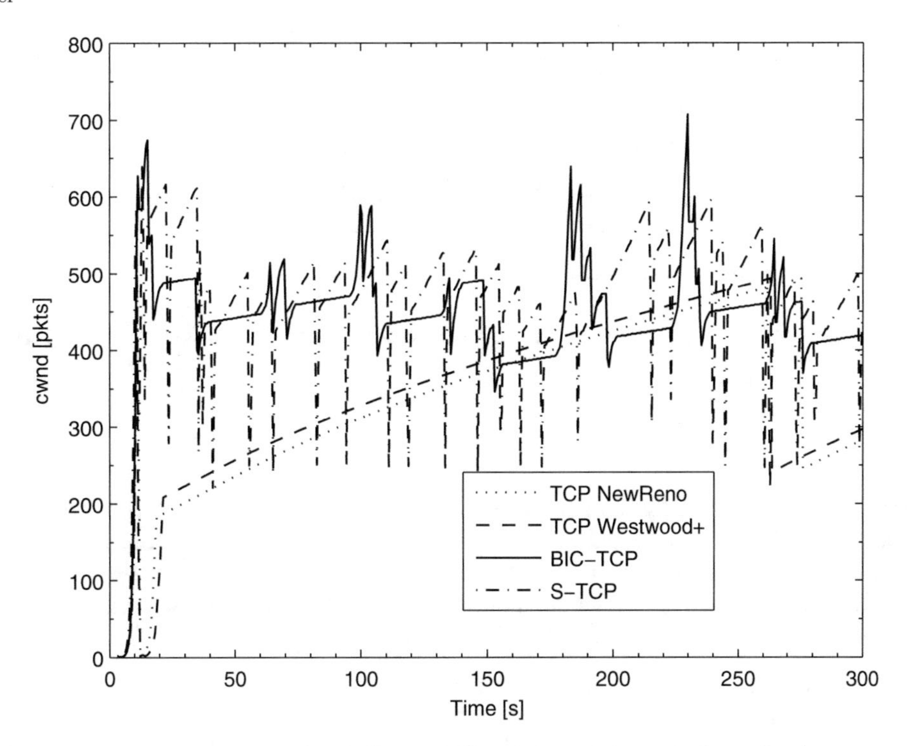

Fig. 2.53: Comparison of cwnd behaviors for different TCP versions (stand-alone flow): we can see that S-TCP and BIC TCP are more aggressive than Westwood+ and NewReno

our assumptions. Hence, the convergence time can be approximated as follows in the TCP NewReno (and Reno) case:

$$T \approx \frac{\mathrm{BDP}}{2}\log_2(\mathrm{BDP})\,[\mathrm{RTT}] \tag{2.39}$$

So far, we have considered fairness issues for "elephant" TCP flows. We note here that there are some fairness problems when elephant and mice TCP flows share a common bottleneck link. Elephant flows operate in TCP congestion avoidance phase, while mice flows primarily operate in the slow-start phase. When elephant and mice TCP flows share a bottleneck link, there can be some fairness issues: mice connections may not be able to exploit a sufficient bandwidth since it is mostly used by elephant connections. Older Web browsers (using HTTP 1.0) would open and close many consecutive short-lived connections to a Web server to fetch all the files ("objects") of a certain Web page. This approach entails that most of the connections used for Web browsing are in the slow-start phase, thus having a poor response time. To avoid this problem, modern Web browsers either reuse one persistent connection (starting from HTTP 1.1) for all the files requested from a particular Web server or open multiple connections simultaneously. In addition to this, the adoption of suitable RED policies at the routers could increase the traffic share of mice TCP flows against elephant TCP flows.

Comparisons of the cwnd Behaviors for Some TCP Versions (LFN Case)

Figure 2.53 shows the cwnd behaviors of some TCP versions, obtained with the NS-2 simulator [63], referring to an LFN scenario with BDP = 250 pkts, buffer size of the bottleneck link $B = 250$ pkts, and initial ssthresh (= initial receiver window) much larger than 600 pkts. With these settings, all the TCP versions have a sudden peak of cwnd at the beginning, well beyond the maximum system capacity of $B + \mathrm{BDP} = 500$ pkts. Hence, there is a massive loss of packets in the initial slow-start phase with the consequent occurrence of RTO expirations, as in NewReno (Impatient version) and Westwood+. Then, NewReno and Westwood+ have a short slow-start phase and a congestion avoidance one to increase slowly cwnd up to the maximum value of $B + \mathrm{BDP}$. On the other hand, BIC TCP and S-TCP exhibit very aggressive behaviors for cwnd.

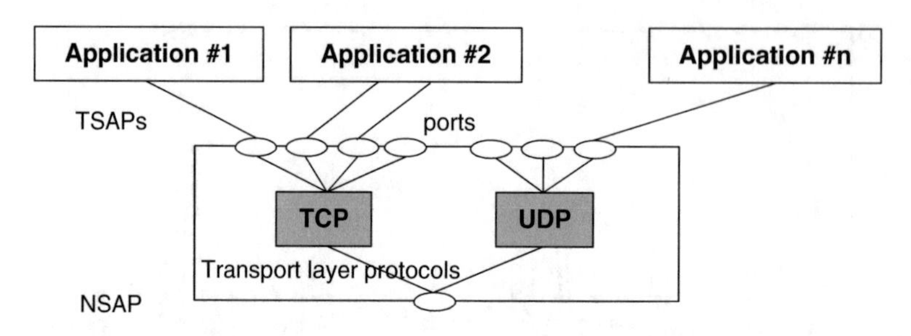

Fig. 2.54: Use of ports to distinguish among different applications running on top of transport layer protocols

2.8.2 UDP

UDP is a connectionless and unreliable transport layer protocol defined in RFC 768 [64]. The UDP protocol is extremely simple. Data from the application layer are handed down to the transport layer, where they are encapsulated into a UDP datagram with a small header of 8 bytes. The datagram is sent to the host without any mechanism to guarantee the safe arrival at the destination. The application program will have the task of checking for errors to try to recover them. UDP provides simple functions beyond those of the IP layer, as described below:

- *Port Numbers.* UDP uses 16-bit port numbers to let multiple processes to use UDP services on the same host. A UDP address is the combination of a 32-bit IP address and a 16-bit port number.
- *Checksum.* Unlike IP, UDP checksums its data and a pseudo-header (as that in Fig. 2.40) to verify their integrity. A packet failing checksum is simply discarded with no further action taken (i.e., no retransmission is requested by UDP).

2.8.3 Port Numbers and Sockets

Different applications run on the same device connected to the Internet. To distinguish among them, 16-bit port numbers have been adopted. These applications (named "services") can run on top of TCP or UDP. Source and destination port numbers are specified in both TCP and UDP headers. The Internet Assigned Numbers Authority (IANA) is responsible for maintaining the official assignments of port numbers for specific uses [65].

A *socket* is an Application Programming Interface (API), usually provided by the operating system, which allows applications to communicate using the protocol stack. Internet sockets are commonly based on the Berkeley sockets standard (BSD socket). A socket address is the binding of an NSAP (IP address) and a TSAP (port number). Several Internet socket types are available; for instance, *connectionless sockets* using UDP and *connection-oriented sockets* using TCP. Local and remote sockets involved in an end-to-end communication are called socket pairs.

TCP and UDP ports are assigned separately since the services provided by TCP and UDP are different. Both TCP and UDP receive requests from higher-layer protocols through TSAPs ports, provide a service, and send requests through the Network SAP (NSAP) for the IP protocol layer. See Fig. 2.54.

Port numbers are divided into three ranges: *System Ports*, also called "well-known" ports (0–1023), *User Ports* (1024–49151), and *Dynamic and/or Private Ports* (49152–65535). RFC 6335 specifies the different uses of these ranges of port numbers. System Ports are assigned by an IETF process for standard protocols. User Ports are assigned by IANA using a review process. Dynamic Ports are not assigned centrally but used only for custom or temporary purposes.

Well-known ports for TCP include
- Echo: 7
- FTP (control): 20
- FTP (data): 21
- Telnet: 23
- SMTP: 25
- HTTP: 80

Well-known ports for UDP include
- DNS: 53
- TFTP: 69
- NTP: 123
- SNMP: 161

Well-known port numbers are reserved across platforms. For instance, a Telnet application on a Windows computer will use TCP port 23 to access a Unix server. Moreover, a Web browser on a terminal will use TCP port 80 on the remote server hosting the desired Web site.

User Ports (1024–49151) and Dynamic and/or Private Ports (49152–65535) can be "ephemeral ports," meaning short-lived ports automatically allocated by the TCP/IP software. Port number 1024 is reserved. Ephemeral ports are adopted by TCP and UDP (as well as by other transport protocols) as the ports used on the client communicating with a well-known port on the server. However, ephemeral ports may also be used on servers. Let us consider two detailed examples.

Example #1: When writing custom client–server applications, port numbers can also be selected in the range 1025–65535. An example of a custom application would be an Internet game that needs to send game update messages to all players. The game server probably would use a port with the number 2000, and the clients would use a port with the number 2001.

Example #2: If two clients operating from the same IP address attach themselves to a server, the server needs to distinguish these two communications. This is achieved by the clients randomly picking two ports with numbers above 1024, say 1025 and 1026. TCP uses this method to multiplex different connections.

2.9 Voice Over IP

Telephone traffic today exploits IP networks. Moreover, end-users are becoming more accustomed to making voice calls through the Internet [66] using client software on computers (softphones), IP phones, and smartphones; this is what is called Voice over IP (VoIP). Today there are many VoIP providers. Some of them have built closed networks for users, offering free calls only within them. However, other VoIP providers have adopted another approach that allows dynamic interconnections between any two domains on the Internet whenever a user wishes to make a call.

The problem of voice transmissions over the Internet is how to reproduce a connection-oriented service in a connectionless IP network. VoIP uses the Real-time Transport Protocol (RTP) together with UDP. RTP/UDP/IP provides end-to-end network transport functions for real-time applications, such as audio and video for unicast and multicast services. A companion protocol, called Real-Time Control Protocol (RTCP), is used to provide out-of-band statistics and control information for an RTP flow. The main task of RTCP is to provide feedbacks on QoS; QoS issues for voice real-time services are addressed later in this section. However, most VoIP applications offer a continuous stream of RTP/UDP/IP packets without taking care of QoS issues.

VoIP calls are not restricted to phones directly served by the IP network (i.e., IP phone-to-IP phone calls) but can also be destined to classical PSTN telephones. In such a case, calls are routed through the IP network to a VoIP/PSTN Media Gateway near the destination telephone. For IP phone-to-IP phone calls, a softswitch is used to connect the calling party with the called party. VoIP softswitches are divided into Class 4 and Class 5 softswitches. Class 4 softswitches are used for routing large volumes of long-distance VoIP calls. Instead, Class 5 softswitches are intended to provide additional services to end-users.

VoIP signaling protocols are divided between Session Control Protocols and Media Control Protocols. Session Control Protocols (e.g., H.323 and SIP) are responsible for the establishment, preservation, tearing down of call sessions, and the negotiation of session parameters such as codecs, bandwidth capabilities, etc. The Media Control Protocols (e.g., MGCP and H.248/Megaco) are used to open and close media connections on VoIP gateways and to process notifications coming from those gateways. The Media Gateways interconnecting IP and PSTN networks are controlled by a Media Gateway Controller using a Media Control Protocol. The two most important protocols of this type are the Media Gateway Control Protocol (MGCP), defined in RFC 3435 and the H.248/Megaco protocol (RFC 3525 and ITU H.248.1). Telephones or gateways involved in setting up a call negotiate which codec to use from a small set of codecs they support. The best option is to code the speech once near the speaker and to decode it once near the listener. Transcoding of speech in the middle of the transmission path degrades the speech quality.

Standard speech codecs are available with output rates in the range from 5 to 64 kbit/s. The choice of a codec varies between different implementations: some implementations rely on narrowband and compressed speech, while others support high-fidelity stereo codecs. Some popular codecs include μ-law (USA, Japan)/A-law (EU) G.711 with Pulse Code Modulation (PCM) at 64 kbit/s, G.722 high-fidelity codec with Adaptive PCM (ADPCM) with a bit-rate up to 64 kbit/s, and the G.729 coded with Conjugate-Structure Algebraic-Code-Excited Linear-Prediction (CS-ACELP) at 8 kbit/s.

The design of a VoIP packet involves a trade-off between payload efficiency (payload size/total packet size) and packetization delay (the time needed for the codec to fill a packet). Since the RTP/UDP/IP header is 40 bytes in IPv4, a payload of 40 bytes would mean efficiency of 50%. Note that 40 bytes are filled in 5 ms at 64 kbit/s and in 40 ms at 8 kbit/s. A packetization delay of 40 ms is significant so that many VoIP systems use 20 ms packets, despite the low payload efficiency of lower bit-rate codecs. Voice coding and packetization entail delays for the voice that are typically larger than those experienced in terrestrial circuit-switched networks.

It is important that VoIP achieves a QoS level comparable to that experienced in the classical PSTN. The main QoS elements are *packet loss rate*, *delay* (latency), and *jitter* (packet delay variation). A large number of factors must be taken into account to achieve a high-quality VoIP call (e.g., codec type, packetization, packet loss rate, delay, jitter, network QoS support, call setup signaling protocol, call admission control, security concerns, and the ability to traverse firewalls). We can make the following general considerations for the QoS management of VoIP traffic.

- IntServ with RSVP is not well suited to VoIP. Firstly, since the bandwidth required for voice traffic is small, the RSVP control traffic would be a significant part of the whole traffic. Secondly, the RSVP router code was not designed to support many thousands of simultaneous connections per router, as expected for the large-scale use of VoIP.
- A better solution for VoIP QoS is to use the EF PHB of DiffServ, which is well suited to achieve scalability. DiffServ relies on a large-capacity network. EF would drop packets at the edge instead of queuing or rerouting them in the presence of network congestion.
- Finally, other possibilities for VoIP QoS are MPLS-TE plus DiffServ (DS-TE).

If VoIP is a small portion of the total traffic, DiffServ or MPLS-TE plus DiffServ may be sufficient. DS-TE promises more efficient use of an IP network carrying a large volume of VoIP traffic but entails greater operational complexity.

ITU-T H.323 [67] and IETF SIP [68] are very important standards for VoIP and advanced telephony services. Both H.323 and SIP adopt RTP to transport the voice. H.323 is quite close to classical telephony protocols (e.g., signaling is based on Q.931 of ISDN) and has attracted industrial interest. In the H.323 architecture, a Multipoint Control Unit (MCU) is responsible for managing video conferences. Interworking between IP network and PSTN is performed by Gateways (GW) for the media stream and by signaling gateways for SS7/IP signaling. The GateKeeper (GK) controls the endpoints (i.e., GW and terminal) of an H.323 domain. Endpoints must register with the GK and perform a call request with consequent call admission. The H.323 standards represent a complete framework, including specific solutions for QoS, security, and mobility support.

Assuming that some QoS mechanisms are supported by the IP network, VoIP platforms (GK or SIP server) must consistently control the transport elements of the network. This means that it is necessary to adopt a Media Gateway Controller (MGC) in H.323 and a Policy Server (PS) in SIP. A protocol such as Megaco/H.248 controls resources and QoS mechanisms (shaping, policing, and tagging) in GWs. In the same way, a PS will control access servers using, for instance, the COPS protocol.

More details on SIP and H.323 are provided in the following sub-sections.

2.9.1 Session Initiation Protocol

Session Initiation Protocol (SIP) is a session-layer transaction protocol that provides advanced signaling and control functionality to establish, modify, and terminate multimedia sessions such as VoIP, as specified in IETF RFC 3261 [68]. The primary SIP functions are location of resources/parties, invitation to service sessions, and session parameters negotiation. SIP can set up and tear down any type of session. With SIP, intelligence is pushed at the network edge, where processing capability is available in desktop computers.

SIP uses a URI (Uniform Resource Identifier) to denote a logical destination, not an IP address. The address could be a nickname, an e-mail address (e.g., sip:rossi@lab.ttl.edu), or a telephone number. A URI is a pointer to a resource, which can generate different responses at different times, depending on the input. A URI typically consists of three parts: the protocol used to communicate with the server (e.g., SIP), the name of the server (e.g., www.labttl.com), and the name of the resource.

While SIP user-agents (i.e., SIP phones) could have peer-to-peer communication without additional intermediaries, SIP servers are used to facilitate end-to-end communication when utilizing SIP as a public service. A *SIP server* (also called *SIP proxy* or *SIP proxy server*) is an intermediary entity that manages the setup of calls between SIP devices and controls call routing. A SIP proxy may also perform authorization, network access control, and some security tasks. SIP uses a request–response client–server transaction model, similar to HTTP. Each transaction starts with a request (in simple text format) invoking a server function (also called "method" that can be INVITE, ACK, BYE, etc.) and ends with a response. Clients send SIP requests, whereas servers accept SIP requests, execute the requested methods, and respond.

SIP can be used both to implement new services and to replicate traditional telephone services. Instant messaging is an example of SIP service. SIP facilitates mobility since a person can use different terminals with the same address and the same services.

The IP Multimedia Subsystem (IMS) is the selected architecture to provide voice services in fourth-generation (4G) and fifth-generation (5G) mobile systems. IMS is the core engine for the 4G VoLTE service and the 5G Voice over New Radio (VoNR).

2.9.2 H.323 Standard

H.323 was originally developed for video teleconferencing on IP networks [67]. The first version of H.323 was released in 1996, while the second version came into effect in January 1998. The current version of H.323 was approved in 2009. The standard encompasses both point-to-point communications and multi-point conferences. Built for packet-based networks, H.323 is well suited to IP networks. The H.323 standard was approved by the world governments as the international standard for voice, video, and data conferencing, defining how devices, such as computers, telephones, smartphones, wireless phones, video conferencing systems, can communicate.

H.323 can integrate with the Internet and the Web and interface with the PSTN to provide a range of applications, such as wholesale transit of voice, prepaid card services, residential voice/video services, and enterprise voice/video services. With H.323, users at remote locations can have a video call and simultaneously edit a document using their personal computers. Not only that, but H.323 allows the users to customize their phones or phone services, locate users, transfer a call, or perform any number of other tasks by using an HTTP interface between the H.323 client and a server in the network. A drawback of H.323 is the call setup time. Since H.323 first establishes a session and only after it negotiates the features and capabilities of that session, the setup of a call on average may take significantly more time than a PSTN call.

H.323 defines several network elements working together to deliver rich multimedia communication services. Those elements are terminals, MCUs, gateways, GKs, and border elements, as detailed below.

- *Multi-point Control Units*: An MCU is a conference bridge connecting the different users of a teleconference. However, H.323 MCUs are also capable of mixing or switching video, in addition to the normal audio mixing.
- *Gateways*: Media Gateways are devices that enable the communication between H.323 networks and other networks, such as legacy PSTN or ISDN networks. Suppose one party in a conversation is utilizing a terminal that is not an H.323 terminal. In that case, the call must pass through a media gateway to enable both parties to communicate.
- *Gatekeepers*: A GK is an optional component in the H.323 network, providing several services to terminals, gateways, and MCU devices. Those services include endpoint registration, address resolution, admission control, user authentication, etc. Among the GK's various functions, address resolution is the most important one since it allows two endpoints to contact each other without knowing the IP address of the other endpoint. GKs may operate in "direct routed" or "gatekeeper routed" mode. The direct routed mode is the most efficient and most widely deployed mode: endpoints utilize the Registration Admission Status (RAS) protocol to learn the IP address of the remote endpoint and a call is established directly with the remote device. Instead, in the

gatekeeper routed mode, call signaling always passes through the gatekeeper, which therefore has full control on the call and the ability to provide supplementary services.

- *Border Elements* and *Peer Elements*: A border element is a signaling entity (signaling gateway), which usually sits at the edge of an administrative domain and communicates with another administrative domain. A border element is used to communicate important things, such as access authorization, call pricing, or other data to enable the communication between two administrative domains.

2.10 New Internet Concepts

The Internet has become very important for everyday life in different areas, such as education, health, defense, commerce, travel, and entertainment. The Internet was not designed for its current level of use. Several critical shortcomings are now appearing in terms of performance, reliability, scalability, security, mobility, and QoS. The approaches towards the future Internet range from small, incremental evolutionary steps to completely new architectural principles (e.g., from the *client–server paradigm*, to cooperative peer structures, to the *publish–subscribe paradigm*).

The evolution of the Internet is based on new techniques such as cloud computing with the remotization of processing, network softwarization that facilitates the distribution and the control of different network functionalities. Today we have hardware with a simplified control layer and remote software in the cloud that can turn this device into a router with different programmable functions. Then, further steps are network virtualization that further facilitates the configuration and the adaptation of the system. Future networks will be loaded for a large percentage by streaming traffic. Therefore, caching techniques adopted in the different nodes will efficiently use network resources, offloading the Internet serves, and reducing the latency. In these new networks, QoS support will be essential. A new approach, called *network slicing*, will also be possible based on different system resources for different traffic classes and application types. A network slice comprises dedicated resources, e.g., processing power, storage, and bandwidth, and has isolation from the other network slices. This will be a fundamental approach for teletraffic engineering.

2.10.1 Core Transport Networks and Internet Hierarchical Architecture

Internet traffic is growing for many reasons, such as increasing content providers, diffusion of cloud applications and social networks, peer-to-peer applications, grid applications, streaming applications, etc. This trend requires the use of broadband geographical networks and IPv6.

The Internet structure is hierarchical, organized in layers or tiers: Tier 1, 2, and 3. A Tier 1 network is an IP network that can exchange traffic with other Tier 1 networks (IP transit or peering) without paying any fees for traffic exchange in either direction. A Tier 1 network or Internet Service Provider (ISP) has a global reach and can connect with all major networks. ISPs at this level (peers) connect to each other and allow traffic to pass for free. These ISPs build infrastructures and make use of Atlantic Internet sea cables. Tier 2 ISPs connect between Tier 1 and Tier 3 ISPs. They have regional or country reach. A Tier 2 network peers for free with some networks but still purchases IP transit or pays for peering to reach at least some Internet portion. Tier 3 ISPs are closest to the end-users in a country and connect to the Internet by charging some money. A Tier 3 network is a network that solely buys transit/peering from other networks to participate in the Internet. A Tier 3 network is a local network in a country. A user connects to the nearest Point-of-Presence (PoP) of an ISP.

A bilateral private peering agreement typically involves a direct physical link between two partners (IP transit). Traffic from one network to the other is then primarily routed through that direct link. Instead, peering between two (or more) networks involves an intermediate peering node, also called Internet Exchange Point (IXP). An IXP is a physical infrastructure through which Internet Service Providers (ISPs) and Content Delivery Networks (CDNs) exchange Internet traffic between their autonomous systems. IXPs reduce the portion of an ISP's traffic delivered via their upstream (higher-tier) transit providers, thereby reducing the average per-bit delivery cost. Furthermore, the increased number of paths available through the IXP improves routing efficiency and fault-tolerance. IXPs are distributed everywhere; for instance, in Italy, we have MIX in Milan, TIX in Florence, and NaMEx in Rome.

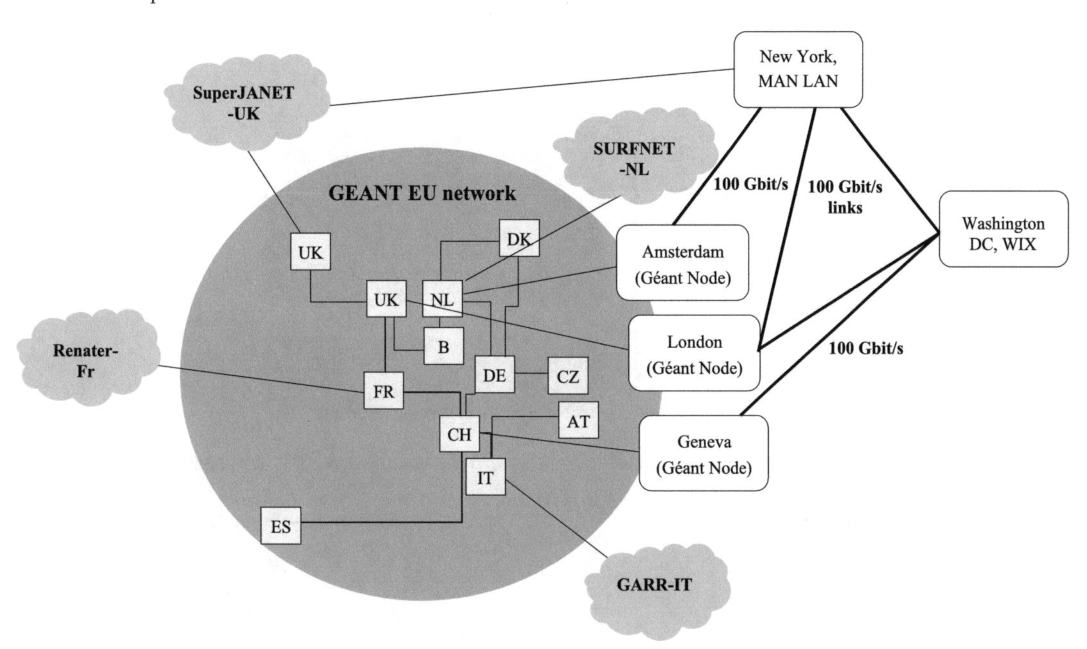

Fig. 2.55: Simplified representation of the GÉANT network (showing some PoPs with Juniper T-series routers and optical fiber links at multiple of 10 Gbit/s, till 100 Gbit/s), connecting NRENs in different European countries, and some transatlantic links to connect to other regions (peering). For instance, in North America, we have peering links with MAN-LAN and WIX to interconnect with the US Energy Sciences Network (ESNet)

An example of a Tier 2 network is represented by the GÉANT multi-gigabit network (GÉANT, a French word meaning "giant"), represents a high-performance pan-European network. GÉANT is the very high-capacity European backbone that interconnects European research and education networks [69, 70]. The most important applications for GÉANT are the sharing of massive amounts of data, distributed data processing, and distributed simulation. These are essential tasks to support worldwide research on physics, astrophysics, medicine, etc. GÉANT has a hierarchical multi-domain infrastructure, interconnecting many National Research and Education Networks (NRENs) of Tear 3, as shown in Fig. 2.55.

GÉANT connects 65 countries, 40 NRENs across Europe, and over 10,000 research and educational institutions in Europe, North America, Africa, and Asian-Pacific Region. Over 1000 terabytes of data are transferred every day via the GÉANT IP backbone (the year 2020). GÉANT is connected with each NREN through access links with capacities ranging from 1 Gbit/s up to 100 Gbit/s. For instance, the Italian NREN, called GARR ("Gruppo per l'Armonizzazione delle Reti della Ricerca"), is connected to GÉANT via 100 Gbit/s links. We have more than 100 PoPs in Italy. At the physical level, PoPs are interconnected using dark fibers. On top of the physical level, the network has two layers for transmission and IP/MPLS. The overall capacity between two backbone PoPs is at least 60 Gbit/s, with 100 Gbit/s on more recently implemented links, particularly those included in the GARR-X Progress project, and 120 Gbit/s between Bologna and Milan. GARR connectivity with MiX and NaMeX exploits multiple links at 100 Gbit/s.

There are basically two fundamental models for interconnecting networks (autonomous systems) serving large communities across geographical areas: the *peering* and the *hierarchical* models. The peering model is based on multiple interconnection agreements with providers with related Internet Exchange Points (IXPs). The hierarchical model is typical of some large communities with common objectives, such as public research communities. GÉANT adopts the hierarchical model. However, GÉANT exploits peering agreements to reach some networks outside Europe through suitable IXPs, as shown in Fig. 2.55.

In the GÉANT network, the access can be according to a Gigabit Ethernet network interface (see Sect. 6.5.1). It is possible to interconnect the different sites using a VLAN over the existing routers. Ethernet is the de facto standard for LANs. Using the same format on wide area networks can prevent Ethernet frames from being encapsulated/decapsulated at the source/destination. This is what is called "Carrier Ethernet."

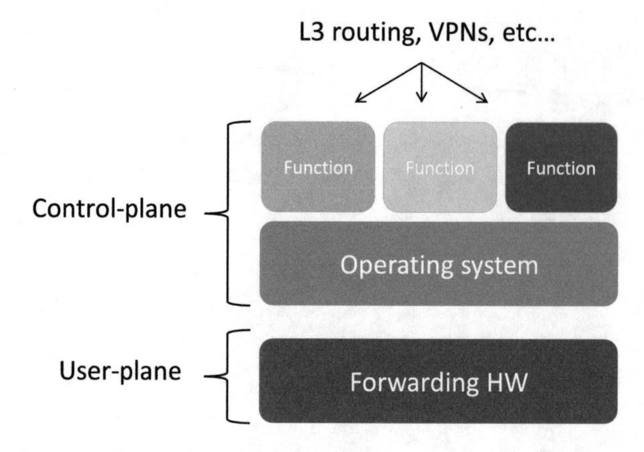

Fig. 2.56: The classical layered protocol stack in network devices: a closed platform

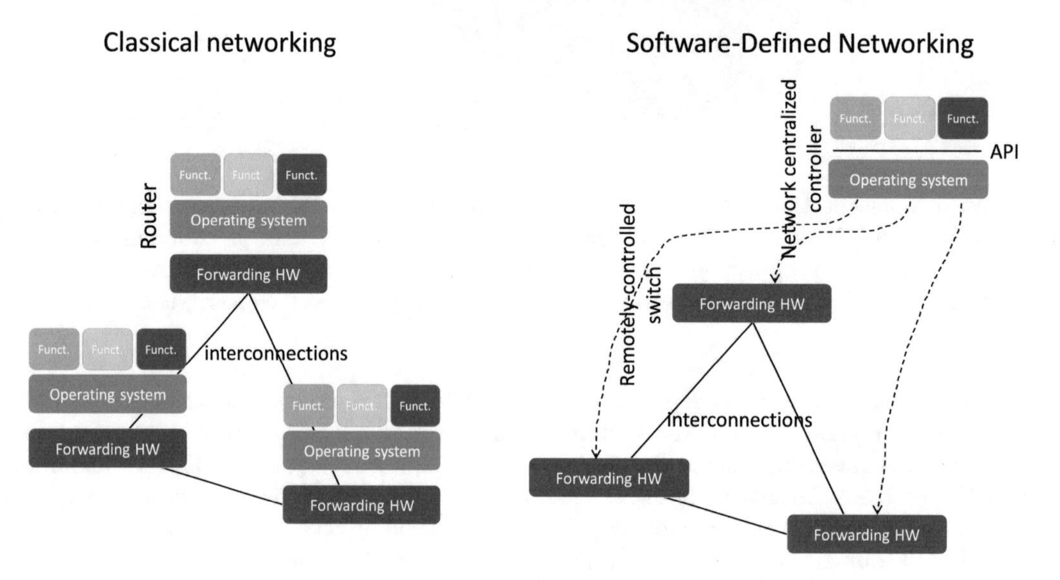

Fig. 2.57: The new paradigm for the networks based on SDN

2.10.2 Network Softwarization

Network protocols should guarantee interoperability, but the problem is due to too many non-interoperable standards. In traditional networking, each hardware has a protocol stack with a data plane and a control plane (see Fig. 2.56). The new paradigm with Software-Defined Networking (SDN) is to have a centralized control plane detached from the single hardware. Moreover, there is an open interface between the forwarding hardware and the centralized control. See Fig. 2.57. The following SDN description is made, having in mind its adoption in fifth-generation (5G) mobile networks [71]. Distributed routing protocols cause adverse effects when there are frequent changes in the network state: they need excessive update messages (large convergence times), thus having possible routing loops while converging. A centralized control could solve these problems.

The SDN technology has the following characteristics [72]: (*i*) decoupling the user plane (part of the network that carries the traffic) from the control plane (part of the network that determines the traffic route and provides the management); (*ii*) centralizing the control functions (network controller) from forwarding nodes: this centralized controller performs functions including routing decision, mobility management, user data management, authentication, and charging, etc.

The user plane is quite simplified with SDN and consists of stateless, distributed forwarding tables performing packet-switching at a very high speed. These tables are defined by a centralized control plane maintaining end-to-end path information and providing advanced functions such as mobility management, policy, and subscription control.

The OpenFlow controller maintains a centralized map of the whole network. Centralized control decisions are sent to the data plane that is responsible for translating them into management actions. Finally, the data plane forwards the packets using flow table match-action protocol.

SDN determines a programmable network infrastructure. By means of the SDN programmability, a network operator can dynamically program data forwarding layer devices to optimize network resource allocation. The network intelligence resides in software-based SDN controllers, and network devices can be configured externally through vendor-independent management software. SDN allows both the centralization of some management functions and the dynamic optimization of the system. The SDN architecture is organized into three different layers:

1. Application layer hosting the applications and communicating with the SDN-enabled controller(s) via a standardized Application Programming Interface (API).
2. Control layer for the definition of several virtualized networks (network slices) and their orchestration. This layer encompasses SDN controllers, network function virtualization manager, and service orchestrator(s) to control and manage both physical and virtualized network functions.
3. Physical network infrastructure layer including all the physical nodes and the transport network.

The SDN controller stores the full network information, including the network topology, dynamic changes of the network status, and global application requirements, such as QoS and security requirements. SDN has many advantages to support Traffic Engineering (TE) due to its distinguishing characteristics, such as isolation of control and forwarding, global centralized control, and network behavior programmability. We can deploy scalable global measurement tasks flexibly in the SDN, collecting real-time network status information and monitoring the traffic centrally in the controller. SDN network measurement parameters mainly include network topology parameters, network traffic parameters, and network performance parameters. The SDN controller detects the current network topology actively using the Link Layer Discovery Protocol (LLDP).

The forwarding equipment has a unified interface to communicate with the SDN controller, which does not depend on different equipment suppliers. The SDN controller can conveniently obtain network status data to schedule network traffic.

There can be multiple paths between the source and destination. The SDN controller maintains the global view of the use of each path in the network using various network measurement technologies. Then, SDN allows for path selection and scheduling of network traffic. In conclusion, SDN can be used to introduce TE techniques in the network, such as traffic load balancing, multipath routing, and QoS scheduling. Traffic application requirements can be considered globally so that flexible, granular traffic scheduling is possible.

OpenFlow is considered one of the first SDN standards [73]. It defines the communication protocol that enables the SDN controller to directly interact with the forwarding plane of network devices such as switches and routers, both physical and virtual, to better adapt to changing business requirements. SDN deals with switch abstraction and matching/action flow table; see Fig. 2.58. OpenFlow allows us to dynamically and centrally control the behavior of switches and routers in the network.

Many applications can be built on top of OpenFlow, and the following are some common examples: MPLS, MPLS-TE (see the previous Sect. 2.7), and VLAN.

OpenFlow has the following tasks:

- It creates a logical representation of the switches in the network so that the controller can process this information.
- It allows the controller to communicate with the switches securely.
- It defines the flow tables and flow entries to change the behavior of a switch and defines how packets are handled (dropped, forwarded, stripped, etc.)

2.10.3 Network Virtualization

The virtualization allows reproducing network functions as logical entities such as logical switches, logical routers, logical gateways, assembled in any topology. The network can be treated as a resource to be managed dynamically. This approach is fundamental to differentiate various data traffic classes, e.g., sensors, city cameras, self-driving cars, real-time communications, and assign resources based on business priorities and SLAs.

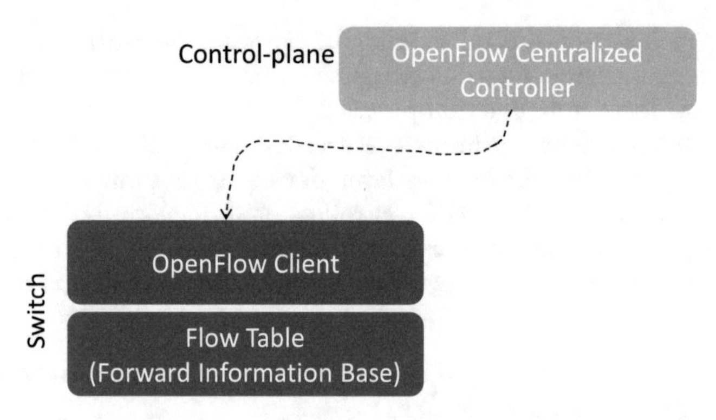

Fig. 2.58: OpenFlow switch abstraction

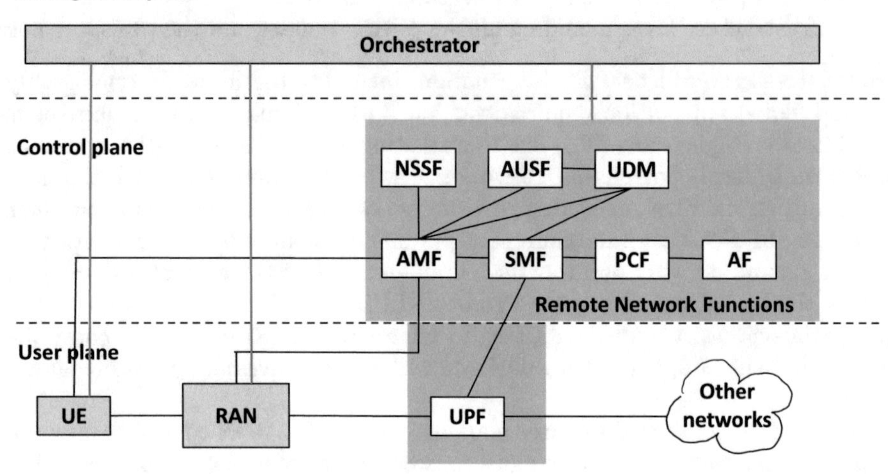

Fig. 2.59: Network virtualization: the case of 5G networks

Network Function Virtualization (NFV) is a framework for the virtualization of network functions (control plane) from layers 2–7, involving application servers, load balancers, routers, switches, databases, etc. The hypervisor is a software controller responsible for all servers' resources in terms of processing, computing, and storage among different virtual machines that are running network functions. These are the so-called Virtual Network Functions (VNFs) [74]. A further coordination among the VNFs is operated by the orchestrator.

5G mobile communication systems adopt the SDN/NFV approach. The organization responsible for the 5G standardization is the Third Generation Partnership Project (3GPP) that has defined the architecture shown in Fig. 2.59 [75]. We can see in this figure many network functions that are virtualized in the control plane, such as Authentication Server Function (AUSF), Access and Mobility Management Function (AMF), Session Management Function (SMF), etc. The acronym UE stands for the User Equipment and represents the mobile terminal/phone.

The complex architecture of upcoming 5G networks calls for an efficient management framework that provides a uniform and coherent orchestration of various resources across the multiple layers of the 5G ecosystem. The solution specified by ETSI is under the name of Management and Orchestration (MANO). It aims to decrease costs and complexity of implementing new services, maintain running services, and manage available resources [76].

The following VNFs are of particular interest for the 5G traffic management that is particularly relevant to this book:

- User Plane Function (UPF). It provides packet routing & forwarding, packet inspection, and QoS handling. UPF can interconnect with multiple networks: it is an anchor point for intra- and inter-radio access network mobility.

- Session Management Function (SMF). It supports session management (session establishment, modification, release), UE IP address allocation & management, DHCP functions, traffic steering configuration for UPF for proper traffic routing, etc.
- Application Function (AF). It is similar to an application server that interacts with the other network functions. AF can be used to influence traffic routing to steer the traffic towards external edge servers.
- Policy Control Function (PCF). It helps to deploy service policies in the 5G networks. PCF can support the following essential tasks: (*i*) 5G QoS policy and charging control function; (*ii*) Policy rules for network slicing, roaming, and mobility management; (*iii*) Collecting subscriber metrics for what concerns usage, applications, and more. The operators analyzing this information can optimize resources and tune their services.

Network virtualization facilitates the obtainment of network slices [77], that is, the generation of virtual subnetworks sharing the same physical infrastructure (e.g., Radio Access Networks, RANs). Network slicing allows the operator to provide customized networks. Each slice entails an independent set of logical network functions optimized to provide the resources for the specific service and traffic that will use the slice. One network can support one or several network slices. Network resources are computing, storage, and networking. The slicing approach allows to dynamically share these resources among the active slices. These slices can have different requirements on functionality (e.g., priority, charging, policy control, security, and mobility), in performance (e.g., latency, mobility, availability, reliability, and data rates), or they can serve only specific users (e.g., public safety users, corporate customers, roamers, or hosting a Mobile Virtual Network Operator). Network slicing facilitates traffic engineering concepts since different traffic classes use different, distinct portions of the resources. A network slice can provide the functionality of a complete network, including radio access network functions and core network functions (e.g., potentially from different vendors).

There are different software platforms to implement virtualization. OpenStack is a free and open-source software platform for cloud computing, mostly deployed as Infrastructure-as-a-Service (IaaS), whereby virtual servers and other resources are made available to customers [78]. The software platform consists of interrelated components that control diverse, multi-vendor hardware pools of processing, storage, and networking resources throughout a data center. Users either manage it through a Web-based dashboard or command-line tools.

2.10.4 Internet-Centric Networking

Globally, IP video traffic will account for 82% of the whole Internet traffic by 2022. With video growth, Internet traffic is evolving since the Internet is increasingly used for information dissemination, rather than for peer-to-peer communications between end hosts.

The basic idea of Information-Centric Networking (ICN) is to enrich network layer functions with content awareness so that routing, forwarding, caching, and data transfer operations are performed on topology-independent content names rather than on IP addresses: addressing is name-based. Differently from the current Internet that assigns hosts/interfaces IP addresses and transfers data packets among endpoints using IP addresses, ICN addresses contents (instead of hosts/interfaces) using names that are independent of their locations [79].

ICN's primary objective is to shift the current host-oriented communication model toward a content-centric model. It relies on location-independent naming, in-network caching, and name-based routing to distribute contents in the network [80]. With ICN, contents are requested directly from the Web, instead of routing to a host/server address that needs to be known *a priori*.

An efficient naming scheme is important for ICN since *data retrieval is based on names assigned to data.* Naming schemes are application-specific and globally unique. There are three different types of naming methods in ICN: Hierarchical, Flat, and Attribute-Value Based names. Hierarchical names of the type "A/B/C/D" provide the benefit of aggregation, potentially leading to a smaller routing table size. Hierarchical names can facilitate the possibility to specify the contents that we like in a more structured way.

On the other hand, the lookup of flat names is more efficient than hierarchical names: a simple hashtable/bloomfilter lookup can determine the next hop for a content request. ICN entails a new type of routers called Content Routers (CR). ICN is based on caching. In particular, in-network caching so that we have distributed caches within the network for the efficient delivery of contents, being them stored closer to the end-users.

NFV and SDN's growing deployment allows for the rapid insertion of new routing techniques such as ICN. ITU-T SG 13 has defined Recommendation Y.3071 on Information-Centric Networking [81].

Different projects have been carried out on ICN that have proposed different possible approaches. Among them, Data-Oriented Network Architecture (DONA), Publish-Subscribe Internet Technology (PURSUIT), and Named Data Networking (NDN). More details are provided in the following sub-sections.

2.10.4.1 DONA

DONA adopts a flat naming scheme and a name resolution scheme using a distributed set of network entities, called Resolution Handlers (RH), with caching capabilities to route requests towards the nearest copies of data. Each information object in DONA belongs to a principal (in what follows also called "publisher"), uniquely identified by a public–private key pair. Names are of the flat type as $P : L$, where P is a hash value of the principal's public key and L is a label attributed by the principal to ensure that names are globally unique in the network. The principal is responsible for organizing the structure of labels when naming the objects that it manages.

The RH is responsible for maintaining information about the availability and the location of contents in the network. Hence, a principal, needing to publish contents, has to send a REGISTER message to its local RH. Upon receiving this message, this RH will store a pointer to its principal for this information. Then, it will propagate the REGISTER message upwards hierarchically as well as to its peers. By doing so, the RH located at the top-most in a given domain are aware of all the registrations inside the network.

Whenever a user wants to receive data, it will send a FIND message to its local RH; this message is propagated hierarchically through RHs until matching is found. Upon a match, there are two possible operations to route the data: coupled and decoupled operation. In coupled operation, the reverse path is used for data routing. Note that the reverse path is created during the forwarding of the FIND message to reach the publisher. Instead, in decoupled operations, once the FIND message reaches the corresponding publisher, the data can be sent directly from the publisher to the user through a regular IP routing/forwarding mechanism. This operation requires maintaining global IP addresses to perform classical routing and forwarding.

There are mainly two types of caching: *on-path caching* and *off-path caching.* If an RH wants to cache a given content with on-path caching, it has to substitute the incoming FIND message IP address with its IP address before forwarding it to the next RH. This ensures that the response data will go through this RH. The RH can thus store the data traversing it. Instead, off-path caching is possible when a given content is available through the local RH, thus utilizing RH's available storage. In this case, a principal can inform its local RH about the availability of the contents.

In DONA, a mobile user changing its location can reissue a new FIND message to the closest RH from its current location. Also, publisher mobility is supported.

2.10.4.2 PURSUIT

From the conceptual point of view, PURSUIT implements three core functionalities of ICN: RendezVous (RV) function, Routing [or Topology Management (TM)], and Forwarding. The RV function acts as a middleman between the subscriber and the publisher. It is responsible for matching the interests (subscriptions) to publications (information objects). The RV nodes can be interconnected via links, forming an RV network. Routing involves determining a path from the publisher to the subscriber, which is used to deliver contents. Here information objects are identified by unique identifiers (IDs): the RV ID and the scope ID. Published information is organized in *scopes.* Scopes can be physical networks, e.g., a university network, or even logical networks. Scope and RV IDs are flat names, meaning they are location independent. Note that information objects can belong to multiple scopes, but they need to carry a unique RV ID within their scope.

PURSUIT adopts the publish/subscribe paradigm. Before publishing contents, publishers have to locate the nearest (local) RV node responsible for managing the desired scope. Once found, a publisher should send a PUBLISH message with the corresponding scope ID. For a subscriber to access contents, he/she must be aware of both its RV ID and scope ID. Then, the subscriber issues a SUBSCRIPTION message towards its local RV. After, the RV will instruct the TM to create a delivery path from the publisher to the subscriber. This route will be sent to the publisher in a START message. Finally, the route will be used by the publisher and by the consecutive intermediate forwarding nodes to forward the content to the desired destination. Along this route, contents may be cached on-path. In case that more than one subscriber requests a specific RV ID, a multicast tree is created to deliver the contents. For off-path caching, a publisher has to advertise the available information to the RV network so that these contents can be taken from local caches.

In this architecture, subscriber mobility is regarded as a two-dimensional problem. The first dimension of the problem concerns the scale of the mobility, which can be local or global. The second dimension of the problem reflects how the mobility is handled by the architecture, that is, statically or dynamically. In a static/local mobility scenario, the mobile terminal needs to issue a new subscription message to its local RV from the new location. This subscription message will ultimately trigger the TM to create a new route. In global subscribe mobility, subscriber mobility is managed by modifying the TM function of the architecture. The publisher mobility is harder than subscriber mobility. It requires that the publisher contacts the new RV and updates the TM about its new position.

2.10.4.3 NDN

In NDN, there are two types of packets. An INTEREST packet that the client (user) sends out to request for content. Then, a DATA packet is sent as a response from the content provider or producer. NDN uses a hierarchical naming scheme that is structured and human-readable. User and content provider must agree upon name construction. Subscribers issue INTEREST packets to retrieve the information they need from the Internet. Once the INTEREST packet reaches an intermediate cache or (at the end) the source that contains the data requested, a DATA packet is sent back. NDN is one of the most common ICN approaches that are available today.

Each NDN node (router) maintains three data structures and a forwarding policy detailed as follows:

- Pending Interest Table (PIT) storing all the INTERESTs that a router has forwarded but not yet satisfied. Each PIT entry records the data name carried in the Interest, together with its incoming and outgoing interface(s).
- Forwarding Information Base (FIB): this is a routing table, mapping name components to interfaces. The FIB itself is populated by a name-prefix based routing protocol and can have multiple output interfaces for each prefix.
- Content Store (CS): a temporary cache of data packets the router has received. Because an NDN packet is meaningful, independently of where it comes from or where it is forwarded, it can be cached to satisfy future INTERESTs. The cache replacement strategy is traditionally Least Recently Used (LRU).
- Forwarding Strategies: a series of policies and rules about forwarding INTEREST and data packets. Note that the Forwarding Strategy may decide to drop an INTEREST packet in certain situations, e.g., if all upstream links are congested, or the INTEREST is suspected to be part of a denial-of-service attack. The default forwarding strategy is the Best Route forwarding strategy.

When an INTEREST packet arrives at an NDN node, i.e., content provider server or an intermediate router, the NDN node checks whether content exists in its cache (i.e., CS) using the content name. If there is a match, a DATA packet is sent back to the user using the same interface from which the INTEREST has arrived. If not, the NDN node checks its PIT table. If the entry exists in the PIT, the INTEREST packet is dropped, and the NDN node records the incoming interface of this INTEREST in the PIT entry. This means that the user already made a content request; hence, this request will not be forwarded again. On the other hand, if it does not exist, a new arrival interface is added to the PIT, and the NDN node determines where to forward the INTEREST packet by checking the FIB table and using the Forwarding Strategy. When the INTEREST packet reaches the source, the DATA packet is generated and sent back to the user on the same path using the reverse path from the PIT. After a certain PIT lifetime, the corresponding PIT entry is removed.

2.10.5 Network Coding

In contrast to source coding, where only data sources can send encoded data, Network Coding (NC) allows nodes in a network to (re-)encode packets before sending them instead of adopting a simple store-and-forward routing approach [82]. NC improves multicast capacity in wired networks, reduces bandwidth requirements for data replications in distributed cache systems, and reduces file-downloading latency in Peer-to-Peer (P2P) file-sharing systems.

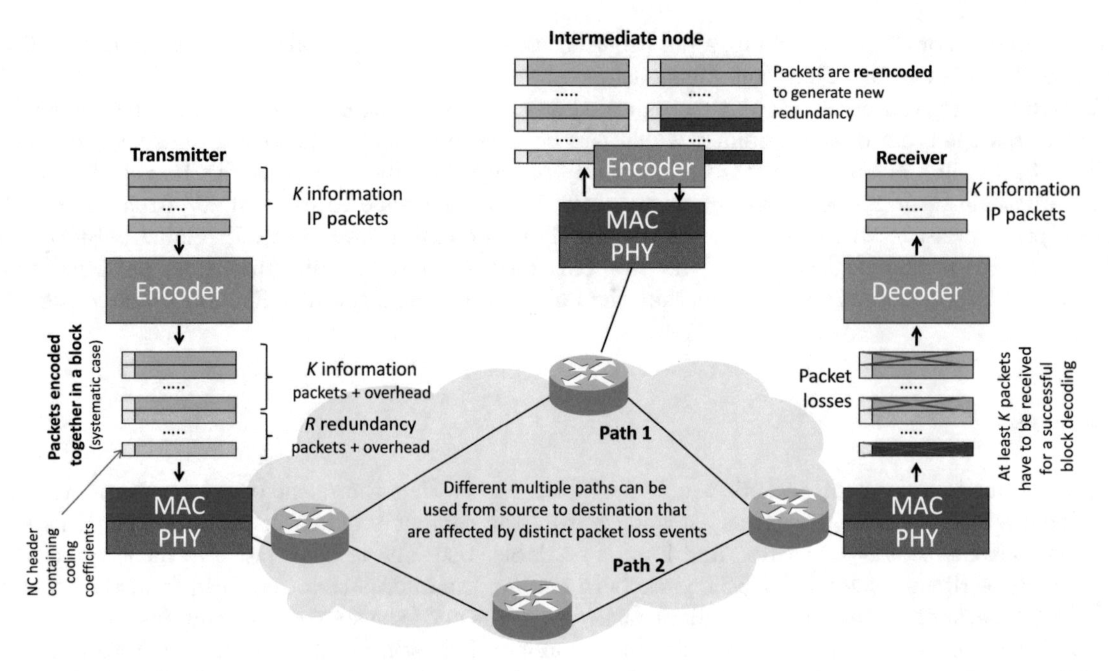

Fig. 2.60: NC scheme inside the network and the possibility to re-encode at intermediate nodes

Let us consider a system that acts as an information relay, such as a router, a node in an ad hoc network, or a node in a peer-to-peer distribution network. Traditionally, when forwarding an information packet destined to some other nodes, we have to repeat the packet, applying the Kirchhoff's law "what goes in, goes out." Instead, network coding allows the output of a node to be a function of the input information: the node can combine several packets it has received into one or several outgoing packets. Moreover, opportunistic routing can improve unicast throughput in wireless mesh networks compared to traditional single-path routing protocols. In opportunistic routing, all neighboring nodes of a transmitter may overhear the data packet and cooperate in forwarding the packets to its destination. The use of network coding in this scenario allows combining the packets reaching the destination from several neighboring nodes without taking care of the exact packet sequencing and being robust to packet losses.

NC operates at the level of packets' binary representation. In particular, encoding can be performed at the level of single bits (XOR operations over Galois Field of size 2) or at the level of groups of q bits, that is, symbols of size 2^q in Galois Fields. When $q = 8$, symbols are bytes, and this is quite convenient for implementations.

NC is different from PHY layer encoding because it deals with the loss of entire packets and not of the errors of bits. NC can be typically applied to packets on top of the MAC layer. Fountain codes are a class of erasure Forward Error Correction (FEC) codes with the property that a potentially infinite sequence of encoded packets can be generated from a given set of source packets (*rateless erasure code*); moreover, the original source packets can ideally be recovered from any subset of the encoded packets of size equal to or only slightly larger than the number of source packets. The term fountain or rateless refers to the fact that these codes do not have a fixed code rate. Random Linear Network Coding (RLNC), Luby Transform (LT) codes, and Raptor/RaptorQ codes are examples of Fountain codes [83]. In particular, Raptor and RaptorQ codes are detailed in RFCs 5053 [84] and 6330 [85], respectively.

NC was initially proposed to achieve reliable data delivery at the maximum data transfer rate in multicast scenarios (single source and multiple receivers) [86]. Using NC techniques, multiple users can recover different lost packets using just one retransmission [87]. NC application areas are:

- Reliability,
- TCP and congestion control,
- Multipath routing,
- Security.

The recent RFC 8975 deals with network coding for satellite systems.

Packets can be lost (erasure) because of congestion or because the MAC layer CRC discards packets with residual errors not recovered by the PHY.

In what follows, we refer to Linear Network Coding (LNC) and actually to RLNC that has unique features, as explained below.

Many characteristics differentiate NC from a common packet-level FEC. In particular, only RLNC can allow the *recoding* of an encoded packet (or a group of packets) at intermediate nodes (see Fig. 2.60). The recoded packet can be used together with first-coded packets to decode the same block. Recoding is performed without decoding the packets. Recoding is particularly suited for both distributed caching and multipath routing. For instance, in the second case, an encoded block is sent via two paths. We can have recoding at intermediate nodes along the paths to add redundancy and cover possible packet losses. This way, the receiver can decode the block if enough encoded packets are received, it does not matter which path they followed.

There are many options on which OSI layer to apply network coding, as:

- Between IP and MAC layers,
- Between transport and IP layers,
- Between application and the lower layers.

These NC alternatives entail different pros and cons and compatibility issues with other protocols, such as IPsec, header compression, etc.

We have intra-flow (or inter-flow) network coding when coding is applied over payloads belonging to the same flow (or to multiple flows). This differentiation also depends on the layer where network coding is applied. We can roughly consider that intra-flow network coding is possible when network coding is applied at higher OSI layers (i.e., transport level and above); instead, inter-flow network coding can be adopted when network coding is applied at lower layers (network-level and below). The decision between these two alternatives is a trade-off between performance gains and operation complexity.

NC can be performed block-by-block or on a sliding window basis [88]:

- In block coding, the original sequence of packets is divided into blocks of a given size, and NC is performed only by applying a transformation block-by-block.
- In sliding window coding, coding blocks are selected based on a sliding window over the stream of packets. Coding blocks are partly overlapping.

In the following description, we refer to unicast traffic applications, even if most of these considerations can also be applied to multicast traffic.

Let us consider encoding a group of packets, where each packet consists of B bits. When the packets to be combined do not have the same size, the shorter ones are padded with trailing 0s. We can interpret q consecutive bits of a packet as a symbol over the Galois Field GF(2^q), with each packet consisting of a vector of $L = B/q$ symbols (assuming B/q is an integer). With LNC, outgoing packets are linear combinations of the original packets, where additions and multiplications are performed over GF(2^q).

Arithmetic in a Galois field (finite field) is different from integer arithmetic. With a Galois field GF(2^q), addition, subtraction, multiplication, and division operations are defined over the numbers 0, 1, ..., $2^q - 1$ in such a way that if a and b are elements of the field, then also $(a + b)$, $(a - b)$, $(a \times b)$, and (a/b) are elements of the field. An element of GF(2^q) can be represented by q bits as $(a_{q-1}, a_{q-2}, \ldots, a_1, a_0)$ and can also be represented in a polynomial form (of degree $q-1$) over GF(2) as $A(x) = a_{q-1}x^{q-1} + a_{q-2}x^{q-2} + \ldots + a_1 x + a_0$. Then, operations can be made using polynomials. Further details on GF(2^q) arithmetic are beyond the scope of the present book.

Let us assume that a block of original packets $P_1, P_2, \ldots, P_K$ has to be sent by one source. In LNC, the ith encoded packet of a block is associated with a sequence of coefficients $\alpha_{i,1}, \ldots, \alpha_{i,K} \in$ GF(2^q) and is equal to $\sum_{j=1}^{K} \alpha_{i,j} P_j$ [89]. This encoded packet also needs a header containing the coefficients $\{\alpha_{i,1}, \ldots, \alpha_{i,K}\}$ that will be used at the receiver to decode the payload, containing the bits of $\sum_{j=1}^{K} \alpha_{i,j} P_j$. The coefficients in the header represent the so-called *encoding vector*. The receiver collects all the encoding vectors of the block of encoded packets and forms the matrix $\mathbf{M} = \{\alpha_{i,j}\}$ of the encoding coefficients.

If the coefficients $\alpha_{i,1}, \ldots, \alpha_{i,K}$ are selected randomly and independently with a uniform distribution over GF(2^q), then we have RLNC.

Let $P_j = \{p_{j,1}, \ldots, p_{j,L}\}$ denote the jth native packet to be sent by the source and $p_{j,\ell} \in$ GF(2^q) be the ℓth symbol of the jth packet. We consider that K source packets are encoded at a node to generate N encoded packet F_i, $i = 1, \ldots, N$ as

$$F_i = \sum_{j=1}^{K} \alpha_{i,j} P_j \quad i = 1, \ldots, N \quad \leftrightarrow \quad \mathbf{F} = \mathbf{M}\,\mathbf{P} \tag{2.40}$$

where $\mathbf{F}$ is the vector of encoded packets (one packet per row) and where $\mathbf{P}$ is the vector of source packets (one per row). Note that all the operations are carried out on symbols belonging to GF(2^q).

We can represent the matrix operation $\mathbf{F} = \mathbf{M}\,\mathbf{P}$ in (2.40) with more details at the level of GF(2^q) symbols as follows:

$$\begin{pmatrix} f_{1,1} & \cdots & f_{1,L} \\ \cdots & \cdots & \cdots \\ f_{N,1} & \cdots & f_{N,L} \end{pmatrix} = \begin{pmatrix} \alpha_{1,1} & \cdots & \alpha_{1,K} \\ \cdots & \cdots & \cdots \\ \alpha_{N,1} & \cdots & \alpha_{N,K} \end{pmatrix} \begin{pmatrix} p_{1,1} & \cdots & p_{1,L} \\ \cdots & \cdots & \cdots \\ p_{K,1} & \cdots & p_{K,L} \end{pmatrix} \tag{2.41}$$

The Gauss–Jordan elimination method can be used to invert matrix $\mathbf{M}$ of the encoding coefficients to decode the block and retrieve the K source packets. Row elimination methods are adopted to remove dependencies among rows. Matrix inversion can be carried out on square matrices. The elimination method applied to the matrix is also meant to remove redundant rows to achieve the matrix square form. Matrix $\mathbf{M}$ is $N \times K$, with $N \geq K$. To decode the original K packets, the rank of matrix $\mathbf{M}$ has to be equal to K, the maximum one. Even working on GF(2^q), the matrix inversion methods are formally the same as in the classical arithmetic. We need $N > K$ if we like that the coding can recover $N - K$ packet losses (erasures). A good choice for the GF size, which works efficiently and provides good performance, is from GF(2^8) to GF(2^{16}). The RLNC decoding technique presented here is referred to, in literature, as batch coding: it needs that the entire batch of encoded packets is received before decoding. A Gaussian elimination variant called "progressive decoding" [90] allows partial decoding after receiving every packet, instead of waiting for all packets of an encoded block to arrive.

RLNC with Gauss–Jordan elimination has decoding complexity of O(K^3). Typical K sizes for RLNC codes are in the order from 10 to 100 packets. Larger block sizes could entail long decoding times that could still be acceptable for some data applications but not compatible with real-time services. To make a comparison with the complexity of packet-level FEC codes, we can consider Raptor and RaptorQ codes that have the advantage that encoding and decoding complexities are O(K).

Consider that for a generation of K source packets, $K+\delta$ coded packets are sent. If more than δ packets are lost, the matrix of received coding coefficients will not be full rank, and no packets can be decoded. In this case, we lose the full block of K packets even if without network coding some of these packets could be delivered correctly at the destination. Hence, in this case, network coding is not so efficient. This problem can be avoided if a *systematic network coding* is adopted [91]. For each generation of K original packets, $K + \delta$ packets are sent, where the first K packets are precisely the original information packets. The packets received of these K will be delivered to the higher layer, irrespectively of the receipt of other packets of the same block since they are already decoded. The last δ packets (redundant packets) are generated using the RLNC scheme so that they can be used to recover the packet losses.

RLNC can be implemented without the use of feedback [92] or with feedback [93]. In this second case, if the decoding of an RLNC block fails, there is the possibility to ask for further redundancy to be transmitted by the sender (feedback scheme); this is possible because RLNC codes are fountain-type. The sender needs to hold in a buffer the block of the packets P_1, P_2, ..., P_K until it receives the confirmation of the correct delivery.

A sub-case of NC is the so-called packet-level FEC, where there is no possibility of recoding the packets at intermediate nodes, but the encoding is only end-to-end. With packet-level FEC, we can use the same types of coding as NC. Let us consider the following study on packet-level FEC performance.

We transmit packets that are composed of n bits on a noisy channel affected by independent errors with Bit Error Rate denoted as BER. Considering that a packet is discarded (erasure) even if it contains a single bit error (use of a CRC code at the packet level), we have to express the probability of successful packet delivery. Let us add a packet-level FEC to improve the successful transmission of packets. This FEC code is operated at the level of packets. We refer to the systematic case where an encoded block is formed of the K original (information) packets plus further R encoded packets (redundancy). We assume to use an "ideal FEC" so that the code can correctly decode the blocks of $K + R$ packets if there are no more than R packet erasures. In a practical code implementation, we could have worse performance due to the real working of the encoder and decoder that operate in a Galois field with specific rules to perform multiplication, addition, subtraction, and division.

A packet is received with errors (and then it is discarded, causing an erasure) with probability p expressed as

$$p = 1 - (1 - BER)^n \tag{2.42}$$

We consider now adding packet-level FEC working on blocks of $K + R$ packets. The number of packet erasures e in a block follows a binomial distribution as follows:

$$\text{Prob}\{e = i\} = \binom{K+R}{i} p^i (1-p)^{K+R-i} \tag{2.43}$$

The mean number of packet losses per block is equal to $(K + R)\,p$. The variance of the number of packet losses per block is $(K + R)\,p\,(1 - p)$ so that the standard deviation results as $\sqrt{(K + R)\,p\,(1 - p)}$.

After FEC decoding, a block is correctly decoded if the number of packet errors is up to R. This occurs according to the following probability P_{Bsuc}:

$$P_{Bsuc} = \sum_{i=0}^{R} \text{Prob}\{e = i\} = \sum_{i=0}^{R} \binom{K+R}{i} p^i (1-p)^{K+R-i} \tag{2.44}$$

If the code is systematic, when the decoding is unsuccessful, we can still deliver some info packets of the block containing erasures according to the packet loss rate p. Then, in this case, we can study the probability of successful packet delivery after FEC decoding, P_{psuc}, conditioning on two cases depending on the comparison of e and R:

$$P_{psuc|\{e \le R\}} = 1 \quad \text{and} \quad P_{psuc|\{e > R\}} = 1 - p \tag{2.45}$$

We remove the conditioning through probability P_{Bsuc} as

$$P_{psuc} = P_{Bsuc} + (1-p)(1 - P_{Bsuc}) = 1 - p(1 - P_{Bsuc}) \tag{2.46}$$

In the case of non-systematic coding, the probability $P_{psuc} \equiv P_{Bsuc}$.

Complete analysis in the RLNC case can be found in [94]. It can be proved that RLNC codes have a performance quite close to the ideal packet-level FEC for enough large q values (for instance $q = 8$ or 16).

We can adopt the following rule-of-thumb to select the redundancy R to protect the codeword adequately from the packet losses. For the sake of simplicity, let us consider a continuous relaxation for the R value. Then, R should be set to cover the mean number of packet losses in a codeword plus some margins due to the standard deviation of the number of packet losses in a codeword. We consider the following condition:

$$R = (K + R)\,p + \psi\sqrt{(K + R)\,p\,(1 - p)} \tag{2.47}$$

where ψ is a coefficient that can be set within 1.5–2 to cover a large percentage of loss configurations.

We can solve this equation for R so that we determine the optimal R value, denoted as R_{opt}. Note that this is a second-order equation for which only the positive solution is acceptable as

$$R_{\text{opt}} \approx \frac{p\left(2K + \psi^2\right) + \psi\sqrt{4Kp + p^2\psi^2}}{2\,(1 - p)} \tag{2.48}$$

where we have to round R to the closest integer.

Let us consider a numerical example with the following settings: $BER = 0.005$ and $n = 100$ so that $p = 0.394$. The behavior of P_{psuc} is shown in Fig. 2.61 for $K = 50$ and 70 and considering both the systematic and the non-systematic cases. We see that we need enough redundancy to reduce the packet loss from the floor level $1 - p$, the successful packet probability we get without encoding. The systematic scheme has better performance than the non-systematic one for lower values of R. These results are in good agreement with the R_{opt} value in (2.48): using $\psi = 1.5$, we have $R_{\text{opt}} = 61$ pkts for $K = 50$ pkts and $R_{\text{opt}} = 82$ pkts for $K = 70$ pkts.

Finally, Fig. 2.62 compares the probability of successful packet delivery after FEC decoding for the ideal packet-level FEC and RLNC with different q values using the same K and R values in the systematic case. The RLNC performance has been obtained according to the analysis in [94]. These results prove that the ideal packet-level FEC provides an upper bound to the RLNC performance. However, the RLNC performance gets closer to the ideal bound as q increases: the RLNC performance is very close to the ideal FEC with $q = 8$.

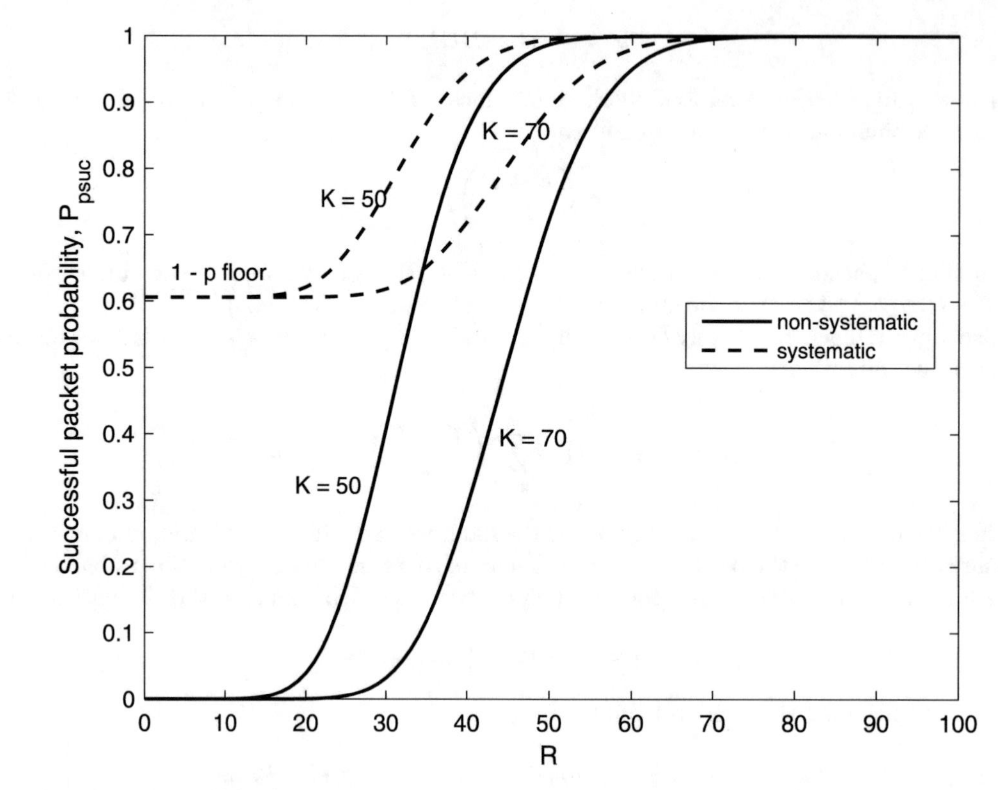

Fig. 2.61: Ideal packet-level FEC performance for $BER = 0.005$ as a function of the redundancy degree R in pkts with $n = 100$ ($p = 0.394$) and different K values

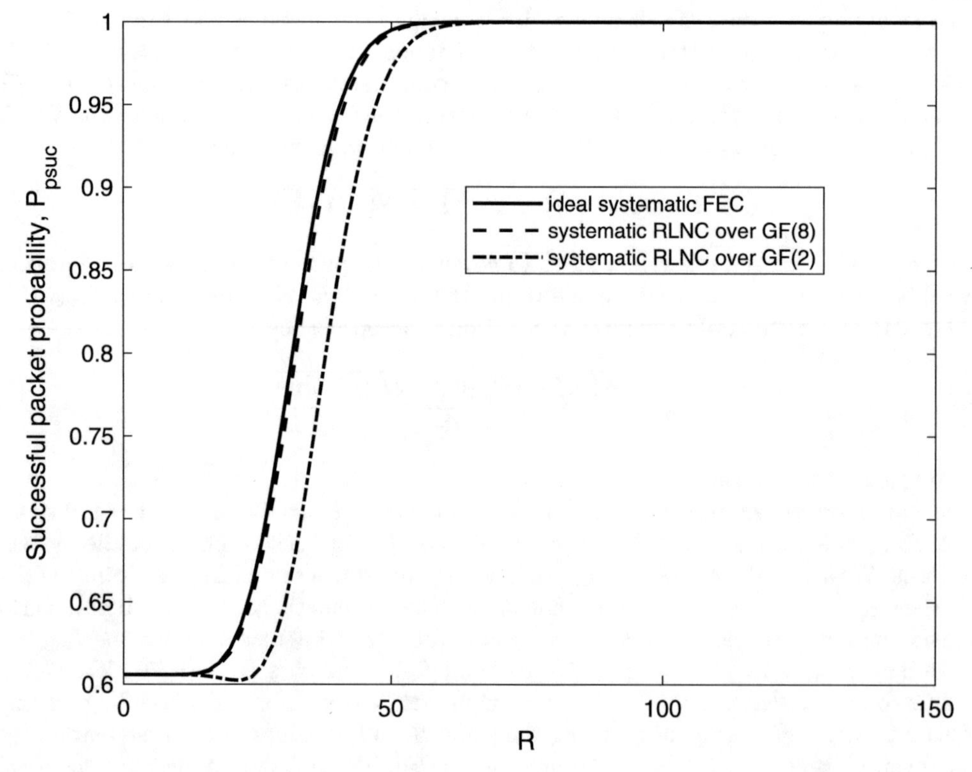

Fig. 2.62: Ideal packet-level FEC performance vs. RLNC for $BER = 0.005$, $n = 100$ ($p = 0.394$), and $K = 50$ pkts

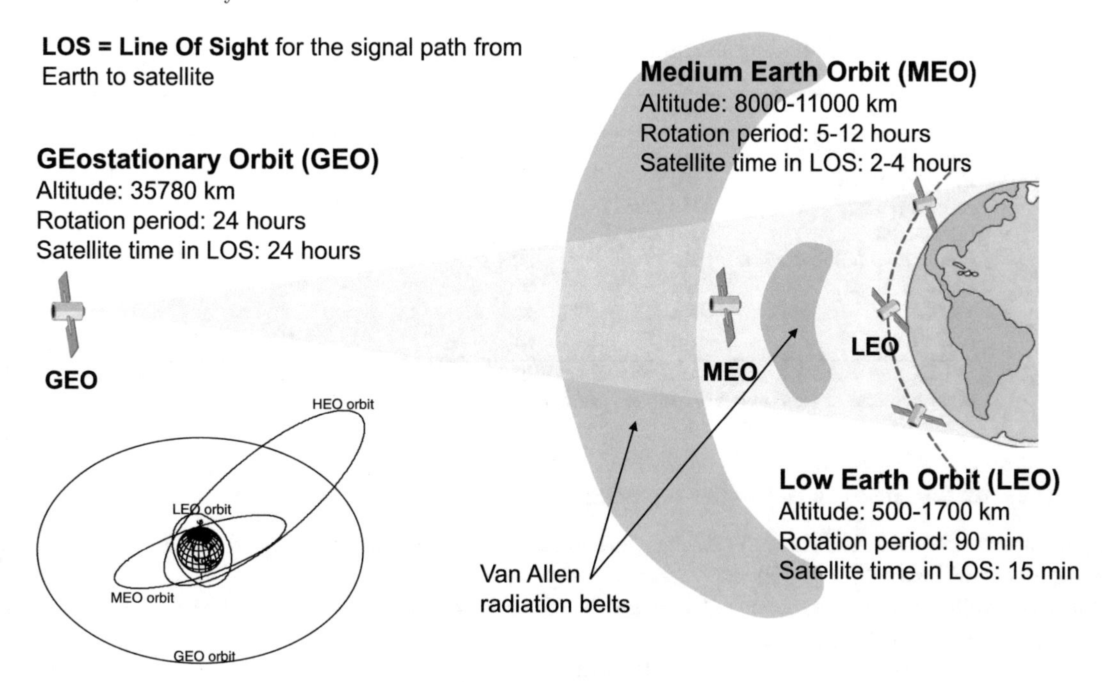

Fig. 2.63: GEO, MEO, and LEO orbital options for satellite communication systems

2.11 Satellite Communication Systems

Satellite communication systems represent an attractive solution to provide high bit-rate services to users in remote areas where the population has no access to high-speed Internet. Nevertheless, when interconnected with local or geographical networks, satellite networks can be the bottleneck of the entire system because of the high propagation delays and throughput limitations. Hence, obtaining the maximum performance from the satellite segment is crucial to reduce the cost of these services. The main advantages of satellite communications are [95]:

- Wide coverage area (a single GEO satellite can provide communication services to almost one-third of the Earth's surface).
- Rapid deployment of new services in broad areas, including developing countries.
- Easy provision of both broadcast and multicast high bit-rate services.
- Integration/Internetworking with terrestrial fixed and wireless networks for joint service support (heterogeneous/hybrid networks).
- Provision of Internet access to people on flights, trains, and ships.
- Provision of backup communications in the presence of emergencies.

Different satellite orbits are available, as detailed below (see Fig. 2.63). GEO satellites are on an equatorial plane at an altitude of about 35,800 km. They have a synchronous motion with respect to the Earth (i.e., 24-h orbital period) so that they are stationary for a user on the Earth. Three GEO satellites are sufficient to cover the whole Earth, except Polar Regions. The RTD values between a GEO satellite and an Earth terminal are about 250 ms when the satellite is at the zenith. This RTD value doubles if the communication is between a terminal and a gateway via satellite. Medium Earth Orbit (MEO) may be circular or elliptical, and its altitude is around 10,000 km. A global system needs a constellation of few tens of MEO satellites. RTD values between the satellite and an Earth terminal are in the range of 85–100 ms for minimum elevation angles greater than 30°. Low Earth Orbit (LEO) satellites are at lower altitudes from 500 to 2000 km. These systems are characterized by constellations of more than 40 LEO satellites with RTD values ranging from 5 to 40 ms for typical minimum elevation angles (from 8° to 40°).

The coverage area of a satellite is divided into many cells (each irradiated by an antenna spot-beam from the satellite) to concentrate the energy on a small area. It is also possible to shape the area served on the Earth. The adoption of multi-beam satellite antennas allows reusing the same frequencies among beams, which are sufficiently spaced, thus increasing the traffic volume carried out by a satellite system.

Table 2.7: Main characteristics of some GEO satellites for communications

Satellite type	Band (user link)	Number of beams
Inmarsat-4 (BGAN)	L	256
HotBird 9, 13C	Ku	1
Amazonas 2	Ku and C	3
Intelsat 23	Ku and C	5
Inmarsat-5 (Global X-press)	Ka	89 + 6
HYLAS 1	Ka	8
ViaSat-1	Ka	72
KaSAT	Ka	82

BGAN and Global X-press are IP-based satellite systems providing mobile services. The other systems in this table are for fixed communication services and TV broadcast services

The following satellite network topologies are possible:

- Transparent satellite star architecture. Terminal-to-terminal communications need two hops via satellite, thus involving a central hub; RTD $\geq$ 500 ms.
- Transparent satellite mesh architecture. Terminal-to-terminal communications need one hop via satellite, being switched on the satellite RTD $\geq$ 250 ms.
- Regenerating satellite mesh architecture. With respect to the previous approach, the satellite is able to decode, correct, and re-encode the signal at the PHY layer; in this case, terminal-to-terminal communications need one hop via satellite.

Transparent satellites are also called bent-pipe satellites.

In satellite networks, traffic flows can adopt the ITU-T Y.1541 QoS classes (see Sect. 2.5). The layer 3 QoS approach currently adopted in satellite networks is based on DiffServ, where queuing and traffic management are for aggregate traffic flows. The DiffServ approach is considered in the ESA Satlab recommendations and is implemented, for instance, by the BGAN (Broadband Global Area Network) satellite system of Inmarsat [96]. According to the satellite industry standpoint, between 4 and 16 queues are manageable to support different traffic classes at the IP level.

To improve the TCP performance in satellite networks (LFN case), Performance Enhancing Proxies (PEP) can be adopted. These are transport layer PEPs, which are interposed between TCP sender and receiver. Other types of PEPs are operating as proxies (local caches) at the application layer. A transport layer PEP typically implements a TCP split technique: the PEP intercepts the segments of a given TCP flow before they reach the satellite link; then, the PEP immediately sends the ACKs back to the sender. This approach reduces the RTT and improves the TCP goodput. Two PEP architectures are possible:

- *Integrated, single PEP*: The PEP is located at the gateway so that the TCP connection, established between the end hosts, is split into two parts. The first connection (between Web server and PEP) uses a standard TCP version and is terminated at the PEP. The second connection (between PEP and end-user) is via satellite and can adopt an enhanced TCP version, compatible with a standard TCP receiver.
- *Distributed PEPs (two PEPs)*: The TCP communication is split into three parts: the PEPs are used to delimit the satellite network where an accelerated TCP version is adopted.

Table 2.7 describes the main characteristics of some GEO satellites [97]. Most of these satellites use Ku and Ka bands, which allow large bandwidths for broadband applications, but suffer from atmospheric events (e.g., clouds, rain events). While fixed services use C and K frequency bands, mobile services are well suited to lower L and S frequency bands assigned at the World Administrative Radio Conference (WARC) 1992.

2.11.1 Satellite Geometry, Number of Satellites per Constellation, Latency Issues

In this sub-section, some analytical tools are provided to describe the distance from users and satellites, the area covered by a satellite under some minimum elevation angle requirement, and the minimum number of satellites to cover the whole Earth under a minimum elevation angle and satellite altitude constraints.

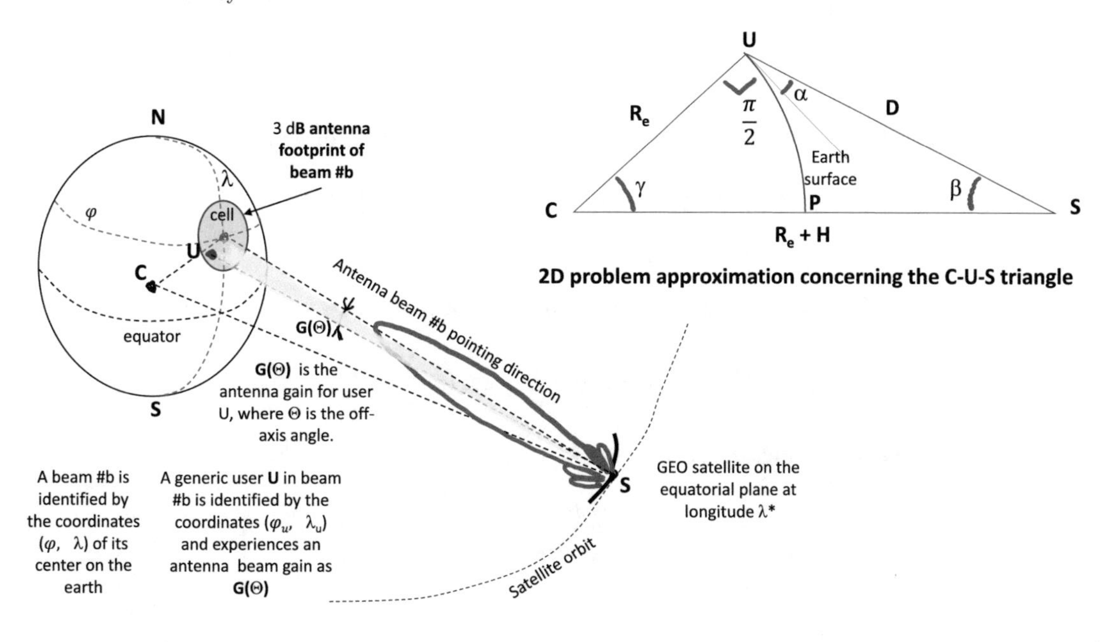

Fig. 2.64: Geometry for the communication from a satellite; the direction of the maximum antenna gain (in this case from the satellite) is called antenna boresight

The system geometry is shown in Fig. 2.64, where we see a satellite S whose antenna beam #b illuminates a cell on the Earth where there is a user U. The cell center generated by the satellite has coordinates with latitude and longitude as (φ, λ), respectively. The user U belonging to this cell has coordinates (φ_u, λ_u). We refer here to a GEO satellite on the equatorial plane with a longitude λ^* and an altitude $H \approx 35{,}800\,\text{km}$. However, many of the following considerations and formulas can also be applied to any satellite with generic altitude H. We also consider the average Earth radius as $R_e = 6378\,\text{km}$. Let C denote the Earth's center. We consider the triangle C-U-S in the space. In this triangle, angles α, β, γ are related to each other according to a spherical-geometry 3D problem; complex formulas are provided in [98].

If we can assume C-U-S on the same plane cutting the Earth along a great circle, the 3D geometrical problem becomes a classical 2D trigonometric problem. This occurs when the GEO satellites have the same longitude as the user: $\lambda^* \equiv \lambda_u$. In this circumstance, the results we get with the simplified 2D problem are correct; otherwise, they are approximate with respect to the actual 3D problem. β is called *nadir angle*; α is called *elevation angle*; γ is called *central angle*. The elevation angle cannot be below about 10° for reliable communications (otherwise, the line-of-sight link to the satellite can be obstructed). Therefore, GEO satellites cannot cover polar regions. This 2D geometry can also be used for MEO or LEO cases if user U is on the orbital plane of the satellite S. The 2D geometry approximation can be used to determine the maximum area covered by a satellite at generical altitude H and with a minimum elevation angle requirement.

Still referring to a generic user (observer) U, we can define one more angle based on the 3D spatial geometry. This is the *azimuth angle*, the angle from the user's North direction, and the direction to point the satellite. Then, user U identifies the position of the satellite S using the satellite elevation and the azimuth angle. Finally, referring to the satellite orbit, the Right Ascension of the Ascending Node (RAAN) is the angle taken on the equator plane from the vernal point to the satellite orbit's ascending node in a counter-clockwise direction.

Let us present some basic results that are applicable to the problem of satellite communication to users in visibility on the Earth.

Sine Laws

We apply the sine law to the triangle C-U-S as

$$\frac{D}{\sin(\gamma)} = \frac{R_e}{\sin(\beta)} = \frac{R_e + H}{\sin\left(\frac{\pi}{2} + \alpha\right)} \tag{2.49}$$

As an elaboration of this formula, we can also achieve the following system of equations:

$$\begin{cases} \frac{R_e+H}{\sin\left(\frac{\pi}{2}+\alpha\right)} = \frac{R_e}{\sin(\beta)} & \Rightarrow \quad \frac{R_e+H}{\cos(\alpha)} = \frac{R_e}{\sin(\beta)} \\ \frac{D(\gamma)}{\sin(\gamma)} = \frac{R_e}{\sin(\beta)} \end{cases} \tag{2.50}$$

We use the first equation in (2.50) to express β as a function of α as follows:

$$\beta = \text{asin}\left\{\frac{R_e}{R_e + H}\cos(\alpha)\right\} \tag{2.51}$$

Otherwise, we can express α as a function of β as follows:

$$\alpha = \text{acos}\left\{\frac{R_e + H}{R_e}\sin(\beta)\right\} \tag{2.52}$$

Sum of Internal Angles of a Triangle

$$\gamma + \beta + \frac{\pi}{2} + \alpha = \pi \quad \Rightarrow \quad \gamma + \beta + \alpha = \frac{\pi}{2} \tag{2.53}$$

Then, solving for γ and using (2.51), we get

$$\gamma = \frac{\pi}{2} - \alpha - \beta = \frac{\pi}{2} - \alpha - \text{asin}\left\{\frac{R_e}{R_e + H}\cos(\alpha)\right\} \tag{2.54}$$

This formula relates the central angle to the elevation angle. The central angle γ is important because it characterizes the side L of the hexagonal deployment of beam footprints on the Earth [99]. The side L on the sphere surface is equal to $L = R_e \times \gamma$.

Cosine Law and the Slant Range

The distance between the satellite in S and the user in U is the so-called *slant range* that can be determined as follows using the cosine law applied to the triangle C-U-S as

$$D = \sqrt{R_e^2 + (R_e + H)^2 - 2R_e(R_e + H)\cos(\gamma)} \tag{2.55}$$

Typically, we know angle β because it can represent half of the satellite beam angular amplitude, characterized as the antenna beam half-power amplitude, θ_{3dB}. We also know R_e and H. The other parameters can be derived as a function of these three main parameters. Note that the maximum β value, $\beta_{\max}$, corresponds to the visible area to a satellite that can be obtained from (2.51) for $\alpha = 0$ (the ideal limit for the satellite visibility, actually the satellite is on the horizon):

$$\beta_{\max} = \text{asin}\left\{\frac{R_e}{R_e + H}\right\} \tag{2.56}$$

Then, β can vary from 0 to $\beta_{\max}$. Correspondingly, the elevation angle α varies from $\pi/2$ (satellite at the zenith) to 0.

Area Covered by a Satellite under a Constraint on the Minimum Elevation Angle

Let us refer to a satellite antenna beam that points right below the satellite. The area covered by the satellite antenna beam with a certain half-power beamwidth θ_{3dB} ($\beta = \theta_{3dB}/2$) can be expressed assuming a hexagonal layout for the beams footprints with side L. On the spherical Earth's surface, the cell generated by this beam has a side $L = R_e \times \gamma_c$. According to (2.54), we have

$$\gamma_c = \frac{\pi}{2} - \frac{\theta_{3dB}}{2} - \text{acos}\left[\frac{R_e + H}{R_e}\sin\left(\frac{\theta_{3dB}}{2}\right)\right] \quad (2.57)$$

where the minimum elevation angle in the cell is $\alpha_{\min} \equiv \text{acos}\left[\frac{R_e+H}{R_e}\sin\left(\frac{\theta_{3dB}}{2}\right)\right]$.

The cell corresponding to the beam footprint on the Earth has an area that can be approximated as that of a planar hexagon: $A_{cell} = \frac{3}{2}\sqrt{3}L^2$.

In the following study, we consider a generic satellite orbital altitude H (not only GEO), and we assume a certain minimum elevation angle $\alpha_{\min} > 0$ (system requirement). Then, each satellite provides the coverage of a hexagonal macro-cell with side $R_e \times \gamma_{\max}$, where $\gamma_{\max}$ is obtained from (2.54) with $\alpha = \alpha_{\min}$:

$$\gamma_{\max} = \frac{\pi}{2} - \alpha_{\min} - \text{asin}\left\{\frac{R_e}{R_e + H}\cos\left(\alpha_{\min}\right)\right\} \quad (2.58)$$

Then, the whole coverage area of a satellite is

$$A_{satellite} = \frac{3}{2}\sqrt{3}\left(\gamma_{\max}R_e\right)^2 \quad (2.59)$$

We can determine the minimum number of satellites for a global coverage dividing the Earth surface $4\pi R_e^2$ by $A_{satellite}$ as

$$Minimum\ number\ of\ satellites\left(\gamma\right) \approx \frac{4\pi R_e^2}{\frac{3}{2}\sqrt{3}\left(\gamma_{\max}R_e\right)^2} = \frac{8\pi}{3\sqrt{3}\gamma_{\max}^2} \quad (2.60)$$

This result approximates the derivations shown in [99], referring to hexagons of coverage on the sphere representing the Earth. The graph in Fig. 2.65 shows the behavior of the minimum number of satellites to cover the whole Earth for a range of satellite altitudes H typical of LEO satellite systems and assuming different possible values of the minimum elevation angle α_{min}. The number of actual satellites can be larger than the minimum in (2.60) if we consider multiple satellite visibility (that is, the coverage areas of the different satellites overlap), as assumed in the mega-LEO constellations that are presented in the next Section.

Maximum Propagation Delay

Let us refer to γ_{max} in (2.61), that is the maximum angle to cover the whole area visible from the satellite at an altitude H and with corresponding minimum elevation angle α_{min}. The maximum distance for the link from the user to the satellite is experienced under the condition with minimum elevation angle α_{min}. Using the γ_{max} value, we adopt the slant range formula in (2.55) to express the maximum distance in the cell from the satellite, $D_{\max}$:

$$D_{\max} = \sqrt{R_e^2 + \left(R_e + H\right)^2 - 2R_e\left(R_e + H\right)\cos\left(\gamma_{\max}\right)} \quad (2.61)$$

Then, the maximum one-way propagation delay $T_{\max}$ from the Earth to the satellite is obtained considering that the signal travels at light speed for the distance $D_{\max}$:

$$T_{\max}\left(H,\ \alpha_{\min}\right) = \frac{D_{\max}\left(H,\ \alpha_{\min}\right)}{c} \quad (2.62)$$

The behavior of $T_{\max}$ in milliseconds as a function of the minimum elevation angle $\alpha_{\min}$ and the satellite altitude H (LEO case) is shown in Fig. 2.66. We can see that $T_{\max}$ decreases with the increase in the minimum elevation

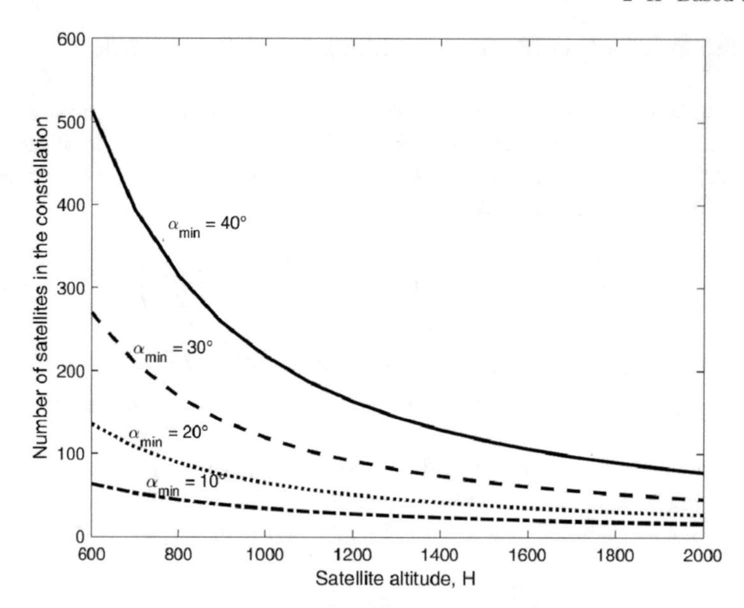

Fig. 2.65: Minimum number of satellites in LEO orbit depending on the constellation altitude and the required minimum elevation angle

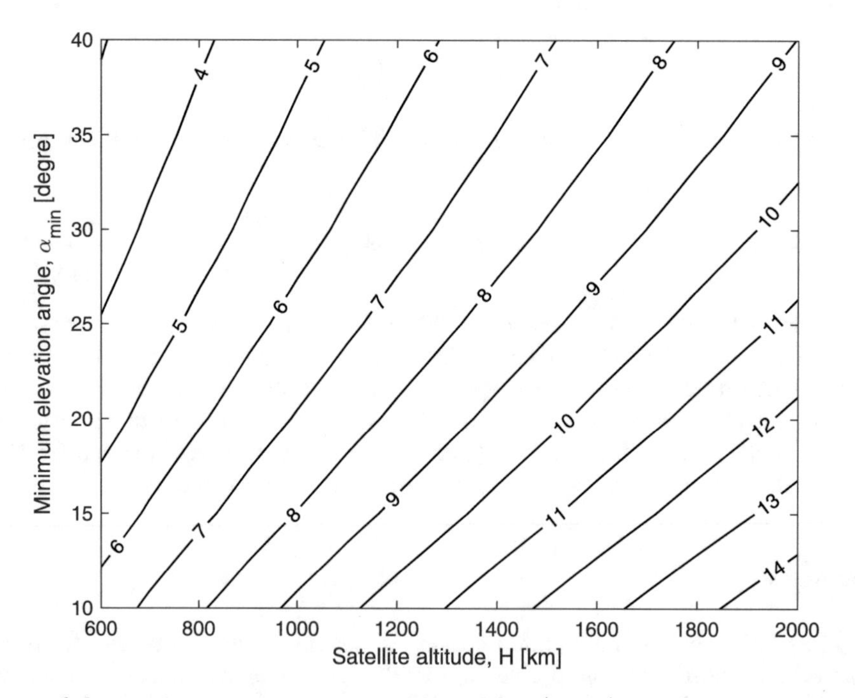

Fig. 2.66: Level curves of the maximum one-way propagation delay (in ms) as a function of the minimum elevation angle and the satellite altitude (LEO case)

angle and that $T_{\max}$ increases with the satellite altitude. Parameter $T_{\max}$ is quite important to characterize the latency experienced by the traffic via the satellite system.

2.11.2 Advanced and Future Satellite Communication Systems

A High Throughput Satellite (HTS) provides a capacity (throughput) that is many times that of a traditional satellite system (as those in Table 2.7) for the same amount of allocated bandwidth in orbit. More than 10 Mbit/s capacity is provided to individual customers. A single HTS GEO satellite can support millions of users. However, there will be HTS not only based on GEO satellites but also based on MEO or LEO satellite orbits. Further evolution of the satellite technology will lead to Very High Throughput Satellite (VHTS) systems with 500 Gbit/s of capacity up to Ultra-High Throughput Satellite (UHTS), achieving capacities larger than 1 Tbit/s, using highly bandwidth-efficient transmission techniques, advanced interference mitigation schemes, and balancing the capacity flexibly among different areas [100].

The following key aspects will determine the total system throughput improvement of HTS/VHTS/UHTS:

- Modulation efficiency (C/B) under the constraint of the Shannon limit $C = B\log_2(1 + SNR)$ bit/s. Note that modulation and coding efficiency should approach the Shannon limit. Gains are limited by the channel non-linearities due to amplifiers.
- Available bandwidth (B) with the expected evolution from current Ku and Ka bands to future Q/V/W bands [millimeter wave domain, also known as the Extremely High Frequency (EHF) spectrum range, from 30 GHz to 300 GHz].
- High reuse of frequencies among the beams.

Adopting a multi-spot-beam antenna on the satellite allows us to reuse the same frequency bandwidth many times in sufficiently separated beams to avoid interference. This approach permits to increase in the traffic load (capacity) supported by a satellite. The number of colors used is the reuse parameter K: the lower is K, the higher is the capacity gain utilizing the reuse. The Signal-to-Interference ratio, S/I, depends on the reuse distance (K) and should be above a certain threshold to allow reliable communications. Typical K values are 3 and 4 but also 1 if physical layer techniques can be adopted to cancel mutual interference, like the pre-coding scheme [101].

Larger frequency bands would allow increasing the feeder link's satellite capacity (from the terrestrial gateway-earth station to the satellite). These frequency bands are available at Q/V (30–75 GHz) and W (75–110 GHz) frequencies in the EHF range. The state-of-the-art design is adopting the Ka band for the user-link (i.e., the link from the satellite to the users), while the feeder link is implemented at higher frequencies, i.e., Q/V band today and W band tomorrow. However, Q/V/W frequency bands suffer more from deep-fading attenuation due to atmospheric events (rain, clouds, etc.) and, of course, larger free space attenuation (combining all the attenuations, we have: $L_{\text{free}} \times L_{atmospheric\_absorption}$). The frequencies around 60 GHz cannot be exploited for satellite transmissions because of Oxygen absorption peaks. The attenuation increases with the frequency for what concerns the water absorption (clouds, rain), reaching a peak around 180 GHz. From these considerations, we can select some favorable EHF bandwidth portions for satellite communications, that is, in Q/V band and W band [102].

Cloud attenuation at W/V bands can contribute significant path loss attenuation (> 10 dB), which is more significant than observed at typical Ka-band frequencies. Moreover, the attenuation due to rain is the dominant loss mechanism when dealing with communications signals above the Ka band. Though rain is a 1% phenomenon (site dependent), its presence could result in unrealistic link margin requirements of 45 dBs for V band (compared to 15 dBs for Ka band and 10 dBs for K band).

2.11.2.1 Smart Gateway Diversity

Services such as teleconference and telemedicine, having high availability requirements, could be critical for the channel conditions experienced in Q/V/W bands. Moreover, reliable applications could experience packet losses and then performance (throughput) degradation. Adaptive Coding Modulation (ACM) alone cannot cope with the deep-fading events and cannot guarantee 99.7% availability. One interesting approach to circumvent these issues is to use architectures with multiple Gateways (GWs): spare GWs enter into service when the currently active N GWs experience adverse meteorological conditions. This is the solution adopted by the Smart Gateway Diversity (SGD) scheme [103]. When the feeder link of a GW experiences a deep atmospheric fading, its traffic is rerouted to another (spare) GW using the terrestrial network segment. The multi-GW architecture achieves the desired availability at the expense of redundant GWs added with respect to the N nominal ones. Hence, SGD typically requires that the GWs are interconnected to each other through a terrestrial network. In particular, a fully centralized architecture is assumed: the switch of a GW towards a spare one and the consequent traffic rerouting are decided by the Network Control Center (NCC).

GWs are located in fixed pre-determined locations. The additional gateways are deployed far away with respect to the active ones to de-correlate the related (atmospheric) fading events. The user terminal has to switch carrier frequency if it has to change the serving GW. If the system is designed to use N GWs, a basic diversity scheme would require $P \equiv N$ redundant GWs to achieve the desired level of availability. The main drawback of this diversity scheme is that we need to double the ground segment of GWs and the underlying costs, which could be unacceptable for future satellite systems. Therefore, it becomes crucial to adopt improved solutions to reduce the number of GWs needed to keep the desired level of reliability. This is the reason why several SGD schemes have been proposed, as detailed below. The following description assumes that a Frequency Division Multiple Access (FDMA) technique is used, where the available satellite spectrum is divided among carriers; carriers are assigned to the GWs according to some rules as shown below.

Three techniques are possible for implementing SGD as detailed below [103].

Smart $N + 0$ Gateways

The basic idea behind the $N + 0$ macro-diversity scheme is that each user beam is served by carriers coming from all the GWs (see Fig. 2.67a). Impairments at a GW will not result in a full outage in some beams but rather in a decrease of throughput in all the user beams to which the GW is connected. Therefore the outage of one of the N GWs will result in a loss of a fraction of $1/N$ of the throughput in each beam (see Fig. 2.67b). There is no GW redundancy (i.e., no additional GWs nor extra bandwidth). This scheme requires a modification at the satellite payload level to filter (separate) carriers and recombine them from different feeder links. This scheme aims at sharing the outage of the GWs among all the involved user beams, thus resulting in a temporarily reduced throughput but without any real outage. In this scheme, there is no terrestrial interconnection among GWs and no extra GWs.

Smart $N + P$ Gateways

This technique allows achieving the desired availability by adding P redundant GWs to the N active ones (see Fig. 2.68a). In case of the outage of one GW, its traffic is rerouted to one of the P spare GWs (see Fig. 2.68b) and restored when the fading event is over. The additional GWs are deployed far away enough from other GWs to ensure proper decorrelation of the rain fading. A terrestrial network must interconnect all GWs to the corresponding backup GWs. This SGD scheme allows obtaining a system feeder link availability larger than 99.9%. Note that each beam is only served by a single GW at a time, either active or backup. This design concept entails no capacity reduction if the number of GWs simultaneously undergoing high fading levels is lower than or equal to P. Onboard the satellite (payload), there is the need for an $N + P : N$ non-blocking switching matrix.

Improved Smart $N + P$ Gateways

This concept is similar to the previous one on smart $N + P$ GWs, except for the fact that in nominal conditions, all the GWs operate concurrently. Let us consider an example with $N = 3$, $P = 1$, and 12 carriers. All the 4 GWs work simultaneously using only 3 carriers (even if they can support 4 carriers). If one GW experiences an outage, its carriers are activated in other GWs. Each of the remaining 3 GWs is fully loaded using 4 carriers that are switched to the same user beams as before through a suitable configuration of the switching matrix onboard the satellite. Thus, we can maintain the same capacity and connectivity as before. This approach improves the system availability at the expense of a more complex satellite payload architecture than the Smart $N + P$ Gateways scheme.

2.11.2.2 Survey on HTS Systems

There has been a 30-times increase in GEO transponders' throughput from 10 Gbit/s in 2006 up to 300 Gbit/s in 2017. This trend is continuing with GEO systems with 500 Gbit/s capacity as well as ambitious mega-LEO constellation projects targeting a capacity of Tbit/s. The frequencies currently adopted are in L, S, K, Ku, and Ka bands, but higher frequency bands (Q/V/W) will also be used in the future. This section provides a survey of today's most relevant satellite systems (either operational or in the design phase).

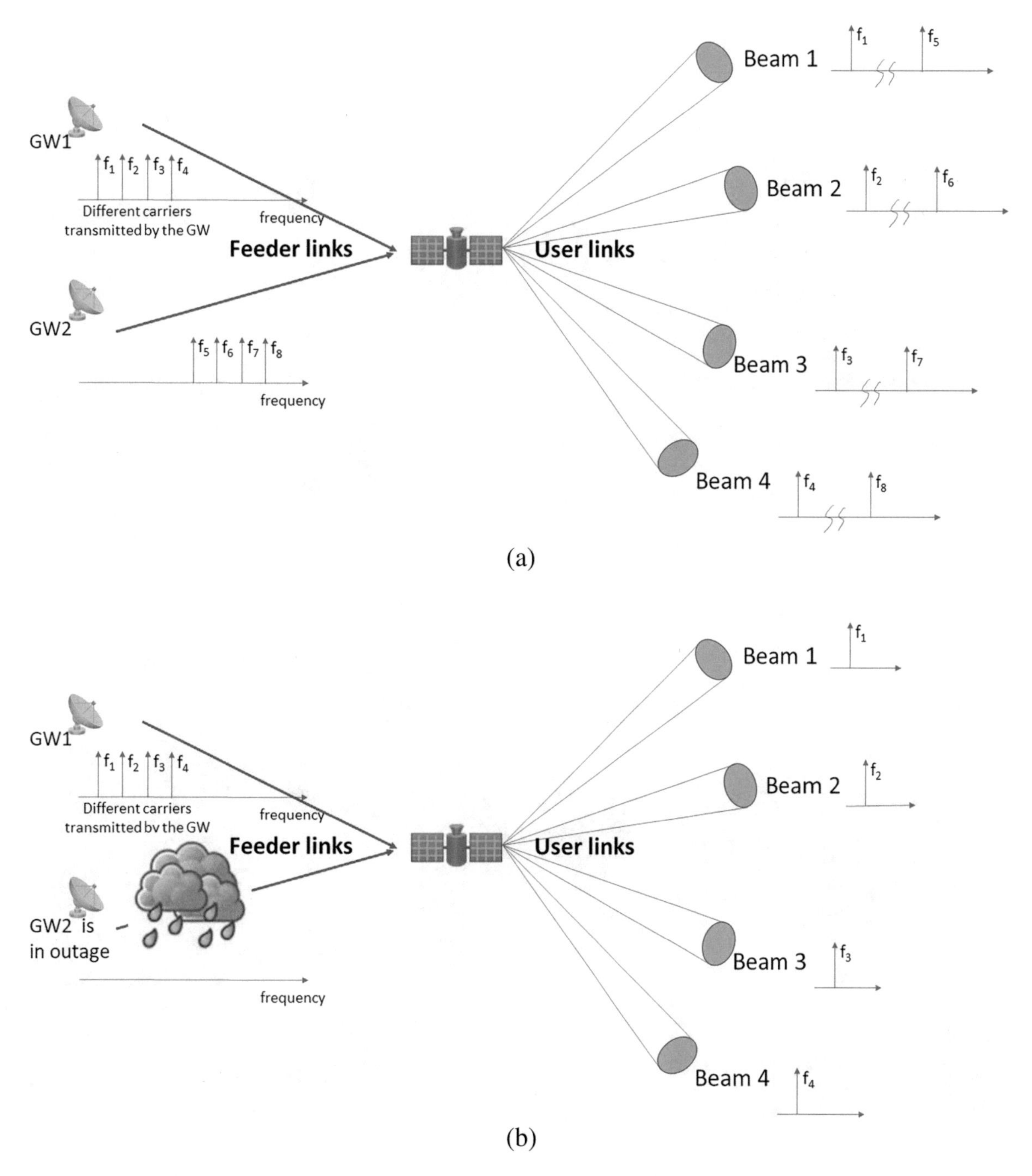

Fig. 2.67: $N + 0$ gateway diversity with $N = 2$ and 8 carriers. (**a**) Clear sky conditions at both GWs. (**b**) GW2 is in outage due to bad weather conditions

GEO HTS

ViaSat-2 (launched in 2017) is a Ka-band HTS GEO satellite that provides more than 300 Gbit/s of total network capacity. ViaSat-2 adopts a dynamic system architecture for auto-shifting traffic among more than 40 GWs. The ViaSat-3 system is an ultra-high capacity satellite platform comprising three ViaSat-3 class satellites (starting operations in the timeframe 2021–2022). Each of them is expected to serve more than 100 GWs to deliver a capacity greater than 1 Tbit/s. The Viasat-3 platform will provide more than 100 Mbit/s residential Internet service, enabling 4K ultra-high-definition video streaming, and will provide up to 1 Gbit/s speed for maritime use.

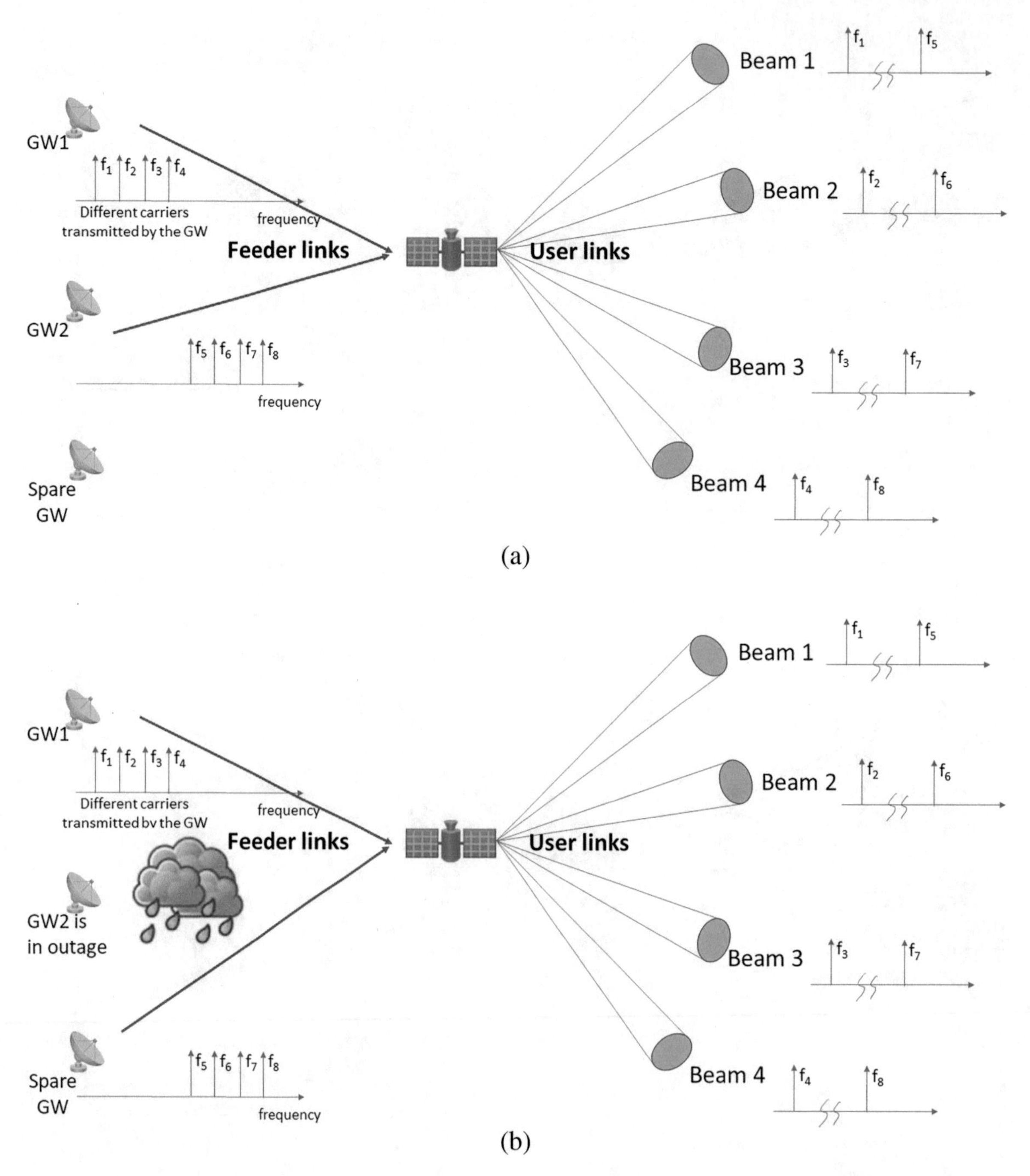

Fig. 2.68: $N + P$ gateway diversity with $N = 2$ and $P = 1$. (**a**) Clear sky conditions at both GWs. (**b**) GW2 is in outage due to bad weather conditions

The Intelsat's EpicNG platform comprising 3 GEO satellites (Intelsat 29e and Intelsat 33e launched in 2016 and Intelsat 32e in 2017) achieves a global coverage (cellular backhaul) with HTS technology based on an open architecture, utilizing C, Ku, and Ka bands with a mixture of wide beams and spot-beams.

The HTS satellites by SES (i.e., SES-12, SES-14, and SES-15) provide coverage over Asia-Pacific, the Americas, and the Atlantic Ocean region. Each satellite comprises a few wide beams and many spot-beams (SES-12: 72 spot-beams, SES-14: 48 spot-beams, SES-15: 46 spot-beams). These beams can use mixed frequencies in Ku and Ka bands. SES is also developing the SES-17 satellite (to be launched in 2021), which will provide 200 Ka-band spot-beams over the Americas and the Atlantic Ocean region.

Eutelsat launched Europe's first GEO HTS, KA-SAT, in 2012, which uses 82 Ka-band spot-beams connected to a network of 10 Earth stations to deliver high-speed Internet at a total throughput of 90 Gbit/s. Eutelsat launched its new HTS, EUTELSAT 172B, in 2017 (the number of the EUTELSAT satellite refers to the longitude of the GEO satellite on the equatorial plane, in this case, 172° East for the Asia-Pacific area), which delivers increased capacity for fast-growing applications in Asia-Pacific, including in-flight and maritime connectivity, cellular backhaul, corporate networks, video, and government services. EUTELSAT 172B has 14 C-band transponders, 36 Ku-band transponders, and a high-throughput Ku-band payload (designed for in-flight broadband services and featuring multiple user spots), delivering an overall throughput of 1.8 Gbit/s.

MEO HTS

The "Other three Billion" (O3b) satellite system owned by SES aims to provide Ka-band broadband trunking connectivity (particularly for Africa and Latin America), using 20 first-generation satellites in a single MEO equatorial ring at about 8000 km of altitude and adopting 9 terrestrial GWs. Each of the first-generation O3b MEO satellites provides 12 fully steerable Ka-band beams (two for GWs and 10 for remote terminals) and has an aggregate capacity of 16 Gbit/s. Each beam illuminates an area of Earth's surface measuring 700 km across. O3b does not use intersatellite links; it relies on the use of GWs to route the traffic received from the satellites. O3b claims a one-way latency of 179 ms for voice communications and an end-to-end round-trip latency of 140 ms for data services. For maritime applications, O3b claims connectivity speeds of over 500 Mbit/s. SES is also working deploying a second-generation of MEO satellites for O3b that are called mPOWER to be operational by 2022 (still on an equatorial ring at 8000 km of altitude) that will provide multiple Terabits of throughput globally to support the digital transformation and the cloud adoption virtually anywhere on the Earth. The upcoming seven O3b mPOWER satellites will provide more than 5000 beams with a smaller ground footprint diameter of approximately 140–280 km. There will be a synergy between the GEO and MEO systems owned by SES, allowing these systems to cooperate in providing services to users. The O3b mPOWER system will route traffic based on applications over GEO, MEO, and terrestrial links; this will be made possible by adopting an SDN approach.

Mega-LEO Constellations

LEO satellites move at extremely fast speeds (several thousand km/h) and complete one orbital revolution in a few hundred minutes. From any specific point on Earth, a satellite is only available for few minutes during each pass. Placing LEO satellites on many orbital planes requires multiple launches with complex launch sequences. The primary input for the constellation design is the geographical areas that need to be covered. The main parameters to plan a constellation of LEO satellites are number of satellites, number of orbital planes, minimum elevation angle, altitude, orbit inclination, plane spacing, and orbit eccentricity. A common approach is adopting a Walker-Delta constellation (also known as Ballard-Rosette constellation) that achieves an almost-global coverage of the Earth's surface using a minimum number of satellites in circular orbits that are inclined with respect to the equator of an angle ζ. Correspondingly, the coverage is for the areas with latitudes λ so that $|\lambda| \leq \zeta$. The satellites are placed on n_1 equally spaced orbital planes (on the equator) with n_2 equally spaced satellites per plane. Hence, $n_1 \times n_2$ satellites are needed in the whole system. Hundreds or thousands of satellites are used in a constellation with many orbital LEO planes to cover the entire Earth; this is why they are called mega-LEO constellations. If there is the need to cover also areas at latitudes larger than ζ, multi-layer constellations are used where different types of orbits are used. For instance, a possibility is to add polar or quasi-polar orbits inclined of about 90° with respect to the equator. Also, constellations made only of quasi-polar orbits are very common (see Fig. 2.69).

Many mega-LEO satellite constellations are in the design or in the implementation phase (beta service phase at the time of writing) [104]. A dedicated terrestrial network of GWs is needed to support the access to the Internet for the traffic coming from the satellites. These satellites typically have onboard processing and switching and, in some cases, Inter-Satellite Links (ISLs) for the direct routing of traffic among satellites in the sky; in this last case, typically, the number of GWs on the Earth is reduced. Necessary details are provided below for some mega-LEO systems.

The Starlink system by SpaceX aims to provide a wide range of broadband communication services for residential, commercial, institutional, government, and professional users worldwide. The satellite payload will support advanced phased-array beam-forming and digital processing technologies in Ku and Ka bands. User terminals will adopt a similar phased-array technology to allow antenna beams to track the LEO satellites. Each Starlink

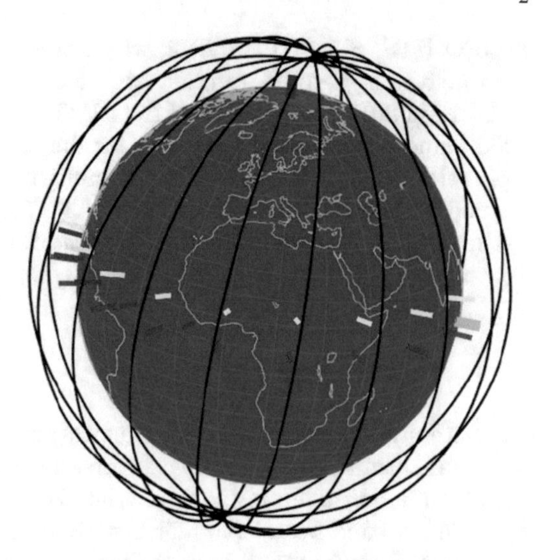

Fig. 2.69: Example of almost polar orbits of a hypothetical LEO constellation at 1000 km of altitude (plotted by the Matlab GUI in [105]—Copyright (c) 2011, Michael Hanchak)

satellite will have 4 optical (laser) ISLs with its neighbors. Thus, these satellites will exchange data for both traffic routing in the space and seamless service continuity (satellite handover). Starlink has an agreement with the US Federal Communications Commission (FCC) to deploy 12,000 satellites in LEO orbits. In particular, the first core (Walker-Delta) constellation to be deployed consists of 1584 satellites, which are evenly distributed on 72 orbital planes at an orbital altitude of 550 km and an inclination of 53°. The huge number of satellites of this system highlights the fact that multiple satellite visibility will be guaranteed. Each satellite in the SpaceX system will carry a high-level digital payload containing phased-array antennas, which will allow each beam to turn and form independently. The minimum elevation of the user terminal is currently planned as 25°. A satellite can cover an area on the Earth with a radius about equal to 940 km; a single antenna spot-beam footprint (cell) has a radius of 8 km. The total throughput of each satellite is estimated to be 17–23 Gbit/s, depending on the characteristics of the user terminal. SpaceX system will use the Ku band for user communications; instead, the GW communications will use the Ka band. We can consider that each Starlink satellite will use about 50 Ku-Band user beams.

The OneWeb system foresees an initial constellation of 650 satellites (the entire plan is for 720 satellites) to provide broadband satellite Internet services to people everywhere [106]. OneWeb is progressively building its constellation of satellites even if this project faced some economic issues. The satellites will operate in circular quasi-polar LEO orbit (86.4°), at approximately 1200 km of altitude. Low mass satellites will be positioned on 12 planes, with 49 satellites per plane and in-orbit spares. The user links will adopt the Ku band, while the GW links will use the Ka band. The satellites will have a bent-pipe architecture with no ISLs. Each OneWeb satellite has 16 Ku-band elliptical beams and 2 Ka band beams. The downlink rate of a single beam can reach 750 Mbit/s, and the upstream rate is 375 Mbit/s.

Like the "Kuiper Belt" representing an area outside the Solar System with a myriad of small asteroids, project Kuiper aims to populate a certain belt around the Earth with many small satellites. In particular, Amazon's Kuiper system will provide high-speed, low-latency Internet broadband services via a fleet of 3236 Ka-band LEO satellites, be organized in different orbital shells: 784 satellites at 590 km altitude, 1296 satellites at around 610 km altitude, and 1156 satellites at 630 km altitude. This system will allow continuous coverage to customers within 56° N and 56° S latitude. The planned minimum elevation angle is 30°. A satellite can cover an area with a radius of about 1091 km. Satellite's lifetime will be 10 years; the satellites are planned to deorbit at the end of their lifetime automatically.

Aerial Platforms

Aerial platforms can be roughly divided into two large groups, that is Low Altitude Platforms (LAP), like Unmanned Aerial Vehicles (UAVs) and drones flying up to 1–2 km of altitude, and High Altitude Platforms (HAPs) in the stratosphere at an altitude of approximately 20 km above the Earth's surface.

HAPs can be balloons, quasi-stationary, and can provide reliable wireless coverage for a quite large geographic area. Depending on their type, HAPs remain almost stationary or fly in a circle. For instance, the Loon LLC company uses high-altitude balloons placed in the stratosphere at an altitude of about 18 km to create an aerial wireless network, providing local 4G coverage. This system was named Project Loon by Google, who found the idea of providing Internet access to the population in remote areas unprecedented and "loony" [107]. In this system, the signal travels through the balloon network from balloon to balloon, then to a ground-based station connected to an Internet Service Provider (ISP), then onto the global Internet. HAPs propagation delay (about 50–85 μs) is significantly lower as compared with LEO satellites (around 2.5 ms), so that HAPs are very appealing for delay-sensitive applications [100]. Because of some economic issues, this project has been shut down recently.

Finally, drones can provide connectivity in a cost-effective and ad hoc way to disaster-struck regions where the terrestrial infrastructure might have been damaged. They can quickly reach the area to be served and transport a light 4G/5G base station on board, thus providing connectivity to mobile terminals on the ground.

Exercises on Part I of the Book

This section contains some exercises about the first part of the book. The main interest is in traffic regulators, Dijkstra routing, deterministic queuing, and cwnd behavior of TCP.

Ex. I.1 We have a Frame Relay network, which applies a policer to control the traffic sources' access. Let us consider a traffic source, which has a periodic ON-OFF bit-rate as a function of time as shown in Fig. 2.70, with parameters b (= burst bit-rate), T (= time length of the source cycle), and l (= xT, burst duration). The policer uses the following assumptions for the measurement interval, T_c, the committed burst size, B_c, and the excess burst size, B_e:

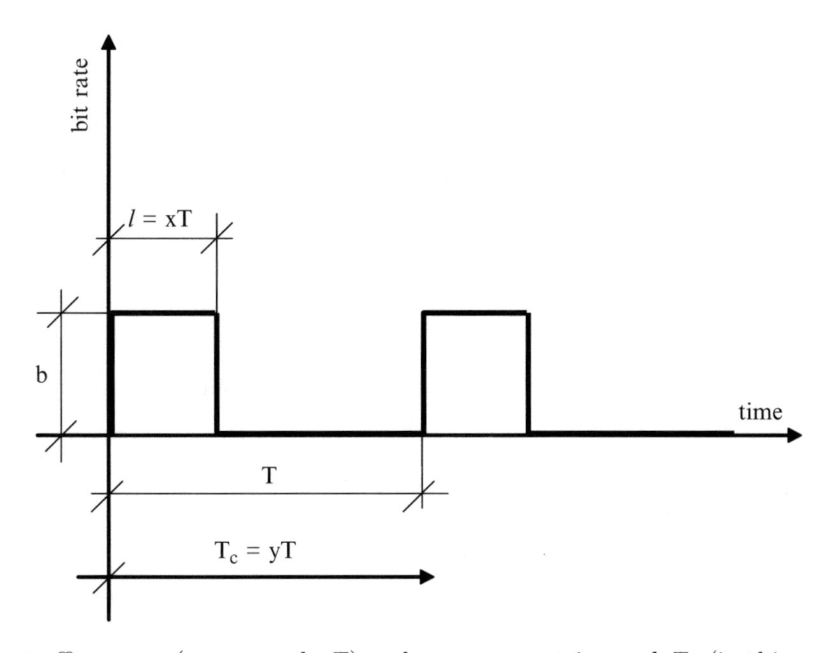

Fig. 2.70: Periodic traffic source (source cycle T) and measurement interval T_c (in this graph $T_c \equiv T$, $y = 1$)

- $B_c/T_c = R_c$, a constant value
- $B_e/T_c = R_e$, a constant value
- $bl/T = bx = R_s$, mean source bit-rate
- A rectangular pulse (burst) represents a single packet in Fig. 2.70
- The measurement interval T_c is applied to the periodic source according to the "phase" shown in Fig. 2.70 so that the source cycle T contains an integer number of measurement intervals T_c ($T_c = yT$, with $y = 1$, 1/2, 1/3, etc.)

ATM switch IN	ATM switch OUT
VPI = 1 VCI = 1	VPI = 2 VCI = 3
VPI = 1 VCI = 2	VPI = 3 VCI = 6
……..	
……..	

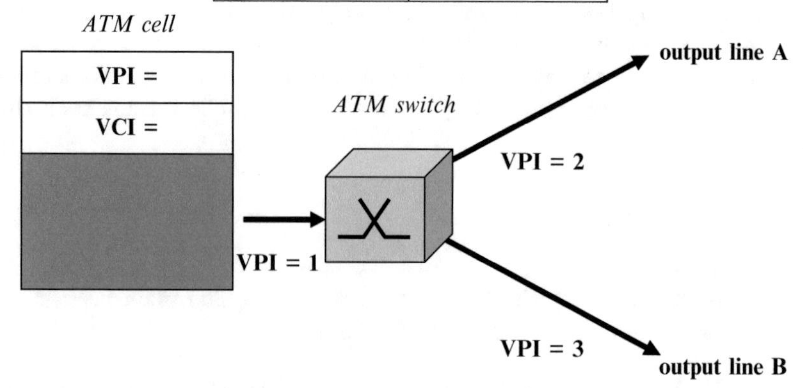

Fig. 2.71: ATM switch and its switching table

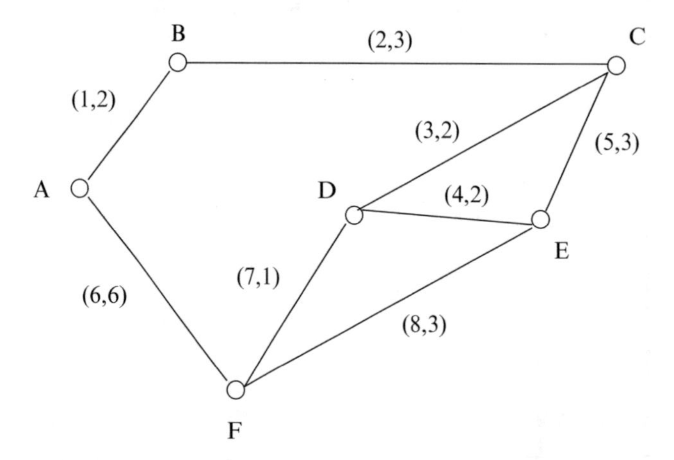

Fig. 2.72: Network with bidirectional links, labeled by (a, c), where "a" denotes the link number and "c" represents the link cost

- Constraints: $bx = R_s \leq R_c$ (so that there is enough capacity to service the traffic source) and $T_c \geq xT \Rightarrow y \geq x$ (the measurement interval is larger than the burst duration)

  It is requested to determine the conditions to have marking or dropping of all generated packets.

Ex. I.2 Let us consider an ATM switch with a switching table, which manages virtual paths and virtual channels, as shown in Fig. 2.71. It is requested to determine the VPI and VCI fields to be used for an input cell if we like that this cell leaves the switch from output line A; in this case, we are also asked to provide the VPI and VCI fields of the corresponding output cell.

Ex. I.3 Let us consider the network depicted in Fig. 2.72: it is requested to determine the sink tree for node A by applying the Dijkstra shortest path routing algorithm.

Ex. I.4 Let us consider an FTP data transfer (TCP "elephant" flow), referring to the network model depicted in Fig. 2.73. We adopt a scenario with IP packets (MTU) of 1500 bytes, Information Bit-Rate (IBR) of the bottleneck link equal to 600 kbit/s, and *physical* Round-Trip Time (RTT) equal to 0.5 s (GEO satellite scenario).

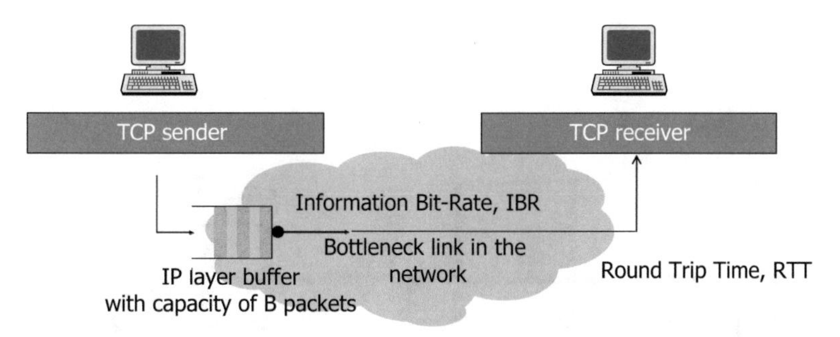

Fig. 2.73: System model for the reliable transfer of data; case of an "elephant" TCP flow (FTP)

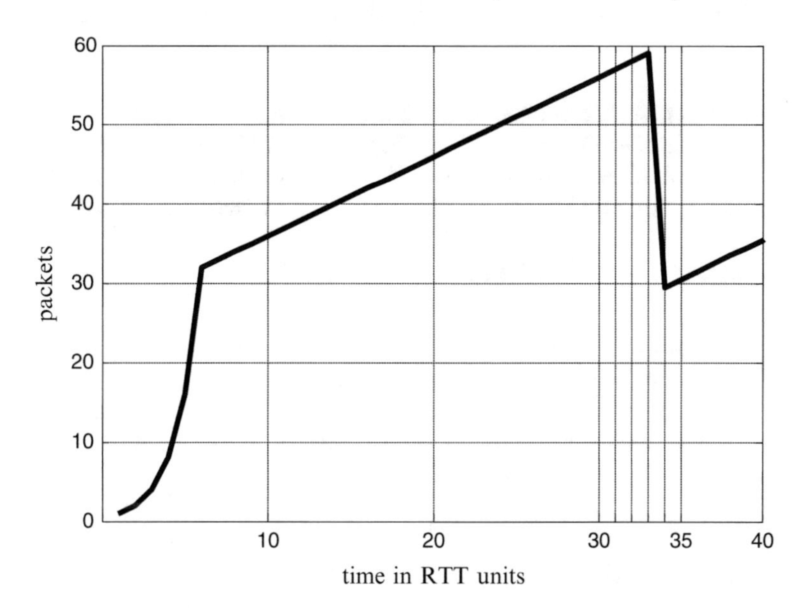

Fig. 2.74: Cwnd behavior for TCP Reno

It is requested to determine the Bandwidth-Delay Product (BDP) and plot the behaviors of both the congestion window (cwnd) and the slow-start threshold (ssthresh) up to 25 RTTs for both TCP Tahoe and TCP NewReno, under the following conditions:

- Bottleneck link buffer capacity $B = 20$ pkts.
- Sockets' buffer size much larger than $B + \text{BDP}$.
- Initial ssthresh value equal to 32 pkts.

Then, it is also requested to show the cwnd behaviors up to 25 RTTs for TCP Tahoe and TCP NewReno with initial ssthresh equal to 64 pkts: what are the differences with respect to the previous case?

Finally, assuming to be able to change the size of the buffer of the bottleneck link, let us determine its optimal size from the TCP throughput standpoint.

Ex. I.5 Let us refer to an FTP transfer (TCP "elephant" flow) on a network characterized by a Bandwidth-Delay Product (BDP) equal to 30 pkts. We are asked to plot cwnd and ssthresh behaviors up to 40 RTTs in the TCP NewReno case under the following conditions:

- Bottleneck link buffer with capacity $B = 10$ pkts.
- Sockets' buffer size much larger than $B + \text{BDP}$.
- Initial ssthresh value equal to 16 pkts.

Ex. I.6 Let us consider a TCP-based traffic flow with the cwnd behavior shown in Fig. 2.74 (the unit of time in abscissa is RTT). Assuming that this cwnd behavior is for the TCP Reno version, it is requested to answer the following questions:

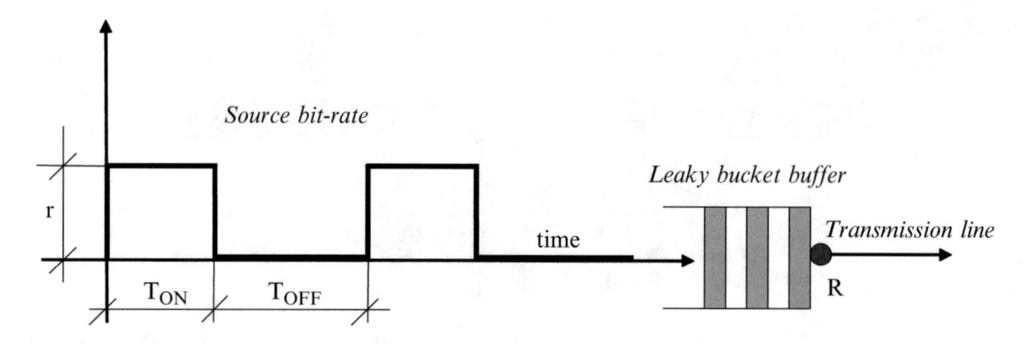

Fig. 2.75: Periodic ON-OFF traffic source at the input of a leaky bucket regulator

- Identify where slow-start and congestion avoidance phases are used in the graph.
- After time 34 RTTs, is the segment loss revealed by three DUPACKs or by an RTO expiration?
- What is the initial ssthresh value? And what is the ssthresh value after time 34 RTTs?
- If we know that the bottleneck link buffer has a capacity of 30 pkts, what is the value of the Bandwidth-Delay-Product (BDP)?
- When is the 63rd TCP segment sent? (RTT interval)

Ex. I.7 Let us consider a network adopting IntServ-Guaranteed Service as quality of service technique. We have a traffic source with a fluid-flow model accessing the network. The traffic source is regulated according to the following T-Spec parameters: $(r, p, b) = (1$ kbit/s, 4 kbit/s, 500 bits) [1 token = 1 bit]. Following the arrival curve approach, it is requested to determine the minimum service rate R to guarantee a delay lower than or equal to $\Delta_{\max} = 150$ ms (let us neglect propagation delays).

Ex. I.8 Referring to the IPv4 address 128.15.10.5, it is requested to determine:

- The class of the IPv4 address and the corresponding network address.
- The most efficient subnet mask for a subnet with 58 hosts.
- An example of IPv4 address of the above subnet.

Ex. I.9 It is requested to determine the classes of the following IPv4 addresses:

(a) 126.12.1.5
(b) 198.15.1.7

How many host addresses are available in the networks corresponding to cases (a) and (b)?

Ex. I.10 Let us consider the ON-OFF periodic traffic source (fluid-flow model) that is feeding a leaky bucket traffic regulator, as shown in Fig. 2.75. Let r denote the rate of the source in the ON state. Let R denote the output rate of the regulator. We assume $r \geq R$. It is requested to determine: (1) the stability condition; (2) the input traffic burstiness; (3) the maximum buffer occupancy; (4) the maximum delay imposed on the traffic by the leaky bucket regulator; (5) the behavior of the regulator buffer occupancy; (6) the output traffic behavior.

Ex. I.11 Let us consider the ON-OFF periodic traffic source (fluid-flow model) shown in Fig. 2.76: during the ON phase, the bit-rate of the source has a linear increase up to R_{max}; during the OFF phase, no traffic is generated. This traffic is feeding a buffer with output rate R. We assume $R_{max}/2 > R$. It is requested to determine: (1) the stability condition; (2) the input traffic burstiness; (3) the maximum buffer occupancy; (4) the maximum delay imposed by the buffer to the input traffic; (5) the buffer size needed to avoid any packet loss.

Ex. I.12 Let us consider the network with the topology in Fig. 2.77. We have to determine the sink tree and the routing table of node A using the Dijkstra shortest path algorithm. Moreover, let us assume that all the links have a capacity of 100 Mbit/s. The traffic from A to C has a mean bit-rate of 60 kbit/s. The traffic flows from A to all the other nodes have a mean bit-rate of 10 kbit/s. We have to determine the network's bottleneck link that is the link with the highest total traffic load because of the Dijkstra routing and the total bit-rate on it.

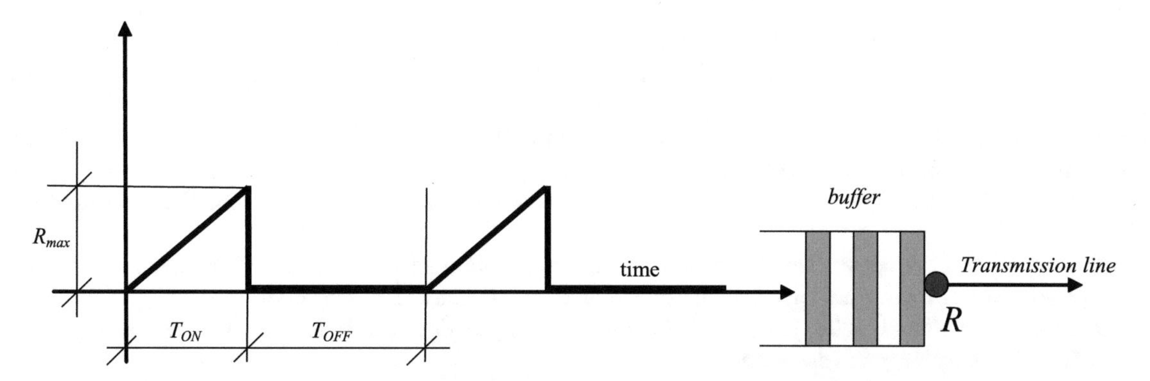

Fig. 2.76: Periodic traffic source offered to a transmission buffer

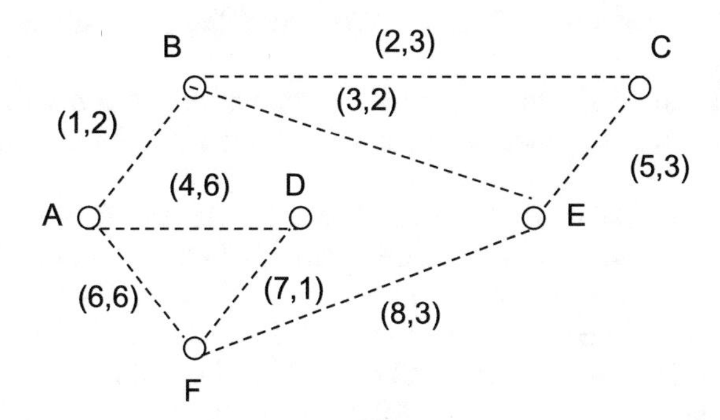

Fig. 2.77: Network with bidirectional links denoted as (a, c), where a is the arc number and c is the cost of the link.

References

1. Cisco Annual Internet Report (2018–2023) White paper networking solutions white paper, March 2020. https://www.cisco.com/c/en/us/solutions/collateral/executive-perspectives/annual-internet-report/white-paper-c11-741490.html
2. Postel J (1972) Telnet protocol. IETF RFC 318
3. Bhushan A (1972) The file transfer protocol. IETF RFC 354
4. Braden R (1973) Comments on file transfer protocol. IETF RFC 430
5. IETF Web site with URL: www.ietf.org/rfc/
6. Postel J (1981) Internet protocol. IETF RFC 791
7. Postel J (1981) Transmission control protocol. IETF RFC 793
8. Feit S (1996) TCP/IP architecture, protocols and implementation with IPv6 and IP security (revised and expanded), 2nd edn. McGraw-Hill, New York
9. Comer DE (2003) Network system design using network processor. Prentice Hall, Upper Saddle River
10. Comer DE (2000) Internetworking with TCP/IP. Principles protocols and architecture, 4th edn. Prentice-Hall, Upper Saddle River
11. Perlman R (1999) Interconnections: bridges, routers, switches, and internetworking protocols. Addison-Wesley, Reading
12. Braden R (1989) Requirements for internet hosts—communication layers. IETF RFC 1122
13. Kirkpatrick S, Stahl M, Recker M (1990) Internet numbers. IETF RFC 1166
14. Reynolds J, Postel J (1994) Assigned numbers. IEFT RFC 1700
15. Deering S, Hinden R (1998) Internet protocol, version 6 (IPv6) specification. RFC 2460
16. Dijkstra EW (1959) A note on two problems in connexion with graphs. Numer Math 1:269–271
17. Cherkassky BV, Goldberg AV, Radzik T (1996) Shortest paths algorithms: theory and experimental evaluation. Math Program A 73(2):129–174
18. Mills DL (1984) Exterior gateway protocol formal specification. IETF RFC 904

19. Rekhter Y, Watson TJ, Li T (1995) A border gateway protocol 4 (BGP-4). IETF RFC 1771
20. Braden R, Clark D, Shenker S (1994) Integrated services in the internet architecture: an overview. IETF RFC 1633
21. Braden R, Zhang L, Berson S, Herzog S, Jamin S (1997) Resource reservation protocol (RSVP) version 1 functional specification. IETF RFC 2205
22. Cruz RL (1991) A calculus for network delay. Part I: network elements in isolation. IEEE Trans Inf Theory 37(1):114–131
23. Parekh AKJ (1992) A generalized processor sharing approach to flow control in integrated service networks. MIT Laboratory for Information and Decision Systems, Report LIDS-TH-2089
24. Le Boudec J-Y, Thiran P (2004) Network calculus: a theory of deterministic queuing systems for the internet, vol 2050, LNCS. Springer, New York
25. Shenker S, Partridge C, Guérin R (1997) Specification of guaranteed quality of service. IETF RFC 2212
26. White PP (1997) RSVP and Integrated Services in the Internet: a tutorial. IEEE Commun Mag 35(5):100–106
27. Blake S, Black D, Carlson M, Davies E, Wang Z, Weiss W (1998) An architecture for differentiated service. IETF RFC 2475
28. Andreadis A, Giambene G (2002) Protocols for high-efficiency wireless networks. Kluwer, New York
29. Floyd S, Jacobson V (1993) Random early detection (RED) gateways for congestion avoidance. IEEE/ACM Trans Netwo 1(4):397–413
30. Bernet Y, Ford P, Yavatkar R, Baker F, Zhang L, Speer M, Braden R, Davie B, Wroclawski J, Felstaine E (2000) A framework for integrated services operation over Diffserv networks. IETF RFC 2998
31. Huston G (2000) Next steps for the IP QoS architecture. IETF RFC 2990
32. Heinanen J (1993) Multiprotocol encapsulation over ATM adaptation layer 5. IETF RFC 1483
33. Laubach M (1994) Classical IP and ARP over ATM. IETF RFC 1577
34. Laubach M, Halpern J (1998) Classical IP and ARP over ATM. IETF RFC 2225
35. Rosen E, Viswanathan A, Callon R (2001) Multiprotocol label switching architecture. IETF RFC 3031
36. Rosen E, Tappan D, Fedorkow G, Rekhter Y, Farinacci D, Li T, Conta A (2001) MPLS label stack encoding. IETF RFC 3032
37. Andersson L, Doolan P, Feldman N, Fredette A, Thomas B (2001) LDP specification 2. IETF RFC 3036
38. Xiao X, Hannan A, Bailey B, Ni L (2000) Traffic engineering with MPLS in the internet. IEEE Netw Mag 14(2):28–33
39. Nichols K, Blake S, Baker F, Black D (1998) Definition of the differentiated services field (DS Field) in the IPv4 and IPv6 headers. IETF RFC 2474
40. Awduche D, Berger L, Gan D, Li T, Srinivasan V, Swallow G (2001) RSVP-TE: extensions to RSVP for LSP tunnels. IETF RFC 3209
41. Jamoussi B et al (2002) Constraint-based LSP setup using LDP. IETF RFC 3212
42. Davie B, Rekhter Y (2000) MPLS: technology and applications. Morgan Kauffmann, San Francisco
43. Mogul J (1990) Path MTU discovery. IETF RFC 1191
44. Allman M, Paxons V, Stevens W (1999) TCP congestion control. IETF RFC 2581
45. Karn P, Partridge C (1987) Improving round-trip time estimates in reliable transport protocols. In: Proc. of ACM SIGCOMM
46. Chang RKC, Chan HY (2000) A throughput deadlock-free TCP for high-speed internet. In: Eighth IEEE International Conference on Networks (ICON‘00)
47. Jain M, Prasad RS, Dovrolis C (2003) The TCP bandwidth-delay product revisited: network buffering, cross traffic, and socket buffer auto-sizing. Technical Report 2003 (GIT-CERCS-03-02). https://smartech.gatech.edu/bitstream/handle/1853/5920/git-cercs-03-02.pdf
48. Jacobson V (1988) Congestion avoidance and control. Comput Commun Rev 18(4):314–329
49. Mathis M, Mahdavi J, Floyd S, Romanow A (1996) TCP selective acknowledgment options. IETF RFC 2018
50. Floyd S, Mahdavi J, Mathis M, Podolsky M (2000) An extension to the selective acknowledgement (SACK) option for TCP. IETF RFC 2883
51. Floyd S, Henderson T, Gurtov A (2004) The NewReno modification to TCP's fast recovery algorithm. IETF RFC 3782
52. Parvez N, Mahanti A, Williamson C (2006) TCP NewReno: slow-but-steady or impatient? In: Proc. of IEEE international conference on communications 2006 (ICC‘06), p 716–722

53. Floyd S (1996) SACK TCP: the sender's congestion control algorithms for the implementation "sack1" in LBNL's "ns" simulator (viewgraphs). Technical report. Presentation to the TCP Large Windows Working Group of the IETF
54. Fall K, Floyd S (1996) Simulation-based comparisons of Tahoe, Reno, and SACK TCP. ACM Comput Commun Rev 26(3):5–21
55. Floyd S (2003) HighSpeed TCP for large congestion windows. IETF RFC 3649
56. Kelly T (2003) Scalable TCP: improving performance in highspeed wide area networks. ACM Comput Commun Rev 33(2):83–91
57. Gopal S, Paul S (2007) TCP dynamics in 802.11 wireless local area networks. In: Proc. of the IEEE International Conference on Communications 2007 (ICC'07)
58. Hoe J (1996) Improving the start-up behavior of a congestion control scheme for TCP. In: Proc. of ACM SIGCOMM
59. Padhye J, Firoiu V, Towsley D, Kurose FJ (2000) Modeling TCP Reno performance: a simple model and its empirical validation. IEEE/ACM Trans Networking 8(2):133–145
60. Gettys J (2012) Bufferbloat: dark buffers in the Internet. IEEE Internet Comput 9(11):95–96
61. Mathis M, Semke J, Mahdavi J, Ott T (1997) The macroscopic behavior of the TCP congestion avoidance algorithm. ACM Comput Commun Rev 27(3):67–82
62. Xu L, Harfoush K, Rhee I (2004) Binary increase congestion control (BIC) for fast long-distance networks. In: IEEE Infocom 2004. Hong Kong
63. NS-2 Network Simulator. http://www.isi.edu/nsnam/ns/
64. Postel J (1980) User datagram protocol. IEFT RFC 768
65. IANA. http://www.iana.org/assignments/service-names-port-numbers/service-names-port-numbers.xml
66. Goode B (2002) Voice over internet protocol (VoIP). IEEE Proc 90(9):1495–1517
67. ITU-T (2000) Packet-based multimedia communications systems. Recommendation H.323v4
68. Rosenberg J et al (2002) SIP: session initiation protocol. IETF RFC 3261
69. FP5 GEANT Project. http://www.geant.net/
70. GEANT2 Project. http://www.geant2.net/
71. Shafi M, Molisch AF, Smith PJ, Haustein T, Zhu P, De Silva P, Tufvesson F, Benjebbour A, Wunder G (2017) 5G: a tutorial overview of standards, trials, challenges, deployment, and Practice. IEEE J Sel Areas Commun 35(6):1201–1221
72. Bianchi G, Capone A (2014) From dumb to smarter switches in software defined networks: an overview of data plane evolution. In: IEEE CAMAD 2014. Athens
73. Open Networking Foundation (2015), OpenFlow v1.5.1. Specification
74. ETSI, Network Functions Virtualisation (NFV) (2017), White paper. http://portal.etsi.org/NFV/NFV_White_Paper_5G.pdf
75. 3GPP. https://www.3gpp.org/
76. ETSI (2014), Network functions virtualisation (NFV); management and orchestration, ETSI GS NFV-MAN 001 V1.1.1
77. Barakabitze AA, Ahmad A, Mijumbi R, Hinesd A (2020) 5G network slicing using SDN and NFV: a survey of taxonomy, architectures and future challenges. Comput Netw 167:1389–1286
78. OpenStack. https://docs.openstack.org/contributors/common/introduction.html
79. D. Kutscher, Eum S, Pentikousis K, Psaras I, Corujo D, Saucez D, Schmidt T, Waehlisch M (2016) Information-centric networking (ICN) research challenges. IETF RFC 7927
80. Bari Md.F, Chowdhury SR, Ahmed R, Boutaba R, Mathieu B (2012) A survey of naming and routing in information-centric networks. IEEE Commun Mag **50**(12):44–53
81. ITUT-T (2017) Data aware networking (information centric networking)—requirements and capabilities. Recommendation Y.3071
82. Ahlswede R, Cai N, Li SYR, Yeung RW (2000) Network information flow. IEEE Trans Inf Theory 46(4):1204–1216
83. Luby M (2002) LT codes. In: Proc. of the IEEE Symposium on the Foundations of Computer Science, p 271–280
84. Luby M, Shokrollahi A, Watson M, Stockhammer T (2007), Raptor forward error correction scheme for object delivery, IETF RFC 5053
85. Luby M, Shokrollahi A, Watson M, Stockhammer T, Minder L (2011) RaptorQ forward error correction scheme for object delivery, IETF RFC 6330

86. Garrammone G, Ninacs T, Erl S (2012) On network coding for reliable multicast via satellite. In Proc. of the 18th Ka and Broadband Communications and 30th AIAA International Communications Satellite Systems Conference (ICSSC), Ottawa
87. Fragouli C, Boudec JYL, Widmer J (2006) Network coding: an instant primer. ACM SIGCOMM Comput Commun Rev 36(1):63–68
88. Firoiu V, Adamson B, Roca V, Adjih C, Bilbao J, Fitzek F, Masucci A, Montpetit M. Network coding taxonomy, Internet draft, draft-irtf-nwcrg-network-coding-taxonomy-00
89. Li SYR, Yeung RW, Cai N (2003) Linear network coding. IEEE Trans Inf Theory 49(2):371–381
90. Shojania H, Li B (2007) Parallelized progressive network coding with hardware acceleration. In: Proc. of the 15th IEEE international workshop on quality of service, p 47–55
91. Prior R, Rodrigues A (2011) Systematic network coding for packet loss concealment in broadcast distribution. In: Proc. of the international conference on information networking 2011 (ICOIN2011), pp. 245–250, Barcelona, 26–28 January 2011
92. Møller JH, Pedersen MV, Fitzek F, Larsen T (2009) Network coding for mobile devices-systematic binary random rateless codes. In: Proc. of the international conference on communications (ICC)
93. Durvy M, Fragouli C, Thiran P (2007) Towards reliable broadcasting using ACKs. In: Proc. IEEE Int. Symp. Inform. Theory
94. Jones AL, Chatzigeorgiou I, Tassi A (2015) Binary systematic network coding for progressive packet decoding. In Proc. of ICC 2015, 8–12 June 2015, London
95. Chini P, Giambene G, Hadzic S (2008) Broadband satellite multimedia networks, chapter XV. In: Cranley N, Murphy L (eds) Handbook of research on wireless multimedia: quality of service and solutions. IGI, Hershey, p 377–397
96. ESA Satlab (2010) SatLabs system recommendations—quality of service specifications
97. Survey on satellite systems. http://space.skyrocket.de/doc_sat/sat-contracts.htm
98. Maral G, Bousquet M, Sun Z (2019) Satellite communications systems: systems, techniques and technology. Wiley, London
99. Werner M, Jahn A, Lutz E, Bottcher A (1995) Analysis of system parameters for LEO/ICO-satellite communication networks. IEEE J Sel Areas Commun 13(2):371–381
100. Kodheli O, Lagunas E et al. (2020) Satellite communications in the new space era: a survey and future challenges. IEEE Commun Surv Tutorials 23(1):70–109
101. Guidotti A, Vanelli-Coralli A (2019) Clustering strategies for multicast precoding in multibeam satellite systems. Int J Satell Commun Netw 38(2):85–104. https://doi.org/10.1002/sat.1312
102. Cianca E, Rossi T, Yahalom A, Pinhasi Y, Farserotu J, Sacchi C (2011) EHF for satellite communications: the new broadband Frontier. Proc IEEE 99(11):1858–1881. https://doi.org/10.1109/JPROC.2011.2158765
103. de Cola T, Ginesi A, Giambene G, Polyzos GC, Siris VA, Fotiou N, Thomas Y (2017) Network and protocol architectures for future satellite systems. Found Trends Netw 12(1–2):1–161
104. del Portillo I., Cameron BG, Crawley EF (2018) A Technical comparison of three low earth orbit satellite constellation systems to provide global broadband. In: 69th international astronautical congress 2018. Bremen
105. Matlab Graphical Interface by Michael Hanchak (2011) Orbits—plot orbits around earth in an interactive manner. https://it.mathworks.com/matlabcentral/fileexchange/31932-orbits-plot-orbits-around-earth-in-an-interactive-manner
106. Xia S, Jiang Q, Zou C, Li G (2019) Beam coverage comparison of LEO satellite systems based on user diversification. IEEE Access 7:181656–181667
107. Web site of the Loon project. https://x.company/projects/loon/

Part II
Queuing Theory and Applications to Networks

Chapter 3
Survey on Probability Theory

Abstract This Chapter deals with the basis of the probability theory, the characterization of random variables using density and distribution functions, most common continuous and discrete random variables, how to deal with sum, difference, product and division of random variables, the use of the probability generating function, characteristic function, and Laplace transform. Some exercises are provided at the end.

Key words: Probability theory, Random variables, Probability Generating Function

3.1 The Notion of Probability and Basic Properties

Probability theory deals with the study of random events [1, 2]. Referring to an *experiment* (e.g., the measure of a quantity, the transmission of a bit or a packet in a telecommunication system, the toss of a dice, etc.), it can be characterized as:

1. The set of all possible results.
2. Classes, i.e., groups of results.
3. The frequencies according to which some classes occur when repeating the same experiment many times.

Let n_{A}^{n} denote the number of times class A occurs when repeating the same experiment n times. The relative frequency for class A, repeating the experiment n times, $f_{\mathrm{A},n}$, is defined as:

$$f_{\mathrm{A},n} = \frac{n_A^n}{n} \tag{3.1}$$

When $n \to \infty$, we expect that $f_{\mathrm{A},n}$ tends to a limit corresponding to a certain form of "regularity" in the statistical behavior of our experiment.

The mathematical model corresponding to the above experiment can be described as:

1. The space S containing all the elementary results: $S = \{\omega\}$.
2. The set of the events (where "event" denotes a group of elementary results). We can perform the typical operations for sets on events $A \subset S$, such as: "union" ($\cup$), "intersection" ($\cap$), "difference" ($\backslash$), "complementation" ($^{-}$).
3. A probabilistic measure representing an application P according to which elementary results in S are mapped into points of the positive real axis $\Re^+$ (see Fig. 3.1):

$$P: \forall A \subset S \to \Re^+$$

The frequency of class A corresponds to $P(A)$ in the limiting case of an increasing number of experiments:

Electronic Supplementary Material The online version contains supplementary material available at (https://doi.org/10.1007/978-3-030-75973-5_3).

G. Giambene, *Queuing Theory and Telecommunications*, Textbooks in Telecommunication Engineering,
https://doi.org/10.1007/978-3-030-75973-5_3

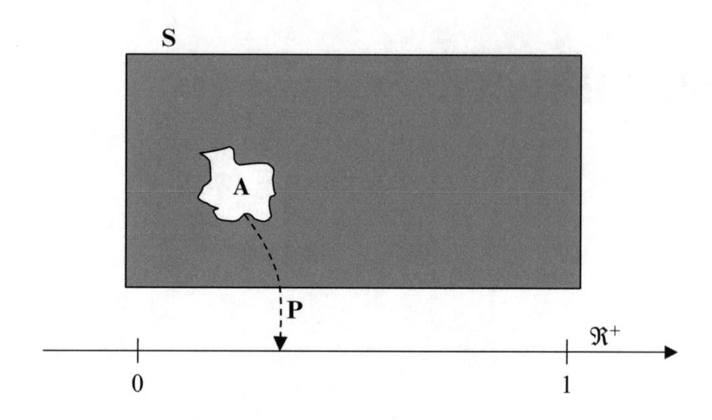

Fig. 3.1: The definition of the probability for an experiment

$$\lim_{n \to +\infty} f_{A,n} = P(A) \tag{3.2}$$

The probabilistic measure of a generic event A, $P(A)$, is axiomatically defined as:

1. $0 \leq P(A) \leq 1, \quad \forall A \subseteq S$.
2. $P(S) = 1$.
3. If $A \cap B = 0$ (i.e., A and B are *disjoint* events), $P(A \cup B) = P(A) + P(B)$.

Typical operations among sets can also be applied to events. For instance:

- "Commutativity": $A \cap B = B \cap A$; $A \cup B = B \cup A$;
- "Associativity": $(A \cap B) \cap C = A \cap (B \cap C)$; $(A \cup B) \cup C = A \cup (B \cup C)$;
- "Distributivity": $A \cap (B \cup C) = (A \cap B) \cup (A \cap C)$.

The intersection of events A and B refers to conditions to be verified by both A and B. Instead, the union of events A and B refers to conditions to be verified by either A or B.

It is easy to show that if A and B are not disjoint events, the following formula holds (and generalizes the previous point number 3):

$$P(A \cup B) = P(A) + P(B) - P(A \cap B) \tag{3.3}$$

Note that we need to include the term $-P(A \cap B)$, otherwise the terms in the intersection of the two sets would be counted twice.

Let us now consider we have to determine the probability of the union of three sets A, B, and C. We have

$$\begin{aligned} P(A \cup B \cup C) =& P(A) + P(B) + P(C) - P(A \cap B) \\ & - P(A \cap C) - P(B \cap C) + P(A \cap B \cap C) \end{aligned} \tag{3.4}$$

We can explain this formula is we look at a diagram showing the sets A, B, C, and their intersections. Then, we start summing the probabilities $P(A) + P(B) + P(C)$ but this is not correct because the elements in the intersections $P(A \cup B)$, $P(A \cup C)$, and $P(B \cup C)$ have been counted too much. Then, we compensate by subtracting these terms. However, we notice that $P(A \cup B \cup C)$ has been removed too much and needs to be added. A further generalization of formula (3.4) is discussed below after the definition of conditional probability and statistical independence.

Let us consider an event B so that $P(B) > 0$. We want to study if there are some implications for the occurrence of event A when event B occurs. Let $P(A|B)$ denote the *conditional probability* of event A given the occurrence of event B. $P(A|B)$ is defined as (Bayes rule):

$$P\left(A \middle| B\right) = \frac{P(A \cap B)}{P(B)} \tag{3.5}$$

Two events A and B are said to be *independent* if and only if the following relation is fulfilled:

$$P(A \cap B) = P(A) \times P(B) \tag{3.6}$$

Three events A, B, and C are independent if three independence conditions are fulfilled for the distinct combinations of two events (i.e., A and B, B and C, A and C) and if the following independence condition is met:

$$P(A \cap B \cap C) = P(A) \times P(B) \times P(C) \tag{3.7}$$

If A and B are independent, then from (3.5) we have: $P(A|B) = P(A)$.

Let us now discuss the generalization of (3.4) through the so-called *inclusion-exclusion principle*, yielding the probability of the union of a generic number n of events A_i for $i = 1, \ldots, n$, as:

$$\begin{aligned} P\left(\bigcup_{i=1}^{n} A_i\right) &= \sum_{i=1}^{n} P(A_i) - \sum_{1 \le i \le j \le n}^{n} P(A_i \cap A_j) \\ &+ \sum_{1 \le i \le j \le k \le n}^{n} P(A_i \cap A_j \cap A_j) - \cdots + (-1)^{n-1} P\left(\bigcap_{i=1}^{n} A_i\right) \end{aligned} \tag{3.8}$$

A relatively simple case is when all the events A_i are independent with the same probability, i.e., $P(A_i) = p$ for any i value. Then, the previous formula simplifies as follows:

$$P\left(\bigcup_{i=1}^{n} A_i\right) = \sum_{k=1}^{n} (-1)^{k-1} \binom{n}{k} p^k = 1 - (1-p)^n \tag{3.9}$$

This formula can be interpreted as follows: the union event is verified when at least one of the n considered events is verified.

Let us define a *complete partition of* S as a set of events A_i, $i = 1, 2, \ldots, n$ so that the two following conditions are fulfilled:

$$\begin{aligned} &S = \bigcup_{i=1}^{n} A_i \\ &A_i \cap A_j = 0, \quad \forall\ i, j \end{aligned} \tag{3.10}$$

Note that in this partition, all events A_i are disjoint.

The *total probability theorem* allows us to express the probability of event B utilizing the conditional probabilities $P(B|A_i)$, where events A_i, $i = 1, 2, \ldots, n$, form a complete partition of S:

$$P(B) = \sum_{i=1}^{n} P(B \cap A_i) = \sum_{i=1}^{n} P\left(B \middle| A_i\right) P(A_i) \tag{3.11}$$

Formula (3.11) can be demonstrated as follows:

$$B = B \cap S = B \cap \bigcup_{i=1}^{n} A_i = \bigcup_{i=1}^{n} (B \cap A_i) \tag{3.12}$$

Since A_i events are disjoint, if we take the probability on both sides of (3.12), we can use the definition of the probabilities to write

$$P(B) = P\left[\bigcup_{i=1}^{n} (B \cap A_i)\right] = \sum_{i=1}^{n} P(B \cap A_i) = \sum_{i=1}^{n} P\left(B \middle| A_i\right) P(A_i)$$

The above result has been obtained since $B \cap A_i$ events are disjoint $\forall i$.

The total probability theorem provides a simple method to deal with the derivation of the probability of complex events B; this approach is based on the derivation of conditional probabilities $P(B|A_i)$, referring to particular conditioning events A_i that constitute a complete partition of S.

Referring to a complete partition A_i of S, we are interested in determining relations between a posteriori conditional probabilities $P(A_i|B)$, and a priori conditional probabilities $P(B|A_i)$:

$$P\left(A_i \middle| B\right) = \frac{P\left(B \middle| A_i\right) P\left(A_i\right)}{\sum_{i=1}^{n} P\left(B \middle| A_i\right) P\left(A_i\right)} \tag{3.13}$$

Let us prove (3.13). From the conditional probability definition, we have

$$P\left(A_i \middle| B\right) = \frac{P\left(A_i \cap B\right)}{P(B)} \quad \text{and} \quad P\left(B \middle| A_i\right) = \frac{P\left(B \cap A_i\right)}{P\left(A_i\right)} \tag{3.14}$$

From both Eqs. (3.14) we obtain $P(A_i \cap B) = P(B \cap A_i)$ so that:

$$P\left(A_i \middle| B\right) P(B) = P\left(A_i \cap B\right) = P\left(B \cap A_i\right) = P\left(B \middle| A_i\right) P\left(A_i\right) \tag{3.15}$$

From (3.15) we can solve $P(A_i|B)$ as:

$$P\left(A_i \middle| B\right) = \frac{P\left(B \middle| A_i\right) P\left(A_i\right)}{P(B)} \tag{3.16}$$

From the total probability theorem, we can express $P(B)$ in (3.16) according to (3.11), thus obtaining (3.13).

3.2 Random Variables: Basic Definitions and Properties

A random variable X is related to an experiment for which a mathematical model has been defined and, in particular, a probability. A random variable X can be defined as an application, which maps elementary results with values on the real axis, $\Re$. Let Ω denote the space of possible values of variable X. Note that Ω can represent a discrete set with finite or infinite values or can represent a continuous set of values. Random variable X can be defined based on the following probability:

$$\text{Prob}\left\{X = x\right\} = P\left\{\omega \in S : X\left(\omega\right) = x\right\} \tag{3.17}$$

where $X(\omega) = x$ denotes the mapping function associated with the random variable definition. $X(\omega)$ maps the experiment outcome $\omega \in \Omega$ onto a certain value $x \in \Re$ on the real axis.

Let us introduce the first example of a random variable as follows. We consider an experiment in which event A occurs with probability $P(A)$ at each trial, independently of the others. Random variable n_A^n denotes the number of times that event A occurs over n trials of the experiment. This random variable is therefore characterized by the following probability: $\text{Prob}\{n_A^n = k\}$, with $k \in \{0,1,\ldots,n\}$. A given "configuration" where event A occurs k times over n trials has a probability of occurring equal to $P(A)^k[1 - P(A)]^{n-k}$. The number of configurations of k events A over n trials is given by the *binomial coefficient*:

$$\binom{n}{k} = \frac{n!}{k!\,(n-k)!} \tag{3.18}$$

In conclusion, the probability distribution function (or probability mass function) of n_A^n is characterized as follows because the different combinations represent disjoint events:

$$\text{Prob}\left\{n_A^n = k\right\} = \binom{n}{k} P(A)^k [1 - P(A)]^{n-k} \tag{3.19}$$

A random variable with a distribution like that shown in (3.19) is binomially distributed.

Other typical examples of random variables are derived from the random experiments related to toss a coin or to roll a dice. For instance, let us refer to the experiment of the dice: the state space is $S = \{1, 2, 3, 4, 5, 6\}$. We can construct different random variables with the outcomes ω of this experiment; for instance, let us consider variables X and Y as follows:

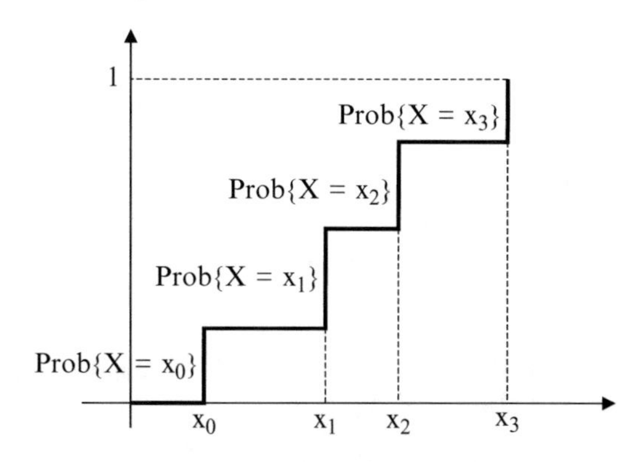

Fig. 3.2: Representation of the PDF for a discrete random variable with values x_0, x_1, x_2, and x_3 (staircase function)

$$X = \begin{cases} 1, \text{if } \omega = 1, \text{with probability } 1/6 \\ 2, \text{if } \omega = 2, \text{with probability } 1/6 \\ 3, \text{if } \omega = 3, \text{with probability } 1/6 \\ 4, \text{if } \omega = 4, \text{with probability } 1/6 \\ 5, \text{if } \omega = 5, \text{with probability } 1/6 \\ 6, \text{if } \omega = 6, \text{with probability } 1/6 \end{cases}$$

or

$$Y = \begin{cases} 1, \text{if } \omega \in \{1, 3, 5\}, \text{with probability } 1/2 \\ 0, \text{if } \omega \in \{2, 4, 6\}, \text{with probability } 1/2 \end{cases}$$

A generic random variable X can be described using the Probability Distribution Function (PDF), $F_X(x)$, as:

$$F_X(x) = \text{Prob}\{X \leq x\} = P\{\omega \in S : X(\omega) \leq x\} \tag{3.20}$$

$F_X(x)$ is a dimensionless quantity since it is a probability.

The typical properties of a PDF function are:

1. $0 \leq F_X(x) \leq 1$.
2. $F_X(x)$ is non-decreasing.
3. $F_X(x)$ tends to 1 for x approaching $+\infty$; $F_X(x)$ tends to 0 for x approaching $-\infty$.
4. $F_X(x)$ is a right-continuous function:

$$\lim_{x \to x_0^+} F_X(x) = F_X(x_0)$$

5. For a discrete random variable X with values x_0, x_1, $x_2, \ldots$, the PDF is a staircase function with steps of amplitudes Prob$\{X = x_i\}$ with $i = 0, 1, \ldots$ corresponding to the points x_i, as shown in Fig. 3.2. In particular, the PDF can be formally expressed as:

$$F_X(x) = \text{Prob}\{X \leq x\} = \sum_{i=0}^{n} \text{Prob}\{X = x_i\} I(x - x_i) \tag{3.21}$$

where $I(\cdot)$ is the *unit step function* (or Heaviside function) defined as follows: $I(x) = 1$ for $x > 0$ and $I(x) = 0$ for $x \leq 0$.

According to the definition of the PDF of a random variable X, we evaluate the probability that X is in a given interval $(x_1, x_2]$, as described below. We start by considering the following event and its equivalent expression:

$$\{X \leq x_2\} = \{X \leq x_1\} \cup \{x_1 < X \leq x_2\}$$

We take the probability of both sides, noticing that we have two disjoint events on the left side:

$$\text{Prob}\{X \leq x_2\} = \text{Prob}\{X \leq x_1\} + \text{Prob}\{x_1 < X \leq x_2\}$$

Then, we solve with respect to $\text{Prob}\{x_1 < X \leq x_2\}$:

$$\begin{aligned}\text{Prob}\{x_1 < X \leq x_2\} &= \text{Prob}\{X \leq x_2\} - \text{Prob}\{X \leq x_1\} \\ &= F_X(x_2) - F_X(x_1)\end{aligned} \tag{3.22}$$

The generic random variable X can be equivalently described in terms of the probability density function (pdf) $f_X(x)$ that is obtained from the corresponding PDF as:

$$f_X(x) = \frac{\mathrm{d}}{\mathrm{d}x} F_X(x) \tag{3.23}$$

Note that $f_X(x)$ has the dimensions of x^{-1}.

The typical properties of a pdf correspond to those of the PDF and can be detailed as follows:

1. $f_X(x) \geq 0$ (non-decreasing PDF function).
2. The PDF function can be obtained according to the following integral:

$$F_X(x) = \int_{-\infty}^{x} f_X(x)\mathrm{d}x \tag{3.24}$$

3. Normalization condition:

$$\int_{-\infty}^{+\infty} f_X(x)\mathrm{d}x = 1 \tag{3.25}$$

For a discrete random variable X, the pdf is characterized by Dirac Delta impulses centered in the points x_i and with amplitudes $\text{Prob}\{X = x_i\}$. Based on (3.23), we obtain the pdf as the derivative of the PDF in (3.21). This derivative must be computed in a general sense by considering that $\mathrm{d}I(x)/\mathrm{d}x = \delta(x)$, the Dirac Delta function centered in the origin. Hence, we have

$$f_X(x) = \sum_{i=0}^{n} \text{Prob}\{X = x_i\}\,\delta(x - x_i) \tag{3.26}$$

Different random variables can be defined for the same experiment. We can consider, for instance, two random variables X and Y, for which we define the following *joint distribution*:

$$F_{XY}(x, y) = \text{Prob}\{X \leq x, Y \leq y\} \tag{3.27}$$

The comma sign "," (to be read as "and") within the "Prob" operator in (3.27) denotes the intersection of events: we study the joint occurrence of $X \leq x$ and $Y \leq y$.

The above joint PDF and the joint pdf are related as follows:

$$f_{XY}(x, y) = \frac{\partial^2}{\partial x \partial y} F_{XY}(x, y) \Longleftrightarrow F_{XY}(x, y) = \int_{-\infty}^{x}\int_{-\infty}^{y} f_{XY}(x, y)\,\mathrm{d}x\,\mathrm{d}y \tag{3.28}$$

If we integrate the joint pdf $f_{XY}(x,y)$ on all possible values of y (or x), we achieve the *marginal* pdf $f_X(x)$ [or $f_Y(y)$]. For instance, we have

$$f_X(x) = \int_{y=-\infty}^{+\infty} f_{XY}(x, y)\,\mathrm{d}y \tag{3.29}$$

Two random variables X and Y are statistically independent if and only if the following condition is fulfilled:

$$f_{XY}(x, y) = f_X(x) \times f_Y(y) \Longleftrightarrow F_{XY}(x, y) = F_X(x) \times F_Y(y) \tag{3.30}$$

In general, n random variables $X_1, X_2, \ldots, X_n$ are statistically independent if and only if the following condition is fulfilled:

$$f_{X_1,X_2,\ldots,X_n}(x_1,x_2,\ldots,x_n) = \prod_{i=1}^{n} f_{X_i}(x) \iff F_{X_1,X_2,\ldots,X_n}(x_1,x_2,\ldots,x_n) = \prod_{i=1}^{n} F_{X_i}(x) \quad (3.31)$$

This condition is simpler than that required for the independence of n events.

Let us consider the conditional PDF of random variable X given the value of another random variable Y:

$$F_{X|Y}(x|y) = \text{Prob}\{X \le x | Y = y\} \quad (3.32)$$

The conditional PDF and the conditional pdf are related as follows:

$$f_{X|Y}(x|y) = \frac{\partial}{\partial x} F_{X|Y}(x|y) = \frac{f_{XY}(x,y)}{f_Y(y)} \quad (3.33)$$

The PDF of variable X can be derived from (3.32) and $f_Y(y)$ as:

$$F_X(x) = \int_{y=-\infty}^{+\infty} F_{X|Y}(x|y) f_Y(y)\mathrm{d}y \quad (3.34)$$

3.2.1 Minimum and Maximum of Random Variables

We have the random variables X and Y for which we know the joint pdf $f_{XY}(x,y)$ and the marginal PDFs. We need to characterize the distribution of the following new variables: $Q = \max\{X, Y\}$, and $W = \min\{X, Y\}$.

The PDF of the maximum, $F_Q(q)$, can be derived as:

$$F_Q(q) = \text{Prob}\{Q \le q\} = \text{Prob}\{X \le q, Y \le q\} \quad (3.35)$$

If X and Y are statistically independent, from (3.35) we have

$$F_Q(q) = \text{Prob}\{X \le q\} \times \text{Prob}\{Y \le q\} = F_X(q) \times F_Y(q) \quad (3.36)$$

The PDF of the minimum, $F_W(w)$, can be obtained as:

$$\begin{aligned} F_W(w) &= \text{Prob}\{W \le w\} = \text{Prob}\{\{X \le w\} \cup \{Y \le w\}\} \\ &= \text{Prob}\{X \le w\} + \text{Prob}\{Y \le w\} - \text{Prob}\{\{X \le w\} \cap \{Y \le w\}\} \end{aligned} \quad (3.37)$$

We can rewrite the result in (3.37) as:

$$F_W(w) = F_X(w) + F_Y(w) - \text{Prob}\{X \le w, Y \le w\} \quad (3.38)$$

If X and Y are statistically independent, from (3.38) we have

$$F_W(w) = F_X(w) + F_Y(w) - F_X(w) \times F_Y(w) \quad (3.39)$$

The corresponding expressions of the pdfs can be obtained through the derivative of $F_Q(q)$ with respect to q and the derivative of $F_W(w)$ with respect to w. Note that random variables Q and W have particular relevance in the field of telecommunications. For instance, let us consider the case where a message is transmitted until its service timeout expires; the effective "message service time" is the minimum between the message transmission time and the service timeout (deadline). Another example is when we have a transmission system with two transmitters that send simultaneously the same information for redundancy: the system's operation is guaranteed for a time, which is the maximum of the lifetimes of the two parts.

3.2.2 Comparisons of Random Variables

We have two random variables X and Y, for which we know PDFs and pdfs. We need to express Prob$\{X > Y\}$. From the definition of conditional probability, we have

$$\text{Prob}\{X > Y\} = \int_Y \text{Prob}\left\{X > y \middle| Y = y\right\} f_Y(y)\text{d}y \tag{3.40}$$

where $f_Y(y)$ denotes the pdf of random variable Y.

If X and Y are statistically independent, we have: Prob$\{X > y | Y = y\} =$ Prob$\{X > y\} = 1 - F_X(y)$. Hence, (3.40) can be elaborated as follows:

$$\begin{aligned}\text{Prob}\{X > Y\} &= \int_Y \text{Prob}\{X > y\} f_Y(y)\text{d}y \\ &= \int_Y [1 - F_X(y)] f_Y(y)\text{d}y = 1 - \int_Y F_X(y) f_Y(y)\text{d}y\end{aligned} \tag{3.41}$$

3.2.3 Moments of Random Variables

The moments are quantities used to characterize the random variables. Their values can be either finite or infinite.

3.2.3.1 Expected Value of a Random Variable

The expected value of random variable X is a statistical mean that can be computed as:

$$E[X] = \begin{cases} \displaystyle\int_{-\infty}^{+\infty} x f_X(x)\text{d}x \text{ for a continuous variable} \\ \displaystyle\sum_i x_i \text{Prob}\{X = x_i\} \text{ for a discrete variable} \end{cases} \tag{3.42}$$

For discrete random variables, we can still adopt the same definition of operator $E[\cdot]$ used for continuous variables but, in this case, the pdf is given by the sum of Dirac Delta functions.

Operator $E[\cdot]$ is linear. For example, let us consider random variable X, and define a new random variable as $aX + b$, where a and b are fixed coefficients. Then, the linearity of operator $E[\cdot]$ can be used as follows:

$$E[aX + b] = aE[X] + b \tag{3.43}$$

If we have two random variables X and Y, we can consider the sum $X + Y$ as a new random variable. The expected value of $X + Y$ can be derived by means of the joint pdf $f_{XY}(x,y)$ as follows:

$$\begin{aligned}E[X+Y] &= \int_{y=-\infty}^{+\infty}\int_{x=-\infty}^{+\infty} (x+y) f_{XY}(x,y)\text{d}x\,\text{d}y \\ &= \int_{y=-\infty}^{+\infty}\int_{x=-\infty}^{+\infty} x f_{XY}(x,y)\text{d}x\,\text{d}y + \int_{y=-\infty}^{+\infty}\int_{x=-\infty}^{+\infty} y f_{XY}(x,y)\text{d}x\,\text{d}y \\ &= \int_{x=-\infty}^{+\infty} x\left[\int_{y=-\infty}^{+\infty} f_{XY}(x,y)\,\text{d}y\right]\text{d}x + \int_{y=-\infty}^{+\infty} y\left[\int_{x=-\infty}^{+\infty} f_{XY}(x,y)\,\text{d}x\right]\text{d}x\end{aligned} \tag{3.44}$$

$$= \int_{x=-\infty}^{+\infty} y f_X(x) \mathrm{d}x + \int_{y=-\infty}^{+\infty} y f_Y(y) \mathrm{d}y = E[X] + E[Y]$$

The important result in (3.44), that is $E[X+Y] = E[X] + E[Y]$, has been obtained without special assumptions (e.g., independence assumption). We will further elaborate on the sum of random variables to characterize the corresponding distribution in Sect. 3.2.7.

If we have two independent random variables X and Y, with joint pdf $f_{XY}(x,y) = f_X(x) \times f_Y(y)$, the random variable given by the product $X \times Y$ has a mean value as follows:

$$\begin{aligned} E[X \times Y] &= \int_{y=-\infty}^{+\infty} \int_{x=-\infty}^{+\infty} (x \times y) f_{XY}(x,y) \mathrm{d}x\, \mathrm{d}y \\ &= \int_{y=-\infty}^{+\infty} y f_Y(y) \mathrm{d}y \times \int_{x=-\infty}^{+\infty} x f_X(x) \mathrm{d}x = E[X] \times E[Y] \end{aligned} \tag{3.45}$$

We will further elaborate the case of the distribution of the product of random variables in Sect. 3.2.7.

3.2.3.2 The *m*th Moment of a Random Variable

The mth moment of random variable X is defined as $E[X^m]$ through operator $E[\cdot]$ shown in (3.42); in particular, for continuous random variables, we have

$$E[X^m] = \int_{-\infty}^{+\infty} x^m f_X(x) \mathrm{d}x \tag{3.46}$$

Of particular relevance is the second moment, representing the mean square value, $E[X^2]$.

3.2.3.3 Variance of a Random Variable

The variance of random variable X is defined through operator $E[\cdot]$ in (3.42) as:

$$\begin{aligned} \mathrm{Var}[X] &= E\left[(X - E[X])^2\right] = E\left[X^2 + \{E[X]\}^2 - 2XE[X]\right] \\ &= \text{using the linearity of operator } E[\cdot] \\ &= E[X^2] + \{E[X]\}^2 - 2\{E[X]\}^2 = E[X^2] - \{E[X]\}^2 \end{aligned} \tag{3.47}$$

If we have two random variables X and Y, we can consider the new random variable given by the sum $X + Y$. The variance of $X + Y$ can be derived using the joint pdf $f_{XY}(x,y)$ as follows:

$$\begin{aligned} \mathrm{Var}[X+Y] &= E\left[(X + Y - E[X] - E[Y])^2\right] \\ &= \int_{y=-\infty}^{+\infty} \int_{x=-\infty}^{+\infty} (X + Y - E[X] - E[Y])^2 f_{XY}(x,y) \mathrm{d}x\, \mathrm{d}y \\ &= E\left[(X - E[X])^2 + (Y - E[Y])^2 + 2(X - E[X])(Y - E[Y])\right] \\ &= \text{using the linearity of operator } E[\cdot] \\ &= E\left[(X - E[X])^2\right] + E\left[(Y - E[Y])^2\right] \\ &\quad + 2E[(X - E[X])(Y - E[Y])] \\ &= \mathrm{Var}[X] + \mathrm{Var}[Y] + 2E[(X - E[X])(Y - E[Y])] \end{aligned} \tag{3.48}$$

The quantity $E[(X - E[X])(Y - E[Y])]$ in (3.48) represents the *covariance* of random variables X and Y. Two random variables with null covariance are said to be *uncorrelated.* If X and Y are statistically independent so that $f_{XY}(x,y) = f_X(x) \times f_Y(y)$, it is easy to show that the covariance is null: $E[(X - E[X])(Y - E[Y])] = 0$. Hence, the statistical independence is a sufficient condition for a null covariance so that the variance of the sum of two random variables becomes

$$\mathrm{Var}\left[X+Y\right]=\mathrm{Var}\left[X\right]+\mathrm{Var}\left[Y\right] \tag{3.49}$$

The square root of the variance of random variable X is the standard deviation σ_X:

$$\sigma_X=\sqrt{\mathrm{Var}\left[X\right]} \tag{3.50}$$

To compare the "randomness" of distributions, we can consider the standard deviation normalized to the mean value. Therefore, the following *coefficient of variation*, C_X, is defined for random variable X:

$$C_X=\frac{\sigma_X}{E\left[X\right]} \tag{3.51}$$

C_X is a dimensionless number. For a deterministic variable, $C_X=0$. More details on the coefficient of variation are provided in Sect. 3.2.6.8.

3.2.3.4 The *m*th Central Moment of a Random Variable

The mth central moment of random variable X is defined as $E[(X-E[X])^m]$ through operator $E[\cdot]$ shown in (3.42); in particular, for continuous random variables, we have

$$E\left[(X-E\left[X\right])^m\right]=\int_{-\infty}^{+\infty}(x-E\left[X\right])^m f_X(x)\mathrm{d}x \tag{3.52}$$

The second central moment is the variance.

3.2.3.5 The *n*th Percentile of a Random Variable

The nth percentile of random variable X [with pdf $f_X(x)$] is a value ζ so that the probability that the values of X are lower than or equal to ζ is equal to n%.

$$\text{The } n\text{th percentile of } X \text{ is a value } \varsigma \text{ so that } \int_{-\infty}^{\varsigma} f_X(x)\mathrm{d}x=\frac{n}{100}$$

3.2.4 Function of a Random Variable

We consider now a random variable X and a function $g(\cdot)$. From random variable X we want to construct a new random variable $Y=g(X)$. Based on the PDF of X, $F_X(x)$, we obtain the PDF of Y, $F_Y(y)$, as:

$$F_Y(y)=\mathrm{Prob}\left\{Y\leq y\right\}=\mathrm{Prob}\left\{g(X)\leq y\right\} \tag{3.53}$$

From (3.53), we have to consider the X values' disjoint intervals where condition $g(X)\leq y$ is fulfilled. Then, we compute the probability as the sum of the probabilities of X belonging to these different intervals. In the simple case where function $g(\cdot)$ is invertible, there is only one interval, and the PDF of Y can be further elaborated. In particular, if $g(X)$ is monotonically increasing with X, we can write

$$F_Y(y)=\mathrm{Prob}\left\{g(X)\leq y\right\}=\mathrm{Prob}\left\{X\leq g^{-1}(y)\right\}=F_X\left[g^{-1}(y)\right] \tag{3.54}$$

Then, the density function of Y can be obtained as:

$$f_Y\left(y\right)=\frac{\mathrm{d}}{\mathrm{d}y}F_X\left[g^{-1}\left(y\right)\right]$$

$$= f_X\left[g^{-1}(y)\right] \times \frac{\mathrm{d}\left[g^{-1}(y)\right]}{\mathrm{d}y} = \frac{f_X\left[g^{-1}(y)\right]}{g'\left[g^{-1}(y)\right]} \tag{3.55}$$

Finally, the mth moment of random variable Y results as:

$$E[Y^m] = \int_y y^m f_Y(y)\,\mathrm{d}y = \int_y g^m(x) \frac{f_X\left[g^{-1}(y)\right]}{g'\left[g^{-1}(y)\right]} \mathrm{d}g(x)$$
$$= \int_x g^m(x) f_X(x)\,\mathrm{d}x \tag{3.56}$$

3.2.5 Discrete Random Variables in the Field of Telecommunications

In this section, basic examples of discrete random variables and the derivation of their principal moments will be shown. In particular, we will consider discrete uniform distribution, geometric distribution, Poisson distribution, and binomial distribution. This section aims to explain some applications of these random variables in the field of telecommunications.

3.2.5.1 The Uniform Distribution (Discrete Case)

A discrete random variable X has a uniform distribution over the discrete set of values $\{1, 2, \ldots, m\}$ when the generic value i is verified with probability $p_i = 1/m$. Let us now derive mean, mean square value, and variance for the discrete uniform random variable:

$$E[X] = \sum_{i=1}^{m} i p_i = \frac{1}{m} \sum_{i=1}^{m} i = \frac{1}{m} \frac{m(m+1)}{2} = \frac{(m+1)}{2} \tag{3.57}$$

$$E\left[X^2\right] = \sum_{i=1}^{m} i^2 p_i = \frac{1}{m} \sum_{i=1}^{m} i^2 = \frac{1}{m} \frac{m(m+1)(2m+1)}{6} = \frac{(m+1)(2m+1)}{6} \tag{3.58}$$

$$\mathrm{Var}[X] = E\left[X^2\right] - E^2[X] = \frac{(m+1)^2}{12} \tag{3.59}$$

3.2.5.2 The Geometric Distribution

A discrete random variable X is geometrically distributed if its probability mass function can be represented as:

$$\mathrm{Prob}\{X = k\} = (1-q)\,q^k, \qquad 0 < q < 1, \qquad k = 0,\ 1,\ 2, \ldots \tag{3.60}$$

where q is a dimensionless parameter and k represents natural numbers.

An example of the use of this random variable is as follows. Let us refer to time-slotted transmissions of packets, where slots are available to transmit packets with probability $1 - q$. The random variable X with the distribution in (3.60) represents the number of slots needed to send one packet by a traffic source.

A variant of the "classical" geometric distribution is the "modified" geometric distribution; in this case, the random variable X has the following probability mass function where the k values start from 1:

$$\mathrm{Prob}\{X = k\} = (1-q)\,q^{k-1}, \qquad 0 < q < 1, \qquad k = 1,\ 2,\ 3, \ldots \tag{3.61}$$

A modified geometric distribution like the above can be used to model the number of transmission attempts to send a packet successfully, if q denotes the probability of a packet loss (or of a packet transmission error).

The following derivations are for the "classical" geometric distribution in (3.60); their adaptation to the "modified" geometric distribution is straightforward.

The normalization condition for the distribution in (3.60) is as follows:

$$\begin{aligned}\sum_{k=0}^{\infty} \text{Prob}\{X=k\} &= \sum_{k=0}^{\infty}(1-q)\,q^k = (1-q)\times\sum_{k=0}^{\infty} q^k \\ &= \text{by invoking the } \textit{geometric} \text{ series} \\ &= (1-q)\times\frac{1}{1-q} = 1\end{aligned} \tag{3.62}$$

The mean value of the distribution in (3.60) results as:

$$\begin{aligned}E[X] &= \sum_{k=0}^{\infty} k\times\text{Prob}\{X=k\} = \sum_{k=0}^{\infty} k\,(1-q)\,q^k = (1-q)\,q\times\sum_{k=0}^{\infty} kq^{k-1} \\ &= \text{note that } kq^{k-1} \text{ is the derivative of } q^k, \\ &\quad \text{thus exchanging sum and derivative} \\ &= (1-q)\,q\times\frac{\mathrm{d}}{\mathrm{d}q}\sum_{k=0}^{\infty} q^k = (1-q)\,q\times\frac{\mathrm{d}}{\mathrm{d}q}\frac{1}{1-q} = \frac{(1-q)\,q}{(1-q)^2} = \frac{q}{1-q}\end{aligned} \tag{3.63}$$

Note that sum and derivative can be exchanged under the assumption of uniform series convergence.

The mean square value of the distribution in (3.60) is

$$\begin{aligned}E\left[X^2\right] &= \sum_{k=0}^{\infty} k^2\times\text{Prob}\{X=k\} = \sum_{k=0}^{\infty} k^2\,(1-q)\,q^k \\ &= (1-q)\,q\times\sum_{k=0}^{\infty} k\left(kq^{k-1}\right) \\ &= \text{note that } kq^{k-1} \text{ is the derivative of } q^k, \\ &\quad \text{thus exchanging sum and derivative} = (1-q)\,q\times\frac{\mathrm{d}}{\mathrm{d}q}\sum_{k=0}^{\infty} kq^k\end{aligned} \tag{3.64}$$

Since from (3.63) we have

$$q\times\sum_{k=0}^{\infty} kq^{k-1} = \sum_{k=0}^{\infty} kq^k = \frac{q}{(1-q)^2},$$

we substitute such expression in (3.64) to obtain $E[X^2]$ as:

$$E\left[X^2\right] = (1-q)\,q\times\frac{\mathrm{d}}{\mathrm{d}q}\frac{q}{(1-q)^2} = (1-q)\,q\times\frac{1+q}{(1-q)^3} = \frac{(1+q)\,q}{(1-q)^2} \tag{3.65}$$

Finally, the variance of the geometric distribution can be obtained as:

$$\text{Var}[X] = E\left[X^2\right] - \{E[X]\}^2 = \frac{(1+q)\,q}{(1-q)^2} - \frac{q^2}{(1-q)^2} = \frac{q}{(1-q)^2} \tag{3.66}$$

3.2.5.3 The Poisson Distribution

A discrete random variable X is Poisson distributed if it has the following probability mass function:

$$\text{Prob}\{X = k\} = \frac{\rho^k}{k!} e^{-\rho}, \quad \rho > 0, \quad k = 0, 1, 2, \ldots \tag{3.67}$$

where ρ is a dimensionless parameter (we will show later that it represents the expected value), and the k values assumed by this random variable are the natural numbers.

An example of the use of this random variable is as follows. Let us assume to count the number k of phone call arrivals at a central office for a generic interval of length t: such a number can be modeled according to a Poisson distribution with a value of a parameter ρ proportional to the duration t through the call arrival rate.

The normalization condition for the distribution in (3.67) is verified as follows:

$$\begin{aligned}\sum_{k=0}^{\infty} \text{Prob}\{X = k\} &= \sum_{k=0}^{\infty} \frac{\rho^k}{k!} e^{-\rho} = e^{-\rho} \times \sum_{k=0}^{\infty} \frac{\rho^k}{k!} \\ &= \text{by invoking the } \textit{exponential} \text{ series} \\ &= e^{-\rho} \times e^{\rho} = 1\end{aligned} \tag{3.68}$$

The mean value of the Poisson distribution (3.67) is as follows:

$$\begin{aligned}E[X] &= \sum_{k=0}^{\infty} k \times \text{Prob}\{X = k\} = \sum_{k=0}^{\infty} k \frac{\rho^k}{k!} e^{-\rho} = e^{-\rho} \times \sum_{k=1}^{\infty} k \frac{\rho^k}{k!} \\ &= \rho\, e^{-\rho} \times \sum_{k=1}^{\infty} \frac{\rho^{k-1}}{(k-1)!} = \rho\, e^{-\rho} \times \sum_{j=0}^{\infty} \frac{\rho^j}{j!} = \rho\, e^{-\rho} \times e^{\rho} = \rho\end{aligned} \tag{3.69}$$

The mean square value of the Poisson distribution (3.67) is as follows:

$$\begin{aligned}E\left[X^2\right] &= \sum_{k=0}^{\infty} k^2 \times \text{Prob}\{X = k\} = \sum_{k=0}^{\infty} k^2 \frac{\rho^k}{k!} e^{-\rho} = e^{-\rho} \times \sum_{k=1}^{\infty} k^2 \frac{\rho^k}{k!} \\ &= \rho\, e^{-\rho} \times \sum_{k=1}^{\infty} k \frac{\rho^{k-1}}{(k-1)!} \\ &= \text{note that } k\rho^{k-1} \text{ is the derivative of } \rho^k, \text{thus exchanging} \\ &\quad \text{sum and derivative} \\ &= \rho\, e^{-\rho} \times \frac{\mathrm{d}}{\mathrm{d}\rho} \left[\rho \sum_{k=1}^{\infty} \frac{\rho^{k-1}}{(k-1)!}\right] = \rho\, e^{-\rho} \times \frac{\mathrm{d}}{\mathrm{d}\rho} \left[\rho \sum_{j=0}^{\infty} \frac{\rho^j}{j!}\right] \\ &= \rho\, e^{-\rho} \times \frac{\mathrm{d}}{\mathrm{d}\rho} \left[\rho\, e^{\rho}\right] = \rho\, e^{-\rho} \times \left[e^{\rho} + \rho\, e^{\rho}\right] = \rho\,(1 + \rho)\end{aligned} \tag{3.70}$$

Finally, the variance of the Poisson distribution can be obtained as:

$$\text{Var}[X] = E\left[X^2\right] - \{E[X]\}^2 = \rho\,(1 + \rho) - \rho^2 = \rho \tag{3.71}$$

Note that mean and variance are equal for the Poisson distribution. This result should not surprise since parameter ρ is dimensionless.

3.2.5.4 The Binomial Distribution

A discrete and finite random variable X is binomially distributed if it has the following probability mass function:

$$\text{Prob}\{X = k\} = \binom{n}{k} p^k (1 - p)^{n-k}, \;\; 0 < p < 1, \;\; k = 0, \; 1, \; 2, \; \ldots, \; n \tag{3.72}$$

This distribution is used to characterize the number of times k a given event with probability p occurs on n different trials of the same experiment. This random variable is particularly useful, for instance, in the case of the transmission of a message with n packets on a channel, which introduces memoryless errors on each packet with probability p. Hence, the number of packet errors per message is binomially distributed as in (3.72).

The normalization condition for the distribution in (3.72) is verified as follows:

$$\sum_{k=0}^{\infty} \text{Prob}\{X=k\} = \sum_{k=0}^{n} \binom{n}{k} p^k (1-p)^{n-k} \\ = \text{by invoking the Newton binomial formula} \\ = [p+1-p]^n = 1 \quad (3.73)$$

The mean value of the binomial distribution in (3.72) is as follows:

$$\begin{aligned} E[X] &= \sum_{k=0}^{n} k \times \text{Prob}\{X=k\} = \sum_{k=1}^{n} k \binom{n}{k} p^k (1-p)^{n-k} \\ &= \sum_{k=1}^{n} \frac{n!}{(k-1)!\,(n-k)!} p^k (1-p)^{n-k} \\ &= np \times \sum_{k=1}^{n} \frac{(n-1)!}{(k-1)!\,[n-1-(k-1)]!} p^{k-1} (1-p)^{n-1-(k-1)} \\ &= np \times \sum_{j=0}^{n-1} \frac{(n-1)!}{j!\,[n-1-j]!} p^j (1-p)^{n-1-j} \\ &= np \times [p+1-p]^{n-1} \\ &= np \end{aligned} \quad (3.74)$$

The mean square value of the binomial distribution in (3.72) is obtained through complex manipulations. We provide below only the final result:

$$\begin{aligned} E\left[X^2\right] &= \sum_{k=0}^{n} k^2 \times \text{Prob}\{X=k\} = \sum_{k=1}^{n} k^2 \binom{n}{k} p^k (1-p)^{n-k} \\ &= np + np^2 (n-1) \end{aligned} \quad (3.75)$$

Finally, the variance of the binomial distribution is

$$\text{Var}[X] = E\left[X^2\right] - \{E[X]\}^2 = np + np^2(n-1) - (np)^2 = np - np^2 \quad (3.76)$$

The graph in Fig. 3.3 compares the behaviors of the probability mass functions for three discrete random variables having the same mean value (=5): the "modified" geometric distribution, the Poisson distribution, and the binomial distribution (with probability $p=0.5$ and maximum value $n=10$).

3.2.6 Continuous Random Variables in the Field of Telecommunications

This section deals with examples and properties of continuous random variables (i.e., uniform, exponential, Erlang, Hyper-exponential, Gaussian, and Pareto distributions) and their applications to the field of telecommunications. Apart from the Gaussian distribution case, these variables are greater than or equal to 0 so that they can represent service times or time intervals in general.

3.2.6.1 Conditional Residual Lifetime

Let us refer to a random variable X representing the duration of a phenomenon started at instant $t=0$. X is a continuous non-negative random variable with pdf $f_X(t)$ and PDF $F_X(t)$. We examine the same phenomenon at time τ, and we assume that it is still active (i.e., $X>\tau$). τ is called the "age" of the phenomenon. We need to determine the PDF of the *conditional residual length* of the event, that is $R_{|\tau} = X - \tau$ with $X>\tau$:

$$F_{R_{|\tau}}(t) = \text{Prob}\left\{X - \tau \le t \middle| X > \tau\right\}$$

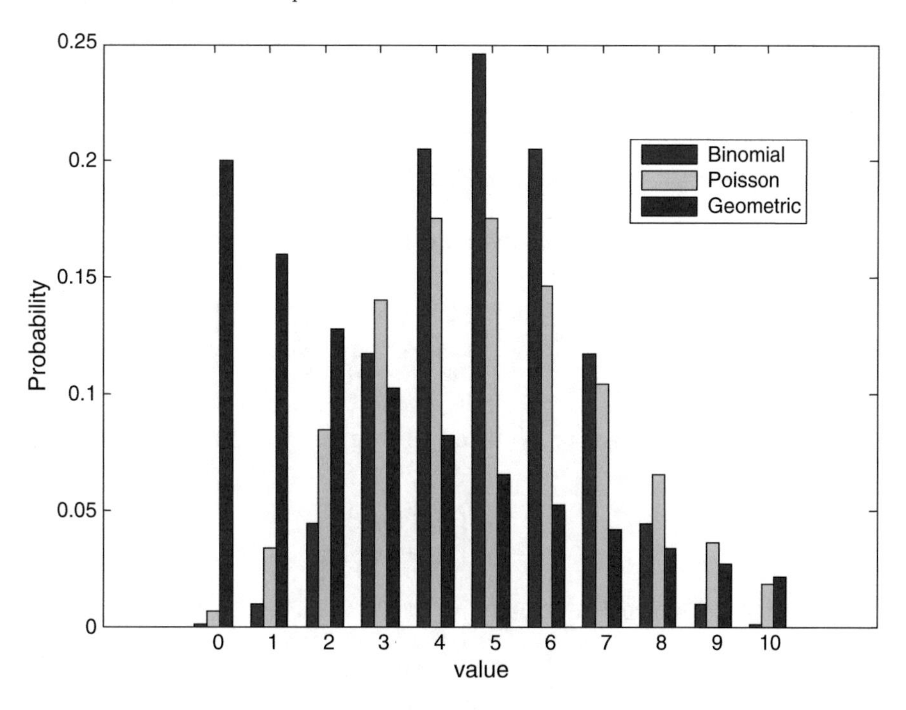

Fig. 3.3: Comparison of discrete random variables with the same mean value (= 5). Note that we have used a bar diagram, which is more appropriate for discrete random variables

$$\begin{aligned}
&= \text{from the definition of } \textit{conditional probability} \\
&= \frac{\text{Prob}\{X-\tau \le t, X > \tau\}}{\text{Prob}\{X > \tau\}} = \frac{\text{Prob}\{\tau < X \le t+\tau\}}{1-\text{Prob}\{X \le \tau\}} \\
&= \frac{\text{Prob}\{X \le t+\tau\} - \text{Prob}\{X \le \tau\}}{1-\text{Prob}\{X \le \tau\}} = \frac{F_X(t+\tau) - F_X(\tau)}{1-F_X(\tau)}
\end{aligned} \tag{3.77}$$

The conditional pdf of the residual length can be obtained by taking the derivative with respect to t of the result in (3.77) as:

$$f_{R_{|\tau}}(t) = \frac{\mathrm{d}}{\mathrm{d}t} F_{R_{|\tau}}(t) = \frac{\mathrm{d}}{\mathrm{d}t}\left[\frac{F_X(t+\tau) - F_X(\tau)}{1-F_X(\tau)}\right] = \frac{f_X(t+\tau)}{1-F_X(\tau)} \tag{3.78}$$

We may note that the PDF and the pdf of the residual length depend on the time τ at which we consider the phenomenon's residual duration.

Let us consider (3.77) and (3.78) for a small value of t that we substitute with Δ in the following notation. We can express the probability that the phenomenon active at instant τ will end in the next small interval Δ as:

$$\text{Prob}\{0 \le X - \tau \le \Delta | X > \tau\} \approx f_{R_{|\tau}}(\Delta)\,\Delta = \frac{f_X(\tau+\Delta)}{1-F_X(\tau)}\Delta \tag{3.79}$$

From (3.79), we define the *hazard rate function* $h_X(\tau)$ as follows [3, 4]:

$$h_X(\tau) = \lim_{\Delta \to 0} f_{R_{|\tau}}(\Delta) = \frac{f_X(\tau)}{1-F_X(\tau)} \tag{3.80}$$

The hazard rate function is not a pdf. It is a non-negative function that is used in the survival analysis to study the lifetimes of devices. The hazard rate function $h_X(\tau)$ represents the instantaneous failure rate of a component (or device) whose lifetime is modeled by random variable X, given that it survived until time τ. The hazard rate represents a dynamic characteristic of a distribution. We can have several behaviors of the hazard rate function, depending on the different random variables; in particular, hazard rate functions $h_X(\tau)$ can be decreasing, constant, or increasing with τ. Some examples are provided in the following Sect. 3.2.6.9 after we have discussed some typical random variables.

3.2.6.2 The Uniform Distribution (Continuous Case)

A continuous random variable X has a uniform distribution in the interval $[a,\ b]$ if it has the following probability density function:

$$f_X(x) = \frac{I(x-b) - I(x-a)}{b-a} = \frac{1}{b-a}\mathrm{rect}\left[\frac{x - \frac{a+b}{2}}{b-a}\right], \quad x \in (-\infty, \quad +\infty) \tag{3.81}$$

where $I(x)$ is the unit step function centered in $x = 0$ and where rect $[x]$ is the rectangular function from $-\frac{1}{2}$ to $\frac{1}{2}$ defined as:

$$\mathrm{rect}\,(x) \triangleq \begin{cases} 1, \text{if } |x| < \frac{1}{2} \\ 0, \text{otherwise} \end{cases} \tag{3.82}$$

The corresponding probability distribution function is

$$F_X\,(x) = \begin{cases} 0, & x < a \\ \frac{x-a}{b-a}, & a \le x < b \\ 1, & x \ge b \end{cases} \tag{3.83}$$

The uniform distribution is used when we have a random phenomenon whose outcomes can span a continuous range of possible values, without any preference.

The normalization condition is straightforwardly verified. Analogously, it is easy to achieve the following results for mean and mean square values:

$$\begin{aligned} E\,[X] &= \int_a^b x \times \frac{1}{b-a}\mathrm{d}x = \frac{b+a}{2} \\ E\left[X^2\right] &= \int_a^b x^2 \times \frac{1}{b-a}\mathrm{d}x = \frac{b^2+ab+a^2}{3} \end{aligned} \tag{3.84}$$

Referring to the conditional residual lifetime discussed in Sect. 3.2.6.1, we consider a phenomenon whose duration X is uniformly distributed from time 0 to time T. We use the distribution defined above in (3.81) and (3.83) where $a = 0$ and $b = T$. Let us assume that this phenomenon started at time $t = 0$ is still active at time τ. We consider $\tau \in [0,\ T]$ to have a possible residual duration of X. Let us use (3.77) to characterize the conditional distribution (PDF) of the residual lifetime $R_{|\tau}$:

$$\begin{aligned} F_{R_{|\tau}}\,(t) &= \frac{F_X\,(t+\tau) - F_X\,(\tau)}{1 - F_X\,(\tau)} = \frac{\frac{t+\tau}{T} - \frac{\tau}{T}}{1 - \frac{\tau}{T}} = \frac{t}{T-\tau} \\ f_{R_{|\tau}}\,(t) &= \frac{1}{T-\tau} \end{aligned} \tag{3.85}$$

where $t \in [0,\ T-\tau]$.

We can see that the conditional distribution of the residual time depends on τ.

3.2.6.3 The Exponential Distribution

A continuous random variable X is exponentially distributed if it has the following probability density function:

$$f_X(x) = \mu\, e^{-\mu x}, \quad x \ge 0 \tag{3.86}$$

where $\mu > 0$ denotes the mean rate with the dimension of time^{-1}.

The corresponding probability distribution function is

$$F_X(x) = 1 - e^{-\mu t}, \quad t \geq 0 \tag{3.87}$$

The exponential distribution is of fundamental importance in the field of queuing systems, as detailed in Chap. 4. In telecommunications, many random phenomena can be described by exponential distributions, such as: the duration of a phone call, the interval between two phone calls arriving at a node of a telecommunication network, the sojourn time in talking or silent phases for a voice traffic source with speech activity detection [5], the interval between the start of two Internet sessions, etc.

The normalization condition for the pdf in (3.86) can be verified as:

$$\int_0^{+\infty} f_X(x)\mathrm{d}x = \int_0^{+\infty} \mu\, e^{-\mu x}\mathrm{d}x = \lim_{a \to +\infty} \left[-e^{-\mu x}\right]_0^a = e^{-\mu 0} - \lim_{a \to +\infty} e^{-\mu a} = 1 \tag{3.88}$$

The mean value of the exponential distribution can be derived from (3.86) as:

$$\begin{aligned} E[X] &= \int_0^{+\infty} x f_X(x)\mathrm{d}x = \int_0^{+\infty} x\mu\, e^{-\mu x}\,\mathrm{d}x \\ &= \text{we use } z = \mu x \\ &= \tfrac{1}{\mu}\int_0^{+\infty} z\, e^{-z}\,\mathrm{d}z \\ &= \text{by employing the rule of the } \textit{integration by parts} \\ &= \tfrac{1}{\mu}\left\{[z \times (-e^{-z})]_0^{+\infty} - \int_0^{+\infty} (-e^{-z})\mathrm{d}z\right\} = \tfrac{1}{\mu}\int_0^{+\infty} e^{-z}\mathrm{d}z = \tfrac{1}{\mu}[-e^{-z}]_0^{+\infty} \\ &= \tfrac{1}{\mu} \end{aligned} \tag{3.89}$$

The mean square value of the exponential distribution can be derived from (3.86) as:

$$\begin{aligned} E\left[X^2\right] &= \int_0^{+\infty} x^2 f_X(x)\mathrm{d}x = \int_0^{+\infty} x^2\mu\, e^{-\mu x}\,\mathrm{d}x \\ &= \text{we use } z = \mu x \\ &= \tfrac{1}{\mu^2}\int_0^{+\infty} z^2\, e^{-z}\,\mathrm{d}z \\ &= \text{by employing the rule of the } \textit{integration by parts} \\ &= \tfrac{1}{\mu^2}\left\{\left[z^2 \times (-e^{-z})\right]_0^{+\infty} - 2\int_0^{+\infty} z\left(-e^{-z}\right)\mathrm{d}z\right\} = \tfrac{2}{\mu^2}\int_0^{+\infty} z\, e^{-z}\,\mathrm{d}z \\ &= \text{from the integrals related to the mean value} = \tfrac{2}{\mu^2} \end{aligned} \tag{3.90}$$

Finally, the variance of the exponential distribution is

$$\mathrm{Var}[X] = E\left[X^2\right] - \{E[X]\}^2 = \frac{2}{\mu^2} - \frac{1}{\mu^2} = \frac{1}{\mu^2} \tag{3.91}$$

Note that in this case, the variance is just equal to the square of the expected value.

The coefficient of variation for a random variable X with exponential distribution is $C_X = 1$. This result gives an interesting method to decide from measurements whether a random variable is exponentially distributed or not. If the measurements on random variable X outcomes indicate that the standard deviation is equal to the expected value, we can characterize X through an exponential distribution.

Memoryless Property of the Exponential Distribution

Referring to the conditional residual lifetime discussed in Sect. 3.2.6.1, let us consider a phenomenon whose duration X is exponentially distributed with a mean rate λ (e.g., the length of a phone call). According to (3.87), the PDF of variable X results as $F_X(t) = 1 - e^{-\mu t}$. From (3.77) and (3.78), we obtain the following interesting results for the distribution of the residual length assuming the phenomenon described by the exponentially distributed variable X is still active at time τ, $R_{|\tau}$:

$$F_{R_{|\tau}}(t) = \frac{F_X(t+\tau) - F_X(\tau)}{1 - F_X(\tau)} = \frac{1 - e^{-\mu(t+\tau)} - [1 - e^{-\mu\tau}]}{1 - [1 - e^{-\mu\tau}]} = 1 - e^{-\mu t}$$
$$f_{R_{|\tau}}(t) = \frac{f_X(t+\tau)}{1 - F_X(\tau)} = \frac{\mu e^{-\mu(t+\tau)}}{1 - [1 - e^{-\mu\tau}]} = \mu e^{-\mu t} \tag{3.92}$$

Hence, the conditional residual length distribution of $R_{|\tau}$ does not depend on τ and is still exponentially distributed with the same mean rate μ of the original length X. Therefore, the exponential distribution is memoryless since its residual length has the same distribution of the whole length. This is a quite powerful property, which will be widely used to analyze Markovian queuing systems. It is possible to prove that the exponential distribution is the sole continuous random variable with such a memoryless property.

Among the discrete random variables, the geometric distribution is the sole random variable with the same memoryless property.

Minimum Between Two Random Variables with Exponential Distributions

We consider two independent, exponentially distributed random variables T_1 and T_2, respectively, with mean rates λ_1 and λ_2. Their PDFs are as follows: $F_{T_1}(t) = 1 - e^{-\lambda_1 t}$ and $F_{T_2}(t) = 1 - e^{-\lambda_2 t}$. We need to characterize the distribution of $T = \min\{T_1, T_2\}$, $F_T(t)$. From (3.39) we have

$$\begin{aligned} F_T(t) &= F_{T_1}(t) + F_{T_2}(t) - F_{T_1}(t) \times F_{T_2}(t) \\ &= 1 - e^{-\lambda_1 t} + 1 - e^{-\lambda_2 t} - \left(1 - e^{-\lambda_1 t}\right) \times \left(1 - e^{-\lambda_2 t}\right) \\ &= 1 - e^{-\lambda_1 t - \lambda_2 t} = 1 - e^{-(\lambda_1 + \lambda_2)t} \end{aligned} \tag{3.93}$$

Hence, random variable T is still exponentially distributed with a mean rate $\lambda_1 + \lambda_2$. In conclusion, the minimum of two independent random variables with exponential distributions is still exponentially distributed with the mean rate given by the sum of the mean rates. Such property can be straightforwardly extended to the minimum of m independent, exponentially distributed random variables.

This property is quite important and allows us to study interesting cases. Let us consider, for instance, a bank with four tellers. A new customer arrives and finds all four tellers occupied by other customers. Assuming that the service time of a customer is exponentially distributed with a mean rate μ, we can determine the distribution of the waiting time experienced by our customer. Using the exponential distribution's memoryless property, the newly arriving customer "knows" that the other currently served customers have a residual service time, which is still exponentially distributed with mean rate μ. Our customer will enter service as soon as the first of the customers currently served completes the service. Therefore, the waiting time is the minimum of four independent, exponentially distributed random variables with a mean rate μ. Consequently, the waiting time is still exponentially distributed with a mean rate 4μ.

Maximum Between Random Variables with Exponential Distributions

We consider n independent exponentially distributed random variables $T_1 \ldots. T_n$, respectively, with mean rates $\lambda_1, \ldots, \lambda_n$. This time we like to study the new random variable obtained as the maximum of $T_1, \ldots, T_n$:

$$Y = \max\{T_1, \ldots, T_n\} \tag{3.94}$$

According to (3.36), the distribution of variable Y can be expressed as:

$$F_Y(y) = \prod_{i=1}^{n} F_{T_i}(y) = \prod_{i=1}^{n} \left(1 - e^{-\lambda_i y}\right) \tag{3.95}$$

Let us now consider the case with T_i iid random variables with exponential distributions and mean rates λ. Then, the distribution function and the probability density functions can be expressed as:

$$F_Y(y) = \left(1 - e^{-\lambda y}\right)^n \quad , \quad f_Y(y) = n\lambda e^{-\lambda y}\left(1 - e^{-\lambda y}\right)^{n-1} \tag{3.96}$$

The distribution of the random variable Y thus obtained is not a classical distribution. For some studies on traffic load, it can be interesting to determine the mean value of variable Y. We provide below the details of its derivation:

$$\begin{aligned} E[Y] &= \int_0^{+\infty} yn\lambda e^{-\lambda y}\left(1 - e^{-\lambda y}\right)^{n-1} dy \\ &= \text{using the Newton Binomial formula} \\ &= \int_0^{+\infty} yn\lambda e^{-\lambda y} \sum_{i=0}^{n-1} \binom{n-1}{i} \left[(-1)^i e^{-\lambda y i}\right] dy \\ &= n \sum_{i=0}^{n-1} \binom{n-1}{i} (-1)^i \int_0^{+\infty} y\lambda e^{-\lambda y(i+1)} dy \\ &= n \sum_{i=0}^{n-1} \binom{n-1}{i} \frac{(-1)^i}{\lambda (i+1)^2} \\ &= \text{setting } i = j-1 \\ &= \frac{1}{\lambda} \sum_{j=1}^{n} \frac{n}{j} \binom{n-1}{j-1} \frac{(-1)^{j-1}}{j} \\ &= \frac{1}{\lambda} \sum_{j=1}^{n} \binom{n}{j} \frac{(-1)^{j-1}}{j} \end{aligned} \tag{3.97}$$

where it is possible to prove that

$$\frac{1}{\lambda} \sum_{j=1}^{n} \binom{n}{j} \frac{(-1)^{j-1}}{j} = \frac{H_n}{\lambda} \tag{3.98}$$

being $H_n \triangleq \sum_{j=1}^{n} \frac{1}{j}$ the nth *harmonic number* [6].

This result can also be achieved in another and more intuitive way by exploiting the memoryless property and the minimum of random variables with exponential distributions. The maximum of n times can be achieved as the sum of the contribution of subsequent transitions among states or *phases*. The first phase is when there is the first timeout of the n exponentially distributed times: this is achieved as the minimum value of n iid exponentially distributed random variables with mean rates λ that is equal to $n\lambda$. In this new state, we have $n-1$ active iid variables of residual lifetimes still exponentially distributed with mean rates λ. Then, because of the exponential distribution's minimum property, the transition from $n-1$ to $n-2$ occurs with a mean rate $(n-1)\lambda$. We have to sum the mean times of all these transitions until we do the last transition from state 1 to state 0 with a mean rate λ. The time to pass from the original state n to state 0 corresponds to variable Y introduced before. Therefore, the mean value of Y, $E[Y]$, is achieved as the sum of the mean values corresponding to the different transitions with mean rates $n\lambda$, $(n-1)\lambda$, ..., λ. Then, we have

$$E[Y] = \sum_{j=1}^{n} \frac{1}{j\lambda} = \frac{H_n}{\lambda} \tag{3.99}$$

and this result is precisely equal to (3.97) and can be expressed in terms of the nth harmonic number, H_n.

Harmonic numbers H_n have been studied since antiquity, and there are many notable formulas involving them. They can roughly approximate the natural logarithm function. For large n values, the following approximation holds for H_n:

$$H_n \approx \ln(n) + \gamma \tag{3.100}$$

where $\gamma = 0.5772156649$ is the Euler–Mascheroni constant.

Comparison Between an Exponentially Distributed Random Variable and Another Variable

Let us consider two independent random variables X and Y. We know that X is exponentially distributed with a mean rate μ. As shown in Sect. 3.2.2 of this chapter, we want to determine Prob$\{X > Y\}$ according to (3.41) as:

$$\begin{aligned}\text{Prob}\{X > Y\} &= \int_0^{+\infty} [1 - F_X(x)]\, f_Y(x) \mathrm{d}x = \int_0^{+\infty} f_Y(x) e^{-\mu x}\, \mathrm{d}x \\ &= \text{LT}\{f_Y(x)\}|_{s=\mu}\end{aligned} \tag{3.101}$$

where LT$\{f_Y(x)\}$ denotes the Laplace transform of the pdf $f_Y(x)$.

If also random variable Y is exponentially distributed with a mean rate λ, we can express Prob$\{X > Y\}$ from (3.101) using the following Laplace transform of the exponential pdf:

$$\text{LT}\{f_Y(x)\} = \text{LT}\left\{\lambda\, e^{-\lambda t}\right\} = \frac{\lambda}{\lambda + s}$$

Hence, we have

$$\text{Prob}\{X > Y\} = \left.\frac{\lambda}{\lambda + s}\right|_{s=\mu} = \frac{\lambda}{\lambda + \mu} \tag{3.102}$$

3.2.6.4 The Erlang Distribution

The *Erlang distribution* is a two-parameter family of continuous probability distributions with support $x \in [0, +\infty)$. The two parameters are: a positive integer k, the "shape," and a positive real number λ, the "rate." Depending on the k values, the distribution is also called 'Erlang-k distribution'. The Erlang-k distribution is obtained as the sum of k iid exponential random variables with the same rate λ. The pdf of the sum is obtained as the convolution of the exponential distribution k times with itself. The resulting pdf can be expressed as:

$$f_X(x) = \frac{\lambda^k x^{k-1} e^{-\lambda x}}{(k-1)!} \tag{3.103}$$

The distribution function $F_X(x)$ is obtained as the integral of the pdf as follows:

$$F_X(x) = \text{Prob}\{X \le x\} = 1 - \sum_{j=0}^{k-1} \frac{(\lambda x)^j}{j!} e^{-\lambda x} \tag{3.104}$$

The mean value is the sum of the mean values (i.e., $1/\lambda$) of the summing random variables. Then, $E[X] = \frac{k}{\lambda}$. Similarly, the variance is the sum of the constituent random variables' variance because they are independent. We have: Var$[X] = \frac{k}{\lambda^2}$.

The coefficient of variation of the Erlang-k distribution is $C_X = \frac{\sigma_X}{E[X]} = \frac{1}{\sqrt{k}} < 1$ (only equal to 1 for $k = 1$, exponential distribution case).

The Laplace transform of the Erlang distribution is just the product of the Laplace transform of the exponential distribution k times with itself that is:

$$\text{LT}\{f_X(X)\} = \text{LT}\left\{\frac{\lambda^k y^{k-1} e^{-\lambda y}}{(k-1)!}\right\} = \left(\frac{\lambda}{\lambda + s}\right)^k \tag{3.105}$$

This distribution and its Laplace transform will be used in Sects. 4.14.2 and 5.1.2 in the next Chapters.

3.2.6.5 The Hyper-Exponential Distribution

A random variable X is said to have a *hyper-exponential distribution* (or a mixture of exponential distributions) when its pdf is given by the weighted sum of the pdfs of independent, exponentially distributed random variables with mean rates λ_i. The pdf of X can be expressed as follows:

$$f_X(x) = \sum_{i=1}^{L} w_i \lambda_i \, e^{-\lambda_i x} \tag{3.106}$$

where the weights $w_i > 0$ for i from 1 to L fulfill the following normalization condition:

$$\sum_{i=1}^{L} w_i = 1$$

The hyper-exponential distribution is an example of mixture density, representing different possible "behaviors" for a given event. From (3.89) and (3.90), it is easy to show that mean and mean square values have the following expressions:

$$E[X] = \sum_{i=1}^{L} \frac{1}{\lambda_i} w_i, \qquad E\left[X^2\right] = \sum_{i=1}^{L} \frac{2}{\lambda_i^2} w_i \tag{3.107}$$

The coefficient of variation of the hyper-exponential random variable X, $C_X = \frac{\sigma_X}{E[X]}$, is greater than 1. This distribution will be used in Sect. 5.1.2.

As a final consideration, it is interesting to note that exponential distributions, hyper-exponential distributions, and Erlang distributions are special cases of *phase-type distributions*, which are related to finite Markov chains (see Chap. 4) with an absorbing state: a phase-type distribution can be seen as the distribution of the time until absorption occurs.

3.2.6.6 The Gaussian Distribution

A continuous random variable X has a Gaussian distribution (or normal distribution) if it has the following probability density function:

$$f_X(x) = \frac{1}{\sqrt{2\pi}\sigma} e^{-\frac{(x-\mu)^2}{2\sigma^2}}, \quad x \in (-\infty, +\infty) \tag{3.108}$$

where μ is a real number and σ is a real positive number.

The corresponding probability distribution function cannot be expressed in a closed form, but we have an integral form as:

$$\begin{aligned}
\text{Prob}\{X \le x\} &= \frac{1}{\sqrt{2\pi}\sigma} \int_{-\infty}^{x} e^{-\frac{(x-\mu)^2}{2\sigma^2}} \mathrm{d}x \\
&= \text{making the substitution } y = \frac{x-\mu}{\sqrt{2}\sigma} \\
&= \frac{1}{\sqrt{\pi}} \int_{-\infty}^{\frac{x-\mu}{\sqrt{2}\sigma}} e^{-y^2} \mathrm{d}y = \frac{1}{\sqrt{\pi}} \int_{-\infty}^{0} e^{-y^2} \mathrm{d}y + \frac{1}{\sqrt{\pi}} \int_{0}^{\frac{x-\mu}{\sqrt{2}\sigma}} e^{-y^2} \mathrm{d}y \\
&= \text{using a result proven later } \int_{-\infty}^{0} e^{-y^2} \mathrm{d}y = \frac{\sqrt{\pi}}{2} \\
&\quad \text{and defining } \frac{2}{\sqrt{\pi}} \int_{0}^{z} e^{-y^2} \mathrm{d}y \triangleq erf\,(z)
\end{aligned} \tag{3.109}$$

$$= \frac{1}{2}\left[1 + erf\left(\frac{x-\mu}{\sqrt{2}\sigma}\right)\right]$$

A Gaussian distribution can be used to characterize the measurements of the resistors of a given production set. Moreover, a random variable sum of n independent identically distributed (iid) random variables[1] tends to have a Gaussian distribution as n tends to infinity because of the *central limit theorem* [2]. In general, data that are influenced by many small and unrelated random effects are approximately normally distributed. For instance, in telecommunications, the sum of elementary traffic sources with iid bit-rates leads to Gaussian traffic. This trend is typical of the Internet, where traffic flow aggregations concentrate an increasing number of traffic sources [7].

Due to the presence of the exponential term

$$e^{-x^2}$$

in the Gaussian pdf, the integrals that allow us to determine the PDF, the mean value, and the variance cannot be expressed in closed forms (using primitives). Suitable methods must be used, as explained below.

Let us consider the normalization condition in the case $\mu = 0$ (the cases with $\mu \neq 0$ are a straightforward extension):

$$\begin{aligned}
\int_{-\infty}^{+\infty} f_X(x)\mathrm{d}x &= \int_{-\infty}^{+\infty} \frac{1}{\sqrt{2\pi}\sigma}\mathrm{e}^{-\frac{x^2}{2\sigma^2}}\mathrm{d}x \\
&= \text{the integrand function is even} \\
&= \frac{1}{\sqrt{\pi}} \times 2 \times \int_0^{+\infty} \mathrm{e}^{-\left(\frac{x}{\sqrt{2}\sigma}\right)^2} d\left(\frac{x}{\sqrt{2}\sigma}\right) \\
&= \frac{2}{\sqrt{\pi}} \times \int_0^{+\infty} \mathrm{e}^{-z^2}\,\mathrm{d}z \qquad (3.110)
\end{aligned}$$

where $\int_0^{+\infty} \mathrm{e}^{-z^2}\,\mathrm{d}z$ can be derived as follows :

$$\left(\int_0^{+\infty} \mathrm{e}^{-z^2}\,\mathrm{d}z\right)^2 = \int_0^{+\infty} \mathrm{e}^{-u^2}\,\mathrm{d}u \times \int_0^{+\infty} \mathrm{e}^{-v^2}\,\mathrm{d}v = \int_0^{+\infty}\int_0^{+\infty} \mathrm{e}^{-(u^2+v^2)}\,\mathrm{d}u\,\mathrm{d}v$$

this double integral will be calculated in *polar coordinates* ρ, θ, as shown below.

We refer to the situation depicted in Fig. 3.4. In particular, we use the following transform of the plane to convert the Cartesian u–v coordinates to polar $\rho - \theta$ coordinates as:

$$\begin{cases} u = \rho\,\cos(\theta) \\ v = \rho\,\sin(\theta) \end{cases} \qquad (3.111)$$

Then, we relate the elementary area $\mathrm{d}u\mathrm{d}v$ in the u–v plane to the elementary area $\mathrm{d}\rho\mathrm{d}\theta$ in the $\rho - \theta$ plane as follows:

$$\mathrm{d}u\mathrm{d}v = J(\rho,\theta)\,\mathrm{d}\rho\mathrm{d}\theta \qquad (3.112)$$

where $J(\rho,\theta)$ is the following Jacobian determinant of the transform in (3.111):

$$J(\rho,\theta) = \begin{vmatrix} \frac{\partial u}{\partial \rho} & \frac{\partial u}{\partial \theta} \\ \frac{\partial v}{\partial \rho} & \frac{\partial v}{\partial \theta} \end{vmatrix} = \frac{\partial u}{\partial \rho}\frac{\partial v}{\partial \theta} - \frac{\partial u}{\partial \theta}\frac{\partial v}{\partial \rho} = \cos^2(\theta)\,\rho + \rho\sin^2(\theta) = \rho \qquad (3.113)$$

Then, using the transform of coordinated and the above relation between $\mathrm{d}u\mathrm{d}v$ and $\mathrm{d}\rho\mathrm{d}\theta$ we can convert the u–v integral in (3.110) to the following one in the $\rho - \theta$ plane:

[1] This property holds for all distributions, except for those "critical cases" where the moments (e.g., the mean) of the random variables do not exist (i.e., are infinite).

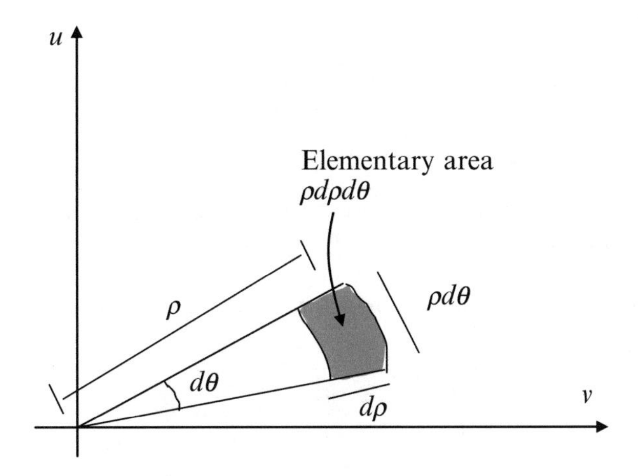

Fig. 3.4: Polar coordinates

$$\begin{aligned}\left(\int_0^{+\infty} e^{-z^2}\,\mathrm{d}z\right)^2 &= \int_0^{+\infty}\int_0^{+\infty} e^{-\left(u^2+v^2\right)}\,\mathrm{d}u\,\mathrm{d}v = \lim_{R\to+\infty}\int_0^{\pi/2}\int_0^{R} e^{-\rho^2}\rho\,\mathrm{d}\rho\,\mathrm{d}\theta \\ &= \int_0^{\pi/2}\mathrm{d}\theta \times \lim_{R\to+\infty}\int_0^{R} e^{-\rho^2}\rho\,\mathrm{d}\rho = \frac{\pi}{2}\times\frac{1}{2}\lim_{R\to+\infty}\int_0^{R} e^{-\rho^2}\,\mathrm{d}\rho^2 \\ &= \frac{\pi}{4}\times\lim_{R\to+\infty}\left[-e^{-\rho^2}\right]_0^R = \frac{\pi}{4}\end{aligned} \tag{3.114}$$

Hence, we have obtained the Euler-Poisson integral result:

$$\int_0^{+\infty} e^{-z^2}\,\mathrm{d}z = \frac{\sqrt{\pi}}{2}$$

In conclusion, by combining (3.110) and the result in (3.114), it is easy to verify the normalization condition of the Gaussian distribution.

Due to the complexity in deriving the expected value and the variance of the Gaussian distribution, we limit the following study to demonstrate that μ is the mean, and σ^2 is the variance.

In particular, we prove the correctness of the following identity for the mean value:

$$\int_{-\infty}^{+\infty} x\frac{1}{\sqrt{2\pi}\sigma}e^{-\frac{(x-\mu)^2}{2\sigma^2}}\,dx \equiv \mu \tag{3.115}$$

From the normalization condition, we have

$$\begin{aligned}&\int_{-\infty}^{+\infty}\frac{1}{\sqrt{2\pi}\sigma}e^{-\frac{(x-\mu)^2}{2\sigma^2}}\,dx = 1\\ &\text{We multiply both sides by } \mu \Rightarrow\\ &\mu\times\int_{-\infty}^{+\infty}\frac{1}{\sqrt{2\pi}\sigma}e^{-\frac{(x-\mu)^2}{2\sigma^2}}\,dx = \mu\end{aligned} \tag{3.116}$$

We sum and subtract μ on the left side of (3.115), where μ can be expressed through an integral expression according to (3.116):

$$\int_{-\infty}^{+\infty} x\frac{1}{\sqrt{2\pi}\sigma}e^{-\frac{(x-\mu)^2}{2\sigma^2}}\,\mathrm{d}x-\mu\times\int_{-\infty}^{+\infty}\frac{1}{\sqrt{2\pi}\sigma}e^{-\frac{(x-\mu)^2}{2\sigma^2}}\,\mathrm{d}x+\mu\equiv\mu$$

This is equivalent to write:

$$\int_{-\infty}^{+\infty}(x-\mu)\frac{1}{\sqrt{2\pi}\sigma}e^{-\frac{(x-\mu)^2}{2\sigma^2}}\,\mathrm{d}x+\mu\equiv\mu \tag{3.117}$$

The integrand is odd with respect to $x-\mu$. Hence, the integral is equal to 0, thus verifying the identity $\mu\equiv\mu$

To prove that σ^2 is the variance of the Gaussian distribution, we manipulate the normalization condition as follows:

$$\int_{-\infty}^{+\infty}e^{-\frac{(x-\mu)^2}{2\sigma^2}}\,\mathrm{d}x=\sqrt{2\pi}\sigma$$

We consider the derivative with respect to σ of both sides of the above expression; by exchanging the integral with the derivative on the left side, we have

$$\int_{-\infty}^{+\infty}\frac{\mathrm{d}}{\mathrm{d}\sigma}e^{-\frac{(x-\mu)^2}{2\sigma^2}}\,\mathrm{d}x=\sqrt{2\pi}\Rightarrow\int_{-\infty}^{+\infty}\frac{(x-\mu)^2}{\sigma^3}e^{-\frac{(x-\mu)^2}{2\sigma^2}}\,\mathrm{d}x=\sqrt{2\pi}$$

We multiply both sides by $\frac{\sigma^2}{\sqrt{2\pi}}$ thus verifying the identity:

$$\int_{-\infty}^{+\infty}(x-\mu)^2\times\frac{1}{\sqrt{2\pi}\sigma}e^{-\frac{(x-\mu)^2}{2\sigma^2}}\,\mathrm{d}x\equiv\sigma^2 \tag{3.118}$$

3.2.6.7 Heavy-Tailed Distributions

A typical characteristic of the traffic in the networks (Internet) is its high variability on a wide range of time scales. This characteristic is at the basis of *self-similar* traffic characteristics [8]: bursts of traffic are not averaged if we aggregate the traffic arrivals on large time intervals. In such circumstances, there is no advantage in terms of bandwidth needs in aggregating (multiplexing) different traffic sources since the aggregate traffic remains bursty. Self-similarity is introduced in the network by "phenomena" modeled by random variables with *heavy-tailed distributions*. A random variable X is said to be heavy-tailed if the following condition is (even definitely) met by its complementary distribution:

$$\mathrm{Prob}\{X>x\}\propto x^{-\alpha},\quad \text{where } 0<\alpha\leq 2 \tag{3.119}$$

From (3.119), we know that the heavy-tailed pdf of X can be characterized as follows:

$$f_X(x)\propto\alpha x^{-(\alpha+1)} \tag{3.120}$$

For the sake of simplicity, let us assume that $x\in[b,+\infty)$. The pdf in (3.120) entails some considerations on the finiteness of mean and variance of a heavy-tailed distribution. In particular, for the expected value, we have

$$E[X]=\int_b^{+\infty}xf_X(x)\mathrm{d}x\propto\alpha\times\int_b^{+\infty}x\times x^{-(\alpha+1)}\,\mathrm{d}x=\alpha\times\int_b^{+\infty}x^{-\alpha}\,\mathrm{d}x \tag{3.121}$$

The integral of the expected value in (3.121) does not converge (i.e., it is infinite) if $\alpha\leq 1$ (the integrand function goes to 0 for $x\to+\infty$ as slowly as or more slowly than $1/x$).

Finally, for the mean square value, we have

$$E\left[X^2\right]=\int_0^{+\infty}x^2f_X(x)\mathrm{d}x\propto\alpha\times\int_0^{+\infty}x^2\times x^{-(\alpha+1)}\,\mathrm{d}x=\alpha\times\int_0^{+\infty}x^{1-\alpha}\,\mathrm{d}x \tag{3.122}$$

The integral of the mean square value in (3.122) does not converge (i.e., it is infinite) if $\alpha-1\leq 1\Rightarrow\alpha\leq 2$.

In conclusion, a heavy-tailed distribution has infinite mean square value and variance, and may have infinite mean value. Heavy-tailed distributions can be used to model the duration of events where the more you wait, the more you have to wait. Further details are provided when determining the hazard rate function of a Pareto-distributed random variable in Sect. 3.2.6.9.

There is evidence for heavy tails in the distributions of the sizes of data objects stored in and transferred via computer systems, such as files in Web servers and files transferred through the Internet.

Today many aspects of the network traffic can be described by heavy-tailed distributions that do not have a finite variance (but with a finite mean). In these cases, sums of independent random variables do not have approximately normal distributions because the central limit theorem is not applicable to variables with heavy-tailed distributions.

The Pareto Distribution

A continuous random variable X has a Pareto distribution if it has the following probability density function:

$$f_X(x) = \frac{\gamma k^\gamma}{x^{\gamma+1}}, \quad x \geq k \tag{3.123}$$

where γ is a real positive number (shape parameter) and k is a positive translation term.

The corresponding PDF can be expressed as:

$$F_X(x) = 1 - \left(\frac{k}{x}\right)^\gamma, \quad x \geq k$$

The Pareto distribution can be used to model, for instance, the duration of an Internet session.

Based on the definition (3.119), the Pareto distribution is heavy-tailed if $0 < \gamma \leq 2$. It is easy to show that the mean value for $\gamma > 1$ is

$$E[X] = \frac{\gamma k}{\gamma - 1} \tag{3.124}$$

Finally, the variance for $\gamma > 2$ results as:

$$\mathrm{Var}[X] = \frac{\gamma k^2}{(\gamma - 1)^2 (\gamma - 2)} \tag{3.125}$$

Depending on the γ value, we can have situations where the variance is infinite (i.e., $\gamma \leq 2$) or even the expected value is infinite (i.e., $\gamma \leq 1$) for the Pareto distribution.

The pdfs of exponential, Gaussian, and Pareto random variables are compared with the same mean value (=10) in the graph in Fig. 3.5.

3.2.6.8 The Histograms of Random Variables

Histograms are experimental tools to characterize the pdfs of random variables. Let us consider a random variable X defined on real numbers. We can divide the real axis (typically, we consider a part of the real axis, where we concentrate our interests) into intervals (also called "bins") with the same size L. Then, based on the definition in (3.1), we repeat n times the experiment characterizing random variable X, recording how many times the outcomes fall into a generic interval; let x_j denote the experiment's outcome at the jth trial. Thus, we can show in a bar graph, called histogram, the number of times n_i that x_j falls into the generic ith bin. If the number of occurrences in each interval is divided by the total number n of trials, we have the relative frequencies $f_i = n_i/n$. As the size L of the bins in abscissa reduces and the number of trials increases (i.e., $n \to \infty$), the *piecewise constant curve* with horizontal segments of length L at height f_i/L tends to be more smoothed and approaches the pdf of X. In Matlab®, the "hist(·)" function can be used to generate histograms, collecting the occurrences of a random variable in each bin.

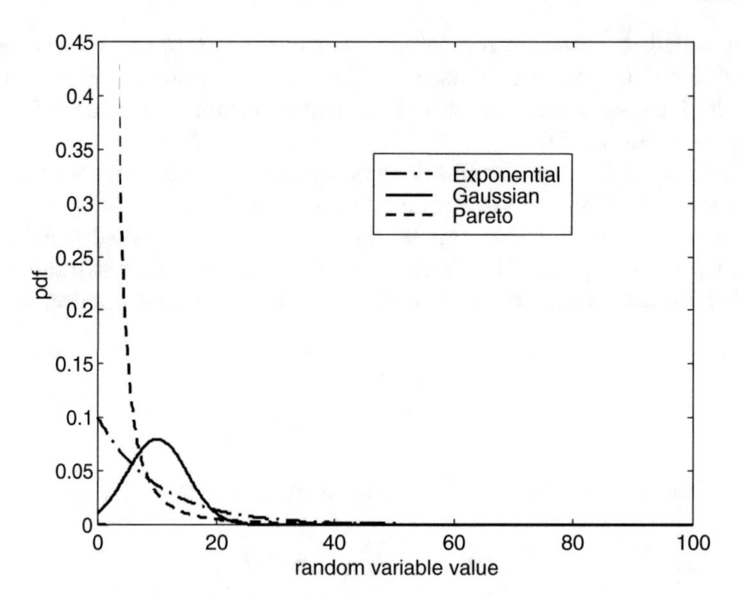

Fig. 3.5: Comparison of the pdfs of continuous random variables with the same mean value (= 10). The Gaussian distribution has standard deviation equal to 5; the Pareto distribution has $\gamma = 1.5$ (so that $k \approx 3.3$)

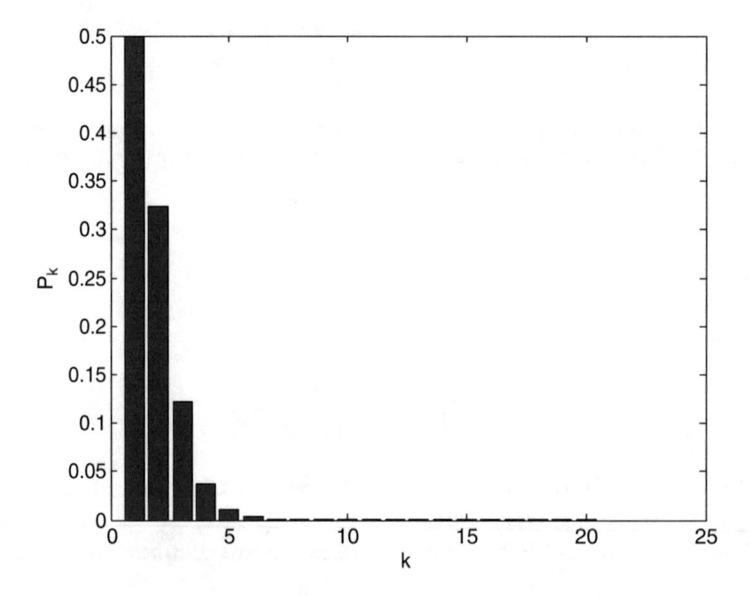

Fig. 3.6: Example of histogram

Figure 3.6 provides an example of a histogram for a distribution that has been plotted using the "bar($\cdot$)" command of Matlab®. To meet the normalization condition, we expect that the histogram has values tending to zero for increasing abscissa values. There are different methods to determine the bins' size to achieve a well-smoothed histogram (i.e., a histogram with reduced fluctuations). The *rule-of-thumb* by Freedman–Diaconis determines the bin size L as a function of the number of trials n as follows [9]:

$$L = 2 \times \text{IQR} \times n^{-1/3}$$

where IQR is the interquartile range, obtained as $\text{IRQ} = Q_3 - Q_1$, where Q_1 is the first and Q_3 is the third quartile of the data: $Q_1 = \text{PDF}^{-1}(0.25)$ and $Q_3 = \text{PDF}^{-1}(0.75)$, being PDF^{-1} the inverse of the PDF.

Note that outcomes x_j ($j = 1, 2, \ldots, n$) can be used to obtain an experimental estimate of both the expected value $E[X]$ and the variance $\text{Var}[X]$ of random variable X through the following formulas [10]:

$$E[X] \approx \frac{\sum_{j=1}^{n} x_j}{n} \quad \text{and} \quad \mathrm{Var}[X] \approx \frac{\sum_{j=1}^{n} (x_j - E[X])^2}{n-1} \tag{3.126}$$

The histogram obtained according to the repeated measurements of a phenomenon can be fitted with some of the distributions we have discussed in the previous sections. We can use the criterion of the coefficient of variation $C_X = \frac{\sigma_X}{E[X]}$ to preselect some most suitable (mathematically defined) distributions [4], as detailed below:

- If $C_X < 1$, we must use a "more regular" distribution than an exponential one; for instance, we can consider an *Erlang distribution*, that is a random variable obtained as sum of k independent, exponentially distributed random variables (k is the so-called shape parameter); for an example, see (4.110) in Chap. 4.
- If $C_X = 1$, we must adopt an exponential distribution.
- If $C_X > 1$, we must use a distribution with "heavier variability" than an exponential one; for instance, we can consider a *hyper-exponential distribution* (as well as other alternatives) having a pdf given by the weighted sum of the pdfs of independent, exponentially distributed random variables [4]. See also Sect. 3.2.6.5.

We consider a mathematically defined distribution having C_X in one of the above ranges as the set of experimental values. Then, we have to fit the distribution parameters to the characteristics of the experimental histogram. Many alternative methods are available to perform this fitting process. The simplest approach is to fit the moments experimentally determined in (3.126). In particular, we fit the mean value for a single-parameter distribution or the mean and variance for a two-parameter distribution. Finally, the goodness-of-fit approach based on the χ^2 ("chi-square") test can be used to see if the fitting achieved with a certain distribution is acceptable or not as well as to compare the goodness of fit of many alternative fitted distributions [11].

3.2.6.9 The Hazard Rate Function for Some Random Variables

As discussed in Sect. 3.2.6.1, the hazard rate function $h_X(\tau)$ is a measure of how fast a phenomenon with total duration X will finish soon after time τ, given that it survived till time τ. The hazard rate function applies only to random variables $X > 0$ that represent time or lifetime. The hazard rate function can give information about the tail of a distribution. If the hazard rate function is decreasing, the distribution has a heavy tail. Instead, if the hazard rate function is increasing, the distribution has a lighter tail.

We have the following hazard rate function for an exponentially distributed random variable X with mean rate μ as in (3.86):

$$h_X(\tau) = \frac{\mu\, e^{-\mu\tau}}{e^{-\mu\tau}} = \mu$$

The exponential distribution is the sole case with a constant hazard rate equal to the mean rate μ. In this particular case, the residual waiting time is unaffected by the waiting time already spent (this is according to the memoryless property of the exponential distribution). The exponential distribution is well suited to model the lifetime of many electronic components as well as the decay of radioactive particles.

We have the following hazard rate function for a Pareto-distributed random variable X as in (3.123):

$$h_X(\tau) = \frac{\frac{\gamma k^{\gamma}}{\tau^{\gamma+1}}}{\left(\frac{k}{\tau}\right)^{\gamma}} = \frac{\gamma}{\tau}$$

Hence, Pareto random variables are characterized by decreasing hazard rate functions. The same is true for hyper-exponential distributions and heavy-tailed distributions. Instead, the Erlang distribution has an increasing hazard rate. For more details on these continuous distributions, the interested reader should refer to [10].

3.2.7 Distribution of Sum, Difference, Product, and Ratio of Two Random Variables

Let us consider X and Y dependent (generic) random variables with joint density function $f_{XY}(x,y)$. Let us determine the pdf of their sum $W = X + Y$, the difference $U = X - Y$, the product $Z = XY$, and the ratio $V = Y/X$, as described below.

Distribution Function of the Sum and of the Difference

$$\begin{aligned}F_W(w) &= \text{Prob}\{W = X + Y \le w\} \\ &= \sum_x \text{Prob}\{Y \le w - x|\, X = x\}\,\text{Prob}\{X = x\} \\ &= \int_{x=-\infty}^{+\infty}\left(\int_{y=-\infty}^{w-x} f_{Y|X}(y|\,x)\,\mathrm{d}y\right) f_X(x)\,\mathrm{d}x \\ &= \int_{x=-\infty}^{+\infty}\int_{y=-\infty}^{w-x} f_{XY}(x,y)\,\mathrm{d}x\mathrm{d}y\end{aligned} \tag{3.127}$$

Taking the derivative with respect to w we obtain the density function of the sum as:

$$\begin{aligned}f_W(w) &= \frac{\mathrm{d}F_W(w)}{\mathrm{d}w} = \frac{\mathrm{d}}{\mathrm{d}w}\int_{x=-\infty}^{+\infty}\left(\int_{y=-\infty}^{w-x} f_{Y|X}(y|\,x)\,\mathrm{d}y\right) f_X(x)\,\mathrm{d}x \\ &= \int_{x=-\infty}^{+\infty}\frac{\mathrm{d}}{\mathrm{d}w}\left(\int_{y=-\infty}^{w-x} f_{Y|X}(y|\,x)\,\mathrm{d}y\right) f_X(x)\,\mathrm{d}x \\ &= \int_{x=-\infty}^{+\infty} f_{Y|X}(w-x|\,x)\,f_X(x)\,\mathrm{d}x \\ &= \int_{x=-\infty}^{+\infty} f_{XY}(x, w-x)\,\mathrm{d}x\end{aligned} \tag{3.128}$$

If X and Y are statistically independent, the conditioned pdf in the expression simplifies as $f_{Y|X}(w-x|\,x) = f_Y(w-x)$. Then, the density function of the sum in the previous expression in (3.128) becomes

$$f_W(w) = \int_{-\infty}^{+\infty} f_Y(w-x)\,f_X(x)\,\mathrm{d}x = f_X(w) \otimes f_Y(w) \tag{3.129}$$

where the symbol $\otimes$ denotes a convolution.

Then, the pdf of the sum W is the convolution of the pdf functions of X and of the pdf Y random variables. A graphical interpretation of this results is shown in Sect. 3.2.7.1.

If we have to study the distribution of the difference, $U = X - Y$, we can reuse the previous results considering it as the sum of X and $-Y$ as: $X + (-Y)$. Then, knowing the pdf of Y, $f_Y(y)$, the pdf of $-Y$ is obtained by overturning the pdf of Y as $f_Y(-y)$; this result will be proven in a next sub-section below. Then, if X and Y are statistically independent, we have

$$f_U(u) = \int_{-\infty}^{+\infty} f_Y(x-u)\,f_X(x)\,\mathrm{d}x = f_X(u) \otimes f_Y(-u) \tag{3.130}$$

If the distribution of Y has a pdf that is symmetrical with respect to the origin (even function), like the zero-mean Gaussian random variable, then $f_Y(-y) = f_Y(y)$ and, hence, $X - Y$ is the same as $X + Y$ in terms of distributions.

Distribution Function of an Overturned Random Variable

We know that random variable X has PDF $F_X(x)$ and pdf $f_X(x)$. We like to determine the distribution of variable $-X$ as follows:

$$F_{-X}(x) = \text{Prob}\{-X \le x\} = \text{Prob}\{X > -x\} = 1 - F_X(-x) \tag{3.131}$$

Then, the corresponding density function can be obtained through a derivative as:

$$f_{-X}(x) = \frac{\mathrm{d}}{\mathrm{d}x}F_{-X}(x) = \frac{\mathrm{d}}{\mathrm{d}x}[1 - F_X(-x)] = f_X(-x) \tag{3.132}$$

Distribution Function of the Product and of the Ratio

We consider the distribution function of the product Z for which we need to split the integral into two parts for $X \ge 0$ and for $X < 0$:

$$\begin{aligned} F_Z(z) &= \text{Prob}\{Z = XY \le z\} \\ &= \int_0^{+\infty} \text{Prob}\left\{Y \le \frac{z}{x}\middle| X = x\right\} f_X(x)\,\mathrm{d}x \\ &\quad + \int_{-\infty}^{0} \text{Prob}\left\{Y \ge \frac{z}{x}\middle| X = x\right\} f_X(x)\,\mathrm{d}x \end{aligned} \tag{3.133}$$

The probability density function of Z is obtained by taking the derivative with respect to Z as:

$$\begin{aligned} f_Z(z) &= \tfrac{\mathrm{d}F_Z(z)}{\mathrm{d}z} = \tfrac{\mathrm{d}}{\mathrm{d}z}\int_0^{+\infty}\text{Prob}\left\{Y \le \tfrac{z}{x}\middle| X = x\right\} f_X(x)\,\mathrm{d}x \\ &\quad + \tfrac{\mathrm{d}}{\mathrm{d}z}\int_{-\infty}^{0}\text{Prob}\left\{Y \ge \tfrac{z}{x}\middle| X = x\right\} f_X(x)\,\mathrm{d}x \\ &= \int_0^{+\infty}\tfrac{\mathrm{d}}{\mathrm{d}z}\text{Prob}\left\{Y \le \tfrac{z}{x}\middle| X = x\right\} f_X(x)\,\mathrm{d}x \\ &\quad + \int_{-\infty}^{0}\tfrac{\mathrm{d}}{\mathrm{d}z}\text{Prob}\left\{Y \ge \tfrac{z}{x}\middle| X = x\right\} f_X(x)\,\mathrm{d}x \\ &= \int_0^{+\infty}\tfrac{1}{x} f_{Y|X}\left(y = \tfrac{z}{x}\middle| x\right) f_X(x)\,\mathrm{d}x - \int_{-\infty}^{0}\tfrac{1}{x} f_{Y|X}\left(y = \tfrac{z}{x}\middle| x\right) f_X(x)\,\mathrm{d}x \\ &= \int_0^{+\infty}\tfrac{1}{x} f_{XY}\left(x,\ \tfrac{z}{x}\right)\mathrm{d}x + \int_{-\infty}^{0}\tfrac{1}{-x} f_{XY}\left(x,\ \tfrac{z}{x}\right)\mathrm{d}x \\ &= \int_{-\infty}^{+\infty}\tfrac{1}{|x|} f_{XY}\left(x,\ \tfrac{z}{x}\right)\mathrm{d}x \end{aligned} \tag{3.134}$$

If variables X and Y are statistically independent, the integrand function becomes $\frac{1}{|x|}f_{XY}\left(x,\ \frac{z}{x}\right) = \frac{1}{|x|}f_X(x)\,f_Y\left(\frac{z}{x}\right)$. Then, we have

$$f_Z(z) = \int_{-\infty}^{+\infty}\frac{1}{|x|}f_X(x)\,f_Y\left(\frac{z}{x}\right)\mathrm{d}x \tag{3.135}$$

Let us determine the mean value of the product random variable Z.

$$\begin{aligned} E[Z] &= \int_{-\infty}^{+\infty} z f_Z(z)\mathrm{d}z = \int_{-\infty}^{+\infty} z\int_{-\infty}^{+\infty}\tfrac{1}{|x|}f_X(x)\,f_Y\left(\tfrac{z}{x}\right)\mathrm{d}x\mathrm{d}z \\ &= \int_{-\infty}^{+\infty}\tfrac{1}{|x|}f_X(x)\left[\int_{-\infty}^{+\infty} z f_Y\left(\tfrac{z}{x}\right)\mathrm{d}z\right]\mathrm{d}x \\ &= \int_{-\infty}^{0}\tfrac{1}{-x}f_X(x)\left[\int_{-\infty}^{+\infty} z f_Y\left(\tfrac{z}{x}\right)\mathrm{d}z\right]\mathrm{d}x + \int_0^{+\infty}\tfrac{1}{x}f_X(x)\left[\int_{-\infty}^{+\infty} z f_Y\left(\tfrac{z}{x}\right)\mathrm{d}z\right]\mathrm{d}x \\ &= \text{let us use } \zeta = \tfrac{z}{x}\text{. Note that the `}-\text{' sign of the first integral is} \\ &\quad\ \text{absorbed by the change of variable} \\ &= \int_{-\infty}^{0}\tfrac{1}{x}f_X(x)\,x^2\left[\int_{-\infty}^{+\infty}\zeta f_Y(\zeta)\,\mathrm{d}\zeta\right]\mathrm{d}x + \int_0^{+\infty}\tfrac{1}{x}f_X(x)\,x^2\left[\int_{-\infty}^{+\infty}\zeta f_Y(\zeta)\,\mathrm{d}\zeta\right]\mathrm{d}x \\ &= \left[\int_{-\infty}^{0} x f_X(x)\,\mathrm{d}x + \int_0^{+\infty} x f_X(x)\mathrm{d}x\right]\left[\int_{-\infty}^{+\infty}\zeta f_Y(\zeta)\,\mathrm{d}\zeta\right] = E[X]\times E[Y] \end{aligned} \tag{3.136}$$

Applying a similar derivation to the above, we can prove that the following result holds for the mean square value of the product of independent random variables:

$$E\left[Z^2\right] = E\left[X^2\right] \times E\left[Y^2\right]. \tag{3.137}$$

Let us, for instance, consider two independent random variables as random variable X with uniform distribution from -10 to $+10$ and random variable Y with uniform distribution from 0 to 2. In this special case, the pdf of the product of the two uniform random variables depends on how their intervals are concatenated on the real axis. Let us express the pdfs of X, $f_X(x)$ and the pdf of Y, $f_Y(y)$ by means of the rectangular function defined in (3.82) as:

$$f_X(x) = \frac{1}{20}\mathrm{rect}\left(\frac{x}{20}\right), \quad f_Y(y) = \frac{1}{2}\mathrm{rect}\left(\frac{y-1}{2}\right) \tag{3.138}$$

The pdf of the product $Z = XY$ can be obtained by applying (3.135) as:

$$f_Z(z) = \frac{1}{40}\int_{-\infty}^{+\infty} \frac{1}{|x|}\mathrm{rect}\left(\frac{x}{20}\right)\mathrm{rect}\left(\frac{z}{2x} - \frac{1}{2}\right)\mathrm{d}x \tag{3.139}$$

To solve this integral, we need to distinguish the case $z \geq 0$ from the case $z < 0$.

- Case $z \geq 0$: the rectangle $\mathrm{rect}\left(\frac{z}{2x} - \frac{1}{2}\right)$ goes for x from $z/2$ to $+\infty$ so that it overlaps with the rectangle $\mathrm{rect}\left(\frac{x}{20}\right)$ only from $z/2$ to 10, assuming that $0 < z < 20$. Then, the integral in (3.139) results as:

$$\begin{aligned} f_Z(z) &= \frac{1}{40}\int_{\frac{z}{2}}^{+10} \frac{1}{x}\mathrm{d}x = \frac{1}{40}\left[\ln(10) - \ln\left(\frac{z}{2}\right)\right] \\ &= \frac{1}{40}\ln\left(\frac{20}{z}\right) \quad \text{for } 0 \leq z \leq 20 \end{aligned} \tag{3.140}$$

- Case $z < 0$: the rectangle $\mathrm{rect}\left(\frac{z}{2x} - \frac{1}{2}\right)$ goes for x from $-\infty$ to $z/2$ so that it overlaps with the rectangle $\mathrm{rect}\left(\frac{x}{20}\right)$ only from -10 to $z/2$, assuming that $-20 < z < 0$. Then, the integral in (3.139) results as:

$$\begin{aligned} f_Z(z) &= \frac{1}{40}\int_{-10}^{\frac{z}{2}} \frac{1}{-x}\mathrm{d}x = \frac{1}{40}\int_{-\frac{z}{2}}^{10} \frac{1}{v}\mathrm{d}v \\ &= \frac{1}{40}\ln\left(\frac{20}{-z}\right) \quad \text{for } -20 \leq z \leq 0 \end{aligned} \tag{3.141}$$

In conclusion, we can combine the two cases in the following notation, representing the pdf of the random variable Z, product of two uniform random variables:

$$f_Z(z) = \begin{cases} \frac{1}{40}\ln\left(\frac{20}{|z|}\right), & \text{for } |z| \leq 20 \\ 0, & \text{otherwise} \end{cases} \tag{3.142}$$

Finally, we may note that the analysis, made here for the pdf of the product of random variables, can also be adapted to the case of the ratio of two random variables, $V = Y/X$, as:

$$\begin{aligned} f_V(v) &= \int_{-\infty}^{+\infty} |x|\, f_{XY}(x,\, vx)\mathrm{d}x \\ &= \text{for independent variables} \\ &= \int_{-\infty}^{+\infty} |x|\, f_X(x)\, f_Y(vx)\mathrm{d}x \end{aligned} \tag{3.143}$$

The problem with the distributions of the ratios is that many times they have not the variance, and, in some cases, they also have not the mean values (meaning that the related integrals do not converge).

The distribution of the ratio of random variables can be useful when we evaluate the time to cover a certain distance, assuming that both the speed and the distance itself are random variables. Let us consider, for instance, that random variables X and Y are independent identically distributed with uniform distribution from 0 to 1.

Then, $f_X(x) = \text{rect}\left(x - \frac{1}{2}\right)$ and $f_Y(y) = \text{rect}\left(y - \frac{1}{2}\right)$. Hence, to determine the distribution of $V = Y/X$ we can apply (3.143), integrating only for $x > 0$, as follows:

$$
\begin{aligned}
f_V(v) &= \int_0^{+\infty} x \,\text{rect}\left(x - \frac{1}{2}\right) \text{rect}\left(vx - \frac{1}{2}\right) \mathrm{d}x \\
&= \int_0^{1} x \,\text{rect}\left(vx - \frac{1}{2}\right) \mathrm{d}x \\
&= \text{where rect}\left(vx - \frac{1}{2}\right) \text{ is equal to 1 from } x = 0 \text{ to } \frac{1}{v} \\
&= \begin{cases} \int_0^1 x \,\text{rect}\left(vx - \frac{1}{2}\right)\mathrm{d}x = \int_0^1 x \,\mathrm{d}x = \frac{1}{2}, \text{for } 0 < v < 1 \\ \int_0^{\frac{1}{v}} x \,\mathrm{d}x = \frac{1}{2v^2}, \text{for } v > 1 \end{cases}
\end{aligned}
\tag{3.144}
$$

Note that this distribution of V has neither the mean value nor the mean square value.

Let us consider another example with an exponential random variable X with rate λ and another independent exponential random variable Y with rate μ. Based on (3.143), we can characterize the pdf of the ratio $V = Y/X$, as follows:

$$
\begin{aligned}
f_V(v) &= \int_0^{+\infty} |x|\, \lambda e^{-\lambda x} \mu e^{-\mu v x} \mathrm{d}x = \lambda\mu \int_0^{+\infty} x e^{-(\lambda+\mu v)x} \mathrm{d}x \\
&= \text{integrating by parts} \\
&= \lambda\mu \left[x \frac{e^{-(\lambda+\mu v)x}}{-(\lambda+\mu v)} \right]_0^{+\infty} - \lambda\mu \int_0^{+\infty} \frac{e^{-(\lambda+\mu v)x}}{-(\lambda+\mu v)} \mathrm{d}x \\
&= \frac{\lambda\mu}{(\lambda+\mu v)^2}
\end{aligned}
\tag{3.145}
$$

for $v \in [0, +\infty)$. Also in this case, we have not mean and mean square values.

3.2.7.1 Graphical Interpretation of the Sum of Independent Random Variables

Let us study the random variable W given by the sum of independent variables X and Y: $W = X + Y$. We consider the joint pdf of X and Y, $f_{XY}(x, y)$ and the marginal pdfs $f_X(x)$ and $f_Y(y)$. The PDF of W can be expressed as follows:

$$
F_W(w) = \text{Prob}\{W \le w\} = \text{Prob}\{X + Y \le w\}
\tag{3.146}
$$

Let us provide a graphical proof of what has been already discussed above in Sect. 3.2.7. Based on the condition in the rightmost term in (3.146), we examine the region in the plane x, y where $x + y \le w$; see the gray region in Fig. 3.7. Hence, we have to compute the following double integral of the joint pdf $f_{XY}(x, y)$ over the gray area:

$$
F_W(w) = \text{Prob}\{X + X \le w\} = \int_{y=-\infty}^{y=+\infty} \int_{x=-\infty}^{x=w-y} f_{XY}(x, y)\,\mathrm{d}x\,\mathrm{d}y
\tag{3.147}
$$

We exploit the statistical independence of X and Y so that $f_{XY}(x, y) = f_X(x) \times f_Y(y)$ can be substituted in (3.147):

$$
\begin{aligned}
F_W(w) &= \int_{y=-\infty}^{y=+\infty} \int_{x=-\infty}^{x=w-y} f_{XY}(x, y)\,\mathrm{d}x\,\mathrm{d}y \\
&= \int_{y=-\infty}^{y=+\infty} f_Y(y) \left[\int_{x=-\infty}^{x=w-y} f_X(x)\,\mathrm{d}x \right] \mathrm{d}y
\end{aligned}
$$

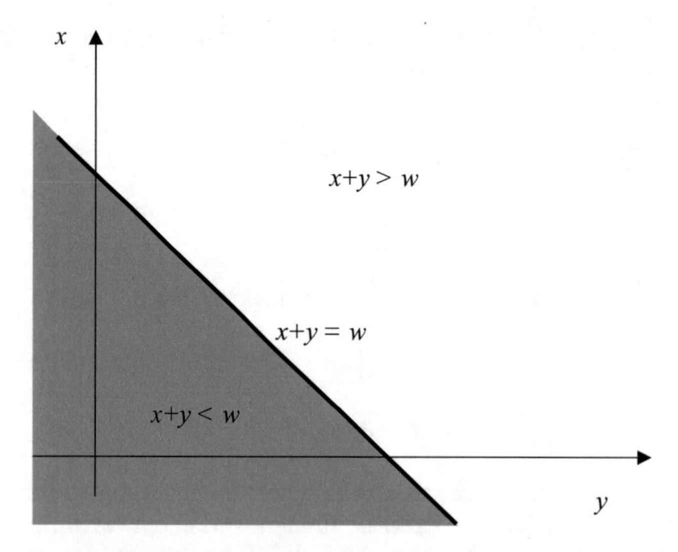

Fig. 3.7: Plane x, y where we perform the double integral

$$= \int_{y=-\infty}^{y=+\infty} f_Y(y) F_X(w-y)\,\mathrm{d}y$$

By taking the derivative of both sides with respect to w, we express the pdf of W, $f_W(w)$, as a function of the pdfs of X and Y as follows:

$$f_W(w) = \frac{\mathrm{d}}{\mathrm{d}w} \int_{y=-\infty}^{+\infty} f_Y(y) F_X(w-y)\,\mathrm{d}y = \int_{y=-\infty}^{+\infty} f_Y(y) f_X(w-y)\,\mathrm{d}y \tag{3.148}$$

From (3.148), we note that the pdf of W is given by the convolution of the pdfs of X and Y: $f_W(w) = f_X(x) \otimes f_Y(y)$. A similar result can be obtained when W is the sum of discrete random variables X and Y: the probability mass function of W is given by the discrete convolution of the probability mass functions of X and Y, as shown below:

$$\mathrm{Prob}\{W=k\} = \sum_i \mathrm{Prob}\{X=i\}\,\mathrm{Prob}\{Y=k-i\} \tag{3.149}$$

The above results for the sum of two independent continuous or discrete random variables can be extended to the sum of n independent random variables.

3.3 Transforms of Random Variables

The transforms applied to the distributions of random variables are powerful tools to characterize these variables' moments. There are transforms for discrete variables and transforms that can be used for both continuous variables and discrete ones. Detailed descriptions are provided in the following sub-sections.

3.3.1 The Probability Generating Function

The Probability Generating Function (PGF) is a transform adopted for non-negative integer-valued[2] discrete random variables for which we know the probability mass function. Let us refer to the generic variable X with

[2] This definition could be extended to random variables having all positive or all negative values.

distribution Prob$\{X=k\}$; its PGF $X(z)$ is a function of the complex variable z that is defined through operator $E[\cdot]$ in (3.42), as follows:

$$X(z) = E\left[z^X\right] = \sum_k z^k \text{Prob}\{X=k\}, \qquad \text{for } |z| \leq 1 \tag{3.150}$$

Note that $X(z)$ is a *power series* with non-negative coefficients. The sum in (3.150) is over all possible values k of random variable X. The PGF defined in (3.150) is a z-transform, except for the change of sign in the exponent of z. Let us examine the basic properties of a PGF:

$$\begin{aligned} &X(z=1) = \sum_k \text{Prob}\{X=k\} = 1 \ (\textit{normalization}) \\ &X(z=0) = \text{Prob}\{X=0\} \leq 1 \ \ (\text{assuming } 0^0 = 1) \end{aligned} \tag{3.151}$$

$$\begin{aligned} &|X(z)| = \left|\sum_k z^k \,\text{Prob}\{X=k\}\right| \\ &\text{we use the } \textit{triangular inequality} \\ &\leq \sum_k \left|z^k \,\text{Prob}\{X=k\}\right| = \sum_k \left|z^k\right| \text{Prob}\{X=k\} \\ &\text{we use the fact that } |z| \leq 1 \\ &\leq \sum_k \text{Prob}\{X=k\} = 1 \\ &\text{In conclusion}: |X(z)| \leq 1 \quad \text{for} \quad |z| \leq 1 \\ &\Rightarrow \text{there cannot be poles in and inside the unit circle} \end{aligned} \tag{3.152}$$

$$\begin{aligned} X'(z) &= \frac{\mathrm{d}}{\mathrm{d}z}\sum_k z^k \,\text{Prob}\{X=k\} \\ &= \text{under the assumption of series } \textit{uniform convergence} \\ &= \sum_k \frac{\mathrm{d}}{\mathrm{d}z} z^k \,\text{Prob}\{X=k\} = \sum_k k z^{k-1} \,\text{Prob}\{X=k\} \\ &\Rightarrow X'(z=1) = \sum_k k \,\text{Prob}\{X=k\} = E[X] \end{aligned} \tag{3.153}$$

$$\begin{aligned} X''(z) &= \frac{\mathrm{d}}{\mathrm{d}z}\sum_k k z^{k-1} \,\text{Prob}\{X=k\} \\ &= \text{under the assumption of series } \textit{uniform convergence} \\ &= \sum_k k \frac{d}{dz} z^{k-1} \,\text{Prob}\{X=k\} = \sum_k k(k-1) z^{k-2} \,\text{Prob}\{X=k\} \\ &\Rightarrow X''(z=1) = \sum_k k^2 \,\text{Prob}\{X=k\} - \sum_k k \,\text{Prob}\{X=k\} \end{aligned} \tag{3.154}$$

Hence, we can express the mean square value by using (3.153):

$$X''(z=1) + X'(z=1) = E\left[X^2\right]$$

A generic complex function (z domain) is characterized by a *radius of convergence* ρ: this complex function is convergent if $|z| < \rho$ and diverges if $|z| > \rho$; on the circle $|z| = \rho$, there is at least one singularity. The radius of convergence is always equal to the distance from the origin of the nearest point where the function has a non-removable singularity.

The PGF of a random variable is a particular case of complex function, i.e., a power series. Based on (3.152), the *radius of convergence* of the PGF must be at least one (a PGF is convergent inside and on the unit disc,

$|z| \leq 1$). In the limiting case of a radius of convergence just equal to 1, the Abel theorem [12] can be used to prove[3] that $N(z)$ has a limit for $z \to 1^-$ and this limit must be equal to 1 because of the normalization condition:

$$\lim_{z \to 1^-} X(z) = 1$$

Let us consider two discrete independent random variables: X with distribution Prob$\{X = k\}$ and PGF $X(z)$ and Y with distribution Prob$\{Y = h\}$ and PGF $Y(z)$. We need to determine the distribution of $W = X + Y$. According to (3.149), it is easy to show that the probability mass function of W is the discrete convolution of the probability mass functions of X and Y:

$$\begin{aligned} \text{Prob}\{W = j\} &= \sum_k \text{Prob}\{X = k\} \text{Prob}\{Y = j - k\} \\ &= \text{Prob}\{X = k\} \otimes_d \text{Prob}\{Y = k\} \end{aligned} \quad (3.155)$$

Equation (3.155) can be transformed in the z-domain, thus yielding the following simple formula in terms of PGFs:

$$W(z) = X(z)Y(z) \quad (3.156)$$

This result can be proved by taking the PGFs of the distributions in (3.155):

$$\begin{aligned} W(z) &= \sum_j z^j \text{Prob}\{W = j\} = \sum_j \sum_k z^j \text{Prob}\{X = k\} \text{Prob}\{Y = j - k\} \\ &= \text{we exchange the sums over } j \text{ and } k \\ &= \sum_k \text{Prob}\{X = k\} \sum_j z^j \text{Prob}\{Y = j - k\} \\ &= \sum_k \text{Prob}\{X = k\} z^k \sum_j z^{j-k} \text{Prob}\{Y = j - k\} = X(z)Y(z) \end{aligned}$$

The previous result can be easily extended to the sum of a generic number of independent random variables. Therefore, the representation in terms of PGFs allows for significant simplifications.

For some derivations in queuing theory (see next chapters) the state probability distribution will be expressed in terms of a PGF. Once solved the system in terms of a PGF $X(z)$, it is essential to derive the relative probability distribution Prob$\{X = k\}$. A simple inversion method can be obtained by looking at the definition of $X(z)$ as a function of the distribution Prob$\{X = k\}$:

$$X(z) = \sum_k z^k \text{Prob}\{X = k\}$$

The above expression can be seen as the Taylor series expansion of $X(z)$ centered in $z = 0$ (i.e., MacLaurin series expansion). Therefore, the coefficients of the expansion of $X(z)$, corresponding to the probability mass function values Prob$\{X = k\}$, can be obtained through successive derivatives computed in $z = 0$ as follows:

$$\text{Prob}\{X = k\} = \frac{1}{k!} \frac{\mathrm{d}^k}{\mathrm{d}z^k} X(z) \bigg|_{z=0} \quad (3.157)$$

The probability mass function of random variable X can be computed based on (3.157) by using the Matlab® symbolic toolbox. We can define the complex variable z as a symbolic one ("syms z") and then express $X(z)$. Thus, we can compute symbolically the ith derivative of $X(z)$ through the "diff(X, i)" command. Finally, we evaluate these derivatives in $z = 0$ using the "eval($\cdot$)" command.

Before concluding this section, it is interesting to consider the following example to compute the PGF of random variable $Y = aX + b$, which is a linear combination of random variable X [having PGF $X(z)$] with coefficients a and b:

[3] The Abel theorem allows to find the limit of a power series (i.e., a PGF in our case) as z approaches 1 from below, even in cases where the radius of convergence of the power series is equal to 1 and we do not know whether this limit is finite or not.

$$Y(z) = E\left\lfloor z^{Y} \right\rfloor = E\left\lfloor z^{aX+b} \right\rfloor = z^{b} E\left\lfloor z^{aX} \right\rfloor = z^{b} X\left(z^{a}\right)$$

3.3.1.1 Sum of a Random Number of Discrete Variables

We consider independent discrete random variables X_i $(i = 1, 2, \dots)$ with probability mass functions $\text{Prob}\{X_i = k\}$ and PGFs $X_i(z)$. We are interested in characterizing the new random variable Y obtained as follows:

$$Y = \sum_{i=1}^{N} X_i \tag{3.158}$$

where N is a discrete random variable with probability mass function $\text{Prob}\{N = j\}$ and PGF $N(z)$.

Conditioning on a given $N = j$ value, the PGF of the sum of X_i for $i = 1, 2, \dots, j$ is obtained according to (3.156) as the product of the $X_i(z)$ functions from $i = 1$ to $i = j$. We remove the conditioning using the distribution of N as:

$$Y(z) = \sum_{j} \left[\prod_{i=1}^{j} X_i(z)\right] \text{Prob}\{N = j\} \tag{3.159}$$

If all the X_i variables are identically distributed so that they have the same PGF $X(z)$, (3.159) becomes

$$Y(z) = \sum_{j} [X(z)]^{j}\, \text{Prob}\{N = j\} = N[X(z)] \tag{3.160}$$

According to (3.160), we get $Y(z)$ if we replace z with $X(z)$ in $N(z)$. We say that random variable Y has a "compound" distribution.

We can derive mean and mean square values of Y by taking the first two derivatives of the PGF in (3.160):

$$\begin{aligned}
E[Y] &= \frac{\mathrm{d}}{\mathrm{d}z} Y(z)\bigg|_{z=1} = \frac{\mathrm{d}}{\mathrm{d}z} N[X(z)]\bigg|_{z=1} \\
&= N'[X(z)] \times X'(z)\big|_{z=1} = N'(1)X'(1) = E[N]\,E[X] \\
E\left[Y^2\right] &= \frac{\mathrm{d}}{\mathrm{d}z} Y(z)\bigg|_{z=1} + \frac{\mathrm{d}^2}{\mathrm{d}z^2} Y(z)\bigg|_{z=1} \\
&= N'(1)X'(1) + \left\{\frac{\mathrm{d}}{\mathrm{d}z} N'[X(z)] \times X'(z)\right\}\bigg|_{z=1} \\
&= N'(1)X'(1) + \left\{N''[X(z)] \times [X'(z)]^2 + N'[X(z)] \times X''(z)\right\}\Big|_{z=1} \\
&= N'(1)X'(1) + N''(1)[X'(1)]^2 + N'(1)X''(1) \\
&= N'(1)\left[X'(1) + X''(1)\right] + N''(1)[X'(1)]^2 \\
&= E[N]\,E\left[X^2\right] + \left\{E\left[N^2\right] - E[N]\right\}\{E[X]\}^2
\end{aligned} \tag{3.161}$$

Based on the above result on $E\left[Y^2\right]$, it is also possible to determine the variance of Y as follows:

$$\begin{aligned}
\text{Var}[Y] &= E\left[Y^2\right] - \{E[Y]\}^2 \\
&= E[N]\,E\left[X^2\right] + \left\{E\left[N^2\right] - E[N]\right\}\{E[X]\}^2 - \{E[X]\}^2\{E[N]\}^2 \\
&= E[N]\,E\left[X^2\right] - E[N]\{E[X]\}^2 + E\left[N^2\right]\{E[X]\}^2 - \{E[X]\}^2\{E[N]\}^2 \\
&= E[N]\,\text{Var}[X] + \{E[X]\}^2\,\text{Var}[N]
\end{aligned} \tag{3.162}$$

This result on $\text{Var}(Y)$ can also be justified by applying the *law of total variance* to variable Y and its conditioned version on N, $Y|N$. Given a random variable Y that can be conditioned on variable N, we can relate as follows the variance of Y and the variance of $Y|N$:

$$\mathrm{Var}\,(Y) = E_N\,[\mathrm{Var}\,[Y|\,N]] + \mathrm{Var}_N\,[E\,[Y|\,N]] \tag{3.163}$$

where $\mathrm{Var}\,[Y|\,N] = N\ \mathrm{Var}\,[X]$ and $E\,[Y|\,N] = N\ E\,[X]$. Note that $E_N\,[\cdot]$ and $\mathrm{Var}_N\,[\cdot]$ represent the expected value and the variance with respect to random variable N as follows:

$$E_N\,[\cdot] \triangleq \sum_i (\cdot)\,\mathrm{Prob}\,\{N=i\}$$

$$\mathrm{Var}_N\,[\cdot] \triangleq \sum_i (\cdot)^2\,\mathrm{Prob}\,\{N=i\} - \left\{\sum_i (\cdot)\,\mathrm{Prob}\,\{N=i\}\right\}^2$$

Note that we cannot simply remove the conditioning on N from $\mathrm{Var}\,[Y|\,N]$ to get $\mathrm{Var}\,(Y)$ because in this way we could get only the first term of the variance of Y in (3.163), but the second term would be missing: the results would be wrong.

3.3.1.2 PGFs of Typical Distributions

This sub-section provides the PGFs of typical discrete random variables.

PGF of a Geometric Distribution

Let us derive the PGF of the "classical" geometric probability mass function in (3.60):

$$X(z) = \sum_{k=0}^{+\infty} z^k \mathrm{Prob}\,\{X=k\} = \sum_{k=0}^{+\infty} (1-q)\,(zq)^k = \frac{1-q}{1-zq} \tag{3.164}$$

The mean and the mean square values of the geometric distribution can be obtained through the derivatives of (3.164) evaluated in $z=1$ as follows:

$$E\,[X] = X'(1) = \left.\frac{\mathrm{d}}{\mathrm{d}z}\frac{1-q}{1-zq}\right|_{z=1} = \left.\frac{q\,(1-q)}{(1-zq)^2}\right|_{z=1} = \frac{q}{1-q} \tag{3.165}$$

$$E\left[X^2\right] = X'(1) + X''(1) = \frac{q}{1-q} + \left.\frac{\mathrm{d}}{\mathrm{d}z}\frac{q\,(1-q)}{(1-zq)^2}\right|_{z=1}$$

$$= \frac{q}{1-q} + \left.\frac{2q^2\,(1-q)}{(1-zq)^3}\right|_{z=1} = \frac{q}{1-q} + \frac{2q^2}{(1-q)^2} = \frac{q\,(1+q)}{(1-q)^2} \tag{3.166}$$

These results are coincident, respectively, with (3.63) and (3.65) that have been obtained according to the classical definitions. We may note that the PGF allows us to derive the random variable's moments more straightforwardly. Finally, the PGF of the "modified" geometric distribution in (3.61) is

$$X(z) = \sum_{k=1}^{+\infty} z^k\,\mathrm{Prob}\,\{X=k\} = \sum_{k=1}^{+\infty} z^k\,(1-q)\,q^{k-1}$$

$$= z\,(1-q)\sum_{k=1}^{+\infty} (zq)^{k-1} = \frac{z\,(1-q)}{1-zq} \tag{3.167}$$

Note that there is a z factor of difference in the PGF from the "classical" geometric distribution in (3.164) and the "modified" one in (3.167). This is because the modified distribution is shifted by one position with respect to the classical one.

PGF of a Poisson Distribution

Let us derive the PGF of the probability mass function in (3.67):

$$\begin{aligned} X(z) &= \sum_{k=0}^{+\infty} z^k \operatorname{Prob}\{X = k\} = \sum_{k=0}^{+\infty} \frac{(z\rho)^k}{k!} e^{-\rho} = e^{-\rho} \sum_{k=0}^{+\infty} \frac{(z\rho)^k}{k!} \\ &= e^{-\rho} \times e^{z\rho} = e^{\rho(z-1)} \end{aligned} \tag{3.168}$$

PGF of a Binomial Distribution

Let us derive the PGF of the probability mass function in (3.72):

$$\begin{aligned} X(z) &= \sum_{k=0}^{n} z^k \operatorname{Prob}\{X = k\} = \sum_{k=0}^{n} \binom{n}{k} (zp)^k (1-p)^{n-k} \\ &= \text{by invoking the Newton binomial formula} \\ &= (1 - p + zp)^n \end{aligned} \tag{3.169}$$

Note that the use of the PGF of a binomial distribution allows a straightforward method for obtaining mean and mean square values with respect to adopting the classical approach in Sect. 3.2.5.4.

Composition of Two Geometrically Distributed Random Variables

Let us consider random variables X_i, independent identically distributed (iid) with "modified" geometric distribution and parameter $1 - q$; see (3.61) and the related PGF in (3.167):

$$\begin{aligned} &\operatorname{Prob}\{X_i = k\} = (1-q)\, q^{k-1}, \quad 0 < q < 1, \quad k = 1,\ 2, \ldots \\ &\text{with PGF}\ \ X(z) = \tfrac{z(1-q)}{1-zq} \end{aligned}$$

We consider the sum of X_i from $i = 1$ to N, where N is a random variable with "modified" geometric distribution (parameter $1 - p$) and probability mass function as:

$$\begin{aligned} &\operatorname{Prob}\{N = j\} = (1-p)\, p^{j-1}, \quad 0 < p < 1, \quad j = 1,\ 2, \ldots \\ &\text{with PGF}\ \ N(z) = \tfrac{z(1-p)}{1-zp} \end{aligned}$$

We are interested in characterizing the random variable Y obtained as:

$$Y = \sum_{i=1}^{N} X_i$$

By exploiting the results shown in Sect. 3.3.1.1, we have that the PGF of Y can be characterized using (3.160) as follows:

$$Y(z) = N[X(z)] = \frac{\frac{z(1-q)}{1-zq}(1-p)}{1 - \frac{z(1-q)}{1-zq} p} = \frac{z(1-q)(1-p)}{1 - z[q + p - pq]}$$

$$= \frac{z(1-q)(1-p)}{1-z[1-(1-q)(1-p)]} \tag{3.170}$$

Hence, the result in (3.170) proves that variable Y still has a "modified" geometric distribution with parameter $(1-q)\times(1-p)$, i.e., the product of the parameters of the distributions composed.

Sum of a Given Number of Independent Bernoulli-Distributed Random Variables

A Bernoulli random variable X is defined as follows:

$$X = \begin{cases} 1, \text{ with probability } p \\ 0, \text{ with probability } 1-p \end{cases} \tag{3.171}$$

The PGF of variable X is: $X(z) = 1 - p + zp$.
Let us consider the sum of n (a given value) iid Bernoulli random variables X_i:

$$Y = \sum_{i=1}^{n} X_i$$

Since X_i are iid, the PGF of Y is obtained as follows:

$$Y(z) = [X(z)]^n = [1 - p + zp]^n \tag{3.172}$$

By comparing (3.172) with the results in Sect. 3.3.1.2 on the "PGF of a Binomial Distribution," we may conclude that Y is binomially distributed. Hence, the sum of Bernoulli random variables yields a binomial random variable.

Sum of a Given Number of Independent Geometrically Distributed Random Variables

Let us consider iid random variables X_i, for $i = 1, 2, \ldots, n$, with modified geometric distribution:

$$\text{Prob}\{X_i = k\} = (1-q)\,q^{k-1}, \quad k = 1, 2, \ldots$$

We study the new random variable Y given by the sum of X_i:

$$Y = \sum_{i=1}^{n} X_i$$

Note that n is a given value, not a random one.
From (3.156) and (3.167), the PGF of Y can be expressed as:

$$Y(z) = [X(z)]^n = \left[\frac{z(1-q)}{1-zq}\right]^n$$

It is possible to show that Y has a "negative" binomial distribution (also called Pascal distribution) with the following probability mass function:

$$\text{Prob}\{Y = j\} = \binom{j-1}{n-1}(1-q)^n q^{j-n}, \quad j = n, n+1, \ldots$$

Composition of Bernoulli and Poisson Random Variables

Let us consider an example similar to the previous one with the sum of N iid variables X_i with Bernoulli distribution, as defined in (3.171). N does not represent a deterministic value but a Poisson-distributed random variable with a mean value ρ. We have to characterize the distribution of the resulting random variable Y:

$$Y = \sum_{i=1}^{N} X_i$$

Random variable N is characterized by the following distribution and PGF:

$$\begin{array}{l} \text{Prob}\{N = j\} = \frac{\rho^j}{j!} e^{-\rho}, \quad \rho > 0, \quad j = 0, 1, 2, \ldots \\ \text{with PGF} \quad N(z) = e^{\rho(z-1)} \end{array}$$

To derive the PGF of random variable Y, first, we condition on a given value of N so that the corresponding $Y(z|N)$ is already given by (3.172). Then, we remove the conditioning utilizing the distribution of N, thus achieving the PGF $Y(z)$ as follows:

$$\begin{aligned} Y(z) &= \sum_{j=0}^{\infty} (1 - p + zp)^j \frac{\rho^j}{j!} e^{-\rho} = e^{-\rho} \sum_{j=0}^{\infty} \frac{[(1 - p + zp)\,\rho]^j}{j!} = e^{-\rho}\, e^{(1-p+zp)\rho} \\ &= e^{p\rho(z-1)} \end{aligned}$$

Hence, we can see that random variable Y has a Poisson distribution with a mean value $p \times \rho$.

3.3.2 The Characteristic Function of a pdf

The characteristic function $\Phi_X(\omega)$ of random variable X [either continuous, for $x \in (-\infty, +\infty)$, or discrete] with pdf $f_X(x)$ is defined as follows for $\omega \in (-\infty, +\infty)$:

$$\Phi_X(\omega) = E\left[e^{j\omega X}\right] = \begin{cases} \int_{-\infty}^{+\infty} f_X(x) e^{j\omega x}\, \mathrm{d}x & \text{for a continuous variable} \\ \sum_k \text{Prob}\{X = k\}\, e^{j\omega k} & \text{for a discrete variable} \end{cases} \tag{3.173}$$

where j denotes the imaginary unit ($j^2 = -1$).

Note that the expression of the characteristic function in the case of a discrete random variable can be obtained from the case of the continuous random variable provided that we use the pdf as a sum of Dirac Delta functions. Then, comparing the resulting expression in (3.173) with the PGF definition in (3.150) we can note that we pass from the characteristic function (ω domain) to the PGF (z domain) with the transform:

$$z = e^{j\omega} \tag{3.174}$$

The characteristic function is similar to a Fourier transform of the pdf. The only difference is the change of the sign in the exponent of the integrand function. The properties of the characteristic function are detailed below, referring to the case of continuous random variables.

$$\Phi_X(\omega = 0) = \int_{-\infty}^{+\infty} f_X(x)\mathrm{d}x = 1 \;\; (normalization) \tag{3.175}$$

$$\begin{aligned} |\Phi_X(\omega)| &= \left| \int_{-\infty}^{+\infty} f_X(x) e^{j\omega x}\, \mathrm{d}x \right| \le \int_{-\infty}^{+\infty} |f_X(x)| \left|e^{j\omega x}\right| \mathrm{d}x \\ &= \text{we use } \left|e^{j\omega x}\right| = 1 \text{ and } f_X(x) \ge 0 \end{aligned} \tag{3.176}$$

$$= \int_{-\infty}^{+\infty} f_X(x)\mathrm{d}x = 1 \Rightarrow |\Phi_X(\omega)| \leq 1$$

We can substitute the series expansion of the exponential term $e^{j\omega x}$ in (3.173) to have

$$\begin{aligned}
\Phi_X(\omega) &= \int_{-\infty}^{+\infty} f_X(x)e^{j\omega x}\,\mathrm{d}x = \int_{-\infty}^{+\infty} f_X(x)\sum_{k=0}^{\infty}\frac{(j\omega x)^k}{k!}\mathrm{d}x \\
&= \int_{-\infty}^{+\infty}\left(1 + j\omega x + \frac{(j\omega x)^2}{2} + \cdots\right) f_X(x)\mathrm{d}x \\
&= 1 + j\omega\int_{-\infty}^{+\infty} x f_X(x)\mathrm{d}x + \frac{(j\omega)^2}{2}\int_{-\infty}^{+\infty} x^2 f_X(x)\mathrm{d}x + \omega^3(\ldots)
\end{aligned} \tag{3.177}$$

By taking the first derivative of (3.177) with respect to ω, we have

$$\begin{aligned}
\Phi'_X(\omega) &= j\int_{-\infty}^{+\infty} x f_X(x)\mathrm{d}x - \omega\int_{-\infty}^{+\infty} x^2 f_X(x)\mathrm{d}x + \omega^2(\ldots) \\
&= jE[X] - \omega E\left[X^2\right] + \omega^2(\ldots)
\end{aligned} \tag{3.178}$$

By taking the second derivative of (3.177) with respect to ω, we have

$$\Phi''_X(\omega) = -\int_{-\infty}^{+\infty} x^2 f_X(x)\mathrm{d}x + \omega(\ldots) = -E\left[X^2\right] + \omega(\ldots) \tag{3.179}$$

Hence, the first two derivatives calculated in $\omega = 0$ allow to obtain mean and mean square value of X, as detailed below:

$$\begin{aligned}
E[X] &= \frac{1}{j}\Phi'_X(\omega = 0) = -j\Phi'_X(\omega = 0) \\
E\left[X^2\right] &= -\Phi'_X(\omega = 0)
\end{aligned} \tag{3.180}$$

In general, we obtain the different moments of random variable X by taking the subsequent derivatives of the characteristic function as follows:

$$E[X^m] = \frac{1}{j^m}\Phi_X^{(m)}(\omega = 0) \tag{3.181}$$

where symbol $\Phi^{(m)}$ denotes the mth derivative of Φ.

Let us consider the sum of the independent random variables X_i, $i = 1, 2, \ldots, n$ characterized by pdfs with characteristic functions $\Phi_{X_i}(\omega)$. We need to determine the characteristic function of the following variable:

$$Y = \sum_{i=1}^{n} X_i$$

We have

$$\begin{aligned}
\Phi_Y(\omega) &= E\left[e^{j\omega\sum_{i=1}^{n} X_i}\right] = E\left[e^{j\omega X_1} \times e^{j\omega X_2} \times \cdots \times e^{j\omega X_n}\right] \\
&= \text{for the statistical independence of } X_i \text{ variables} \\
&= E\left[e^{j\omega X_1}\right] \times E\left[e^{j\omega X_2}\right] \times \cdots \times E\left[e^{j\omega X_n}\right] = \prod_{i=1}^{n}\Phi_{X_i}(\omega)
\end{aligned} \tag{3.182}$$

Hence, the characteristic function of the sum of independent random variables is the product of the characteristic functions of these variables.

3.3.2.1 Sum of a Random Number of Continuous Random Variables

Similarly to what was done in Sect. 3.3.1.1, we consider independent continuous random variables X_i $(i = 1, 2, \dots)$ with pdfs that have characteristic functions $\Phi_{Xi}(\omega)$. We are interested in studying the new random variable Y obtained as follows:

$$Y = \sum_{i=1}^{N} X_i \tag{3.183}$$

where N is a discrete random variable with probability mass function Prob$\{N = j\}$ and PGF $N(z)$.

Conditioning on a given $N = n$ value, we use the characteristic function of Y, as shown in (3.182). Then, we remove the conditioning through the probability mass function of N:

$$\Phi_Y(\omega) = \sum_{j=1}^{\infty} \left[\prod_{i=1}^{n} \Phi_{X_i}(\omega) \right] \text{Prob}\{N = j\} \tag{3.184}$$

In the case where variables X_i are iid with characteristic function $\Phi_X(\omega)$, the result in (3.184) can be further elaborated as:

$$\Phi_Y(\omega) = \sum_{j=1}^{\infty} \Phi_X^j(\omega)\, \text{Prob}\{N = j\} = N[\Phi_X(\omega)] \tag{3.185}$$

Therefore, the characteristic function is obtained using the PGF of N and replacing z with the characteristic function $\Phi_X(\omega)$. The final result in (3.185) makes it possible to evaluate the "compound distribution" moments quickly through the derivatives of $\Phi_Y(\omega)$. In particular, we have

$$\begin{aligned}
E[Y] &= -j\Phi_Y'(\omega = 0) = -j\frac{\mathrm{d}}{\mathrm{d}\omega} N[\Phi_X(\omega)]\Big|_{\omega=0} \\
&= -j\{N'[\Phi_X(\omega)] \times \Phi_X'(\omega)\}|_{\omega=0} = N'[1] \times [-j\Phi_X'(0)] \\
&= E[N] \times E[X] \\
E[Y^2] &= -\Phi_Y''(\omega = 0) = -\left\{\frac{\mathrm{d}}{\mathrm{d}\omega} N'[\Phi_X(\omega)] \times \Phi_X'(\omega)\right\}\Big|_{\omega=0} \\
&= \left\{-N''[\Phi_X(\omega)] \times [\Phi_X'(\omega)]^2 - N'[\Phi_X(\omega)] \times \Phi_X''(\omega)\right\}\Big|_{\omega=0} \\
&= N''[1] \times [-j\Phi_X'(0)]^2 + N'[1] \times [-\Phi_X''(0)] \\
&= \{E[N^2] - E[N]\}\{E[X]\}^2 + E[N]\,E[X^2]
\end{aligned} \tag{3.186}$$

These results fully agree with those shown in Sect. 3.3.1.1.

3.3.2.2 The Characteristic Function of an Exponential Random Variable

Let X be exponentially distributed with mean rate μ. Hence, its characteristic function is obtained as:

$$\Phi_X(\omega) = \int_0^{+\infty} \mu\, e^{-\mu x}\, e^{j\omega x}\, \mathrm{d}x = \frac{\mu}{\mu - j\omega} \tag{3.187}$$

Composition of Exponential and Modified Geometric Random Variables

Let us consider iid variables X_i, with exponential distribution and mean rate μ. We refer to the distribution in (3.86) and the corresponding characteristic function in (3.187):

$$f_X(x) = \mu\, e^{-\mu x}, \quad x \geq 0$$
$$\text{with characteristic function } \Phi_X(\omega) = \frac{\mu}{\mu - j\omega}$$

We sum random variables X_i from $i = 1$ to N, where N has a modified geometric distribution with parameter $1 - p$ as:

$$\text{Prob}\{N = j\} = (1-p)\,p^{j-1}, \quad 0 < p < 1, \quad j = 1,\ 2, \ldots$$
$$\text{with PGF } \ N(z) = \frac{z(1-p)}{1-zp}$$

We are interested in characterizing the random variable Y obtained as:

$$Y = \sum_{i=1}^{N} X_i$$

Conditioning on a given N value, the corresponding random variable has a characteristic function $\Phi_Y(\omega|N)$ obtained as:

$$\Phi_Y\left(\omega \middle| N\right) = \left[\frac{\mu}{\mu - j\omega}\right]^N$$

We remove the conditioning on N employing its distribution:

$$\Phi_Y(\omega) = \sum_{j=1}^{\infty} \left[\frac{\mu}{\mu - j\omega}\right]^j (1-p)\,p^{j-1} = N\left(z = \frac{\mu}{\mu - j\omega}\right) = \frac{\frac{\mu}{\mu - j\omega}(1-p)}{1 - \frac{\mu}{\mu - j\omega}p}$$
$$= \frac{\mu(1-p)}{\mu(1-p) - j\omega}$$

Hence, we note that the above characteristic function $\Phi_Y(\omega)$ corresponds to an exponentially distributed random variable with a mean rate $\mu(1-p)$.

3.3.2.3 The Characteristic Function of a Gaussian Random Variable

Let X be a Gaussian random variable with mean value μ and standard deviation σ. Its characteristic function can be derived as:

$$\begin{aligned}\Phi_X(\omega) &= \int_{-\infty}^{+\infty} \frac{1}{\sqrt{2\pi}\sigma} e^{-\frac{(x-\mu)^2}{2\sigma^2}}\, e^{j\omega x}\, dx \\ &= \text{we make the substitution } u = \frac{x-\mu}{\sqrt{2}\sigma} \\ &= \frac{e^{j\omega\mu}}{\sqrt{\pi}} \int_{-\infty}^{+\infty} e^{-u^2}\, e^{j\omega\sqrt{2}\sigma u}\, du\end{aligned} \tag{3.188}$$

By using the Residue theorem for the integration in the complex domain [13], it is possible to prove that the Fourier transform of the exponential impulse is as follows:

$$\int_{-\infty}^{+\infty} e^{-u^2}\, e^{-j\omega u}\, du = \sqrt{\pi} e^{-\frac{\omega^2}{4}} \tag{3.189}$$

In (3.189) we make the substitution

$$\omega \to -\omega\sqrt{2}\sigma \Rightarrow \int_{-\infty}^{+\infty} e^{-u^2} e^{j\omega\sqrt{2}\sigma u} \, du = \sqrt{\pi} e^{-\frac{(\omega\sigma)^2}{2}}$$

and we substitute the previous result into (3.188) to express the characteristic function of a Gaussian random variable:

$$\begin{aligned}\Phi_X(\omega) &= \frac{e^{j\omega\mu}}{\sqrt{\pi}} \int_{-\infty}^{+\infty} e^{-u^2} e^{j\omega\sqrt{2}\sigma u} \, du = \frac{e^{j\omega\mu}}{\sqrt{\pi}} \times \sqrt{\pi}\, e^{-\frac{(\omega\sigma)^2}{2}} \\ &= e^{-\frac{(\omega\sigma)^2}{2} + j\omega\mu}\end{aligned} \tag{3.190}$$

3.3.2.4 Inversion of a Characteristic Function

Let us consider a random variable X with the characteristic function $\Phi_X(\omega)$. We can obtain the pdf $f_X(x)$ of X using the following anti-transform formula:

$$f_X(x) = \frac{1}{2\pi} \int_{-\infty}^{+\infty} \Phi_X(\omega)\, e^{-j\omega x} \, d\omega \tag{3.191}$$

where the integral is in the sense of the Cauchy *principal value*.

3.3.3 The Laplace Transform of a pdf

For random variable X [either continuous for $x \in [0, +\infty)$, or discrete for $x \geq 0$] we can use the Laplace transform $X(s)$ of its pdf $f_X(x)$ according to the following definition:

$$X(s) = E\left[e^{-sX}\right] = \begin{cases} \displaystyle\int_0^{+\infty} f_X(x) e^{-sx} \, dx \text{ for a continuous variable} \\ \displaystyle\sum_i \text{Prob}\{X = k\}\, e^{-sk} \text{ for a discrete variable} \end{cases} \tag{3.192}$$

If we compare the characteristic function of random variable X, $\Phi_X(\omega)$, in (3.173) with its Laplace transform, $X(s)$, in (3.192), we note that we pass from the s domain to the ω one by using the following transform:

$$s = -j\omega \tag{3.193}$$

Note that $X(s = 0) = 1$ represents the normalization condition. Following the same approach as for the characteristic function, we can obtain the moments of random variable X as functions of the derivatives of $X(s)$ computed in $s = 0$:

$$E[X^m] = (-1)^m X^{(m)}(s = 0) \tag{3.194}$$

If random variable X has finite moments of all orders, then the Laplace transform $X(s)$ is an analytic function for all values of s with real part, Re$\{s\}$, greater than 0 [12, 13].

The anti-transform of $X(s)$ is carried out by means of the classical methods for Laplace transforms. The Laplace transforms of the pdfs of random variables will be particularly important for analyzing M/G/1 queuing systems in Chap. 5.

3.3.3.1 The Laplace Transform of the pdf of an Exponential Random Variable

Let X denote an exponentially distributed random variable with mean rate μ. The Laplace transform $X(s)$ of this pdf can be obtained by substituting $s = -j\omega$ in (3.187). Hence, we have:

$$X(s) = \int_0^{+\infty} \mu\, e^{-\mu x}\, e^{-sx}\, \mathrm{d}x = \frac{\mu}{\mu + s} \tag{3.195}$$

3.3.3.2 The Laplace Transform of the pdf of a Pareto Random Variable

Let X denote a random variable with Pareto pdf as in (3.123):

$$f_X(x) = \frac{\gamma k^{\gamma}}{x^{\gamma+1}}, \quad x \geq k$$

The Laplace transform $X(s)$ of this pdf can be derived as described below:

$$X(s) = \int_k^{+\infty} \frac{\gamma k^{\gamma}}{x^{\gamma+1}} e^{-sx}\, \mathrm{d}x = (\gamma k^{\gamma}) \times \int_k^{+\infty} x^{-\gamma-1}\, e^{-sx}\, \mathrm{d}x \tag{3.196}$$

The difficulty in obtaining this Laplace transform is that, in general, exponent γ is a real positive number (not an integer number). Hence, $X(s)$ cannot be expressed in a closed form through elementary functions. We resort to using the (upper) incomplete Gamma function $\Gamma(a, y)$ defined below [14]:

$$\Gamma(a, y) = \int_y^{+\infty} e^{-t} t^{a-1}\, \mathrm{d}t \tag{3.197}$$

We compute function $\Gamma(a, y)$ for $a = -\gamma$ and $y = s \times k$:

$$\begin{aligned} \Gamma(-\gamma, sk) &= \int_{sk}^{+\infty} e^{-t} t^{-\gamma-1}\, \mathrm{d}t \\ &= \text{we make the substitution } t = sx \\ &= s^{-\gamma} \int_k^{+\infty} e^{-sx} x^{-\gamma-1}\, \mathrm{d}x \end{aligned} \tag{3.198}$$

If we compare the expression of $\Gamma(-\gamma, sk)$ in (3.198) with the $X(s)$ expression in (3.196), we obtain the following result for the Laplace transform of a Pareto pdf:

$$\begin{aligned} X(s) &= (\gamma k^{\gamma}) \times \int_k^{+\infty} x^{-\gamma-1}\, e^{-sx}\, \mathrm{d}x = (\gamma k^{\gamma}) \times s^{\gamma} \Gamma(-\gamma, sk) \\ &= \gamma (sk)^{\gamma} \Gamma(-\gamma, sk) \end{aligned} \tag{3.199}$$

3.4 Some Considerations on Combinatorics and the Urn Theory

Many probability problems can be modeled as balls to be laid in urns (or balls to be drawn from urns) with different conditions (e.g., balls of different colors or balls indistinguishable, etc.) and rules on how to fill the urns [15]. Combinatorial problems as well as urn models can be used to model real phenomena in diverse areas such as physics, genetics, economics, clinical trials, modeling of networks, and many others. Combinatorics is a sector of mathematics that deals with counting how many configurations are possible using sets of objects. Classical combinatorial problems concern *combinations* and *permutations*. For instance, choosing r objects, having for each of them n different types, leads to n^r different combinations. On the other hand, when studying how many configurations we have in choosing k objects out of n (differently from permutations, here the order of the objects does not count), we can have $\binom{n}{k}$ combinations. Finally, n distinct objects can be arranged in $n!$ different ways.

Back to the urn theory basics, we are concerned about the problem of distributing n balls into m urns, under some rules to lay the balls in the urns. A typical problem is to count how many combinations are possible. Another problem is to study the occupancy of the urns in terms of the probability, that is the probability distribution of the number of urns with r balls. Many different problems are possible depending on distinguishable or indistinguishable balls and the condition of distinguishable or indistinguishable urns. Moreover, the distribution of balls can take place either with exclusion or without exclusion; the term "with exclusion" means that no box can contain more than one ball, and the term "without exclusion" means that a box may contain more than one ball. We assume that there is no limit to the number of balls the urns may contain.

For instance, there are m^n different ways to distribute n distinguishable balls into m distinguishable urns. The cases with indistinguishable urns are more complex and are not considered here. In what follows, we are interested in the cases with indistinguishable balls and distinguishable urns. Then, we have the following number of combinations for n balls into m urns:

$$\binom{n+m-1}{n} = \binom{n+m-1}{m-1} \tag{3.200}$$

Balls fill the urns according to some rules. A simple case is when each ball can be put in the m urns with the same probability $1/m$. In this case, we can easily determine the distribution of the number of urns with exactly r balls as follows.

The most important assumption is the independence of the urns. The state of an urn is independent of the state of the other urns. We consider that the generic ith ball of the n corresponds to a random variable X_i representing the condition that this ball goes into an urn; this is according to the following Bernoulli random variable:

$$X_i = \begin{cases} 1, & \text{with probability } \frac{1}{m} \\ 0, & \text{with probability } 1 - \frac{1}{m} \end{cases} \tag{3.201}$$

Since we have n balls, for every urn, we have to sum n iid random variables as X_i to characterize the occupancy of an urn. We obtain a random variable Y with binomial distribution as shown below:

$$Y = \sum_{i=1}^{n} X_i \qquad \Rightarrow \text{Prob}\{Y = \ell\} = \binom{n}{\ell}\left(\frac{1}{m}\right)^{\ell}\left(1 - \frac{1}{m}\right)^{n-\ell} \tag{3.202}$$

Considering that an iid random variable Y represents each urn (approximation for a large number of balls), we can determine the distribution of the random variable M_k of the number of urns containing exactly k balls as:

$$\text{Prob}\{M_k = j\} = \binom{m}{j} \text{Prob}\{Y = k\}^j \left[1 - \text{Prob}\{Y = k\}\right]^{m-j}. \tag{3.203}$$

where $k \le n$.

Then, the mean number of urns with k balls can be expressed as follows:

$$E[M_k] = m\binom{n}{k}\left(\frac{1}{m}\right)^{k}\left(1 - \frac{1}{m}\right)^{n-k} \tag{3.204}$$

This analysis can be useful for slotted-based random-access MAC protocols (as considered in Chap. 6) to determine how many slots (= urns) contain a single packet transmission (= ball) since multiple packet transmissions in the same slot cause collisions and the destruction of all these packets. This is a case of distinguishable urns since the m slots available for the transmissions of the packets can be numbered. In this case, we have to consider $\ell = 1$ in (3.202) and $k = 1$ in both (3.203) and (3.204). Then, the success probability for a transmission attempt (given n attempts and m urns), P_s, can be obtained by dividing $E[M_1]$ from (3.204) by n, the number of packet transmissions:

$$P_s = \frac{E[M_1]}{n} = \left(1 - \frac{1}{m}\right)^{n-1} \tag{3.205}$$

In the previous study, we have assumed that every ball can be laid with the same probability in every urn. Different results can be achieved for the distribution of M_k if we consider different assumptions for the distribution of the balls in the urns. For instance, another interesting case is due to Bose-Einstein and considers that each configuration for the indistinguishable balls in the urns has the same probability of occurring; in this case, balls are not uniformly distributed in the urns so that $\text{Prob}\{M_k = j\}$ changes with respect to (3.203). More details on this interesting case can be found in [16].

3.5 Methods for the Generation of Random Variables

A typical approach to obtain samples of random variables is to generate samples of a basic random variable and then using a transform to obtain the samples of the desired random variable. Let us refer to the two following techniques:

- Method of the inverse of the PDF (the quantile function).
- Method of the transform.

Let us describe both techniques through some examples.

3.5.1 Method of the Inverse of the Distribution Function

We consider a random variable U with uniform distribution from 0 to 1 with samples obtained through a pseudo-random number generator. We need to generate a random variable X with PDF $F_X(x)$. Assuming that there is the inverse function of $F_X(x)$, $F_X^{-1}(y)$, we can obtain samples of random variable X from the samples of the uniform random variable U using the following formula:

$$X = F_X^{-1}(U) \tag{3.206}$$

3.5.1.1 Generation of an Exponentially Distributed Random Variable

By means of the method of the inverse of the PDF, we are interested in generating samples of an exponentially distributed random variable X with a mean rate λ. The corresponding PDF is quite simple and is invertible on the whole domain, as shown below:

$$\begin{aligned} F_X(x) &= 1 - e^{-\lambda x} \\ F_X^{-1}(y) &= -\frac{\ln(1-y)}{\lambda} \end{aligned} \tag{3.207}$$

Hence, samples of X can be obtained by generating samples of the uniformly distributed random variable U and then by computing the formula $F_X^{-1}(U) = -\ln(1-U)/\lambda$ according to (3.207).

3.5.2 Method of the Transform

A random variable X with PDF $F_X(x)$ can be used to obtain a new random variable $Y = g(X)$, as shown in Sect. 3.2.4.

The PDF of Y can be easily derived according to (3.53) and (3.54), if function $g(\cdot)$ is invertible. Note that in this case X has a generic distribution, not uniform.

3.5.2.1 Generation of a Pareto-Distributed Random Variable

The transform method can be easily used to obtain a Pareto-distributed random variable Y starting from an exponential one X, as follows. Let us consider random variable X with exponential distribution and mean rate γ. Then, $Y = k \times e^X$ is Pareto-distributed[4] with pdf $f_Y(y)$ as shown in (3.123).

[4] Another approach to generate a Pareto-distributed random variable with the pdf shown in (3.123) is to use the previous method of the inversion of the PDF. Accordingly, we have that $Y = k \times (1-U)^{-1/\gamma}$ is Pareto-distributed with pdf given by (3.123), where U (and hence $1-U$) is a random variable uniformly distributed from 0 to 1.

3.6 Exercises

This section contains some exercises that involve the derivation of distributions, PGFs, Laplace transforms, and characteristic functions.

Ex. 3.1 We know that a telecommunication device experiences failures after an exponentially distributed time with a mean value $1/\lambda$ (=*Mean Time Between Failures*, MTBF). Let us assume that a central control system monitors the equipment's status at regular intervals of length T to verify whether there is a failure. We have to determine the probability mass function of variable N = number of checks to be made to find a failure ($N = 1, 2, \ldots$) and the mean time to find a failure, T_f.

Ex. 3.2 We consider a telephone private branch exchange with a single output line. At time $t = 0$, a data transfer (modem) starts that uses the output line for a duration U, modeled according to a uniform distribution in $[0, T]$. Let us assume that at time $\tau > 0$, a call arrives and finds a busy output line due to the previous data transfer. It is requested to determine the distribution of the time W the call has to wait before obtaining a free output line.

Ex. 3.3 A phone user A makes a call at time t_0 through a private branch exchange with a single output line. It finds a busy output line due to another call started from an indefinite time (the duration of calls is exponentially distributed with a mean value $1/\lambda = 3$ min). We have to determine the probability according to which user A finds a busy output line if he/she tries again to call at time $t_0 + \tau$, where τ is exponentially distributed with a mean value $1/\mu = 2$ min.

Ex. 3.4 Two transmitters simultaneously send the same information flow for redundancy reasons. Each transmitter has a failure after a time with exponential distribution and mean value T. Let us refer to the system at time $t = 0$ when both transmitters are working correctly from an indefinite time. Let us determine:

- The mean waiting time for the first failure, $E[t_m]$.
- The pdf of the time t_M to have that both transmitters do not work.
- The mean value of t_M.

Please explain whether we need to adopt the exponential distribution's memoryless property when answering the above questions.

Ex. 3.5 A private branch exchange has 4 output lines. Let us assume that a phone call arrives when three output lines are busy due to preexisting calls so that this call uses the latest available output line from the exchange. Assuming that no other call arrives at the exchange, we have to determine the mean time T from the arrival of the last call to the instant when all four calls are over. In this study, we consider that the duration of each call is exponentially distributed with a mean value $1/\mu$.

Ex. 3.6 We have the following PGF $X(z)$ of a discrete random variable X:

$$X(z) = z^2(1 - p + zp)^N$$

We have to determine the following quantities:

- The mean value of X.
- The mean square value of X.
- The distribution of X.
- The minimum value of X.

Ex. 3.7 Let us consider a packet of N bits, containing a code able to correct t bit errors. Bit errors are independent (due to the use of interleaving) and occur with probability BER. It is requested to determine the packet error probability after decoding.

Ex. 3.8 Let us consider the PGF $M(z)$ shown below for the random variable M:

$$M(z) = \frac{z^2 p - z^2}{z^2 p - 1}$$

It is requested to determine:

- The probability mass function of M.
- The minimum value of M.
- The mean and the mean square value of M.
- The probability that $M > 4$.

Ex. 3.9 Let us consider the following function of complex variable z:

$$X(z) = \frac{1}{2z - 1}$$

May this function be the PGF of a discrete random variable?

Ex. 3.10 Let us consider a mobile phone operator that sells phone services according to two possible charging schemes:

1. The cost of a phone call increases by a fixed amount at regular intervals (units); each charge is made in advance for the corresponding interval; the cost is c_1 euros/interval, and each interval lasts 1 min.
2. The cost of a phone call depends on the actual call duration according to a rate of c_2 euros/min.

Assuming that the call duration is exponentially distributed with mean rate μ in min^{-1}, it is requested to compare the two charging schemes in terms of average expenditure per call to find the most convenient one.

Ex. 3.11 Let us consider the following functions of complex variable z:

$$\frac{1}{2z - 1}, \quad \frac{z}{2}, \quad \frac{z(z+1)}{2}, \quad z\left[\frac{z+1}{2}\right]^5, \quad \frac{z(z+1)}{4 - z(z+1)}$$

For each case, we have to verify if it is a PGF and, if yes, it is requested to invert the function to obtain the corresponding probability mass function.

Ex. 3.12 Let us consider a random variable X with probability distribution function $F_X(x)$ and probability density function $f_X(x)$ for $-\infty \leq x \leq +\infty$. We are requested to determine the new random variable distribution obtained by taking only the positive values of X (truncated distribution).

Ex. 3.13 Let us consider a mobile user traveling in a radio coverage cell with speed V having a uniform distribution from 0 to v_{max}. If the distance R traveled in the cell from border to border is uniformly distributed from 0 to d, we have to determine the pdf of the time spent by the user in the cell.

References

1. Papoulis A, Pillai SU (2001) Probability, random variables and stochastic processes. McGraw Hill, Avenel, NJ
2. Feller W (1971) Probability theory and its applications, vol. II, 2nd edn. Wiley, New York
3. Rinne H (2014) The hazard rate – Theory and inference. Justus–Liebig–University. http://geb.uni-giessen.de/geb/volltexte/2014/10793/
4. Gross D, Harris CM (1974) Fundamentals of queueing theory. Wiley, New York
5. Nanda S (1994) Stability evaluation and design of the PRMA joint voice data system. IEEE Trans Commun 42(3):2092–2104
6. Donald K (1997). The art of computer programming. Volume 1: Fundamental algorithms, 3rd edn. Addison-Wesley, pp 75–79. ISBN 978-0-201-89683-1
7. Addie RG, Zuckerman M, Neame TD (1998) Broadband traffic modeling: simple solutions to hard problems. IEEE Commun Mag 36(8):88–95
8. Willinger W, Taqqu MS, Sherman R, Wilson DV (1997) Self-similarity through high-variability: statistical analysis of Ethernet LAN Traffic at the source level. IEEE/ACM Trans Netw 5(1):71–86

9. Freedman D, Diaconis P (1981) On the histogram as a density estimator: L2 theory. Probability theory and related fields, vol. 57, No. 4. Springer, Berlin, pp 453–476. ISSN 0178-8051
10. Spiegel MR (1975) Schaum's outline of theory and problems of probability and statistics. Schaum's outline series. McGraw-Hill, New York
11. Snedecor GW, Cochran WG (1989) Statistical methods, 8th edn. Iowa State University Press, Ames
12. Priestley HA (2003) Introduction to complex analysis. Oxford University Press, Oxford
13. Spiegel MR (1968) Schaum's outline of complex variables. Schaum's outline series. McGraw-Hill, New York
14. Abramovitz M, Stegun I (1970) Handbook of mathematical functions. Dover, New York
15. Normam LJ, Kotz S (1977) Urn models and their application. Wiley, New York
16. Menon VV, Indira NK (1990) A Poisson approximation in an urn model with indistinguishable balls. J Statist Plann Inference 26(1):93–101

Chapter 4
Markov Chains and Queuing Theory

Abstract This chapter deals with the basic aspects of queuing theory as stochastic processes and then addresses the Markov queues showing how they can be solved and the most important performance parameters derived. In particular, the following queuing systems are solved as: M/M/1, M/M/S, M/M/S/S, M/M/S/S/P. The Erlang-B analysis is provided and applied to a general service time. An approach is presented to solve multi-dimensional Markov chains and related blocking probabilities. This section ends with a comprehensive list of exercises.

Key words: Stochastic processes, Markov chains, Markov queues, Erlang-B, PASTA

4.1 Queues and Stochastic Processes

Telecommunication systems are characterized by the transmission of data on wired or wireless links. In these cases, we have that different "messages" share the use of the same transmission resources. Typical examples can be as follows:

- Different phone calls arrive at a switching node and must be routed on a limited set of output links.
- Different packets need to be sent on the same link.

Transmission requests can be different instances of the same process or be generated by concurrent (and uncoordinated) processes, sharing the same transmission resources. All these cases involve the queuing of either different packets or different calls if there are not enough resources for their simultaneous transmissions. In telecommunication networks, the following ones are typical examples of problems that can be tackled by queuing theory:

- Performance analysis for the transmission on links and corresponding buffer dimensioning
- Network planning (i.e., planning of the capacity needed to interconnect the different nodes of a telecommunication network)
- Performance evaluation of access protocols where different "users" contend for the same resources

A *queue* is characterized by an *arrival process* of service requests, a *waiting list* of the requests to be processed, a *discipline* according to which the requests in the queue are selected to be served, and a *service process*. Queues are special cases of stochastic processes that are represented by a state $X(t)$, denoting the number of service requests or "entities" or "customers" queued at time t. In this chapter, several cases will be considered to understand the theory for Markovian queues.

A stochastic process $X(t)$ is identified by a different distribution of random variable X at different instants t. Let $f_{X(t)}(\tau)$ denote the pdf of process X at time τ. A stochastic process can be characterized as follows [1]:

- The *state-space*, that is the set of all the possible values, which can be taken by $X(t)$. Such space can be continuous or discrete (if the state-space is discrete, the stochastic process is called a *chain*).

Electronic Supplementary Material The online version contains supplementary material available at (https://doi.org/10.1007/978-3-030-75973-5_4).

G. Giambene, *Queuing Theory and Telecommunications*, Textbooks in Telecommunication Engineering,
https://doi.org/10.1007/978-3-030-75973-5_4

- *Time variable*: t can belong to a continuous set or a discrete one.
- *Correlation characteristics* among $X(t)$ random variables at different instants t.

In order to account for the process correlation, we describe $X(t)$ in terms of its joint probability distribution function, sampling the process at different instants $\mathbf{t} = \{t_1, t_2, \ldots, t_n\}$ for any n:

$$\mathrm{PDF}_{\mathrm{X}}(\mathbf{x}, \mathbf{t}) = \mathrm{Prob}\{X(t_1) \le x_1, X(t_2) \le x_2, \ldots, X(t_n) \le x_n\} \tag{4.1}$$

where we consider vector $\mathbf{x} = \{x_1, x_2, \ldots, x_n\}$.

The expected value $E[X(t)]$ and the autocorrelation $R(t_1, t_2)$ of process $X(t)$ can be expressed as

$$E[X(t)] = \int_{-\infty}^{+\infty} \tau f_{X(t)}(\tau)\,\mathrm{d}\tau, \quad R(t_1, t_2) = E[X(t_2)X(t_1)]$$

where the computation of $E[X(t_2)X(t_1)]$ requires a double integral and the use of the joint probability density function of variables $X(t_1)$ and $X(t_2)$.

A process $X(t)$ is *strict-sense stationary* if the following equality holds for any n and $\mathbf{t}$ (i.e., distribution $\mathrm{PDF}_{\mathrm{X}}(\mathbf{x}, \mathbf{t})$ is invariant to time shifts):

$$\mathrm{PDF}_{\mathrm{X}}(\mathbf{x}, \mathbf{t} + \tau) = \mathrm{PDF}_{\mathrm{X}}(\mathbf{x}, \mathbf{t}) \tag{4.2}$$

Moreover, a process $X(t)$ is *wide-sense stationary* if its expected value $E[X(t)]$ and its autocorrelation $R(t, t+\tau) = E[X(t)X(t+\tau)]$ are independent of t: $E[X(t)] = \mu$ and $R(t, t+\tau) = R(\tau)$. Of course condition (4.2) implies the wide-sense stationarity.

A process is *independent* if we have for any n and $\mathbf{t}$:

$$\mathrm{PDF}_{\mathrm{X}}(\mathbf{x}, \mathbf{t}) = \mathrm{Prob}\{X(t_1) \le x_1\}\,\mathrm{Prob}\{X(t_2) \le x_2\} \cdots \mathrm{Prob}\{X(t_n) \le x_n\} \tag{4.3}$$

The same relation in (4.3) holds in terms of probability density functions (we take the partial derivatives $\partial x_1, \ldots, \partial x_n$ on the left side and we take the total derivatives of the single distributions on the right side). In the case of an independent process $X(t)$, the random variables at the different instants, $X(t_i)$, are completely uncorrelated.

A particular type of stochastic process is a chain that evolves in time by making transitions between states, i.e., discrete values of $X(t)$. These transitions can occur at any instant in continuous-time chains or at specific instants in discrete-time chains. A *Markov chain* ([1]) is characterized by the fact that its state value at instant t_{n+1}, $X(t_{n+1})$, depends only on its state value at the previous instant t_n, $X(t_n)$ [2]. The formal definition of a Markov chain $X(t)$ is

$$\begin{aligned}\mathrm{Prob}\Big\{X(t_{n+1}) = x_{n+1} \Big| X(t_n) = x_n, X(t_{n-1}) = x_{n-1}, \ldots, X(t_1) = x_1\Big\} \\ = \mathrm{Prob}\Big\{X(t_{n+1}) = x_{n+1} \Big| X(t_n) = x_n\Big\}\end{aligned} \tag{4.4}$$

The evolution of a Markov chain does not depend on how long the chain is in the current state. This *memoryless characteristic* implies that state sojourn times are exponentially distributed for a continuous-time chain or geometrically distributed for a discrete-time chain. In what follows, we refer mainly to continuous-time Markov chains, where the transitions from one state to another are characterized by mean rates.

There are Markovian chains, called *birth–death Markov chains*, where the transitions from the generic state $X = i$ are only towards state $X = i - 1$ or towards state $X = i + 1$. These chains will be used to model some queues (denoted as M/M/...) that will be described later in this chapter.

Other important types of chains are as follows:

- *Renewal processes*: These are "point" processes (i.e., *arrival processes* or only-birth processes), like the arrival of points on the time axis. Intervals between adjacent arrivals (points) are iid, according to a general distribution. A generic arrival process can be equivalently characterized by the process $N(t)$ of the number of arrivals in a generic interval t or the distribution of the interarrival times. A special case of renewal processes is the Poisson arrival process, where interarrival times are exponentially distributed with a constant rate; see Sect. 4.2.

[1] These chains are named after the Russian mathematician Andrey Markov.

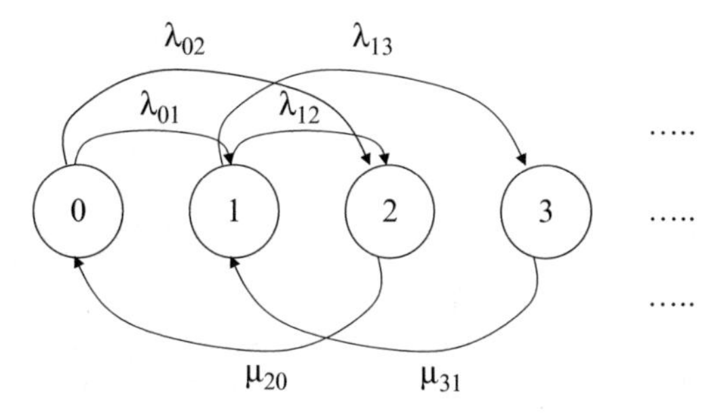

Fig. 4.1: Example of continuous-time Markov chain with mean transition rates between states

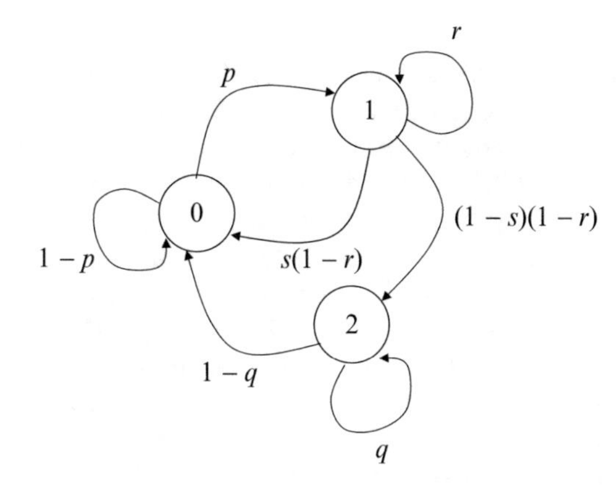

Fig. 4.2: Example of discrete-time Markov chain with transitional probabilities between states

- *Semi-Markov chains*: These chains have a general distribution of the state sojourn time. We can study a chain of this type at the state transition time so that we obtain an *imbedded Markov chain.* Semi-Markov chains will be used to model M/G/1 queues, as described in Chap. 5.

Markov chains are characterized by diagrams with *states* (represented by circles) and *transitions* (represented by directed arcs). In the case of a continuous-time chain, transitions may occur at any time and are characterized by exponentially distributed intervals with mean rates shown above the arcs of the transitions (see the example in Fig. 4.1). Instead, transitions can only occur at given instants for discrete-time chains; probabilities are used to characterize the transitions that correspond to geometrically distributed intervals. In the discrete-time case, states may have transitions into themselves (see the example in Fig. 4.2). The sum of all the transitional probabilities leaving a state must be equal to 1. A Markov chain is said to be *irreducible* if it is possible to get to any state from any state. A state i has period k if any return to state i must occur in multiples of k steps. If $k = 1$, then the state is said to be aperiodic. A Markov chain is *aperiodic* if every state is aperiodic. More details on the analysis of discrete-time Markov chains can be found in [1, 3].

4.1.1 Compound Arrival Processes and Implications

Let us consider the case where each arrival carries multiple "service requests" or "objects": for instance, the arrival of a message that carries multiple packets simultaneously (this could be case of an IP packet fragmented into many layer 2 packets arriving at a MAC layer queue). This group arrival case can have different names in the literature, such as *bulk* arrival process, *batched* arrival process, and *compound* arrival process. These names will be used interchangeably in this book.

However, there is a difference in the compound arrival processes between continuous-time cases and discrete-time ones, as shown in Fig. 4.3:

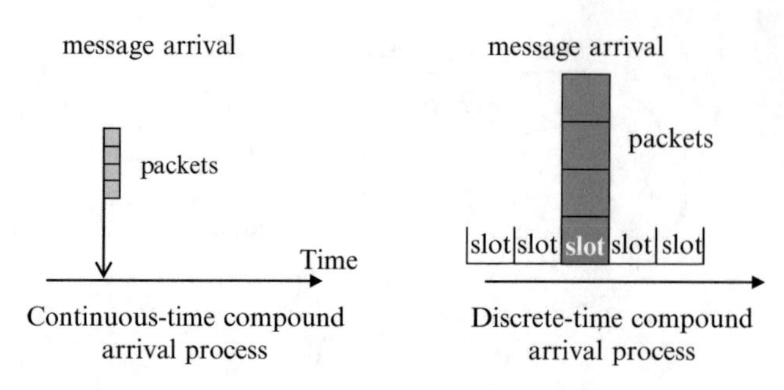

Fig. 4.3: Continuous-time and discrete-time compound arrival processes

- In the continuous-time cases, all the "objects" (i.e., packets) of a group arrive simultaneously at a queuing system. This is possible since we consider that all of these "objects" are generated by the operating system at an extremely faster speed than the service rate of the queue. An example of this arrival process is the compound Poisson process, as described in Sect. 4.2.3. However, there can also be some special compound processes where the different "objects" corresponding to a given arrival do not arrive simultaneously but are generated at a constant rate (see, for instance, Exercise 4.14 and the solution manual).
- In the discrete-time cases, compound arrivals are synchronized with time slots: a message arrival needs a slot to become available to the queue. A message contains multiple "objects" (e.g., packets). This is consistent with the store-and-forward model and refers to the case where messages are transmitted from a network node and arrive at another node, propagating in a communication link: a message must first be stored in the queue of this node to be processed and be ready for the transmission towards the next node. An example of this arrival process is the slot-based binomial packet arrival process, as detailed in Sect. 5.6.

Note that the service can also be done in batches (in groups), if many "objects" are served together in a given frame time.

4.2 Poisson Arrival Process

A Poisson process can be used to describe the number of arrivals $N(t)$ (or equivalently N_t) for any interval of duration t. We have a Poisson arrival process if the following condition holds:

$$\text{Prob}\{N_t = k\} = \frac{(\lambda t)^k}{k!} e^{-\lambda t} \tag{4.5}$$

where λ is the mean arrival rate.

The PGF of the number of arrivals in an interval of duration t, $N_t(z)$, is as follows:

$$N_t(z) = \sum_{k=0}^{\infty} z^k \frac{(\lambda t)^k}{k!} e^{-\lambda t} = e^{-\lambda t} \sum_{k=0}^{\infty} \frac{(z\lambda t)^k}{k!} = e^{-\lambda t} \times e^{zt\lambda} = e^{\lambda t(z-1)} \tag{4.6}$$

The mean number of arrivals in an interval of duration t, $E[N_t]$, and the mean square value of the number of arrivals in t, $E[N_t^2]$, are obtained as follows:

$$\begin{aligned} E[N_t] &= \left.\frac{\mathrm{d}N(z)}{dz}\right|_{z=1} = \left.\lambda t e^{\lambda t(z-1)}\right|_{z=1} = \lambda t \\ E\left[N_t^2\right] &= \left.\frac{\mathrm{d}^2 N(z)}{\mathrm{d}z^2}\right|_{z=1} + \left.\frac{\mathrm{d}N(z)}{\mathrm{d}z}\right|_{z=1} \\ &= \left.(\lambda t)^2 e^{\lambda t(z-1)}\right|_{z=1} + \left.\lambda t e^{\lambda t(z-1)}\right|_{z=1} = \lambda^2 t^2 + \lambda t \end{aligned} \tag{4.7}$$

On the basis of $E[N_t]$ in (4.7), it is evident that λ represents the mean arrival rate of the process. Note that the variance of N_t, $\mathrm{Var}[N_t]$, is equal to its expected value: this is a special characteristic of Poisson processes.

The number of Poisson arrivals in disjoint intervals is statistically independent; instead, *the number of Poisson arrivals in overlapped intervals is not independent.* Hence, N_t and N_s, where t and s are generic instants, are not independent variables. However, even if N_t and N_s are not independent variables, $N_t - N_s$, and N_s are independent variables if $t > s$. The autocorrelation function of a Poisson process can be obtained as follows, referring to a case with $t > s$:

$$\begin{aligned} R(t,s) &= E[N_t \times N_s] = E\left[(N_t - N_s) N_s + N_s^2\right] \\ &= E[N_t - N_s] \times E[N_s] + E\left[N_s^2\right] = \lambda^2 ts + \lambda s \end{aligned}$$

We note that $R(t, s) \neq R(t - s)$, meaning that in this case $R(t, s)$ does not depend on t and s by means of their difference $t - s$. This result together with the fact that $E[N_t] = \lambda t$ (the average value depends on time) allows us to state that *the Poisson process is not wide-sense stationary.* Nevertheless, increments of Poisson processes are stationary (for instance, $N_t - N_s$).

The autocovariance of the Poisson process can be obtained according to the following definition and considering the previous result for $R(t, s)$ with $t > s$:

$$\begin{aligned} C_{NN}(t,s) &= E[(N_t - E[N_t]) \times (N_s - E[N_s])] \\ &= E[N_t \times N_s] - E[N_t] \times E[N_s] = \lambda s \end{aligned}$$

Let us define the Index of Dispersion for Counts (IDC) for a generic arrival process (or point process) as the ratio between the variance of the number of arrivals in a given interval t and the mean number of arrivals in the same interval:

$$\mathrm{IDC}_t = \frac{\mathrm{Var}[N_t]}{E[N_t]} \tag{4.8}$$

For a Poisson process $\mathrm{IDC}_t \equiv 1$, $\forall t$. In general, for a renewal process, $\mathrm{IDC}_t \neq 1$. An arrival process is *peaked* if $\mathrm{IDC}_t > 1$; an arrival process is *smoothed* if $\mathrm{IDC}_t < 1$. If IDC reduces, arrivals are spaced in time more regularly. The limiting case is when $\mathrm{IDC} = 0$, so that the arrival process is deterministic: arrivals occur at fixed, regular intervals. Conversely, when $\mathrm{IDC} > 1$, arrivals tend to occur in bursts (i.e., bursty arrival process). Bursty arrival processes cause the sudden queuing of requests in queuing systems and consequently high delays. For given resources and mean arrival rate, the mean queuing delay increases with IDC. An alternative way to characterize the burstiness of an arrival process will be described in Sect. 4.15 based on the peakedness parameter.

Note that a Poisson arrival process is characterized by only one parameter, i.e., the mean rate λ. From measurements on traffic traces, we can consider having a Poisson process when the mean and variance of the number of arrivals in intervals of length t are equal; correspondingly, we derive λ as the ratio of the mean number of arrivals in an interval of length t and time t itself.

Let us study the statistics of interarrival times t_a for the Poisson process. Let $t = 0$ denote the instant of the last arrival. We determine the probability that the next arrival occurs at a generic instant $t > 0$; this is equivalent to consider the probability that there is no Poisson arrival in the interval $(0, t)$, which is equal to $e^{-\lambda t}$. We have thus obtained the complementary distribution of t_a as

$$\begin{aligned} \mathrm{Prob}\{t_a > t\} = e^{-\lambda t} &\Longleftrightarrow \mathrm{Prob}\{t_a \leq t\} = 1 - e^{-\lambda t} \\ &\Longleftrightarrow \mathrm{pdf}_{t_a}(t) = \lambda e^{-\lambda t}, \quad t \geq 0 \end{aligned}$$

Hence, t_a is exponentially distributed with mean rate λ. Interarrival times are iid. It is possible to prove that we have a Poisson arrival process with mean rate λ if and only if interarrival times are exponentially distributed with mean rate λ (mean value $1/\lambda$).

Poisson processes are quite important in the field of telecommunications since they may model the arrival of several types of events, such as:

- The arrival of new calls at a node of a telephone network (see Fig. 4.4)
- The start of Web browsing sessions for a given user
- The arrival of email messages in a packet data network

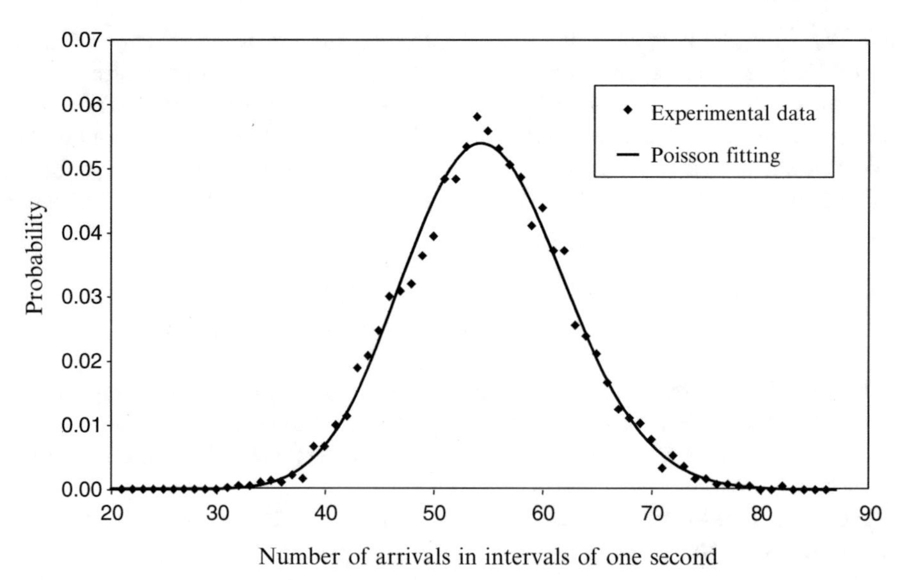

Fig. 4.4: Histogram of the arrivals at a switching node in a telephone network and Poisson fitting

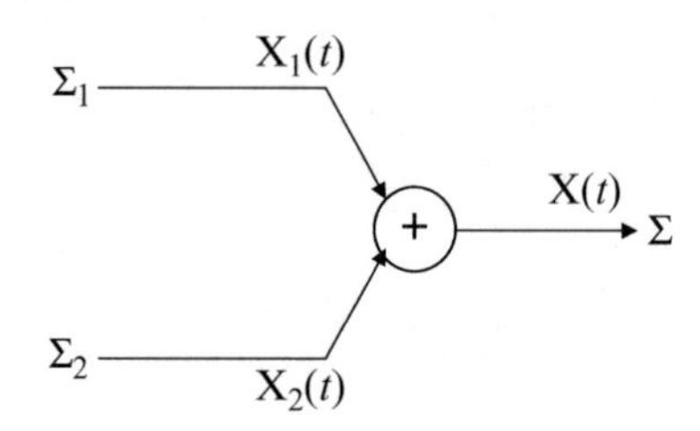

Fig. 4.5: Sum of independent Poisson processes

- The arrival of packets in access networks, as studied in Chap. 6.

As a final remark, it is important to point out that the wide adoption of exponential distributions and Poisson arrival processes is not completely linked to the empirical evidence (measurements), but rather to the ease of conditioning with the help of the memoryless property [4].

In the following sections, we examine important properties of the Poisson arrival processes.

4.2.1 Sum of Independent Poisson Processes

Let us consider two independent sources of Poisson arrivals Σ_1 and Σ_2 with related mean rates λ_1 and λ_2. The arrivals of the two sources are added together to form another source $\Sigma = \Sigma_1 + \Sigma_2$. Let us characterize the process of Σ. We denote with $X_1(t)$ $[X_2(t)]$ the number of arrivals from Σ_1 (Σ_2) in a given interval of duration t. We want to characterize the sum process $X(t) = X_1(t) + X_2(t)$ (Fig. 4.5).

Since $X_1(t)$ and $X_2(t)$ are independent, the PGF of $X(t)$, $X(z)$, is given by the product of the PGF of $X_1(t)$, $X_1(z)$, and the PGF of $X_2(t)$, $X_2(z)$. Considering the Poisson characteristic of both $X_1(t)$ and $X_2(t)$, we have

$$X(z) = X_1(z)X_2(z) = e^{\lambda_1 t(z-1)} e^{\lambda_2 t(z-1)} = e^{(\lambda_1+\lambda_2)t(z-1)} \tag{4.9}$$

From (4.9), we note that the PGF $X(z)$ corresponds to that of a Poisson process with mean rate $\lambda_1 + \lambda_2$. In conclusion, *the process sum of two independent Poisson processes is still a Poisson process with the mean rate given by the sum of the mean rates of the processes.* This is an important property in telecommunication networks since nodes can receive Poisson arrivals of messages from different and independent sources. Another typical example is given by a private branch exchange that collects call arrivals from different phone users.

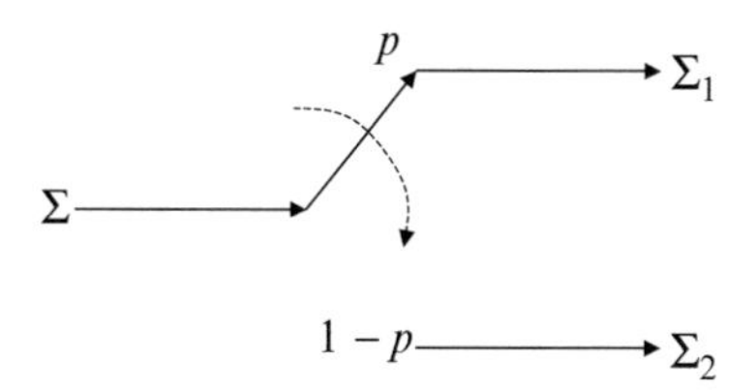

Fig. 4.6: Random splitting of a Poisson process

4.2.2 Random Splitting of a Poisson Process

We consider a Poisson process (mean rate λ), whose arrivals are randomly switched on two output lines: an arrival is sent to line #1 with probability p or to line #2 with probability $1-p$ (Fig. 4.6).

Let us characterize the output process from line #1 (corresponding to source Σ_1) using the statistics of the interarrival times, t_{a_1}. We need to express the distribution of t_{a_1}, knowing the distribution of the interarrival times t_a of the Poisson input process Σ: t_a is exponentially distributed with mean value $1/\lambda$ and Laplace transform of its pdf as $T_a(s) = \lambda/(\lambda+s)$. We refer to a given instant $t=0$ where an arrival from Σ finds the switch in position #1 so that it is forwarded to line #1; then, t_{a_1} denotes the next instant at which an arrival from Σ is switched to line #1. We determine the distribution of t_{a_1} conditioned on the number of arrivals generated by Σ, k, in order to have the next arrival at line #1. In particular:

- $k=1$ with probability p, so that t_{a_1} is equal to t_a
- $k=2$ with probability $p(1-p)$, so that t_{a_1} is the sum of two iid variables with the same distribution as t_a
- $k=3$ with probability $p(1-p)^2$, so that t_{a_1} is the sum of three iid variables with the same distribution as t_a

Therefore, operating in terms of Laplace transforms of pdfs and removing the conditioning on k (using a modified geometric distribution with parameter p), we have

$$T_{a_1}(s) = \sum_{k=1}^{\infty} [T_a(s)]^k p(1-p)^{k-1} = \frac{pT_a(s)}{1-T_a(s)(1-p)} \tag{4.10}$$

It is easy to note that the distribution of t_{a_1} is the composition of the exponential distribution of t_a and the modified geometric distribution of k (see "Composition of Exponential and Modified Geometric Random Variables" in Sect. 3.3.2.2 of Chap. 3). By substituting the expression of $T_a(s)$ in (4.10), we have

$$T_{a_1}(s) = \frac{p\frac{\lambda}{\lambda+s}}{1-\frac{\lambda}{\lambda+s}(1-p)} = \frac{p\lambda}{\lambda p+s} \tag{4.11}$$

Hence, variable t_{a_1} is also exponentially distributed with mean rate $p\lambda$ and the output process from line #1 is still Poisson with mean rate $p\lambda$. Analogously, we can prove that the output process from line #2 (corresponding to source Σ_2) is Poisson with mean rate $(1-p)\lambda$.

The random splitting of Poisson arrivals can be adopted to model the routing of traffic in a network as a "macroscopic" stochastic process.

4.2.3 Compound Poisson Processes

We consider a Poisson arrival process with mean rate λ, where each arrival does not convey a single "object" (or "service request"), but a group of "objects." The lengths of these arrivals in "objects" are iid with generic distribution and the corresponding PGF denoted by $M(z)$. We know that the number of arrivals in an interval of duration t is according to the distribution in (4.5) with the PGF $N_t(z)$ in (4.6). Therefore, the PGF of the number of "objects" arrived in the interval t, $N_{tc}(z)$, can be obtained by conditioning on the number k of groups arrived in t: $N_{tc|k}(z) = M^k(z)$. Then, we derive $N_{tc}(z)$ by means of the Poisson distribution of k:

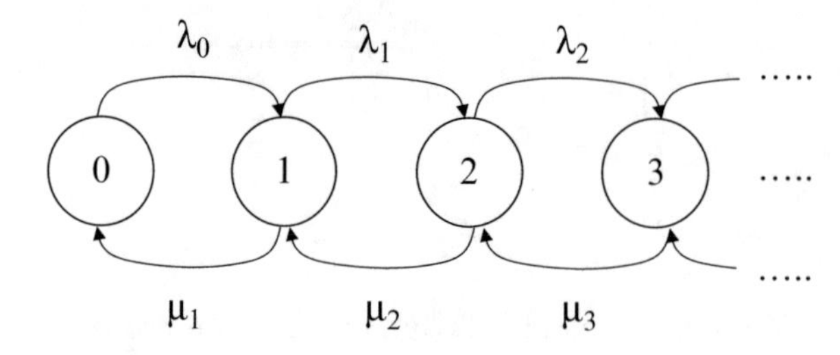

Fig. 4.7: Birth–death Markov chain

$$N_{tc}(z) = \sum_{k=0}^{\infty} M^k(z) \frac{(\lambda t)^k}{k!} e^{-\lambda t} = N_t\left[M(z)\right] = e^{\lambda t[M(z)-1]}$$

In conclusion, the distribution of the number of "objects" arrived in the interval of duration t is obtained as the composition of the variable number of group arrivals in t and the variable number of "objects" per arrival.

This compound arrival process is particularly suited to model the arrival of layer 3 packets fragmented into layer 2 packets at a layer 2 transmission buffer.

4.3 Birth–Death Markov Chains

We study here continuous-time Markov chains describing the behavior of a "population" with states representing the natural numbers $\{0, 1, 2, \dots\}$. For a generic state k, only transitions to states $k-1$ and $k+1$ are allowed. Let us denote:

- λ_i, the *mean birth rate* from state i to state $i+1$
- μ_m, the *mean death (or completion) rate* from state m to state $m-1$
- $P_n(t)$, the probability of state n that, in general, depends on the time

A generic example of a Markov chain is shown in Fig. 4.7, where we assume an infinite number of states.

When the arrival process does not depend on the state, so that $\lambda_i = \lambda\ \forall i$, the arrival process (as well as the Markov chain) is said to be *homogeneous.*

The Kolmogorov–Chapman equations describe the time behavior of this chain. Their analysis is beyond the scope of this book. The interested reader may refer to [1, 3] for more details. We study the chain at equilibrium (assuming that there is an equilibrium). A sufficient condition to have a *steady-state behavior* is the following *ergodicity condition*:

$$\exists \ \text{an index } k_0 \text{ so that} \quad \lambda_k/\mu_k < 1 \quad \forall\ k \geq k_0 \tag{4.12}$$

This condition implies that there is a state beyond which the birth rate is lower than the death rate. In what follows, we will consider that *a queue is stable* if the ergodicity condition (4.12) is met.

Assuming that (4.12) is fulfilled, there is *regime condition* where state probabilities P_n do not depend on time. Hence, we can study the chain in Fig. 4.7 at equilibrium by imposing the balance of the "fluxes" across any closed curve surrounding states in the diagram. In particular, these curves can be *circles* around states or *cuts* that intercept transitions between two states. The most straightforward approach is to make cuts between any pair of states, as shown in Fig. 4.8 and write the corresponding balance equations in sequence as described below. Basically we adopt a recursive approach where the solution of P_i allows us to solve P_{i+1}.

$$\text{cut 1 balance}: \lambda_0 P_0 = \mu_1 P_1 \ \Rightarrow \ P_1 = \frac{\lambda_0}{\mu_1} P_0$$

$$\text{cut 2 balance}: \lambda_1 P_1 = \mu_2 P_2 \ \Rightarrow \ P_2 = \frac{\lambda_1}{\mu_2} P_1 = \frac{\lambda_1}{\mu_2}\frac{\lambda_0}{\mu_1} P_0$$

$$\vdots$$

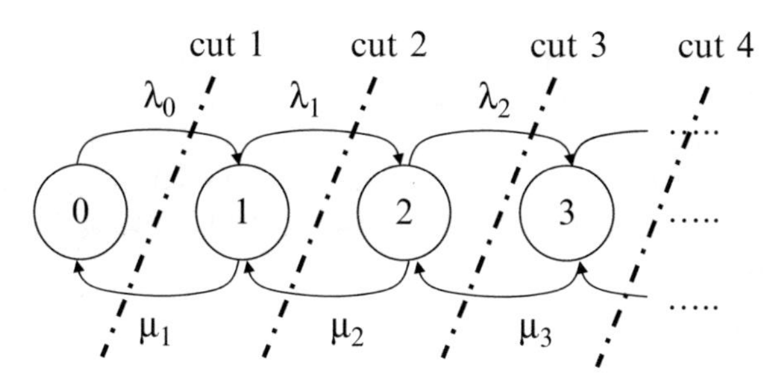

Fig. 4.8: Cuts for the balance equations at equilibrium

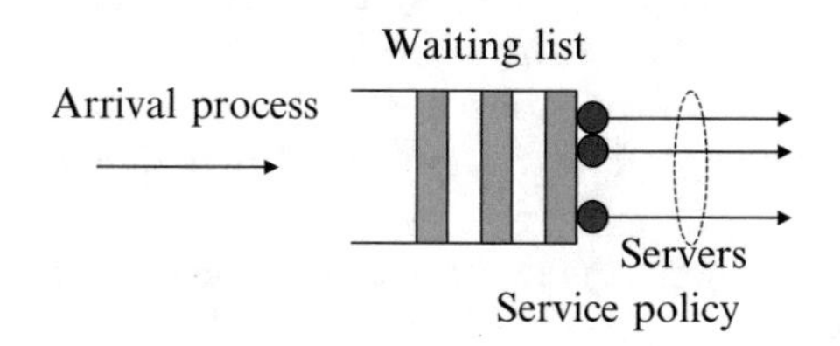

Fig. 4.9: Representation of a generic queue

$$\text{cut } i \text{ balance} : \lambda_{i-1}P_{i-1} = \mu_i P_i \;\Rightarrow\; P_i = \frac{\lambda_{i-1}}{\mu_i}P_{i-1} = P_0\prod_{n=1}^{i}\frac{\lambda_{n-1}}{\mu_n} \quad \forall i \geq 1 \tag{4.13}$$

All state probabilities are expressed as functions of both transition rates and the probability of state "0," P_0. Therefore, we impose the following normalization condition to determine P_0:

$$\sum_{i=0}^{\infty} P_i = 1 \Rightarrow P_0\sum_{i=0}^{\infty}\frac{P_i}{P_0} = 1 \Rightarrow P_0\left(1+\sum_{i=1}^{\infty}\prod_{n=1}^{i}\frac{\lambda_{n-1}}{\mu_n}\right) = 1$$
$$\Rightarrow P_0 = \frac{1}{1+\sum_{i=1}^{\infty}\prod_{n=1}^{i}\frac{\lambda_{n-1}}{\mu_n}} \tag{4.14}$$

We will show later in this chapter how to use birth–death Markov chains to model some queuing systems.

4.4 Notations for Queuing Systems

As shown in Fig. 4.9, a queue can be characterized as follows:

- Arrival process of requests
- List of requests waiting for service
- Policy adopted for the service of the different requests in the list
- Number of servers characterizing the maximum number of requests that can be processed simultaneously
- Statistics of the service duration of each request

To describe the above aspects, the following notation has been introduced by David George Kendall in 1953 [5] (Kendall was the English mathematician who first used the term "queuing system" in his 1951 paper [6]):

$$A/B/S/\Delta/E$$

where "A" denotes the type of the arrival process (e.g., $A=\text{M}$ for a Poisson process; $A=\text{GI}$ for a renewal arrival process; $A=\text{D}$ for a deterministic process). "B" represents the statistics of the service time of a request (e.g., $B=\text{M}$ for an exponentially distributed service duration; $B=\text{G}$ for a generally distributed service process; $B=\text{D}$

for a deterministic service time). "S" indicates the number of servers (i.e., S can be a given integer value or even infinite). "Δ" denotes the number of rooms for the requests in the queuing system, including the currently served request(s): Δ can be an integer value or infinite (in this case, Δ is omitted in the notation); $\Delta \geq S$. Finally, "E" specifies how many sources can produce requests of service: E can be an integer value or infinite (in this case E is omitted). Many service policies have been proposed in the literature; among them we can consider:

- First Input First Output (FIFO)
- Last Input First Output (LIFO)
- Random
- Round-Robin (RR) if the queue is shared by different traffic sources

In the case of a compound arrival process, the queuing system may admit different models. For instance, studying the system at the level of each single "object" of the arrivals, we have models of the type $A^{[\text{comp}]}/B/S\ldots$, where the apex "[comp]" denotes the distribution of the group of arrivals; instead, studying the system at the macroscopic level of each arrival, the model is of the type $A/B/S\ldots$ where B now depends on the distribution "[comp]" of the group of arrivals.

In the case of batched services, the generic queue notation becomes $A/B^{[b]}/S\ldots$, where apex "[b]" represents the number of requests served together.

A queue is said to be *work-conserving* if its server is not empty as long as the queue contains requests ("work") to be served. If the service discipline is not work-conserving, the server can sometimes be "on vacation" so that there is work in the system, but the server is not processing it for a while. In general, vacations lead to an additional delay contribution to the latency experienced by a request in the queue.

In the field of telecommunications, the arrival process typically corresponds to the occurrence of phone calls or messages or packets (i.e., service requests) that have to be transmitted (i.e., served) through a suitable link. Arrival and service processes characterize the traffic. Let λ denote the mean arrival rate and $E[X]$ the mean service duration. A simple way to describe the traffic is given by the *traffic intensity*, ρ:

$$\rho = \lambda E\left[X\right] \tag{4.15}$$

Traffic intensity ρ is measured in "Erlangs," also shortened as "Erl."[1]

4.5 Little Theorem and Insensitivity Property

A queue can be characterized by: (1) the mean number of requests, N, in the queue, including those in service and (2) the mean system delay, T, from the entrance of a request in the queue until the end of its service. We consider here an important result, which allows us to relate T to N in the most general cases of queuing systems. This is the Little's law that was first guessed by Little and then rigorously demonstrated (in the form of a theorem) in a paper and following works [7]. There are many proofs in the literature and all of them are based on very general assumptions. In particular, we can consider the following hypotheses made by Little referring to a generic $G/G/S/\Delta$ queuing system (the queuing system is like a "black box"):

- *Boundary condition*: The queue must become empty at some time instants (this is assured if the queue is stable, as we consider below).
- *Conservation of customers*: All arriving customers (i.e., requests entering the system) will eventually complete their service and will leave the system.

In addition to the above, we consider that the queuing system admits a *steady-state* and is described by an *ergodic process* (time averages are equal to the corresponding statistical averages). Let λ denote the mean arrival rate of customers offered to the queue. Then, the Little theorem states that $N = \lambda T$. A proof of the Little theorem [7] is provided in the following sub-section. Note that alternative hypotheses are considered in [8] to prove the Little theorem.

[1] Agner Krarup Erlang was a Danish engineer who worked for the Copenhagen Telephone Company. He was a pioneer of the queuing theory with his paper published in 1909. Note that the traffic intensity is a dimensionless quantity, but CCIF (a predecessor of ITU-T) decided in 1946 to adopt the "Erlang" as the unit of measurement of the traffic intensity in honor of Erlang's work.

We consider below the *insensitivity property*, according to which the distribution of the number of customers in the system (and thus the mean delay using the Little theorem) is independent of the queuing discipline (i.e., the service order). This property can be asserted under the following general assumptions [9]:

- *The service policy is independent of the service time.*
- *The service policy is work-conserving.*

On the basis of the insensitivity property, many queuing disciplines (e.g., FIFO, LIFO, Random, as well as PS[2]) are characterized by the same mean queuing delay T. In contrast, other moments of the delay do depend on the queuing discipline. We can prove that the following conditions are valid: $E(T_{\text{FIFO}}) = E(T_{\text{Random}}) = E(T_{\text{LIFO}})$ and $\text{Var}(T_{\text{FIFO}}) < \text{Var}(T_{\text{Random}}) < \text{Var}(T_{\text{LIFO}})$.

Note that the assumptions of the insensitivity property exclude the cases where the service order depends on the service time; this would be, for instance, the case of the Shortest Processing Time (SPT) policy, where the request in the queue with the shortest service time is served first. The insensitivity result can be demonstrated by using the Kleinrock *conservation law* in [10, 11] for the mean delays of the different traffic classes in a priority queue.

4.5.1 Proof of the Little Theorem

We consider that the queue is empty at time $t = 0$. Let us denote:

- $\alpha(t)$ = arrival curve, i.e., number of requests arrived in the interval $(0,\ t)$
- $\beta(t)$ = departure curve, i.e., number of requests completing their service in the interval $(0,\ t)$
- t_i = arrival instant of the ith request
- t'_i = departure instant (i.e., service completion) of the ith request

We neglect the cases of multiple arrivals (or departures) at the same instant. Therefore, both $\alpha(t)$ and $\beta(t)$ increase of +1 at arrival and departure instants, respectively. The arrival instants are obviously ordered in time as: $t_1 < t_2 < t_3 < \ldots$. Instead, the ordering of the departure instants in time $t'_1,\ t'_2,\ t'_3, \ldots$ depends on the queuing policy adopted (e.g., in the FIFO case, $t'_1 < t'_2 < t'_3 < \ldots$).

The proof of the Little theorem is carried out under general assumptions on the service policy. The following relations are used:

- $T_i = t'_i - t_i$ represents the time spent in the system (delay) by the ith request.
- $N(t) = \alpha(t) - \beta(t)$ is the number of requests in the queue at the instant $t \geq 0$.

Let us consider a generic instant $t = H$, where $\alpha(H) = \beta(H)$, so that the system is empty (i.e., $N(H) = 0$). The interval from instant t_1 to instant H is called "busy period," i.e., the interval during which the system is non-empty. If the queue is stable (i.e., admits a steady-state), there must be eventually an instant H in which the system becomes empty. For instance, let us refer to the diagram of arrivals and departures in Fig. 4.10. The time average of the delay experienced by a request arrived at the queue in the interval $(0,\ H)$ is

$$\overline{T_H} = \frac{\sum_{i=1}^{\alpha(H)} T_i}{\alpha(H)} = \frac{\sum_{i=1}^{\alpha(H)} (t'_i - t_i)}{\alpha(H)} = \frac{\sum_{i=1}^{\alpha(H)} t'_i - \sum_{i=1}^{\alpha(H)} t_i}{\alpha(H)} \tag{4.16}$$

If we consider the right-side equality in (4.16), we notice that the term $\sum_{i=1}^{\alpha(H)} t'_i$ represents the area between curve $\alpha(t)$ and the ordinate axis in Fig. 4.10. Similarly, $\sum_{i=1}^{\alpha(H)} t_i$ represents the area between curve $\beta(t)$ and the ordinate

[2] Processor Sharing (PS) is an *ideal* service discipline where the server is equally shared among all customers in the queue. Let us consider a single-server queue of the M/M/1−PS type. Surprisingly, it is possible to show that even in this special case, the mean delay T is insensitive to the service time distribution. Hence, the following queuing systems are "equivalent" in terms of mean delay T and mean number of requests N: M/M/1−FIFO, M/M/1−LIFO, M/M/1−PS.

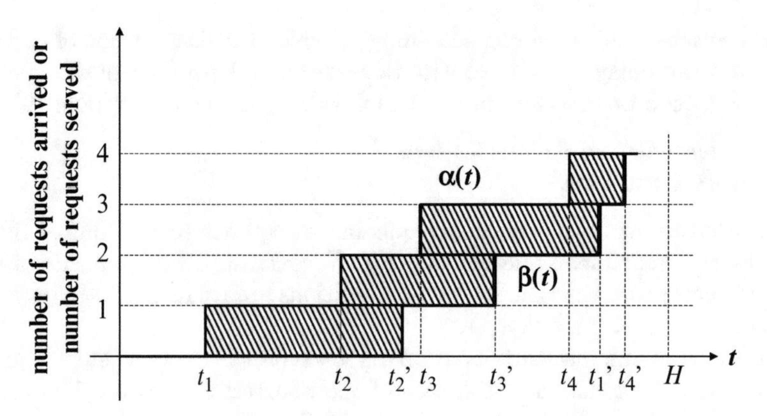

Fig. 4.10: Diagram of arrivals and departures for the queue. The area comprised between $\alpha(t)$ and $\beta(t)$ curves has been *highlighted*. The service policy for the requests has been assumed random

axis. Hence, the difference $\sum_{i=1}^{\alpha(H)} t_i' - \sum_{i=1}^{\alpha(H)} t_i$ is the highlighted area in Fig. 4.10, which can also be expressed as: $\int_0^H [\alpha(t) - \beta(t)]\,dt = \int_0^H N(t)dt$. Since $\overline{N_H} = \frac{1}{H}\int_0^H N(t)dt$ represents the time average of the number of requests in the queue in the interval (0, H), and $\overline{\lambda_H} = \alpha(H)/H$ represents the average arrival rate in the interval (0, H), we can elaborate (4.16) as follows:

$$\overline{T_H} = \frac{\int_0^H N(t)dt}{\alpha(H)} = \frac{H}{\alpha(H)} \times \frac{1}{H}\int_0^H N(t)dt = \frac{\overline{N_H}}{\overline{\lambda_H}} \tag{4.17}$$

Through the ergodicity assumption, *time averages* $\overline{T_H}$, $\overline{N_H}$ and $\overline{\lambda_H}$ are equal to the corresponding *statistical averages*, denoted here as T, N and λ, respectively. The above proof can also be extended to a generic instant H where $\alpha(H) > \beta(H)$, but still referring to a stable queue. Therefore, we can express the Little theorem result [7] by means of the following equality, which relates the mean number of requests in the queue, N, to the mean system delay, T:

$$T = \frac{N}{\lambda} \quad \Longleftrightarrow \quad N = \lambda T \tag{4.18}$$

This formula can also be applied to queues where arriving customers can be blocked with some probability. However, in these cases, λ has to be substituted in (4.18) by the mean rate of the requests entering the queue, λ_s.

Formula (4.18) can be utilized to study the two different parts of a queue: the service part and the waiting list. Let us introduce the following notations:

- $E[X]$, the mean service time of a request
- $E[W]$, the mean time spent in the queue waiting for service
- N_Q, the mean number of requests in the waiting list
- N_S, the mean number of requests in service

We can write the following mean delay balance:

$$T = E[X] + E[W] \tag{4.19}$$

By multiplying both sides of (4.19) by λ (the mean arrival rate) and applying the Little theorem twice (i.e., both to the whole queue and to its different parts), we have

$$\lambda T = \lambda E[X] + \lambda E[W] \Rightarrow N = N_S + N_Q \tag{4.20}$$

Based on (4.15), the mean number of requests in service N_S is equal to the input traffic intensity $\rho = \lambda E[X]$; if there are rejected requests from the queue, we have to consider λ_s in (4.20), so that $N_S = \lambda_s E[X]$ denotes the intensity of the input traffic accepted in the queue. The utilization factor of a server, φ, is given by N_S/S, where S denotes the number of servers of a queue. Of course, $\varphi \in [0, 1)$.

Note that a packet-switched network consists of nodes and links and each node can be modeled as a set of buffers for the transmission on the corresponding output links. The Little theorem has a quite general validity and can also be applied to a whole packet-switched network. In particular, this theorem can be used to relate the mean delay experienced by a message (or packet) from input to output of the network, T, to the mean number of messages (or packets) in the whole network, N, through the total mean arrival rate λ of messages entering the network. Also in this case, we use the formula $T = N/\lambda$.

4.6 M/M/1 Queue Analysis

Let us consider a queue with Poisson arrivals of requests (mean rate λ), exponentially distributed service times (mean rate μ), single server, infinite rooms, and an infinite population of users. This is an M/M/1 queue, according to the Kendall notation. Considering that the state of the system is given by the number of requests in the queue (including the one served), we can model the M/M/1 queue as a birth–death Markov chain with $\lambda_i \equiv \lambda$ and $\mu_i \equiv \mu$, as shown in Fig. 4.11.

The intensity of the input traffic is $\rho = \lambda/\mu$. The ergodicity condition for queue stability is met if the traffic intensity $\rho = \lambda/\mu < 1$ Erlang. Then, the M/M/1 queue can be solved using (4.13) and (4.14). We have

$$\begin{aligned} P_i &= P_0\left(\frac{\lambda}{\mu}\right)^i = P_0\rho^i \\ P_0 &= \frac{1}{1+\sum_{i=1}^{\infty}\rho^i} = \frac{1}{\sum_{i=0}^{\infty}\rho^i} = 1-\rho \quad \text{(normalization)} \end{aligned} \tag{4.21}$$

From (4.21), we note that the state probability is geometrically distributed: $P_i = (1-\rho)\rho^i$. The stability (ergodicity) condition and (4.21) lead to $P_0 > 0$: *if the queue is stable, it must occasionally become empty.*

The PGF of the state probability distribution is obtained as follows:

$$P(z) = \sum_{i=0}^{\infty}(1-\rho)\rho^i z^i = \frac{1-\rho}{1-z\rho} \tag{4.22}$$

The mean number of requests in the system, N, can be obtained by means of the first derivative of the PGF of the state distribution:

$$N = \sum_{i=0}^{\infty} i(1-\rho)\rho^i = \left.\frac{\mathrm{d}P(z)}{\mathrm{d}z}\right|_{z=1} = \frac{\rho}{1-\rho} \tag{4.23}$$

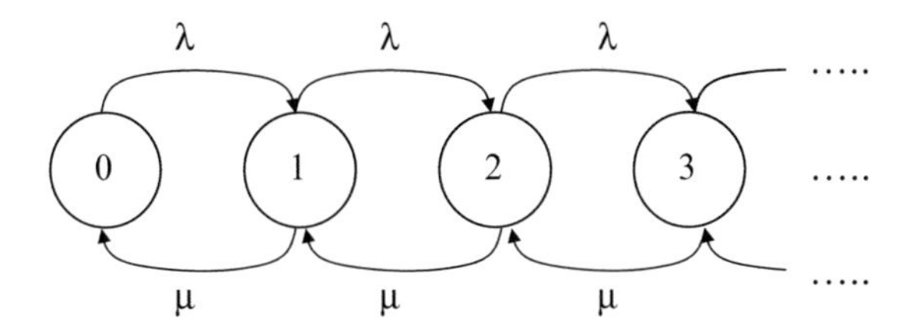

Fig. 4.11: Continuous-time Markov chain modeling an M/M/1 queue

The mean delay from the arrival of a request to its completion, T, is obtained by applying the Little theorem to (4.23):

$$T = \frac{N}{\lambda} = \frac{1}{\mu - \lambda} \tag{4.24}$$

These results on N and T, describing the queue behavior at the "first order," do not depend on the queuing discipline according to the insensitivity property shown in Sect. 4.5.[3]

The ergodicity condition (stability) implies that both N and T have finite values. As $\rho \to 1$ Erlang (or, equivalently $\lambda \to \mu$), the queue becomes congested so that both N and T increase and tend to infinity.

The traffic carried out by the queue (i.e., the throughput), γ, is given by

$$\gamma = \sum_{i=1}^{\infty} \mu (1-\rho)\rho^i = \mu \sum_{i=1}^{\infty} (1-\rho)\rho^i = \mu (1 - P_0) \tag{4.25}$$

In stability conditions, γ is coincident with λ, so that (4.25) can be modified as $\lambda = \mu(1 - P_0)$ or, equivalently, $\rho = 1 - P_0$; this *result is valid in general for any G/G/1 queue* and is equivalent to the P_0 formula in the normalization condition in (4.21).

Finally, the probability that the M/M/1 queue contains more than i requests is obtained as follows:

$$\begin{aligned} \text{Prob}\{n > i\} &= 1 - \sum_{j=0}^{i} (1-\rho)\,\rho^j \\ &= 1 - (1-\rho)\frac{1-\rho^{i+1}}{1-\rho} = \rho^{i+1} \end{aligned} \tag{4.26}$$

4.7 M/M/1/K Queue Analysis

We consider a special case of the M/M/1 queue, where there are only K rooms for requests: the possible states belong to the finite set $\{0, 1, 2, \ldots, K\}$. The notation is M/M/1/K. We use a birth–death Markov chain with $\lambda_i \equiv \lambda$ for $i < K$ and $\mu_i \equiv \mu$ for $i \le K$. If a new arrival finds the queue in the state $i = K$, the new arrival is blocked (i.e., refused from the queue), as described in Fig. 4.12. This queuing model can be adopted to describe a private branch exchange with many input lines, just one output line, and able to queue up to $K - 1$ calls if they find a busy output line.

The intensity of the arrival process (offered traffic) is $\rho = \lambda/\mu$. This queue can be solved by means of (4.13) and a modified version of (4.14), thus obtaining:

$$P_i = P_0 \left(\frac{\lambda}{\mu}\right)^i = P_0 \rho^i$$

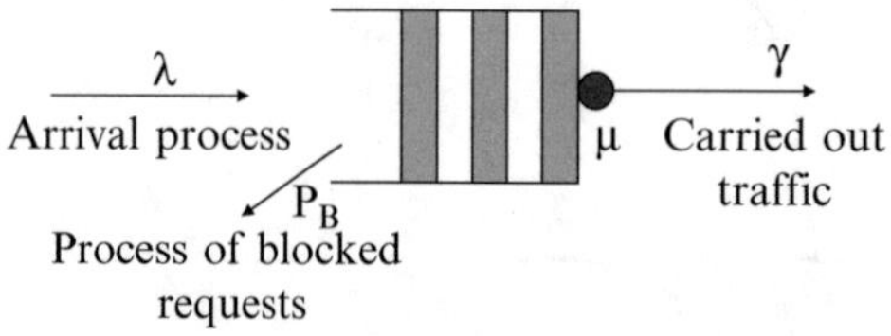

Fig. 4.12: M/M/1/K queue with the process of blocked requests

[3] The insensitivity is lost for some special queue disciplines; this happens when, for instance, the service order is determined by the duration of the service itself, as in the SPT case [12].

$$P_0 = \frac{1}{1+\sum_{i=1}^{K}\rho^i} = \frac{1}{\sum_{i=0}^{K}\rho^i} = \frac{1-\rho}{1-\rho^{K+1}} \quad \text{(normalization)} \tag{4.27}$$

The state probability distribution in (4.27) is obtained from the distribution (4.21) truncated to $i = K$. P_0 is decreasing to 0 with ρ. The limit of P_0 for $\rho \to 1$ Erlang is equal to $1/(K+1)$ using the Hôpital rule. Note that P_0 is positive (tending to 0) for $\rho > 1$ Erlang.

In this special case, the ergodicity condition for the queue stability is met for any ρ value (even if $\rho > 1$ Erlang): since the number of states is finite, there is a state $i = k$ starting from which $\lambda_i = 0$ so that these λ_i values are for sure lower than μ_i.

By means of the Poisson Arrivals See Times Averages (PASTA) property (see Sect. 4.7.1), the probability of state K coincides with the blocking probability experienced by new arrivals, P_B:

$$P_B \equiv P_K = \frac{1-\rho}{1-\rho^{K+1}}\rho^K \tag{4.28}$$

Throughput γ is obtained as

$$\gamma = \sum_{i=1}^{K} \mu P_i = \mu(1-P_0) \tag{4.29}$$

Since the system is stable, γ must be equal to the mean arrival rate accepted in the queue, $\lambda_s = \lambda - \lambda P_B$:

$$\lambda - \lambda P_B = \mu(1-P_0) \quad \Rightarrow \quad \rho(1-P_B) = 1-P_0 \tag{4.30}$$

The PGF of the state probability distribution results as

$$P(z) = \sum_{i=0}^{K} \frac{1-\rho}{1-\rho^{K+1}}\rho^i z^i = \frac{1-\rho}{1-\rho^{K+1}} \frac{1-(\rho z)^{K+1}}{1-\rho z} \tag{4.31}$$

The average number of requests in the queue is obtained as

$$N = \left.\frac{\mathrm{d}P(z)}{\mathrm{d}z}\right|_{z=1} = \frac{\rho}{1-\rho} - \frac{(K+1)\rho^{K+1}}{1-\rho^{K+1}} \tag{4.32}$$

N is equal to zero for $\rho = 0$ and tends asymptotically to K as ρ goes to infinity (the singularity in $\rho = 1$ Erlang can be removed, thus yielding $N = K/2$).

The mean delay can be obtained using the Little theorem as follows:

$$T = \frac{N}{\lambda - \lambda P_B} = \frac{1}{\lambda}\frac{\rho}{1-\rho}\frac{1-\rho^{K+1}}{1-\rho^K} - \frac{1}{\lambda}\frac{(K+1)\rho^{K+1}}{1-\rho^K} \tag{4.33}$$

4.7.1 PASTA Property

In the case of a Poisson arrival process with a constant rate, independent of the state (homogeneous case), the probability that an arrival finds the queue in a given state i coincides with the probability of that state, P_i. This is due to the Poisson Arrivals See Times Averages (PASTA) property defined by R. W. Wolff in 1982 [13]. For M/–/–/– queues where the arrival process is Poisson, the state probabilities as seen at the random instants of new arrivals are the same as the state probability as seen by a random outside observer; these steady-state probabilities are coincident with the percentages of time for which the states occur, referring to ergodic processes. In other words, the fraction of arrivals finding the system in a given state i is equal to the fraction of time the system is in state i and, then, is equal to P_i.

We provide a simplified proof of the PASTA property as follows [14]. We consider the queuing system at regime so that all the probabilities related to the states are not varying with the time. Let $A_n(t)$ denote the probability

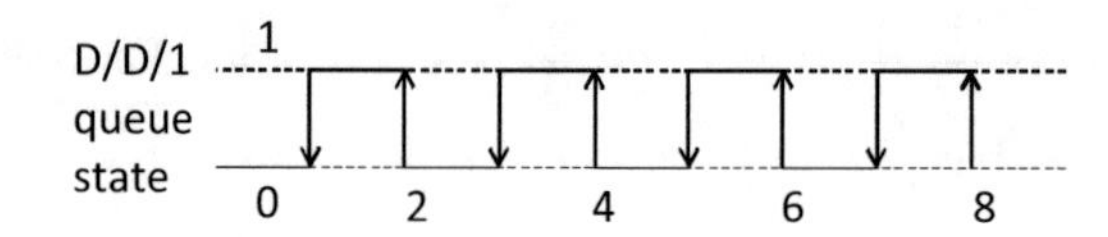

Fig. 4.13: D/D/1 queue example

that an arrival occurring at the generic time t finds the system in state n (i.e., with other n requests inside). At regime, these probabilities do not depend on time t, so that we simply use probabilities A_n.

Let us recall that P_n denotes the probability that the system is in state n. We like to prove that $A_n = P_n$ for $n \geq 0$.

Let $N(t)$ denote the number of requests in the queuing system at the generic time t. In regime conditions, $\text{Prob}\{N(t) = n\} = P_n$. Let $A(t_1, t_2)$ denote the number of Poisson arrivals in the interval from time t_1 to time t_2.

We characterize $A_n(t)$, the probability of one arrival at time t finding the system in state n, as the probability of being in state $N(t) = n$ at time t and that soon after time t (in an elementary interval of amplitude δ) we have a new arrival. We have

$$A_n(t) = \lim_{\delta \to 0} \text{Prob}\{N(t) = n | A(t, t+\delta) = 1\} \tag{4.34}$$

We apply the Bayes rule to remove the conditioning as follows:

$$\begin{aligned} A_n(t) &= \lim_{\delta \to 0} \frac{\text{Prob}\{N(t) = n, A(t, t+\delta) = 1\}}{\text{Prob}\{A(t, t+\delta) = 1\}} \\ &= \lim_{\delta \to 0} \frac{\text{Prob}\{A(t, t+\delta) = 1 | N(t) = n\} \text{Prob}\{N(t) = n\}}{\text{Prob}\{A(t, t+\delta) = 1\}} \\ &= \text{since the Poisson arrival process is memoryless} \\ &= \lim_{\delta \to 0} \frac{\text{Prob}\{A(t, t+\delta) = 1\} \text{Prob}\{N(t) = n\}}{\text{Prob}\{A(t, t+\delta) = 1\}} = \text{Prob}\{N(t) = n\} \end{aligned} \tag{4.35}$$

This completes the proof. Then, we can conclude that in a queue with Poisson arrivals, the probability that an arrival finds the system in state n is the same as the probability that the system is in state n.

The PASTA property does not apply to state-dependent Poisson arrival processes or non-Poisson arrival processes. For instance, let us consider a D/D/1 queuing system, which is empty at time 0, with periodic arrivals at times 1, 3, 5, ... s and with service times of 1 s (see Fig. 4.13): new arrivals always find an empty system, so for them it is as if $P_0 = 1$ (100%). Nevertheless, the queue is empty for 50% of the time, thus yielding $P_0 = 0.5$.

In discrete-time systems, where the arrival process is slot-based, the equivalent of the PASTA property is the BASTA (Bernoulli Arrivals See Times Averages) property.

4.8 M/M/S Queue Analysis

We consider a queue with a Poisson arrival process (mean rate λ), exponentially distributed service times (mean rate μ), and S servers. This is an M/M/S queue, according to the Kendall notation. The birth rate is always equal to λ (i.e., $\lambda_i \equiv \lambda \ \forall i$) and the death rate depends on the state. In the case of a generic state with $i \leq S$, there are i requests served simultaneously; by invoking the memoryless property of the exponential distribution, each served request has a residual lifetime, which is exponentially distributed with mean rate μ. Therefore, the time needed for the transition from state i to state $i-1$ is the minimum among i times exponentially distributed (each with mean rate μ); this minimum is still exponentially distributed with mean rate $\mu_i = i\mu$ (see "Minimum Between Two Random Variables with Exponential Distribution" in Sect. 3.2.6.3 of Chap. 3). For a generic state with $i > S$, the mean completion rate μ_i is equal to $S\mu$. The Markov chain modeling this queue is shown in Fig. 4.14.

The intensity of the arrival process is $\rho = \lambda/\mu$. The ergodicity condition for the stability of the queue requires that $\lambda/(S\mu) < 1$ (i.e., the M/M/S queue can support a traffic intensity ρ up to S Erlangs). The M/M/S queue

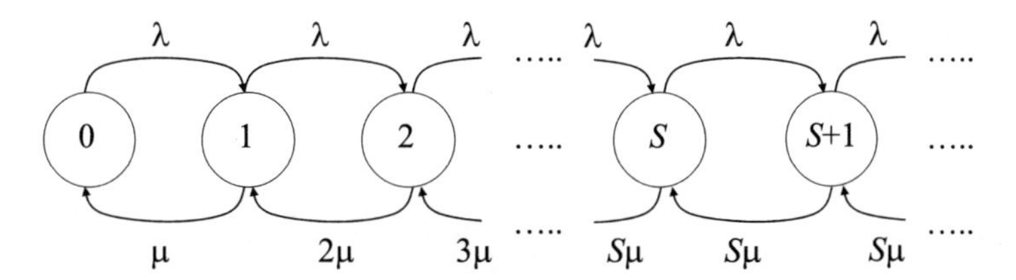

Fig. 4.14: Continuous-time Markov chain modeling an M/M/S queue

can be solved using (4.13) and (4.14), thus obtaining:

$$\begin{aligned} &\text{cut 1 balance}: \ \lambda P_0 = \mu P_1 \ \Rightarrow \ P_1 = \frac{\lambda}{\mu} P_0 = \rho P_0 \\ &\text{cut 2 balance}: \ \lambda P_1 = 2\mu P_2 \ \Rightarrow \ P_2 = \frac{\lambda}{2\mu} P_1 = \frac{\rho^2}{2} P_0 \\ &\dots \\ &\text{cut } S \text{ balance}: \ \lambda P_{S-1} = S\mu P_S \ \Rightarrow \ P_S = \frac{\lambda}{S\mu} P_{S-1} = \frac{\rho^s}{S!} P_0 \\ &\text{cut } S+1 \text{ balance}: \ \lambda P_S = S\mu P_{S+1} \ \Rightarrow \ P_{S+1} = \frac{\lambda}{S\mu} P_S = \frac{\rho^{s+1}}{S\,S!} P_0 \\ &\dots \end{aligned} \tag{4.36}$$

Probability P_0 is obtained by means of the normalization condition as follows:

$$P_0 = \frac{1}{1 + \sum_{i=1}^{\infty} \prod_{n=1}^{i} \frac{\lambda_{n-1}}{\mu_n}} = \frac{1}{\sum_{i=0}^{S-1} \frac{\rho^i}{i!} + \sum_{i=S}^{\infty} \frac{\rho^i}{S!\ S^{i-S}}} = \frac{1}{\sum_{i=0}^{S-1} \frac{\rho^i}{i!} + \frac{S\rho^S}{S!(S-\rho)}} \tag{4.37}$$

It is interesting to note that state probabilities P_n in (4.36) need to be calculated by means of an iterative method, because of the presence of factorial terms and the ratios of very high numbers when n increases. The recursive process starts by computing P_1/P_0; this result is used to compute $P_2/P_0 = (\rho/2) \times P_1/P_0$, and so on. Simultaneously, we sum all these values of P_n/P_0 to obtain P_0 according to (4.37). We can truncate this process for a sufficiently high value of n so that the corresponding terms P_n/P_0 add negligible contributions. Similar considerations can be applied to other queuing systems as well.

The probability that a new arrival finds all the servers busy (so that it is queued), P_C, is given by

$$P_C = \sum_{i=S}^{\infty} P_i = P_0 \sum_{i=S}^{\infty} \frac{P_i}{P_0} = \frac{S\rho^S}{S!\,(S-\rho)} P_0 = \frac{\frac{S\rho^S}{S!(S-\rho)}}{\sum_{i=0}^{S-1} \frac{\rho^i}{i!} + \frac{S\rho^S}{S!(S-\rho)}} \tag{4.38}$$

This is the well-known *Erlang-C formula*, typically used to design the number of servers S to achieve a reasonable queuing probability (e.g., $P_C \leq 1\%$).

Let us now concentrate on the mean queue length, L_q, that can be expressed as follows:

$$\begin{aligned} L_q &= \sum_{i=S+1}^{\infty} (i-S)\, P_i = P_0 \sum_{i=S+1}^{\infty} (i-S) \frac{\rho^i}{S!S^{i-S}} \\ &\text{let us set } \ j = i - S - 1 \\ &= P_0 \frac{\rho^{S+1}}{S!S} \sum_{j=0}^{\infty} (j+1) \left(\frac{\rho}{S}\right)^j = P_0 \frac{\rho^{S+1}}{S!S} \sum_{j=0}^{\infty} S \frac{\mathrm{d}}{\mathrm{d}\rho} \left(\frac{\rho}{S}\right)^{j+1} \\ &= P_0 \frac{\rho^{S+1}}{S!} \frac{\mathrm{d}}{\mathrm{d}\rho} \frac{\rho}{S} \frac{1}{1-\frac{\rho}{S}} = \frac{P_0 \rho^{S+1} S}{S!\,(S-\rho)^2} \ \ \text{[pkts]} \end{aligned} \tag{4.39}$$

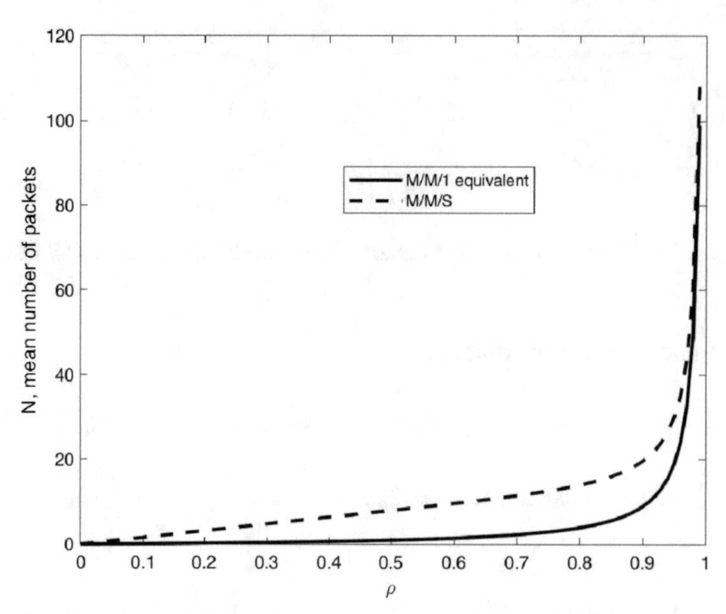

Fig. 4.15: Comparison of the M/M/S queue and the M/M/1 equivalent one with the same capacity ($S = 16$)

Moreover, the mean number of objects in the queue, N, is obtained by summing the mean number of requests in service, $\frac{\lambda}{\mu}$, and the mean number of requests in the queue, L_q, as:

$$N = S\rho + L_q = S\rho + \frac{P_0 \rho^{S+1} S}{S!\,(S-\rho)^2} \quad \text{[pkts]} \tag{4.40}$$

Then, the mean delay T can be obtained by means of the Little theorem, dividing N by λ or, equivalently, by summing the mean service time $\frac{1}{\mu}$ to the mean queuing delay that, according to the Little theorem, is obtained as $\frac{L_q}{\lambda}$. We have

$$T = \frac{1}{\mu} + \frac{L_q}{\lambda} = \frac{1}{\mu}\left[1 + \frac{P_0 \rho^S S}{S!\,(S-\rho)^2}\right] \quad \text{[s]} \tag{4.41}$$

Let us now compare the above M/M/S system with mean arrival rate λ and mean service rate μ with a corresponding queuing system that concentrates all the capacity in a single server: this is an M/M/1 queue with the same mean arrival rate λ, but with the mean service rate equal to $S\mu$. We would like to confront the mean number of requests of these two cases, that is (4.40) with the corresponding mean number of requests in the M/M/1 case below:

$$N_{M/M/1\_eq} = \frac{\rho}{S-\rho} \quad \text{[pkts]} \tag{4.42}$$

The results of this comparison are shown in Fig. 4.15 for $S = 16$. We can see that the equivalent M/M/1 queue yields much better results in terms of requests in the queue and then, using the Little theorem, also in terms of mean delay. Then, we may conclude that aggregating the capacity in a single server is much better than using multiple parallel servers of lower capacity.

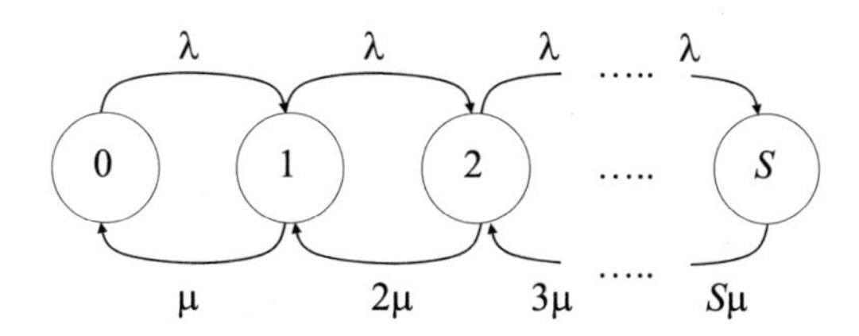

Fig. 4.16: Continuous-time Markov chain modeling an M/M/S/S queue

4.9 M/M/S/S Queue Analysis

This queue has $S+1$ states for i from 0 to S, as shown in Fig. 4.16. The mean birth rate is $\lambda_i \equiv \lambda \ \forall i$ and the mean death rate is $\mu_i = i\mu \ \forall i$. The ergodicity condition for the queue stability is always fulfilled since there is a finite number of states.

By exploiting the derivations made in the M/M/S case, we achieve the following state probability distribution for the M/M/S/S queue:

$$P_i = \frac{\rho^i}{i!} P_0, \quad \text{for } i = 1, 2, \ldots, S$$
$$\text{where} \quad P_0 = \frac{1}{\sum_{i=0}^{S} \frac{\rho^i}{i!}} \tag{4.43}$$

Since the arrival process is Poisson (the mean arrival rate does not depend on the state), the probability that a new request is blocked and lost because of the unavailability of rooms in the queue, P_B, is obtained as the probability that the queue is in the state S, P_S (PASTA property):

$$P_B(S, \rho) \equiv P_S = \frac{\rho^S}{S! \sum_{i=0}^{S} \frac{\rho^i}{i!}} \tag{4.44}$$

This is the well-known *Erlang-B formula*. It is commonly assumed that blocked calls are lost (not reattempted); in practice, rejected calls are reattempted after a certain amount of time so that they can be considered as uncorrelated to the previous ones (and included in the mean arrival rate λ).

The Erlang-B formula is particularly useful for dimensioning circuit-switched networks; for instance, the number of output links S from a node for which we know the input traffic intensity. In particular, the Erlang-B formula can be applied to *typical problems*, like the following one:

> given the traffic intensity $\rho = 12$ Erlangs, we have to determine the minimum number of "servers" S of a switching center to guarantee that the blocking probability $P_B \le 5\%$.

Note that the Erlang-B formula cannot be calculated directly if the number of servers, S, is high due to the presence of the factorial terms. Therefore, the following recursive approach is adopted to compute the Erlang-B formula $P_B(S, \rho)$ with S servers and input traffic intensity ρ. By setting $P_B(0, \rho) = 1$, we obtain $P_B(S, \rho)$ recursively computing the following formula:

$$\frac{1}{P_B(i, \rho)} = 1 + \frac{i}{\rho P_B(i-1, \rho)} \tag{4.45}$$

The recursive approach has been adopted to generate the Erlang-B tabulation in Table 4.1 that can be used to solve the above problem. We consider the column labeled with 5% (blocking probability), and starting from the top, we stop at the first entry greater than or equal to 12 Erlangs, i.e., 12.5. Correspondingly, we read the value of S equal to 17 servers in the leftmost column.

The utilization factor of a server is $\varphi = \lambda(1 - P_B)/\mu/S$. For a given $P_B = P_B^*$ and ρ, we can determine the smallest $S = S^*$ integer value so that $P_B(S^*, \rho = \lambda/\mu) \le P_B^*$. Then, the utilization results as $\varphi = \rho[1 - P_B(S^*,$

Table 4.1: Erlang-B table

S	1%	2%	3%	5%	7%
1	0.0101	0.0204	0.0309	0.0526	0.0753
2	0.153	0.223	0.282	0.381	0.470
3	0.455	0.602	0.715	0.899	1.06
4	0.869	1.09	1.26	1.52	1.75
5	1.36	1.66	1.88	2.22	2.50
6	1.91	2.28	2.54	2.96	3.30
7	2.50	2.94	3.25	3.74	4.14
8	3.13	3.63	3.99	4.54	5.00
9	3.78	4.34	4.75	5.37	5.88
10	4.46	5.08	5.53	6.22	6.78
11	5.16	5.84	6.33	7.08	7.69
12	5.88	6.61	7.14	7.95	8.61
13	6.61	7.40	7.97	8.83	9.54
14	7.35	8.20	8.80	9.73	10.5
15	8.11	9.01	9.65	10.6	11.4
16	8.88	9.83	10.5	11.5	12.4
17	9.65	10.7	11.4	12.5	13.4
18	10.4	11.5	12.2	13.4	14.3
19	11.2	12.3	13.1	14.3	15.3
20	12.0	13.2	14.0	15.2	16.3
21	12.8	14.0	14.9	16.2	17.3
22	13.7	14.9	15.8	17.1	18.2
23	14.5	15.8	16.7	18.1	19.2
24	15.3	16.6	17.6	19.0	20.2
25	16.1	17.5	18.5	20.0	21.2
26	17.0	18.4	19.4	20.9	22.2
27	17.8	19.3	20.3	21.9	23.2
28	18.6	20.2	21.2	22.9	24.2
29	19.5	21.0	22.1	23.8	25.2
30	20.3	21.9	23.1	24.8	26.2

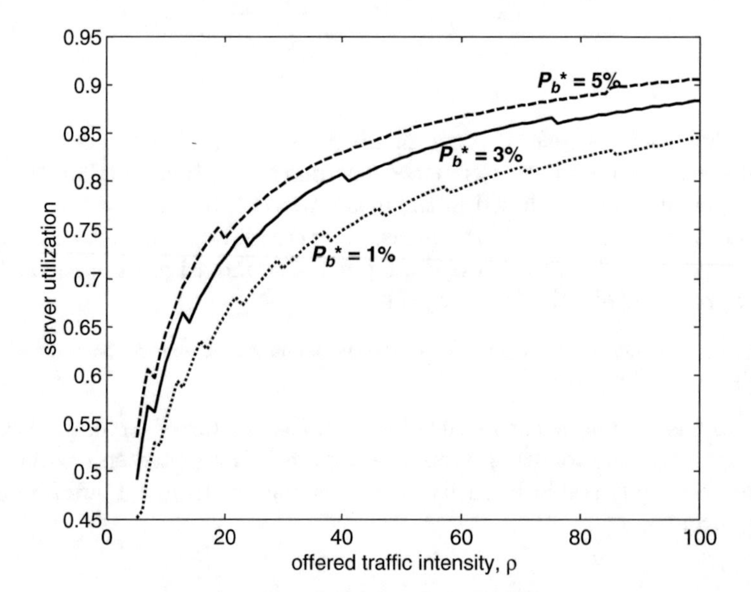

Fig. 4.17: Server utilization versus input traffic intensity for an M/M/S/S queue

$\rho = \lambda/\mu)]/S^*$. In Fig. 4.17, we have plotted φ versus ρ. The steps in the graph are due to the granularity of the integer values of S^* that are used to fulfill the P_{B} constraint. We can note that the utilization of servers φ increases with the input traffic intensity ρ for a given blocking probability P_{B}^*; this result is consistent with the multiplexing effect. Of course, φ increases if higher values of P_{B}^* are permitted.

The mean number of requests N in an M/M/S/S system can be derived as

$$N = \sum_{i=0}^{S} iP_i = P_0 \sum_{i=1}^{S} i\frac{P_i}{P_0} = P_0 \sum_{i=1}^{S} \frac{i\rho^i}{i!} = \rho P_0 \sum_{i=1}^{S} \frac{\rho^{i-1}}{(i-1)!} = \rho\,(1 - P_S) \tag{4.46}$$

The mean arrival rate accepted in the system is

$$\lambda_s = \overline{\lambda} = \sum_{i=0}^{S-1} \lambda_i P_i = \lambda \sum_{i=0}^{S-1} P_i = \lambda\,(1 - P_S) \tag{4.47}$$

Due to system stability, the mean arrival rate accepted in the system λ_s is also equivalent to the mean traffic carried by the system, γ. Note that the mean arrival rate refused by the system is given by λP_B. Even if the input process is Poisson, the process of refused requests and the process of accepted requests are not Poisson. In particular, *the refused traffic is a peaked process and the carried traffic is a smoothed process*, as discussed in the previous Sect. 4.2.

By means of the Little theorem, we can derive the mean delay experienced by a request accepted in the system, T, as follows:

$$T = \frac{N}{\lambda\,(1 - P_S)} = \frac{1}{\mu} \tag{4.48}$$

Since there is not a waiting phase in the M/M/S/S queue, all carried requests experience a (mean) delay equal to their (mean) service time, as expressed by (4.48).

It is possible to prove that the M/M/S/S state distribution is also valid for an M/G/S/S queue with the same input traffic intensity: the state probability distribution has *the property of insensitivity* to the statistics of the service time (only the mean value has impact through the input traffic intensity) [15, 16]. Therefore, the Erlang-B formula (that only depends on the number of servers, the mean arrival rate, and the mean service duration) can also be adopted in the general case of M/G/S/S queues. This generalization of the Erlang-B formula is quite important since current service times are no longer exponentially distributed; for instance, Web browsing sessions typically have durations modeled by Pareto distributions.

4.10 The M/M/S/S/P Queue Analysis

The Erlang-B formula is derived under the assumption of Poisson arrivals, which is an arrival process with a constant rate. However, arrivals are actually phone calls made by users. Each user has on ON–OFF behavior, meaning that phone call intervals (ON times) are separated by idle (OFF) times; both intervals are exponentially distributed with mean rates μ and λ, respectively. Let ρ denote the maximum traffic intensity contributed by each user: $\rho = \frac{\lambda}{\mu}$. Each user admits a two-state Markov chain model with activity probability (probability to be in the ON state) equal to $P_{ON} = \frac{\lambda}{\lambda+\mu} = \frac{\rho}{1+\rho}$.

In the case of a finite (discrete) number of users, P, the arrival process of calls to a switch is not pure Poisson. We consider $P > S$ that corresponds to the case of a loss queuing system (there is no blocking if $P \leq S$). In this case, we can still adopt a Markovian model of the system, but now the arrival rate $\lambda_i = (P - i)\lambda$ depends on the state i (i.e., the number of calls already in progress), as shown in Fig. 4.18. This figure also shows the hypothetical transition that corresponds to the arrivals that are blocked. Note that this Markov chain is not homogeneous. Consequently, the PASTA property is not applicable: the call blocking probability P_B is not equal to the probability of being in the state where all lines are busy, P_S. In this case, we need to differentiate between "call congestion" and "time congestion" thus using a more general approach to determine the blocking probability. Let us consider the following definitions:

- *Time congestion* or time blocking, E, is the fraction of time during which all the servers are busy; this is the probability that the system is in the state S, P_S.
- *Call congestion* or call blocking, P_B, is the probability that an incoming call finds all the servers busy and it is blocked; this is the call blocking probability in our queuing system.

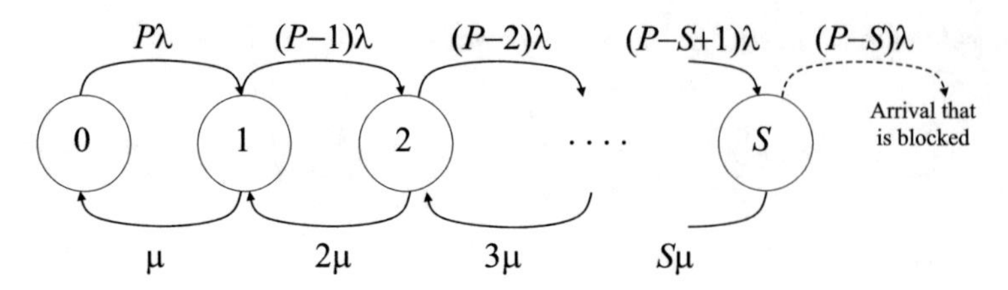

Fig. 4.18: Markov chain model for the M/M/S/S/P queue

For a general arrival process, E is different from $P_{\rm B}$. Only for a Poisson arrival process, we have $E \equiv P_{\rm B}$ (PASTA property).

We can solve the chain in Fig. 4.18 to determine the state probability distribution P_0, P_1, ..., P_S through the cut equilibrium conditions and the normalization condition as follows:

$$\begin{aligned} P\lambda P_0 = \mu P_1 \;&\Rightarrow\; P_1 = \frac{P\lambda}{\mu} P_0 = P\rho P_0 \\ (P-1)\lambda P_1 = 2\mu P_2 \;&\Rightarrow\; P_2 = \frac{(P-1)\lambda}{2!\mu} P_1 = \binom{P}{2}\rho^2 P_0 \\ &\ldots \\ (P-S+1)\lambda P_{S-1} = S\mu P_S \;&\Rightarrow\; P_S = \frac{(P-S+1)\lambda}{S\mu} P_{S-1} = \binom{P}{i}\rho^i P_0 \end{aligned} \tag{4.49}$$

The normalization condition to express P_0 is

$$P_0 = \frac{1}{1 + \sum_{i=1}^{S} {}^{P_i}/_{P_0}} = \frac{1}{\sum_{i=0}^{S} \binom{P}{i}\rho^i} \tag{4.50}$$

Solving these equations, we get the following expression for the state probabilities P_i for $i = 0, \ldots, S$:

$$P_i = \frac{\binom{P}{i}\rho^i}{\sum_{i=0}^{S} \binom{P}{i}\rho^i} \tag{4.51}$$

The state probability distribution P_i in (4.51) can be further elaborated considering the previous relation between P_{ON} and ρ so that $\rho = \frac{P_{ON}}{1-P_{ON}}$. We substitute this expression in (4.51). Then, we multiply both numerator and denominator by $(1-P_{ON})^P$. We obtain the following new expression of the state probability distribution:

$$P_i = \frac{\binom{P}{i}\rho^i}{\sum_{i=0}^{S} \binom{P}{i}\rho^i} = \frac{\binom{P}{i} P_{ON}^i (1-P_{ON})^{P-i}}{\sum_{i=0}^{S} \binom{P}{i} P_{ON}^i (1-P_{ON})^{P-i}} \tag{4.52}$$

This new expression of P_i can be justified because we have i independent users (with i from 0 to S) out of P that are active and admitted in the system, where each of the P users can be active with probability P_{ON}. Then, we need to consider a binomial distribution with i active users out of P that is truncated because only up to $S < P$ active users are admitted. The probability P_i contains the binomial distribution with the normalization term due to the truncation at the denominator.

Like in the M/M/S/S loss system, the state probability distribution is insensitive to the form of the distribution of the call duration. The results shown here are also valid for an M/G/S/S/P queuing system.

The percentage of time that the system is completely full (time congestion) is equal to $P_{full} = P_S$, as

$$E = P_S = \frac{\binom{P}{S}\rho^S}{\sum_{i=0}^{S}\binom{P}{i}\rho^i} \tag{4.53}$$

However, P_{full} is not equal to the blocking probability of new arrivals, P_B, because PASTA cannot be applied!

To compute P_B (arrival congestion, call blocking probability) we need a new approach, considering the arrival rate averaged over all the states λ_{avg} and the blocking rate averaged over all the states (in this model, only state S entails blocking), λ_{block}, obtained as follows:

$$\lambda_{avg} = \sum_{n=0}^{S}(P-n)\lambda P_n = (P-N)\lambda\,, \quad \lambda_{block} = (P-S)\,\lambda P_S \tag{4.54}$$

where N denotes the mean number of users active in the system, obtained as

$$N = \frac{\sum_{i=1}^{S} i\binom{P}{i}\rho^i}{\sum_{i=0}^{S}\binom{P}{i}\rho^i} \tag{4.55}$$

Then, the blocking probability P_B can be obtained by means of the following formula:

$$P_B = \frac{\lambda_{block}}{\lambda_{avg}} \tag{4.56}$$

Using the definition in (4.56) and the expressions for λ_{avg} and λ_{block} in (4.54), through some algebraic manipulations, we get the following result for P_B, known as *Engset's formula*:

$$P_B = P_S\frac{P-S}{P-N} = \frac{\binom{P-1}{S}\rho^S}{\sum_{i=0}^{S}\binom{P-1}{i}\rho^i} \tag{4.57}$$

From this result, we can see that $P_B \neq P_S$.

The mean arrival rate accepted in this system is

$$\lambda_s = \overline{\lambda} = \sum_{i=0}^{S-1}(P-i)\,\lambda P_i = (P-N)\,\lambda - (P-S)\,\lambda P_S = \lambda_{avg} - \lambda_{block} \tag{4.58}$$

Then, applying Little, the mean delay T is

$$T = \frac{N}{\overline{\lambda}} = \frac{\sum_{i=1}^{S} i\binom{P}{i}\rho^i}{\lambda\sum_{i=0}^{S-1}(P-i)\binom{P}{i}\rho^i} = \frac{\rho\sum_{i=1}^{S} P\binom{P-1}{i-1}\rho^{i-1}}{\lambda\sum_{i=0}^{S-1} P\binom{P-1}{i}\rho^i} = \frac{1}{\mu} \tag{4.59}$$

This result is consistent with the fact that there is no waiting part in this queue.

A conservative approach to estimate the blocking probability is P_B obtained using the Erlang-B formula with $\lambda_i = P\lambda$, for $\forall\, i$ value. In this way, we approximate the M/M/S/S/P system as M/M/S/S with peak traffic intensity $\rho_{\max} = P\rho$ so that the Erlang-B formula, in this case, yields an upper bound of P_B in (4.57) as

$$P_B = \frac{\rho_{\max}^S}{S!\sum_{i=0}^{S}\frac{\rho_{\max}^i}{i}} \tag{4.60}$$

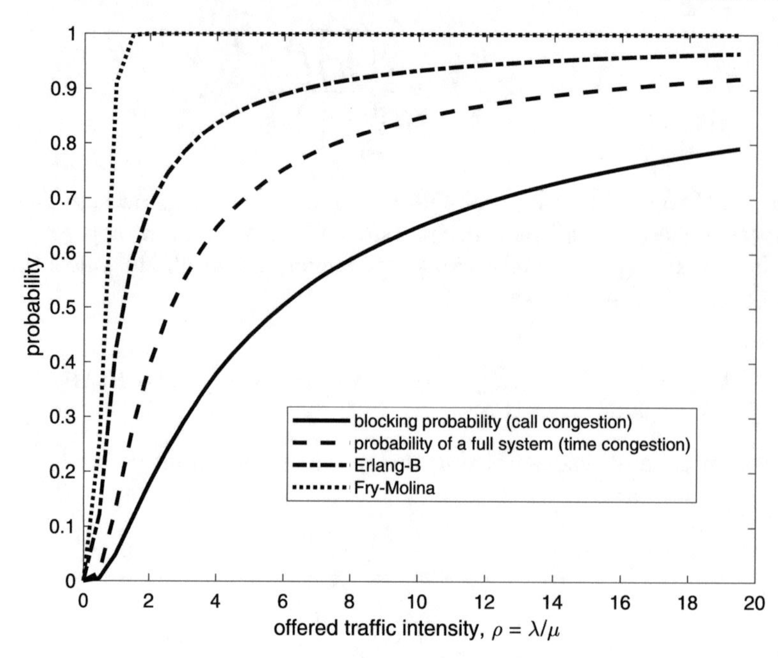

Fig. 4.19: Comparison of approaches for the blocking probability of an M/M/8/8/12 system (the Erlang-B case and the Fry–Molina one use the peak traffic load as 12ρ)

Figure 4.19 shows the blocking probability comparison for $S = 8$ and $P = 12$ among Engset's formula, the $P_{full} = P_S$ probability of the M/M/S/S/P system, and the upper bound of the Erlang-B formula in (4.60). In this graph, another approach is shown using the Fry–Molina formula instead of the Erlang-B one, as it will be explained in Sect. 4.11. We can see that for our system P_B is different from P_S and that the Erlang-B approach as well as the Fry–Molina one gives an upper bound that is not tight.

If we apply the same method in (4.56) for P_B to the M/M/S/S queue (homogeneous case, where the PASTA property can be applied), we get the classical Erlang-B formula.

4.11 The M/M/∞ Queue Analysis

This is a limiting case of both M/M/S and M/M/S/S queues for $S \to \infty$. In particular, the arrival process is Poisson with a mean rate λ. Each request has a service time exponentially distributed with a mean rate μ and there are infinite servers so that there is no waiting phase: any request always finds a free server. Consequently, we use a Markov chain model with $\lambda_i \equiv \lambda \ \forall\, i \geq 0$ and $\mu_i = i\mu \ \forall\, i \geq 1$. Let $\rho = \lambda/\mu$ denote the intensity of the arrival process. By using the cut equilibrium equations and the normalization condition, similarly to (4.43), we have

$$\begin{aligned} P_i &= \frac{\rho^i}{i!} P_0, \quad \text{for } i = 1, 2, \ldots, \infty \\ \text{where } P_0 &= \frac{1}{\sum_{i=0}^{\infty} \frac{\rho^i}{i!}} = e^{-\rho} \end{aligned} \tag{4.61}$$

Hence, the state probability is Poisson distributed with PGF expressed as

$$P(z) = \sum_{i=0}^{\infty} \frac{z^i \rho^i}{i!} e^{-\rho} = e^{-\rho} \sum_{i=0}^{\infty} \frac{(z\rho)^i}{i!} = e^{-\rho} \times e^{z\rho} = e^{\rho(z-1)} \tag{4.62}$$

The mean number of requests in the system can be obtained by means of the first derivative of $P(z)$:

$$N = \sum_{i=0}^{\infty} i \frac{\rho^i}{i!} e^{-\rho} = \left. \frac{\mathrm{d}P(z)}{\mathrm{d}z} \right|_{z=1} = \left. \rho e^{\rho(z-1)} \right|_{z=1} = \rho \tag{4.63}$$

From (4.63) we have that, as expected, the mean number of requests in the system is equal to the mean number of requests in service. According to the Little theorem, the mean system delay is $T = N/\lambda = 1/\mu$ (i.e., the mean service time).

It is possible to prove that the state probability distribution of the M/M/∞ system in (4.61) can also be applied to the M/G/∞ case [16]. This is an important result to study some particular traffic sources, such as the M/Pareto one [17]. Moreover, the M/G/∞ theory can be adopted to solve the M/D/∞ queue that models Aloha-like access protocols, as discussed in Chap. 6.

It is interesting to note that the M/M/∞ queue is adopted to describe another model for blocking systems where new blocked arrivals are held for a certain time. This is the case of the Fry–Molina blocked calls held model [18]: a new call arriving at the system when all the servers are busy will be kept in the system for a further time, which is equal to the service time it would have obtained, if it was accepted. If a server becomes empty during this time, the call will occupy the server for its remaining lifetime that in the case of exponentially distributed service times (as assumed here) is equal to the unencumbered call duration. Then, this system admits an M/M/∞ where the completion rate from the nth state is $n\mu$. The blocking probability corresponds in this case to the probability of states from $n = S$ to ∞, according to the PASTA property. For a system with traffic intensity ρ and S servers we apply the M/M/∞ model with same ρ (the state probability distribution if Poisson) and we express the Fry–Molina blocking probability as

$$P_B = \sum_{n=S}^{\infty} P_n = \sum_{n=S}^{\infty} \frac{\rho^n}{n!} e^{-\rho} = e^{-\rho} \sum_{n=S}^{\infty} \frac{\rho^n}{n!} = 1 - e^{-\rho} \sum_{n=0}^{S-1} \frac{\rho^n}{n!} \tag{4.64}$$

The Fry–Molina blocking probability is higher than the corresponding Erlang-B one [19], as shown in the previous Fig. 4.19. The Fry–Molina approach is better suited to represent those cases when blocked attempts are reattempted and not completely cleared as assumed in the Erlang-B analysis.

4.12 Solution of Markovian Queues Directly in the z-Domain

It is possible to solve the recursive equations obtained from the equilibrium balance conditions of Markov chains directly in the z-domain. This approach can be particularly useful when the balance conditions are more complex than those of birth–death Markov chains. To understand this method, let us first refer to the simple case of the solution of the M/M/1 chain in Fig. 4.11. In particular, let us consider the cut equilibrium conditions in (4.13) applied to this case as follows:

$$\lambda P_{i-1} = \mu P_i \,, \quad \forall i \geq 1 \tag{4.65}$$

This is a recursive equation that we will now solve in the z-domain, multiplying both sides by z^i and summing all these conditions for i from 1 to $+\infty$ as

$$\begin{aligned} &\sum_{i=1}^{+\infty} \lambda P_{i-1} z^i = \sum_{i=1}^{+\infty} \mu P_i z^i \quad \Rightarrow \quad \lambda z \sum_{i=1}^{+\infty} P_{i-1} z^{i-1} = \mu \sum_{i=1}^{+\infty} P_i z^i \\ &\Rightarrow \quad \lambda z P(z) = \mu \left[P(z) - P_0 \right] \quad \Rightarrow \quad P(z) = \frac{P_0}{1-\rho z} \end{aligned} \tag{4.66}$$

where $\rho = \frac{\lambda}{\mu}$ and P_0 is still obtained by means of the normalization condition that is now applied using the $P(z)$ expression in (4.66) as

$$P(z=1) = 1 \quad \Rightarrow \quad P_0 = 1 - \rho \tag{4.67}$$

We have thus reobtained the $P(z)$ solution for an M/M/1 queue, as shown in (4.22).

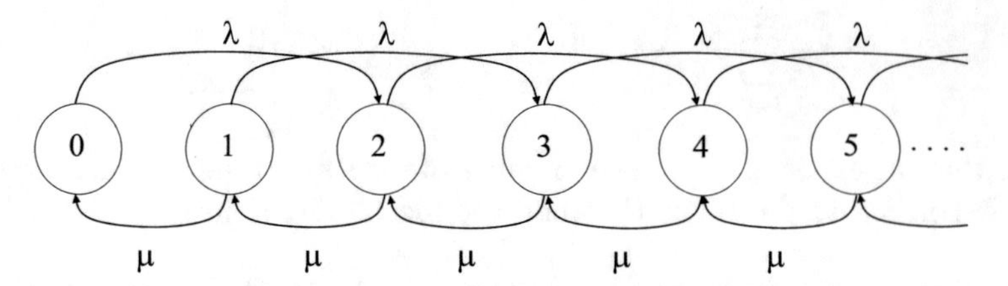

Fig. 4.20: Markov chain model for the queue with multiple simultaneous arrivals

This method in the z-domain is not needed in the classical M/M/1 case but is really useful in the following new example where the recursive relations characterizing the state probabilities are more complex.

Let us consider a queuing system with a single server where each arrival event, modeled with Poisson distribution (mean rate λ), entails the simultaneous arrival of two customers at the system (batched arrivals). Each customer requires an exponentially distributed service time with a mean rate μ. The state diagram is shown in Fig. 4.20. We adopt the following generic flow balance conditions for the cut equilibriums that have a recursive form:

$$\begin{cases} \lambda P_0 = \mu P_1, & \text{for } i = 1 \\ \lambda P_{i-2} + \lambda P_{i-1} = \mu P_i, & \forall i \geq 2 \end{cases} \tag{4.68}$$

We need to transform these recursive formulas in the z-domain by multiplying both sides of the ith equation by z^i and summing all these conditions for i from 1 to $+\infty$ as

$$\begin{aligned} &\lambda P_0 z + \sum_{i=2}^{+\infty} \left(\lambda P_{i-2} z^i + \lambda P_{i-1} z^i\right) = \sum_{i=1}^{+\infty} \mu P_i z^i \quad \Rightarrow \\ &\lambda P_0 z + \lambda \sum_{i=2}^{+\infty} P_{i-2} z^i + \lambda \sum_{i=2}^{+\infty} P_{i-1} z^i = \sum_{i=1}^{+\infty} \mu P_i z^i \quad \Rightarrow \\ &\lambda P_0 z + \lambda z^2 P(z) + \lambda z \left[P(z) - P_0\right] = \mu \left[P(z) - P_0\right] \quad \Rightarrow \end{aligned} \tag{4.69}$$

$$P(z) = \frac{P_0}{1 - \rho z^2 - \rho z}$$

where $\rho = \frac{\lambda}{\mu}$ and P_0 is still obtained by means of the normalization condition that is now applied as $P(z=1) = 1$, so that we have

$$P(z=1) = 1 \quad \Rightarrow \quad P_0 = 1 - 2\rho \tag{4.70}$$

The stability condition $P_0 > 0$ entails that $2\rho < 1$ Erl (actually, the traffic intensity is 2ρ).

We can invert the PGF $P(z)$ by applying the same method as shown in Sect. 5.1.2 based on the expansion in partial fractions. Otherwise, we can also determine directly P_n by applying the classical methods to solve recursions to (4.68). The poles of $P(z)$ are the inverse of the roots of the characteristic equation to solve the recursions in (4.68).

The mean number of requests in the system, N, is obtained as the first derivative of $P(z)$ computed in $z = 1$ as

$$P'(z) = \frac{(1-2\rho)(2\rho z + \rho)}{\left(1 - \rho z^2 - \rho z\right)^2} \quad \Rightarrow \quad P'(z=1) = N = \frac{3\rho}{1-2\rho} \tag{4.71}$$

The mean delay is obtained by applying Little and dividing N by 2λ since we have two arrivals per arrival event:

$$T = \frac{N}{2\lambda} = \frac{1}{2\lambda}\frac{3\rho}{1-2\rho} = \frac{3}{2(\mu - 2\lambda)} \tag{4.72}$$

If we consider the corresponding M/M/1 system with mean arrival rate 2λ and mean service rate μ, we get a mean delay $T = \frac{1}{\mu - 2\lambda}$. This M/M/1 system has the same stability limit of our system. However, the case with coupled arrivals has a higher mean delay because the concentrated arrivals cause greater congestion of the queue.

Finally, a similar z-domain conversion of the recursive formulas of the state probability can also be adopted to solve the case with batched service, where a service time is suitable to complete simultaneously the service of two requests.

4.13 Multi-Dimensional Erlang-B Cases

We consider 2 independent Poisson arrival processes with mean rates λ_1 and λ_2. The service times of these two arrivals are exponentially distributed with mean rates μ_1 and μ_2, respectively, with $\mu_1 \neq \mu_2$. There are no waiting places. These arrivals of requests share the resources from a common pool of capacity C. Each type 1 arrival needs a capacity C_1; each type 2 arrival needs a capacity C_2. A new arrival is blocked and lost if it finds the system when not enough capacity is available, that is the residual capacity is lower than C_1 for type 1 arrivals and the residual capacity is lower than C_2 for type 2.

To study this system, we have to use a two-dimensional (2D) Markov chain where the generic state (i, j) represents how many requests of type 1 and how many requests of type 2 are simultaneously present in the system.

Because of the capacity limit C, not all the states (i, j) are possible, but only those of the set Ω defined below:

$$\Omega = \{ (i, j) \mid iC_1 + jC_2 \leq C \} \tag{4.73}$$

Let us denote the traffic intensity due to type 1 arrivals as $\rho_1 = {}^{\lambda_1}/_{\mu_1}$. Moreover, the traffic intensity due to type 2 arrivals is $\rho_2 = {}^{\lambda_2}/_{\mu_2}$.

Figure 4.21 shows an example of the 2D Markov chain modeling the system for $C_1 = 1$ and $C_2 = 2$ units of capacity with $C = 4$ units of capacity. This chain has 9 states. For numerical results we will also use $\rho_1 = 2$ Erl and $\rho_2 = 1$ Erl. Note that the system would become mono-dimensional if $\mu_1 = \mu_2$.

We can solve the state probabilities by writing equilibrium conditions to balance the flows around each state. We obtain a linear system with 9 equations, where one of them is redundant so that we need to add the normalization condition to express $P(0, 0)$. Each of the following equations represents an equilibrium condition related to the state for which the probability is on the left side.

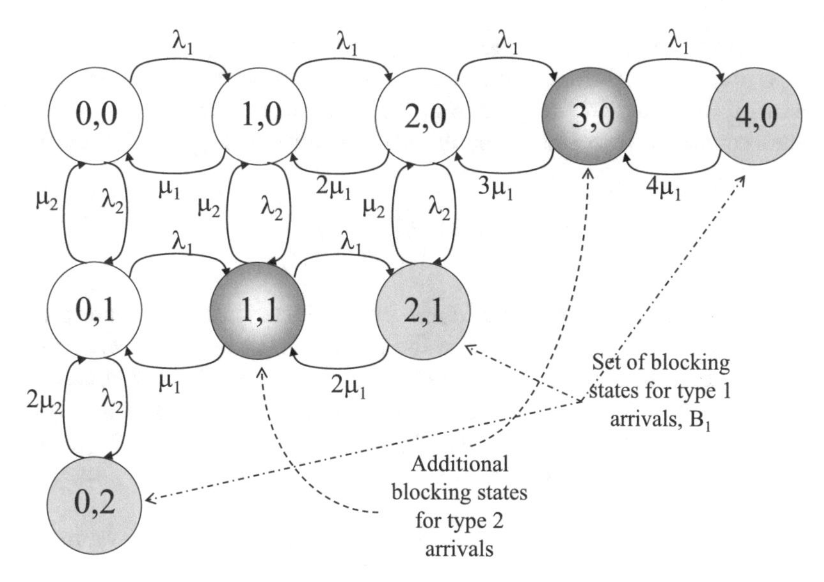

Fig. 4.21: Example of two-dimensional Markov Chain with limit capacity $C = 4$, $C_1 = 1$, and $C_2 = 2$ units of capacity

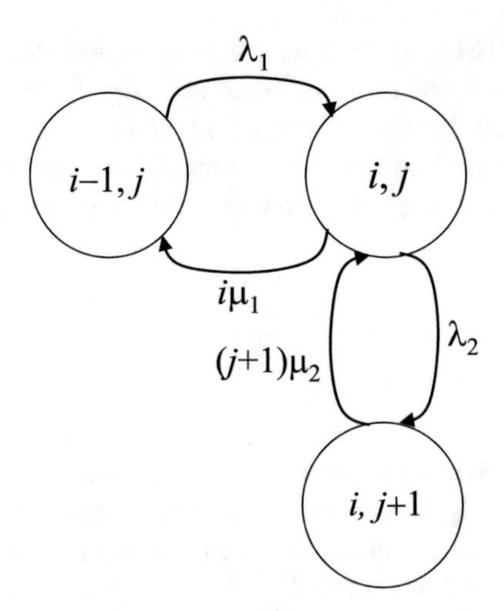

Fig. 4.22: Local equilibrium cases

$$\begin{aligned}(\lambda_1+\lambda_2)\,P(0,0)&=\mu_1 P(1,0)+\mu_2 P(0,1)\\(\lambda_1+\lambda_2+\mu_1)\,P(1,0)&=2\mu_1 P(2,0)+\mu_2 P(1,1)+\lambda_1 P(0,0)\\(\lambda_1+\lambda_2+\mu_2)\,P(0,1)&=\mu_1 P(1,1)+2\mu_2 P(0,2)\\(\lambda_2+\mu_1+\mu_2)\,P(1,1)&=\lambda_1 P(0,1)+2\mu_1 P(2,1)+\lambda_2 P(1,0)\\(\lambda_1+\lambda_2+2\mu_1)\,P(2,0)&=3\mu_1 P(3,0)+\mu_2 P(2,1)+\lambda_1 P(1,0)\\(2\mu_1+\mu_2)\,P(2,1)&=\lambda_2 P(2,0)+\lambda_1 P(1,1)\\(\lambda_1+3\mu_1)\,P(3,0)&=\lambda_1 P(2,0)+4\mu_1 P(4,0)\\2\mu_2 P(0,2)&=\lambda_2 P(0,1)\\4\mu_1 P(4,0)&=\lambda_1 P(3,0)\end{aligned}\tag{4.74}$$

As in the case of one-dimensional (1D) Markov queues, one of these equations is redundant. For instance, we can see that the 3rd equation can be obtained as a linear combination of the other equations. Then, we need to add the normalization condition stating that the sum of the probabilities over all the states yields one. This condition will be useful to determine $P(0,0)$.

Even if this is a linear system, we need a numerical approach with matrices to solve it. In these cases, it has been shown in [20] that writing equilibrium around nodes is equivalent to state local equilibriums among the flows between two adjacent states. Referring to the generic group of adjacent states in Fig. 4.22, we have the following cases for the local balance equations (like the cut equilibrium conditions for 1D Markov chains):

$$\begin{aligned}\lambda_1 P(i-1,j)&=i\mu_1 P(i,j)\\\lambda_2 P(i,j)&=(j+1)\,\mu_2 P(i,j+1)\end{aligned}\tag{4.75}$$

This type of solution is possible because this 2D Markov chain is *reversible* [21]. For a continuous-time Markov chain, the necessary and sufficient condition for reversibility (*Kolmogorov cycle condition*) is that the "circulation flow" (i.e., the product of streams parameters) among any four neighboring states in a square equals zero. That means that the clockwise flow is equal to the counterclockwise flow.[4] For instance, considering the states $(1,0)$, $(2,0)$, $(1,1)$, and $(2,1)$ in Fig. 4.21, we have $\lambda_1\lambda_2 2\mu_1\mu_2$ (clockwise) $=$ $\lambda_1\lambda_2 2\mu_1\mu_2$ (counterclockwise). The reversibility property leads to local balance equations between any two neighboring states of the process as shown in (4.75).

Then, if we progressively solve the first type of equations in (4.75) along a row, we have

$$P(i,j)=\frac{\lambda_1}{i\mu_1}P(i-1,j)=\ldots=\frac{\rho_1^i}{i!}P(0,j)\tag{4.76}$$

[4] It is possible to show that all the birth–death (1D) Markov chains are reversible.

Analogously, if we progressively solve the second type of equations in (4.75) along a column, we have

$$P(i,j) = \frac{\lambda_2}{j\mu_2} P(i,j-1) = \ldots = \frac{\rho_2^j}{j!} P(i,0) \tag{4.77}$$

In conclusion, if we use (4.77) for $i = 0$ in (4.76), we obtain the following product form expression for the probability of the generic state $P(i,j)$ as a function of the probability $P(0,0)$ [22]:

$$P(i,j) = \frac{\rho_1^i \rho_2^j}{i!j!} P(0,0) \tag{4.78}$$

where $P(0,0) = 1/G(\Omega)$ is obtained as a normalization term summing contributions all over the state-space Ω as

$$G(\Omega) = \sum_{(i,j)\in\Omega} \frac{\rho_1^i \rho_2^j}{i!j!} \tag{4.79}$$

According to Fig. 4.21, the set Ω of all the possible states is

$$\Omega = \left\{ \begin{array}{l} (0,0)\ (1,0)\ (2,0)\ (3,0)\ (4,0) \\ (0,1)\ (1,1)\ (2,1) \\ (0,2) \end{array} \right\} \tag{4.80}$$

Let B_1 and B_2 denote the sets of states where type 1 and type 2 arrivals are blocked, respectively. According to Fig. 4.21, these blocking states are

$$\begin{aligned} B_1 &= \{ (0,2)\ (2,1)\ (4,0) \} \\ B_2 &= B_1 \cup \{ (3,0)\ (1,1) \} \end{aligned} \tag{4.81}$$

This Markov chain is homogeneous because we have the same arrival processes in all the states with the total mean rate $\lambda_1 + \lambda_2$. Then, we can apply the PASTA property. Hence, the probability that a type 1 request is blocked, P_{B1}, and the probability that a type 2 request is blocked, P_{B2}, can be obtained by summing the probabilities of all the states where these type 1 or 2 arrivals are blocked, denoted as sets B_1 and B_2, respectively.

$$P_{B1} = \sum_{(i,j)\in B_1} P(i,j) \quad , \quad P_{B2} = \sum_{(i,j)\in B_2} P(i,j) \tag{4.82}$$

Note that P_{B1} and P_{B1} are actually conditional probabilities on an arrival of type 1 or 2, respectively. If we need to obtain an overall blocking probability, P_B, we have to average P_{B1} and P_{B2} by the probabilities that a new arrival is a type 1 arrival and that a new arrival is a type 2 arrival:

$$P_B = \frac{\lambda_1}{\lambda_1 + \lambda_2} P_{B1} + \frac{\lambda_2}{\lambda_1 + \lambda_2} P_{B2} \tag{4.83}$$

Referring to our numerical example, we get: $P_{B1} = 0.2533$ and $P_{B2} = 0.52$. We can notice that in this case where $C_2 = 2 \times C_1$, also $P_{B2} \approx 2 \times P_{B1}$.

It is interesting to note that the "product form solution" for the state probability distribution in (4.78)–(4.79) is also valid adopting other conditions to delimit the states $(i,j) \in \Omega$ for dimensions i (type 1) and j (type 2). For instance, we can consider the system with C servers and the state-space delimited by the following conditions: $0 \le i \le K, i + j \le C, K < C$ (case of a system with two service classes with preferential service for j-class).

As a concluding remark on 2D Markov chains, it is essential to note that there is not a single solution method for all these chains. We have shown a simple method that is suitable for loss queuing systems with different service rates and truncated state-space. It is easier to find a solution if at least one dimension is limited or both, as in our example. There are algorithmic methods, which are known as "matrix geometric methods" that allow us to efficiently analyze a 2D Markov chain provided that the Markov chain has a certain block structure suitable for service times that are combinations of exponential random variables (i.e., phase-type distributions of which exponential, Erlang-n, and hyper-exponential are just some examples).

Still referring to our example with state Ω specified in (4.73), we consider the special case when $C_1 = C_2$ so that the Markov chain in Fig. 4.21 becomes rectangular and type 1 and type 2 arrivals experience both the same blocking conditions and then the same blocking probability P_B, as described below. Let ℓ denote the total maximum number of arrivals that can be admitted in the system so that $\ell = {}^{C}/_{C_1} = {}^{C}/_{C_2}$ Then, the state-space Ω can be expressed as

$$\Omega = \{(i, j)|\ \ i + j \leq \ell\} \tag{4.84}$$

where i and j are greater than or equal to 0.

State probabilities can still be expressed by (4.78) and (4.79).

The blocking states for type 1 arrivals (B_1) and type 2 arrivals (B_2) are the same and denoted using the following set B:

$$B = \{(i, j)|\ \ i + j = \ell\} \tag{4.85}$$

According to the PASTA property, the blocking probability P_B is the probability that the system is in one of the states of set B and can be expressed as follows through the Newton binomial formula:

$$\begin{aligned} P_B &= \sum_{B} P(i, j) = \sum_{i=0}^{\ell} P(i, \ell - i) = \frac{1}{G(\Omega)} \sum_{i=0}^{\ell} \frac{\rho_1^i}{i!} \frac{\rho_2^{\ell-i}}{(\ell - i)!} \\ &= \frac{1}{G(\Omega)} \sum_{i=0}^{\ell} \binom{\ell}{i} \frac{\rho_1^i \rho_2^{\ell-i}}{\ell!} = \frac{1}{G(\Omega)} \frac{(\rho_1 + \rho_2)^{\ell}}{\ell!} \end{aligned} \tag{4.86}$$

where $G(\Omega)$ can be expressed according to (4.79) through some derivations as

$$\begin{aligned} G(\Omega) &= \sum_{i=0}^{\ell} \sum_{j=0}^{\ell-i} \frac{\rho_1^i}{i!} \frac{\rho_2^j}{j!} = \begin{array}{c}\text{using}\\ k = i + j\end{array} = \sum_{i=0}^{\ell} \sum_{k=i}^{\ell} \frac{\rho_1^i}{i!} \frac{\rho_2^{k-i}}{(k-i)!} = \begin{array}{c}\text{reorganizing the}\\ \text{terms of the double}\\ \text{sum in a different way}\end{array} \\ &= \sum_{k=0}^{\ell} \sum_{i=0}^{k} \frac{\rho_1^i}{i!} \frac{\rho_2^{k-i}}{(k-i)!} = \sum_{k=0}^{\ell} \frac{\rho_2^k}{k!} \sum_{i=0}^{k} \binom{k}{i} \left(\frac{\rho_1}{\rho_2}\right)^i = \begin{array}{c}\text{using the Newton}\\ \text{binomial formula}\end{array} \\ &= \sum_{k=0}^{\ell} \frac{\rho_2^k}{k!} \left(1 + \frac{\rho_1}{\rho_2}\right)^k = \sum_{k=0}^{\ell} \frac{(\rho_2 + \rho_1)^k}{k!} \end{aligned} \tag{4.87}$$

Combining (4.86) and (4.87), we achieve the following formula for the blocking probability of both calls (this is the blocking probability conditioned on an arrival of type 1 or 2 that is also equal to the overall blocking probability):

$$P_B = \frac{(\rho_1 + \rho_2)^{\ell}}{\ell! \sum_{k=0}^{\ell} \frac{(\rho_2 + \rho_1)^k}{k!}} \tag{4.88}$$

If we refer to the total traffic load as $\rho_1 + \rho_2 = \rho$, we can see that equation (4.88) is exactly the classical Erlang-B formula; this is correct since in this case the Markov chain becomes 1D and of the M/M/ℓ/ℓ type.

In conclusion, the same method described above can also be applied to solve a loss queuing system with more than two classes. For instance, we can consider three classes and the most general case where arrivals of type 1 need a capacity C_1, arrivals of type 2 need a capacity C_2, and arrivals of type 3 need a capacity C_3 out of the total capacity C. Using the following constraint to delimit the set of states Ω

$$\Omega = \{(i, j, k) \mid iC_1 + jC_2 + kC_3 \leq C\} \tag{4.89}$$

the state probability distribution results as

$$P(i, j, k) = \frac{1}{G(\Omega)} \frac{\rho_1^i \rho_2^j \rho_3^k}{i! j! k!} \tag{4.90}$$

where

$$G(\Omega) = \sum_{(i,j,k)\in\Omega} \frac{\rho_1^i \rho_2^j \rho_3^k}{i!j!k!} \quad (4.91)$$

There is an iterative method by Kaufman and Robert [22, 23] that provides an efficient numerical approach to compute the blocking probabilities for a multi-rate system with a generic number of users/classes but under the constraint that the total link capacity C and the capacity needs for the different classes $C_1, C_2, \ldots, C_N$ are positive integers. Let n_i (with i from 1 to N) represent how many class-i services are in progress, sharing the link capacity C. The state-space of the system is characterized using the vector of n_i values, $\mathbf{n} = \{n_i\}$, and the vector of capacity values $\mathbf{b} = \{C_i\}$. Let $\mathbf{n} \cdot \mathbf{b}$ denote the scalar product of the vectors $\mathbf{n}$ and $\mathbf{b}$ as

$$\mathbf{n} \cdot \mathbf{b} = \sum_{i=1}^{N} n_i C_i \quad (4.92)$$

Then, the state-space Ω in this general case can be represented as

$$\Omega = \{\mathbf{n} : \mathbf{n} \cdot \mathbf{b} \le C\} \quad (4.93)$$

Following the same notation as before, we have the following expression for the normalization constant:

$$G(\Omega) = \sum_{\mathbf{n}\in\Omega} \frac{\rho_1^{n_1}}{n_1!} \frac{\rho_2^{n_2}}{n_2!} \cdots \frac{\rho_N^{n_N}}{n_N!} \quad (4.94)$$

where $\rho_i = \lambda_i/\mu_i$ denotes the intensity of the ith class.

Let B_i denote the set of blocking states for the ith class. These states are those where the available capacity is lower than C_i, that is where the total allocated capacity $\mathbf{n} \cdot \mathbf{b}$ is from $C - C_i + 1$ to C. We use the following notations:

$$\begin{aligned}
\Omega(c) &= \{\mathbf{n} : \mathbf{n} \cdot \mathbf{b} = c\}, \text{for } c = 0, \ldots, C \\
G(c) = G(\Omega(c)) &= \sum_{\mathbf{n}\in\Omega(c)} \frac{\rho_1^{n_1}}{n_1!} \frac{\rho_2^{n_2}}{n_2!} \cdots \frac{\rho_N^{n_N}}{n_N!} \\
\Omega &= \bigcup_{c=0}^{C} \Omega(c) \\
B_i &= \bigcup_{c=C-C_i+1}^{C} \Omega(c)
\end{aligned} \quad (4.95)$$

where $G(c)$ denotes the "unnormalized" probability that the system is in one of the states belonging to $\Omega(c)$.
Note that the sets of states $\Omega(c)$ and $\Omega(c+j)$ are disjoint and correspond to the unnormalized probabilities $G(c)$ and $G(c+j)$, respectively. Therefore, sets B_i are given by the union of disjoint sets so that their probabilities can be obtained as the sum of the corresponding probabilities $\Omega(c)$. Then, we have $G(\Omega) = \sum_{c=0}^{C} G(c)$.

Then, the blocking probability for the generic ith class can be determined as follows:

$$P_{B_i} = \frac{\sum_{c=C-C_i+1}^{C} G(c)}{\sum_{c=0}^{C} G(c)}, \text{for } i = 0, \ldots, N \quad (4.96)$$

Then, the Kaufman–Robert algorithm adopts the following iterations for c from 1 to C:

$$G(c) = \sum_{i=1}^{N} \frac{\rho_i C_i}{c} G(c - C_i) \quad (4.97)$$

with $G(0) = 1$ ([5]) and $G(c) = 0$ for $c < 0$.

With these recursions, we can obtain all the $G(c)$ terms to compute the blocking probabilities in (4.96).

Let us reconsider the example at the beginning of this section and apply the Kaufman–Robert algorithm to recompute the blocking probabilities. In this case, we have two traffic classes for which: $C_1 = 1$ unit, $C_2 = 2$ unit, $C = 4$ units, $\rho_1 = 2$ Erl, and $\rho_2 = 1$ Erl. We perform the following iterations for c from 0 to 4 using (4.97) with $G(0) = 1$ and $G(c) = 0$ for $c < 0$. We have

$$G(c) = \frac{2}{c}G(c-1) + \frac{2}{c}G(c-2) \tag{4.98}$$

We obtain the following results:

$$\begin{aligned} G(1) &= 2G(0) + 2G(-1) = 2 \\ G(2) &= G(1) + G(0) = 3 \\ G(3) &= \frac{2}{3}G(2) + \frac{2}{3}G(1) = \frac{10}{3} \\ G(4) &= \frac{1}{2}G(3) + \frac{1}{2}G(2) = \frac{19}{6} \end{aligned} \tag{4.99}$$

Finally, the blocking probabilities for the two classes, P_{B_1} and P_{B_2}, result as

$$\begin{aligned} P_{B_1} &= \frac{G(4)}{G(0)+G(1)+G(2)+G(3)+G(4)} = \frac{19}{75} \approx 0.253 \\ P_{B_2} &= \frac{G(3)+G(4)}{G(0)+G(1)+G(2)+G(3)+G(4)} = 0.52 \end{aligned} \tag{4.100}$$

and these results are coincident with those obtained at the beginning.

4.14 Distribution of the Queuing Delays in the FIFO Case

In this section, we focus on the pdf of the queuing delay for the FIFO service discipline. We will determine this pdf using its Laplace transform.

4.14.1 M/M/1 Case

Let $f_D(t)$ denote the pdf of T_D, the queuing delay that a request experiences from the entrance in the M/M/1 queue to the completion of its service. Moreover, we denote the Laplace transform of $f_D(t)$ with $T_D(s)$.

Due to the FIFO policy, the n requests left in the system when the service of a given request A completes, are those arrived at the queue according to the Poisson input process during the queuing delay T_D of request A. Then, the probability distribution of n, P_n, can be expressed as

$$\begin{aligned} P_n &= \int_0^{+\infty} \text{Prob}\left\{ n \text{ Poisson arrivals in } t \middle| T_D = t \right\} f_D(t)dt \\ &= \int_0^{+\infty} \frac{(\lambda t)^n}{n!} e^{-\lambda t} f_D(t)dt \end{aligned} \tag{4.101}$$

For queuing systems where the state changes at most by +1 or −1, the probability distribution of n, P_n, at the service completion instants is equivalent to the probability distribution of n at arrival instants (Kleinrock principle [1]). Moreover, because of the PASTA property, the probability distribution of n at arrival instants coincides with the state probability distribution of the M/M/1 queue. Hence, P_n is also the state probability distribution of the

[5] In this case with $c = 0$, we have just the case where $n_1 = n_2 = \ldots = n_N = 0$ so that $G(c)$ in (4.95) yields 1.

queue: the PGF of n is therefore equal to the $P(z)$ function in (4.22) for the M/M/1 case. On the other hand, the PGF of n can also be computed by using the expressions of probabilities P_n in (4.101) as

$$\begin{aligned} P(z) &= \sum_{n=0}^{\infty} z^n \int_0^{+\infty} \frac{(\lambda t)^n}{n!} e^{-\lambda t} f_{\mathrm{D}}(t)\mathrm{d}t = \int_0^{+\infty} e^{-\lambda t} f_{\mathrm{D}}(t) \sum_{n=0}^{\infty} \frac{(\lambda t z)^n}{n!}\mathrm{d}t \\ &= \int_0^{+\infty} e^{-\lambda t} f_{\mathrm{D}}(t) e^{\lambda t z}\mathrm{d}t = \int_0^{+\infty} f_{\mathrm{D}}(t) e^{-\lambda t(1-z)}\mathrm{d}t = T_{\mathrm{D}}(s)|_{s=\lambda(1-z)} \end{aligned} \tag{4.102}$$

Note that in (4.102) we have exchanged the integral with the sum. We have therefore obtained a useful relation between the PGF of the state probability distribution of the M/M/1 queue, $P(z)$, and the Laplace transform of the pdf of the queuing delay, $T_{\mathrm{D}}(s)$, by using $s = \lambda(1-z)$. Since $P(z)$ is also given by (4.22), we obtain $T_{\mathrm{D}}(s)$ from (4.102) by using the substitution $z = 1 - s/\lambda$ (this is the inverse of $s = \lambda(1-z)$):

$$T_{\mathrm{D}}(s) = P(z)\Big|_{z=1-s/\lambda} = \frac{1-\rho}{1-z\rho}\Big|_{z=1-s/\lambda} = \frac{1-\rho}{1-\rho+\rho s/\lambda} = \frac{\mu-\lambda}{\mu-\lambda+s} \tag{4.103}$$

The inversion of the Laplace transform in (4.103) allows us to prove that T_{D} is exponentially distributed with mean rate $\mu - \lambda$ (> 0 under the ergodicity condition):

$$f_{\mathrm{D}}(t) = (\mu-\lambda)\, e^{-(\mu-\lambda)t}, \quad t \geq 0 \tag{4.104}$$

Let T_{Sv} denote the service time, with the corresponding exponential pdf $f_{\mathrm{Sv}}(t)$ that has Laplace transform equal to $T_{\mathrm{Sv}}(s) = \mu/(\mu+s)$.

We consider another method to derive $T_{\mathrm{D}}(s)$. Due to the PASTA property, the state probability P_n also represents the probability that an arrival finds n requests in the queue and, due to the FIFO policy, must wait for their completion before being served. Conditioning on an arrival that finds n requests in the system, the related pdf of the delay, $T_{\mathrm{D}|n}$, has a Laplace transform equal to ${T_{\mathrm{Sv}}}^{n+1}(s)$. We remove the conditioning by means of the P_n distribution as

$$T_{\mathrm{D}}(s) = \sum_{n=0}^{\infty} T_{\mathrm{Sv}}^{n+1}(s) P_n = T_{\mathrm{Sv}}(s) \sum_{n=0}^{\infty} T_{\mathrm{Sv}}^{n}(s) P_n = T_{\mathrm{Sv}}(s) \times P(z)|_{z=T_{\mathrm{Sv}}(s)} \tag{4.105}$$

It is easy to verify that (4.105) is equivalent to (4.103).

The above procedure in (4.103) can also be applied to equate the PGF of the number of requests in the waiting list to the Laplace transform (computed for $s = \lambda(1-z)$) of the pdf of the time spent in the queue waiting for service, T_{W}. However, we express $T_{\mathrm{W}}(s)$ using another approach recalling that T_{W} and T_{Sv} are independent random variables. Since $T_{\mathrm{D}} = T_{\mathrm{W}} + T_{\mathrm{Sv}}$, we use the following product formula with the Laplace transforms of the pdfs:

$$T_{\mathrm{D}}(s) = T_{\mathrm{Sv}}(s) \times T_{\mathrm{W}}(s) \tag{4.106}$$

On the basis of the previous expression of $T_{\mathrm{D}}(s)$ and $T_{\mathrm{Sv}}(s)$ we can express $T_{\mathrm{W}}(s)$ as

$$\begin{aligned} T_{\mathrm{W}}(s) &= \frac{T_{\mathrm{D}}(s)}{T_{\mathrm{Sv}}(s)} = \frac{(\mu-\lambda)(\mu+s)}{(\mu-\lambda+s)\,\mu} = \frac{(\mu-\lambda)(\mu-\lambda+\lambda+s)}{(\mu-\lambda+s)\,\mu} \\ &= \frac{(\mu-\lambda)}{\mu} + \frac{\lambda}{\mu}\frac{(\mu-\lambda)}{(\mu-\lambda+s)} \end{aligned} \tag{4.107}$$

Note that the PGF of the number of requests waiting in the queue is obtained by making the substitution $s = \lambda(1-z)$ in (4.107).

Moreover, $T_{\mathrm{W}}(s)$ in (4.107) can be anti-transformed by considering that the first term is a constant with anti-transform proportional to the Dirac Delta function, $\delta(t)$, and the second term is proportional to $T_{\mathrm{D}}(s)$ with anti-transform given by (4.104). Hence, the pdf of the waiting time, $f_{\mathrm{W}}(t)$, results as

$$f_{\mathrm{w}}(t) = \frac{\mu-\lambda}{\mu}\delta(t) + \frac{\lambda}{\mu}(\mu-\lambda)\, e^{-(\mu-\lambda)t}, \quad t \geq 0 \tag{4.108}$$

This formula can be interpreted as follows, by noticing that $\rho=\lambda/\mu$ represents the probability that the server of the queue is busy ($\rho=1-P_0$):

- If a newly arriving call finds the queue empty (with probability $P_0=1-\rho=(\mu-\lambda)/\mu$), there is no wait and the pdf of T_W (in this case $T_{W|\text{empty}}$) coincides with $\delta(t)$.
- If a newly arriving call finds the queue busy (with probability $1-P_0=\rho$), there is a waiting time corresponding to the residual duration of the currently served request plus the delay due to the requests that the new arrival finds in the queue. Due to the memoryless property of the exponential distribution, the residual lifetime of the currently served request is still exponentially distributed with a mean rate μ. Following the reasoning given below, we can prove that in this case, the Laplace transform of the waiting time (denoted by $T_{W|\text{non-empty}}$) is equal to (4.103) so that the corresponding pdf of the waiting time is equal to $f_D(t)$.

Similarly to (4.105), $T_{W|\text{non-empty}}(s)$ can be determined conditioning on the number of arrivals $n>0$ found in the system (i.e., the sum of n iid T_{Sv} times) and then removing the conditioning through the distribution of n for $n>0$: $P_n/(1-P_0)$.

$$T_{W|\text{non-empty}}(s)=\sum_{n=1}^{\infty}T_{Sv}^{n}(s)\frac{P_n}{1-P_0}\equiv T_D(s)$$

We note that $T_{W|\text{non-empty}}(s)$ has the same pdf as that in (4.104) for T_D and can be found in the second term on the right side of (4.108).

4.14.2 M/M/S Case

We focus here on the distribution of the waiting time in the M/M/S queue with FIFO discipline. We adopt an approach similar to that used for obtaining (4.105) but now working in the time domain rather than in the Laplace one. We refer to a newly arriving call that finds n requests in the system. Let P_n denote the probability of state n. Correspondingly, we determine the pdf of the waiting time, $f_{W|n}(t)$, as

$$f_{W|n}(t)=\begin{cases}\delta(t), & n<S\\ f_S(t)\otimes f_S(t)\cdots\otimes f_S(t), & n\geq S\end{cases},\quad t\geq 0 \tag{4.109}$$

where symbol $\otimes$ denotes the convolution and where for $n\geq S$ we consider the $n-S+1$-fold convolution of the pdf $f_S(t)$ of the completion time of the first request among S in service; this time is exponentially distributed with mean rate $S\mu$: $f_S(t)=S\mu e^{-S\mu t}$, for $t\geq 0$ (for $S=1$, $f_S(t)$ has been previously denoted by $f_{Sv}(t)$).

Equation (4.109) can be explained as follows:

- If a new arrival finds $n<S$ requests already in the system, it is immediately served (i.e., there is no waiting time) so that the pdf of the waiting time is equal to $\delta(t)$.
- If the new arrival finds $n\geq S$ requests already in the system, the waiting time corresponds to the time to have $n-S+1$ service completions. Since these time intervals are iid, the pdf of their sum is equal to the $n-S+1$-fold convolution of $f_S(t)$.

The $n-S+1$-fold convolution of $f_S(t)$, each exponentially distributed with mean rate $S\mu$, is given by the Erlang distribution with "shape parameter" equal to $n-S+1$ and "rate" equal to $S\mu$ (see Chap. 3, Sect. 3.103 where $\lambda=S\mu$ and $k=n-S+1$):

$$f_{W|n\geq S}(t)=(S\mu)^{n-S+1}\frac{t^{n-S}}{(n-S)!}e^{-S\mu t},\quad t\geq 0 \tag{4.110}$$

We can now determine the complementary distribution of the waiting time:

$$1-F_W(t)=\text{Prob}\{W>t\}=\sum_{n=S}^{\infty}\text{Prob}\left\{W>t\middle|n\right\}P_n$$

$$= \sum_{n=S}^{\infty} \left[1 - \text{Prob}\left\{W \leq t \middle| n\right\}\right] P_n = \sum_{n=S}^{\infty} P_n - \sum_{n=S}^{\infty} \int_0^t f_{\mathrm{W}|n}(t) \mathrm{d}t P_n$$
$$= P_{\mathrm{C}} - P_{\mathrm{C}} \left[1 - e^{\mu(\rho - S)t}\right] = P_{\mathrm{C}} e^{\mu(\rho - S)t}$$

where P_{C} denotes the Erlang-C formula given in (4.38) and $\rho = \lambda/\mu$ is the input traffic intensity.

The above formula allows us to express the PDF $F_{\mathrm{W}}(t)$ and the corresponding pdf $f_{\mathrm{W}}(t)$ as follows:

$$\begin{aligned} F_{\mathrm{W}}(t) &= 1 - P_{\mathrm{C}} e^{-\mu(S-\rho)t} = (1 - P_{\mathrm{C}}) + P_{\mathrm{C}} \left[1 - e^{-\mu(S-\rho)t}\right] \Longleftrightarrow \\ f_{\mathrm{W}}(t) &= (1 - P_{\mathrm{C}})\, \delta(t) + P_{\mathrm{C}} \mu\, (S - \rho)\, e^{-\mu(S-\rho)t} \\ &= (1 - P_{\mathrm{C}})\, \delta(t) + P_{\mathrm{C}}\, (\mu S - \lambda)\, e^{-(\mu S - \lambda)t} \end{aligned} \tag{4.111}$$

If $S = 1$, the pdf $f_{\mathrm{W}}(t)$ obtained in (4.111) for an M/M/S queue becomes equal to (4.108) for an M/M/1 queue. Finally, we can take the Laplace transform of the pdf in (4.111) to obtain $T_{\mathrm{W}}(s)$ as

$$T_{\mathrm{W}}(s) = (1 - P_{\mathrm{C}}) + P_{\mathrm{C}} \frac{\mu S - \lambda}{\mu S - \lambda + s} = \frac{\mu S - \lambda + s\,(1 - P_{\mathrm{C}})}{\mu S - \lambda + s} \tag{4.112}$$

Similarly to (4.106), the Laplace transform of the pdf of the whole queuing system delay T_{D} is obtained as

$$T_{\mathrm{D}}(s) = T_{\mathrm{W}}(s) \times T_{\mathrm{Sv}}(s) = (1 - P_{\mathrm{C}}) \frac{\mu}{\mu + s} + P_{\mathrm{C}} \frac{(\mu S - \lambda)\, \mu}{(\mu S - \lambda + s)\, (\mu + s)} \tag{4.113}$$

4.15 Erlang-B Generalization for Non-Poisson Arrivals

We describe here some approximate approaches to extend the use of the Erlang-B formula to loss queuing systems with general renewal arrival processes. Before analyzing these cases, we study the properties of the traffic types involved in an M/M/S/S queuing system.

4.15.1 The Traffic Types in the M/M/S/S Queue

Let us study the characteristics of both the *refused traffic* (at the input of the queue) and the *carried traffic* (at the output of the queue) for an M/M/S/S queuing system with a mean arrival rate λ and a mean completion rate μ. These processes will be described using mean and variance.

The *peakedness parameter* z of a given traffic process (with a certain arrival process and a particular service time distribution) is defined as if this traffic was at the input of a fictitious queue with infinite servers (a G/G/∞ queue where all the requests are in service):

$$z = \frac{\text{Var}\,[n]}{E\,[n]} \tag{4.114}$$

where $E[n]$ and Var$[n]$ represent the mean and variance of the state probability distribution P_n, referring to the number of requests in the G/G/∞ queue. In what follows, we will use the following notations: $A = E[n]$ (A is equal to the traffic intensity according to (4.20) with $W = 0$ because the system has an infinite number of servers) and $V = \text{Var}[n]$.

The peakedness parameter z provides a better description of the traffic than the IDC parameter introduced in Sect. 4.2, since the value of z depends on both the arrival process and the service process, whereas IDC only depends on the arrival process. The peakedness parameter z has the same dimensions as n.

If $z < 1$, the traffic is said to be *smoothed*; if $z = 1$, the traffic has Poisson arrivals (the fictitious queuing system is of the M/G/∞ type); if $z > 1$, the traffic is said to be *peaked*. The traffic with Poisson arrivals is the boundary case between more regular arrivals (i.e., smoothed traffic) and more bursty arrivals (i.e., peaked traffic).

Let us go back to the characterization of the different traffic types involved in an M/M/S/S queuing system. As for the carried traffic (also referred to as the traffic of accepted requests), we have to consider the mean and variance of the number n of busy servers. Since there are no waiting rooms, the state probability distribution P_n in (4.43) is actually the distribution of the number of busy servers. Hence, the mean and variance of the carried traffic can be easily obtained as [24, 25]

$$E[n] = A_\mathrm{C} = \rho(1 - P_\mathrm{B}), \quad \mathrm{Var}[n] = V_\mathrm{C} = A_\mathrm{C} - \rho P_\mathrm{B}(S - A_\mathrm{C}) \tag{4.115}$$

where P_B is the Erlang-B blocking probability (4.44), depending on the number of servers S and the input traffic intensity $\rho = \lambda/\mu = A$.

The peakedness of the carried traffic is

$$z_\mathrm{C} = \frac{V_\mathrm{C}}{A_\mathrm{C}} = \frac{A_\mathrm{C}\left[1 - \rho P_\mathrm{B}\left(\frac{S}{A_\mathrm{C}} - 1\right)\right]}{A_\mathrm{C}} = 1 - \rho P_\mathrm{B}\left(\frac{S}{A_\mathrm{C}} - 1\right) \tag{4.116}$$

It is possible to show that z_C in (4.116) is lower than 1: the carried traffic is non-Poisson and smoothed.

The traffic of requests that are blocked and not accepted into the system due to congestion (i.e., refused traffic) can be studied as detailed in [24, 25], thus obtaining mean and variance as follows:

$$A_\mathrm{B} = \rho P_\mathrm{B}, \quad V_\mathrm{B} = A_\mathrm{B}\left(1 - A_\mathrm{B} + \frac{\rho}{S - \rho + A_\mathrm{B} + 1}\right) \tag{4.117}$$

where the above V_B formula is also known as the Riordan formula.

Hence, the peakedness of the refused traffic is obtained as

$$z_\mathrm{B} = \frac{V_\mathrm{B}}{A_\mathrm{B}} = 1 - A_\mathrm{B} + \frac{\rho}{S - \rho + A_\mathrm{B} + 1} \tag{4.118}$$

It is possible to prove that z_B in (4.118) is greater than 1: the refused traffic is non-Poisson and peaked [24]. The arrival process corresponding to the refused traffic is an Interrupted Poisson Process (IPP).

Note that for given values of S and ρ, we can compute the moments of the refused traffic according to (4.117). Instead, given "generic" values of A_B and V_B, it is possible to invert the system of non-linear equations in (4.117) to find the corresponding values of S and ρ. Since non-integer S values could be needed, the solution can be obtained only by means of the Erlang-B extension to real positive values of the number of servers (*Fortet representation*) [25]:

$$P_\mathrm{B}^*(S, \rho) = \frac{1}{\rho \int_0^{+\infty} e^{-\rho y}(1 + y)^S \mathrm{d}y} = \frac{\rho^S e^{-\rho}}{\Gamma(S + 1, \rho)}, \quad S \geq 0 \tag{4.119}$$

where $\Gamma(a, b)$ denotes the incomplete Gamma function, already defined in Chap. 3, equation (3.197).

Numerical methods have to be adopted to invert (4.117), where P_B is given by (4.119).

4.15.2 Blocking Probability for Non-Poisson Arrivals

Let us consider a generic traffic with mean (intensity) A and variance V (hence, peakedness factor $z = V/A$). This traffic is at the input of a general loss queuing system with S servers and no waiting rooms: G/G/S/S system.

We are interested in deriving the probability that an arrival (here also named "call," referring to the classical telephony for which this theory was developed) finds all the servers busy so that it experiences a blocking event. This is an interesting generalization of the Erlang-B problem. We will describe briefly below two approaches [25]: the Wilkinson method [24] for peaked traffic and the Fredericks method [26] for both peaked and smoothed traffic. Other useful considerations on moment matching techniques can be found in [27].

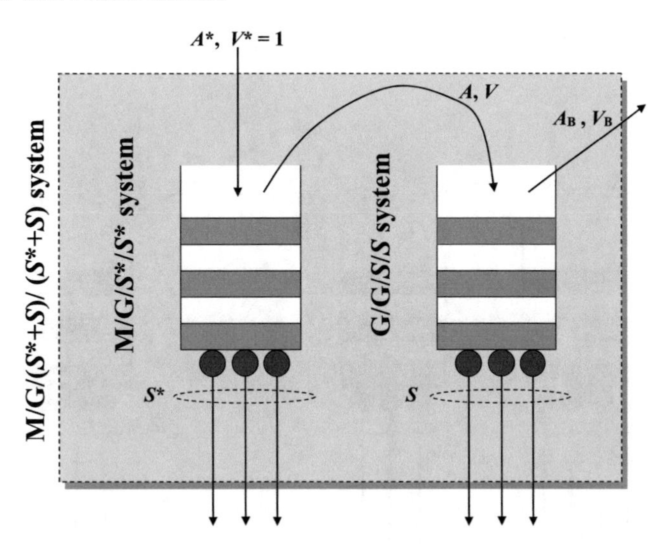

Fig. 4.23: Model of the loss queuing system ("approximate equivalent system") according to the ERT–Wilkinson approach (queues have no waiting rooms)

4.15.2.1 Equivalent Random Theory (ERT), Also Known as Wilkinson Method

Let us assume that the input traffic with moments A and V ($V > A$, for peaked traffic) can be obtained as the traffic refused by a certain M/G/S^*/S^* loss queuing system with unknown A^* (input traffic intensity) and S^* (number of servers) and Poisson input traffic. Note that A^* and S^* can be obtained by inverting (4.117), where $P_B{}^*$ is given by the extended Erlang-B formula in (4.119). In particular, we have

$$\begin{cases} A = A^* P_B^* \left(S^*, A^*\right) \\ V = A\left(1 - A + \frac{A^*}{S^* - A^* + A + 1}\right) \end{cases} \tag{4.120}$$

An approximate solution can be obtained by considering that

$$A^* = V + 3z\left(z - 1\right)$$

with $z = V/A$. Moreover, we can obtain S^* by inverting the following equation according to (4.120):

$$A = A^* P_B^* \left(S^*, A^*\right)$$

Analogously to what is shown in Sect. 4.10 to determine the general formula of the call congestion for non-Poisson arrival processes, the blocking probability of a G/G/S/S queue with S servers and input traffic with moments A and V can be computed as $P_B = A_B/A$, where A_B can be seen as the traffic rejected by a fictitious (equivalent) loss queuing system of the M/G/$(S + S^*)$/$(S + S^*)$ type with Poisson input process and intensity A^* (see Fig. 4.23). From (4.117), we have $A_B = A^* P_B{}^*(S + S^*, A^*)$, where $P_B{}^*(S + S^*, A^*)$ is obtained from (4.119) due to the possible non-integer value of $S + S^*$. Note that due to the insensitivity property, the results obtained in Sect. 4.15.1 for an M/M/S/S queue can also be reapplied here to a loss queuing system with general service distribution. Hence, the blocking probability P_B experienced by the traffic A, V due to the G/G/S/S queue can be approximated as [24]

$$P_B = \frac{A_B}{A} = \frac{A^* P_B^* \left(S + S^*, A^*\right)}{A} \tag{4.121}$$

This method is approximated since the input traffic A, V cannot be obtained in general as the traffic refused by a loss queuing system with Poisson input traffic. This method can be extended to the case where the input traffic is the sum of independent traffic contributions with mean A_i and variance V_i; in fact, we employ the above formulas with $A = \Sigma A_i$ and $V = \Sigma V_i$ [27].

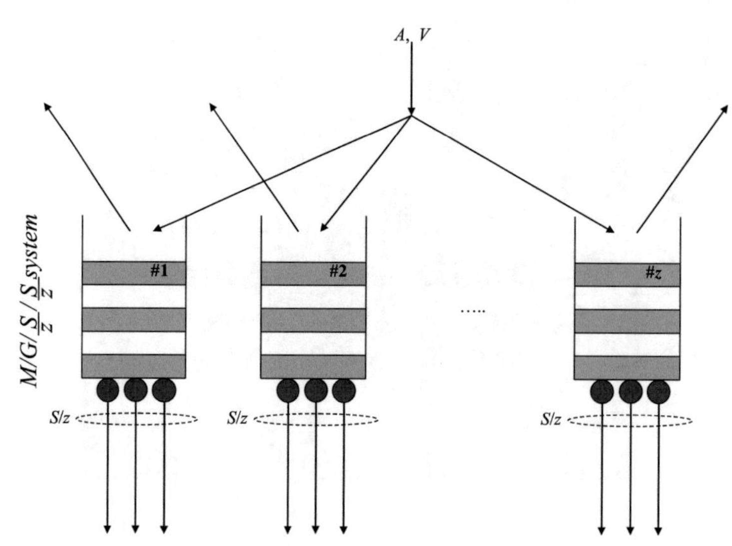

Fig. 4.24: Model of the loss queuing system ("approximate equivalent system") according to the Fredericks approach (queues have no waiting rooms)

4.15.2.2 Fredericks Method

This method in general can be applied to both smoothed and peaked traffic cases, even if it is described here for an input peaked traffic with A and V values so that $V > A$. This method is based on an approximate equivalent system detailed as follows:

- The arrivals are considered to occur in groups of fixed size $z = V/A$.
- The groups arrive according to a Poisson process.

It is possible to show that this arrival process has the same mean A and variance V of the original input process [26]. For the sake of simplicity, we assume that both z and S/z are integer values. Each of the z arrivals of a group is sent to a different sub-queue with S/z servers; there are z different sub-queues of this type. This equivalent system is depicted in Fig. 4.24.

The blocking probability of the G/G/S/S queue with S servers and input traffic with moments A, V is approximated by the blocking probability of a loss queuing system with S/z servers and Poisson input traffic with intensity A/z:

$$P_{\mathrm{B}} = P_{\mathrm{B}}\left(\frac{S}{z}, \frac{A}{z}\right) \tag{4.122}$$

where the function P_{B} on the right side of (4.122) is computed in general by means of the Erlang-B extension in (4.119).

Let us compare the $P_{\mathrm{B}}(S/z,\ A/z)$ value obtained by (4.122) for a generic traffic (A, V) and S servers with the blocking probability $P_{\mathrm{B,Poisson}}(S,\ A)$ of an M/G/S/S system with the same intensity A (i.e., the classical Erlang-B formula). We have that:

- If $V > A$ (peaked traffic), $P_{\mathrm{B}}(S/z,\ A/z) > P_{\mathrm{B,Poisson}}(S,\ A)$.
- If $V \leq A$ (smoothed traffic), $P_{\mathrm{B}}(S/z,\ A/z) \leq P_{\mathrm{B,Poisson}}(S,\ A)$.

Figure 4.25 shows a comparison between the Wilkinson method and the Fredericks one for the blocking probability experienced by a G/G/S/S system for increasing peakedness factor values ($z > 1$, peaked traffic) with input traffic $A = 5$ Erlangs and $S = 8$ servers. This graph also shows the value of the blocking probability for Poisson arrivals (Erlang-B formula) for the same A and S values, which is 7%. We may note that as z increases, the blocking probability for peaked traffic increases with respect to the Poisson case. Moreover, the Wilkinson method gives a slightly higher blocking probability value than the Fredericks one. The difference of the blocking probability values of these methods with respect to the Poisson case (i.e., Erlang-B value) is noticeable.

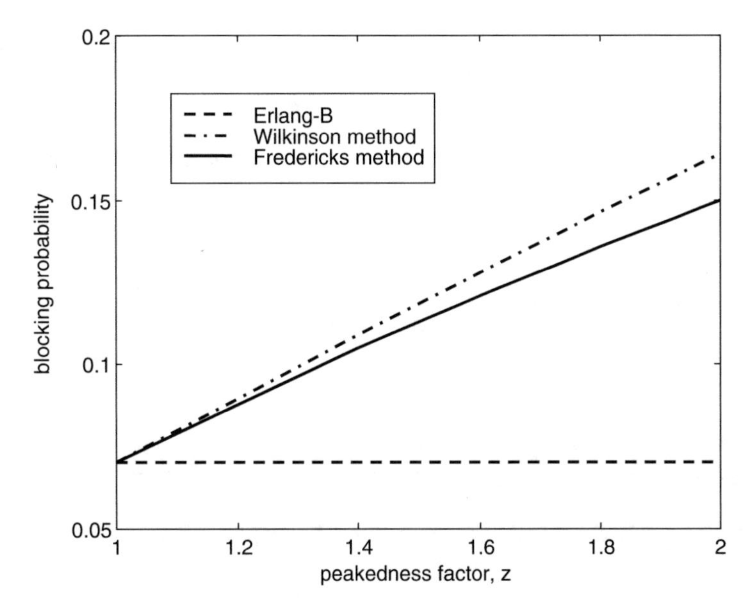

Fig. 4.25: Comparison between the Wilkinson method and the Fredericks one as a function of the z value

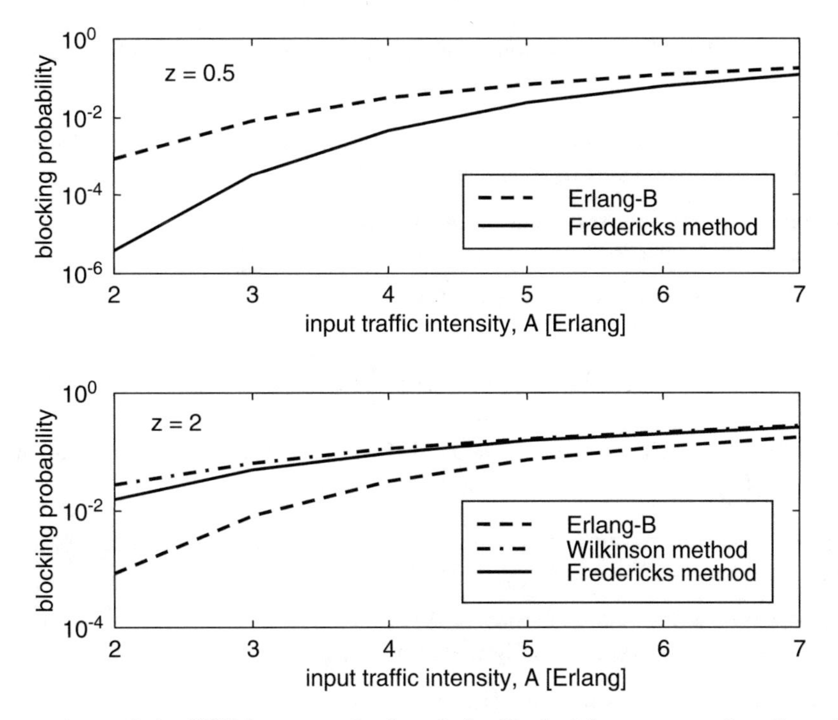

Fig. 4.26: Comparison of the Wilkinson method and the Fredericks one as a function of the A value

Finally, Fig. 4.26 compares Wilkinson, Fredericks, and Erlang-B blocking probabilities as functions of the input traffic intensity A, for both $z = 0.5$ (smoothed traffic) and $z = 2$ (peaked traffic) with $S = 8$ servers. Of course, the Erlang-B curve does not depend on the z value ($z = 1$ for Poisson traffic). Moreover, in the case $z = 0.5$, the Wilkinson method cannot be applied. Based on these results, we can conclude that *the Erlang-B formula overestimates the blocking probability with smoothed traffic and underestimates the blocking probability with peaked traffic.*

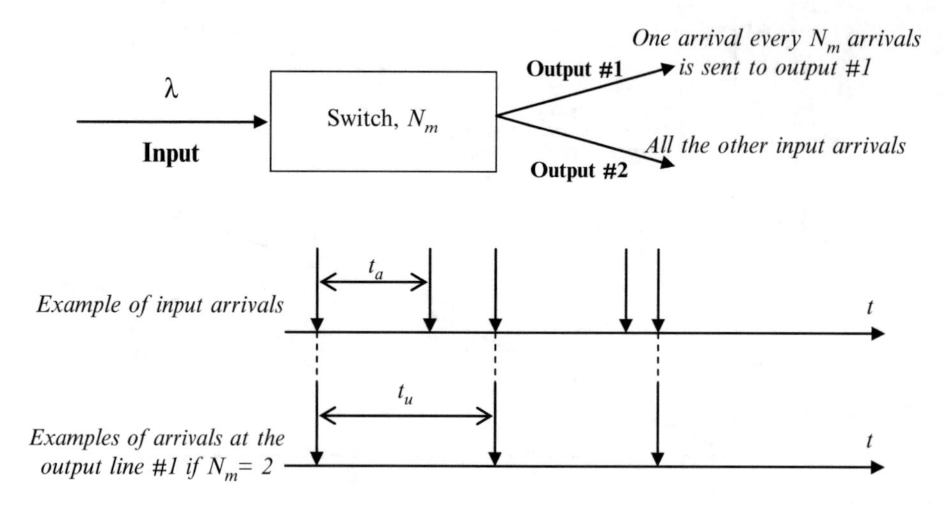

Fig. 4.27: Switch that divides arrivals between the two output lines based on a stochastic choice

4.16 Exercises

This section contains a collection of exercises where Markovian queues are adopted to model telecommunication systems.

Ex. 4.1 We consider a Poisson arrival process with a mean rate λ at the input of a switch, as shown in Fig. 4.27. Arrivals are distributed between the two output lines as follows:

- Output line #1 receives one arrival every N_m input arrivals.
- Output line #2 receives all arrivals not sent to output line #1.

Let us assume that N_m is a random variable with distribution:

$$\text{Prob}\{N_m = l\} = \frac{1}{L}\left(1 - \frac{1}{L}\right)^{l-1}, \quad \text{for } l \geq 1$$

We have to evaluate the probability density function of the interarrival times for output line #1 to characterize the arrival process at this line and the corresponding mean arrival rate.

Ex. 4.2 We consider a buffer that receives messages to be sent. Two modems are available to transmit messages from the buffer; modems operate at the same speed. We know that:

- The message arrival process is Poisson with mean rate λ.
- The message transmission time is exponentially distributed with mean value $E[X]$.

It is requested to determine the following quantities:

- The traffic intensity offered to the buffer in Erlangs
- The mean number of messages in the buffer
- The mean delay from a message arrival at the buffer until it is transmitted completely
- Could the buffer support input traffic with $\lambda = 10$ msgs/s and $E[X] = 2$ s?

Ex. 4.3 An *Internet Service Provider* (ISP) has to dimension a *Point-of-Presence* (POP) in the territory, which can handle up to S simultaneous Internet connections (due to a limited number of available IP addresses and/or because of a limited processing capability). If a new Internet connection is requested to the POP by a user when there are other S connections already in progress, the new connection request is blocked. We have to determine S, guaranteeing that the blocking probability $P_B < 3\%$. We know that:

- Each subscriber generates Internet sessions according to a Poisson process with a mean rate λ.
- Internet sessions have a duration that is generally distributed.

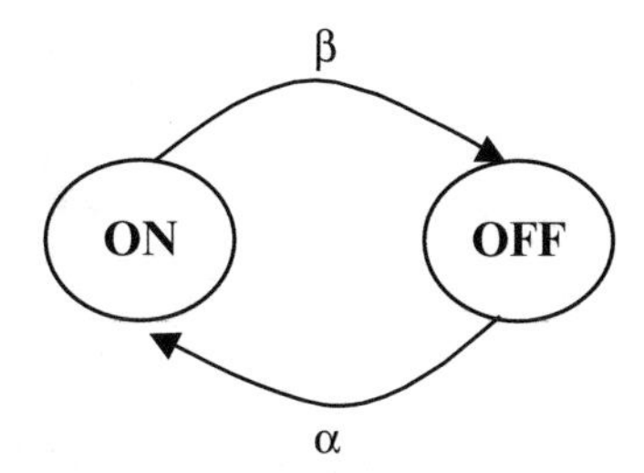

Fig. 4.28: Model of a voice source with activity detection

- Each subscriber is connected to the POP on average 1 h per day, thus contributing a traffic load of about 41 mErlangs.
- We have $U = 100$ subscribers/POP.

Ex. 4.4 We consider a traffic regulator that manages the arrivals of messages at a transmission line buffer. Messages arrive according to exponentially distributed interarrival times with mean rate l. The traffic regulator acts as follows: a newly arriving message is sent to the transmission buffer with probability q; otherwise, a newly arriving message is blocked with probability $1 - q$. The message transmission time has an exponential distribution with a mean rate μ. It is requested to determine:

- A suitable model for the buffer
- The stability condition of the buffer
- The mean delay from the arrival of a message at the buffer until its complete transmission

Ex. 4.5 We consider a multiplexer, which collects messages arriving according to exponentially distributed interarrival times. The multiplexer is composed of a buffer and a transmission line. We make the following approximation: the transmission time of a message is exponentially distributed with mean value $E[X] = 10$ ms. From measurements on the state of the buffer, we know that the empty buffer probability is $P_0 = 0.8$. It is requested to determine the mean message delay.

Ex. 4.6 We consider a private branch exchange, which collects phone calls generated in a company with 1,000 phone users, each contributing Poisson traffic of 30 mErlangs. We have to design the number S of output lines from the private branch exchange to the public network's central office to guarantee a blocking probability for new calls lower than or equal to 3%. What is the increase in the number of output lines if the number of users becomes equal to 1,300, still requiring a blocking probability of 3%? It is requested to compare the percentage of traffic increase $\Delta\rho\%$ corresponding to the percentage of increase in the number of output lines $\Delta S\%$.

Ex. 4.7 We have a packet-switched telecommunication system where N simultaneous phone conversations with speech activity detection are managed by a central office. A Markov chain with ON and OFF states is adopted to model the behavior of the traffic of each voice source, as shown in Fig. 4.28. In the ON state, a voice source generates a bit-rate R_{ON}; in the OFF state, no bit-rate is produced.We have to determine the statistical distribution of the total bit-rate generated by the N sources that produce traffic at the central office.

Ex. 4.8 Referring to the network of queues in Fig. 4.29, we need to determine the mean number of messages in all queues of the network and the total mean delay of a message from input to output of the network.

Ex. 4.9 We have a buffer for the transmission of messages, which arrive according to exponentially distributed interarrival times with mean value $E[X]$. The transmission time of a message is according to an exponential distribution with mean value $E[T]$. The buffer adopts a self-regulation technique for the input traffic: when the number of messages in the buffer is greater than or equal to S, any new arrival can be rejected with probability $1 - p$. It is requested to model this system to identify the buffer's stability condition and evaluate the probability that new arrivals are blocked and refused.

Ex. 4.10 A link uses two parallel transmitters at 5 Mbit/s. Each transmitter has a buffer with an infinite capacity to store messages. Messages arrive at the link according to a Poisson process with a mean rate $\lambda = 20$ msgs/s and

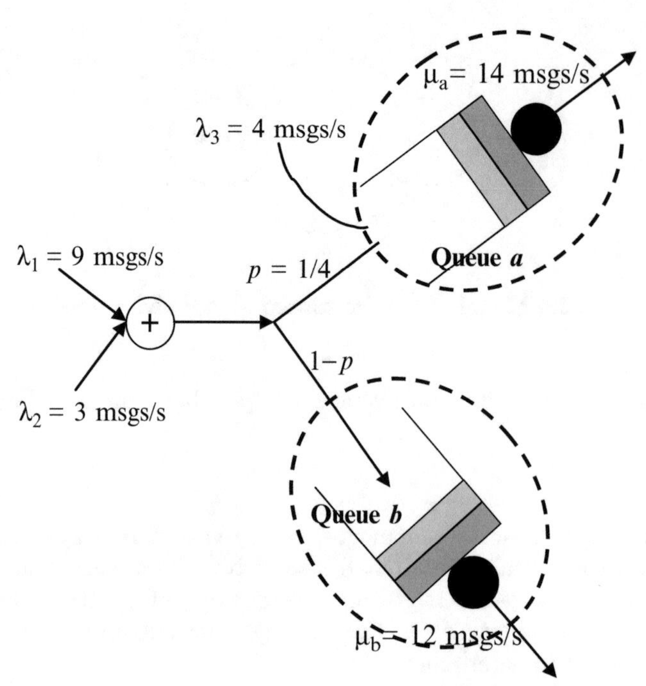

Fig. 4.29: System composed of two queues

have a mean length of 100 kbits. A switch at the input divides the messages between the two transmitters with equal probability.

- We have to evaluate the mean delay T from message arrival to transmission completion.
- We assume that the operator substitutes the two transmitters with a single transmitter having a rate of 10 Mbit/s; we have to evaluate the mean message delay and compare this result with that obtained in the previous case.

Ex. 4.11 We have an M/M/1 queue with mean arrival rate λ, mean service time μ, and FIFO service discipline. It is requested to obtain the Laplace transform of the probability density function of the delay. What is the probability that a generic arrival finds an empty queue?

Ex. 4.12 A radio link adopts four parallel transmitters for redundancy reasons. The operational characteristics of the transmitters require that each of them is switched off (for maintenance or recovery actions) according to a Poisson process with a mean interarrival time of 1 month. A technician performing maintenance and recovery actions needs an exponentially distributed time with a mean duration of 12 hours to fix a problem. Two technicians are available. We have to address the following issues:

1. To define a suitable model for the system
2. To determine the probability distribution of the number of transmitters switched off at a generic instant
3. To derive the probability that no transmitter is working for this radio link
4. To determine the mean delay from when a transmitter is switched off to when it is reactivated

Ex. 4.13 A transmission system for messages (composed of packets) is characterized as follows:

- The probability distribution of the number of messages in the system can be approximated by an M/M/1 system, which is empty with probability $P_0 = 0.5$.
- Each message is composed of a random number of packets according to the following distribution:

$$\text{Prob}\,\{\text{num. of packets in a message} = k\} = q(1-q)^{k-1}, k \in 1, 2, \ldots$$

We have to determine the probability distribution of the total number of packets in the queuing system. Moreover, let us consider the transmission system at a given instant: assuming that we have started to count

10 packets in the queue and that there are other packets, what is the distribution of the number of packets remaining in the queue?

Ex. 4.14 We consider a traffic source, which generates traffic as follows:

- Message arrivals occur according to a Poisson process with a mean rate λ.
- Each arrival triggers the generation of the packets of a message. A message has a length in packets according to a modified geometric distribution with a mean value L. The packets of a message are not generated instantaneously but are generated at a constant rate of r pkts/s.

We have to determine the distribution of the number of packets generated simultaneously by the traffic source at a generic instant.

Ex. 4.15 We have a transmission line sending the messages arriving at a buffer. Each message can wait for service for a maximum time (deadline). A message not served within its deadline is discarded from the buffer. We model the maximum waiting time of a message as a random variable with exponential distribution and a mean rate γ. Messages arrive according to a Poisson process with a mean rate λ, and their transmission time is exponentially distributed with a mean rate μ. We need to determine:

1. A suitable queuing model for the system
2. The mean number of messages in the transmission buffer

Ex. 4.16 We have a private branch exchange with two output lines and no waiting part. This exchange can receive two different types of calls, with corresponding independent arrival processes:

- A *type #1 phone call* requiring one output line. This arrival process is Poisson with mean rate λ_1 and the call length is exponentially distributed with mean rate μ_1.
- A *type #2 phone call* requiring two output lines. This arrival process is Poisson with mean rate λ_2 and the call length is exponentially distributed with mean rate μ_2.

A new call arriving at the exchange is blocked and lost if it needs a number of output lines greater than those available. We have to model this system and determine the blocking probability for both type #1 and type #2 calls.

Ex. 4.17 We consider a Private Branch eXchange (PBX) with a single output line. Calls arrive according to exponentially distributed interarrival times with a mean rate α. The length of each call is according to an exponential distribution with a mean rate γ. We have to analyze two different cases:

- *Case #1*: The PBX can put new calls on a waiting list if they find a busy output line. It is requested to model this system and to express the probability that an incoming call is put on the waiting list, P_C.
- *Case #2*: The PBX has no waiting list: if an incoming call finds a busy output line, the call is blocked and lost. It is requested to model this system and to express the call blocking probability P_B. What is the maximum traffic intensity in Erlangs that can be supported with a blocking probability lower than 1%?

Finally, we have to compare the stability limits of these two different cases.

Ex. 4.18 We have a Time Division Multiplexing (TDM) transmission line. The arrival process from this line is characterized as follows on a slot basis (duration T_s):

- The slot-based arrival process is memoryless.
- A slot carries a message (containing a random number of packets) with probability p and is empty with probability $1 - p$.
- The lengths of messages in packets are iid; let $L(z)$ denote the PGF of the message length in packets.

The messages coming from the TDM line are switched on a slot basis on two different output lines, A and B, as detailed in Fig. 4.30. The switching process is random and memoryless from slot to slot: a message is addressed towards line A with probability q; instead, a message is addressed towards line B with probability $1 - q$.

It is requested: (1) to characterize the arrival process of messages at line A on a slot basis; (2) to determine the PGF of the number of packets arrived at line A on a slot basis.

Ex. 4.19 Let us refer to a node of a telecommunication network receiving packet-based traffic as follows:

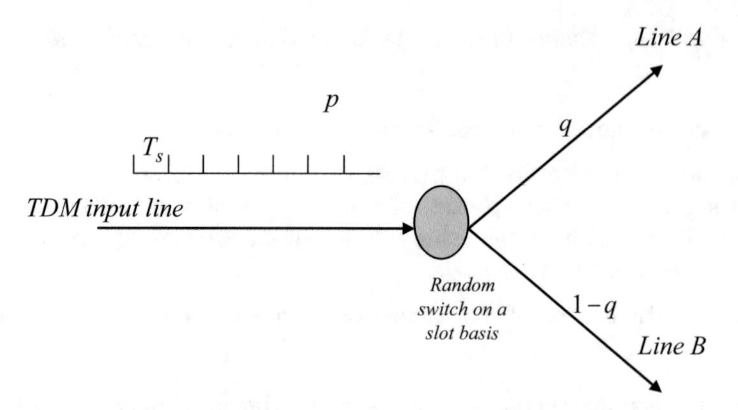

Fig. 4.30: TDM line with random splitting

- Messages arrive according to exponentially distributed interarrival times with mean value T_a.
- Each message is composed of a binomially distributed number of packets with mean value M[6].
- The maximum length of a message is equal to L pkts.

We need to derive:

- The PGF of the number of packets arrived in a generic time interval T
- The mean number of packets arrived in T

Ex. 4.20 We have a group of modems that receive dial-up Internet connections calls (circuit-switched calls) from a very large number of different users according to exponentially distributed interarrival times with a mean value of 10 s. We have to determine:

1. The PGF of the number of calls arrived in a generic interval T
2. The probability that, starting from a generic instant, more than 20 s are needed to receive the third call
3. The PGF $A_c(z)$ of the number of calls arrived in the time interval of the duration of a call, which is exponentially distributed with mean rate μ

Ex. 4.21 We have m independent Poisson arrival processes of messages, each with a mean rate λ. Messages arrive at a transmission system, which has a total transmission capacity C. Each message requires an exponentially distributed time to be sent (service time). It is requested to compare the mean delay experienced by a message in two different cases for what concerns the sharing of capacity C:

1. We use a *separate queue for each traffic flow* (*deterministic multiplexing*): each queue has a transmission capacity C/m, corresponding to a mean message transmission time equal to $1/\mu$.
2. We use a *single queue collecting all traffic flows* (*statistical multiplexing*), with a transmission capacity equal to C and a corresponding mean message transmission time equal to $1/(m\mu)$.

Ex. 4.22 Let us consider a buffer of a transmission system (= queuing system), where packets arrive according to an ON–OFF traffic source (see Fig. 4.31). Sojourn times in ON and OFF states are exponentially distributed, with mean rates μ_{ON} and μ_{OFF}, respectively. When the source is in the OFF state, no packet is generated. When the source is in the ON state, packets are generated at a constant rate of r pkts/s. Considering that the system needs a time T to transmit a packet, we have to determine:

1. *First case*:

    (a) The burstiness index of the traffic source as a function of parameters μ_{ON} and μ_{OFF}
    (b) The traffic intensity offered to the system in Erlangs

[6] For the sake of simplicity, let us assume here that it is possible to receive an empty message (i.e., a message without packets). Otherwise, we should rescale the binomial distribution to exclude the empty message case. Of course, the solution method of this exercise does not depend on the distribution adopted for the message length.

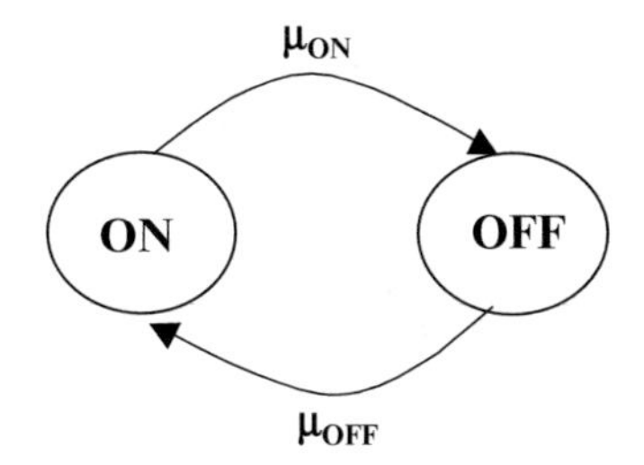

Fig. 4.31: Model of the traffic source

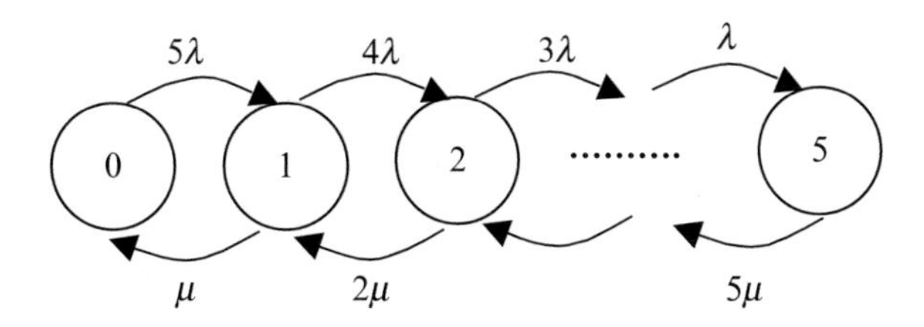

Fig. 4.32: Markov chain modulating the bit-rate generated by the video traffic source (fluid-flow model)

(c) The mean number of packets in the system (buffer), N, if the mean delay in transmitting a packet is equal to $5T$

2. *Second case*: If $\mu_{ON} = 1\ s^{-1}$, $\mu_{OFF} = 1/3\ s^{-1}$, $r = 4$ pkts/s, and $T = 1$ s is the transmission system stable or not?

Ex. 4.23 We have a variable bit-rate video traffic source whose bit-rate (fluid-flow traffic model) is characterized by the continuous-time Markov chain shown in Fig. 4.32 with parameters λ and μ (see also reference [28]). The source can be in one of the six states, $i \in \{0, 1, \ldots, 5\}$. When the traffic source is in state i, the traffic is generated according to a constant bit-rate equal to iV bit/s. We have to determine: (1) the state probability distribution of the chain modulating the traffic generation as a function of λ and μ; (2) the mean bit-rate and the burstiness of the traffic produced by the source as a function of λ, μ, and V; (3) the traffic intensity produced by this source if its bits are sent on a transmission line with a capacity of R bit/s.

Ex. 4.24 We have to dimension the communication part of an Automatic Teller Machine (ATM) system. We know that customers arrive at the ATM machine according to a Poisson process with a mean rate λ (proportional to the service area). We consider that a customer is blocked and refused if the ATM machine is busy when it arrives: the customer should try again after some time. Then, the ATM machine can be modeled as a loss queuing system with a single server and no waiting part. Let us assume that the service time of a customer is according to a general distribution with mean value T. We have to study this system and determine the blocking probability P_B that a generic customer experiences because the ATM machine is busy. Finally, it is requested to determine the maximum value of the customer arrival rate λ to guarantee $P_B < 1\%$.

Ex. 4.25 Let us consider two independent traffic sources whose traffic is at the input of a multiplexer. The following Markov-modulated fluid-flow models characterize the two traffic sources.

The traffic source #1 has the model shown below: in the state "0" no traffic is generated, while in the state "1" a constant bit-rate V is generated.

The traffic source #2 has the model shown below: in the state "0" no traffic is generated, while in the state "1" a constant bit-rate R is generated.

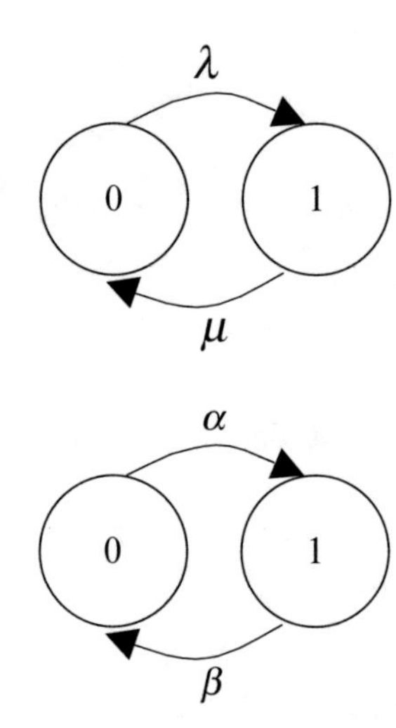

It is requested to determine the bit-rate distribution of the aggregate traffic.

Ex. 4.26 Let us consider a Next Generation Network (NGN) supporting VoIP calls, each needing a bandwidth BW_{call} to guarantee acceptable voice quality. Let $BW_{gateway}$ denote the capacity of the output link from the VoIP gateway. We assume that VoIP calls arrive at the gateway according to a Poisson process and have a generally distributed length. From measurements, we know the maximum arrival rate of VoIP calls at the gateway in the busy hour; this is denoted by parameter BHCA (Busy Hour Call Attempts). Still, from measurements, we know the Mean Call Duration, denoted by parameter MCD. If a new call arrives at the VoIP gateway and does not find an available bandwidth equal to BW_{call}, it is blocked and refused by some Call Admission Control (CAC) functionality. What is the grade of service provided by the gateway to the VoIP traffic? In particular, it is requested to analyze the blocking probability of VoIP calls due to CAC.

Ex. 4.27 Let us consider a node with one transmitter; when the number of messages exceeds a certain threshold, K, a second dial-up transmitter is activated to reduce the congestion at the node. We assume that the messages arrive according to a Poisson process with a mean rate λ and that the messages have an exponentially distributed transmission time with mean rate μ. We have to determine the mean number of messages in the system N and the mean message delay T.

Ex. 4.28 A link consists of a buffer and two transmitters with exponential service times with different mean rates μ_1 and μ_2, where $1/\mu_1 < 1/\mu_2$. A message entering an empty system is routed to the faster of the two servers, i.e., to the server with mean rate μ_1, where it remains until its service completion. The system can hold one message in addition to those in service. We suppose that a message can switch to a faster server if available. It is requested to write down the state diagram and to solve the state probability distribution assuming a Poisson message arrival process with a mean rate λ. Please motivate the values of the departures from the states of the model adequately.

Ex. 4.29 In a dormitory kitchen, there are 2 ovens for cooking and 3 chairs. In the dormitory, there are 8 students and each of them goes to the kitchen on average every $1/\alpha$ time to cook (if he/she is not cooking already); the interarrival time is exponentially distributed. When a new student arrives and finds both the ovens occupied, the student looks for a place on the chairs. If all the chairs are occupied as well, the student goes back to his/her room and tries again later. Students spend an exponentially distributed time for cooking with a mean value $1/\beta$. It is requested to:

- Draw the state transition diagram
- Determine the state probability distribution as a function of α and β and express the mean number of students in the kitchen
- Explain if we can or we cannot apply the PASTA property to study the probability that a student cannot enter the kitchen because it is full

Ex. 4.30 We consider two types of calls arrivals to a cell of a mobile telephone network: "new calls" that originate in the cell and "handover calls" that arrive in the cell because the corresponding user with an active call is entering the cell from an adjacent one. The cell has $S = 10$ channels to serve the calls; each call needs one channel. To prioritize handover calls, $K = 2$ channels out of S are reserved in the cell only for handover calls, while the rest of the channels are available to both types of calls. New calls arrive in the cell according to a Poisson process with mean rate $\lambda_{nc} = 125$ calls/hour. Handover calls arrive in the cell according to an independent Poisson process with mean rate $\lambda_{ho} = 50$ calls/hour. Both new calls and handover calls have exponentially distributed durations with mean value $1/\mu = 2$ minutes.

- Draw the state diagram of the number of occupied channels in the cell.
- Express the blocking probability in the cell of new calls, P_{Bnc}, and the blocking probability of handover calls, P_{Bho}.

Ex. 4.31 Network slicing is a network virtualization technique that splits a single physical infrastructure into multiple virtual networks, called slices. We consider that slice requests are received by the system orchestrator that has to allocate resources of the radio access network from a shared pool. There are slice requests of two types:

- Type #1, requesting capacity C_1 occurring with a Poisson process with mean rate λ_1 and lasting for a mean time $1/\mu_1$
- Type #2, requesting capacity $C_2 \neq C_1$ occurring with a Poisson process with mean rate λ_2 and lasting for a mean time $1/\mu_2$

Let us consider that the total system capacity is C_{tot}. A new slice request is refused if there is not enough capacity to support it. We need to determine the blocking probability experienced by type #1 requests and by type #2 requests, knowing that $C_1 = 1$ Gbit/s, $C_2 = 2$ Gbit/s, and $C_{tot} = 4$ Gbit/s.

Ex. 4.32 Let us consider a network node that can store and serve 1 packet at most. Packets arrive at the node according to a Poisson process. The service of a packet involves two independent tasks that are performed in sequence: the error check and the packet transmission to the output link. Each task requires an exponentially distributed time with an average $1/\mu = 30$ ms. We observe that the node is empty for 60% of the time. We have to model this system through a Markov chain and determine the system's average number of packets.

References

1. Kleinrock L (1976) Queuing systems. Wiley, New York
2. Markov AA (1907) Extension of the limit theorems of probability theory to a sum of variables connected in a chain. The notes of the Imperial Academy of Sciences of St. Petersburg VIII Series, Physio-Mathematical College, vol XXII, No. 9, December 5, 1907
3. Hayes JF (1986) Modeling and analysis of computer communication networks. Plenum Press, New York
4. Paxson V, Floyd S (1995) Wide-area traffic: the failure of Poisson modelling. IEEE/ACM Trans Network 3(3):226–244
5. Kendall DG (1953) Stochastic processes occurring in the theory of queues and their analysis by the method of imbedded Markov chain. Ann Math Stat 24:338–354
6. Kendall DG (1951) Some problems in the theory of queues. J Roy Stat Soc B 13:151–185
7. Little JDC (1961) A proof of the queueing formula L = λW. Oper Res 9:383–387
8. Jewell WS (1967) A simple proof of L = λW. Oper Res 15(6):109–116
9. Schrage LE, Miller LW (1966) The queue M/G/1 with the shortest remaining processing time. Oper Res 14:670–684
10. Kleinrock L, Scholl M (1980) Packet switching in radio channels: new conflict-free multiple access schemes. IEEE Trans Commun 28(7):1015–1029
11. Scholl M, Kleinrock L (1983) On the M/G/1 queue with rest periods and certain service-independent queueing disciplines. Oper Res 31(4):705–719
12. Hyytiä E, Penttinen A, Aalto S (2012) Size and state-aware dispatching problem with queue-specific job sizes. Eur J Oper Res 217(2):357–370

13. Wolff RW (1982) Poisson arrivals see time averages. Oper Res 30(2):223–231
14. Morro G (2018) Queueing theory. Thesis on Mathematics, Universitat de Barcelona, June 26, 2018.
15. Erlang AK (1918) Solutions of some problems in the theory of probabilities of significance in automatic telephone exchanges. POEEJ 10:189–197 (translated from the paper in Danish appeared on Elektroteknikeren, vol 13, 1917)
16. Ross SM (1983) Stochastic processes. Wiley, New York
17. Addie RG, Zuckerman M, Neame TD (1998) Broadband traffic modeling: simple solutions to hard problems. IEEE Commun Mag 36(8):88–95
18. Iversen, VB (2015) Teletraffic engineering and network planning. DTU Fotonik
19. Myskja A (1995) Teletraffic, Telektronikk, vol 91(2/3)
20. Takagi H, Walke BH (eds) (2008) Spectrum requirement planning in wireless communications: model and methodology for IMT-Advanced. Wiley, New York
21. Chang J (2007) Stochastic processes
22. Kaufman JS (1981) Blocking in a shared resource environment. IEEE Trans Commun COM-29(10):1474–1481
23. Roberts JW (1981) A service system with heterogeneous user requirements. In: Pujolle G (ed) Perf. of data commun. systems and their applications. North-Holland Publishing, pp 423–431
24. Wilkinson RI (1956) Theories for toll traffic engineering in the U.S.A. Bell Syst Techn J 35:421–514
25. Buttò M, Colombo G, Tofoni T, Tonietti A (1991) Ingegneria del traffico nelle reti di telecomunicazioni. L'Aquila, Italy
26. Fredericks AA (1980) Congestion in blocking systems—a simple approximation technique. Bell Syst Tech J 59(6):805–827
27. Delbrouck LEN (1991) A unified approximate evaluation of congestion functions for smooth and peaky traffics. IEEE Trans Commun 29(2):85–91
28. Blondia C, Casals O (1992) Performance analysis of statistical multiplexing of VBR sources. In: Proc. of INFOCOM'92, pp 828–838

Chapter 5
M/G/1 Queuing Theory and Applications

Abstract The Markovian chains and the related queues have the advantage of a simple solution method to determine the state probability distribution. Unfortunately, there are many cases where the service time has not an exponential distribution. This section presents the analytical method to solve the M/G/1 queue in the transform domain in terms of probability generating functions. The theoretical approach is based on imbedded Markov chains. Special cases of batched services and differentiated service times are addressed. Many application examples are provided to study the implications of different imbedding options.

Key words: M/G/1 analysis, Imbedded Markov chain, Differentiated service times, Batched service, Kleinrock principle

5.1 The M/G/1 Queuing Theory

The M/G/1 theory is a powerful tool, generalizing Markovian queues' solution to the case of general service time distributions. There are many applications of the M/G/1 theory in telecommunications; for instance, it can be used to study the queuing of fixed-size packets transmitted on a given link (i.e., M/D/1 case). Moreover, this theory yields results, which are compatible with the M/M/1 theory, based on birth–death Markov chains.

In the M/G/1 theory, the arrival process is Poisson with mean arrival rate λ, but, in general, the service time is not exponentially distributed. Hence, the service process has a certain memory: if there is a request in service at a given instant, its *residual service time* has a distribution depending on the time elapsed since the beginning of its service. Let us refer to a generic instant t. The system is described by a *two-dimensional* state $S(t)$, as follows:

- Number of requests in the system at instant t, $n(t)$.
- Elapsed time from the beginning of the service of the currently served request, $\tau(t)$. This parameter $\tau(t)$ is also called *age* of the currently served request. Note that in the Markovian M/M/1 case, the residual service time's pdf does not depend on $\tau(t)$ because of the memoryless property of the exponential distribution.

Hence, $S(t) = \{n(t),\ \tau(t)\}$. To characterize these queues, we study their behaviors at specific time instants ζ_i where we obtain a mono-dimensional simplification of state $S(\zeta_i)$. The M/G/1 queue is studied at specific imbedding instants, where we obtain a Markovian system again; this is a so-called *imbedded Markov chain* [1, 2]. Different alternatives are available to select instants ζ_i. There is no need that instants ζ_i be equally spaced in time. Typical choices for ζ_i instants are:

1. Service completion instants
2. Arrival instants (this is the case adopted in [3] to study the waiting times in G/M/1 queues)
3. Regularly spaced instants for cases with service based on time slots

---

Electronic Supplementary Material The online version contains supplementary material available at (https://doi.org/10.1007/978-3-030-75973-5_5).

G. Giambene, *Queuing Theory and Telecommunications*, Textbooks in Telecommunication Engineering,
https://doi.org/10.1007/978-3-030-75973-5_5

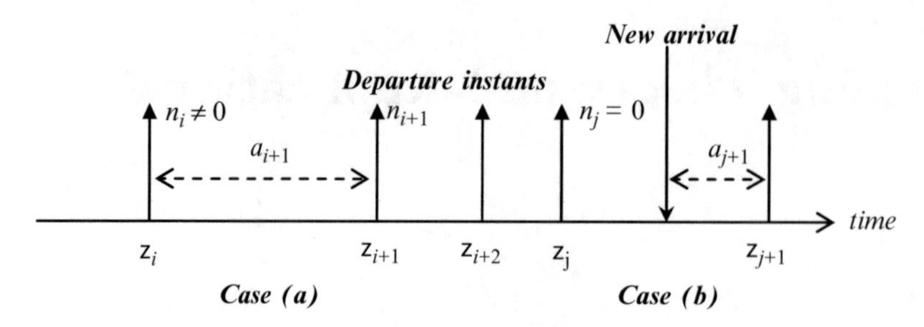

Fig. 5.1: Time diagram of service completion events and new arrivals

It makes a difference in selecting the imbedding points; different imbedding options applied to the same system in general do not yield the same results. In this chapter, let us refer to the first type of imbedding points: let ζ_i denote the service completion instant of the ith request arrived at the queue. We have that $\tau(\zeta_i) \equiv 0 \ \forall i$ since at instant ζ_i a request has completed its service, and no new request has yet started its service. Hence, at these instants ζ_i the state becomes mono-dimensional: $S(\zeta_i) \equiv n(\zeta_i) = n_i$, where n_i denotes the number of requests in the queue soon after the service completion of the ith request. Let a_i indicate the number of requests arrived at the queue during the service time of the ith request (ending at instant ζ_i). Note that n_i and a_i random variables are also used with different imbedding points, but the distributions of both n_i and a_i depend on the imbedding instants selected.

Let us refer to the situation depicted in Fig. 5.1. If $n_i \neq 0$ (i.e., case (a) in Fig. 5.1), the following balance is valid at the next service completion instant: $n_{i+1} = n_i - 1 + a_{i+1}$. Instead, if $n_i = 0$ (i.e., case (b) in Fig. 5.1), we have to wait for the next arrival, which is served immediately so that at the next completion instant ζ_{i+1} the system contains only the arrivals occurred during the service time of the last request; this number is still represented by variable a_{i+1}. Hence, we have: $n_{i+1} = a_{i+1}$.

Let us recall that the indicator (Heaviside) function is defined as: $I(x) = 1$ for $x > 0$; $I(x) = 0$ for $x \leq 0$. By means of function $I(x)$, we can represent n_{i+1} with an expression, which is valid for both $n_i \neq 0$ and $n_i = 0$, as shown below:

$$n_{i+1} = n_i - I(n_i) + a_{i+1} \tag{5.1}$$

Alternative notations to the above are detailed below:

$$n_{i+1} = \max\{n_i - 1, 0\} + a_{i+1}$$
$$n_{i+1} = (n_i - 1)^+ + a_{i+1}$$

The difference equation (5.1) describes the behavior of the M/G/1 queue at imbedding instants. Since the variables at the instant ζ_{i+1} depend only on the variables at instant ζ_i, equation (5.1) characterizes the M/G/1 system through a discrete-time Markov chain (or, more correctly, an imbedded Markov chain). Note that the method of imbedding instants is quite general and has also been applied to study G/M/1 queues (general iid interarrival times; exponentially distributed service times; one server), where the chain is imbedded at the arrival instants of the input process [3].

Let $G(t)$ denote the PDF of the service time, X: $G(t) = \text{Prob}\{X \leq t\}$. Let $g(t)$ denote the pdf of the service time: $g(t) = \mathrm{d}G(t)/\mathrm{d}t$. The mean service time is indicated as $E[X]$.

Let us assume that the M/G/1 queue admits a steady-state with P_n denoting the probability (at regime) to have n requests in the queue at imbedding instants:

$$\lim_{i \to \infty} P_{n_{i+1}} = \lim_{i \to \infty} P_{n_i} = P_n$$

Hence, we have

$$\lim_{i \to \infty} E[n_{i+1}] = \lim_{i \to \infty} E[n_i] = E[n], \text{where } E[n] \text{ denotes the regime value.}$$

By taking the expected values at both sides of (5.1), we have

$$E\left[n_{i+1}\right]=E\left[n_i\right]-E\left[I\left(n_i\right)\right]+E\left[a_{i+1}\right] \tag{5.2}$$

Hence, if we take the limit of both sides for $i \to \infty$, we obtain regime values as

$$E[n]=E[n]-E[I(n)]+E[a] \Rightarrow E[a]=E[I(n)]$$

We can evaluate $E[I(n)]$ through the state probability distribution as

$$E[I(n)]=\sum_{n=0}^{\infty} I(n) P_n=\sum_{n=1}^{\infty} P_n=1-P_0 \tag{5.3}$$

By using (5.3) and the expression at regime corresponding to (5.2), we can obtain probability P_0 as

$$P_0=1-E[a] \tag{5.4}$$

Let us recall that based on the PASTA property P_0 (or $1-P_0$) is the probability that a new Poisson arrival finds an empty (or a non-empty) M/G/1 queue.

The mean number of arrivals during the service time of a request, $E[a]$, can be obtained as the mean number of Poisson arrivals conditioned on a given service time $X=t$, $E[a|X=t]=\lambda t$, and, then, removing the conditioning using the pdf $g(t)$ of X:

$$E[a]=\int_0^{\infty} E\left[a \middle| X=t\right] g(t) \mathrm{d}t=\lambda \int_0^{\infty} t g(t) \mathrm{d}t=\lambda E[X] \tag{5.5}$$

From (5.5), we note that $E[a]$ corresponds to the traffic intensity expressed in Erlangs, ρ. The M/G/1 queue is stable if $P_0>0$ or, equivalently on the basis of (5.4) and (5.5), if $\rho<1$ Erlang.

We focus here on the solution of the difference equation (5.1) in the z-domain by means of PGFs. First of all, we consider the equality obtained by taking the exponentiation with base z on both sides of (5.1) for any index i value:

$$z^{n_{i+1}}=z^{n_i-I(n_i)+a_{i+1}}, \forall i$$

Then, we multiply both sides by the joint distribution $\text{Prob}\{n_{i+1}=h,\ n_i=k,\ a_{i+1}=j\}$ and we sum over h, k, j. The summations on k and j can be removed on the left side; moreover, the summation on h can be removed on the right side. Details are as follows:

$$\begin{aligned}
&\text{left side}: \\
&\sum_h \sum_k \sum_j z^{n_{i+1}} P_{n_{i+1}, n_i, a_{i+1}}=\sum_h z^{n_{i+1}} \sum_k \sum_j P_{n_{i+1}, n_i, a_{i+1}}=\sum_h z^{n_{i+1}} P_{n_{i+1}} \\
&\text{right side}: \\
&\sum_h \sum_k \sum_j z^{n_i-I(n_i)+a_{i+1}} P_{n_{i+1}, n_i, a_{i+1}}=\sum_k \sum_j z^{n_i-I(n_i)+a_{i+1}} \sum_h P_{n_{i+1}, n_i, a_{i+1}} \\
&\quad=\sum_k \sum_j z^{n_i-I(n_i)+a_{i+1}} P_{n_i, a_{i+1}}
\end{aligned}$$

By equating the two expressions above, we obtain

$$\sum_h z^{n_{i+1}} P_{n_{i+1}}=\sum_k \sum_j z^{n_i-I(n_i)+a_{i+1}} P_{n_i, a_{i+1}} \tag{5.6}$$

In order to solve the imbedded Markov chain, we make the following assumptions:

1. Memoryless arrival process[1]

---

[1] In the case of continuous-time processes, we have to consider Poisson (or compound Poisson) processes. Instead, in the case of discrete-time processes, we have to consider Bernoulli or Binomial arrival processes on a slot basis (in this respect, symbol M used to denote the arrival process at the queue has to be considered in a wider sense and as such it will be substituted by "M").

2. Arrival process independent of the number of requests in the queue: n_i and a_{i+1} are independent variables[2]

The above assumptions are quite general and can be met by many systems. We do not impose any request on the service discipline (e.g., FIFO, LIFO, random). In particular, they are verified in the special case of Poisson arrivals and general service time, which are both independent of the queue state.

Under the previous assumption #2, $\text{Prob}\{n_i = k,\ a_{i+1} = j\} = \text{Prob}\{n_i = k\} \times \text{Prob}\{a_{i+1} = j\}$. Therefore, the left side in (5.6) can be rewritten as

$$\sum_h z^{n_{i+1}} P_{n_{i+1}} = \sum_k z^{n_i - I(n_i)} P_{n_i} \sum_j z^{a_{i+1}} P_{a_{i+1}} \tag{5.7}$$

Let $P(z)$ denote the PGF at regime of the state probability distribution at the imbedding instants. Let $A(z)$ denote the PGF at regime of the number of arrivals during the service time of a request. Let us work on a term of the right side of (5.7).

Moreover, note that

$$\begin{aligned} \sum_{k=0}^{\infty} z^{n_i - I(n_i)} P_{n_i} &= P_{0i} + \sum_{k=1}^{\infty} z^{n_i - 1} P_{n_i} = P_{0i} + z^{-1} \sum_{k=1}^{\infty} z^{n_i} P_{n_i} \\ &= P_{0i} + z^{-1} \left\{ \sum_{k=0}^{\infty} z^{n_i} P_{n_i} - P_{0i} \right\} \end{aligned} \tag{5.8}$$

By considering the situation at regime (i.e., for $i \to \infty$), we can eliminate subscript i in (5.7) and (5.8). Then, we substitute (5.8) in (5.7) where we use the PGFs $P(z)$ and $A(z)$:

$$P(z) = \left\{ P_0 + z^{-1} \left[P(z) - P_0 \right] \right\} A(z) \tag{5.9}$$

Finally, we can solve $P(z)$ in (5.9):

$$P(z)\,[z - A(z)] = P_0\,(z-1)\,A(z) \Rightarrow P(z) = P_0 \frac{(z-1)\,A(z)}{z - A(z)} \tag{5.10}$$

The PGF of the state probability distribution in (5.10) represents a general formula, which can be applied to all the imbedded Markov chains fulfilling (5.1) under the previous assumptions #1 and #2. In particular, the PGF in (5.10) is valid for any service policy, provided that the conditions of the insensitivity property are fulfilled (see Sect. 4.5).

Since P_0 is determined from (5.4), the PGF of the state probability distribution depends only on the PGF $A(z)$, which, in turn, depends on both the arrival process and the imbedding instants. The state probability distribution can be obtained by inverting (5.10). This is not an easy task since there may not be a closed-form solution: the PGF in (5.10) typically does not correspond to a classical distribution. A possible approach to invert $P(z)$ is to adopt the Taylor series expansion centered in $z = 0$, as shown in Sect. 3.3.1: the coefficients of the expansion represent the state probability distribution. This approach can be easily implemented using a numerical method based on the Matlab® symbolic toolbox. Another method to invert (5.10) will be described in Sect. 5.5.

By means of (5.4), the *stability condition* $P_0 > 0$ can be expressed as follows, noticing that $E[a] = A'(z=1)$:

$$P_0 = 1 - A'(1) > 0 \Rightarrow A'(1) < 1 \quad \text{[Erlang]}$$

Under the assumption of Poisson arrivals and imbedding at the service completion instants, $A(z)$ can be derived considering the PGF of the number of arrivals in a given interval $X = t$, $A(z|X = t) = e^{\lambda t(z-1)}$ and then removing the conditioning through the pdf of the service time X, $g(t)$:

$$A(z) = \int_0^{+\infty} e^{\lambda t(z-1)} g(t) dt = \Gamma\left[s = -\lambda\,(z-1)\right] \tag{5.11}$$

[2] Note that it is also possible to solve (5.6) by removing such assumption: we obtain a recursive formula to determine the state probabilities P_n at imbedding instants. More details are provided in the following Sect. 5.5.

where $\Gamma(s)$ denotes the Laplace transform of the pdf $g(t)$. On the basis of the expression of $A(z)$ in (5.11) we can evaluate $A'(1)$ and $A''(1)$ as follows:

$$\left.\frac{\mathrm{d}A(z)}{\mathrm{d}z}\right|_{z=1} = -\lambda\Gamma'\left[-\lambda\left(z-1\right)\right]\big|_{z=1} = \lambda\left[-\Gamma'(0)\right] = \lambda E\left[X\right] \tag{5.12}$$

$$\begin{aligned}\left.\frac{\mathrm{d}^2A(z)}{\mathrm{d}z^2}\right|_{z=1} &= \left.\frac{\mathrm{d}}{\mathrm{d}z}\left\{-\lambda\Gamma'\left[-\lambda\left(z-1\right)\right]\right\}\right|_{z=1} = \lambda^2\Gamma''\left[-\lambda\left(z-1\right)\right]\big|_{z=1} \\ &= \lambda^2\Gamma''(0) = \lambda^2 E\left[X^2\right]\end{aligned} \tag{5.13}$$

Note that (5.12) is equivalent to (5.5).

The PGF in (5.10) has a singularity in $z=1$ (a removable singularity according to the Abel theorem), which causes some problems for both the normalization condition and the derivation of the moments of the distribution. Of course, we can use the Hôpital theorem to prove that $P(z=1)=1$ (normalization). Moreover, the moments of the state probability distribution can be obtained by taking subsequent derivatives on both sides of the leftmost expression in (5.10). With the first derivative, we have

$$P'(z)\left[z-A(z)\right] + P(z)\left[1-A'(z)\right] = P_0A(z) + P_0\left(z-1\right)A'(z) \tag{5.14}$$

If we evaluate (5.14) in $z=1$, we obtain $P_0=1-A'(1)$; this is the same expression as in (5.4).

If we derive again (5.14) on both sides with respect to z, we obtain

$$\begin{aligned}&P''(z)\left[z-A(z)\right] + 2P'(z)\left[1-A'(z)\right] + P(z)\left[-A''(z)\right] \\ &= 2P_0A'(z) + P_0\left(z-1\right)A''(z)\end{aligned} \tag{5.15}$$

If we evaluate (5.15) in $z=1$ and we use (5.4) for P_0, we have

$$\begin{aligned}&2P'(1)\left[1-A'(1)\right] - A''(z) = 2P_0A'(1) \\ &\Rightarrow\ N = P'(1) = A'(1) + \frac{A''(z)}{2\left[1-A'(1)\right]}\end{aligned} \tag{5.16}$$

The mean number of requests in the queue at imbedding instants, N, depends on the first two derivatives of $A(z)$ computed in $z=1$. Let us recall that the stability condition is met if $1-A'(1)>0$, i.e., traffic intensity is lower than 1 Erlang. Note that (5.16) is a general expression, which could also be applied to memoryless arrival processes different from the Poisson one provided that the imbedded system is characterized by (5.1). If we refer to Poisson arrivals (i.e., the classical M/G/1 queue) and imbedding points at service completion instants, we can substitute (5.12) and (5.13) in (5.16), thus yielding

$$N = \lambda E\left[X\right] + \frac{\lambda^2 E\left[X^2\right]}{2\left[1-\lambda E\left[X\right]\right]} \tag{5.17}$$

We can derive the mean delay for crossing the queuing system, T, by applying the Little theorem to (5.16) for the more general case or to (5.17) for the Poisson arrival case. In particular, referring to (5.17), we obtain the well-known Pollaczek–Khinchin formula for the mean queuing delay [1, 2, 4]:

$$T = \frac{N}{\lambda} = E\left[X\right] + \frac{\lambda E\left[X^2\right]}{2\left[1-\lambda E\left[X\right]\right]} \tag{5.18}$$

Note that in (5.18) the first contribution to the mean delay is $E[X]$, i.e., the *mean service time*; instead, the second contribution $\lambda E[X^2]/\{2[1-\lambda E[X]]\}$ represents the *mean waiting time*. The mean queuing delay is related to the second moment of the service time distribution. In particular, the mean waiting time increases with the service time variance, considering a certain fixed mean service time. If the input arrival process's traffic intensity $\lambda E[X]$ tends to 1 Erlang (stability limit), the mean delay tends to infinity.

In the case of exponentially distributed service times (mean rate μ), the above formulas (5.17) and (5.18) yield the same expressions of the M/M/1 queue as shown in Chap. 4. In this case, we have $\Gamma(s)=\mu/(\mu+s)$, $E[X]=1/\mu$, and $E[X^2]=2/\mu^2$. As shown in [1, 2], this result permits to conjecture that the state probability distribution obtained for an M/G/1 system at the imbedding instants is also valid at any instants. These considerations can be supported more formally by introducing the Kleinrock principle [1]: for queuing systems where the state changes at most by +1 or −1 (we refer here to actual changes in the number of requests in the queue and not to what happens only at imbedding instants), the state distribution as seen by an arriving customer is the same as that seen by a departing customer. Hence, the state probability distribution at departure instants is equal to the state probability distribution at arrival instants. Moreover, by applying the PASTA property (in the Poisson arrival case), the state probability distribution at arrival instants is also valid at generic instants (random observer). Hence, using both the Kleinrock principle and the PASTA property, we can conclude that the state probability distribution at service completion instants D_k coincides with the state probability distribution at arrival instants Q_k and then with the distribution of the continuous-time system (random observer), P_k: $D_k = Q_k = P_k$. As for discrete-time (Markov) systems, the equivalent BASTA property can be adopted to determine the probability that an arrival finds the queue in a certain state as the probability of the corresponding state.

5.1.1 Alternative M/G/1 Approach Based on the Residual Life

In this section, an alternative method is provided to study the waiting time of an M/G/1 queue under the assumption of Poisson arrivals. Under PASTA and the Kleinrock principle, we know that $D_k = Q_k = P_k$ and we can refer to the state probability distribution at arrival instants Q_k. Let X denote the random variable of the service time and let R denote the random variable of the *residual service time* (this is different from τ, representing the age of the currently served request). An arriving request experiences a waiting time W due to the residual lifetime R plus the delay due to the service of the other requests found in the queue (we refer to a FIFO policy in any case the mean delay is invariant under the insensitivity conditions to apply the Little theorem) [5]. We have

$$
\begin{aligned}
E[W] &= \sum_{k=1}^{\infty} [E[R] + (k-1)E[X]]\,Q_k \\
&= E[R](1-P_0) + E[X]\sum_{k=1}^{\infty}(k-1)P_k \\
&= E[R]\rho + E[X]E[Q]
\end{aligned}
\tag{5.19}
$$

where $\rho = \lambda E[X]$ and $E[Q]$ denotes the mean number of requests in the queue that according to the Little theorem results as $E[Q] = \lambda E[W]$. Then, we can solve (5.19) with respect to $E[W]$ as

$$
E[W] = \frac{\rho E[R]}{1-\rho} \quad [s] \tag{5.20}
$$

Now we have to express the mean value of the residual life of the service of a request.

We know that a request arrives when there is a currently served request with service time X with density function $g(t)$ according to our notations. To characterize the distribution of R we use the *residual life theorem* (or *excess life theorem*) of the renewal theory.[3] We have the renewal process of service times X_i and jump times (or renewal epochs) at the end of the service ξ_i. Times X_i are iid. We denote with $R(t)$ the residual life at time t. The residual life is defined as the interval from t until the next renewal epoch.

It is possible to prove that the limiting pdf of the residual time can be expressed as [6]

$$
f_R(t) = \frac{1 - PDF_X(t)}{E[X]} \tag{5.21}
$$

where $PDF_X(t)$ is the probability distribution of the service time X, which is the integral of the pdf $g(t)$.

This residual life pdf can be useful for instance when studying the handover process from one cell to another in a cellular system where the time spent by a user in subsequent cells can be represented by a renewal process. The residual lifetime is the remaining time spent in cell where the user starts or receives a new call.

[3] Let us recall that a renewal process is an arrival process in which the interarrival intervals are positive iid random variables.

Then, we can prove the following result:

$$E[R] = \int_0^{+\infty} t f_R(t)\,\mathrm{d}t = \frac{1}{E[X]} \int_0^{+\infty} t\left[1 - PDF_X(t)\right]\mathrm{d}t = \frac{E\left[X^2\right]}{2E[X]} \tag{5.22}$$

where the integral can be solved by applying the rule of the integration by parts.

In conclusion, substituting (5.23) in (5.20), we obtain the following result that is consistent with the mean waiting time contribution in (5.18):

$$E[W] = \frac{\rho}{1-\rho}\frac{E\left[X^2\right]}{2E[X]} = \frac{\lambda E\left[X^2\right]}{2(1-\rho)} \quad [s] \tag{5.23}$$

To compute the (unconditional) mean residual service time $E[R]$, we can also consider an alternative approach to what has been shown before [7]. We consider the process $\{R(t), t \geq 0\}$ where $R(t)$ is the residual service time of the customer in service at time t. We refer to a very long time interval $[0,\ T]$ (we will take after the limit for $T \to +\infty$). We have

$$E[R] = \frac{1}{T}\int_0^T R(t)\,\mathrm{d}t \tag{5.24}$$

Let us see how can we compute this integral as sum of rectangles. Let $a(T)$ be the number of service completions by time T and X_i the ith service time. Notice that the function $R(t)$ takes the value zero during times when there is no customer in service and jumps to the value of X_i at the point of time the ith service time commences. During a service time it linearly decreases with rate of one and reaches zero at the end of a service time. Therefore, the area under the curve $R(t)$ is equal to the sum of the areas of $a(T)$ isosceles right triangles where the side of the ith triangle is X_i. Therefore, for large T, we can ignore the last possibly incomplete triangle, so we obtain

$$E[R] = E[R] = \frac{1}{T}\sum_{i=1}^{a(T)} \frac{1}{2}X_i^2 = \frac{1}{2}\frac{a(T)}{T}\frac{1}{a(T)}\sum_{i=1}^{a(T)} X_i^2 \tag{5.25}$$

Taking the limit of the previous $E[R]$ expression as T goes to $+\infty$, we have that $\frac{a(T)}{T} \to \frac{1}{E[X]}$ because counter $a(T)$ is related to departures (a similar counter is used in Sect. 4.5.1 to deal with the number of arrivals in T so that in that case it is related to the mean interarrival time). Moreover, $\frac{1}{a(T)}\sum_{i=1}^{a(T)} X_i^2 \to E\left[X^2\right]$. Then, we get the following result:

$$E[R] = \frac{E\left[X^2\right]}{2E[X]} \tag{5.26}$$

This is the same $E[R]$ we have achieved before thus proving the validity of this alternative approach. A further proof of this result can be found on [6] referring to the renewal reward processes.

5.1.2 Some Special Cases for the Inversion of the M/G/1 PGF

We have solved the M/G/1 system in the z transform domain in terms of the PGF $P(z)$ of the state probability distribution (5.9). If we need to obtain the state probability distribution P_n, we have to invert $P(z)$. In general, it is quite complex to invert the $P(z)$ to determine P_n and, apart from some special cases that are discussed below, numerical methods are needed.

Let us refer to Poisson arrivals with mean rate λ, imbedding at the service completion instants. We use (5.10) and (5.11) so that we have

$$P(z) = P_0 \frac{(z-1)\Gamma(\lambda - \lambda z)}{z - \Gamma(\lambda - \lambda z)} \tag{5.27}$$

Suppose the Laplace transform $\Gamma(s)$ of the pdf $g(t)$ of the service time is a rational function (i.e., the ratio of polynomials) in s. In that case, the right-hand side of (5.27) can be decomposed into *partial fractions* to obtain the inverse transform easily. This happens when the service time has a distribution of the following types (Chap. 3): exponential, Erlang-n, and hyper-exponential. The Laplace transforms of the corresponding pdfs are:

- Exponential random variable: $\Gamma(s) = \frac{\mu}{\mu+s}$
- Erlang-n random variable (sum of n exponential iid random variables so that the pdf is the convolution of the exponential pdf n times with itself): $\Gamma(s) = \left(\frac{\mu}{\mu+s}\right)^n$
- Hyper-exponential random variable (random combination of exponential random variables): $\Gamma(s) = \sum_{i=1}^{n} p_i \frac{\mu_i}{\mu_i+s}$, where $\sum_{i=1}^{n} p_i = 1$

For all the other cases, numerical inversion methods are needed; an example of a numerical inversion method to determine the pdf of the delay is shown in Sect. 5.3.

Let us show this inversion method referring to the second case with Erlang-n distributed random variable and $n = 2$.

$$P(z) = P_0 \frac{(z-1)\left(\frac{\mu}{\mu+\lambda-\lambda z}\right)^2}{z - \left(\frac{\mu}{\mu+\lambda-\lambda z}\right)^2} \tag{5.28}$$

where $P_0 = 1 - \rho$ and $\rho = \lambda/\mu$.

Through some algebraic manipulations, we get

$$P(z) = \frac{(1-\rho)(z-1)\frac{4}{\rho^2}}{z^3 - \frac{4}{\rho}\left(1+\frac{\rho}{2}\right)z^2 + \left(1+\frac{\rho}{2}\right)^2 \frac{4}{\rho^2} z - \frac{4}{\rho^2}} = \frac{(1-\rho)(z-1)\frac{4}{\rho^2}}{(z-1)(z-z_1)(z-z_2)} \tag{5.29}$$

where we have expanded the denominator using its poles in 1, z_1, and z_2. The first pole in $z = 1$ is canceled by the corresponding zero of the numerator (in the M/G/1 $P(z)$ formula, the singularity in $z = 1$ is always removable). The last two poles z_1 and z_2 have to be determined. We can directly solve the 3rd-degree equation of the denominator's zeros in (5.29) or by following the method below that takes advantage on the fact that we already know that a solution is in $z = 1$.

To factorize the denominator as shown in (5.29), let us consider the following intermediate equality:

$$z^3 - \frac{4}{\rho}\left(1+\frac{\rho}{2}\right)z^2 + \left(1+\frac{\rho}{2}\right)^2 \frac{4}{\rho^2} z - \frac{4}{\rho^2} = (z-1)\left(z^2 - bz + c\right) \tag{5.30}$$

Let us expand the right side as follows:

$$(z-1)\left(z^2 - bz + c\right) = z^3 - (1+b)z^2 + (b+c)z - c \tag{5.31}$$

Then, we equate the left-side term in (5.30) with the right-side term in (5.31), and applying the identity principle of polynomials, we can set the following linear system of equations to determine the values of b and c (note that we omit one redundant equation) as:

$$\begin{cases} \frac{4}{\rho}\left(1+\frac{\rho}{2}\right) = 1 + b \\ \frac{4}{\rho^2} = c \end{cases} \Rightarrow \begin{cases} b = \frac{4+\rho}{\rho} \\ c = \frac{4}{\rho^2} \end{cases} \tag{5.32}$$

Then, z_1 and z_2 are the solutions of the following second-order equation:

$$\left(z^2 - bz + c\right) = 0 \quad \Rightarrow z^2 - \left(\frac{4+\rho}{\rho}\right) z + \frac{4}{\rho^2} = 0 \tag{5.33}$$

We obtain the following expressions for z_1 and z_2, solving this equation:

$$z_1 = \frac{4+\rho+\sqrt{\rho(\rho+8)}}{2\rho}, \quad \text{and} \quad z_2 = \frac{4+\rho-\sqrt{\rho(\rho+8)}}{2\rho} \tag{5.34}$$

We can verify that both z_1 and z_2 are outside the unit disk in the complex plane, that is: $|z_1| > 1$ and $|z_2| > 1$. This condition is needed to verify the feasibility of $P(z)$.

The expression in (5.29), where we canceled the pole and the zero in 1 and we expressed the denominator with the known z_1 and z_2 values, can be expanded in partial fractions to facilitate its inversion as

$$P(z) = (1-\rho)\frac{4}{\rho^2}\left\{\frac{1}{(z-z_1)(z-z_2)}\right\} \tag{5.35}$$

$$= (1-\rho)\frac{4}{\rho^2}\left\{\frac{A}{(z-z_1)}+\frac{B}{(z-z_2)}\right\} \tag{5.36}$$

where coefficients A and B can be determined as

$$A = \left.\frac{1}{(z-z_2)}\right|_{z=z_1} = \frac{\rho}{\sqrt{\rho(\rho+8)}}, \quad B = \left.\frac{1}{(z-z_1)}\right|_{z=z_2} = -A \tag{5.37}$$

Let us express $P(z)$ in (5.36) as follows, further elaborating the notation with partial fractions as

$$P(z) = (1-\rho)\frac{4}{\rho^2}A\left\{-\frac{\frac{1}{z_1}}{\left(1-\frac{z}{z_1}\right)}+\frac{\frac{1}{z_2}}{\left(1-\frac{z}{z_2}\right)}\right\} \tag{5.38}$$

It is relatively easy to invert $P(z)$ using this last expression because the terms $\frac{1}{\left(1-\frac{z}{a}\right)}$ (apart from a normalization coefficient that we do not consider here) can be anti-transformed as $\left(\frac{1}{a}\right)^n$. Then, applying this rule to (5.38), we achieve the following state probability distribution that is the combination of two geometric terms:

$$P_n = \frac{4(1-\rho)}{\rho\sqrt{\rho(\rho+8)}}\left\{\frac{1}{z_2}\left(\frac{1}{z_2}\right)^n-\frac{1}{z_1}\left(\frac{1}{z_1}\right)^n\right\} \tag{5.39}$$

5.1.3 The M/D/1 Case

In this system, the requests have a fixed, constant service time, τ. For instance, this is the case of the transmission of packets of a given size on a link with constant capacity. Note the change of notation since we will use τ to represent fixed service times from here on (previously, τ was used to represent the age of the currently served request). Therefore, the pdf of the service time becomes $g(t)=\delta(t-\tau)$, where $\delta(\cdot)$ denotes the Dirac Delta function. The corresponding Laplace transform is $\Gamma(s)=e^{-\tau s}$. By using (5.11), we have $A(z)=\Gamma(s)|_{s=-\lambda(z-1)}=e^{\tau\lambda(z-1)}$. Note that $\lambda\tau$ is the intensity of the input traffic in Erlangs. By substituting this expression of $A(z)$ in (5.10), we obtain $P(z)$ with imbedding points at the service completion instants as

$$P(z) = (1-\lambda\tau)\frac{(z-1)\,e^{\lambda\tau(z-1)}}{z-e^{\lambda\tau(z-1)}} \tag{5.40}$$

Note that the PGF of an M/D/1 queue in (5.40) cannot be anti-transformed in a closed form so that a numerical method (as discussed in Sect. 3.3.1) is needed to obtain the state probability distribution.

Finally, the mean number of requests in the queue N can be expressed according to (5.17) as

$$N = \lambda\tau + \frac{\lambda^2\tau^2}{2[1-\lambda\tau]} = \frac{\lambda\tau}{1-\lambda\tau} - \frac{\lambda^2\tau^2}{2[1-\lambda\tau]} \tag{5.41}$$

Hence, in the second expression of N in (5.41), there is a contribution corresponding to an M/M/1 queue (with the same mean arrival rate and the same mean service time) minus a positive term. Hence, the congestion of an M/D/1 queue is lower than that of the corresponding M/M/1 queue. The same relation holds for the mean system delay given by (5.18). This is consistent with the fact that the exponential distribution has a mean square value two times larger than that of a deterministic distribution for the same mean service time.

5.2 M/G/1 System Delay Distribution in the FIFO Case

This section provides an extension of the study made in Sect. 4.14.1 to the case of general service times. As long as possible, we keep the same notations as those used in Sect. 4.14.1. Let us refer to a queue with FIFO discipline, Poisson arrivals, general service time, and system imbedded at service completion instants. The n requests left in the system at the service completion instant are those arrived during the system delay T_D experienced by the request just served; see Fig. 5.2.

The probability distribution for the n requests in the system at the service completion instants coincides with the state probability distribution with PGF $P(z)$ in (5.10). This PGF of a random variable n can also be obtained referring to the fact that these n requests are the arrivals at the system during the system delay T_D, with the corresponding pdf $f_D(t)$ (note that $f_D(t)$ is the unknown distribution that we need to characterize). Let us first condition our study on a given system delay $T_D = t$ so that the PGF of the number of Poisson arrivals in this interval is $P(z|T_D = t) = e^{\lambda t(z-1)}$. Then, we remove the conditioning using the following integral with the pdf $f_D(t)$ as:

$$P(z) = \int_0^{+\infty} e^{\lambda t(z-1)} f_D(t) dt = T_D\left[s = -\lambda(z-1)\right] \tag{5.42}$$

where $T_D(s)$ is the Laplace transform of the pdf $f_D(t)$.

Note that formula (5.42), i.e., $P(z) = T_D[s = -\lambda(z-1)]$, is a sort of "generalization" of the Little theorem in the FIFO case. If we take the derivatives of both sides with respect to z and we evaluate them in $z = 1$, we obtain the mean number of requests in the system N as a function of the mean arrival rate λ and the mean system delay $T = E[T_D]$:

$$\begin{aligned} N = P'(z)\big|_{z=1} &= \frac{d}{dz} T_D\left[s = -\lambda(z-1)\right]\bigg|_{z=1} \\ &= -\lambda T_D'\left[s = -\lambda(z-1)\right]\big|_{z=1} = \lambda E\left[T_D\right] = \lambda T \end{aligned}$$

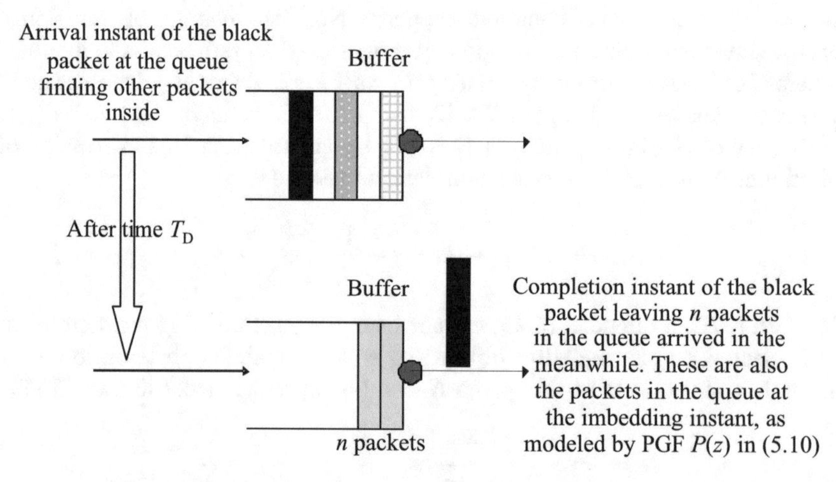

Fig. 5.2: Relation between random variable T_D of the queuing delay and the PGF $P(z)$ of the number of requests n in the queue at imbedding instants

The generalization is in the sense that (5.42) permits to relate the moments of the number of requests in the system and the moments of the system delay.

It is important to stress that we have adopted the transform $s = -\lambda(z-1)$ from the Laplace "s"-domain to the PGF "z"-domain in both (5.11) and (5.42). Vice versa, from the z-domain to the s-domain, we use the inverse transform $z = 1 - s/\lambda$. In particular, since we know the expression of $P(z)$ from (5.10), we can use (5.42), and the inverse transform to obtain the Laplace transform of $f_{\mathrm{D}}(t)$, where $A(z)|_{z=1-s/\lambda}$ is replaced by $\Gamma(s)$ according to (5.11):

$$T_{\mathrm{D}}(s) = P(z)|_{z=1-s/\lambda} = P_0 \frac{(z-1)\,A(z)}{z - A(z)}\bigg|_{z=1-s/\lambda} = P_0 \frac{s\Gamma(s)}{s - \lambda + \lambda\Gamma(s)} \tag{5.43}$$

The system delay T_{D} is the sum of two independent contributions: the service time (with Laplace transform of the pdf denoted as $\Gamma(s)$) and the waiting time (with Laplace transform of the pdf denoted as $T_{\mathrm{W}}(s)$). Hence, the Laplace transform of the pdf of the waiting time can be obtained from (5.43) as

$$T_{\mathrm{W}}(s) = \frac{T_{\mathrm{D}}(s)}{\Gamma(s)} = P_0 \frac{s}{s - \lambda + \lambda\Gamma(s)} \tag{5.44}$$

For bulk (compound) Poisson arrivals with PGF $M(z)$ of the length of each group, formula (5.42) can be generalized as

$$P(z) = T_{\mathrm{D}}\left[s = -\lambda\left(M(z) - 1\right)\right] \tag{5.45}$$

Then, the inverse transform $z = z(s)$ now becomes

$$z = M^{-1}\left(1 - \frac{s}{\lambda}\right) \tag{5.46}$$

assuming that $M(z)$ is an invertible function.

In conclusion, we can determine the Laplace transform of the pdf of the system delay, $T_{\mathrm{D}}(s)$, as

$$T_D\left(s\right) = P\left[z = M^{-1}\left(1 - \frac{s}{\lambda}\right)\right] \tag{5.47}$$

Note that the above approach in (5.45) and (5.46) can also be applied to the M/M/1 case with compound arrivals (i.e., $\mathrm{M}^{[\mathrm{comp}]}/\mathrm{M}/1$), thus further extending the study made in Sect. 4.14.1.

5.3 Numerical Inversion Method of the Laplace Transform of the Delay

This section provides a numerical method to invert the Laplace transform $\Pi(s)$ of the pdf $\pi(t)$ of a certain random variable in time; in particular, we know $\Pi(s)$ and we would like to characterize $\pi(t)$. This chapter will be particularly useful when $\Pi(s)$ has a complex expression, which does not allow the inversion in terms of elementary functions. This is for instance the typical case that happens when we need to invert the Laplace transform $T_{\mathrm{D}}(s)$ in (5.43) to obtain $f_{\mathrm{D}}(t)$.

Let us focus on the inversion of $\Pi(s)$ to obtain $\pi(t)$. We start by changing the variable from the Laplace "s"-domain to the frequency "f" one so that Laplace transforms become Fourier transforms: we use $s = j2\pi f$, where j is the imaginary unit ($j^2 = -1$). Then, we take the samples in the frequency domain $\Pi(s = j2\pi f_n)$ with interval f_{c} (see below) and we apply an inverse Fourier transform algorithm, by considering scaled samples in the frequency domain by $1/T_{\mathrm{c}}$, where T_{c} denotes the sampling interval in the time domain. Matlab® supports the efficient Inverse Fast Fourier Transform (IFFT) algorithm.

We make the approximation that the Fourier components $\Pi(s = j2\pi f)$ are negligible for $f > f_{\max}$. We determine $f_{\max}$ so that $T_{\mathrm{c}} = 1/(2f_{\max})$ is the time granularity (sampling interval) that we consider for the pdf $\pi(t)$. Moreover, the number of samples, N_{s}, is determined by adopting the approximation that the pdf $\pi(t)$ is equal to zero for $t > N_{\mathrm{s}}T_{\mathrm{c}}$. Since the pdf $\pi(t)$ is unknown, we will use suitable values for $N_{\mathrm{s}}T_{\mathrm{c}}$ and we will verify *a posteriori* whether the pdf obtained has negligible values for $t > N_s T_c$. Finally, knowing $f_{\max}$ and N_s we can determine f_c based on the relation $N_s f_c = 2f_{\max}$.

This method needs that the values of parameters T_c, $f_{\max}$, N_s, and f_c be determined. According to the above, we adopt the following method:

1. We use the relation $T_c = 1/(2f_{\max})$ and we choose a T_c value to obtain the corresponding $f_{\max}$ value so that $\Pi(s = j2\pi f)$ is negligible for $f > f_{\max}$. Otherwise, we can directly choose the value of $f_{\max}$ so that $\Pi(s = j2\pi f)$ is negligible for $f > f_{\max}$ and determine the corresponding T_c value.
2. Knowing $f_{\max}$ we can use the following relation: $N_s f_c = 2f_{\max}$. Hence, we can choose the N_s value (possibly, a power of 2 to use the IFFT algorithm) to obtain the corresponding f_c value.
3. We apply the IFFT algorithm to the N_s samples $\Pi(s = j2\pi f_n)$, where $f_n = n \times f_c$, for $n = 0, \ldots, N_s - 1$. This allows us to determine the N_s samples $\pi(t = t_n)$, where $t_n = n \times T_c$, for $n = 0, \ldots, N_s - 1$.
4. Finally, we verify *a posteriori* that the obtained samples of the pdf $\pi(t)$ tend to zero for t approaching the $N_s T_c$ value.

We adopt the above approach to invert the Laplace transform $T_D(s)$ in (5.43) to obtain the pdf $f_D(t)$ of the delay of an M/G/1 FIFO queue. We consider the following example referring to a buffer for the transmission on a link. Transmission time is slotted. Each slot is used to transmit one packet. Packets arrive in groups, named messages. The message arrival process is Poisson. We apply the previous method to determine the message delay's pdf, $f_D(t)$. In particular, we compare the pdf of the message delay in three cases with different message length distributions. In particular, the queuing system is characterized as

- The traffic offered to the buffer is generated by M_d independent terminals.
- Each terminal generates message arrivals according to a Poisson process with mean rate λ.
- Messages have a random length X in packets according to the three different distributions below (in all these cases, the mean message length is 6 pkts):

  Deterministic distribution:

$$\text{Prob}\{X = k\} = \begin{cases} 1, & \text{for } k = 6 \\ 0, & \text{otherwise} \end{cases}$$

  "Modified" geometric distribution:

$$\text{Prob}\{X = k\} = (1 - q)\, q^{k-1}, \quad \text{with } q = 5/6, \quad k = 1, 2, 3, \ldots$$

  "Truncated Pareto" distribution [8]:

$$\text{Prob}\{X = k\} = \begin{cases} 1 - \left(\dfrac{h}{kL_p + 1}\right)^{v}, & k = L_{w,\min} \\ \left(\frac{h}{(k-1)L_p+1}\right)^{v} - \left(\frac{h}{kL_p+1}\right)^{v}, & L_{w,\min} < k < L_{w,\max} \\ \left(\frac{h}{(k-1)L_p+1}\right)^{v}, & k = L_{w,\max} \end{cases}$$

  where $L_{w,\min} = \lceil h/L_p \rceil$, $L_{w,\max} = \lceil m/L_p \rceil$, and symbol $\lceil . \rceil$ denotes the *ceiling function*. The selected numeric values are: $v = 1.565$, $m = 5{,}000$ bytes, $h = 50$ bytes, and $L_p = 30$ bytes.

  The mean square values are equal to 36 pkts^2, 66 pkts^2, 88.91 pkts^2, respectively, for deterministic, geometric, and truncated-Pareto cases.

- Each packet requires a time T_s to be transmitted.
- Output transmissions are time-slotted with slot duration equal to T_s.
- Messages are served according to a FIFO policy.

We study the above M/D/1 system by imbedding the chain at the message transmission completion instants. Hence, the PGF $P(z)$ of the number of messages in the system is given by (5.10), where $A(z) = L[A_s(z)]$, $L(z)$ is the PGF of the message length in packets, and $A_s(z)$ denotes the PGF of the number of message arrivals in a time slot T_s (= packet transmission time): $A_s(z) = e^{\lambda M_d T_s (z-1)}$. We substitute $z = 1 - s/\lambda$ in (5.10), so that textitP$(z = 1 - s/\lambda) = T_D(s)$ yields the Laplace transform of the pdf of message delay $f_D(t)$. In applying the above Laplace transform inversion method, we select T_c to be coincident with the minimum possible message

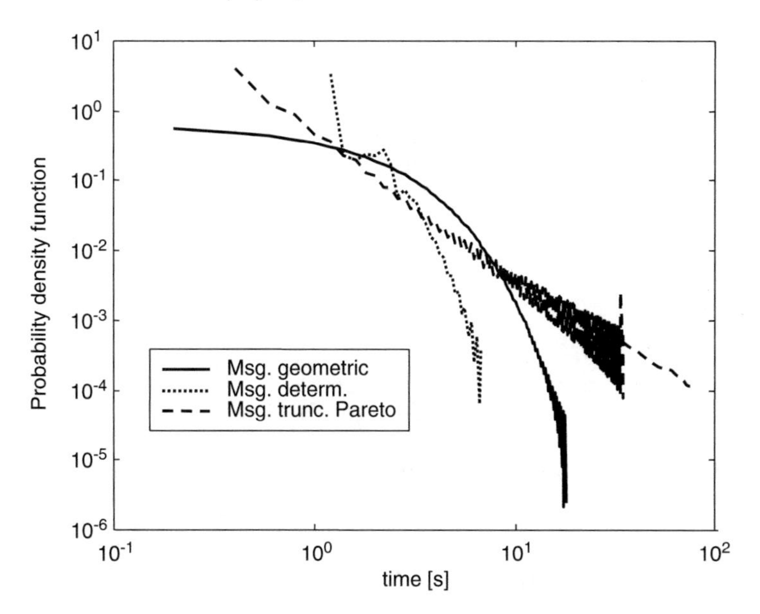

Fig. 5.3: Graph of the pdfs of the M/G/1 delay $f_D(t)$ obtained using the proposed numerical inversion method for different message length distributions

delay, that is T_s, considering that the minimum message length is one packet (this is true only in the geometric case; the other cases have larger values of the minimum message length).

We make the following numerical assumptions: $M_d = 6$ terminals, $\lambda = 0.05$ msgs/s/terminal, $T_c \equiv T_s = 0.2$ s. Since the mean message length is equal to 6 pkts/message, the total traffic intensity offered to the transmission buffer (queue) is 0.36 Erlangs.

On the basis of the previous point No. 1, we use $f_{max} = 1/(2T_s) = 2.5$ Hz. Then, according to the above point No. 2, we select $N_s = 4{,}096$ so that the frequency sampling interval becomes $f_c = 2f_{max}/N_s \approx 0.0012$ Hz. This method has been implemented in Matlab® by means of its IFFT algorithm (see the previous point No. 3). The obtained pdfs for the message delay in the three different cases are shown in Fig. 5.3. We can note that the pdfs have negligible values for times greater than 10 s ($<< N_s T_c$, according to point No. 4). Moreover, the truncated Pareto case, having the greatest mean square value of the message length, entails the heaviest tail in the pdf of the message delay, $f_D(t)$.

5.4 Impact of the Service Time Distribution on M/G/1 Queue

The mean number of requests in an M/G/1 queue depends on mean and mean square values $\{E[X]$ and $E[X^2]\}$ of the message service time. $E[X^2]$ has an impact on the waiting part, as shown in (5.17). We consider the coefficient of variation C_v of the M/G/1 queue's service time distribution, referring to the definition in (3.51) and the comments in Sect. 3.2.6.3; the definition of C_v^2 is provided below:

$$C_v^2 = \frac{\mathrm{Var}[X]}{E^2[X]} = \frac{E[X^2]}{E^2[X]} - 1 \Rightarrow E[X^2] = E^2[X](C_v^2 + 1) \tag{5.48}$$

Let us recall that C_v is equal to zero 0 for a deterministic service time, is below 1 for Erlang-n distributions, is 1 for exponential distributions, is greater than 1 for hyper-exponential distributions, and is $+\infty$ for heavy-tailed distributions (e.g., case of the Pareto random variable, as shown in the next section).

We compare the mean number of requests in our M/G/1 queue with those in an M/M/1 queue, having the same traffic intensity ρ. Let λ denote the mean arrival rate and let $E[X]$ denote the mean service time: $\rho = \lambda E[X] < 1$ Erlang. Let us recall that the mean number of requests in the M/M/1 queue, $N_{M/M/1}$, is given by (4.23) as

$$N_{\mathrm{M/M/1}} = \frac{\lambda E[X]}{1-\lambda E[X]} \tag{5.49}$$

Moreover, the mean number of requests in our M/G/1 queue in (5.17) can be rewritten as follows using the coefficient of variation C_v of the service time distribution:

$$N_{\mathrm{M/G/1}} = \lambda E[X] + \frac{\lambda^2 E\left[X^2\right]}{2\left[1-\lambda E[X]\right]} = N_{\mathrm{M/M/1}}\left[1+\rho\left(\frac{C_\mathrm{v}^2-1}{2}\right)\right] \tag{5.50}$$

Based on (5.50), we have that

$$\begin{aligned} N_{\mathrm{M/M/1}} &< N_{\mathrm{M/G/1}}, \text{ if } C_\mathrm{v} > 1 \\ N_{\mathrm{M/M/1}} &> N_{\mathrm{M/G/1}}, \text{ if } C_\mathrm{v} < 1 \end{aligned} \tag{5.51}$$

An interesting case for our M/G/1 queue is represented by the Weibull-distributed service time. This distribution depends on two parameters $\beta > 0$ (*scale parameter*) and $k > 0$ (*shape parameter*), as expressed below:

$$f_{\beta,k}(t) = \frac{k}{\beta}\left(\frac{t}{\beta}\right)^{k-1} e^{-\left(\frac{t}{\beta}\right)^k}, \quad t \geq 0 \tag{5.52}$$

Depending on the values of the two parameters β and k, the Weibull distribution can represent a family of distributions with different C_v values. In particular, the distribution becomes exponential for $k=1$. Moreover, we have a Rayleigh distribution for $k=2$. Finally, the distribution becomes deterministic for $k \to \infty$. The moments of the Weibull distribution can be expressed as a function of the Gamma $\Gamma(\cdot)$, as defined in (3.197) for $y=0$; in particular, the square value of the coefficient of variation can be determined as

$$C_\mathrm{v}^2(k) + 1 = \frac{E\left[X^2\right]}{\{E[X]\}^2} = \frac{\Gamma\left(1+\frac{2}{k}\right)}{\Gamma^2\left(1+\frac{1}{k}\right)} \tag{5.53}$$

Correspondingly, we note that the coefficient of variation C_v varies from $+\infty$ to 0 as k goes from 0 to $+\infty$. The traffic intensity ρ can be expressed as

$$\rho(\lambda, \beta, k) = \lambda E[X] = \lambda\beta\Gamma\left(1+\frac{1}{k}\right) \tag{5.54}$$

The graph in Fig. 5.4 of the mean number of requests N in the M/G/1 queue for Weibull-distributed service times has been obtained as a function of the traffic intensity ρ for different ${C_\mathrm{v}}^2$ (i.e., k) values and a fixed mean service time $E[X]$. We can note that N increases significantly with ${C_\mathrm{v}}^2$ for a given ρ value.

Note that Weibull and Pareto distributions are difficult to use for the service time of an M/G/1 queue because their Laplace transforms cannot be expressed in closed forms. For instance, the Laplace transform of a Pareto pdf can be defined in terms of the incomplete Gamma function, as shown in Sect. 3.3.3.2

5.4.1 The Case of the M/P/1 Queue

According to data traffic measurements, many random variables associated with Internet traffic are heavy-tailed [9]. For example, the file length at the application layer follows a Pareto (heavy-tailed) distribution. At the session layer, session durations also have a Pareto distribution. At the network layer, the packet interarrival times are according to heavy-tailed distributions. These phenomena with heavy tails are responsible for introducing self-similarity (fractal property) in the traffic, meaning that the traffic profile (bit-rate as a function of time) is the same at different time scales of aggregation. While non-fractal traffic tends to smooth for progressive time scale aggregation, this is not the case for self-similar traffic. This is a common situation in today's network and causes congestions that are difficult to manage (long-range dependence).

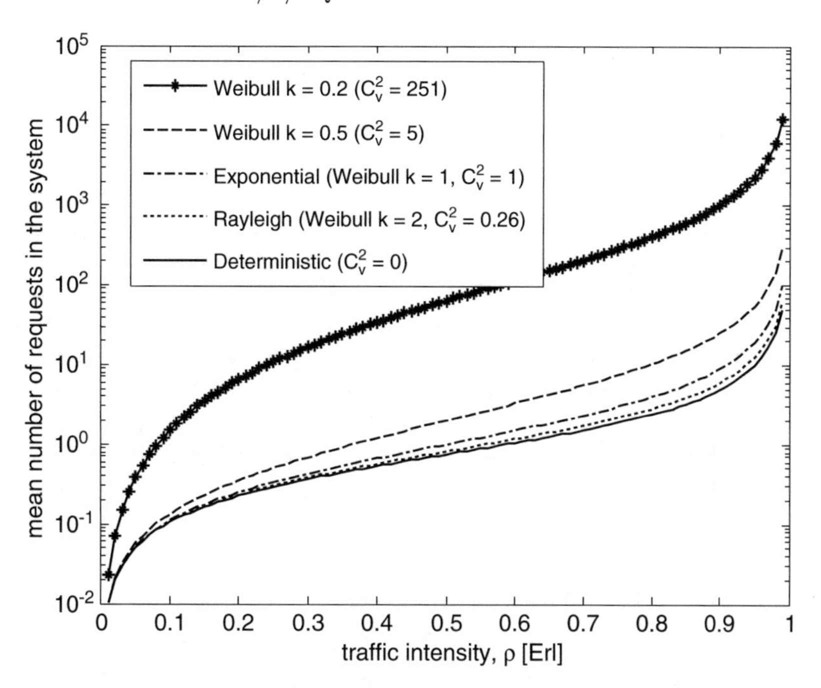

Fig. 5.4: Mean number of requests in the M/G/1 queue as a function of the traffic intensity ρ for different $C_v{}^2$ values of the service time distribution of the Weibull type

We refer in this sub-section to the case of service time with heavy-tailed Pareto distributions. In common random variables, there is a very low probability of extremely high values. Instead, heavy-tailed random variables have a non-trivial probability of being extremely large.

Let us recall that the Pareto probability density function is defined as

$$f_X(x) = \frac{\gamma k^\gamma}{x^{\gamma+1}}, \quad x \geq k \tag{5.55}$$

where k and γ are real positive terms.

The existence of the moments of the Pareto random variable depends on the value of the shape parameter γ. In particular, the variance is infinite and the mean value is finite for $1 < \gamma \leq 2$; both variance and mean values are infinite for $\gamma \leq 1$. The Pareto distribution is heavy-tailed if $\gamma \leq 2$, as shown in Sect. 3.2.6.7.

A very special case in the M/G/1 queue study is when *the service time has a heavy-tailed distribution where the variance does not exist* (i.e., they are infinite). In particular, let us refer to the Pareto case with $1 < \gamma \leq 2$ so that the mean value is finite and the variance (as well as the mean square value) is infinite so that the coefficient of variation C_v is infinite. The service time has a heavy-tailed distribution. Let us denote this queuing system as M/P/1 queue, where "P" stands for "Pareto." Let us recall that the Pollaczek–Khinchin formula (5.18) can be applied to determine the mean delay of an M/G/1 queue if the mean service time is finite due to the stability condition. This entails a sort of *paradox*: the M/P/1 queue is stable (there exists the state probability distribution, as well as the distribution of the delay, and the probability of an empty queue is $P_0 > 0$ [10]), but its mean delay is infinite because the mean square value of the service time is infinity (actually, it does not exist); this is a very special case where an infinite mean delay does not imply the queue instability!

A difficulty in analyzing these queues is that heavy-tailed distributions do not generally have closed-form Laplace transforms so that the PGF of the state probability distribution can be determined only using numerical methods. For Poisson arrivals, the PGF of the number of arrivals in the Pareto service time can be expressed by means of the incomplete Gamma function, $\Gamma(a, y)$, as follows:

$$A(z) = \sum_{n=0}^{\infty} z^n \int_k^{+\infty} \frac{(\lambda t)^n}{n!} e^{-\lambda t} \frac{\gamma k^\gamma}{t^{\gamma+1}} \mathrm{d}t = \gamma k^\gamma \int_k^{+\infty} e^{-\gamma t(1-z)} t^{-\gamma-1} \mathrm{d}t = \tag{5.56}$$
$$= \gamma \left[\lambda k (1-z)\right]^\gamma \Gamma\left[-\gamma, \lambda k (1-z)\right]$$

where

$$\Gamma(a, y) = \int_{y}^{+\infty} e^{-t} t^{a-1} \mathrm{d}t \tag{5.57}$$

Moreover, the delay distribution in the FIFO case can be determined by approximating the Laplace transform of the service time and then inverting numerically the resulting Laplace transform in (5.43), as shown in Sect. 5.3. However, in these queuing systems with heavy-tailed distributions of the service time, the interest is not on the study of the mean delay, but instead on the queue overflow probability with finite rooms in the queue.

There are also problems in simulating M/P/1 queues as γ approaches 2 from above; the simulation can be extremely slow to approach the regime condition (i.e., we need very long simulation runs and many repetitions to achieve reliable results). Practically, M/P/1 queues (with $1 < \gamma \leq 2$ or even with γ close to 2 from above) are studied by truncating the Pareto distribution, as considered in Sect. 5.3 [8], thus avoiding the infinite variance caused by the distribution tail.

5.5 M/G/1 Theory with State-Dependent Arrival Process

To solve the M/G/1 queue at imbedding instants, we have written the difference equation (5.1):

$$n_{i+1} = n_i - I(n_i) + a_{i+1}$$

Then, we have expressed the PGF of the number of requests in the system, $P(z)$, under the assumption that a_{i+1} and n_i are independent. If this is not the case (i.e., a_{i+1} depends on n_i), the difference equation (5.1) can be solved by considering that it represents a discrete-time Markov chain with state-dependent transitions, as described in Fig. 5.5: each transition corresponds to a_n arrivals during the imbedding interval, where subscript n is not connected to the time evolution of the system (we are at regime), but to the originating state n. We refer here to the case of imbedding at the end of the service completion instants (see Fig. 5.5). Prob$\{a_n = j\}$ denotes the transition probability due to j arrivals in the state n. Transition probabilities must satisfy the following condition:

$$\sum_{j=0}^{\infty} \text{Prob}\{a_n = j\} = 1, \quad \text{for } \forall n \tag{5.58}$$

Note that we can have infinite possible transitions leaving a state.

In general, the derivation of the state probability distribution P_n of the discrete-time Markov chain requires either a matrix geometric approach or cut equilibrium conditions solved progressively (recursions). Let us refer to

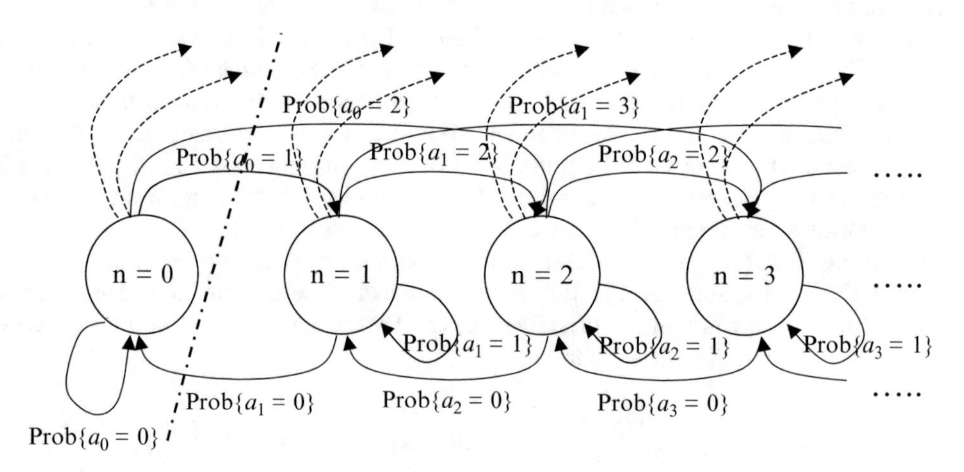

Fig. 5.5: State diagram of a “general” M/G/1 queue at imbedding instants (discrete-time system)

this second case; for example, we can study the equilibrium for the cut shown in Fig. 5.5:

$$P_0\left[1-\text{Prob}\{a_0=0\}\right]=P_1\text{Prob}\{a_1=0\}\Rightarrow$$
$$P_1=\frac{1-\text{Prob}\{a_0=0\}}{\text{Prob}\{a_1=0\}}P_0 \tag{5.59}$$

Then, infinite cut equilibrium conditions should be used to determine the state probability distribution as a function of P_0, according to a recursive approach (i.e., solving P_1, then P_2, then P_3, etc.). P_0 is determined by imposing the normalization condition.

Another (and more formal) solution approach is to use the general expression in (5.6) that is reproduced below by omitting the subscripts, referring to the regime conditions:

$$\sum_k z^k P_k=\sum_n\sum_a z^{n-I(n)+a}P_{n,a} \tag{5.60}$$

where term a in the exponent of the right term is actually $a=a(n)$.

Now the arrival process depends on the state n. Then, the joint probability $P_{n,a}$ is not the product of the marginal probabilities. We need to use the conditional probability so that we can write $P_{n,a}=P_{a|n}\times P_n$ and use it in the right term of (5.60):

$$\sum_k z^k P_k=\sum_n\sum_a z^{n-I(n)+a}P_{a|n}P_n \tag{5.61}$$

where $P_{a|n}$ is the probability of a arrivals when the chain is in state n; $\mathrm{P}_{a|n}$ characterizes the state-dependent arrival process.

Equation (5.61) can be used to determine the state probabilities P_n. Since we have polynomials in z on both sides of (5.61), we use the identity principle of polynomials: we equate the coefficients of the same z^k terms, $k\geq 0$, appearing on both sides of (5.61). We have

$$P_k=\sum_{n-I(n)+a=k}P_{a|n}P_n \tag{5.62}$$

where the sum on the right side is for all the combinations of $n\geq 0$ and $a\geq 0$ that satisfy the condition $n-I(n)+a=k$. Due to the term $I(n)$, we distinguish the case $n=0$ from the cases $n>0$ in (5.62):

$$P_k=P_{a=k|n=0}P_{n=0}+\sum_{\substack{n-1+a=k\\ n>0}}P_{a|n}P_n$$
$$\Longleftrightarrow P_k=P_{k|0}P_0+\sum_{a=0}^{k}P_{a|k-a+1}P_{k-a+1} \tag{5.63}$$

where the notation of conditional arrival probabilities has been simplified as follows: $P_{a=i|n=j}=P_{i|j}$.

Note that the right term in (5.63) is a linear combination of state probabilities P_0, P_1, ... P_k, P_{k+1}. Hence, we can express P_{k+1} as a function of P_0, P_1, ... P_k as follows:

$$P_k=P_{k|0}P_0+P_{0|k+1}P_{k+1}+\sum_{a=1}^{k}P_{a|k-a+1}P_{k-a+1}\Rightarrow$$
$$P_{k+1}=\frac{P_k-P_{k|0}P_0-\sum_{a=1}^{k}P_{a|k-a+1}P_{k-a+1}}{P_{0|k+1}} \tag{5.64}$$

We have thus obtained a *recursive approach*: given the conditional arrival probabilities and state probabilities P_0, P_1, ... P_k, we can obtain P_{k+1}. Then, the "normalized" state probabilities P_{k+1}/P_0 can be obtained iteratively as follows for k = 0, 1, ...:

$$\frac{P_{k+1}}{P_0} = \frac{\frac{P_k}{P_0} - P_{k|0} - \sum_{a=1}^{k} P_{a|k+1-a} \frac{P_{k+1-a}}{P_0}}{P_{0|k+1}} \quad (5.65)$$

Once the P_{k+1}/P_0 values have been obtained recursively from (5.65), probability P_0 is given by the normalization condition as

$$P_0 = \frac{1}{1 + \sum_{k=0}^{\infty} \frac{P_{k+1}}{P_0}} \quad (5.66)$$

For practical numerical evaluations, we can truncate the state probability distribution for $k > k_0$ so that P_{k+1}/P_0 values are below a given threshold; then, we obtain P_0 from the normalization condition. Note that (5.63) (or (5.64)) computed for $k=0$ yields (5.59), which has been already obtained using a cut equilibrium condition.

Of course, the M/G/1 solutions (5.65) and (5.66) for a state-dependent arrival process are also valid for a state-independent arrival process; it is sufficient to omit the conditioning in probabilities $P_{a|n} = P_a$ (this distribution corresponds to the PGF $A(z)$ used in Sect. 5.1). Therefore, (5.65) and (5.66) also allow us to invert the PGF $P(z)$ of the state probability distribution in (5.10).

5.6 Applications of the M/G/1 Analysis to Slotted-Based Arrivals and Departures

We study a case with slotted-based arrivals and departures for which we can apply an "M"/G/1 model, where "M" stands for memoryless arrival process (not Poisson, but Bernoulli or binomial process). Most of the considerations made here can be applied to Time Division Multiplexing transmission technologies.

We have a multiplexer receiving N_ℓ synchronous input time division flows of traffic from distinct input lines. The arriving packets are stored in a buffer with infinite rooms. There is only one output flow. Input and output lines are synchronized with the same slot duration, τ. One slot allows conveying one packet (i.e., input and output lines have the same speed). This system can be modeled as shown in Fig. 5.6.

We consider that each slot of an input line transports a packet with probability p. This behavior is memoryless from slot to slot. Each input line then contributes a (simple) Bernoulli arrival process of packets on a slot basis. Hence, the number of packets arriving at the multiplexer on a slot basis is given by the sum of N_ℓ independent Bernoulli processes; this is a binomial process with the distribution detailed below:

$$\text{Prob}\{n \text{ packets arrived in a slot}\} = \binom{N_\ell}{n} p^n (1-p)^{N_\ell - n} \quad (5.67)$$

The transmission time of each packet is fixed and equal to T and there is only one output line. This system evolves at discrete-time instants. Hence, the multiplexer can be described according to a ΣBernoulli/D/1 (or Binomial/D/1) queuing model, as analyzed below.

We select the imbedding instants at the end of the slots of the output transmission line, ξ_i. In this case, n_i denotes the number of packets at the end of the ith slot of the output line (instant ξ_i^+) and a_i denotes the number of packets arrived at the buffer during the ith slot (these arrivals complete at instants ξ_i^- because of the synchronization assumption).

We consider that *a packet needs the time of one slot to arrive at the buffer; one packet must arrive completely before being counted in n_i and before being available for transmission* (store-and-forward model). If $n_i \neq 0$, at the $(i+1)$-th imbedding instant, we have $n_{i+1} = n_i - 1 + a_{i+1}$ since one packet has been transmitted from the buffer and a_{i+1} packets have arrived in the $(i+1)$-th slot; see Fig. 5.7a. Instead, if $n_i = 0$ and there are a_{i+1} packet arrivals during the $(i+1)$-th slot, this is also the number of packets in the buffer at the end of the $(i+1)$-th slot: $n_{i+1} = a_{i+1}$;[4] see Fig. 5.7b. In conclusion, we can write the following balance:

[4] At instant ξ_i^+, the queue is empty, $n_i = 0$. Hence, during the $(i+1)$-th slot no packet is transmitted and at the end of this slot (instant $\xi_{i+1}{}^-$) the system contains the new requests a_{i+1} arrived. There is no service completion at instant ξ_{i+1}.

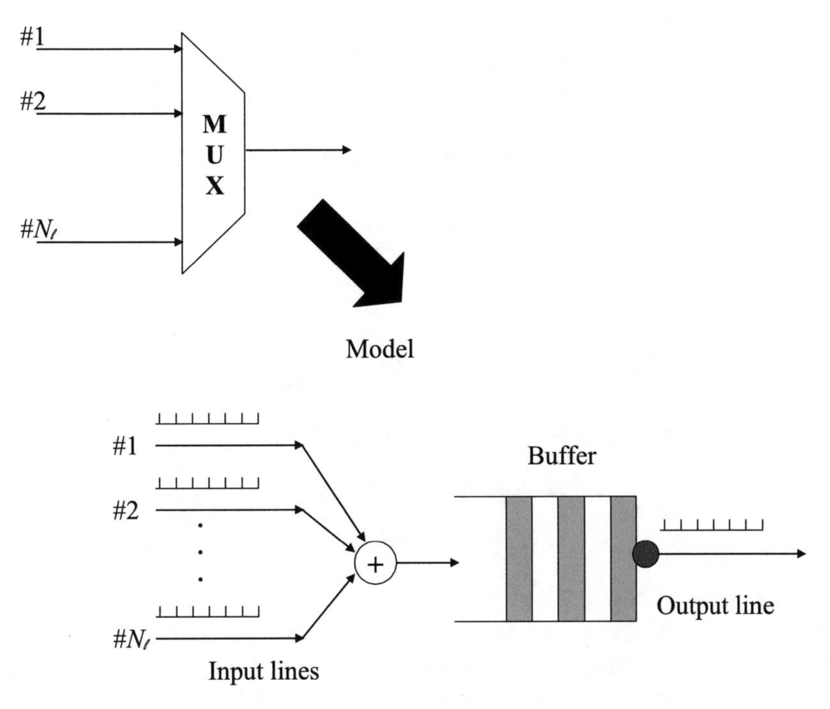

Fig. 5.6: Multiplexer with infinite rooms for packets

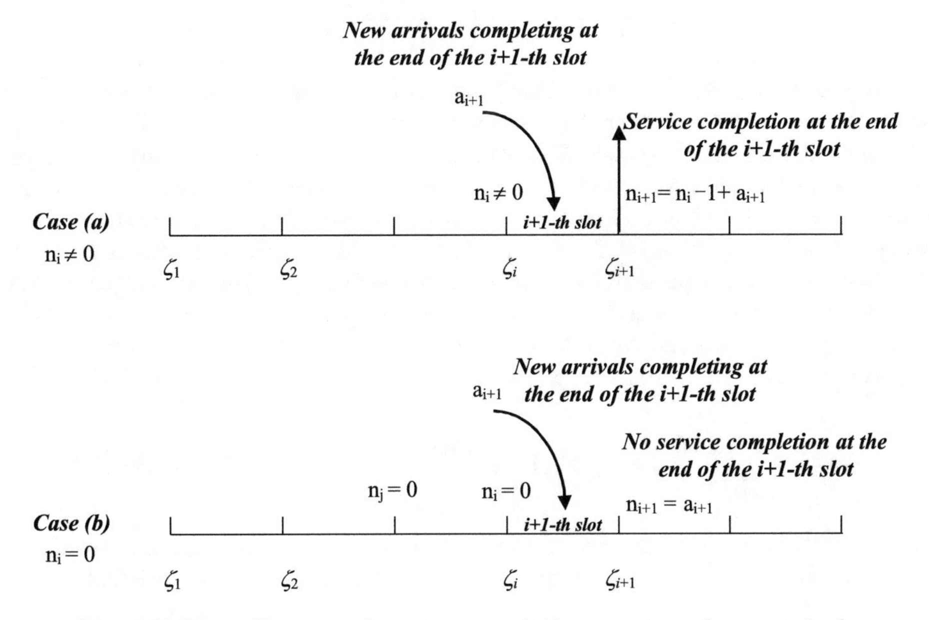

Fig. 5.7: Time diagram of service completion events and new arrivals

$$n_{i+1} = n_i - 1 + a_{i+1} \text{ for } n_i > 0 \text{ and } n_{i+1} = a_{i+1} \text{ for } n_i = 0$$

Interestingly, the above difference equation corresponds to (5.1), which has been derived for a classical M/G/1 queue imbedded at the service completion instants. In this case, we have a memoryless arrival process that is independent of the state (= number of packets in the buffer at imbedding instants). Hence, the PGF of the state probability distribution is given by (5.10), where $A(z)$ represents the PGF of the number of packets arrived in a slot according to the binomial process (5.67):

$$A(z) = \sum_{n=0}^{N_\ell} \binom{N_\ell}{n} z^n p^n (1-p)^{N_\ell - n} = (1 - p + zp)^{N_\ell} \tag{5.68}$$

Finally, the mean number of packets in the multiplexer, N_p, is obtained from (5.16) as

$$N_\mathrm{p} = A'(z=1) + \frac{A''(z=1)}{2[1 - A'(z=1)]} \text{ [packets in the buffer]} \tag{5.69}$$

where $A'(z=1)$ and $A''(z=1)$ are given as

$$\begin{aligned} A'(z)|_{z=1} &= N_\ell (1-p+zp)^{N_\ell - 1} p\Big|_{z=1} = N_\ell p \\ A''(z)|_{z=1} &= \frac{\mathrm{d}}{\mathrm{d}z} N_\ell (1-p+zp)^{N_\ell - 1} p\Big|_{z=1} \\ &= N_\ell p \,(N_\ell - 1)\,(1-p+zp)^{N_\ell - 2} p\Big|_{z=1} \\ &= N_\ell \,(N_\ell - 1)\, p^2 \end{aligned}$$

The queue stability condition is $N_\ell p < 1$ pkts/slot (or Erlang) so that the denominator of N_p is positive and finite. The mean delay T_p experienced by a packet can be obtained from N_p by means of the Little theorem. We need to know the mean rate according to which packets arrive at the buffer; this value is given by $A'(z=1)$, representing the mean number of packets generated in a slot and, hence, the mean packet arrival rate in pkts/slot.[5]

$$T_\mathrm{p} = \frac{N_\mathrm{p}}{A'(z=1)} = 1 + \frac{(N_\ell - 1)\,p}{2\,(1 - N_\ell p)} \text{ [slots]} \tag{5.70}$$

It is interesting to note that if $N_\ell = 1$, the waiting part in (5.70) vanishes: in this case, we have an input line with the same speed as the output line so that messages do not experience any wait for the service.

Note that we cannot use here the Pollaczek–Khinchin formula (5.18) because the arrival process is not Poisson.

This example can be generalized as follows by referring to a compound arrival process on a slot basis. In particular, we consider that each slot of an input line carries with probability p one message (OSI layer 3) formed by a random number of packets (OSI layer 2) with related PGF $L(z)$. Hence, in this case, distribution (5.67) is associated with the number of messages arrived at the multiplexer in a slot. We still apply the M/G/1 theory (5.1) by selecting the imbedding points at the end of the slots of the output line. Such choice permits both to study the buffer congestion at the packet level and to evaluate the mean delay experienced by packets. Therefore, the PGF of the number of packets arrived in a time slot, $A_c(z)$, is obtained by composing (5.68) with $L(z)$ as follows:

$$A_c(z) = \sum_{n=0}^{N_\ell} \binom{N_\ell}{n} L(z)^n p^n (1-p)^{N_\ell - n} = [1 - p + L(z)p]^{N_\ell} = A[L(z)] \tag{5.71}$$

In the above sum, $L(z)^0 = 1$ for $n = 0$ represents the PGF of the deterministic number 0 and corresponds to the case with no packet arrivals in the slot. The stability condition becomes $A_c'(z=1) < 1$ Erlang, $\Rightarrow N_\ell p L'(z=1) < 1$ pkts/slot.

The mean number of packets in the system, N_p, can be obtained by substituting the derivatives of $A_c(z)$ computed in $z = 1$ in (5.16). Finally, T_p in slot units can be obtained by applying the Little theorem, dividing N_p by $A_c'(z=1)$.

Note that if we consider a buffer of finite capacity (let us say $C_{\max}$ packets, including the packet served), the classical PGF approach (5.1)–(5.10) cannot be adopted to study the multiplexer. We have to adopt an approach similar to that in Sect. 5.5, based on the state diagram in Fig. 5.5. The classical approach assumes that the number of arrivals a_{i+1} is independent of n_i. Such assumption is no longer valid in the presence of a finite buffer since a new packet arrival is rejected if it finds the system in the state $n_i = C_{\max}$ (i.e., the arrivals admitted in the system do depend on the current n_i value) [2]. Two different options are available for the imbedding instants:

1. The instants of packet transmission completions as in (5.1) and in Fig. 5.5

[5] In the case of time-slotted systems, the application of the Little theorem entails to divide the mean number of requests in the queue by the mean number of packets generated per slot, which typically corresponds to $A'(z=1)$, also representing the traffic intensity.

2. The instants of slot end of the output line since we have time-slotted transmissions (as considered so far in this section)

In both cases, the state diagram in Fig. 5.5 should be truncated: since the number of packets in the system cannot be greater than C_{max}: some state transitions in Fig. 5.5 have to be merged (with the corresponding packet loss events). For instance, considering the current state $n_i \neq 0$, the next state can be only $n_{i+1} = n_i - 1$ a+ a^*_{i+1}, where $a^*_{i+1} = \min(a_{i+1}, C_{max} - n_i + 1)$. Correspondingly, we have transitions merged and packet loss events when $a_{i+1} > C_{max} - n_i + 1$.

5.7 Advanced M/G/1 Cases

We are interested in studying advanced "M"/G/1 cases, modeled by difference equations that are generalizations of (5.1), as shown below:

$$n_{i+1} = \begin{cases} \max\{n_i - B, 0\} + a_{i+1}, & \text{if } n_i \geq 1 \\ a_{i+1}, & \text{if } n_i = 0 \end{cases} \tag{5.72a}$$

or

$$n_{i+1} = \begin{cases} n_i - 1 + a_{i+1}, & \text{if } n_i \geq 1 \\ a_{i+1} + \Delta_i, & \text{if } n_i = 0 \end{cases} \tag{5.72b}$$

Case (5.72*a*) with $B > 1$ (being B a deterministic value or a random variable) is used in the presence of batched service per imbedding interval (e.g., there is a frame-based service with many slots per frame for the transmission of packets). Case (5.72*b*) with $\Delta > 0$ (being Δ a random variable) is used when the arrivals at an empty buffer experience a sort of synchronization delay before their service can start (this is the "service differentiation"). The solution of case (5.72*a*) will be analyzed in Sect. 5.7.2 using the Rouché theorem [2, 11]; instead, the solution of case (5.72*b*) will be studied in Sect. 5.7.1. In general, we can say that the difference equations in (5.72*a*, 5.72*b*) can be solved in the z-domain in terms of the PGF of the state probability distribution by reapplying the same method presented in Sect. 5.1.

5.7.1 M/G/1 Theory with Differentiated Service Times

There are many ways to generalize the difference equation (5.1). We consider here the cases where the service time distribution of a request arriving at an empty buffer is different from the service time distribution of a request served after a waiting time in the queue. In particular, the arrivals at an empty buffer experience an additional delay to be served because of a sort of *rest period* (or *vacation time*). This system with "differentiated service times" is the subject of this section and can be seen as a particular case of the M/G/1 queue with server vacations [1, 2].

We imbed the queue at the instants of service completion. Let X denote the "normal service time," and X^* denote the "differentiated service time" for arrivals that occur when the buffer is empty. Let n_i represent the number of requests in the queue when completion of the ith request. Let a_i denote the number of requests arrived at the queue during the service time X of the ith request arrived at a non-empty buffer. Due to the differentiation, we denote with $a_i{}^*$ the number of requests arrived at the queue during the service time X^* of the ith request arrived at an empty buffer. We can write the following difference equation, which describes system behavior [2, 12]:

- $n_{i+1} = n_i - 1 + a_{i+1}$, if $n_i \neq 0$
- $n_{i+1} = a^*_{i+1}$, if $n_i = 0$

For instance, we can adopt this theory for those transmission systems, where a message arriving at an empty buffer requires a synchronization time before its transmission can start.

The method to solve the above difference equation is analogous to that used in the classical M/G/1 case in Sect. 5.1. System stability is assured if $A'(z = 1) < 1$ Erlang (note that service differentiation has no impact on system stability). Under the assumption of the arrival process independent of the system state, we obtain the following results for the empty state probability, the PGF of the state probability distribution $P(z)$, and the mean number of requests in the system, N:

$$
\begin{aligned}
P_0 &= \frac{1-E[a]}{1-E[a]+E[a^*]} = \frac{1-A'(1)}{1-A'(1)+A^{*\prime}(1)} \\
P(z) &= P_0\frac{A(z)-zA^*(z)}{A(z)-z} \\
N &= P_0\frac{2A^{*'}(1)+A^{*''}(1)-A''(1)}{2[1-A'(1)]} + \frac{A''(1)}{2[1-A'(1)]}
\end{aligned}
\tag{5.73}
$$

where $A(z)$ is the PGF of the number of arrivals during a normal service time, $A^*(z)$ is the PGF of the number of arrivals during a differentiated service time, $E[a] = A'(z=1)$, and $E[a^*] = A^{*\prime}(z=1)$. It is easy to show that $A'(z=1) = \lambda E[X]$ and $A^{*\prime}(z=1) = \lambda E[X^*]$ in the case of a Poisson arrival process of requests with mean rate λ.

In general, if $A^*(z) = W(z)A(z)$, we can use (5.73) to express the mean number of requests in the system as

$$
N = \frac{2W'(1)+W''(1)}{2[1+W'(1)]} + A'(1) + \frac{A''(1)}{2[1-A'(1)]}
\tag{5.74}
$$

$A(z)$ and $W(z)$ depend on the arrival process. In the above formula, the stability limit still depends on $A'(1)$: $A'(1) < 1$. Moreover, there is an additional term in N with respect to (5.16) that only depends on the PGF $W(z)$.

The mean system delay T is obtained through the Little theorem as $T = N/\lambda$. In the FIFO case, the Laplace transform of the system delay pdf is obtained by substituting $z = 1 - s/\lambda$ in the $P(z)$ in (5.73).

The difference equation described here would be suitable to address the M/G/1 case where the server becomes vacant when the system becomes empty (case of multiple vacations till a new arrival) for a time V according to a specific distribution. However, arrivals in $W(z)$ are not the arrivals in the whole vacation time, but just those in the residual vacation time as seen by a Poisson arrival at an empty system. Then, the classical M/G/1 mean delay needs an additional term depending on the mean residual life of the vacation interval V, that is $\frac{E[V^2]}{2E[V]}$, as shown in Sect. 5.1.1 [13].

5.7.2 M/D$^{[b]}$/1 Theory with Batched Service

We have a multiplexer receiving traffic from N_ℓ input synchronous Time Division Multiplexing (TDM) lines. Each input slot has a duration τ and may convey one packet with probability $p < 1$, uncorrelated from slot to slot and from line to line; with probability $1-p$ the slot is occupied by other higher priority traffic and cannot be used. The TDM line at the multiplexer's output is synchronized with the input TDM lines and has a packet transmission time equal to $\tau/2$: the "speeds" of the output line are double as compared to the input lines.

We study this system by considering an M/D$^{[2]}$/1 model with batched service, imbedding the chain at the instants of slot ends of input lines. Let n_i denote the number of packets in the multiplexer at the end of the ith slot; let a_i denote the number of packets arrived at the multiplexer during the ith slot from the N_ℓ input lines. The arrival process is independent from slot to slot and from input line to input line. Making considerations similar to those in Sects. 5.6 and 5.8.3.2, we write the following difference equation where we have taken the different speeds of input and output lines into due account:

$$
n_{i+1} = \begin{cases} n_i - 2 + a_{i+1}, & n_i \geq 2 \\ n_i - 1 + a_{i+1}, & n_i = 1 \\ a_{i+1}, & n_i = 0 \end{cases}
\tag{5.75}
$$

For instance, if $n_i \geq 2$, at the end of the next input slot, two packets in the multiplexer can be transmitted so that $n_{i+1} = n_i - 2 + a_{i+1}$. Utilizing the indicator function, we can rewrite the above balance in a more compact form as

$$n_{i+1} = n_i - I(n_i) - I(n_i - 1) + a_{i+1} \tag{5.76}$$

This is the difference equation modeling the behavior of the system. Assuming that there is a regime, we can find the PGF $P(z)$ of the state probability distribution (i.e., the probability mass function of the number of packets in the multiplexer) by adopting a similar approach to that in (5.6), under the assumption that n_i and a_{i+1} are independent and that the arrival process is memoryless. We have

$$\sum_{n_{i+1}=0}^{\infty} z^{n_{i+1}} P_{n_{i+1}} = \sum_{n_i=0}^{\infty} z^{n_i - I(n_i) - I(n_i-1)} P_{n_i} \times \sum_{a_{i+1}=0}^{\infty} z^{a_{i+1}} P_{a_{i+1}} \tag{5.77}$$

Referring to a regime condition, we can omit subscript i in the above expression so that we obtain

$$P(z) = \left\{ P_0 + P_1 + \sum_{n=2}^{\infty} z^{n-2} P_n \right\} \times A(z) \tag{5.78}$$

where $A(z)$ is the PGF of the number of packets arrived in a slot from the input lines. We note that each input line contributes a packet with probability p; hence, $A(z)$ is the PGF of a binomially distributed random variable as in Sect. 5.6:

$$\text{Prob}\{a = l\} = \binom{N_\ell}{l} p^l (1-p)^{N_\ell - l} \iff A(z) = (1 - p + zp)^{N_\ell} \tag{5.79}$$

The Eq. (5.78) in $P(z)$ can be further manipulated as follows:

$$P(z) = \{P_0 + P_1 + z^{-2}[P(z) - P_0 - zP_1]\} \times A(z) \iff$$

$$P(z) = \frac{\sum_{i=0}^{1} \left(z^2 - z^i\right) P_i}{z^2 - A(z)} A(z) \tag{5.80}$$

For deriving the mean number of packets in the buffer, we use the following expression in $P(z)$, as explained later:

$$P(z)\left[z^2 - A(z)\right] = A(z) \sum_{i=0}^{1} \left(z^2 - z^i\right) P_i \tag{5.81}$$

In the above $P(z)$ expression, we have two unknown terms: the probability of no packets in the multiplexer, P_0, and the probability of one packet in the multiplexer, P_1. These terms are derived as described below. However, before going on, we need to establish the system stability condition:

$$\text{(mean packet arrival rate)} \times \text{(mean packet transmission time)} < 1 \text{ Erlang}$$
$$\iff \left(\frac{N_\ell p}{T}\right) \times \left(\frac{T}{2}\right) < 1 \text{ Erlang} \iff N_\ell p < 2$$

Note that this stability condition corresponds to $A'(1) < 2$.

Under the stability assumption, we know that the PGF of the state probability distribution, $P(z)$, must fulfill the condition $|P(z)| \leq 1$ for $|z| \leq 1$. Hence, $P(z)$ cannot have poles on and inside the unit circle in the complex plane. Based on the Rouché theorem [2, 11] we can prove that $z^2 - A(z) = 0$ (i.e., the denominator of $P(z)$ in (5.80)) has *two distinct* solutions within the circle $|z| \leq 1$ for any $A(z)$ expression if $A'(1) < 2$; one solution is for $z = z_0 = 1$ and the other is denoted by z_1 (i.e., $z_1^2 - A(z_1) = 0$, $|z_1| \leq 1$, $z_1 \neq 1$).

The standard application of the Rouché theorem requires that $A(z)$ has a radius of convergence strictly larger than one (we have seen in Chap. 3 that the radius of convergence of a PGF must be at least one and then this holds for $A(z)$) and this commonly is assured by the stability condition. In general, the Rouché theorem is applied to the equation $z^S - A(z) = 0$ under the condition $A'(1) < S$ to guarantee the existence of S solutions within the circle $|z| \leq 1$ with one solution being $z = 1$.

The expression of $P(z)$ in (5.80) represents a PGF if the poles of $P(z)$ for $|z| \le 1$ due to $z^2 - A(z) = 0$ are canceled by the zeros of the numerator. Hence, P_0 and P_1 are determined by imposing the normalization condition $P(z=1) = 1$ (this is equivalent to canceling the pole in $z = 1$) and the pole cancellation in $z = z_1$:

$$\begin{cases} \lim_{z \to 1^-} \dfrac{\sum_{i=0}^{1} \left(z^2 - z^i\right) P_i}{z^2 - A(z)} A(z) = 1 \\ \sum_{i=0}^{1} \left(z_1^2 - z_1^i\right) P_i = 0 \end{cases} \tag{5.82}$$

Note that factor $A(z)$ on the left term in (5.81) cannot contribute to the cancellation in $z = 1$ because $A(1) = 1$. Hence, $A(z)$ can be removed from the first condition (i.e., in the limit term) in (5.82).

In the above system, the limit is indeterminate and can be solved by means of the Hôpital rule:

$$\begin{cases} \lim_{z \to 1^-} \frac{(2z-1)P_1 + 2zP_0}{2z - A'(z)} = 1 \\ \sum_{i=0}^{1} \left(z_1^2 - z_1^i\right) P_i = 0 \end{cases} \Rightarrow \begin{cases} \frac{P_1 + 2P_0}{2 - A'(1)} = 1 \\ \left(z_1^2 - 1\right) P_0 + \left(z_1^2 - z_1\right) P_1 = 0 \end{cases}$$

$$\Rightarrow \begin{cases} P_0 = \frac{z_1}{z_1 - 1} \left[2 - A'(1)\right] \\ P_1 = -\frac{z_1 + 1}{z_1 - 1} \left[2 - A'(1)\right] \end{cases}$$

We have thus obtained P_0 and P_1 as functions of the solution $z_1 \neq 1$ of the equation $z^2 - A(z) = 0$ in the complex domain, where $A(z)$ is given by (5.79).

By differentiating twice the non-fractional expression of $P(z)$ in (5.81) and by using the above formula $(P_1 + 2P_0)/[2 - A'(1)] = 1$, we can quickly obtain the following result for the mean number of packets in the multiplexer:

$$\begin{aligned} N_p &= A'(1) + \frac{P_0 + P_1 - 1}{2 - A'(1)} + \frac{A''(1)}{2\left[2 - A'(1)\right]} \\ &= A'(1) + \frac{1}{1 - z_1} + \frac{A''(1) - 2}{2\left[2 - A'(1)\right]} \text{ [pkts]} \end{aligned} \tag{5.83}$$

The mean packet delay can be obtained by means of the Little theorem, dividing N_p by $A'(1)$. This method, however, causes some inconsistency because the input slot is twice than the output slot (= packet transmission time) and we embed to the end of the slot of the input lines, thus not being possible to detect lower times than τ (i.e., the service time is not well represented). Moreover, the Kleinrock principle is not applicable here. Then, we achieve the following coarse result for T_p as

$$T_p = 1 + \frac{1}{(1 - z_1) A'(1)} + \frac{A''(1) - 2}{2A'(1)\left[2 - A'(1)\right]} \text{ [input slots]} \tag{5.84}$$

We have a simple case for the state probability distribution when $N_\ell = 2$. In fact, $A(z) = (1 - p + zp)^2$ and z_1 can be obtained by solving the following equation:

$$z^2 - (1 - p + zp)^2 = 0 \iff (z - 1 + p - zp) \times (z + 1 - p + zp) = 0$$

$$\Rightarrow \begin{cases} (z - 1 + p - zp) = 0 \\ (z + 1 - p + zp) = 0 \end{cases} \Rightarrow \begin{cases} z_0 = 1 \\ z_1 = -\frac{1-p}{1+p} < 0 \end{cases}$$

In this special case, P_0 and P_1 result as

$$\begin{cases} P_0 = \dfrac{z_1}{z_1 - 1} \left[2 - A'(1)\right] \\ P_1 = -\frac{z_1 + 1}{z_1 - 1} \left[2 - A'(1)\right] \end{cases} \Rightarrow \begin{cases} P_0 = (1 - p)^2 \\ P_1 = 2p(1 - p) \end{cases}$$

Then, the mean number of packets in the multiplexer results as

$$N_{\mathrm{p}} = 2p + \frac{1+p}{2} + \frac{p^2-1}{2(1-p)} = 2p \,[\text{pkts}] \tag{5.85}$$

As we can see in this special case, we have no stability limit for $p = 1$ because in this case we have two input lines but also the output line has a double capacity (we have not stability issues, and the queue has only the service part).

Note that this study can be generalized to the case of an output slot of duration τ/b (= packet transmission time), where b is an integer number greater than or equal to 2.

5.8 Different Imbedding Options for M/G/1

We are interested in studying advanced "M"/G/1 cases with compound arrivals (both discrete-time or continuous-time) that may experience synchronization issues because of a time division transmission (TDM system) with slots of duration τ. Each arrival contains a random number $\ell \geq 1$ of packets with PGF equal to $L(z)$, bulk arrivals. The packet transmission time is equal to τ. In the following study, we do not make any assumption on the service discipline (unless differently stated) apart from the fulfillment of the insensitivity conditions to apply the Little theorem. In the cases we investigate here, we can have different imbedding alternatives at the message level, at the packet level, and at the end of the output slot (if the output is time-slotted). In particular, we consider the three different cases, as follows (see Fig. 5.8):

(A) Compound Poisson arrivals and continuous-time service
(B) Poisson (or Poisson compound) arrivals and slotted service times
(C) Time-slotted and memoryless (Bernoulli/binomial) arrival process and slotted service times

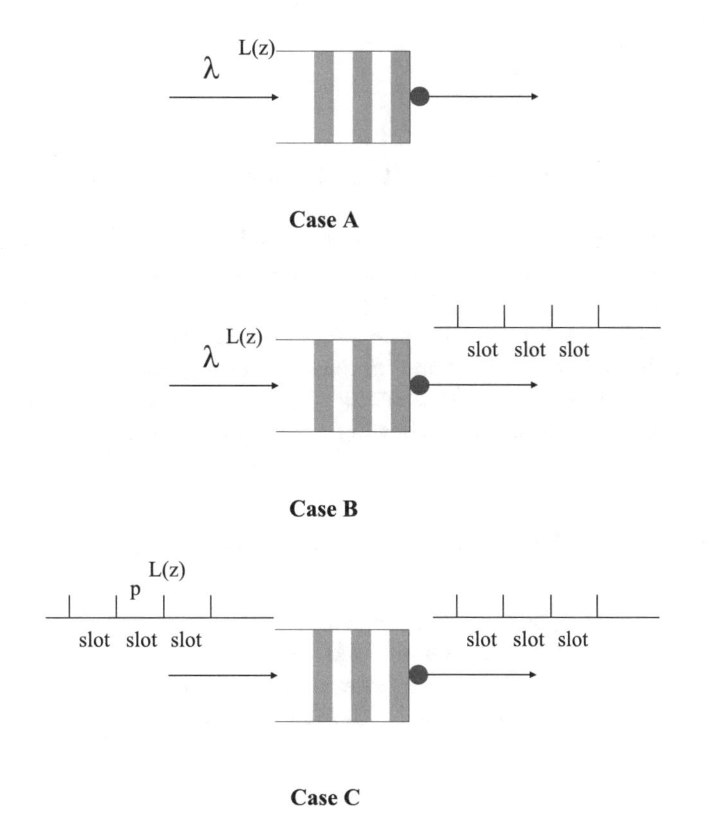

Fig. 5.8: Cases to be investigated for advanced M/G/1 systems

As sub-cases of both B and C, we have the situations where the service is organized according to a TDMA frame so that multiple slots (say them b) can also be assigned to the service of the queue, thus leading to the case with multiple service completions in the imbedding interval. In the "Case C," we have to assume a sort of synchronization between input and output slots and that output slots contain an integer number of input slots (typically, they have the same length). The available imbedding options for these three cases are as follows:

Case A

- To the packet transmission end
- To the message transmission end

Cases B and C

- To the packet transmission end
- To the end of the output slot
- To the end of the TDMA frame (in case of TDMA transmission frame)
- To the message transmission end

Each message is composed of a random number of packets, each requiring a time τ to be transmitted. Note that in this chapter, all packets of the same message arrive simultaneously. The PGF of the message length in packets, $L(z)$, corresponds to the PGF of the message transmission time in time units τ. Subsequent messages have iid lengths.

This study will be first carried out under general assumptions for $L(z)$ and, then, particularized for the case of messages having a length in packets ℓ according to a modified geometric distribution:

$$\text{Prob}\{\text{message length} = k \text{ pkts}\} = \frac{1}{L}\left(1 - \frac{1}{L}\right)^{k-1}, \quad k > 0$$

$$\Rightarrow L(z) = \frac{z/L}{1 - z\left(1 - \frac{1}{L}\right)} \tag{5.86}$$

In this special case, we have $L'(1) = L$ pkts (i.e., the mean message length in packets) and $L''(1) = 2[L'(1)]^2 - 2L'(1) = 2L(L-1)$.

As explained before, the same system can admit different imbedding options and can be described by different queuing models. However, *not all the imbedding options give the same results*. Studying the system at the packet level, the arrival process is compound (i.e., each message arrival simultaneously carries many packets) so that we cannot apply PASTA and the Kleinrock principle. The solution at the imbedding instants cannot be generalized at any instants. We have distinct solutions for the different imbedding instants in these cases. Nevertheless, we expect to have the *same stability condition, independently of the imbedding instants selected.*

The following sub-sections analyze these cases, showing the solutions with the different imbedding options.

5.8.1 Case A: Continuous-Time, Compound Poisson Arrivals

We consider a transmission line with a buffer, where messages arrive according to a Poisson process with a mean arrival rate λ. The arrival process and the transmission one are not time-slotted, but continuous-time in this case. We study this system at the level of both packets and messages, considering an M/G/1 queuing model. Therefore, we have to solve it in two different ways by selecting the imbedding instants, as detailed in the following sub-sections.

5.8.1.1 Imbedding at Packet Transmission Completion

Let n_i denote the number of packets in the buffer at the end of the transmission of the ith packet; let a_i denote the number of packets arrived at the buffer during the service time of the ith packet. We can write the classical M/G/1 difference equation (5.1) at the selected imbedding instants as

$$n_{i+1} = n_i - I(n_i) + a_{i+1}$$

However, the use of Eq. (5.1) in this case entails an *approximation* (see Sect. 5.8.1.2) for $n_i = 0$ due to the presence of bulk arrivals (i.e., packets arrive in messages of variable length): when $n_i = 0$, we have to wait for the next message arrival, and the service completion of the first packet arrived to obtain n_{i+1}. Therefore, in n_{i+1} we will have not only the a_{i+1} packets arrived during the service time of a packet (whose message arrived at an empty buffer), but also the residual number of packets of its message, $\ell - 1$. Hence, the exact formula in the case $n_i = 0$ would be: $n_{i+1} = a_{i+1} + \ell - 1$. In this section, however, we make the approximation to neglect the $\ell - 1$ term so that $n_{i+1} = a_{i+1}$ for $n_i = 0$ as in (5.1). We can remove such approximation by adopting an approach similar to that used for the "differentiated service time" in Sect. 5.7.1 that will be discussed in the next sub-section.

The PGF of the number of packets in the system is given by (5.10) where $A(z)$ is the PGF of the number of packets arrived at the buffer during the service time τ of one packet. Note that the PGF of the number of messages arrived in the service time τ of a packet related to the Poisson arrival process is $e^{\lambda\tau(z-1)}$. If there is a single message arrival in τ, the PGF of the number of packets arrived is $L(z)$; if there are two message arrivals in τ, the PGF of the number of packets arrived is $L(z)^2$. By removing the conditioning on the number of messages arrived in τ, we have

$$A(z) = \sum_{n=0}^{\infty} [L(z)]^n \text{Prob}\{n \text{ message arrivals in } \tau\} = e^{\lambda\tau[L(z)-1]} \tag{5.87}$$

From (5.16) the mean number of packets in the buffer N_p is

$$N_\text{p} = A'(1) + \frac{A''(1)}{2[1 - A'(1)]} \text{ [pkts]} \tag{5.88}$$

where

$$\begin{aligned}
A'(z=1) &= \frac{\mathrm{d}}{\mathrm{d}z} e^{\lambda\tau[L(z)-1]}\Big|_{z=1} = e^{\lambda\tau[L(z)-1]} \times \lambda\tau L'(z)\Big|_{z=1} = \lambda\tau L'(1) \\
A''(z=1) &= \frac{\mathrm{d}}{\mathrm{d}z} e^{\lambda\tau[L(z)-1]} \times \lambda\tau L'(z)\Big|_{z=1} \\
&= e^{\lambda\tau[L(z)-1]} \times [\lambda\tau L'(z)]^2\Big|_{z=1} \\
&\quad + e^{\lambda\tau[L(z)-1]} \times \lambda\tau L''(z)\Big|_{z=1} \\
&= [\lambda\tau L'(1)]^2 + \lambda\tau L''(1)
\end{aligned} \tag{5.89}$$

The stability of the buffer is assured if $\lambda\tau L'(1) < 1$ Erlang. By substituting (5.89) in (5.88), we obtain the following expression for the mean number of packets in the buffer:

$$N_\text{p} = \lambda\tau L'(1) + \frac{[\lambda\tau L'(1)]^2 + \lambda\tau L''(1)}{2[1 - \lambda\tau L'(1)]} \text{ [pkts]} \tag{5.90}$$

The mean packet delay, T_p, is obtained from (5.90), dividing by the mean packet arrival rate of $\lambda L'(1)$ pkts/s according to the Little theorem:

$$T_\text{p} = \frac{N_\text{p}}{\lambda L'(1)} = \tau + \frac{\lambda[\tau]^2 L'(1) + \tau^{L''(1)} \Big/ L'(1)}{2[1 - \lambda\tau L'(1)]} \text{ [s]} \tag{5.91}$$

When $\lambda \to 0$, T_p according to (5.91) results as

$$T_\text{p} = \tau + \tau\frac{L''(1)}{2L'(1)} \tag{5.92}$$

In this formula, the first term is the packet transmission time; the second term is a waiting time that we also have without traffic because the arrivals occur in groups and the mean delay, in this case, is the delay of a generic packet of this group. Following the same method as shown in [3], we can prove that the term $\tau \frac{L''(1)}{2L'(1)} = \tau \left\{ \frac{E[\ell^2]}{2E[\ell]} - \frac{1}{2} \right\}$ is related to the wait (delay) because our packet is just one generic packet of the group with PGF $L(z)$.

Let us consider the delay experienced by a packet because it is part of a message with length in packets according to the PGF $L(z)$. Let us condition on a message of length ℓ. A packet in this message experiences a wait equal to 0, 1, ... $\ell - 1$ transmission units τ. Then, the mean wait $E[W|\ell]$ is obtained as

$$E[W|\ell] = \frac{\tau}{\ell}\sum_{i=0}^{\ell-1} i = \frac{\tau}{\ell}\frac{\ell(\ell-1)}{2} = \tau\frac{(\ell-1)}{2} \tag{5.93}$$

We need to remove the conditioning from ℓ in the mean value $E[W|\ell]$. We can do this by considering the probability p_ℓ that a random packet belongs to a message of size ℓ. We would be tempted to use for p_ℓ the probability of a message with length ℓ, denoted here as q_ℓ, but this is not correct. We need a sort of biased distribution and we follow here the reasoning made in [3]. Let us consider an arbitrarily long time interval u. Messages of size ℓ arrive at a rate λq_ℓ; then, a time u at which we have a mean number of $\lambda q_\ell \ell u$ packets from these messages. In total, we have the following mean number of messages considering the different possible lengths of these messages as $\sum_{k=1}^{\infty} \lambda q_k k u$. Then, the probability that a random packet belongs to a message of size ℓ can be approximated as

$$p_\ell = \lim_{u\to\infty} \frac{\lambda q_\ell \ell u}{\sum_{k=1}^{\infty} \lambda q_k k u} = \frac{q_\ell \ell}{\sum_{k=1}^{\infty} k q_k} \tag{5.94}$$

We can now remove the conditioning on ℓ in $E[W|\ell]$ to determine the mean waiting time w experienced by a packet because part of a message

$$E[W] = \sum_{\ell=1}^{\infty} E[W|\ell]\, p_\ell = \frac{\sum_{\ell=1}^{\infty} \tau \frac{(\ell-1)}{2} q_\ell \ell}{\sum_{k=1}^{\infty} k q_k} = \tau \frac{L''(1)}{2\,L'(1)} \tag{5.95}$$

This result completes the proof of the meaning of the term $\frac{L''(1)}{2\,L'(1)}$.

In the particular case of the modified geometric distribution, we use the formula $L''(1) = 2[L'(1)]^2 - 2L'(1)$. We have

$$T_\mathrm{p} = \tau + \frac{\lambda[\tau]^2 L'(1) + 2\tau L'(1) - 2\tau}{2\left[1 - \lambda\tau L'(1)\right]} \text{ [s]} \tag{5.96}$$

In the FIFO case with a bulk arrival process and PGF of the message length as $L(z)$, we use (5.45) to relate the PGF of the number of packets in the buffer, $P(z)$, to the Laplace transform of the pdf of the packet delay, $T_{\mathrm{Dp}}(s)$, computed for $s = \lambda[1 - L(z)]$[6] [3]. Hence, referring to a message length with modified geometric distribution, we use the $L(z)$ expression in (5.86) and obtain $s = s(z)$ as

$$s = \lambda\left[1 - \frac{z/L}{1 - z\left(1 - \frac{1}{L}\right)}\right]$$

We can invert the above relation to obtain $z = z(s)$ as

$$z = L^{-1}\left(1 - \frac{s}{\lambda}\right) = \frac{s - \lambda}{s\left(1 - \frac{1}{L}\right) - \lambda} \tag{5.97}$$

where $L^{-1}(\cdot)$ is the inverse function of $L(\cdot)$.

[6] We have to use this expression for $s = s(z)$ because of the compound arrival process and the imbedding instants at the level of packets. However, we should use the more simple formula $s = \lambda(1 - z)$ if the imbedding points are at message transmission completion instants.

Since the PGF $P(z)$ of the number of packets in the buffer is given by (5.10) with $A(z)$ as in (5.87), we substitute to z the expression in (5.97) to obtain the Laplace transform of the pdf of the packet delay, $T_{\mathrm{Dp}}(s)$:

$$T_{\mathrm{Dp}}(s) = [1 - \lambda\tau L'(1)] \frac{\left[\dfrac{s-\lambda}{s\left(1-\frac{1}{L}\right)-\lambda} - 1\right] e^{\lambda\tau\left\{L\left[\dfrac{s-\lambda}{s\left(1-\frac{1}{L}\right)-\lambda}\right]-1\right\}}}{\dfrac{s-\lambda}{s\left(1-\frac{1}{L}\right)-\lambda} - e^{\lambda\tau\left\{L\left[\dfrac{s-\lambda}{s\left(1-\frac{1}{L}\right)-\lambda}\right]-1\right\}}} \tag{5.98}$$

where $L(z)$ is given by (5.86) and where we have terms of the following type at the exponent:

$$L\left[\frac{s-\lambda}{s\left(1-\frac{1}{L}\right)-\lambda}\right] = L\left[L^{-1}\left(1-\frac{s}{\lambda}\right)\right] = 1 - \frac{s}{\lambda}$$

Consequently, $T_{\mathrm{Dp}}(s)$ can also be expressed in the following form:

$$T_{\mathrm{Dp}}(s) = [1 - \lambda\tau L'(1)] \frac{s \times e^{-\lambda\tau\, s/\lambda}}{(s-\lambda)\, L - [s\,(L-1) - \lambda L] \times e^{-\lambda\tau\, s/\lambda}} \tag{5.99}$$

Due to the complexity of this Laplace transform, its inversion can be obtained only through a numerical method, as shown in Sect. 5.3.

5.8.1.2 The Differentiated Theory Applied to Compound Arrivals

In this section, we study the same queuing system as in Sect. 5.8.1.1 with imbedding points at the end of packet transmission instants, but we remove the approximation to always consider messages of one packet for those arrivals occurring at an empty buffer. We apply the differentiated theory. Variables n_i and a_i are defined as in Sect. 5.8.1.1. Moreover, we also have to consider random variable $a_i{}^*$ denoting the number of packets arrived during the service time of the first packet of a group arrived at an empty buffer. Let ℓ denote the random length of a message in packets; the corresponding PGF is denoted by $L(z)$. Based on Sect. 5.7.1, the exact difference equation modeling this system results as

- $n_{i+1} = n_i - 1 + a_{i+1}$, if $n_i \neq 0$
- $n_{i+1} = a^*_{i+1} = \ell - 1 + a_{i+1}$, if $n_i = 0$

Hence, $A^*(z) = A(z) \times L(z)/z$ and $A(z) = e^{\lambda T[L(z)-1]}$, as shown in (5.87). Thus, substituting the $A^*(z)$ expression in (5.73), we obtain the empty buffer probability P_0 and the mean number of packets in the queue N_{p} as

$$\begin{aligned} P_0 &= \frac{1 - A'(1)}{L'(1)} \\ P(z) &= P_0 \frac{A(z)\,[1 - L(z)]}{A(z) - z} \\ N_p &= \frac{L''(1)}{2L'(1)} + A'(1) + \frac{A''(1)}{2\,[1 - A'(1)]} \text{ [pkts]} \end{aligned} \tag{5.100}$$

Note that these results in (5.100) are valid for any $A(z)$ of a compound arrival process and any $L(z)$; however, we consider $A(z) = e^{\lambda\tau[L(z)-1]}$ in this chapter.

The first term in N_{p}, that is $\frac{L''(1)}{2L'(1)}$, is due to differentiation depending on the message length statistics; this term disappears if messages are formed of a single packet $L(z) = z$. The second part in N_{p} is the classical M/G/1 formula, as in the approximated model in Sect. 5.8.1.1; this also implies the same stability limit in both cases. We note that the N_p first term in (5.100) is missing in the corresponding formula in (5.90). This term is zero only if $L''(1) = 0$: this is true only if messages have a fixed length of one packet.

The Little theorem cannot be applied here [3] to N_p in (5.100) since N_p contains some packets (i.e., $\frac{L''(1)}{2L'(1)}$) at imbedding instants that do not depend on the Poisson arrival process (there are these packets also if $\lambda \to 0$). In the case of a modified geometric distribution of the message length in packets, we have $L''(1)/[2 \times L'(1)] = L'(1) - 1$.

Note also that the Kleinrock principle cannot be applied here due to the compound process: the results at completion instants cannot be extended to other instants.

5.8.1.3 Imbedding at Message Transmission Completion

In this case, n_i represents the number of messages in the buffer at the end of the ith message's transmission; moreover, a_i denotes the number of messages arrived at the buffer during the ith message's service time. We can write the classical M/G/1 difference equation (5.1) at the imbedding instants selected. Therefore, the PGF $P(z)$ of the number of messages in the system is given by (5.10), where $A(z)$ is the PGF of the number of messages arrived at the buffer during the service time of a message. Let us condition on the message service time. First, we consider the service time of a message of one packet (= service time τ) so that $A(z|\tau) = e^{\lambda\tau(z-1)}$. Then, we consider the PGF $A(z|n\,\tau) = [A(z|\tau)]^n$ for the service time of a generic message of n packets since message arrivals are independent in subsequent τ units. We remove the conditioning on n using the message length distribution with PGF $L(z)$. We have

$$A(z) = \sum_{n=1}^{\infty} \left[A\left(z\middle|T\right)\right]^n \text{Prob}\{\text{message with } n \text{ pkts}\}$$
$$= L\left[A\left(z\middle|T\right)\right] = L\left[e^{\lambda T(z-1)}\right] \quad (5.101)$$

With this imbedding option, $A(z)$ is obtained by composing $L(z)$ and $e^{\lambda\tau(z-1)}$ in the opposite way with respect to what is done in (5.87), where imbedding instants are at the end of the packet transmission.

The derivatives of $A(z)$ computed in $z = 1$ result as

$$\begin{aligned}
A'(z=1) &= \frac{d}{dz} L\left[e^{\lambda\tau(z-1)}\right]\Big|_{z=1} = L'\left[e^{\lambda\tau(z-1)}\right] \times e^{\lambda\tau(z-1)}\lambda\tau\Big|_{z=1} \\
&= L'(1)\lambda\tau \\
A''(z=1) &= \frac{d}{dz} L'\left[e^{\lambda\tau(z-1)}\right] \times e^{\lambda\tau(z-1)}\lambda\tau\Big|_{z=1} \\
&= L''\left[e^{\lambda\tau(z-1)}\right] \times \left[e^{\lambda\tau(z-1)}\lambda\tau\right]^2\Big|_{z=1} \\
&\quad + L'\left[e^{\lambda\tau(z-1)}\right] \times (\lambda\tau)^2 e^{\lambda\tau(z-1)}\Big|_{z=1} \\
&= [\lambda\tau]^2 L''(1) + [\lambda\tau]^2 L'(1) \\
&= [\lambda\tau]^2 \left[L''(1) + L'(1)\right]
\end{aligned} \quad (5.102)$$

As in the previous study related to packets, the stability condition is $\lambda\tau L'(1) < 1$ Erlang. The mean number of messages in the buffer, N_m, is obtained as

$$\begin{aligned}
N_m &= A'(1) + \frac{A''(1)}{2\left[1 - A'(1)\right]} \\
&= L'(1)\lambda\tau + \frac{[\lambda\tau]^2\left[L''(1) + L'(1)\right]}{2\left[1 - L'(1)\lambda\tau\right]} \text{ [msgs]}
\end{aligned} \quad (5.103)$$

Since the mean arrival rate of messages is λ, we apply the Little theorem to (5.103) to derive the mean message delay T_m as

$$T_\mathrm{m}=\frac{N_\mathrm{m}}{\lambda}=L'(1)\tau+\frac{\lambda[\tau]^2\,[L''(1)+L'(1)]}{2\,[1-L'(1)\lambda\tau]}\ [\mathrm{s}] \tag{5.104}$$

Note that (5.103) and (5.104) could be easily derived by applying the Pollaczek–Khinchin formula (5.17) and (5.18) without the need to use $A(z)$.

Under the assumption of messages with modified geometric distribution, we can substitute the following formula in (5.104): $L''(1)=2[L'(1)]^2-2L'(1)$. We have

$$T_\mathrm{m}=L'(1)\tau+\frac{\lambda[\tau]^2\left[2L'(1)^2-L'(1)\right]}{2\,[1-L'(1)\lambda\tau]}\ [\mathrm{s}] \tag{5.105}$$

Through algebraic manipulations, we can prove that (5.96) and (5.105) are equal, $T_\mathrm{p}\equiv T_\mathrm{m}$: in the presence of a geometrically distributed message length and Poisson arrivals, the mean packet delay is coincident with the mean message delay. This result is valid for any service policy of the packets or of the messages in the buffer, provided that the conditions of the insensitivity property are met (see Sect. 4.5). However, it is essential to remark that this surprising result of $T_\mathrm{p}\equiv T_\mathrm{m}$ is valid only as a first approximation because we have derived T_p by considering that message arrivals at an empty buffer always contain one packet.

In the FIFO case, we make the substitution in (5.43),

$$z=1-\frac{s}{\lambda}$$

in the PGF $P(z)$ in (5.10) of the number of messages in the buffer with $A(z)$ given by (5.101); we thus obtain the Laplace transform of the pdf of the message delay, $T_\mathrm{Dm}(s)$, as

$$T_\mathrm{Dm}(s)=\left[1-\lambda\tau L'(1)\right]\frac{\left(-\frac{s}{\lambda}\right)L\left[e^{-s\tau}\right]}{1-\frac{s}{\lambda}-L\left[e^{-s\tau}\right]} \tag{5.106}$$

where $L\left[e^{-s\tau}\right]$ is the Laplace transform of the message transmission time.

Under the assumption of geometrically distributed messages, we use in (5.106) the $L(z)$ expression given in (5.86) so that the resulting formula from (5.106) is equal[7] to that in (5.99). In conclusion, we have obtained the very interesting result that, in the FIFO case with a geometrically distributed message length, the delay distribution for packets and messages is equal: $T_\mathrm{Dp}(s)\equiv T_\mathrm{Dm}(s)$. This very special condition, only possible with geometrically distributed messages, is a generalization of the equality already obtained for the mean values (without the FIFO assumption), i.e., (5.96) and (5.105) are equal.[7]

5.8.2 Case B: Compound Poisson Arrivals and Slotted Service Times

Let us analyze in detail the implications for the different imbedding options of "Case B":

1. *Imbedding at the end of the packet transmission time* (see Fig. 5.9a) to study the statistics of the buffer occupancy in terms of packets (like MAC layer performance). This chapter requires to adopt a "differentiation approach" (see Sect. 5.7.1) since continuous-time arrivals at an empty buffer experience extra delays due to the wait for the right transmission synchronism (TDM). Notation: $\mathrm{M}^{[\mathrm{GI}]}/\mathrm{D}/1$. Let δ denote the synchronization delay of the arrival at an empty buffer with respect to the beginning of the next slot. Moreover, Δ denotes the number of packet arrivals according to the compound process in time δ.
2. *Imbedding the system at the end of the output TDM slot* (see Fig. 5.9b), to have a classical $\mathrm{M}^{[\mathrm{GI}]}/\mathrm{D}/1$ case (the same notation as before) without any service differentiation issue.
3. *Imbedding at the end of the TDMA frame* assuming that there are b slots available for the service of the queue in the frame duration, thus having a batched service since we can serve up to b packets per frame. Notation: $\mathrm{M}^{[\mathrm{GI}]}/\mathrm{D}^{[b]}/1$ (see Fig. 5.10a).
4. *Imbedding at the end of the message transmission time* (see Fig. 5.10b) to study the message delay distribution (like layer 3 performance). We have to account for the synchronism of the arrival at an empty buffer. Neglecting

[7] Let us recall that this is true under the approximation concerning message arrivals at an empty buffer.

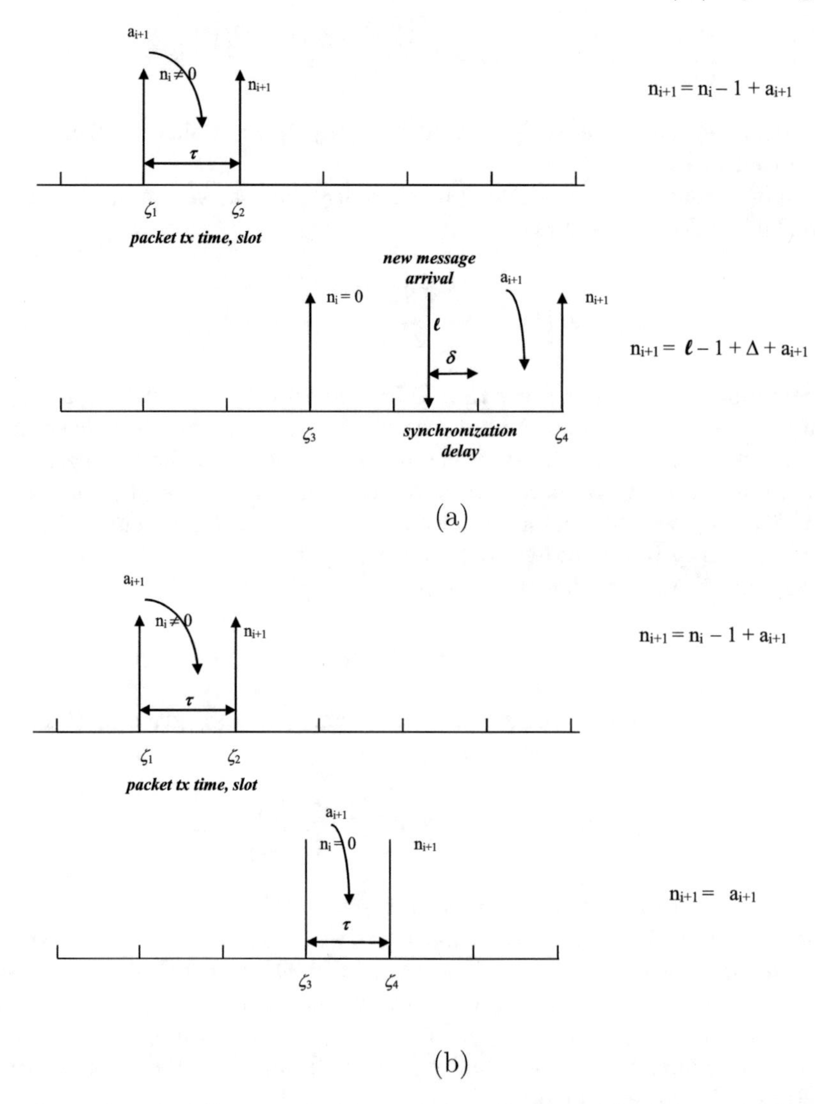

Fig. 5.9: Solution of "Case B" with different imbedding options (Part I): (**a**) at the end of the packet transmission; (**b**) at the end of the output slot. Note that n_i and a_i can have different meanings in these different sub-cases

the synchronization issues, we could simply apply the Pollaczek–Khinchin formula in this case. Notation: M/G/1.

So far, we have considered that the transmission of one packet requires just one output slot. However, "Case B" could be made more complex considering that an output slot is unavailable to carry the traffic of the queue with probability p and available with probability $1-p$; this mechanism is memoryless from slot to slot. This way, the service time of a packet has a geometric modified distribution with PGF and derivatives as detailed below:

$$T_s(z) = \frac{(1-p)\,z}{1-zp}, \quad T'(1) = \frac{1}{1-p}, \quad T''(1) = \frac{2p}{(1-p)^2} \tag{5.107}$$

5.8.2.1 Imbedding at the End of the Packet Transmission

According to Fig. 5.9a, we have to solve the following difference equations:

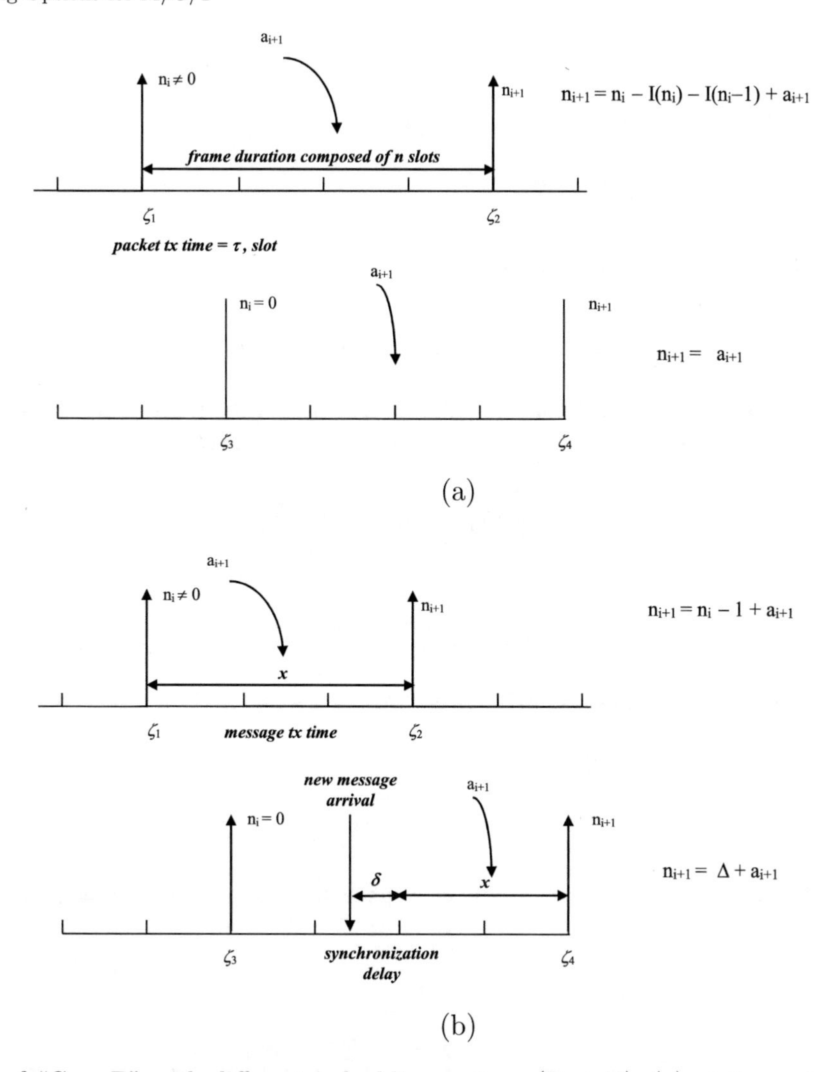

Fig. 5.10: Solution of "Case B" with different imbedding options (Part II): (**a**) at the end of the output frame; (**b**) at the end of the message transmission. Note that n_i and a_i can have different meanings in these different sub-cases

$$
\begin{aligned}
n_{i+1} &= n_i - 1 + a_{i+1}, \quad if\ n_i > 0 \\
n_{i+1} &= \ell - 1 + \Delta + a_{i+1}, \quad if\ n_i = 0
\end{aligned}
\tag{5.108}
$$

For these difference equations, we use the same solution approach as for the case with differentiated service times in Sect. 5.7.1 for which we have to express the PGF of the sum $\ell - 1 + \Delta$. Moreover, the PGF $A(z)$ corresponds at regime to the random variables a_{i+1} and represents the PGF of the number of arrivals in a slot time τ due to the compound Poisson process: $A(z) = e^{\lambda\tau[L(z)-1]}$. Note that Δ is the PGF of the number of packet arrivals due to the compound Poisson process in the synchronization delay δ. We need to derive the pdf of the random variable δ as follows and then the PGF of Δ, denoted as $\Delta(z)$.

δ is the remaining time in a slot after a Poisson arrival in this slot (residual time). This arrival occurs at a time t_0 after the beginning of the slot considered to start at time $t = 0$ for the sake of simplicity. Because of the Poisson interarrival time distribution, time t_0 has an exponential distribution with mean rate λ. Then, the residual time in the slot results as $\delta = \tau - t_0$ for which we can express the probability distribution function as follows:

$$\mathrm{Prob}\{\delta \le t\} = \mathrm{Prob}\{\tau - t_0 \le t | t_0 \le \tau\} = \frac{\mathrm{Prob}\{\tau - t \le t_0 \le \tau\}}{\mathrm{Prob}\{t_0 \le \tau\}}$$
$$= \frac{e^{-\lambda\tau}\left(e^{\lambda t} - 1\right)}{1 - e^{-\lambda\tau}} \tag{5.109}$$

The corresponding density function of δ can be obtained by taking the derivative with respect to t as

$$f_\delta(t) = \frac{e^{-\lambda\tau}\lambda e^{\lambda t}}{1 - e^{-\lambda\tau}}, \quad for\ t \in [0,\ \tau] \tag{5.110}$$

Finally, the PGF $\Delta(z)$ is obtained considering that in a generic time t the PGF of the number of packet arrivals is $e^{\lambda t[L(z)-1]}$. Then, removing the dependency on t by means of the pdf $f_\delta(t)$, we have

$$\Delta(z) = \int_0^\tau e^{\lambda t[L(z)-1]} f_\delta(t)\,\mathrm{d}t = \frac{\lambda e^{-\lambda\tau}}{1 - e^{-\lambda\tau}} \int_0^\tau e^{\lambda t L(z)}\mathrm{d}t$$
$$= \frac{\lambda e^{-\lambda\tau}}{1 - e^{-\lambda\tau}} \left[\frac{e^{\lambda t L(z)}}{\lambda L(z)}\right]_0^\tau = \frac{e^{\lambda\tau[L(z)-1]} - e^{-\lambda\tau}}{\left(1 - e^{-\lambda\tau}\right) L(z)} \tag{5.111}$$

Let us use the derivations at the end of Sect. 5.7.1; accordingly, we adopt the following definitions:

$$A^*(z) = W(z)\,A(z)$$
$$W(z) = \frac{L(z)}{z}\Delta(z) = \frac{e^{\lambda\tau[L(z)-1]} - e^{-\lambda\tau}}{\left(1 - e^{-\lambda\tau}\right) z} \tag{5.112}$$

Then, we can use (5.74) to express the mean number of packets using the first and second derivatives of $W(z)$ in $z = 1$ as

$$W'(1) = \frac{\lambda\tau L'(1)}{\left(1 - e^{-\lambda\tau}\right)} - 1$$
$$W''(1) = \frac{\lambda\tau L''(1) + \lambda^2\tau^2 L'^2(1) - 2\lambda\tau L'(1)}{\left(1 - e^{-\lambda\tau}\right)} + 2 \tag{5.113}$$

Moreover, the derivatives of $A(z)$ are

$$A'(1) = \lambda\tau L'(1)$$
$$A''(1) = \lambda\tau L''(1) + \left[\lambda\tau L'(1)\right]^2 \tag{5.114}$$

Substituting these expressions in (5.74) we get

$$N_\mathrm{p} = \frac{L''(1)}{2L'(1)} + \frac{\lambda\tau}{2} L'(1) + \lambda\tau L'(1) + \frac{\lambda\tau L''(1) + \left[\lambda\tau L'(1)\right]^2}{2\left[1 - \lambda\tau L'(1)\right]}\ \text{[pkts]} \tag{5.115}$$

We can see that the mean number of packets N_p is given by four contributions that can be justified as follows:

(a) $L''(1)/[2L'(1)]$, term due to the message arrival at an empty buffer
(b) $\frac{\lambda\tau}{2}L'(1)$, term due to the synchronization delay due to the wait for the start of the output slot
(c) $\lambda\tau L'(1)$, the classical M/G/1 service part
(d) $\frac{\lambda\tau L''(1) + \left[\lambda\tau L'(1)\right]^2}{2[1 - \lambda\tau L'(1)]}$, the classical M/G/1 waiting part

The stability limit is $A'(1) < 1 \Rightarrow \lambda\tau L'(1) < 1$ Erl (or pkts/slot).

In general, we cannot apply the Little theorem to (5.115) because of the first contribution in (5.115). On the other hand, if we neglect the effect of the message length for arrivals at an empty buffer, we can apply the Little theorem only to the above terms b, c, and d. To do so, we need to divide N by the mean arrival rate in packets, that is $\lambda L'(1)$:

$$T_{\mathrm{p}} \approx \frac{\tau}{2} + \tau + \frac{\tau \frac{L''(1)}{L'(1)} + \lambda \tau^2 L'(1)}{2\left[1 - \lambda \tau L'(1)\right]} \text{ [s]} \tag{5.116}$$

Let us now consider the different case that the output slot can be unavailable with probability p; we can still apply the difference equation (5.108), but now $A(z)$ is the PGF of the number of packet arrivals not in a single slot τ, but in a number of slots according to PGF $T_s(z)$. In particular, we have $A(z) = T_s\left(e^{\lambda\tau[L(z)-1]}\right)$ for which the first and second derivatives for $z = 1$ result as

$$\begin{aligned} A'(1) &= \lambda \tau L'(1) T_s'(1) \\ A''(1) &= T_s''(1)\left[\lambda \tau L'(1)\right]^2 + T_s'(1)\left\{\lambda \tau L''(1) + \left[\lambda \tau L'(1)\right]^2\right\} \\ &= \left[\lambda \tau L'(1)\right]^2 \left[T_s''(1) + T_s'(1)\right] + \lambda \tau L''(1) T_s'(1) \end{aligned} \tag{5.117}$$

The stability limit is now different because the unavailability of the output slot with probability p has the effect of extending the service time of each packet: $\lambda \tau L'(1) T_s'(1) < 1$ Erl (or pkts/slot). With these new expressions for $A'(1)$ and for $A''(1)$ we can reapply Eq. (5.74) using the same $W'(1)$ and $W''(1)$ as in (5.113) to express the mean number of packets in the system. Then, we can elaborate (5.115), using the new expressions for $A'(1)$ and $A''(1)$ as

$$\begin{aligned} N_{\mathrm{p}} = &\frac{L''(1)}{2L'(1)} + \frac{\lambda\tau}{2} L'(1) + \lambda \tau L'(1) T_s'(1) \\ &+ \frac{\lambda \tau L''(1) T_s'(1) + \left[T_s'(1) + T_s''(1)\right]\left[\lambda \tau L''(1)\right]^2}{2\left[1 - \lambda \tau L'(1) T_s'(1)\right]} \text{ [pkts]} \end{aligned} \tag{5.118}$$

5.8.2.2 Imbedding at the Slot End

In this case, referring to Fig. 5.9b, we have to solve the following difference equations:

$$\begin{aligned} &n_{i+1} = n_i - 1 + a_{i+1}, \quad if\ n_i > 0 \\ &n_{i+1} = a_{i+1}, \quad if\ n_i = 0 \end{aligned} \tag{5.119}$$

These are the classical M/G/1 difference equations in (5.1) for which the mean number of packets in the queue can be expressed using (5.16) where $A(z) = e^{\lambda\tau[L(z)-1]}$ and its derivatives computed for $z = 1$ are expressed as

$$\begin{aligned} A'(1) &= \lambda \tau L'(1) \\ A''(1) &= \lambda \tau L''(1) + \left[\lambda \tau L'(1)\right]^2 \end{aligned} \tag{5.120}$$

Then, the mean number of packets results as

$$N_{\mathrm{p}} = \lambda \tau L'(1) + \frac{\lambda \tau L''(1) + \left[\lambda \tau L'(1)\right]^2}{2\left[1 - \lambda \tau L'(1)\right]} \text{ [pkts]} \tag{5.121}$$

The mean packet delay can be obtained by applying to (5.121) the Little theorem dividing N_{p} by $\lambda L'(1)$, the packet arrival rate:

$$T_{\mathrm{p}} = \tau + \frac{\tau \frac{L''(1)}{L'(1)} + \lambda \tau^2 L'(1)}{2\left[1 - \lambda \tau L'(1)\right]} \text{ [s]} \tag{5.122}$$

With this imbedding option, we have different results for N_{p} and T_{p} with respect to the previous section with imbedding instants at the end of each slot, even if the differences are minor. In both cases, we have however the same stability condition $\lambda \tau L'(1) < 1$ Erl.

Under the condition that the output slot can be unavailable with probability p, we cannot apply the difference equation (5.119) because if $n_i \neq 0$, we have to consider that we cannot have service in a slot if this slot is unavailable. Then, the difference equations result as

$$\begin{aligned} n_{i+1} &= n_i - m + a_{i+1}, \quad if\ n_i > 0 \\ n_{i+1} &= a_{i+1}, \quad if\ n_i = 0 \end{aligned} \tag{5.123}$$

where m is a Bernoulli random variable that is equal to 1 with probability $1-p$ and is equal to 0 with probability p; then the corresponding PGF is $M(z) = p + z(1-p)$. Note that $A(z)$, representing the PGF of the number of packets arrived at the buffer in a slot, is the same as before: $A(z) = e^{\lambda\tau[L(z)-1]}$.

The difference equation in (5.123) is slightly different from that in (5.1) because now the service is probabilistic. Our system admits an M/G/1 model with a randomly available server. By following the same approach in (5.2)–(5.7), we transform the difference equation in the z-domain, and we achieve the following expression of the PGF of the number of packets in the queue:

$$P(z) = P_0 \frac{(1-p)(z-1)A(z)}{z - [1-p+zp]A(z)} \tag{5.124}$$

where

$$P_0 = \frac{1-p-A'(z=1)}{1-p} \tag{5.125}$$

Since $P(z)$ has a singularity in $z=1$, we can derive the mean number of packets in the buffer, N_p, by multiplying both sides of (5.124) by the denominator and by differentiating twice and computing in $z=1$. The final result is

$$N_\mathrm{p} = P'(1) = A'(1) + \frac{pA'(1)}{1-p-A'(1)} + \frac{A''(1)}{2[1-p-A'(1)]} \text{ [pkts]} \tag{5.126}$$

where $A'(1)$ and $A''(1)$ are given in (5.120).

Note that (5.126) for $p = 0$ becomes the classical mean number of requests of the M/G/1 system in (5.16). Similarly, $P(z)$ and P_0 for $p = 0$ also reduce to those of the classical M/G/1 system.

The system is stable under the condition that $p + \lambda\tau L'(1) < 1$ pkts/slot.

We can apply the Little theorem to derive the mean packet delay, T_p, by considering that $A'(1)$ denotes the mean number of packets arrived at the buffer in a time slot; hence, $A'(1)$ is the mean packet arrival rate in pkts/slot.

$$\begin{aligned} T_\mathrm{p} &= \frac{N_\mathrm{p}}{A'(1)} = 1 + \frac{p}{1-p-A'(1)} + \frac{A''(1)\big/A'(1)}{2[1-p-A'(1)]} \\ &= 1 + \frac{p}{1-p-\lambda\tau L'(1)} + \frac{L''(1)/L'(1) + \lambda\tau L'(1)}{2[1-p-\lambda\tau L'(1)]} \text{ [slots]} \end{aligned} \tag{5.127}$$

5.8.2.3 Imbedding at the End of the Frame in a TDMA Case

Let us consider a TDMA system where there are N_s slots per frame and it is possible to allocate multiple service opportunities, say them b, per frame to serve our queue and the related traffic (see Fig. 5.10a): we can transmit up to b packets per frame from our queue. We have Poisson arrivals of messages with length $\ell \geq 1$ and PGF $L(z)$. Each packet needs one slot to be transmitted. We can imbed the system at the end of each frame. We can solve this system by applying the considerations made in Sect. 5.7.2. Then, we have bulk arrivals and batched service. In this case, the allocations of slots to a traffic source are on a frame basis. According to the previous notations, this queuing system can be denoted as $\mathrm{M}^{[L(z)]}/\mathrm{D}^{[b]}/1$. The difference equation (5.1) has to be modified when $n_i \neq 0$, thus obtaining the following expression:

$$n_{i+1} = \max\{n_i - b,\ 0\} + a_{i+1} \tag{5.128}$$

where a_i denotes the number of packets arrived at the queue in a frame interval.

The arrival process has a PGF obtained as the nth power exponent of the PGF of the arrivals in a single output slot, that is $A(z) = e^{\lambda \tau N_s [L(z)-1]}$. The derivatives of this PGF are

$$A'(1) = \lambda \tau N_s L'(1)$$
$$A''(1) = [\lambda \tau N_s L'(1)]^2 + \lambda \tau N_s L''(1) \tag{5.129}$$

For the case $b = 2$, this system can be solved by re-applying the same method shown in Sect. 5.7.2 but with a different $A(z)$ function. According to (5.83), we have

$$N_p = A'(1) + \frac{1}{1-z_1} + \frac{A''(1) - 2}{2[2 - A'(1)]}$$
$$= \lambda \tau N_s L'(1) + \frac{1}{1-z_1} + \frac{[\lambda \tau N_s L'(1)]^2 + \lambda \tau N_s L''(1) - 2}{2[2 - \lambda \tau N_s L'(1)]} \quad \text{[pkts]} \tag{5.130}$$

Using the Rouché theorem, z_1 is the solution different from 1 inside the unit circle of the equation $z^2 - e^{\lambda \tau N_s [L(z)-1]} = 0$. Following the same approach as in [2], z_1 can be obtained as the solution of the following equation:

$$z_1 = e^{j\pi + \frac{\lambda \tau N_s}{2}[L(z_1)-1]} \quad \Rightarrow \quad z_1 = -e^{\frac{\lambda \tau N_s}{2}[L(z_1)-1]} \tag{5.131}$$

We need a numerical method to solve this equation and to approximate z_1. The stability condition is $\lambda \tau N_s L'(1) < 2$ (pkts/frame).

The mean delay can be determined by applying the Little theorem, dividing N_p by the mean arrival rate of packets $\lambda \tau N_s L'(1)$ in pkts/frame. The mean packet delay is measured with granularity at the level of the embedding interval, that is, at the frame level; this is a coarse granularity for the delay.

$$T_p = 1 + \frac{1}{\lambda \tau N_s L'(1)(1-z_1)} +$$
$$+ \frac{\lambda \tau N_s L'(1) + {}^{L''(1)}\!/_{L'(1)} - {}^{2}\!/_{[\lambda \tau N_s L'(1)]}}{2[2 - \lambda \tau N_s L'(1)]} \quad \text{[frames]} \tag{5.132}$$

We would like to provide a heuristic proof that there is no singularity for $\lambda \to 0$ in the mean delay T_p in (5.132). In general, we see that the solution different from 1 of the equation $z^2 - e^{\lambda \tau N_s [L(z)-1]} = 0$ tends to -1 for $\lambda \to 0$ so that $z_1 \to -1$. Then, if we like to study the singularity of T_p for $\lambda \to 0$, we need the solution around -1 for the Eq. (5.131). We perform this study in the simple case of modified geometrical distribution of the message length with mean value $L'(1)$. Then, we consider

$$z_1 = -e^{\frac{\lambda \tau N_s}{2}[L(z_1)-1]}, \quad \text{where } L(z_1) = \frac{\frac{z_1}{L'(1)}}{1 - z_1\left(1 - \frac{1}{L'(1)}\right)} \tag{5.133}$$

We can have a rough estimate of the solution z_1 if we take the zero-order expansion of the left term of (5.133) centered in $z = -1$. We obtain the following result:

$$z_1 \approx -e^{\frac{\lambda \tau N_s L'(1)}{1 - 2L'(1)}} \tag{5.134}$$

We can verify that the z_1 value given in (5.134) tends to -1 for $\lambda \to 0$. In general, for $\lambda \to 0$, we can see that the following terms of $\frac{1}{1-z_1} - \frac{1}{2 - \lambda \tau N_s L'(1)} \to \frac{1}{2} - \frac{1}{2}$ tend to 0. Now if we divide this quantity that tends to 0 by $\lambda \tau N_s L'(1)$ to apply the Little theorem, we have an indeterminate case of the type 0/0 that can be analyzed through the Hôpital rule. We can then verify that this term tends to a finite value different from 0 when $\lambda \to 0$. In conclusion, we have

$$T_p(\lambda \to 0) \approx L'(1) + \frac{1}{2}\left(\frac{1 - L'(1)}{2L'(1) - 1}\right) \tag{5.135}$$

Note that the results we achieve in this case are not directly comparable with the others within "Case B" because now the organization of the transmission is different, and only b slots out of N_s are available to serve the traffic of our queue.

5.8.2.4 Imbedding at the End of the Message Transmission

In this case, referring to Fig. 5.10b, we have to solve the following difference equations:

$$\begin{aligned} n_{i+1} &= n_i - 1 + a_{i+1}, \quad if\ n_i > 0 \\ n_{i+1} &= \Delta + a_{i+1}, \quad if\ n_i = 0 \end{aligned} \tag{5.136}$$

where n_i, a_i, and Δ have a different meaning than before in Sect. 5.8.2.1 since they refer to the number of message arrivals in an imbedding interval, representing the message's service time with length in slots given by the PGF $L(z)$. The PGFs $\Delta(z)$ and $A(z)$ can be expressed as follows:

$$\begin{aligned} A(z) &= L\left[e^{\lambda\tau(z-1)}\right] \\ \Delta(z) &= \int_0^{\tau} e^{\lambda t(z-1)} f_\delta(t)\,\mathrm{d}t = \frac{e^{-\lambda\tau}}{(1-e^{-\lambda\tau})}\frac{e^{\lambda\tau z}-1}{z} \end{aligned} \tag{5.137}$$

where $f_\delta(t)$ is the same pdf of the time δ expressed in (5.110) and where $\Delta(z)$ can be obtained from $W(z)$ in (5.112), considering $L(z) = z$.

To determine the mean number of messages in the queue N_m we need to apply the solution formula in (5.74), and then we compute the first and second derivatives of both $\Delta(z) = W(z)$ and $A(z)$ for $z = 1$ as follows:

$$\begin{aligned} A'(1) &= \lambda\tau L'(1) \\ A''(1) &= (\lambda\tau)^2\left[L''(1) + L'(1)\right] \\ \Delta'(1) &= \frac{\lambda\tau}{(1-e^{-\lambda\tau})} - 1 \\ \Delta''(1) &= \frac{\lambda^2\tau^2 - 2\lambda\tau}{(1-e^{-\lambda\tau})} + 2 \end{aligned}$$

where $\Delta'(1)$ and $\Delta''(1)$ can also be obtained from (5.113), considering $L(z) = z$ so that $L'(1) = 1$ and $L''(1) = 0$.

Then, the mean number of messages N_m can be obtained using Eq. (5.74) for the case of differentiated service times as

$$N_\mathrm{m} = \frac{\lambda\tau}{2} + \lambda\tau L'(1) + \frac{(\lambda\tau)^2\left[L''(1) + L'(1)\right]}{2\left[1 - \lambda\tau L'(1)\right]}\ [\text{pkts}] \tag{5.138}$$

We can apply the Little theorem to (5.138) dividing N_m by λ (the mean message arrival rate) as

$$T_\mathrm{m} = \frac{\tau}{2} + \tau L'(1) + \frac{\lambda\tau^2\left[L''(1) + L'(1)\right]}{2\left[1 - \lambda\tau L'(1)\right]}\ [\text{s}] \tag{5.139}$$

We can see that the mean message delay has a classical M/G/1 term plus an additional delay of $\frac{\tau}{2}$ to account for the synchronization delay for the arrival of messages at an empty buffer. We can also see that there is no proportionality relationship between T_m in (5.139) and T_p in (5.116). Finally, the stability condition here is $\lambda\tau L'(1) < 1$ Erl that is the same as the previous cases with imbedding at the packet level.

Under the condition that the output slot can be unavailable with probability p, we can still apply the difference equation (5.136), but now $A(z)$ is the PGF of the number of message arrivals not in a number of slots τ according to the PGF $L(z)$, but in a number of slots according to the composition of the PGFs $L(z)$ and $T_s(z)$. In particular, since each packet of the message needs a service time expressed by the PGF $T_s(z)$, for a message with ℓ packets

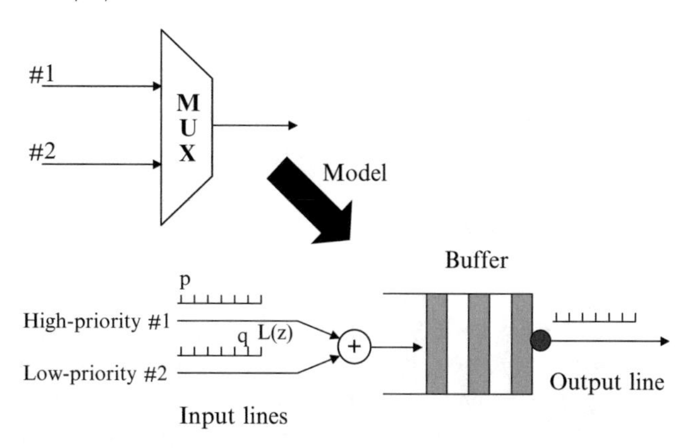

Fig. 5.11: Multiplexer with different priorities for the input lines and no room limitations for the packets

we have a PGF of the service time as $T_s^\ell(z)$. We remove the conditioning on ℓ using the probability distribution corresponding to the PGF $L(z)$. Then, the PGF of the message service time in slots is $L[T_s(z)]$. In each slot, we have a PGF of the number of message arrivals as $e^{\lambda\tau(z-1)}$, and then, the new $A(z)$, PGF of the number of message arrivals in the service time of a message, results as

$$A(z) = L\left[T_s\left(e^{\lambda\tau(z-1)}\right)\right] \tag{5.140}$$

To determine the mean number of messages in the queue N_{m} we need to apply the solution formula (5.74) again with the new $A(z)$ but using the same $W(z)$ function as before. Then, we need to compute the first and second derivatives of the new $A(z)$ for $z = 1$ as

$$\begin{aligned} A'(1) &= \lambda\tau L'(1)\,T_s'(1) \\ A''(1) &= (\lambda\tau)^2\left\{L''(1)\,[T_s'(1)]^2 + L'(1)\,T_s''(1) + L'(1)\,T_s'(1)\right\} \end{aligned} \tag{5.141}$$

In conclusion, the mean message delay, in this case, results as

$$T_{\mathrm{m}} = \frac{\tau}{2} + \tau L'(1)\,T_s'(1) + \frac{\lambda\tau^2\left\{L''(1)\,[T_s'(1)]^2 + L'(1)\,T_s''(1) + L'(1)\,T_s'(1)\right\}}{2\,[1-\lambda\tau L'(1)\,T_s'(1)]}\ [\mathrm{s}] \tag{5.142}$$

5.8.3 Case C: Time-Slotted and Memoryless Arrival Process and Slotted Service Times

As for discrete-time systems with memoryless arrival processes (i.e., Bernoulli or binomial processes) of packets, we can reapply the above considerations and the different imbedding options as in Sect. 5.8.2 (i.e., "Case B").

For a Binomial arrival process (as a Bernoulli compound process), the generalization of M/G/1 results from imbedding instants to any instant (random observer) is not possible because we cannot apply the Kleinrock principle and the BASTA property. Hence, different imbedding options result in different mean values (still having the same stability limit); however, we can reasonably expect that the two cases should not differ significantly from a numerical standpoint.

We consider here a multiplexer receiving two synchronous time division input traffic flows (see Fig. 5.11). These two input lines have different priorities:

- Each slot of the high-priority line conveys a packet with probability p.
- Each slot of the low-priority line conveys one message with probability q; each message is composed of a random number of packets, $\ell \geq 1$, according to the PGF $L(z)$.

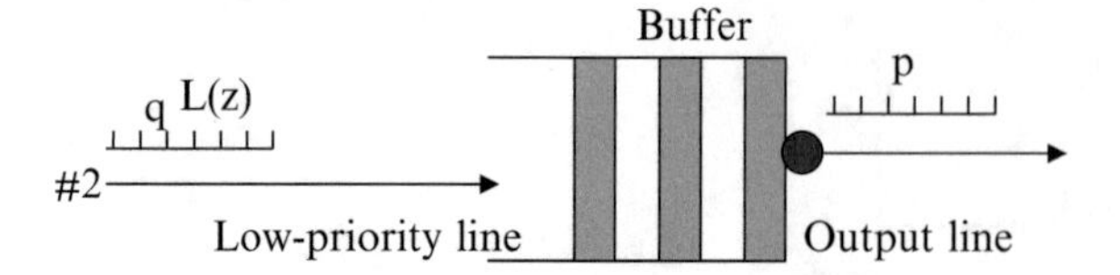

Fig. 5.12: Equivalent model for low-priority traffic

This is a generalization of "Case C" in Sect. 5.8 with only the low-priority traffic and $p = 0$. We assume that output and input lines are synchronous and have the same slot duration, τ, suitable for the transmission of one packet. The multiplexer stores the packets before transmission in a buffer of infinite capacity. We have to study the queuing phenomena experienced by the low-priority line due to the presence of the packets of the high-priority line.

High-priority traffic only sees itself: for its correct behavior, we have to consider $p < 1$ (stability). There is no waiting part in the queue for high-priority traffic since input and output lines have the same speed. We will further discuss and prove this aspect later, adapting the same analysis of the low-priority traffic but considering a simple Bernoulli arrival process and the prioritization in these packets' service.

Due to high-priority traffic, the slots of the output line are available for the low-priority packets with probability $1 - p$ and unavailable with probability p. Hence, the equivalent service model for low-priority traffic is shown in Fig. 5.12.

Figure 5.12 describes the system for low-priority traffic, evolving at discrete-time instants. As explained in Sect. 5.8, we can consider three different imbedding options to solve this discrete-time system as an "M"/G/1 queue. These different possibilities correspond to different meanings for n_i and a_i and entail distinct expressions for the difference equations modeling the system. More details are as follows.

5.8.3.1 Imbedding at Transmission End of Low-Priority Packets

The service time of a low-priority packet is according to a modified geometric distribution with parameter p, depending on the availability of output slots. In this case, n_i denotes the number of packets in the buffer from the low-priority line at the end of the transmission of the ith low-priority packet; moreover, a_i denotes the number of packets from the low-priority line arrived at the buffer during the service time of the ith low-priority packet.

We have to solve the following difference equations similarly to (5.108) for the case with differentiated service times:

$$\begin{aligned} n_{i+1} &= n_i - 1 + a_{i+1}, \quad if\ n_i > 0 \\ n_{i+1} &= \ell - 1 + \Delta + a_{i+1}, \quad if\ n_i = 0 \end{aligned} \tag{5.143}$$

where $\Delta = 0$ since the slot arrival process is synchronized with the slotted-based packet service.

Then, we can solve this problem using the "differentiated service time" case in Sect. 5.7.1. In particular, referring to the notation in Sect. 5.7.1, we have that $W(z) = L(z)/z$ and $A(z)$ is the PGF of the number of low-priority packets arrived at the buffer during the service time of a low-priority packet.

Let us derive $A(z)$. The transmission time t of a low-priority packet is according to a modified geometric distribution, as

$$\text{Prob}\left\{\begin{array}{l}\text{number of slots needed to} \\ \text{transmit a low-priority packet} = n\end{array}\right\} = (1-p)\,p^{n-1} \tag{5.144}$$

The corresponding PGF is

$$T_s(z) = \sum_{n=1}^{\infty} z^n\,(1-p)\,p^{n-1} = \frac{z\,(1-p)}{1 - zp} \tag{5.145}$$

The PGF of the number of low-priority packets arrived at the buffer in a time slot is $A(z|\text{slot}) = 1 - q + qL(z)$. Since arrivals are uncorrelated from slot to slot, we have $A(z|n \text{ slots}) = [1 - q + qL(z)]^n$. We remove the conditioning on n using distribution (5.144) as

$$A(z) = \sum_{n=1}^{\infty} [1-q+qL(z)]^n (1-p)\, p^{n-1} = T_s\left[A\left(z\middle|\text{slot}\right)\right]$$
$$= \frac{[1-q+qL(z)]\,(1-p)}{1-[1-q+qL(z)]\,p} \tag{5.146}$$

Note that the expression of $A(z)$ is different from that of the next case in Sect. 5.8.3.2 (see (5.154)). The derivatives of $A(z)$ computed in $z=1$ can be obtained as

$$A'(z=1) = \frac{\mathrm{d}}{\mathrm{d}z} T_s\left[A\left(z\middle|\text{slot}\right)\right]\Big|_{z=1} = T_s'\left[A\left(z\middle|\text{slot}\right)\right] \times A'\left(z\middle|\text{slot}\right)\Big|_{z=1}$$
$$= T_s'[1] \times A'\left(1\middle|\text{slot}\right) = \frac{qL'(1)}{1-p}$$
$$A''(z=1) = \frac{\mathrm{d}}{\mathrm{d}z} T_s{}'\left[A\left(z\middle|\text{slot}\right)\right] \times A'\left(z\middle|\text{slot}\right)\Big|_{z=1}$$
$$= T_s{}''\left[A\left(z\middle|\text{slot}\right)\right] \times \left[A'\left(z\middle|\text{slot}\right)\right]^2\Big|_{z=1}$$
$$+ T_s{}'\left[A\left(z\middle|\text{slot}\right)\right] \times A''\left(z\middle|\text{slot}\right)\Big|_{z=1}$$
$$= T_s{}''[1] \times \left[A'\left(1\middle|\text{slot}\right)\right]^2 + T_s{}'[1] \times A''\left(1\middle|\text{slot}\right)$$
$$= \frac{2p}{(1-p)^2}[qL'(1)]^2 + \frac{1}{1-p} qL''(1)$$

The buffer stability condition is $A'(z=1) < 1$ pkts/slot $\Rightarrow \frac{qL'(1)}{(1-p)} < 1$ pkts/slot $\Rightarrow qL'(1) + p < 1$ pkts/slot. Note that this is the same stability condition derived in the previous case with different imbedding instants (see Sect. 5.8.3.2).

The derivatives of $W(z)$ are

$$W'(1) = L'(1) - 1$$
$$W''(1) = L''(1) - 2L'(1) + 2 \tag{5.147}$$

Then, we can determine the mean number of packets N_p applying (5.74) as

$$N_\text{p} = \frac{L''(1)}{2L'(1)} + \frac{qL'(1)}{1-p} + \frac{\frac{2p}{(1-p)^2}[qL'(1)]^2 + \frac{1}{1-p} qL''(1)}{2\left[1-\frac{qL'(1)}{1-p}\right]} \quad \text{[pkts]} \tag{5.148}$$

The mean packet delay could be obtained from N_p in (5.148) dividing by the mean packet arrival rate of $qL'(z=1)$ pkts/slot according to the Little theorem. However, the problem is that we cannot apply the Little theorem to the term $\frac{L''(1)}{2L'(1)}$ in (5.148) because it does not correspond to an arrival process [3]. Then, as a first approximation, we could apply the Little theorem to the other terms in (5.148), as if we adopt the following approximation for the difference equations when $n_i = 0$: $n_{i+1} = a_{i+1}$. Then, we achieve the following result for the mean packet delay T_p:

$$T_\text{p} \approx \frac{1}{1-p} + \frac{\frac{2pqL'(1)}{(1-p)^2} + \frac{L''(1)\big/L'(1)}{1-p}}{2\left[1-\frac{qL'(1)}{1-p}\right]} \quad \text{[slots]} \tag{5.149}$$

The results in (5.148) and (5.149) can be adapted to the pure "Case C" if we consider $p = 0$ so that the low-priority traffic is the only traffic that is present and we can determine N_p and T_p for a slotted input arrival process (Bernoulli compound) and slotted-based service time. We have

$$N_{\mathrm{p}} = \frac{L''(1)}{2L'(1)} + qL'(1) + \frac{qL''(1)}{2[1 - qL'(1)]} \quad \text{[pkts]} \tag{5.150}$$

$$T_{\mathrm{p}} \approx 1 + \frac{L''(1)/L'(1)}{2[1 - qL'(1)]} \quad \text{[slots]} \tag{5.151}$$

5.8.3.2 Imbedding at Slot End of the Output Line

Let n_i denote the number of packets in the buffer from the low-priority line at the end of the ith slot of the output line; let a_i denote the number of packets arrived at the buffer from the low-priority line during the ith slot. We can write the following balance:

$$n_{i+1} = \begin{cases} n_i - m + a_{i+1}, & n_i > 0 \\ a_{i+1}, & n_i = 0 \end{cases} \tag{5.152}$$

where m is a Bernoulli random variable characterized below:

$$m = \begin{cases} 1, & \text{with prob. } 1 - p \\ 0, & \text{with prob. } p \end{cases} \tag{5.153}$$

To justify the above Eq. (5.152), we can refer to the same considerations made in Sect. 5.6. Moreover, we can solve the difference equation in (5.152), considering that it is formally the same as (5.123) in Sect. 5.8.2.2. The only difference is the definitions of n_i and a_i so that now we have to change the PGF $A(z)$ because of the slotted-based arrival process as

$$A(z) = 1 - q + qL(z) \tag{5.154}$$

This $A(z)$ is the PGF of the number of packets arrived at the buffer in a slot from the low-priority line. Then, the PGF of the number of packets in the queue (from the low-priority line) is still given by (5.124) and (5.125). The mean number of packets in the buffer from the low-priority line can be obtained from (5.126), where we use the derivatives of $A(z)$ in (5.154), that is $A'(z=1) = qL'(z=1)$ and $A''(z=1) = qL''(z=1)$. The system is stable under the condition that $P_0 > 0 \Rightarrow 1 - p - A'(z=1) > 0 \Rightarrow p + qL'(z=1) < 1$ pkts/slot. We achieve the following result for N_{p}:

$$N_{\mathrm{p}} = qL'(1) + \frac{pqL'(1)}{1 - p - qL'(1)} + \frac{qL''(1)}{2[1 - p - qL'(1)]} \quad \text{[pkts]} \tag{5.155}$$

We can apply the Little theorem to derive the mean packet delay experienced by the packets of the low-priority line, T_{p}, by considering that $A'(z=1) = qL'(1)$ denotes the mean number of packets from the low-priority line arrived at the buffer in a time slot; hence, $A'(z=1)$ is the mean packet arrival rate in pkts/slot.

$$T_{\mathrm{p}} = 1 + \frac{p}{1 - p - qL'(1)} + \frac{L''(1)/L'(1)}{2[1 - p - qL'(1)]} \quad \text{[slots]} \tag{5.156}$$

Through some algebraic manipulations, we can easily prove that the approximate result in terms of T_{p} in (5.149) is equal to (5.156), which refers to different imbedding instants. On the other hand, the results in terms of N_{p} are different because in any case we cannot apply BASTA and the Kleinrock principle due to bulk arrivals.

In conclusion, we can state that imbedding the queue at the end of the output slot is much simpler than imbedding at the end of the packet transmission time.

The results in (5.155) and (5.156) achieved here can be adapted to the pure "Case C" if we consider $p = 0$ so that the low-priority traffic is the only traffic that is present and we can determine N_{p} and T_{p} under these conditions with a slotted input arrival (Bernoulli compound) process and slotted-based service time. We have

$$N_{\mathrm{p}} = qL'(1) + \frac{qL''(1)}{2[1 - qL'(1)]} \quad \text{[pkts]} \tag{5.157}$$

$$T_{\mathrm{p}} = 1 + \frac{L''(1)/L'(1)}{2\left[1 - qL'(1)\right]} \text{ [slots]} \tag{5.158}$$

Note that we can also analyze the high-priority line still imbedding at the end of the output slot. In this case, the server is always available (i.e., $m = 1$) and arrivals are of single packets (i.e., $L(z) = z$). Then, the arrival process is Bernoulli so that the PGF of the number of high-priority packets on a slot basis is $A(z) = 1 - p + pz$. We can adopt the classical M/G/1 difference equations (5.1) to analyze high-priority traffic. The stability limit is $A'(1) = p < 1$. Then, the mean number of packets from (5.16) is $N_{\mathrm{p}} = p$ [packets] and the mean packet delay applying the Little theorem is $T_{\mathrm{p}} = 1$ [slot]: there is no waiting time for high-priority packets (see also Exercise 5.14 at the end of this chapter). Since we can apply BASTA and the Kleinrock principle, these results are valid at any time instant. These results are also valid if we imbed the system at the end of the transmission of high-priority packets.

5.8.3.3 Imbedding at the Transmission End of the Low-Priority Messages

In this case, n_i represents the number of messages in the buffer (from the low-priority line) at the instant of transmission completion of the ith low-priority message, whereas a_i is the number of messages (from the low-priority line) arrived at the buffer during the service time of the ith message. Such service time depends on two random phenomena: (1) the availability of output slots for low-priority traffic with probability $1 - p$; (2) the number of packets arrived per message. In this case, we can apply the classical M/G/1 theory and write the same balance equation as in (5.1) to model the system. Hence, the mean number of messages N_{m} is given by (5.16), where $A(z)$ is the PGF of the number of low-priority messages arrived at the buffer during the service time of a low-priority message.

Let us derive $A(z)$ conditioning on the service time of a message with ℓ packets (low-priority line): $A(z|\ell \text{ pkts})$. If $\ell = 1$, $A(z|\text{one pkt})$ denotes the PGF of the number of messages arrived during the service time of one packet: $A(z|\text{one pkt}) = T_s[1 - q + zq]$, where $T_s(z)$ is given by (5.145) and $1 - q + zq$ denotes the PGF of the number of messages arrived in a slot. Then, $A(z|\ell \text{ pkts})$ corresponds to the sum of the arrivals on ℓ slots. Since the arrivals in different slots are independent, $A(z|\ell \text{ pkts}) = [A(z|\text{one pkt})]^{\ell}$. We remove the conditioning using the distribution of ℓ, which is characterized by the PGF $L(z)$:

$$\begin{aligned} A(z) &= \sum_{\ell=1}^{\infty} \left[A\left(z\middle|\text{one pkt}\right)\right]^{\ell} \text{Prob}\left\{\text{message with } \ell \text{ pkts}\right\} \\ &= L\left[T_s\left(1 - q + zq\right)\right] = L\left[\frac{\left(1 - q + zq\right)\left(1 - p\right)}{1 - \left(1 - q + zq\right)p}\right] \end{aligned} \tag{5.159}$$

It is easy to obtain $A(z)$ derivatives to express the mean number of messages N_{m} in the queue according to (5.16). In particular, we have

$$\begin{aligned} A'\left(z = 1\right) &= L'(1)\frac{1}{1 - p}q \\ A''\left(z = 1\right) &= \left(\frac{q}{1 - p}\right)^2 \left[2pL'(1) + L''(1)\right] \end{aligned} \tag{5.160}$$

The stability condition is now $A'(z=1) < 1 \Rightarrow \frac{qL'(z=1)}{(1-p)} < 1$ pkts/slot, the same condition as in all other imbedding cases for this system.

In this case, we cannot apply the Pollaczek–Khinchin formula (5.18) to express the mean message delay T_{m} since the arrival process is compound Poisson. Then, T_{m} is obtained by dividing the mean number of messages in the buffer (5.16) where $A'(z=1)$ and $A''(z=1)$ are given in (5.160) by the mean message arrival rate, corresponding to q messages/slot:

$$T_{\mathrm{m}} = \frac{N_{\mathrm{m}}}{q} = \frac{L'(1)}{1 - p} + \frac{\frac{q}{(1-p)^2}\left[2pL'(1) + L''(1)\right]}{2\left[1 - \frac{qL'(1)}{1-p}\right]} \text{ [slots]} \tag{5.161}$$

Note that (5.161) is not proportional to (5.149) through the mean message length in packets (i.e., $L'(z=1) \times T_\text{p} \neq T_\text{m}$), due to the queuing part of the formula, as already commented in Sect. 5.8.2.4.

For the sake of completeness, in the case $p = 0$, Eq. (5.161) yields the following result for the mean message delay:

$$T_\text{m} = L'(1) + \frac{qL''(1)}{2\left[1 - qL'(1)\right]} \quad \text{[slots]} \tag{5.162}$$

As a final comment to the examples shown in Sect. 5.8.3, it is important to remark that we imbed the chain at the packet transmission completion instants to study the statistical parameters related to the packets (i.e., mean number of packets in the buffer and mean packet delay); instead, we imbed the chain at the message transmission completion instants to evaluate the performance at the message level (i.e., the mean number of messages in the buffer and mean message delay).

5.9 Exercises

This section contains exercises on the M/G/1 theory with some applications to the packet-based traffic, to the Automatic ReQuest repeat (ARQ) protocol, etc.

Ex. 5.1 We consider a transmission system with a buffer. The transmitter is used to send packets on a radio channel. We know that:

- Packets arrive in groups of messages.
- Messages arrive according to exponentially distributed interarrival times with a mean value equal to T_a seconds.
- The length l_m of a message in packets is according to the following distribution (memoryless from message to message):

$$\text{Prob}\left\{l_\text{m} = n \text{ pkts}\right\} = q(1-q)^{n-1}, \quad n \in \{1, 2, \ \ldots\}$$

- The buffer has infinite capacity.
- The radio channel causes a packet to be received with errors with probability p; packet errors are memoryless from packet to packet.
- An ARQ scheme is adopted to manage the retransmissions of the packets received with errors.
- Round-trip propagation delays in receiving ACKs (to notify the correct receipt of a packet) are negligible with respect to the deterministic packet transmission time, T.
- A packet remains in the buffer until its ACK is received.

We have to determine the mean number of packets in the buffer and the mean delay that a packet experiences from its arrival at the buffer to its last (and successful) transmission.

Ex. 5.2 Messages arrive at a node of a telecommunication network to be transmitted on an output line. From measurements, we know that the arrival process and the service process are characterized as follows:

- Interarrival times υ are distributed so that $E[\upsilon^2] \approx 2E[\upsilon]^2$.
- The message service time τ has a distribution so that $E[\tau^2] \approx E[\tau]^2$.

A suitable queuing model should be envisaged for this system to determine the mean delay experienced by a message to cross the node.

Ex. 5.3 We consider a Time Division Multiplexing (TDM) transmission line with a buffer receiving regulated input traffic from U sources. The TDM slot duration coincides with the packet transmission time. The regulation of each traffic source operates as follows: (1) a source generates one packet in a slot with probability g; (2) a source generating one packet does not generate further packets until the previous one has been transmitted. Considering a generic number n of packets in the buffer, the packet arrival process on a slot basis is characterized by the following conditional probability:

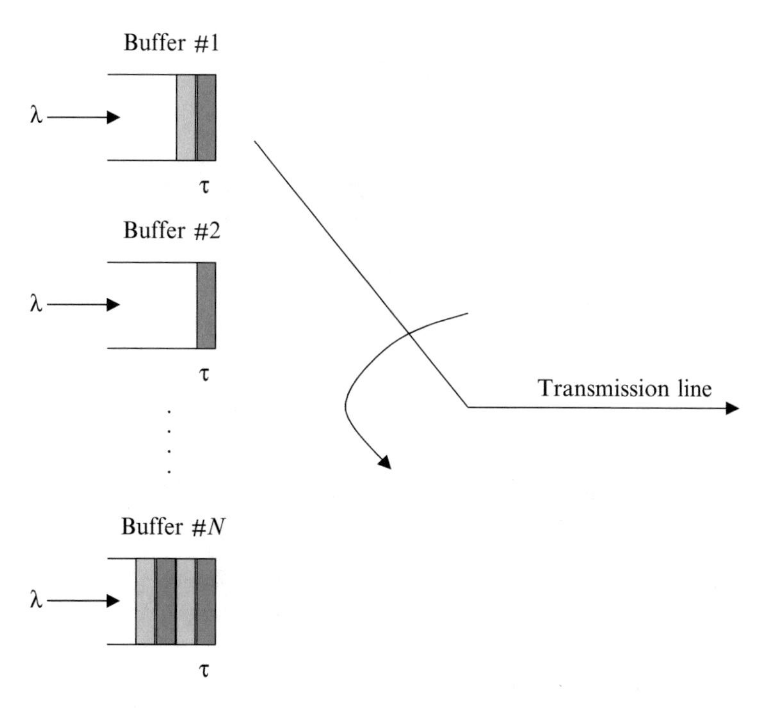

Fig. 5.13: Transmission system on a shared line

$$\begin{aligned}&\text{Prob}\left\{a = l \text{ pkts arrived in a slot}\middle| n\right\}\\&= \text{Prob}\{a_n = l\}\\&= \begin{cases}\binom{U-n}{l} g^l (1-g)^{U-n-l}, \text{ for } 0 \le l \le U-n\\ 0, \text{otherwise}\end{cases}\end{aligned}$$

for $n \in \{0, 1, 2, \ldots, U\}$ and $l \in \{0, \ldots, U-n\}$.

It is requested to model this system in the case $U = 2$ and derive the mean number of packets in the buffer as a function of g.

Ex. 5.4 We have a buffer of a transmission line that receives messages coming from two independent processes:

- *First traffic*: Poisson message arrival process with mean rate λ_1 and exponentially distributed service time with mean rate μ_1
- *Second traffic*: Poisson message arrival process with mean rate λ_2 and exponentially distributed service time with mean rate μ_2

Assuming $\mu_1 \neq \mu_2$, we have to determine the mean delay from the message arrival at the buffer (from one of the two input processes) to the message transmission completion.

Ex. 5.5 Let us consider a traffic source that generates packets according to a Poisson arrival process with a mean rate λ arrivals/s. This traffic is controlled by a *leaky bucket regulator*, allowing packets to be transmitted at a rate of 1 packet every T_c s. It is requested to model this system and determine the following quantities: the probability that an arriving packet finds an empty leaky bucket and the mean waiting time experienced by a packet in the leaky bucket regulator before starting its transmission.

Ex. 5.6 We consider the transmission system outlined in Fig. 5.13: we have N input traffic flows (each characterized by an independent Poisson arrival process of packets with mean rate λ), which correspond to distinct buffers served by a shared transmission line. Let τ denote the packet transmission time.The transmission line serves the different buffers cyclically: it transmits a packet from a buffer (if it is not empty) and then switches instantaneously to

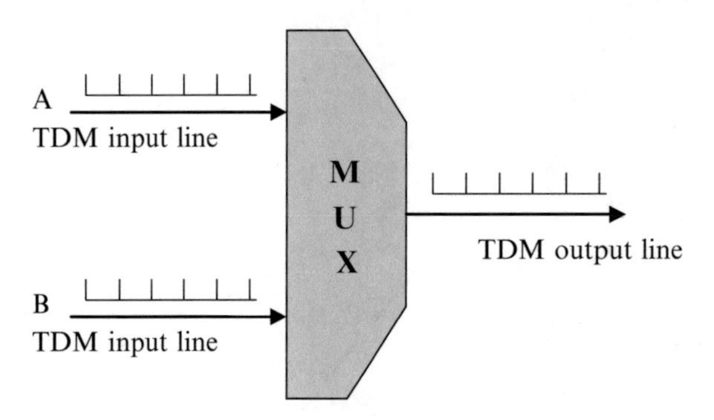

Fig. 5.14: Multiplexer with two input lines and one output line

service the next buffer according to a fixed service cycle.[8] We have to determine the mean delay experienced by a packet from its arrival at the system to its departure.

Ex. 5.7 A multiplexer receives traffic from two Time Division Multiplexing (TDM) input lines (line A and line B) and has a single TDM output line, as shown in Fig. 5.14.Let us assume:

- A time slot, T_c, of all TDM lines can convey one packet.
- Input and output TDM lines are synchronized.
- The number of packets n_A arrived from line A at the multiplexer in T_c is according to a Poisson distribution with mean value λT_c.
- The number of packets n_B arrived from line B at the multiplexer in T_c is according to the following distribution:

$$\text{Prob}\{n_B = k \text{ pkts}\} = q(1-q)^k, \quad k \in \{0, 1, \dots\}$$

- Both arrival processes from the two input lines are discrete-time, independent, and memoryless from slot to slot.

We have to determine: (1) the stability condition for the buffer of the multiplexer; (2) the mean number of packets in the buffer; and (3) the mean packet delay from the arrival of a packet at the multiplexer (from one of the two input lines) to its transmission on the output line.

Ex. 5.8 We refer to a leaky bucket regulator that "filters" the packets generated by a traffic source. This regulator can send a packet every time T. We have a Time Division Multiplexing (TDM) line at the regulator's input and output. These lines are synchronous with slot duration T. The packet arrival process (input line) is characterized as follows:

- A slot carries a message with probability q; otherwise, it is empty.
- Each message comprises a random number of packets with PGF $L(z)$; note that a message has a maximum length of L_{max} packets.

  It is requested to evaluate the following quantities:

- The mean delay experienced by a packet from input to output of the regulator
- The burstiness of the output traffic to be compared with that of the input traffic

Ex. 5.9 We have a transmission buffer where messages arrive according to a Poisson process (mean rate λ) and have a general service time distribution with pdf $g(t)$. We need to characterize the message service completion process for this M/G/1 system.

Ex. 5.10 Let us consider a multiplexer receiving input traffic due to many elementary contributions. Packets arrive at the multiplexer according to a Poisson process with a mean rate λ. An output Time Division Multiplexing

[8] This is a special case of the round-robin scheme with a threshold, which can be studied on the basis of what is written in Sects. 6.3.1 and 6.3.3. Other schemes could also be considered here, like the exhaustive service or the gated service, as explained in Sects. 6.3.1 and 6.3.3. These aspects are however beyond the scope of the present exercise.

(TDM) line is used. A packet is transmitted in a slot of duration T. It is requested to determine the mean number of packets in the buffer and the mean delay experienced by a packet from its arrival at the buffer to its transmission completion.

Ex. 5.11 Let us consider a multiplexer receiving input traffic from N synchronous Time Division Multiplexing (TDM) lines. Each input slot has a duration T and may convey one packet with probability p, uncorrelated from slot to slot and from line to line. At the multiplexer's output, we consider two synchronous lines, each requiring a time T to transmit a packet. We have to model this system and determine the probability generating function of the number of packets in the multiplexer.

Ex. 5.12 We have a traffic source, which injects packets into the network according to a token bucket regulator. Packets arrive at the regulator's buffer according to a Poisson process with mean interarrival times equal to T. The effect of the regulator on the transmission of the packets is modeled as follows: a packet arriving at the head of the buffer finds an available token for its immediate transmission with probability p; otherwise (i.e., with probability $1-p$), the packet has to wait for a token according to an exponentially distributed time with mean rate μ. For the sake of simplicity, we neglect the transmission time for a packet that has received its token. We have to evaluate the mean delay experienced by a packet to be injected into the network.

Ex. 5.13 We consider the data traffic flow generated by a given user (host); this flow first crosses an IP layer queue and then a MAC layer (tandem) queue. IP packets arriving at the layer 2 queue are fragmented to generate fixed-length layer 2 packets (padding is used), whose transmission time is τ. The length of an IP packet in layer 2 (MAC) packets is modeled using a random variable with modified geometric distribution and mean value L. Let us assume that the arrival process of IP packets at the layer 2 queue is Poisson with mean interarrival time T. We have to determine the mean delay experienced by a layer 2 packet from the arrival instant at the layer 2 queue to its complete transmission.

Ex. 5.14 Let us consider a queuing system of the "M"/G/1 type modeling a transmission buffer. Referring to imbedding points at service completion instants, the queue is characterized by the classical difference equation: $n_{i+1} = n_i - I(n_i) + a_{i+1}$. We know that a_i is independent of n_i and that the arrival process is memoryless. We have to verify whether the following probability generating function of random variable a_i,

$$A(z) = [1 + (c-1)\,z]\,/c \quad (\text{where } c > 1 \text{ is a constant}),$$

allows an empty waiting list in the buffer (i.e., a request arriving at the system is served immediately).

Ex. 5.15 We have to investigate a queuing system with feedback as follows: (1) message arrivals occur according to a Poisson process with a mean rate λ; (2) the service time of a message is exponentially distributed with a mean rate μ; (3) when the service of a message completes, the message can be fed back to the queue with probability p or leaves the system with probability $1-p$; and (4) a given message always has the same service length every time it crosses the queue. We have to determine the mean delay experienced by a message from its first arrival at the system to the instant when it leaves the system definitively. Here, we have to solve this exercise by applying the M/G/1 theory. However, this exercise can also be solved (with some approximation) using the Jackson theorem, as shown in Chap. 7.

Ex. 5.16 Let us consider a network node where packets arrive from an input TDM line with slot duration equal to T. The slot-based packet arrival process is described by a random interarrival time t_a with the following distribution: $\text{Prob}\{t_a = k \text{ slots}\} = q(1-q)^k$. When a packet arrives at the node, it is routed internally to the node either towards queue #1 with probability p or towards queue #2 with probability $1-p$. Queue #1 has a slotted service process with slot length equal to T (as the input slot). Queue #2 has a slotted service process with a slot length equal to $2T$ (twice the input slot length). The model of the node is described in Fig. 5.15. It is requested to determine: the mean packet delay T_1 that a packet experiences to cross queue #1, the mean packet delay T_2 that a packet experiences to cross queue #2, and the total mean packet delay T from node input to output.

Ex. 5.17 Let us consider a node where messages arrive according to a Poisson process with a mean rate of 10 messages per hour. Each message to be served needs a processing time plus the actual transmission time. These

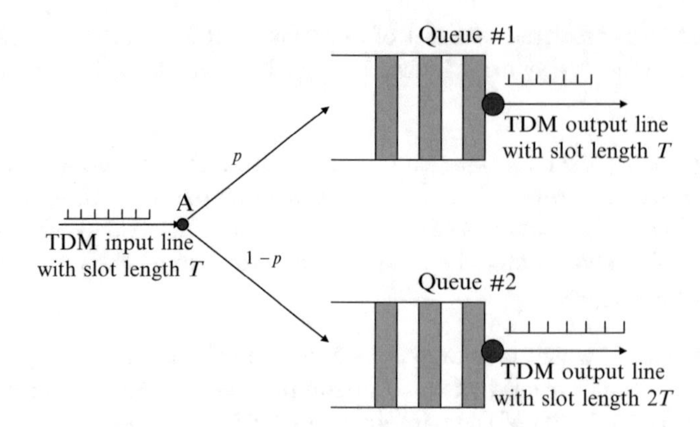

Fig. 5.15: Model of the node considered. Input and output slots are synchronized

two times have independent exponential distributions with the same mean value of 1 min. We have to determine the mean number of messages in the system and the mean message delay.

Ex. 5.18 Let us consider an M/G/1 queue, where the server successfully completes a service time with probability p. If a service is not completed successfully, it is soon repeated until it is successful. Let λ denote the mean message arrival rate and $f(t)$ the pdf of the generic service time X_i. We consider two different cases:

i The repeated service times of a customer are identical: $X_i = X_j = X$.
ii The repeated service times of a customer are different ($X_i \neq X_j$), but iid at each pass according to the distribution with pdf $f(t)$.

We have to characterize the "total service time" Y of a customer (i.e., the time needed until the customer leaves the system) and the mean delay T experienced by a customer in both cases. Are the two cases admitting the same solution? This exercise will be useful to understand the feedback queues in Chap. 7 and the so-called Kleinrock assumption.

{**Hint**: Let N denote the random number of passes through the queue by a customer and let X denote the customer service time of a single pass. X has pdf given by $f(t)$. We assume that N and X are independent. **In case i**, the total service time is $Y = N \times X$: $E[Y] = E[N \times X] = E[N] \times E[X]$ and $E[Y^2] = E[(N \times X)^2] = E[N^2] \times E[X^2]$. **In case ii**, we can derive $E[Y|N = n]$ and $E[Y^2|N = n]$ and then remove the conditioning to obtain $E[Y]$ and $E[Y^2]$.

References

1. Kleinrock L (1976) Queuing systems. Wiley, New York
2. Hayes JF (1986) Modeling and analysis of computer communication networks. Plenum Press, New York
3. Kleinrock L, Gail R (1996) Queuing systems. Wiley, New York
4. Gross D, Harris CM (1974) Fundamentals of queueing theory. Wiley, New York
5. Sztrik J (2016), Basic queueing theory. GlobeEdit, OmniScriptum GmbH & Co, KG, Saarbrucken (2016), ISBN 978-3-639-73471-3
6. Ross SM (1997) Introduction to probability models. Academic, Cambridge
7. Zukerman M (2020) Introduction to Queueing Theory and Stochastic Teletraffic Models. arXiv:1307.2968
8. Andreadis A, Giambene G (2002) Protocols for high-efficiency wireless networks. Kluwer, New York
9. Shortle J, Gross D, Fischer MJ, Masi DMB (2003) Numerical methods for analyzing queues with heavy-tailed distributions. Telecommunications Network Design and Management. Springer, Berlin, pp193–206.
10. Ramsay CM (2007) Exact waiting time and queue size distributions for equilibrium M/G/1 queues with Pareto service. J Queueing Syst 57(4):147–155

11. Giambene G, Hadzic-Puzovic S (2013) Downlink performance analysis for broadband wireless systems with multiple packet sizes. Perform Eval 70(5):364–386
12. Heines TS (1979) Buffer behavior in computer communication systems. IEEE Trans Comput c-28(8):573–576
13. Kawasaki N, Takagi H, Takahashi Y, Hong SJ, Hasegawa T (2000) Waiting time analysis of $M^X/G/1$ queues with&without vacations under random order of service discipline. J Oper Res Soc Japan 43(4):455–468

Chapter 6
Local Area Networks and Analysis

Abstract Local area networks make use of a data link protocol to connect the various hosts in them. This chapter provides a background on Medium Access Protocol (MAC), including Aloha and CSMA types. Methods are presented for the performance analysis of the most common protocols. A special approach, called Equilibrium Point Analysis, has also been discussed for the analysis of random access protocols. Finally, a survey is provided on both IEEE 802.3 (Ethernet) and IEEE 802.11 (WiFi) standards and their evolutions towards higher capacity.

Key words: Medium access protocols, Aloha, CSMA, Ethernet, WiFi

6.1 Introduction

Traditional networks adopt point-to-point links with dedicated resources between couples of nodes. There is no interference among these channels: the transmission between a pair of (source/destination) nodes does not affect the communications of other pairs of nodes. However, point-to-point links require the topology to be fixed, determined during the network design phase. When point-to-point links are not economical, not available, or when dynamic topologies are required, *broadcast channels* can be used. Broadcast channels are characterized by the fact that more than one destination can receive every transmitted message. Radio, television, satellite, wireless systems, and some Local Area Networks (LANs) use broadcast channels. They have both advantages and disadvantages. For instance, if a message is destined to a large number of nodes, then a broadcast channel is the best solution. However, transmissions over a broadcast channel interfere with one another: one transmission coinciding in time with another may cause *interference* (i.e., collisions) so that none of them is received correctly. In other words, the success of a transmission between a pair of nodes is no longer independent of other transmissions. The interference must be avoided or at least controlled to achieve a successful transmission. Moreover, there is the need to manage the retransmissions of lost packets due to collisions.

A broadcast medium is the common choice of LANs, intending with this term short-range networks allowing many terminals to access shared transmission facilities. An important problem in telecommunication systems arises when different terminals need to access a broadcast transmission medium (e.g., to exchange data with a central controller). This is the typical task of Medium Access Control (MAC) protocols, which belong to layer 2 of the OSI reference model [1, 2]. Above the MAC protocol level, the Logical Link Control (LLC) layer can provide a reliable transmission medium to upper layers. Forward Error Correction (FEC) and Automatic reQuest Repeat (ARQ) techniques are implemented at this level to correct packet errors and request the retransmission of packets if the FEC correction has been unable to recover the errors. LLC can provide three different service types to higher layers: (1) connectionless service, (2) connection-oriented service, and (3) end-to-end acknowledged service for connectionless traffic.

Electronic Supplementary Material The online version contains supplementary material available at (https://doi.org/10.1007/978-3-030-75973-5_6).

G. Giambene, *Queuing Theory and Telecommunications*, Textbooks in Telecommunication Engineering,
https://doi.org/10.1007/978-3-030-75973-5_6

MAC protocols were born with the advent of computer networks and the adoption of packet-switched transmissions to allow different traffic flows sharing the same transmission medium. The development of MAC protocols can be related to the diffusion of the Internet. MAC protocol characteristics depend on LAN topology, type of shared medium, and traffic types. A MAC protocol taxonomy is described below based on the scheme according to which resources are assigned to terminals.

1. *Fixed access protocols* granting permission to transmit only to one terminal at a time to avoid collisions of messages on the shared medium. Access rights are statically defined for the terminals. This MAC protocol class encompasses classical techniques to differentiate user transmissions in time, frequency, or code domains (note that hybrid cases are possible). Correspondingly, we have Time Division Multiple Access (TDMA), Frequency Division Multiple Access (FDMA), and Code Division Multiple Access (CDMA).
2. *Contention-based protocols* may give transmission rights to several terminals at the same time. These MAC protocols may cause two or more terminals to transmit simultaneously and their messages to collide on the shared medium. Hence, suitable collision resolution schemes have to be used; these schemes introduce a random delay before attempting the transmission of a collided packet again. This MAC protocol class encompasses Pure Aloha, Slotted Aloha, and Carrier Sense Multiple Access (CSMA).
3. *Demand-assignment protocols* grant access to the network based on requests made by the terminals. Resources used to send requests are distinct from those used for information traffic. Demand-assignment schemes represent a large family of MAC protocols that can belong to one of the following three types: *polling scheme*, *token-based approach*, and *Reservation-Aloha*. There is only one terminal in a polling scheme to transmit at a time through a specific poll request containing the terminal address. This approach can be adopted in broadcast media (e.g., tree or bus) to select one terminal at a time. The token-based approach is used in ring or bus networks: a terminal is enabled to transmit using a token (a generic enabling message, without any address) that is circulated among the terminals of the LAN. Finally, the Reservation-Aloha protocol can be used in radio systems: it adopts a contention phase (signaling) on resources separated in time from those used for information traffic. Terminals can send transmission requests during the contention phase. Once the central controller successfully receives a request, it allocates specific transmission resources to the corresponding terminal.

The problem with many different MAC protocols is that they are suitable for some applications (and corresponding traffic types) but often are not suited to the characteristics of other applications. Fixed access schemes are suitable for constant traffic sources but not efficient for bursty traffic sources. Contention-based techniques are adequate for bursty and sporadic traffic sources. Finally, demand-assignment protocols are appropriate in the presence of bursty sources, generating heavy (almost constant) traffic for sufficiently long time intervals.

Several MAC schemes have been proposed in the literature. The aim is to design an efficient MAC protocol to guarantee differentiated Quality of Service (QoS) levels for many traffic classes. In particular, a MAC protocol should meet the following requirements:

- *Priority*: Managing different traffic classes with suitable priority levels to support differentiated QoS requirements.
- *Fairness*: Fair sharing of resources among the traffic sources of a given traffic class.
- *Latency*: Guaranteeing prompt access to resources for real-time and interactive traffic flows.
- *Efficiency*: Allowing high utilization of resources.
- *Stability*: Guaranteeing the correct management of transmission requests so that the mean rate of packets generated by terminals equals the mean rate of packets successfully delivered to their destinations.

The typical parameters for evaluating the performance of a MAC protocol in a LAN are the *system throughput* (i.e., the degree of utilization of shared transmission resources) and the *mean delay* experienced by a packet or a message.

Various topologies and distinct media can be used in the LANs, as described in Fig. 6.1 [3].

There is a single path from a central controller to each host (terminal) in a tree network, and the adopted medium can be a coaxial cable or an optical fiber. In a bus network, the shared medium is of the broadcast type (typically, a coaxial cable or a twisted pair) so that when a host transmits, all others receive. In the ring case, the transmission is typically point to point using an optical fiber; there is a transmission direction on the ring. Finally, a star network can be obtained both in case #1 of a wireless medium from the hosts to a central controller (broadcast medium) and in case #2 of a wired link with hosts interconnected to a switching device (point-to-point medium). In the above network architectures, the MAC protocol can be centrally coordinated by a controller.

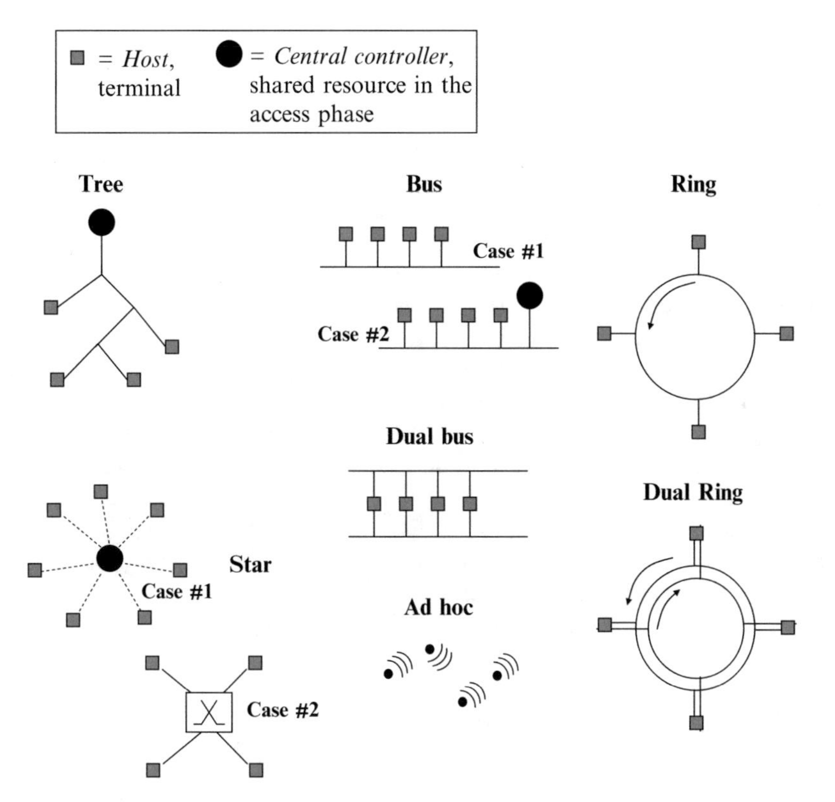

Fig. 6.1: Different topologies and shared media in Local Area Networks (LANs). The presence of a controller denotes a centralized MAC scheme; otherwise, a decentralized protocol is used

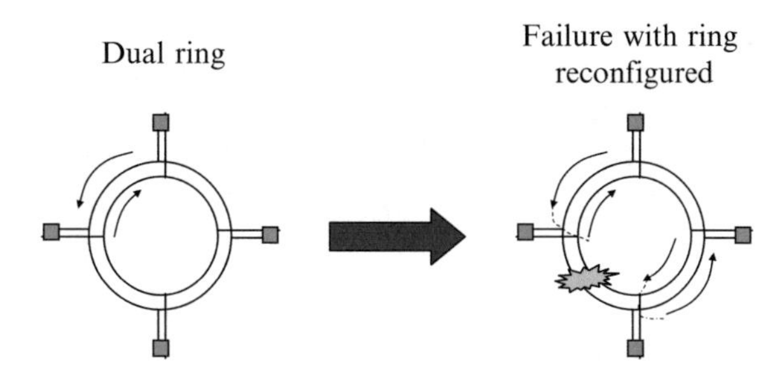

Fig. 6.2: Robustness to failures of the double ring topology

Otherwise, the MAC protocol is decentralized, and each host (i.e., its MAC layer) can decide autonomously when a transmission is permitted.

The tree topology is typical of cable television networks, which use an analog coaxial bus for broadcast transmissions with large available bandwidth. In addition to this, tree topology is also adopted by optical fiber technologies such as HFC (Hybrid Fiber/Cox), PON (Passive Optical Network), and GPON (Gigabit-PON) where different users (leaves) have to be reached from a central office (root). A PON can also support other network topologies, such as star, bus, ring, and mesh.

The bus topology is typical of small-scale LANs for offices, organizations, and campuses. Both single bus (e.g., Ethernet) and dual bus (e.g., Distributed Queue Dual Bus, DQDB, metropolitan area network) architectures are possible.

The ring topology is well suited to large-size LANs and metropolitan area networks. On the ring, there is a well-defined direction for the transmission of data. In the Fiber Distributed Data Interface (FDDI) technology (ANSI X3T9.5 standard), two rings (primary and secondary rings) are used with opposite transmission directions. In common operating conditions, practically one of the two rings is not used. In case of a failure, the closest stations reveal the problem and switch the ring, as described in Fig. 6.2 so that the topology becomes a dual bus.

The radio star topology is typical of wireless and cellular systems, where several mobile terminals exchange data with an access point or a central base station, providing coverage to a particular area. We can also have the radio star topology in the case where different terminals communicate through a satellite. Finally, a wired star topology is obtained when terminals are connected to a central switch using cables (e.g., switched Ethernet case).

In the "ad hoc" wireless topology, there are no base stations (or access points): mobile nodes communicate directly with one another within the range of the radio link. These wireless networks are called "ad hoc" because the topology of the communication links between nodes is dynamic, based on the communication ranges of the nodes and their positions. Mobile ad hoc networks, recognized in the literature as MANETs, are self-configuring networks of mobile routers. A subclass of these networks is represented by Vehicular Ad hoc Networks (VANETs), which allow mobile vehicles to communicate with one another as well as with roadside devices.

Future MAC protocols will adopt innovative approaches based on novel protocol architectures, especially suited to wireless transmissions. In particular, the ISO/OSI reference protocol stack should also be enriched with interfaces (and the exchange of related information) between non-adjacent OSI layers. The *cross-layer protocol architecture* envisages interfaces between and beyond adjacent layers. Although interfaces between adjacent layers are preferable, future systems should support *new interfaces between non-adjacent layers.* This allows higher-layer (lower-layer) protocols to make better decisions, depending on direct knowledge of lower-layer (higher-layer) conditions. Hence, for instance, the MAC protocols of the air interface can be *adaptive*, based on signaling information coming from physical, network, transport, and application layers.

This chapter aims to review the most important MAC protocols and analyze their performance.

6.1.1 Standards for Local Area Networks

The IEEE institute has defined many working groups for LAN standards that are characterized by the numbers 802.xx. In particular, we can consider the list below [4]:

- 802.1 Internetworking
- 802.2 Logical Link Control (LLC)
- 802.3 Ethernet, based initially on Carrier Sense Multiple Access/Collision Detection (CSMA/CD) and after amended by new supplements to extend the capacity such as 802.3u for Fast Ethernet at 100 Mbit/s, 802.3z, 802.3ab, 802.3ah for Gigabit Ethernet, 802.3ae for 10 Gigabit Ethernet, and 802.3ba and 802.3bm for 40 and 100 Gigabit Ethernet.
- 802.4 Token Bus
- 802.5 Token Ring
- 802.6 DQDB
- 802.7 Broadband technical advisory group
- 802.8 Fiber-optic technical advisory group
- 802.9 Integrated Voice/Data Networks
- 802.10 Network Security
- 802.11 Wireless LAN standards (e.g., 802.11, 802.11a, 802.11b, 802.11e, 802.11g, 802.11h, 802.11i, 802.11j, 802.11n, 802.11ac, etc.). In particular, the following standards are commercially identified under the name of Wi-Fi (Wireless Fidelity): 802.11, 802.11a, 802.11b, 802.11g, 802.11n, 802.11ac, 802.11ax
- 802.12 Demand Priority Access LAN, 100BaseVG-AnyLAN
- 802.13 Not used
- 802.14 Data over cable TV (cable modems, hybrid fiber/coax)
- 802.15 Wireless Personal Area Networks (WPANs)
- 802.16 Broadband wireless access (wireless metropolitan area network), commercialized under the name of WiMAX (Worldwide Interoperability for Microwave Access)
- 802.17 Resilient Packet Ring (RPR)
- 802.18 Radio regulatory technical advisory group
- 802.19 Coexistence technical advisory group
- 802.20 Mobile Broadband Wireless Access Systems
- 802.21 Media Independent Handover and Interoperability
- 802.22 Wireless Regional Area Networks
- 802.23 Emergency Services Working Group

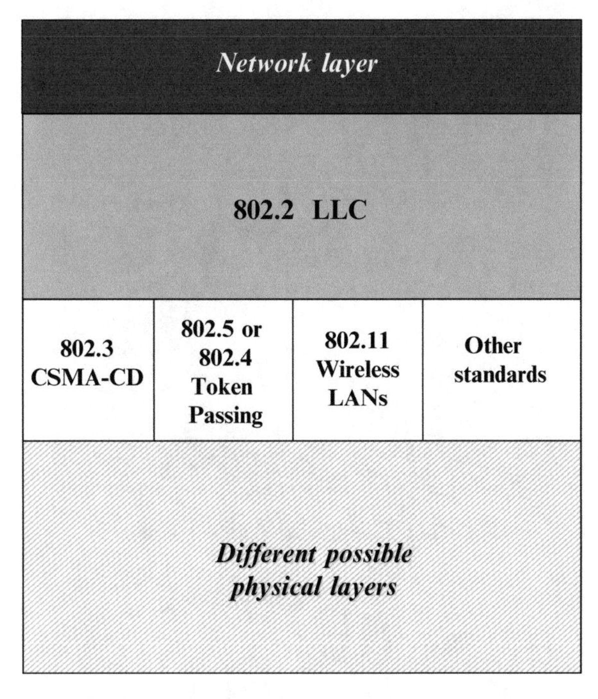

Fig. 6.3: LAN protocol stack

- 802.24 Smart Grid TAG
- 892.25 Omni-Range Area Networks.

The IEEE 802 reference protocol stack is shown in Fig. 6.3.

The MAC layer protocol and the frame format are specific to the different LANs. Some fundamental fields are present in all the frames, in particular, the addresses of both source and destination. MAC layer addresses (also known as physical addresses) of the IEEE 802 format are composed of 2 or 6 bytes. In the first case (2-byte address), a manager is requested, which assigns the addresses to the different stations connected to the LAN. To avoid this task, addresses are globally defined and assigned according to the 6-byte format (classical MAC address). This is the prevailing solution today: each manufacturer of LAN stations buys a block of addresses (globally administered by IEEE), unique to the network terminals produced. MAC addresses are typically burned into the adapter ROM (e.g., network card). Such an approach allows a plug-and-play method for setting a LAN.

The LLC protocol (IEEE 802.2) is above the MAC layer and is common to all the different IEEE access technologies (e.g., 802.3, 802.4, 802.5, 802.11, etc.). Typical LLC functions control the data flow and support recovery actions when data are received with errors. The LLC protocol data unit is encapsulated into the MAC protocol data unit (frame). The LLC protocol data unit has a header containing the SAP addresses of both source and destination. These SAPs make it possible to identify the network layer processes that generate data or that need to receive data. SAP addresses are one byte long and belong to two broad categories: IEEE-administered and manufacturer-implemented.

6.2 Contention-Based MAC Protocols

6.2.1 Aloha Protocol

In 1970, Norman Manuel Abramson defined and implemented a local packet-based wireless transmission system. It was an experimental system operating in the UHF band to connect computers on various campuses in the Hawaiian islands using the radio medium [5]. A central controller of the network at the University of Hawaii in Honolulu was used to broadcast data packets to the different terminals. A reverse procedure was necessary to allow the various terminals, spread on many Hawaiian islands, to transmit to the central host in Honolulu (see Fig. 6.4). The idea of Abramson was to allow terminals to send packets randomly without any form of coordination made by the central controller. This is a straightforward access scheme, named Aloha protocol, where terminals try to transmit as soon as they have packets ready to be sent. This type of access protocol is particularly suitable for those cases where

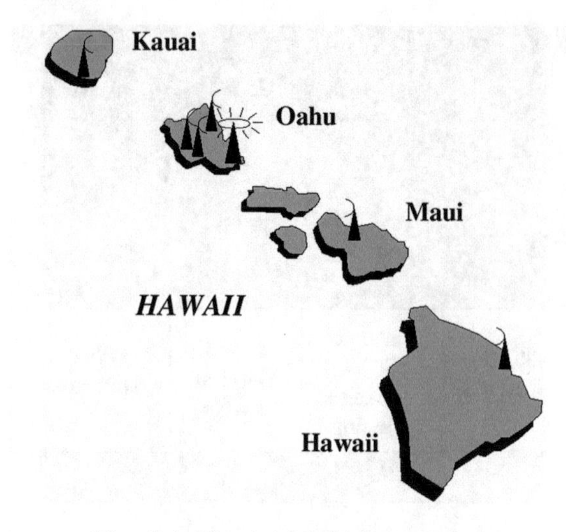

Fig. 6.4: The initial Aloha network

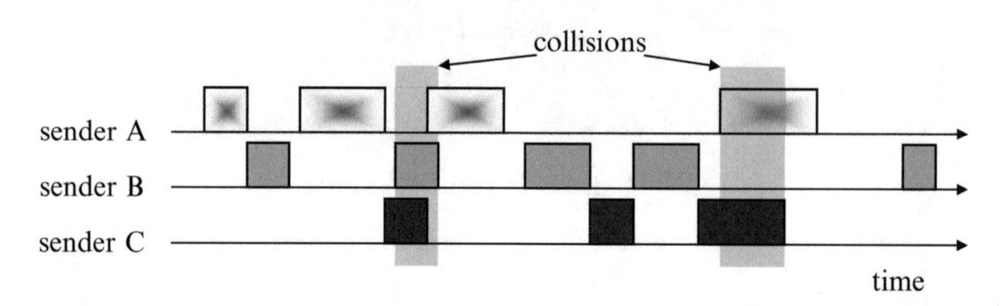

Fig. 6.5: Collisions with the Aloha protocol. The transmissions of the different terminals, shown here on separate time axes, are concurrent and use a shared channel

the packet transmission time is much shorter than the propagation delay from remote terminals to the central controller. Such a protocol, originally experimented as a ground system, was later (1973) also implemented via a GEO satellite to cover broad areas; this system was named AlohaNET. Today, the Aloha protocol is still important in the Internet of Things (IoT) case, where many sensors need to transmit small message reports to a sink. For instance, this is the case of both SigFox and LoRa sensors that can use multiple Aloha channels to send packet data. These systems are characterized by low-power transmissions and a high density of sensors [6].

Transmissions are organized in packets. If more packets are received simultaneously by the central controller, there is a *collision*, which typically leads to the destruction of all the colliding packets[1] (see Fig. 6.5). Each transmitting terminal recognizes that its packet has not been received correctly if it does not receive an acknowledgment within a certain timeout. Otherwise, the central controller may broadcast everything it receives on another frequency so that every terminal can realize whether its transmission was successful or not by hearing the broadcast channel. The Aloha protocol is a reliable scheme: if a packet has collided, it is retransmitted after a random delay (*backoff algorithm*) to avoid repeated collisions. Retransmission attempts are carried out until the packet is successfully received. In some practical implementations, there is a maximum number of retransmissions after which the packet is discarded, thus accepting a small packet loss probability; this could be useful for protocol stability reasons.

The Aloha protocol's behavior is unaffected by the propagation delays from the central controller to remote terminals. Therefore, it is also well suited to packet-based access to satellites.

The Aloha protocol is analyzed here under the assumption that all transmitted packets have the same length. New packet arrivals occur according to a Poisson process with mean rate λ. Each packet requires a time T to be

[1] There could be some cases in which one packet is received at a level significantly higher than the others so that it can be correctly decoded even in the case of a collision. This is the so-called capture effect, not considered here for a conservative analysis of the Aloha protocol. Another possibility to reduce the number of collisions would be to adopt Successive Interference Cancellation (SIC) techniques [7]. In these schemes, particularly used in satellite networks, a new packet #i is transmitted many times: the first successful transmission of packet #i allows us to cancel the collisions of packet #i with other packet transmissions by means of iterative interference cancellation.

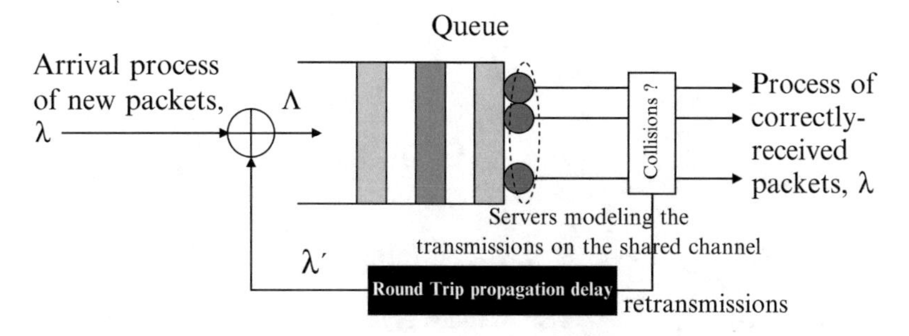

Fig. 6.6: Model of the Aloha protocol through a queuing system. When a (new or retransmitted) packet arrives at the queue, it is sent immediately so that the queue has infinite servers and no waiting part: M/D/∞ queue

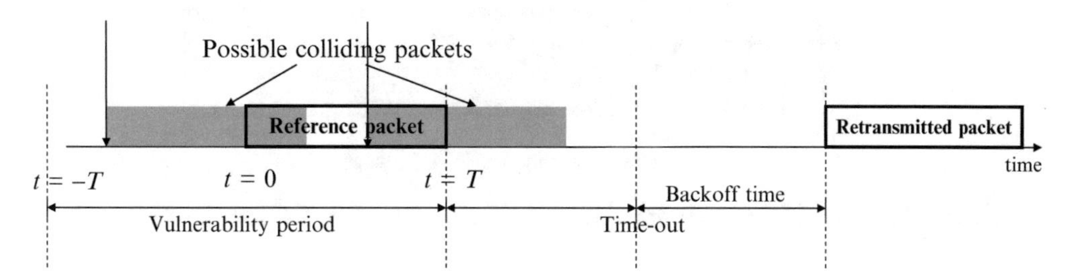

Fig. 6.7: Possible collision examples with our reference packet transmitted

transmitted. We consider an infinite number of elementary traffic sources in the network so that the mean arrival rate of new packets is insensitive to the number of packet transmissions in progress. The total arrival process (new packets plus retransmitted ones) is not Poisson but peaked because of collisions and retransmissions. However, for the sake of simplicity, we consider here that the retransmission process is Poisson with a mean rate λ' so that the total arrival process is still Poisson with total arrival rate $\Lambda = \lambda + \lambda'$. The Aloha protocol can be described through the model shown in Fig. 6.6. The feedback scheme included in Fig. 6.6 is due to collisions and retransmissions. This is a positive feedback system, thus highlighting the possible risks of protocol instability.

The intensity of the traffic offered to the network is $S = \lambda T$ in Erlangs. The intensity of the whole traffic circulating in the network due to new packet arrivals and retransmissions is $G = (\lambda' + \lambda)T = \Lambda T$ in Erlangs. If the access protocol has a *stable behavior*, we expect that the mean rate of packets entering the system equals the mean rate of correctly delivered packets leaving the system. Hence, under the stability assumption, S also denotes the intensity of the carried traffic or *throughput*. Consequently, the ratio S/G represents the probability of successful packet transmission, P_s, at each attempt with the Aloha protocol:

$$\frac{S}{G} = P_s \tag{6.1}$$

We need to derive the success probability P_s for a packet transmission starting at instant $t = 0$ and ending at instant $t = T$. The situation is depicted in Fig. 6.7, where we can also note the presence of some possible colliding packets with our reference packet transmitted.

Collisions with our packet are caused by other packets generated in the *vulnerability period* starting at time $t = -T$ and ending at time $t = T$. Therefore, our packet transmission is successful if there is no packet generation (according to the Poisson process with mean rate Λ) in the vulnerability period of duration $2T$; this occurs with probability $e^{-2\Lambda T}$:

$$\frac{S}{G} = P_s = e^{-2\Lambda T} \tag{6.2}$$

Since, $\Lambda T = G$, we can rewrite (6.2) as

$$S = G\, e^{-2G} \tag{6.3}$$

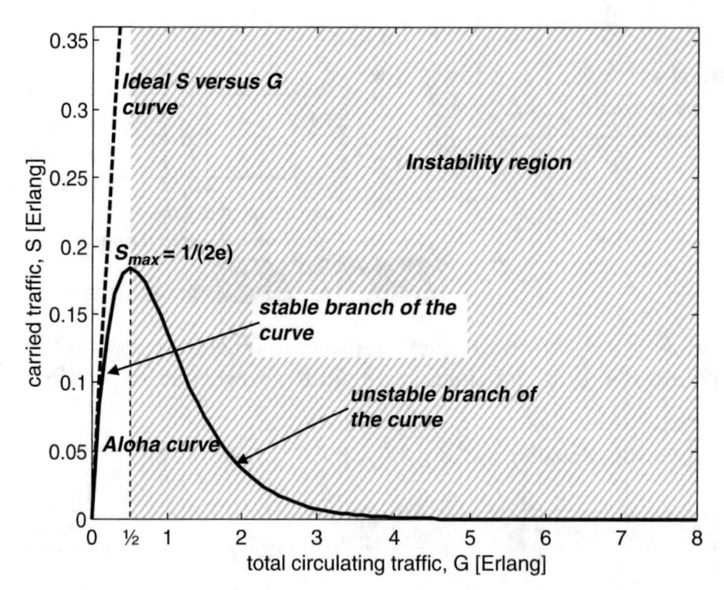

Fig. 6.8: Carried traffic (throughput) versus total circulating traffic for the Aloha protocol

Equation (6.3) relates the carried traffic intensity S and the total circulating traffic intensity G. If $G \to 0$, $S \to 0$. If $G \to \infty$, $S \to 0$. Hence, S as a function of G has an extreme point. In particular, this extreme is a maximum, which can be obtained by equating the derivative of (6.3) to 0:

$$\frac{\mathrm{d}S}{\mathrm{d}G} = e^{-2G} + G\,e^{-2G}\,(-2) = 0 \Rightarrow e^{-2G}\,(1-2G) = 0 \Rightarrow G = \frac{1}{2} \tag{6.4}$$

S has a maximum for $G = 1/2$ Erlangs, and this value is $S_{\max} = 1/(2e) \approx 0.18$ Erlangs according to (6.3): the maximum achievable throughput of an Aloha access system is 18%. Therefore, this protocol results in very low transmission medium utilization (the maximum possible utilization would be 100%, corresponding to $S = 1$ Erlang).

The behavior of S as a function of G is shown in Fig. 6.8. Note that we should consider S as an independent variable and $G = G(S)$ as a dependent variable so that the graph in Fig. 6.8 should more appropriately be turned over to have S in abscissa and G in ordinates. The function $G = G(s)$ that is the inversion of (6.3) is commonly known in the literature as Lambert W function.

We can note that for a given value of S we have two different possible values of G. This particular situation can be explained by considering that *the Aloha protocol is inherently unstable on the basis of the current modeling assumptions*. In real systems, instability can be avoided by considering a finite number of terminals and a suitably large retransmission interval of the backoff algorithm used after a collision. In Fig. 6.8, the part of the curve for $G > 1/2$ Erlangs corresponds to the unstable behavior of the protocol (if G increases, S decreases to 0: no traffic is delivered correctly). Through a suitable design of the backoff algorithm, we can consider with some approximation that for a given value of $S \leq S_{\max} = 1/(2e)$ Erlangs, there is only one solution for G for the real protocol. This is a stable solution that approximately corresponds to the solution for $G \leq 1/2$ Erlangs in the graph in Fig. 6.8 (obtained under ideal assumptions!). If we try to increase S beyond $S = S_{\max}$, the Aloha protocol experiences instability: S approaches 0 very rapidly, and the channel is saturated by collisions [2]. In conclusion, for $S \leq 0.18$ Erlangs, we numerically solve (6.3) by considering the solution of G lower than or equal to $1/2$ Erlangs: this is the only valid solution for system stability (see also the considerations made in Sect. 6.2.2).

The Aloha protocol is quite simple to implement, but it wastes the capacity of the shared medium. The maximum utilization is 18%.

Ideally, if the coordination for the transmission instants of the different packets were perfect (i.e., one transmission at once, no collisions: $S = G$ until $G = 1$ Erlang), the access protocol would admit an M/D/1 model with mean packet arrival rate λ and packet transmission time T. However, the Aloha system is characterized by an M/D/∞ model, since packets (new arrivals and retransmissions) arrive according to a Poisson process with a mean rate Λ, a packet transmission requires a deterministic time T, and infinite packet transmissions (due to the presence of an ideally infinite number of traffic sources) can be carried out simultaneously. We can solve this chain on the basis of the method shown in Sect. 4.11. Hence, the state probability distribution (i.e., the distribution of the number N of simultaneously transmitted packets) is Poisson with mean value $2\Lambda T$:

$$\text{Prob}\{N = k\} = \frac{(2\Lambda T)^k}{k!}e^{-2\Lambda T} = \frac{(2G)^k}{k!}e^{-2G} \quad (6.5)$$

where G is determined by numerically solving (6.3) for a given S value (under the stability assumption, only the solution with $G \leq 1/2$ Erlangs is valid).

The number of transmission attempts to successfully send a packet, L, has a modified geometric distribution with parameter $P_s = e^{-2\Lambda T} = e^{-2G}$:

$$\text{Prob}\{L = j\} = P_s(1 - P_s)^{j-1} = e^{-2G}\left(1 - e^{-2G}\right)^{j-1}, \qquad j = 1, 2, \ldots \quad (6.6)$$

The mean number of attempts for successful packet transmission is equal to $1/P_s$. Hence, the mean access delay for the successful transmission of a packet $E[T_p]$ with the Aloha protocol (i.e., the mean time from the beginning of the first packet transmission attempt until the reception of the acknowledgment for the correctly delivered packet) is

$$\begin{aligned} E[T_p] &= \left(\frac{1}{P_s} - 1\right)\{T + \Delta + E[R]\} + T + \Delta \\ &= \left(e^{2G} - 1\right)\{T + \Delta + E[R]\} + T + \Delta, \qquad \text{for } G \leq 1/2 \end{aligned} \quad (6.7)$$

where Δ denotes the round-trip propagation delay (from the remote terminal to the central controller and, then, back to the remote terminal; we consider here that all terminals experience the same Δ value), and $E[R]$ denotes the mean waiting time before attempting a packet retransmission according to the backoff algorithm. Note that G is determined by numerically solving (6.3) for a given S value. Moreover, we have neglected the ACK transmission/reception time (or included this time in Δ).

Note that the mean packet delay $E[T_p]$ could also be defined up to the time when the packet is successfully delivered; in this case, the last Δ symbol on the left side of (6.7) has to be replaced by $\Delta/2$.

Equations (6.7) and (6.3) together allow determining the mean packet delay $E[T_p]$ as a function of S. Figure 6.9 shows the graph of $E[T_p]$ in T units versus S for the Aloha protocol with $E[R] = 4$ [T units] and $\Delta = 10$ [T units], referring only to the stable part of the protocol (a complete curve is shown in [2], where there are two possible $E[T_p]$ values for each S value). Note that $E[T_p]$ has an infinite value for $S > S_{\max}$ because the Aloha channel is full of collisions, and no packet is delivered successfully.

Equation (6.1) is quite general and can be used to model many Aloha-like protocols. The only differences in these cases will be the expressions of P_s. For instance, we can consider that even a non-collided packet needs retransmissions due to errors caused by the radio channel with probability P_e (memoryless channel). In this case, the carried traffic versus total circulating traffic equation (6.3) should be modified as follows:[2]

$$\frac{S}{G} = P_s = (1 - P_e) \times e^{-2\Lambda T} \quad (6.8)$$

In this case, the curve of S versus G is similar to that shown in Fig. 6.8. The maximum value of S is obtained for $G = 1/2$ Erlangs and is equal to $(1 - P_e)/(2e)$ Erlangs. As expected, the presence of packet errors reduces the traffic intensity supported by the protocol under stability conditions.

The approach represented by (6.1) can also be used to study two important protocol variants described in the following sections: the Slotted-Aloha protocol and the Aloha scheme with ideal capture effect.

6.2.2 Slotted-Aloha Protocol

Because of the Aloha protocol's low throughput, it was soon understood the need for some improvements. In 1972, Roberts described a method for doubling the capacity of Aloha by dividing the time into slots, each corresponding

[2] We neglect here the possibility of errors on the feedback channel, which is used to send the acknowledgments of correctly received packets.

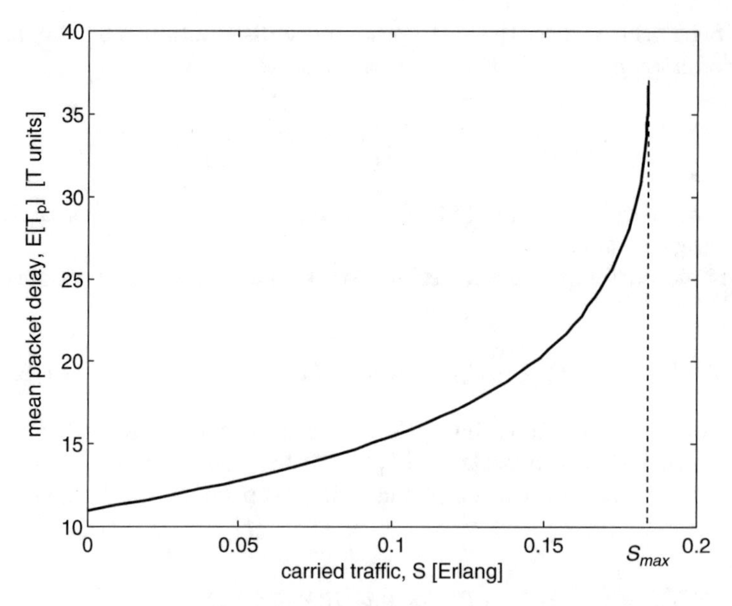

Fig. 6.9: Mean packet transmission delay $E[T_\mathrm{p}]$ versus S for the Aloha protocol ($E[R]=4$ [T units] and $\Delta=10$ [T units]). In this graph, we have shown only the part of the curve of $E[T_\mathrm{p}]$ versus S, which corresponds to a stable protocol behavior. Otherwise, two $E[T_\mathrm{p}]$ values correspond to the same S value for $S \in [0,\, S_{\max}]$

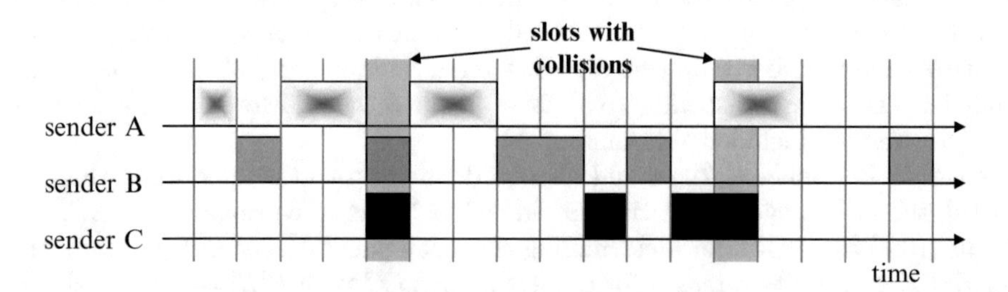

Fig. 6.10: Collisions with the Slotted-Aloha protocol. The transmissions of the terminals, shown on separate time axes, are concurrent and use a shared channel

to the transmission of one packet [8]: packet transmissions are performed by remote terminals so that the central controller receives packets only at predetermined instants of time. Since time is slotted, this protocol has been named Slotted-Aloha (S-Aloha). T denotes the duration of a slot and also the packet transmission time. To achieve global synchronization, the central controller broadcasts synchronization pulses. Because of the synchronization, a packet arriving within one slot is transmitted by the corresponding terminal at the beginning of the next slot.

Since a remote terminal can only transmit at predetermined instants of time, collisions can occur only with other packets sent in the same time slot (see Fig. 6.10). Terminals experiencing a collision reschedule their attempts after a random retransmission delay according to a backoff algorithm.

For this protocol study, we adopt the same approach as the Aloha case with Poisson arrivals (infinite elementary users) of fixed-length packets. Hence, we use (6.1), where we express P_s by considering that the vulnerability period is now equal to one slot, T. The transmission of a packet is successful if there is no packet arrival (according to the Poisson process with total mean rate Λ) in the slot before that in which our reference packet has been transmitted: $P_s = e^{-\Lambda T}$.

$$\frac{S}{G} = P_s = e^{-\Lambda T} \Rightarrow S = Ge^{-G} \tag{6.9}$$

S has a maximum value as a function of G. This extreme can be obtained by equating the derivative of (6.9) to 0:

$$\frac{\mathrm{d}S}{\mathrm{d}G} = e^{-G} + G\,e^{-G}(-1) = 0 \Rightarrow e^{-G}(1-G) = 0 \Rightarrow G = 1 \tag{6.10}$$

S has a maximum for $G = 1$ Erlang and its value is $S_{\max} = 1/e \approx 0.36$ Erlangs. The maximum achievable throughput of a Slotted-Aloha protocol is twice that of a classical Aloha scheme. This is the advantage obtained by adopting time synchronization. For $S > S_{\max}$ (or, equivalently, $G > 1$ Erlang), the Slotted-Aloha protocol is unstable. The "stable branch" of the S versus G curve is that for $G < 1$ Erlang, as explained below.

A more detailed analysis of this protocol, including stability issues, is carried out in [9], where terminals are considered to be either "thinking" or "backlogged" (i.e., terminals that have one packet to be transmitted) according to a Markov chain model. Thus, it is possible to characterize the conditions under which there is a single stable solution. However, if the number of terminals is quite high, there is the risk of saturated solutions or multiple stable solutions: the protocol behavior is not good in these conditions. The system can be stabilized by adequately enlarging the randomization interval used for the retransmission of collided packets. In [9], it is shown that *the branch of* (6.9) *for* $G < 1$ *Erlang with an infinite number of terminals represents a good approximation of the stable solution obtained with the refined model for the same traffic intensity S.* More details on this refined model are provided in Sect. 6.2.4.

By using the same notations adopted for the analysis of the classical Aloha protocol, the mean packet (access) delay $E[T_{\mathrm{p}}]$ with Slotted-Aloha results as

$$\begin{aligned} E\left[T_{\mathrm{p}}\right] &= \frac{T}{2} + \left(\frac{1}{P_s} - 1\right)\{T + \Delta + E\left[R\right]\} + T + \Delta \\ &= \frac{T}{2} + \left(e^{G} - 1\right)\{T + \Delta + E\left[R\right]\} + T + \Delta, \qquad \text{for } G \leq 1 \end{aligned} \tag{6.11}$$

Also, in this case, we have neglected the ACK transmission/reception time. The additional delay term $T/2$ with respect to the classical Aloha scheme is necessary to account for the fact that packet arrivals occurring according to a Poisson process have to wait for a mean time $T/2$ to start their transmissions: $T/2$ is the mean time from the Poisson arrival within a slot to the starting instant of the next slot (this result is intuitive since there are no privileged instants within a slot for Poisson arrivals); for a more formal proof of this mean time $T/2$, the interested reader should refer to Chap. 5 and, in particular, to Sect. 5.8.2.1.

6.2.2.1 Slotted-Aloha Protocol with a Finite Number of Terminals

In this study, we consider a finite number N of independent users so that the arrival process is binomial on a slot basis [10]. Let us denote

- S_i the probability of transmitting a packet successfully on a slot for the ith user and
- G_i the total probability of transmitting a packet on a slot for the ith user.

We assume that all the users generate the same traffic load. Hence, the total carried traffic S on a slot and the total circulating traffic G on a slot can be expressed as

$$S = \sum_{i=1}^{N} S_i = NS_i \quad \text{and} \quad G = \sum_{i=1}^{N} G_i = NG_i \tag{6.12}$$

G also represents the total average number of packets transmitted on a slot. The probability of a successful transmission on a slot by the ith user, $S_i = S/N$, can be expressed as the joint probability of independent events so that it is the product of the probability that the ith user transmits on the slot, $G_i = G/N$, and the probability that no other user transmits on the same slot, $\Pi_j(1 - G_j) = (1 - G_j)^{N-1} = (1 - G/N)^{N-1}$:

$$\frac{S}{N} = \frac{G}{N}\left(1 - \frac{G}{N}\right)^{N-1} \Rightarrow S = G\left(1 - \frac{G}{N}\right)^{N-1} \tag{6.13}$$

In conclusion, (6.13) relates the total carried traffic S and the total circulating traffic G for the Slotted-Aloha system with N terminals. The maximum throughput is still obtained by considering the null-derivative condition for (6.13):

$$\frac{\mathrm{d}S}{\mathrm{d}G} = \left(1 - \frac{G}{N}\right)^{N-1} + G\,(N-1)\left(1 - \frac{G}{N}\right)^{N-2}\left(-\frac{1}{N}\right) = 0$$

$$\Rightarrow \left(1-\frac{G}{N}\right)^{N-2}\left[1-\frac{G}{N}-\frac{G}{N}(N-1)\right]=0 \tag{6.14}$$

The above equation is fulfilled for $G=1$ Erlang; correspondingly, we have the maximum throughput $S_{\max}=(1-1/N)^{N-1}$ Erlangs (note that the above derivative is also equal to 0 for $G=N$; this is a trivial case of no interest since we have that all stations simultaneously transmit and collide: the throughput is zero).

For $N\to\infty$ (case of infinite, independent, elementary sources), (6.13) can be expressed through the following notable limit:

$$\lim_{N\to\infty}\left(1-\frac{G}{N}\right)^{N-1}=e^{-G} \tag{6.15}$$

Hence, we obtain

$$S=G\,e^{-G}$$

i.e., the same result as in (6.9), obtained for an infinite number of terminals.

6.2.3 The Aloha Protocol with Ideal Capture Effect

We refer here to a Slotted-Aloha case (but these considerations can also be applied to the classical Aloha scheme), and we assume that when there are n colliding packets with our reference packet, the central controller is always able to receive one packet (ideal capture effect) correctly. We consider that the success probability is uniform over all the $n+1$ colliding packets. Hence, one packet out of $n+1$ is correctly received with probability $1/(n+1)$. We refer to the case of infinite users and we adopt (6.1), where P_s is derived as follows using the Poisson assumption for new arrivals (mean rate λ) and new arrivals plus retransmissions (mean rate Λ). P_s is obtained as the weighted sum over the sub-cases corresponding to the different n values (i.e., the number of colliding packets generated in T, slot duration, by the Poisson process with mean rate Λ); weights are given by the probability of n arrivals in T because of the Poisson process with mean rate Λ.

$$\begin{aligned}\frac{S}{G}=P_s&=\sum_{n=0}^{\infty}\frac{1}{n+1}P_n=\sum_{n=0}^{\infty}\frac{1}{n+1}\frac{(\Lambda T)^n}{n!}e^{-\Lambda T}\\&=\frac{e^{-\Lambda T}}{\Lambda T}\sum_{i=1}^{\infty}\frac{(\Lambda T)^i}{i!}=\frac{e^{-\Lambda T}}{\Lambda T}\left[\sum_{i=0}^{\infty}\frac{(\Lambda T)^i}{i!}-1\right]\\&=\frac{e^{-\Lambda T}}{\Lambda T}\left[e^{\Lambda T}-1\right]=\frac{1-e^{-\Lambda T}}{\Lambda T}=\frac{1-e^{-G}}{G}\qquad\Rightarrow S=1-e^{-G}\end{aligned} \tag{6.16}$$

In this protocol, the carried traffic S is a monotonically increasing function of G. For G close to 0, S is close to 0 as well. When G increases, S tends to 1 Erlang, the maximum achievable throughput. This access protocol has no stability issues, but of course it is ideal. The capture effect is possible in practice but depends on the relative powers of the packets received at the central controller: one packet among the colliding ones can be successfully received only in special cases, depending on the signal-to-noise ratio.

Referring to the ideal capture case, we invert the formula $S=S(G)$ in (6.16) so that $G=-\ln(1-S)$. Hence, $P_s=S/G=-S/\ln(1-S)$.

By using the same notations adopted for the analysis of the classical Aloha protocol, the mean packet (access) delay $E[T_{\mathrm{p}}]$ with Slotted-Aloha and ideal capture results as

$$\begin{aligned}E\left[T_{\mathrm{p}}\right]&=\frac{T}{2}+\left(\frac{1}{P_s}-1\right)\{T+\Delta+E\left[R\right]\}+T+\Delta\\&=\frac{T}{2}+\left(\frac{G}{1-e^{-G}}-1\right)\{T+\Delta+E\left[R\right]\}+T+\Delta,\quad\text{for}\quad\forall G>0\end{aligned}$$

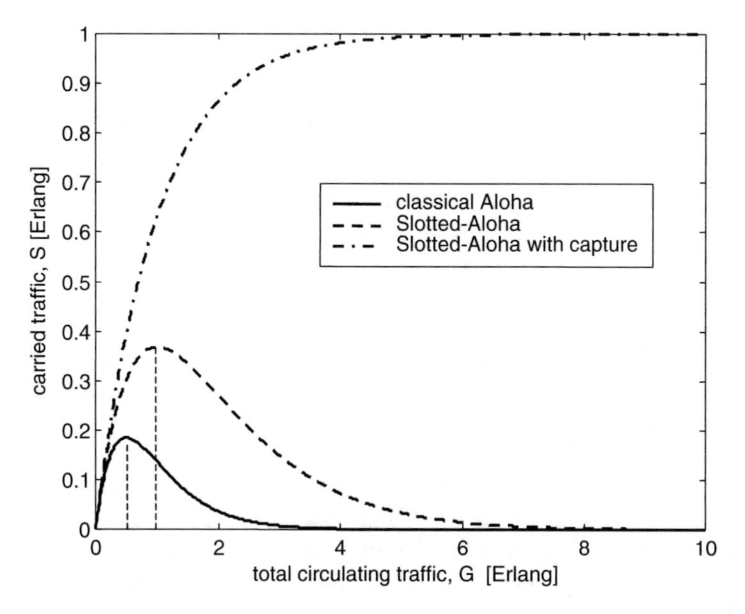

Fig. 6.11: Comparison of various Aloha versions in terms of carried traffic versus total circulating traffic

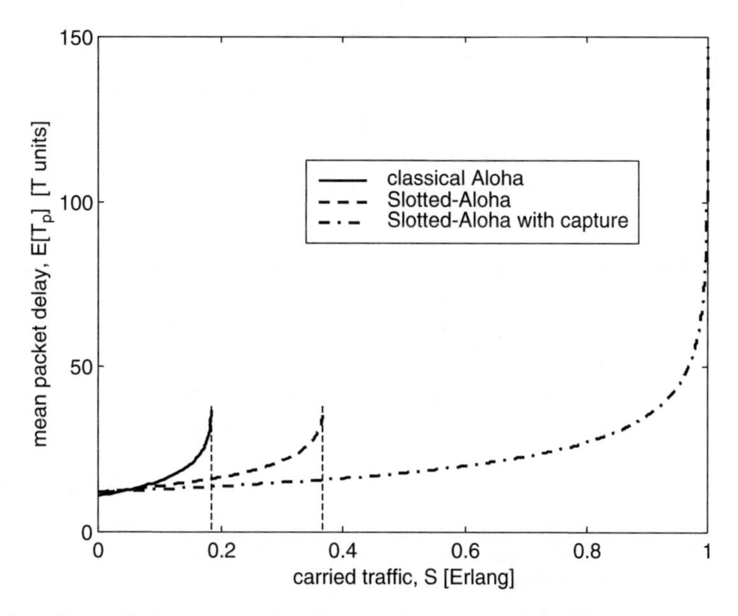

Fig. 6.12: Comparison of various Aloha versions in terms of mean packet delay for $\Delta = 10$ [T units] and $E[R] = 4$ [T units]. We have shown only the stable parts of the curves in the Aloha and Slotted-Aloha cases

$$= \frac{T}{2} + \left[-\frac{S}{\ln(1-S)} - 1 \right] \{T + \Delta + E[R]\} + T + \Delta, \quad \text{for } 0 \le S < 1 \tag{6.17}$$

Note that there would be no need for a retransmission delay for collided packets in an ideal capture system (in the above formula, we could even consider $E[R] = 0$).

Figures 6.11 and 6.12 compare the different variants of the Aloha protocol (i.e., the classical version, the slotted version, and the ideal capture version) in terms of both throughput S and mean packet delay $E[T_\mathrm{p}]$.

We can note that the throughput curves in Fig. 6.11 for Aloha and Slotted-Aloha have a maximum, denoting that these protocols have a stability limit. Instead, there is no maximum in the Slotted-Aloha case with ideal capture so that this protocol is always stable.

Referring to Fig. 6.12, the Aloha (Slotted-Aloha) protocol has infinite $E[T_\mathrm{p}]$ values for $S > 0.18$ Erlangs (for $S > 0.36$ Erlangs). Instead, the Slotted-Aloha protocol with capture has infinite $E[T_\mathrm{p}]$ values for $S > 1$ Erlang.

6.2.4 Alternative Approaches for Aloha Protocol Analysis

Three different approaches can be adopted to analyze the performance of random access protocols (and, in particular, of Aloha protocols):

- *S–G analysis* as done so far for the Aloha protocols under the assumption of Poisson arrivals and infinite elementary traffic sources. This is a simplified approach for a finite or infinite number of terminals. There is, however, no consideration of queuing issues due to the buffers of terminals. This approach is suitable for analyzing the mean access delay $E[T_p]$ and the throughput S.
- *Markov chain method* [9], where the behavior of the whole system (or of a single terminal) is modeled by a discrete-time Markov chain with a set of states, representing, for instance, the number of backlogged terminals (i.e., terminals having packets ready to be transmitted). This method is applied to a finite number M of terminals; it is typically not only the most accurate but also the most complex approach since transition probabilities have to be determined between states. The access protocol stability is analyzed based on the throughput behavior. In some cases (e.g., analysis of the MAC for WiFi), the study is typically carried out in *saturated conditions*, assuming that all terminals are always backlogged [11]. Saturated models are suitable for analyzing the mean access delay $E[T_p]$ and the throughput S. Instead, *non-saturated models* can be used to characterize the mean packet delay (queue plus access) and the stability limit, combining queue stability and access protocol stability effects.
- *Equilibrium Point Analysis* (*EPA*) [12–15], where the behavior of each terminal is modeled by a state diagram. This method is applied to a finite number M of terminals. State transitions are characterized by suitable probabilities. Equilibrium conditions are written for each state, considering that it is populated by an average (equilibrium) number of terminals. Stability conditions are determined considering that EPA equations represent the null-gradient condition of a potential function (equilibrium point) in the state variables. The *catastrophe theory* is used to study the behavior of dynamic systems under the effect of some control parameters. In this case, it can be used to characterize the system parameter values that allow stable equilibrium points of the random access protocols [16].

For instance, we can consider the Markov chain model of the Slotted-Aloha protocol, defined in Sect. 6.2.2, following the method shown in [9].

The number of backlogged (contending) terminals n (out of M) represents the Markov chain state. This is a sort of non-saturated model, but there is no queuing of packets at the terminals due to the following assumption on traffic generation: when a terminal has generated a packet (i.e., a packet is ready for transmission) on a certain slot with probability σ, it immediately sends this packet but cannot generate a new packet before the current one has been correctly received.

After a collision, a terminal reschedules a new transmission in the next slot with probability p (this assumption is slightly different from what is considered so far in the Slotted-Aloha case) to have a geometric distribution of the retransmission time. The adoption of this memoryless distribution allows us to adopt a Markov chain model for the protocol. The classical Slotted-Aloha S–G analysis shown in Sect. 6.2.2 is considered equivalent to this Markov chain approach if $1/p = \{\Delta + E[R]\}/T$, where Δ, $E[R]$ and T are expressed here in seconds. In this model, the arrival process of new packets is according to a binomial distribution depending on M (the total number of terminals) and σ. Note that if M tends to infinity (infinite population of users) so that $M\sigma = S$, the packet arrival process is Poisson with intensity S (mean rate in pkts/slot). This system is modeled by the discrete-time Markov chain in Fig. 6.13 (note that the packet arrival process is similar to that considered in Exercise 5.3), where the state is the number of backlogged terminals and where P_{ij} denotes the transition probability from state $n = \mathrm{i}$ to state $n = j$. Probabilities P_{ij} with $i < j$ are related to new packet arrivals (i.e., new terminals becoming backlogged out of $M - i$) on a slot. Note that $P_{ij} = 0$ for $i = 0$ and $j = 1$: a new packet arrival occurring in state 0 is soon transmitted without collisions, thus contributing to the transition $0 \to 0$ (this transition occurs with probability P_{00} given by two contributions: the probability of no arrival and the probability of one arrival in a slot). Probability P_{ij} with $i = j + 1$ entails a successful transmission and no new arrivals on a slot ($P_{ij} = 0$ for $i - j \geq 2$). More details on probabilities P_{ij} can be found in [9].

This Markov chain (see Fig. 6.13) can be solved to determine the state probability distribution P_n for $n =$ 0, ..., M, using an approach similar to that shown in Sect. 5.5. The throughput is determined conditioned on the state n, and then the average throughput S_{out} is obtained by removing the conditioning summing over the state-space with probability P_n. The average number of backlogged packets in the system (corresponding to the

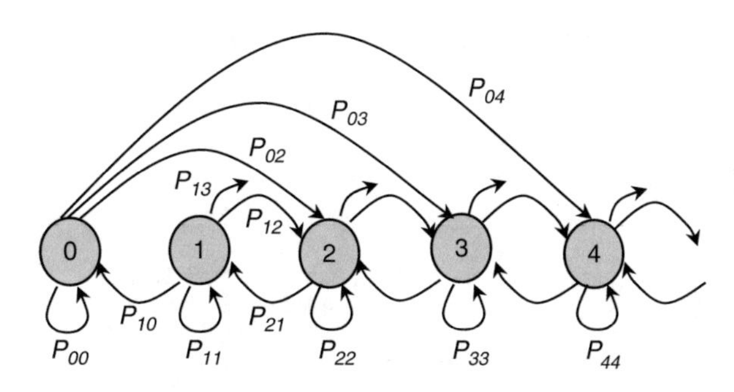

Fig. 6.13: Markov chain model of the Slotted-Aloha protocol

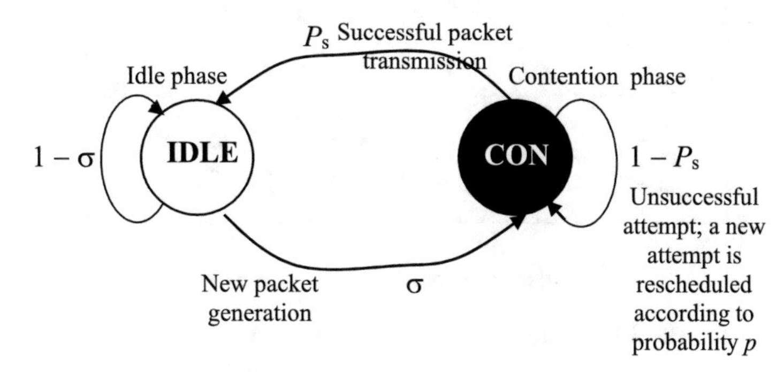

Fig. 6.14: Terminal state diagram with the EPA approach for the Slotted-Aloha protocol, p-persistent case

average number of backlogged terminals) is $N = \sum_{n=0}^{M} nP_n$. Then, we can determine the mean packet delay dividing N by the throughput S_{out}, according to the Little theorem. However, we have also to add T and Δ to take the successful transmission and the propagation delay into account. Hence, we have $E[T_{\text{p}}] = N/S_{\text{out}} + 1 + \Delta/T$ [T units].

An alternative graphical approach is carried out on the plane $(S,\ n)$ to determine the system throughput and study the system stability. It is based on the intersection of the *throughput equilibrium contour* [locus of the points on the plane $(S,\ n)$ where throughput S_{out} is equal to input traffic S] and a *load line* depending on M and σ parameters. It is shown in [9] that the protocol is stable in the case of a single intersection. Given the average retransmission interval $E[R]$ and the packet generation probability on a slot σ, there is a maximum number of terminals, $M_{\max}$, which can be supported with stable behavior. In the case of multiple intersections (two or three solutions), one solution is stable and another is unstable: the system can operate around the stable condition only for a finite time interval; statistical fluctuations may lead the system towards saturation. Interestingly, it has been shown in [9] that the throughput obtained averaging over the state-space is well approximated by the solution on the stable branch of the S–G approach in (6.9).

Let us now consider the EPA analysis of the Slotted-Aloha protocol under the following generic assumptions: Poisson arrival process of packets with mean rate λ and transmission attempts on a slot basis (length T) according to probability p. The behavior of each terminal is modeled using a two-state diagram, as shown in Fig. 6.14 with idle (IDLE) and contention (CON) states, where we assume that (1) the outcome of the transmission attempt on a slot (i.e., collision or not collision) is known within the end of the same slot as in the case of negligible propagation delays and (2) a terminal generating a packet (according to a Poisson process) enters the contention phase and cannot generate a new packet until the previous one is received successfully.[3]

Let s and c denote the equilibrium number of terminals in IDLE and CON states, respectively. In Fig. 6.14, we use the following symbols: σ represents the probability of a new Poisson arrival on a slot ($\sigma = 1 - e^{-\lambda T}$); P_s is the probability at equilibrium that our backlogged terminal has a successful transmission on a slot [$P_s = p(1-p)^{c-1}$]. Finally, M denotes the total number of terminals. We can write the following EPA balance equations representing a non-linear system:

[3] Removing this approximation, the (mean) packet delay can be characterized using an M/G/1 queuing model, but a more refined terminal state diagram is needed with respect to that in Fig. 6.14 (i.e., the queue occupancy status has to be included in the model).

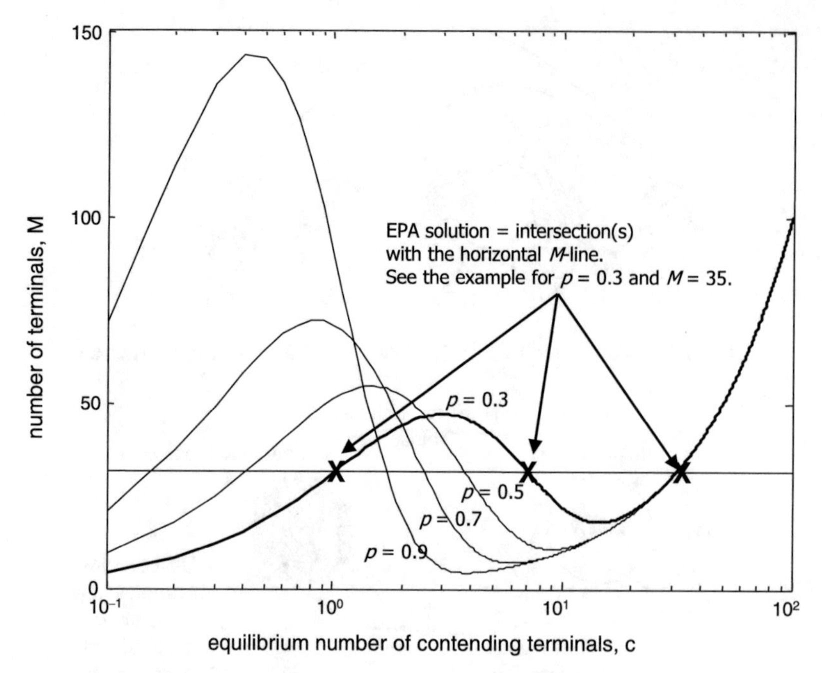

Fig. 6.15: Graphical method for the EPA solution of the Slotted-Aloha p-persistent protocol with $1/\sigma = 100$ and different p values. The EPA solutions in the case with $M = 35$ terminals and $p = 0.3$ are denoted by "X"

$$\begin{cases} s\sigma = cP_s \\ s + c = M \end{cases} \Rightarrow c + c\frac{p}{\sigma}(1-p)^{c-1} = M \tag{6.18}$$

Note that $S = \lambda T \approx \sigma$ (for low traffic intensity values). Thus, we have obtained a single EPA equation in the unknown term c, which can be solved numerically or graphically. Figure 6.15 shows the behavior of this EPA equation as a function of c with $1/\sigma = 100$ and for different p values: the EPA solutions are given by the intersections of the curve with the horizontal line for an ordinate value equal to M. Actually, the graph in Fig. 6.15 has some similarities with the S versus n graph and load curve considered in the previous Markovian approach [9], even if the EPA method is different.

The EPA equation must have a single solution to have a stable protocol behavior. However, depending on the p value, in Fig. 6.15, we have situations with a single EPA solution (stability) and situations with three EPA solutions (bistability). In particular, with $1/\sigma = 100$, there is a single and stable EPA solution up to $M_{\max} = 18$ terminals for $p = 0.3$, up to $M_{\max} = 11$ terminals for $p = 0.5$, and up to $M_{\max} = 8$ terminals for $p = 0.7$. Hence, by increasing the p value, the protocol is more aggressive (i.e., more collisions occur) so that it can support a lower number of terminals. The protocol performance depends on M and "control parameters" σ and p. The possible change in the behavior of the protocol (from a stable situation to an unstable one) can be characterized according to the *catastrophe theory* [16]. In particular, there is a "cusp point" (c_{cusp}, σ_{cusp}, M_{cusp}), which can be determined as a function p by a system composed of the EPA equation, the first and the second derivatives of the EPA equation, as shown in [15, 16]. It is possible to prove that $M_{\text{cusp}}(p) = -4/\ln(1-p)$. Therefore, if $M < M_{\text{cusp}}(p)$, there is a single and stable EPA solution for $\forall\sigma$ value. Instead, if $M \geq M_{\text{cusp}}(p)$, there are three EPA solutions for σ in a certain interval $[\sigma_1(p), \sigma_2(p)]$ "centered" around the cusp value $\sigma_{\text{cusp}}(p) = p(1-p)^{M\text{cusp}/2-1}$ and a single EPA solution for σ outside that interval.

There is a relation between probability p and the retransmission delay of the real protocol: similarly to what was suggested by Kleinrock in [9], we consider $1/p = \{\Delta + E[R]\}/T$. Thus, the EPA approach allows us to identify a stability condition relating M_{cusp} and $E[R]$ as

$$M_{\text{cusp}} = -\frac{4}{\ln\left(1 - \frac{1}{\Delta/T + E[R]/T}\right)}.$$

If we use a certain $E[R]$ value, M has to be lower than M_{cusp} if we like to have a stable protocol behavior.

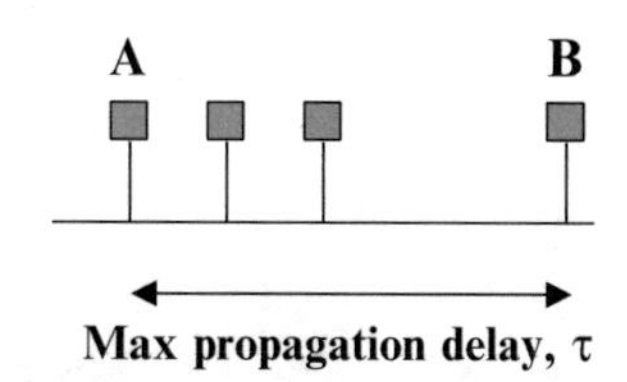

Fig. 6.16: Bus topology and maximum propagation delay from the farthest terminals

Assuming a stable protocol configuration, from the equilibrium EPA solution c, we can derive the distribution of the terminals in the different states [12–14]; a typical assumption is to consider the number of contending terminals according to a geometric distribution with mean value equal to c, the equilibrium number of contending terminals (case of a single EPA solution). Moreover, according to this model, the mean packet delay in slots is equal to $E[T_\mathrm{p}] = 1/P_s = 1/[p(1-p)^{c-1}]$. This performance parameter cannot go to infinity but tends to saturate to the value $E[T_\mathrm{p}]_{|\mathrm{sat}} = 1/[p(1-p)^{M-1}]$ as S tends to 1 Erlang due to the assumed model with a finite number of terminals.

6.2.5 CSMA Protocols

There are some random access schemes that allow us to improve the throughput performance of Aloha-type protocols if the packet transmission time, T, is much lower than the maximum (one-way) propagation delay in the LAN, τ (see Fig. 6.16). Note that in the Aloha cases, we used parameter Δ with a value corresponding to 2τ adopted here. This new class of random access protocols typically uses a broadcast physical medium (e.g., a single bus) so that a remote station (listening to the physical medium) recognizes whether another transmission is in progress or not (*carrier sensing*). If another transmission is revealed, the remote station refrains from transmitting to avoid collisions. The protocols of this type are called Carrier Sense Multiple Access (CSMA) and are detailed in [1, 2, 17]. CSMA schemes are based on decentralized control. Both slotted and unslotted options are possible for each version of the CSMA protocol. Since the performance difference between slotted and unslotted versions of the same protocol is relatively small (the throughput improvement of the slotted version is much lower than that of the Aloha protocol), we will refer basically to unslotted cases.

A special line coding must be used to achieve carrier sensing. This is needed to avoid that a bit "0" corresponds to a 0-V level for all the bit duration. To solve this issue, the Ethernet standard (IEEE 802.3), based on a variant of the CSMA protocol, was standardized to use Manchester encoding (see Sect. 6.5.1). Moreover, since the medium is of the broadcast type, a transmitting terminal cannot simultaneously receive a signal; otherwise, there is a collision event. Hence, half-duplex transmissions are typical of CSMA protocols.

Collisions may still occur with the CSMA protocol since a terminal recognizes that another terminal is using the medium only after a (maximum) delay τ. Referring to the typical situation in Fig. 6.16, we consider that station A starts transmitting at time $t=0$; this signal reaches station B at time $t=\tau$ (worst case). If station B generates a new packet at instant $t=\tau-\varepsilon$ (where ε denotes an elementary positive value), station B can transmit this packet, thus causing a collision. Based on this example, we can state that time interval τ is the *vulnerability period* of this protocol. The efficiency of the carrier sensing approach improves as the following parameter a reduces:

$$a = \frac{\tau}{T} \tag{6.19}$$

The slotted versions of the CSMA protocols adopt a time slot of duration τ, even if the CSMA scheme implemented by the Ethernet standard uses a time slot of length 2τ (see Sect. 6.5.1). Carrier sensing should avoid any collision in the ideal case with $\tau=0$.

The collision phenomenon with CSMA is described in Fig. 6.17, where two stations activate transmissions within time τ. This representation is useful to highlight the time wasted due to a collision.

When a terminal recognizes that its packet has collided, the packet is retransmitted after a random waiting time. A truncated Binary Exponential Backoff (BEB) algorithm is used: the retransmission delay is randomized within a time window, which grows exponentially (up to a maximum value) after each collision of the same packet. Such an approach entails a sort of Last Input First Output (LIFO) effect: the terminal (among the colliding ones)

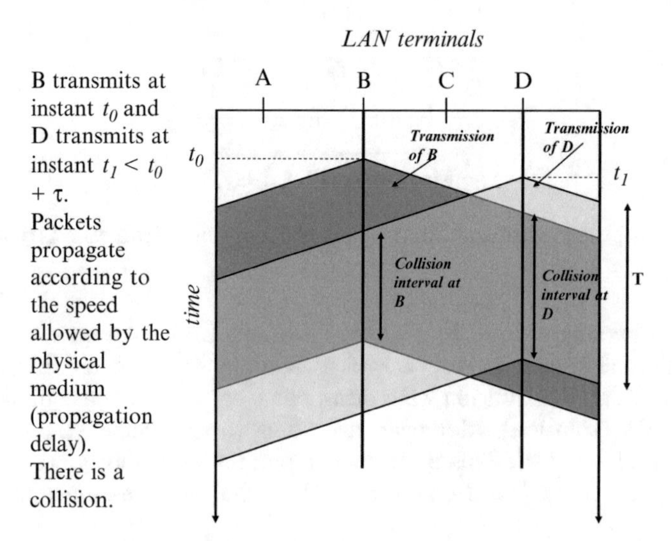
LAN terminals
A
B
C
D
B transmits at instant t_0 and D transmits at instant $t_1 < t_0 + \tau$. Packets propagate according to the speed allowed by the physical medium (propagation delay). There is a collision.
t_0
t_1
time
Transmission of B
Transmission of D
Collision interval at B
Collision interval at D
T

Fig. 6.17: Collision with CSMA: the entire packet transmission time is wasted

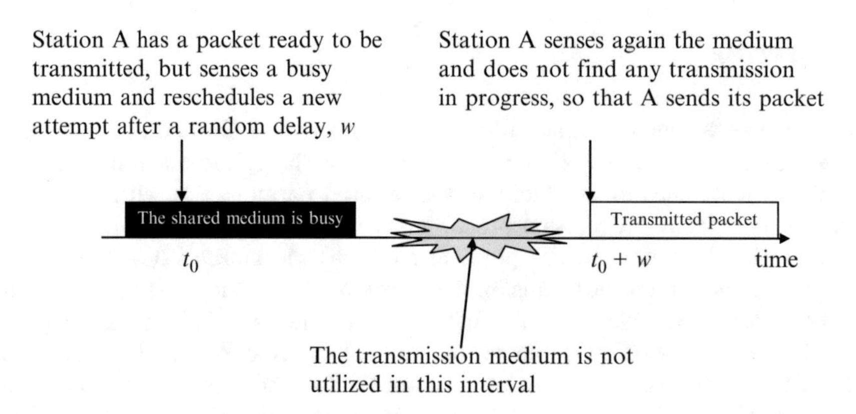
Station A has a packet ready to be transmitted, but senses a busy medium and reschedules a new attempt after a random delay, w
Station A senses again the medium and does not find any transmission in progress, so that A sends its packet
The shared medium is busy
Transmitted packet
t_0
$t_0 + w$
time
The transmission medium is not utilized in this interval

Fig. 6.18: Transmission after a busy line period with non-persistent CSMA

selecting the lowest retransmission delay has the highest probability to be successful in the packet transmission. This is more likely to happen at the first attempt.

For the analysis of CSMA protocols, we will assume a Poisson arrival process of new packets with a mean rate λ (i.e., an infinite population of users). Hence, the offered traffic (= carried traffic, throughput, under stability assumption) is $S = \lambda T$ Erlangs, whereas the total circulating traffic (new arrivals plus retransmissions due to collisions, with total mean rate Λ) is $G = \Lambda T$ Erlangs.

In the following sub-sections, the different variants of the CSMA protocol are described.

6.2.5.1 Unslotted, Non-persistent CSMA

When a terminal is ready to send its packet, it senses the broadcast medium and acts as follows [2, 17]:

- If no transmission is revealed (i.e., the channel is free), the terminal transmits its packet.
- If a transmission is revealed, the terminal goes back to the previous point after a random delay (i.e., the same delay adopted to reschedule transmissions after a collision).

An example of the protocol behavior is shown in Fig. 6.18.

A packet transmission may be successful or not due to a collision. An acknowledgment scheme with a timeout is used to notify a collision.

Figure 6.19 shows the behavior of the offered traffic S as a function of the total circulating traffic G for non-persistent CSMA protocol with different a values based on the analysis carried out in Sect. 6.2.5.6. If $a > 0$, the

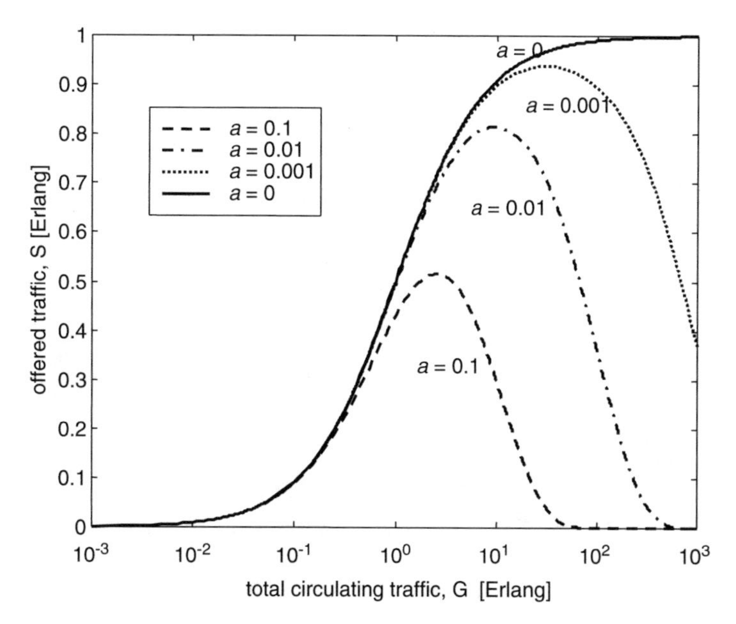

Fig. 6.19: Throughput behavior of unslotted non-persistent CSMA

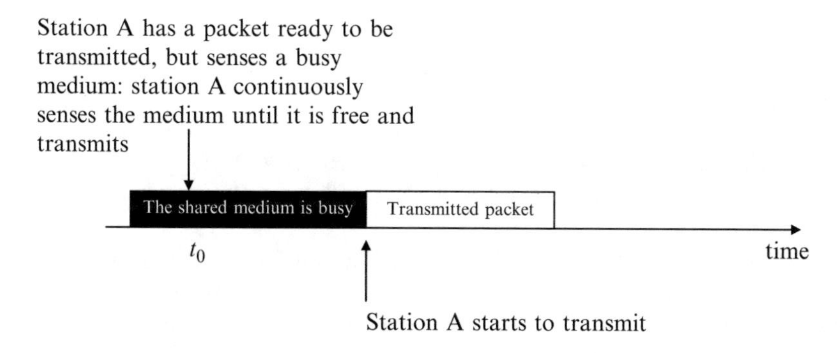

Fig. 6.20: Transmissions after a busy line period with 1-persistent CSMA

throughput of non-persistent CSMA has a maximum that denotes a boundary condition beyond which there is instability. If $a = 0$, we have an ideal situation with no collisions so that throughput S tends monotonically to 1 Erlang as G increases; there is no instability for $a = 0$. Hence, the a value has a significant impact on the protocol behavior.

However, the non-persistent CSMA protocol does not allow us to exploit resources efficiently; this is because a packet is not transmitted immediately as soon as the medium is free. This is the reason why the following 1-persistent CSMA protocol has been proposed.

6.2.5.2 Unslotted, 1-Persistent CSMA

When a terminal is ready to send its packet, it senses the broadcast medium and acts as follows [2, 17]:

- If no transmission is revealed (i.e., the channel is free), the terminal immediately sends its packet.
- If a transmission is revealed, the terminal waits and transmits its packet as soon as the medium is sensed free.

An example of the protocol behavior is shown in Fig. 6.20.

However, this protocol has more collisions than the non-persistent case due to the following situation. Let us consider two stations, A and B, which need to transmit a new packet that arrived during the transmission of another packet by station C. Both A and B will wait for the end of the previous transmission and will start to transmit as soon as they sense a free channel due to the completion of the service of C. Consequently, A and B will

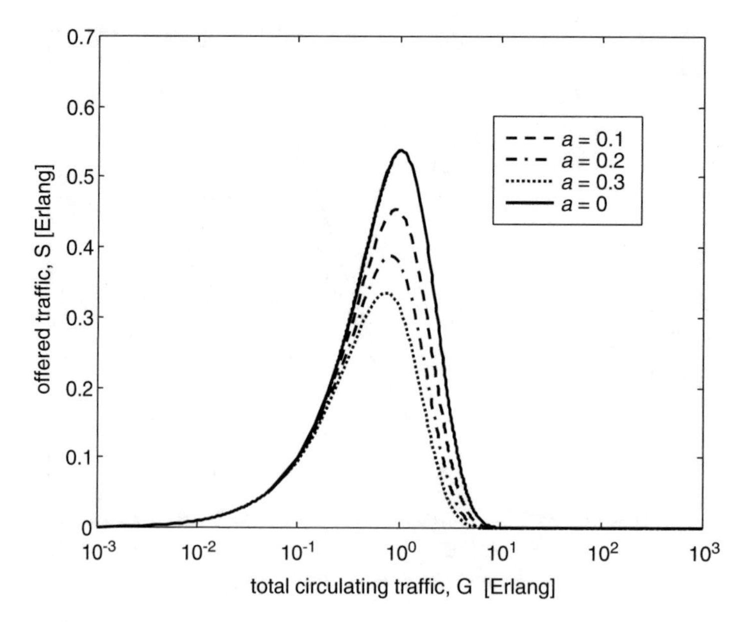

Fig. 6.21: Throughput behavior for the unslotted version of 1-persistent CSMA

start to transmit almost simultaneously, thus causing a collision. With this scheme, collisions are possible even if $a = 0$, i.e., the propagation delay $\tau = 0$.

The throughput behavior of 1-persistent CSMA is shown in Fig. 6.21. The curve always has a maximum, which denotes a boundary condition for protocol stability. The performance strongly depends on the a value. Even if $a = 0$, the peak of the throughput is significantly lower than 1 Erlang. The peak of the throughput reduces if a tends to 1.

Practically, 1-persistent CSMA does not provide any throughput improvement with respect to non-persistent CSMA. It is important to randomize the starting instant of a packet transmission after a busy period on the shared medium to improve the 1-persistent CSMA protocol efficiency. Such an improvement is accomplished by the following p-persistent protocol.

6.2.5.3 Slotted, p-Persistent CSMA

When a terminal is ready to send its packet, it senses the broadcast medium and acts as follows [2, 17]:

(a) If the medium is free, then transmit.
(b) If the medium is busy, then wait until it is free.
(c) As soon as the medium becomes free, a slotted transmission scheme is adopted, being τ the slot duration.

  1. At the new slot, the terminal transmits with probability p and does not transmit with probability $1 - p$, thus performing the next step.
  2. If the channel is free at the new slot, the above point #1 is performed (probabilistic transmission scheme); otherwise, a random waiting time (as in the case of a collision) is used, and then the algorithm is restarted from the above point a.

Note that for the p-persistent CSMA case, we have not slotted and unslotted cases, but just the slotted version described above. An example of the access phase is described in Fig. 6.22.

The performance of the p-persistent CSMA protocol (in terms of mean packet delay and throughput) depends on both a and p values. The analysis of the p-persistent CSMA throughput is a quite complex task, as detailed in [17]. A closed-form approximate expression of S as a function of G, a, and p is available only for small p values, $p < 0.1$. The corresponding graph has been shown in Fig. 6.23. We can note that the maximum throughput of the p-persistent CSMA scheme is sensitive to the a value: lower a values guarantee a higher peak of the throughput. Considering higher values of p than those in Fig. 6.23, the throughput peak of p-persistent CSMA improves as

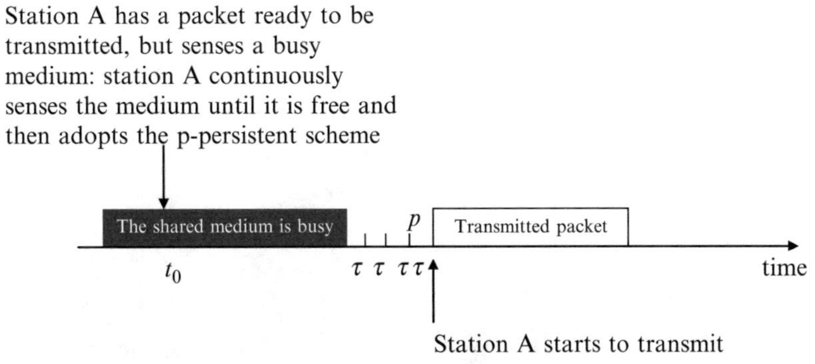

Station A has a packet ready to be transmitted, but senses a busy medium: station A continuously senses the medium until it is free and then adopts the p-persistent scheme

The shared medium is busy

p

Transmitted packet

t_0

$\tau\ \tau\ \tau\ \tau$

time

Station A starts to transmit

Fig. 6.22: Transmission after a busy line period with p-persistent CSMA

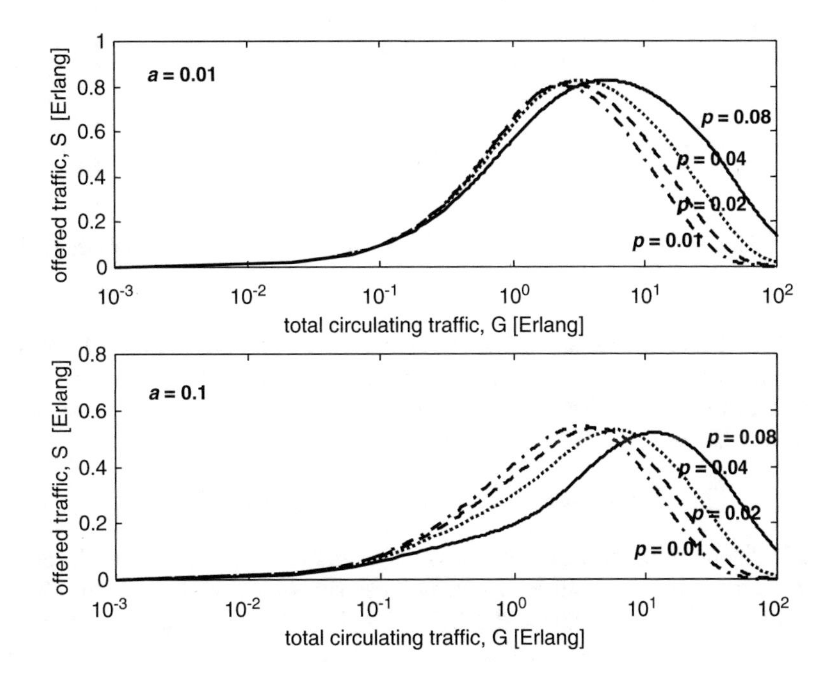

Fig. 6.23: S versus G for slotted p-persistent CSMA with different p values for both $a = 0.01$ and $a = 0.1$

p reduces; however, packets experience higher delays as p decreases. Then, a trade-off has to be adopted for the selection of the p value.

6.2.5.4 CSMA with Collision Detection

When a collision occurs, there is a loss of efficiency with CSMA, because we waste the resources of the shared medium for the whole packet transmission time T. In order to overcome this problem, the Collision Detection (CD) mechanism has been considered [1, 2, 17]. As soon as a certain terminal B realizes that its packet transmission has undergone a collision, terminal B immediately stops the packet transmission and sends a special *jam message*.[4] The terminals receiving the jam signal discard the received packet. Then, terminal B waits for a random time (according to the backoff algorithm for collision resolution) and then returns to the initial phase of carrier sensing to verify whether the physical medium is free or not.

A terminal listens before and while talking with the CD variant of the protocol. The CD scheme requires that a terminal reads what it is transmitting: if there are differences, the terminal realizes that a collision is occurring.

---

[4] CSMA/CD does not require an acknowledgment scheme with a timeout to detect collisions.

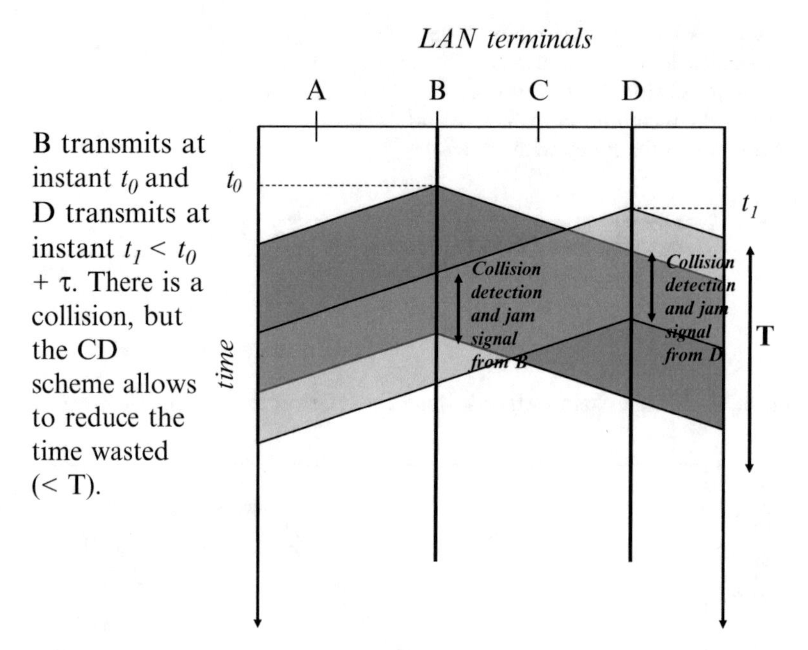

Fig. 6.24: Collisions with CSMA/CD: the waste of transmission time is shorter than that with CSMA

To verify whether a packet is transmitted without collisions, a host must detect a collision before it finishes transmitting a packet. Hence, the application of the CD scheme imposes a *constraint* on the minimum transmission time of a packet in relation to the maximum round-trip propagation delay 2τ of the network; this constraint also depends on the transmission bit-rate R_b adopted in the network as follows:

$$(\text{Minimum packet length in bits})\,/R_b \geq 2\tau \tag{6.20}$$

In the Ethernet standard, the minimum packet length is 64 bytes. More details are in Sect. 6.5.1.

The CSMA/CD scheme manages the collisions and reduces the waste of time because of collisions, as we can see by comparing Fig. 6.17 with Fig. 6.24. Then, the CD algorithm allows increasing the throughput of LANs.

The CD approach can be used with non-persistent, 1-persistent, or p-persistent variants of CSMA, both slotted and unslotted.

6.2.5.5 Comparison Among Random Access Protocols

The throughput comparison of the various random access schemes is shown in Fig. 6.25 for $a = 0.01$ [17]. All the curves have a maximum, highlighting a boundary condition for protocol stability. The throughput values of CSMA protocols are better than those of Aloha ones. We may notice that the performance of p-persistent CSMA schemes for low p values is equivalent to that of 1-persistent techniques for $G \leq 1$ Erlang. Non-persistent schemes are stable for higher values of G and achieve higher maximum values of S. Finally, the 1-persistent CSMA scheme achieves a good enough throughput performance, provided that $G < 1$ Erlang; since in these conditions, also the mean packet delay performance of 1-persistent CSMA is good, we can conclude that 1-persistent CSMA can represent a good solution for LANs.

Figure 6.26 compares the different random access protocols in terms of the maximum of S as a function of the a value [17]. When parameter a increases, the maximum throughputs of CSMA protocols decrease; instead, the maximum throughputs of Aloha protocols are constant. As a goes beyond 1, the maximum throughputs achieved by CSMA protocols reduce significantly below those of Aloha protocols. These results confirm that CSMA protocols achieve a good efficiency only in the presence of low propagation delays with respect to the packet transmission time (i.e., $\tau < T$). Finally, as expected, non-persistent CSMA/CD outperforms non-persistent CSMA for any value of a, thus confirming the good impact of the CD scheme.

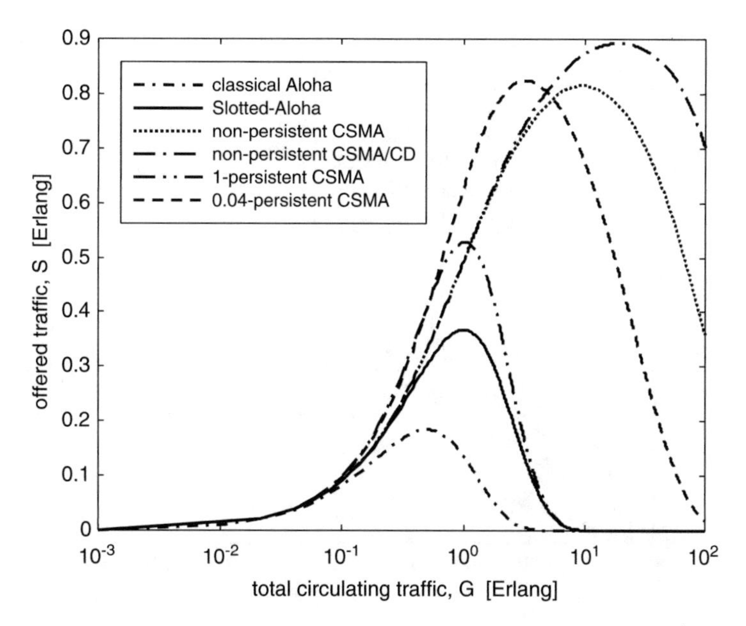

Fig. 6.25: Offered traffic versus total circulating traffic for different random access protocols for $a = 0.01$

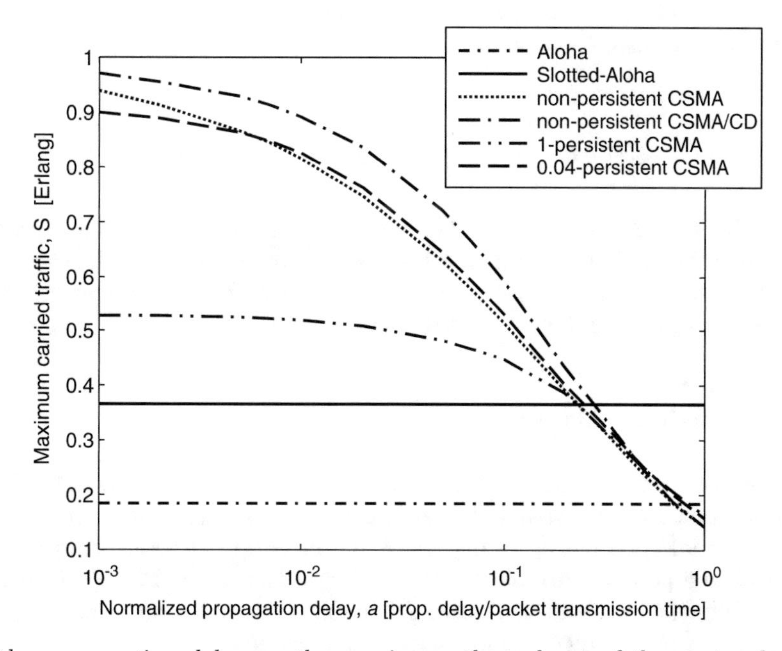

Fig. 6.26: Impact of the propagation delay on the maximum throughput of the protocols. In the non-persistent CSMA/CD case, the normalized jam message (to the packet length) has been considered equal to 0.2 [T units]

The p value of the p-persistent scheme could be selected optimally for the different a values (i.e., the p value that maximizes the carried traffic).

Figure 6.27 compares the different protocols in terms of the mean packet delay $E[T_p]$ for $a = 0.01$ {corresponding to $\Delta = 2a = 0.02$ [T units] according to the definition made for (6.7)}, $E[R] = 4$ [T units], and jam message duration $= 0.2$ [T units].[5] This graph has been obtained in all cases by applying a formula equivalent to (6.7) as [17]:

[5] In the IEEE 802.3 standard, a packet has a minimum length of 64 bytes and a maximum length of 1518 bytes. The length of the jam message is 32 or 48 bits. Hence, the assumption made here of a jam message equal to 0.2 in [T units] is a conservative choice.

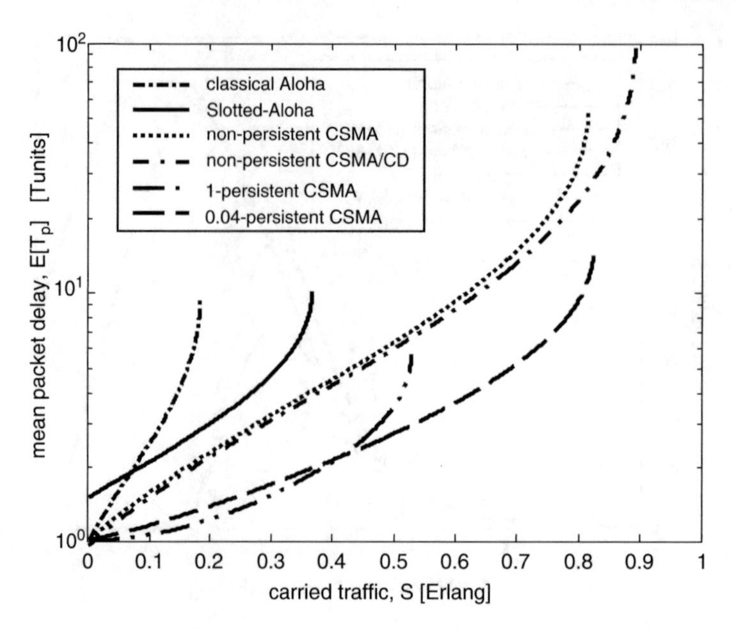

Fig. 6.27: Comparison of the random access protocols in terms of mean packet delay versus carried traffic S for $a = 0.01$ (corresponding to $\Delta = 0.02$ [T units]), $E[R] = 4$ [T units], and jam message duration = 0.2 [T units]. Only the stable branches of the curves have been shown in this graph

$$E\left[T_{\mathrm{p}}\right] \approx \left(\frac{G}{S} - 1\right)\{T + \Delta + E\left[R\right]\} + T + \Delta$$

This formula is approximated in CSMA cases. The interesting result is that non-persistent CSMA protocols allow supporting more traffic than Aloha schemes, but the mean packet delay of CSMA schemes can be much higher as S increases. The 1-persistent scheme and p-persistent protocols with low p values achieve similar throughput performance for low S values and attain low mean packet delay values.

6.2.5.6 CSMA S–G Throughput Analysis

This section aims to describe an analytical approach for studying the throughput of CSMA protocols and, in particular, of unslotted non-persistent CSMA. The shared medium alternates busy periods, B, during which there are packet transmissions and idle periods, I, during which the medium is unused. A cycle is composed of a busy period and the subsequent idle period. Variable U is the interval in a cycle during which the channel is used to transmit a packet successfully (i.e., without collisions). This study is carried out assuming no buffering at nodes. The channel throughput S can be obtained using the following formula [2, 17]:

$$S = \frac{E\left[U\right]}{E\left[B\right] + E\left[I\right]} \tag{6.21}$$

In the case of successful packet transmission, we have a busy period with a single packet transmission. Hence, variable B is equal to $T + \tau$, since the packet transmission time is T and a further time τ is necessary to have that a free channel condition is perceived by all terminals. Instead, in a busy period with multiple packet transmissions, there are collisions and B is greater than $T + \tau$.

Let us compute the mean value of the idle period. Let us remark that arrivals are according to a Poisson process with a total mean rate Λ (new arrival plus retransmissions). We can invoke the memoryless assumption, considering that after a packet transmission interval has ended, a new arrival will need a residual time of a packet interarrival time, which is still exponentially distributed with a mean rate Λ. Hence, the mean duration of an idle period after a packet transmission is

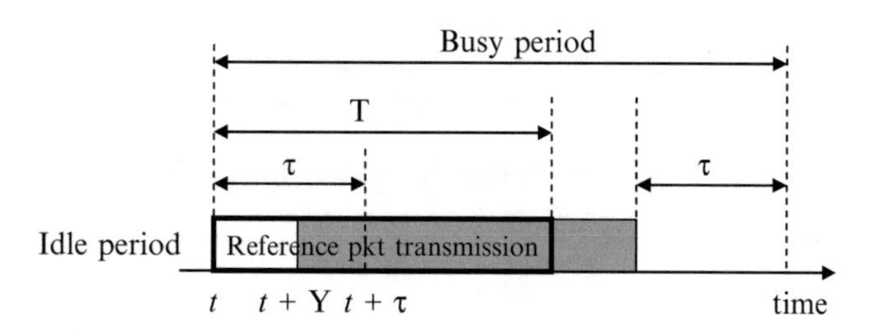

Fig. 6.28: Characterization of the busy period in a general case

$$E[I] = \frac{1}{\Lambda} \tag{6.22}$$

The mean time during a cycle for which the channel is used to transmit a packet successfully, $E[U]$, can be obtained by considering that the transmission of a packet (time T) is successful if no other generation occurs in the vulnerability window τ at the beginning of the packet transmission; this occurs with the probability of no arrival of the total Poisson process with mean rate Λ in τ:

$$U = \begin{cases} T, & \text{with probability } e^{-\Lambda\tau} \\ 0, & \text{otherwise} \end{cases} \quad \Rightarrow \quad E[U] = T\,e^{-\Lambda\tau} \tag{6.23}$$

We need to analyze variable B and its expected value. Let us refer to Fig. 6.28. We consider that an idle period ends with the beginning of a new (reference) packet transmission at time t. Collisions may occur because of packet generations made by other stations with mean rate Λ during the vulnerability interval from instant t to instant $t+\tau$. Let $t+Y$ denote the arrival instant of the last packet colliding with the reference packet. We must have $0 \le Y < \tau$. Please note that the limiting case with $Y=0$ implies that no collision occurs with our reference packet: this is the case where the busy period is successfully used for the transmission of a packet (i.e., $B = T+\tau$).

Referring to Fig. 6.28, the busy period starts at time t and ends at time $t+Y+T+\tau$. Hence, $B = Y+T+\tau$. We characterize the PDF of variable Y, $F_Y(x)$, in the interval $[0, \tau]$ as

$$\begin{aligned} F_Y(x) &= \text{Prob}\{Y \le x\} \\ &= \text{Prob}\{\text{no arrival in the interval of length } \tau - x\} = e^{-\Lambda(\tau - x)} \end{aligned} \tag{6.24}$$

The corresponding pdf $f_Y(x)$ is

$$f_Y(x) = \frac{\mathrm{d}}{\mathrm{d}x} F_Y(x) = \Lambda\, e^{-\Lambda(\tau - x)} \tag{6.25}$$

Hence, we can derive the expected value of Y from (6.25) as follows:

$$\begin{aligned} E[Y] &= \int_0^\tau x f_Y(x) \mathrm{d}x = \int_0^\tau x \Lambda e^{-\Lambda(\tau-x)}\, \mathrm{d}x \\ &= \frac{e^{-\Lambda\tau}}{\Lambda} \int_0^\tau x\Lambda\, e^{\Lambda x}\, \mathrm{d}\Lambda x = \text{we integrate by parts} \\ &= \frac{e^{-\Lambda\tau}}{\Lambda} \left[x\Lambda\, e^{\Lambda x} - e^{\Lambda x}\right]_0^\tau = \tau + \frac{e^{-\Lambda\tau} - 1}{\Lambda} \end{aligned} \tag{6.26}$$

We can express the expected value of B as

$$E[B] = E[Y] + T + \tau = 2\tau + \frac{e^{-\Lambda\tau} - 1}{\Lambda} + T \tag{6.27}$$

Using (6.21)–(6.23), and (6.27), we can achieve the following throughput result by considering that $\tau/T = a$ and $\Lambda\tau = Ga$:

$$S = \frac{E[U]}{E[B]+E[I]} = \frac{Te^{-\Lambda\tau}}{2\tau + \frac{e^{-\Lambda\tau}-1}{\Lambda} + T + \frac{1}{\Lambda}} = \frac{G\,e^{-Ga}}{G(1+2a)+e^{-Ga}} \tag{6.28}$$

In the limiting (and ideal) case for $a \to 0$, we obtain the following simple result:

$$S = \frac{G}{G+1} \quad \text{[Erlangs]} \tag{6.29}$$

The behavior of (6.28) has already been shown in Fig. 6.19 for different a values, including the ideal case $a=0$. The peak of the throughput increases if a decreases, since collisions are less likely.

Finally, the mean packet delay, $E[T_p]$, can be derived using an approach similar to (6.7). By neglecting the ACK transmission/receive time (or including this time in Δ), we have [17]

$$\begin{aligned} E[T_p] &\approx \left(\frac{G}{S}-1\right)\{T+\Delta+E[R]\}+T+\Delta \\ &= G\,e^{Ga}(1+2a)\{T+2\tau+E[R]\}+T+2\tau \end{aligned} \tag{6.30}$$

Note that formula (6.30) has already been used for drawing the non-persistent CSMA curve in Fig. 6.27.

6.3 Demand-Assignment Protocols

Pure random access protocols do not guarantee fairness or bounded access delays for real-time traffics. This is the reason why other access protocols have been proposed, which allow more regulated access of the terminals to the shared medium. This section considers polling protocols, token-based schemes, and Reservation-Aloha [1–3].

6.3.1 Polling Protocols

These schemes are based on a cyclic authorization according to which terminals are enabled to transmit. The following broadcast topologies are suitable to support polling access protocols: tree, bus, and wireless star. In the case of a tree topology with centralized control, we have the classical roll-call polling. In the bus topology case, we may have a decentralized control scheme, called hub polling. In the following description, we refer to the case with a centralized controller. In particular, we consider that the central controller enables the transmissions of a remote terminal by sending a special broadcast signal, named poll message, which contains the address of the remote terminal to enable its transmissions. The polled terminal (recognizing its address in the polling message) is enabled to transmit the contents of its buffer to the central controller according to three different techniques [3]:

- *Gated technique*: A terminal sends only the packets, which were in its buffer at the arrival instant of the authorization to transmit.
- *Exhaustive technique*: A terminal sends all packets in its buffer (i.e., a terminal releases the control only when its buffer is empty).
- *Exhaustive limited technique*: A terminal sends up to T_{max} packets, regardless of whether these packets had arrived before or during the service interval of the terminal.

At the end of the terminal transmission interval, the control is returned to the central controller, which starts to poll the next remote terminal according to a cyclic service scheme. A classical *round-robin* technique can be adopted; a *weighted round-robin* scheme should be used if remote terminals contribute unequal traffic loads [18].

According to the polling approach, a remote terminal is polled even if it has no message to send to the central controller. Therefore, this access scheme is efficient only if remote terminals have regular traffic to be sent to the central controller. The *protocol overhead* to interrogate remote terminals reduces the efficiency. There is a certain waste of time (i.e., time not used to convey information traffic) to poll a terminal and allow the terminal to return the control to the central controller.

In a decentralized control scheme, the different terminals directly exchange the polling message, and this is similar to the token passing schemes described below.

Bytes	1	1	1	2 or 6	2 or 6	Variable	4	1	1
	Start delimiter	**Access control**	Frame control	Dest. Address	Source Address	Data	CRC	**EOT**	Frame status

Fig. 6.29: IEEE 802.5 frame format with the length of the different fields. *Shaded fields* come from the token

6.3.2 Token Passing Protocols

Unlike Ethernet, token networks allow a single terminal to transmit at a time, i.e., the terminal with the token. A typical ring topology (either physical or logical) is used for these LANs. The token rotates around the ring and arrives in turn at each node. A ring terminal copies all data and tokens (received on the input interface) and repeats them along the ring (output interface) by adding a delay of 1-bit time; a terminal needs to buffer just one bit. When a terminal wishes to transmit packet(s), it grabs the token when it passes and appends its data to the data train. When the transmission completes, the terminal releases the token and sends it on its way. A token ring network has a star topology with a central wiring center (a sort of hub): the logical topology is different from the physical topology.

There are two variants of the token ring protocol, depending on the policy to release a token on behalf of a terminal, which has completed its transmission.

- Release After Reception (RAR): A terminal captures the token, transmits data, waits for data to propagate successfully around the ring, and then releases the token. Such an approach enables the terminals to detect erroneous frames and to retransmit them.
- Release After Transmission (RAT): A terminal captures the token, transmits data, and then releases the token so that the next terminal on the ring can use the token after a short propagation delay.

Each terminal in the network can be served according to one of the schemes already described for the polling protocol (e.g., gated and exhaustive techniques).

FDDI was an American technology for Metropolitan Area Networks (MANs) operating at 100 Mbit/s with optical fiber as the physical medium (year 1980). FDDI was based on a dual-ring topology. FDDI adopted a token ring protocol with RAT policy and a limit to the token holding time on behalf of a terminal. In the case of a ring failure, the terminals closer to the failure point switch the rings to achieve a virtual bus topology, as shown in Fig. 6.2.

The IEEE standards for token protocols are IEEE 802.4 for a bus topology (token bus standard) and IEEE 802.5 for a ring topology (token ring standard) with RAR approach.

The IEEE 802.5 token ring standard is based on an IBM proposal. The transmission of bits requires a differential Manchester line encoding.[6] The token is a small packet circulating around the ring or included in the transmitted frame. The token is composed of a *token delimiter* (1 byte, where the encoding scheme is violated to distinguish such byte from the rest of the frame), an *access control field* (1 byte), and an *end of token* (1 byte). A "free token" is a 3-byte message used to release the control to the next station according to the cycle order. If a terminal receiving the token has no information to send, it passes the token to the next terminal of the cycle. If a terminal receiving the token has information to transmit, it seizes the token, alters 1 bit of the token access control field (so that the initial part of the token becomes the initial part of a packet), appends the information that it wants to transmit, and sends this information packet to the next terminal on the ring.

Each station can hold the token for a maximum time, called Token Holding Time (THT). The Token Rotation Time (TRT) denotes the time taken by a token to traverse the ring.

While the packet travels along the ring, no token can be on the network, which means that other stations needing to transmit must wait. In the transmitted packet, both token delimiter and access control field are at the beginning of the packet, while the end delimiter (End of Token, EOT) is at the end of the packet, as shown in Fig. 6.29. The packet travels along the ring until it reaches the destination station, which copies this information; then, the packet continues to travel along the ring until it is finally removed by the sending terminal (RAR approach), which checks the returning packet to see whether the destination has copied this packet. The maximum number of terminals on the ring is 250 in the IEEE 802.5 standard.

[6] With differential Manchester line encoding, there is always a level transition in the middle of a bit. In the case of bit "1" (or "0") transmission, we have the first half of the signal equal (or complemented with respect) to the last part of the previous bit.

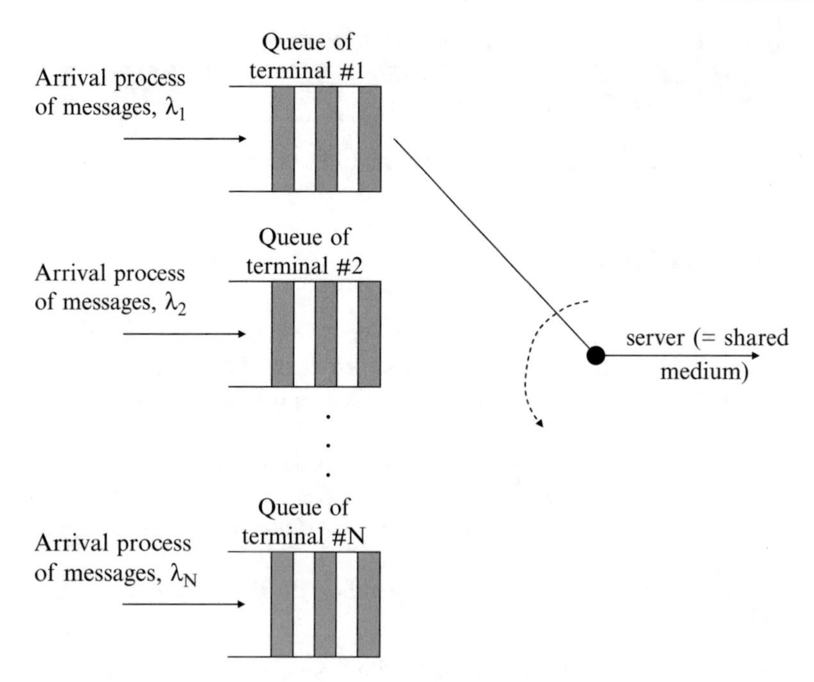

Fig. 6.30: Model of the statistical multiplexer used for both token-based and polling-based schemes

The access control field of the token (one byte) is structured as follows: the three most significant bits contain a priority field (i.e., a priority level from 0 to 7), then there is a token bit (used to differentiate a token from a data/command packet), a monitor bit (used by the active monitor to determine whether a packet is traveling endlessly in the ring), and the three least significant bits contain the reservation field.

IEEE 802.5 adopts a sophisticated priority system that allows some user-designated, high-priority terminals to use the network more frequently than other terminals. The access control field has two subfields to control the priority: the priority field and the reservation field. Only the terminals with a priority equal to or higher than the priority value set in a token can seize that token. After the token is seized and changed to an information frame, only the terminals with a priority value higher than that of the transmitting terminal can reserve the token for the next round in the network: when the next token is generated, it includes the higher priority of the reserving terminal. Terminals raising the token priority level must reset the previous priority level after their transmissions are completed.

Token ring networks employ three different types of cabling:

- UTP cables of categories 3, 4, and 5 for 4 Mbit/s and UTP cables of categories 4 and 5 for 16 Mbit/s
- STP cables of type 9
- Optical fibers.

The maximum frame length L is determined by the transmission bit-rate R on the ring (i.e., 4 or 16 Mbit/s) and the THT value (i.e., the maximum time for which a terminal can seize the token, which is about equal to 10 ms): $L = \text{THT} \times R$. Hence, the maximum frame length is constrained to 5000 bytes and 20,000 bytes, respectively, for 4 and 16 Mbit/s bit-rate transmissions on the ring.

6.3.3 Analysis of Token and Polling Schemes

In the following study, we analyze the performance of polling-based and token-based schemes using a common model with as many transmission queues as the number N of the terminals sharing the LAN and with one server cyclically assigned to the various queues, as shown in Fig. 6.30 [2]. The cyclic assignment allows the statistical multiplexing of the traffic flows of different terminals on the output line.

Cyclic resource assignment schemes (polling-based or token-based) have bounded access delays. Let us consider that each terminal (when enabled) may transmit for a maximum time T_{max}, according to an exhaustive limited scheme. Hence, if the network has N terminals, the maximum access delay for a terminal is $T_{\max} \times (N-1)$, if we neglect the times to switch the control from one terminal to another.

The generic ith queue (ith terminal) has an input process characterized by a mean message arrival rate λ_i. Each message has a random length in packets l_i that, in general, may have a different distribution from queue to queue. Let T_p denote the packet transmission time. Let T_i denote the service time for the ith queue. Let δ_i denote the overhead time to switch the service from the ith queue to the next $(i+1)$th queue according to the service cycle. The overhead time depends on the adopted protocol and on the LAN topology. For instance, in a token ring network, δ_i is the propagation delay from terminal i to terminal $i+1$, including a synchronization time for terminal $i+1$. Instead, in a tree network with roll-call polling, δ_i is the round-trip propagation delay between the central controller and terminal i plus the synchronization time of terminal i plus the time to send the address of the polled terminal i. Note that δ_i are deterministic values.

We are interested in characterizing the cycle time T_c, i.e., the time interval from the instant when the server starts to serve a generic queue (terminal) to the instant when the server returns to serve the same queue (after completing a cycle). The following formula is valid:

$$T_\mathrm{c} = \sum_{i=1}^{N} (T_i + \delta_i) \tag{6.31}$$

Note that random variables T_i for $i=1,\ldots,N$ are not statistically independent; for example, if a queue uses the server for a long time (according to the exhaustive discipline), we may expect that the other queues experience congestion and, hence, high delays. Therefore, there is a correlation in the service times of the different queues.

We can take the expected value of both sides of (6.31); we exploit the linearity of the $E[\cdot]$ operator even in the presence of correlated random variables. The following result is obtained:

$$E[T_\mathrm{c}] = E\left[\sum_{i=1}^{N} (T_i + \delta_i)\right] \Rightarrow E[T_\mathrm{c}] = \sum_{i=1}^{N} \{E[T_i] + \delta_i\} \tag{6.32}$$

The mean service time of the ith queue in the cycle, $E[T_i]$, is the mean time to send the packets arrived with mean rate λ_i during a cycle with mean duration $E[T_\mathrm{c}]$:

$$E[T_i] = \lambda_i E[T_\mathrm{c}] E[l_i] T_\mathrm{p} \tag{6.33}$$

By substituting (6.33) in (6.32), we obtain an equation in $E[T_\mathrm{c}]$ as

$$\begin{aligned} E[T_\mathrm{c}] &= \sum_{i=1}^{N} \{\lambda_i E[T_\mathrm{c}] E[l_i] T_\mathrm{p} + \delta_i\} \Rightarrow \\ E[T_\mathrm{c}] &\left\{1 - \sum_{i=1}^{N} \lambda_i E[l_i] T_\mathrm{p}\right\} = \sum_{i=1}^{N} \delta_i \Rightarrow \\ E[T_\mathrm{c}] &= \frac{\sum_{i=1}^{N} \delta_i}{1 - \sum_{i=1}^{N} \lambda_i E[l_i] T_\mathrm{p}} \end{aligned} \tag{6.34}$$

Note that the above considerations are valid only in the case that the total *protocol overhead* also representing the *minimum total latency* $\Sigma\delta_i > 0$; otherwise, (6.32) becomes an identity and the proposed approach cannot allow us to determine the mean cycle duration. $\Sigma\delta_i = 0$ is the case of a round-robin scheduler where there is one queue that can be divided into many different (virtual) sub-queues as the number of traffic flows or users sharing the same transmission resources. The polling systems with zero switch-over periods have been analyzed in [19].

The value of $E[T_\mathrm{c}]$ in (6.34) is finite if the following stability condition is fulfilled:

$$\rho_\mathrm{tot} = \sum_{i=1}^{N} \lambda_i E[l_i] T_\mathrm{p} < 1 \quad \text{[Erlang]} \tag{6.35}$$

Equation (6.35) shows that the total traffic load ρ_{tot} offered by the N terminals is equal to the sum of the traffic loads contributed by each terminal. If $\rho_{tot} \to 1$ Erlang, the system becomes congested.

$E[T_c]/2$ is the mean delay a packet arriving at an empty queue must wait for the arrival of the server.

Let us consider that the arrival processes at the different queues are Poisson and independent. Let us assume that the buffers have infinite capacity. If overhead times δ_i are negligible, the queuing of messages in the whole system depicted in Fig. 6.30 can be modeled using an M/G/1 global queue (with a special service policy). Then, the mean message delay is given by the well-known Pollaczek–Khinchin formula [see Chap. 5, (5.18)].

$$T = E[X] + \frac{\lambda_{tot} E[X^2]}{2[1 - \lambda_{tot} E[X]]} = \sum_{i=1}^{N} \frac{\lambda_i}{\lambda_{tot}} E[l_i] T_p + \frac{\sum_{i=1}^{N} \lambda_i E[l_i^2] T_p^2}{2\left\{1 - \sum_{i=1}^{N} \lambda_i E[l_i] T_p\right\}} \tag{6.36}$$

where λ_{tot} denotes the sum of all λ_i values from $i = 1$ to N.

The result in (6.36) is consistent with the fact that the mean queuing delay does not depend on the service discipline, provided that the insensitivity property conditions are met, as detailed in Sect. 4.5.

If overhead times δ_i are not negligible, an additional term must be included in the mean message delay in (6.36) to take the wasted times into account. In particular, we consider that this additional term is equal to $E[T_c]/2$ [3]. More accurate results have been obtained (distinguishing gated and exhaustive techniques) in the case of constant overhead times ($\delta_i = \delta$) with all N terminals having the same traffic characteristics ($\lambda_i = \lambda$, $l_i = l$) [2]. In particular, the following results have been obtained, considering an M/G/1 *model with server vacations* due to overhead times [20]:

$$T = E[X] + \frac{N\lambda E[X^2]}{2[1 - N\lambda E[X]]} + \frac{E[T_c]}{2} \times \begin{cases} (1 - \lambda E[X]), & \text{exhaustive} \\ (1 + \lambda E[X]), & \text{gated} \end{cases} \tag{6.37}$$

where $E[T_c] = N\delta/[1 - N\lambda E[X]]$, $E[X] = E[l] T_p$ and $E[X^2] = E[l^2] T_p^2$.

Let us consider the mean transfer time, T_{transf}, i.e., the mean delay from the message arrival at a given terminal (queue) to its completed delivery to another terminal in the case of a ring topology. The mean transfer time can be obtained by adding a mean ring propagation delay to (6.37) as

$$T_{transf} = T + \frac{1}{2} \sum_{i=1}^{N} \delta_i \tag{6.38}$$

Note that a fundamental term for the characterization of polling and token protocols is the derivation of the total latency $\Sigma\delta_i$. We describe below some interesting cases taken from reference [3].

Roll-call polling:

$$\delta_i = t_p + t_s + \tau_i \quad \Rightarrow \quad \sum_{i=1}^{N} \delta_i = N t_p + N t_s + \sum_{i=1}^{N} \tau_i \tag{6.39}$$

where t_p is the transmission time of the polling message (containing the address of the ith remote terminal), t_s is the synchronization time of the ith remote terminal, and τ_i is the round-trip propagation delay between the central controller and the ith terminal.

Token ring or token bus:

$$\sum_{i=1}^{N} \delta_i = N t_s + \tau \tag{6.40}$$

where τ denotes the propagation delay on the entire network (in the derivation of $\Sigma\delta_i$, we do not consider the token transmission time since it is practically included in the frame time).

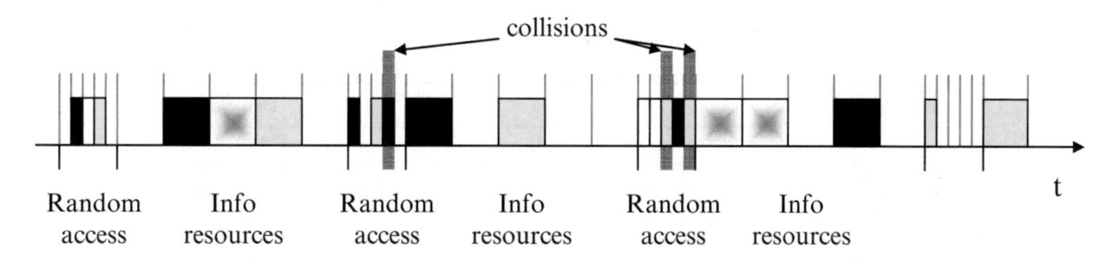

Fig. 6.31: The R-Aloha protocol with access phase and transmission one

The ring must be long enough to hold the entire token. Hence, the minimum ring latency $Nt_s + \tau$ has to be greater than the token transmission time (i.e., the transmission of 3 bytes at the bit-rate R of the ring):

$$Nt_s + \tau > \frac{24}{R} \tag{6.41}$$

Note that $t_s = 1/R$ because of the delay of one-bit time introduced by a ring terminal. Condition (6.41) provides a constraint that relates the minimum number of stations on the ring, N, to the ring bit-rate R and the ring length (proportional to τ). A special station, called ring monitor, can be added to the network to introduce additional delays so that the minimum number of stations is lowered. Finally, let us notice that the maximum ring latency results as $N \times \mathrm{THT} + \tau$.

6.3.4 Reservation-Aloha (R-Aloha) Protocol

The R-Aloha protocol was proposed by Roberts in 1973 [21]. It is based on a frame of length T_f, divided into N slots. The initial slots of the frame are minislotted to allow the transmission of minipackets to request the reservation of transmission resources according to a contention scheme (request phase). The terminals use the other slots of the frame to transmit data packets using the reservation acquired (grant).

Let m denote the number of minislots of the access phase. Transmission attempts are randomized with equal priority over all the m minislots of the access phase. If there is a collision, a terminal retries a new transmission attempt in one of the minislots of the next frame in the case of a persistency level 1. If the number of minislots is low with respect to the potential number of attempting terminals, a persistency probability p_pers lower than 1 must be used to avoid instability problems, as with Aloha-like protocols. An attempt is carried out by selecting a frame based on probability p_pers and then randomly using one minislot within this frame.

As shown in Fig. 6.31, information slots are distinct from access slots, used in the frame to send transmission requests (minipackets sent on minislots). A remote terminal needing to transmit a message generated in the middle of a frame must send a request (minipacket) in the access phase at the beginning of the next frame selected based on the p_pers mechanism. Once a request has been received correctly, resources are assigned (when available) according to a reservation scheme.

A reservation guarantees the use of one or more slots per frame. A terminal retains the reservation until its buffer is empty. Each message has a random length in packets, where the packet transmission time coincides with the slot. This access scheme is particularly efficient and suited to a traffic source with ON–OFF behavior, producing a constant data rate only during the ON phase.

We study the R-Aloha access phase by analyzing the throughput of successfully carried access requests per frame in the 1-persistent case. We consider an R-Aloha frame formed of one initial minislotted slot and then $N-1$ information slots. Let m denote the number of minislots in the initial contention slot of the frame. We refer to a Poisson arrival process of new requests with a mean rate λ. Let Λ denote the total mean arrival rate, including new transmission requests and retransmissions of collided requests. The number of transmission requests N_r managed in the access phase of a frame is Poisson distributed as

$$\mathrm{Prob}\{N_\mathrm{r} = k\} = P_k = \frac{(\Lambda T_\mathrm{f})^k}{k!} e^{-\Lambda T_\mathrm{f}} \tag{6.42}$$

Table 6.1: Distribution of the access delay D for 1-persistent R-Aloha

Access delay D in slots	Probability
$N/2+L$	P_s
$N/2+N+L$	$(1-P_s)P_s$
...	...
$N/2+(k-1)N+L$, where k denotes the kth transmission attempt	$(1-P_s)^{k-1}P_s$

The N_r requests generated in the current frame are transmitted in the minislots at the beginning of the next frame.

Let $S=\lambda T_\mathrm{f}$ denote the mean input traffic of new transmission requests per frame. Let $G=\Lambda T_\mathrm{f}$ denote the total mean traffic of transmission requests per frame. Like the Aloha protocols, we can write the following formula:

$$\frac{S}{G}=P_s \tag{6.43}$$

where P_s denotes the probability of successfully transmitting a request (i.e., no collision).

We derive P_s assuming that a target user has transmitted its request in a given minislot; then, we condition the study on the number of terminals making a transmission attempt in the same access phase according to distribution P_k in (6.42). In particular, if there is no other attempt in the access phase (with probability P_0), $P_{s|0}$ is equal to 1. Moreover, if there is another transmission attempt (with probability P_1), $P_{s|1}$ is equal to $1-1/m$, since we have to exclude the possibility that this attempt is made on the same minislot of our tagged transmission. In general, if there are other k transmission attempts (with probability P_k), $P_{s|k}$ is equal to $(1-1/m)^k$, since no other attempt must be made on the same slot of our tagged transmission. Note that this result on $P_{s|k}$ can be justified using the urn theory, as described in Chap. 3, Sect. 3.4.

In conclusion, probability P_s is obtained removing the conditioning on k as

$$\begin{aligned}P_s &= \sum_{k=0}^{\infty}\left(1-\frac{1}{m}\right)^k P_k = e^{-\Lambda T_\mathrm{f}} \times \sum_{k=0}^{\infty}\frac{\left[\Lambda T_f\left(1-\frac{1}{m}\right)\right]^k}{k!} \\ &= e^{-\Lambda T_\mathrm{f}} \times e^{\Lambda T_\mathrm{f}\left(1-\frac{1}{m}\right)} = e^{-\frac{\Lambda T_\mathrm{f}}{m}} = e^{-\frac{G}{m}}\end{aligned} \tag{6.44}$$

By considering (6.43) and (6.44), we achieve the following expression to relate S and G:

$$S = G\,e^{-\frac{G}{m}} \tag{6.45}$$

$S=S(G)$ has a maximum for $G=m$, which is equal to $S_{\max}=m/e$ Erlangs. This is a reasonable result, since the R-Aloha access phase is similar to an S-Aloha system with m parallel channels. The R-Aloha access phase is stable if $S=\lambda T_\mathrm{f}\le m/e$ transmission requests per frame. Due to the traffic capacity of the R-Aloha frame, we have also to consider the following traffic stability constraint considering that each arrival carries a single packet: $\lambda T_\mathrm{f} < N-1$ packets per frame. Thus, considering together the access phase stability and the traffic stability limit, we have the following constraint: $\lambda T_\mathrm{f} < \min\{m/e, N-1\}$. It is a good design choice that m/e and $N-1$ are almost equal; otherwise, the minislots or the information slots are not used efficiently.

We can determine the distribution of the *access delay*, D, from the arrival of a packet to the successful transmission of its minipacket in an access phase. We consider in general that there are L minislotted slots per access phase (i.e., $L\times m$ minislots per frame) and that a terminal knows the outcome of a transmission attempt in the access phase of a frame within the end of the same frame (i.e., the frame duration T_f has to be longer than the round-trip propagation delay). A 1-persistent scheme has been considered here. The access delay D has the modified geometric distribution shown in Table 6.1. Note that delay D has to account for the initial delay (with a mean value of $N/2$ slots) to wait for the start of the next contention phase. Moreover, at each unsuccessful attempt, an entire frame time is lost, i.e., N slots.

Hence, the mean access delay, $E[D]$, results as

$$E\left[D\right]=\frac{N}{2}+\left(\frac{1}{P_s}-1\right)N+L=\frac{N}{2}+\left(\frac{G}{S}-1\right)N+L\ \ \text{[slots]} \tag{6.46}$$

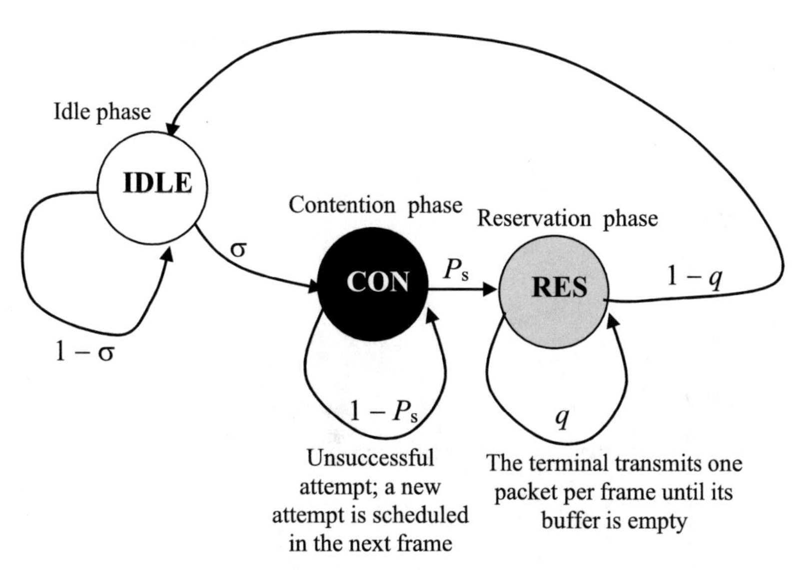

Fig. 6.32: Terminal state diagram for 1-persistent R-Aloha, where we consider the case of the reservation of just one slot per frame

where G as a function of S can be derived numerically from (6.45).

In this study, parameter $G = \Lambda T_f$ also represents the mean number of contending terminals per frame (1-persistent case).

Let us now consider the following additional assumption that a terminal cannot generate a new packet until the previous one has been successfully transmitted. Hence, there is no more than one packet in the buffer of each terminal. Then, the mean packet transmission delay $E[T_p]$ is obtained by summing the mean packet service time to the mean access delay $E[D]$. The mean packet service time can be characterized using an M/G/1 queuing model, as detailed in Chap. 5.

An alternative approach to study the access phase (still considering that no more than one resource can be assigned to a terminal per frame) is to adopt the EPA approach [12], referring to cases with a finite number M of terminals. In this study, we consider a more general assumption than before: messages are not formed of a single packet but by more packets according to a modified geometric distribution with parameter $1-q$. The behavior of each terminal is described by the three-state diagram in Fig. 6.32, where state transitions occur at the end of each frame. We use the following symbols: σ is the probability of a new message arrival at an empty buffer on a frame basis $\left(\sigma = 1 - e^{-\lambda T_f}\right)$; P_s is the success probability for an attempt on a minislot $[P_s = (1-1/m)^{c-1}$, where c denotes the equilibrium number of terminals in the CON state]; and q is the parameter used to model the length of a message according to a modified geometric distribution (if $q=0$, we have messages composed of a single packet, similarly to what has been considered before as well as in the Slotted-Aloha study [9]).

Let s, c, and r denote the equilibrium number of terminals in IDLE, CON, and RES states, respectively. We can write the following EPA balance equations (non-linear system):

$$\begin{cases} s\sigma = cP_s \\ r(1-q) = s\sigma \\ s+c+r = M \end{cases} \Rightarrow c + c\left(\frac{1}{\sigma} + \frac{1}{1-q}\right)\left(1-\frac{1}{m}\right)^{c-1} = M \tag{6.47}$$

We have thus obtained a single EPA equation in the unknown term c, which can be solved numerically or graphically. In Fig. 6.33, we show the behavior of this equation as a function of c in the case that $1/\sigma + 1/(1-q) = 100$ and for different m values: the solution is given by the intersection of the curve with the horizontal line for ordinate value equal to M. To have a stable protocol behavior, the EPA system must have a single solution. There is a single and stable EPA solution up to $M_{\max} = 12$ terminals for $m = 2$ minislots/frame, up to $M_{\max} = 19$ terminals for $m = 3$ minislots/frame, and up to $M_{\max} = 27$ terminals for $m = 4$ minislots/frame. Hence, if we increase the m value, the access protocol can support a higher number of terminals (stable behavior). Also in the R-Aloha case, we can determine the cusp point as shown in Sect. 6.2.4.

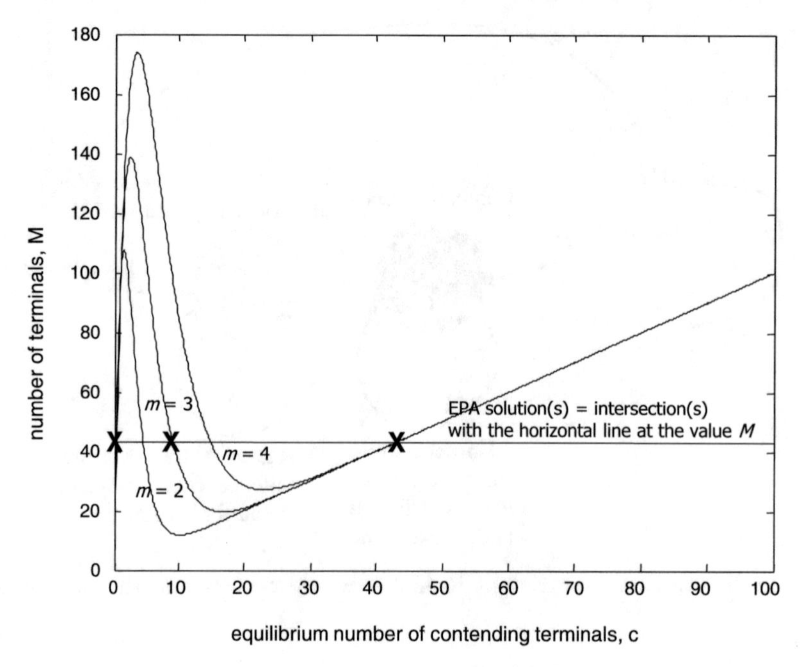

Fig. 6.33: EPA graphical solution method for 1-persistent R-Aloha with different m values and $1/\sigma + 1/(1-q) = 100$. We have denoted by "X" the EPA solutions in the case with $M = 44$ terminals and $m = 3$ minislots/frame

6.3.5 Packet Reservation Multiple Access (PRMA) Protocol

Packet Reservation Multiple Access (PRMA) is a MAC protocol proposed for terrestrial micro-cellular systems. It is based on a TDMA air interface (see also Sect. 6.4.2). A PRMA carrier has a frame structure of length T_f with N slots. Each slot can be used by a terminal to transmit a packet. Differently from R-Aloha, there is no distinction in the frame between information and access slots: all the slots can be used for both functions.

This access protocol is conceived to improve the efficiency of managing voice traffic sources, which are equipped with Speech Activity Detection (SAD): a voice source produces packets at regular intervals only during a talking phase; otherwise, no traffic is generated (silent phase). The traffic model of a voice source with SAD is Markovian with two states: talking and silent phases. The time intervals spent in talking and silent phases are exponentially distributed with mean values $t_1 = 1$ s and $t_2 = 1.35$ s, respectively, [12, 14].

As soon as a voice source starts a talkspurt, it performs the transmission of the first packet on an available slot according to a Slotted-Aloha scheme with permission probability, p_v. Terminals attempting simultaneously collide and no reservation is achieved. If the transmission attempt is successful, it represents the implicit request to reserve the same slot in subsequent frames. Each packet has a header containing the address of the sending terminal and the other control fields. The terminal waits for receiving the outcome of its reservation attempt from the cell controller (base station). The acknowledgment is practically instantaneous in terrestrial micro-cellular systems. In the case of a collision, a new attempt is performed according to the permission probability scheme. As soon as a transmission is successful, a terminal acquires the exclusive use of one slot per frame. A voice source generates packets at regular intervals in the talking phase. Hence, it is essential that the packets generation interval of the codec coincides with the frame length T_f.

A terminal releases a reservation at the end of a talkspurt by setting an End-of-Transmission (EoT) flag in the header of the last packet to be sent. Otherwise, there is also the possibility to release a reservation by inserting an empty packet at the end of a talkspurt; however, this approach may cause ambiguities because the radio channel may attenuate a packet, thus causing a false channel release request.

PRMA can manage different traffic classes using differentiated permission probabilities. As soon as a traffic source of a generic class needs to transmit, it acquires a reservation using the same random access scheme but with a suitable permission probability value.

During the access phase, many transmission attempts could be needed to acquire a reservation. Hence, a terminal may experience access delays, which become longer as the number of terminals sharing the same resources increases. In the case of voice packets (real-time traffic), there is a maximum delay, D_{vmax} ($= 32\,\text{ms}$), within which a packet must be successfully transmitted, and otherwise the packet is dropped and the terminal attempts the transmission of the next packet. With PRMA, only the first packets of a talkspurt may experience packet dropping; this phenomenon is named *front-end clipping.* The PRMA protocol has to be designed to control the access delay experienced by voice packets so that the packet dropping probability is lower than 1%. This entails basically a limit to the capacity of voice terminals.

The PRMA protocol, initially conceived for micro-cellular systems with low propagation delays with respect to the packet transmission time, has also been extended to the case of systems with much higher propagation delays, requiring only that these delays are lower than or equal to the frame length (T_{f} in the range of 10–40 ms). It has been proved that PRMA-like schemes can support different traffic classes in Low Earth Orbit (LEO) satellite networks, characterized by propagation delays much lower than those of GEO satellites, as explained in Sect. 2.11 [13, 15].

The performance of the PRMA protocol has been analyzed in the literature using the EPA approach, as shown in [12, 14].

6.3.6 Efficiency Comparison: CSMA/CD vs. Token Protocols

This section is aimed at comparing the CSMA/CD random access scheme (IEEE 802.3) with the token ring protocol (IEEE 802.5) in terms of *efficiency* η, i.e., the percentage of time that LAN resources are used to transmit data successfully [22]. Under the stability assumption, the efficiency corresponds to the *traffic intensity* S or the *throughput* carried out by the LAN.

The following study is performed under simplifying assumptions. In particular, we consider that there are N terminals always backlogged in the system with one packet of fixed length T to be transmitted. Hence, this study is carried out in saturated conditions. The maximum one-way propagation delay between any two terminals of the LAN is denoted by τ, and the normalized maximum propagation delay is $a = \tau / T$.

6.3.6.1 CSMA/CD Efficiency Analysis

The efficiency analysis is carried out considering that the time on the transmission medium is divided between intervals spent to transmit data successfully (*useful intervals*) and intervals spent to contend for the use of resources on the broadcast medium (*contention intervals*). Useful intervals have a length equal to T; instead, contention intervals have an average duration $E[C]$ that we are going to derive below.

Using the collision detection scheme, a terminal knows whether its transmission was successful or not within a time 2τ from the starting instant of its transmission. Hence, we can model the contention interval after a packet transmission as organized in minislots of length 2τ (this is the typical slot duration adopted for the BEB phase in the IEEE 802.3 standard).

In this model, every contending terminal may decide to transmit (according to its backoff algorithm) with probability q at each slot and knows the result (success or collision) within the end of the same slot. With probability $1 - q$, the above procedure is rescheduled for the next slot. The peculiarity of this model (and related analytical approach) is that the slot-based probabilistic transmission scheme is adopted for both new transmission attempts and retransmissions; see Fig. 6.34.

We consider that after a successful transmission phase, there are still N terminals, which contend for the transmission of their packets (saturation condition). The number of terminals transmitting on the same slot is binomially distributed with parameters N and q. Hence, the transmission attempt on a slot is successful with the probability $P_s(N, q)$ that only one terminal transmits on that slot:

$$P_s\left(N,q\right) = Nq(1-q)^{N-1} \tag{6.48}$$

$P_s(N, q)$ is equal to 0 for both $q = 0$ and $q = 1$. In the following analysis, we refer to the value of q maximizing $P_s(N, q)$ to evaluate the CSMA/CD efficiency in the best conditions. We have

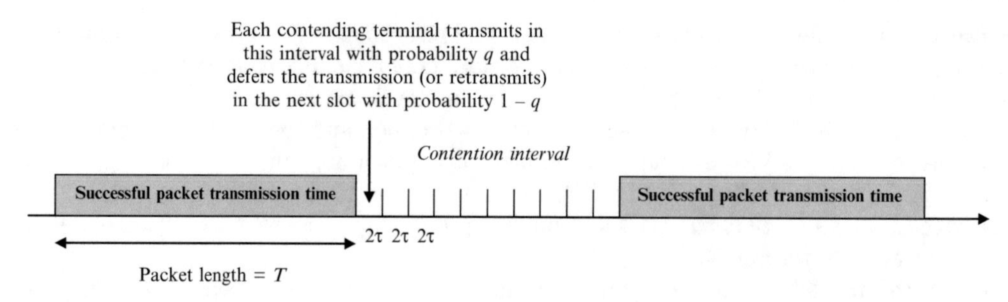

Fig. 6.34: CSMA/CD model for analysis

$$\frac{\mathrm{d}}{\mathrm{d}q}P_s(N,q) = N(1-q)^{N-2}(1-qN) = 0 \iff q = \frac{1}{N} \tag{6.49}$$

Hence, considering the optimum q value, we achieve the following expression for $P_{s,\mathrm{opt}}(N,\, q) = P_{s,\mathrm{opt}}(N)$:

$$P_{s,\mathrm{opt}}(N) = \left(1 - \frac{1}{N}\right)^{N-1} \tag{6.50}$$

Note that $P_{s,\mathrm{opt}}(N=2) = 1/2$. $P_{s,\mathrm{opt}}(N)$ decreases as N increases. The limit of $P_{s,\mathrm{opt}}(N)$ for $N \to \infty$ (practically, many elementary terminals in the LAN) is equal to $e^{-1} \approx 0.36$, the maximum traffic load for the Slotted-Aloha protocol applied to the minislot access.

The time in slots for the first successful transmission (by one of the N terminals) is according to a modified geometric distribution (in number of slots) with parameter $P_{s,\mathrm{opt}}(N)$. Therefore, the mean number of slots for the first successful transmission, $E[n_{\mathrm{slot}}]$, results as

$$E\left[n_{\mathrm{slot}}\right] = \frac{1}{P_{s,\mathrm{opt}}(N)} = \left(1 - \frac{1}{N}\right)^{1-N} \tag{6.51}$$

The mean length of the contention phase is $E[C] = 2\tau E[n_{\mathrm{slot}}] - 2\tau$, since we have to exclude the last slot in which the packet is transmitted successfully.

We can therefore express the CSMA/CD (IEEE 802.3) efficiency (actually an upper bound) as follows similarly to (6.21) [1]:

$$\begin{aligned}\eta_{\mathrm{CSMA/CD}}(N) &= \frac{T}{T + E[C]} = \frac{T}{T + 2\tau\left[\left(1-\frac{1}{N}\right)^{1-N} - 1\right]} \\ &= \frac{1}{1 + 2a\left[\left(1-\frac{1}{N}\right)^{1-N} - 1\right]}\end{aligned} \tag{6.52}$$

Note that the above $\eta_{\mathrm{CSMA/CD}}$ can be considered equivalent to the maximum possible throughput in Erlangs that the protocol can support under an ideally optimal collision resolution algorithm.

Finally, the limiting value of $\eta_{\mathrm{CSMA/CD}}$ for $N \to \infty$ (i.e., the minimum value of $\eta_{\mathrm{CSMA/CD}}$) is as follows:

$$\lim_{N\to\infty} \eta_{\mathrm{CSMA/CD}}(N) = \frac{1}{1 + 2a\,[e-1]} \approx \frac{1}{1 + 3.43a} \tag{6.53}$$

According to (6.53), the higher the propagation delay (i.e., a), the longer the contention interval, and the lower the efficiency. Moreover, the efficiency decreases with N up to the limit in (6.53).

6.3.6.2 Token Ring Efficiency Analysis

We study the efficiency of a token ring scheme with RAR policy as in IEEE 802.5: if a terminal transmits a frame, it releases the token when it receives the transmitted frame, which has propagated along the entire ring. In this analysis, we assume that when a terminal acquires the token, it always has one packet (fixed-length T) to transmit. We also assume that there are N terminals at a regular distance on the ring.

Ring resources are used according to a periodic sequence of packet transmission time, including the propagation time back to the originating terminal to notify the release of the token (*busy line interval*), B, and the time to propagate the token to the next terminal (*protocol overhead interval*), O_N. Hence, the efficiency of the token ring protocol can be expressed [similarly to (6.21)] as

$$\eta_{\text{token ring}}(N) = \frac{T}{B + O_N} = \frac{1}{\frac{B}{T} + \frac{O_N}{T}} \tag{6.54}$$

Referring to the RAR policy, two different cases are possible, depending on the value of the normalized propagation delay on the ring: $a = \tau/T$, where τ denotes the propagation delay on the entire ring.

Case with $a < 1$ (i.e., $\tau < T$): a reference terminal receives the token at time $t = 0$ and starts to transmit a packet. At time $t = a \times T$, this terminal starts to receive the packet that has propagated along the ring. At time $t = T$, the transmission of the packet of our terminal ends and the terminal releases the token. The released token reaches the next terminal in the ring after a time τ/N. Hence, $B/T = 1$ and $O_N/T = a/N$ so that the efficiency can be expressed as

$$\eta_{\text{token ring}, a<1}(N) = \frac{1}{1 + \frac{a}{N}} \tag{6.55}$$

Case with $a > 1$ (i.e., $\tau > T$): a reference terminal receives the token at time $t = 0$ and starts to transmit a packet. At time $t = T$, the transmission of the packet of the terminal ends. At time $t = a \times T$, the terminal starts to receive the packet, which has propagated along the ring and releases the token. The released token reaches the next terminal in the ring after a time τ/N. Hence, $B/T = a$ and $O_N/T = a/N$ so that the efficiency can be expressed as

$$\eta_{\text{token ring}, a>1}(N) = \frac{1}{a + \frac{a}{N}} \tag{6.56}$$

In conclusion, both (6.55) and (6.56) can be summarized in the following token ring (IEEE 802.5) efficiency expression (we can consider that the efficiency is equivalent to the throughput expressed in Erlangs):

$$\eta_{\text{token ring}}(N) = \frac{1}{\max(1, a) + \frac{a}{N}} = \frac{N}{\max(N, aN) + a} \tag{6.57}$$

Finally, the limiting value of $\eta_{\text{token ring}}$ for $N \to \infty$ (i.e., the maximum value) is as follows:

$$\lim_{N\to\infty} \eta_{\text{token ring}}(N) = \frac{1}{\max(1, a)} \tag{6.58}$$

6.3.6.3 Efficiency Comparisons

The graphs in Fig. 6.35 compare the optimal efficiency of CSMA/CD from (6.52) with the efficiency of the token ring protocol from (6.57) as a function of both the number of terminals N and the normalized maximum propagation delay a (note that the a value depends on the physical length of the LAN, the transmission bit-rate, and the frame size).

We can note that the token ring efficiency (or, equivalently, the maximum traffic intensity S) increases with N due to the reduction in the time to send the token to the next terminal. In contrast, the CSMA/CD efficiency decreases with N due to increased collision rate. Moreover, the efficiencies of both CSMA/CD and token ring decrease with a (CSMA/CD efficiency decreases more significantly with a). The above S values could also be compared with the maximum S values of Aloha and Slotted-Aloha schemes; for instance, for $N \to \infty$, $\eta_{\text{Aloha}} \to 1/(2e) \approx 0.18$ Erlangs and $\eta_{\text{S-Aloha}} \to 1/e \approx 0.36$ Erlangs; both values are independent of a.

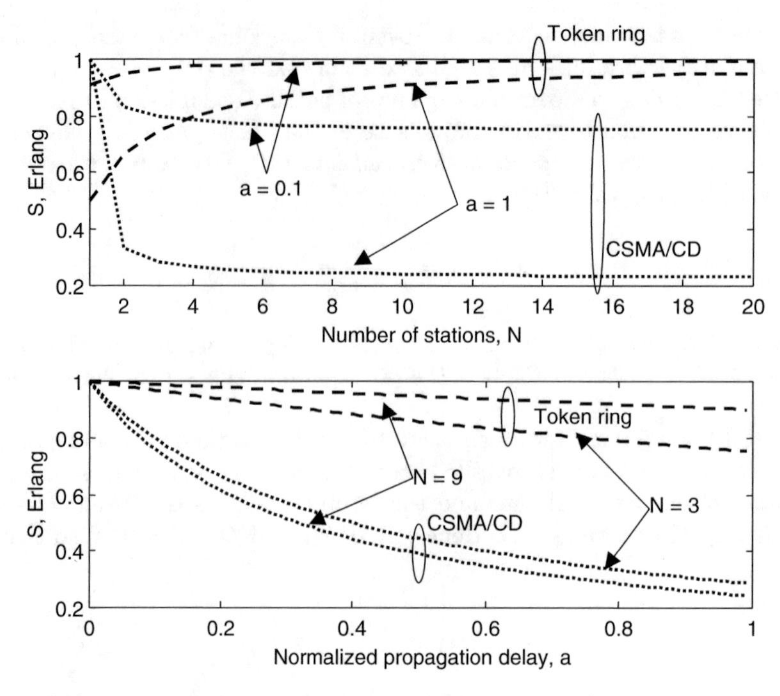

Fig. 6.35: Comparisons of the access protocols in terms of efficiency η (corresponding to the maximum traffic intensity S in Erlangs); *dotted curves* are for CSMA/CD cases and *dashed curves* are for token ring cases

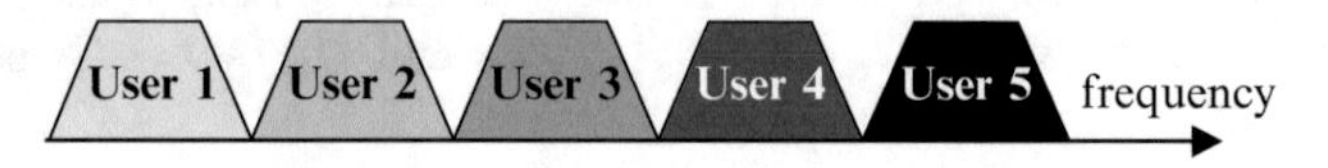

Fig. 6.36: FDMA technique: bandwidth partitioned into channels

6.4 Fixed Assignment Protocols

This section is devoted to the description of the access schemes with a rigid assignment of resources to the various terminals. These schemes are suitable to support continuous and fixed traffic patterns [13]. The following techniques have quite general applications, but we refer to radio transmissions of cellular communications in most cases.

6.4.1 Frequency Division Multiple Access (FDMA)

The frequency band available to the system is divided into separate portions, each of them used for a given channel (Fig. 6.36); the different channels are distributed among terminals. Adjacent bands have guard spaces to avoid inter-channel interference.

One disadvantage of FDMA is the lack of flexibility for the support of variable bit-rate transmissions, an essential prerequisite for multimedia communication systems.

6.4.2 Time Division Multiple Access (TDMA)

In this scheme, each terminal can transmit on the whole bandwidth of a carrier, but only for a short interval of time (slot), which is repeated periodically according to a frame structure. Transmissions are organized into frames, each of them containing a given number of slot intervals, N_s, to transmit packets, as shown in Fig. 6.37.

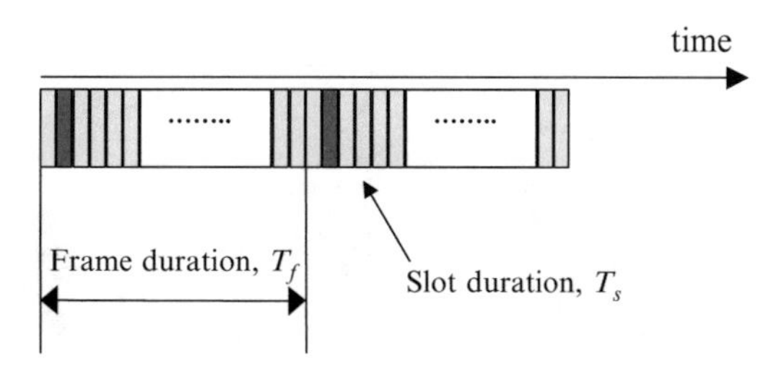

Fig. 6.37: TDMA frame with slots

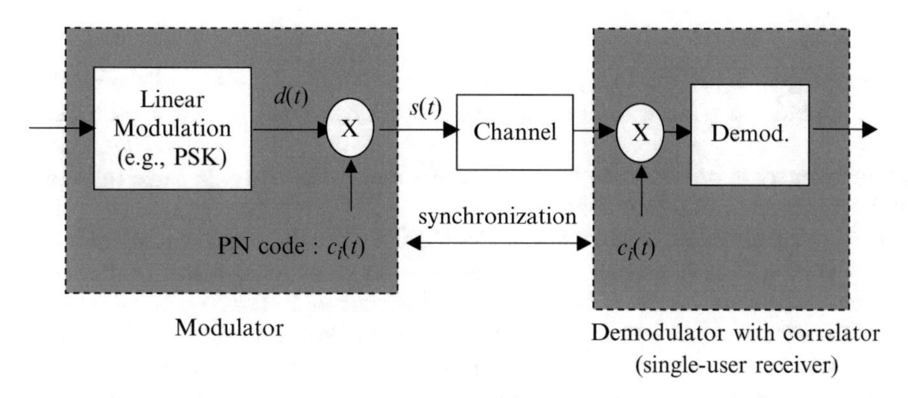

Fig. 6.38: Spreading and de-spreading processes for the ith DS-CDMA user; the product by the PN code with $c_i(t)$ at the modulator is used to widen the bandwidth occupied by the signal

The main disadvantage of TDMA is the high peak transmit power, which is required to send packets in the assigned slots. Moreover, a fine synchronization must be achieved to align packet transmissions with the time slot in the frame. Finally, a rigid resource allocation is allowed by classical TDMA: for instance, a voice traffic source has assigned one slot per frame also during silent periods among talkspurts.

6.4.3 Code Division Multiple Access (CDMA)

The basic concept of CDMA is to spread the transmitted signal over a larger bandwidth (Spread Spectrum, SS). Such a technique was developed as a jamming countermeasure for military applications in 1950s. Accordingly, the signal is spread over a bandwidth PG times larger than the original one using a suitable modulation based on a PseudoNoise (PN) code[7] [23–26]. PG is the so-called Processing Gain. The higher PG, the higher the spreading bandwidth, and the greater the system capacity. Distinct codes must be used to distinguish the different simultaneous transmissions in the same band. The receiver must use a code sequence synchronous with that of the received signal to correctly de-spread the signal.

There are two different techniques to obtain spread-spectrum transmissions:

- Direct Sequence (DS), where the user binary signal is multiplied by the PN code with bits (named *chips*), whose length is PG times shorter than that of the original bits. This spreading scheme is well suited to Phase Shift Keying (PSK) and Quadrature Phase Shift Keying (QPSK) modulations (see Fig. 6.38).
- Frequency Hopping (FH), where the PN code is used to change the frequency of the transmitted symbols (see Fig. 6.39). We have a fast hopping if the frequency is changed at each new symbol; instead, we have a slow hopping if the frequency varies after a given number of symbols. Frequency Shift Keying (FSK) modulation is well suited to the FH scheme.

[7] PN codes are cyclic codes (e.g., gold codes) that well approximate the generation of random bits 0 and 1. These codes must have a high peak for the autocorrelation (synchronization purposes) and very low cross-correlation values (orthogonality of different users).

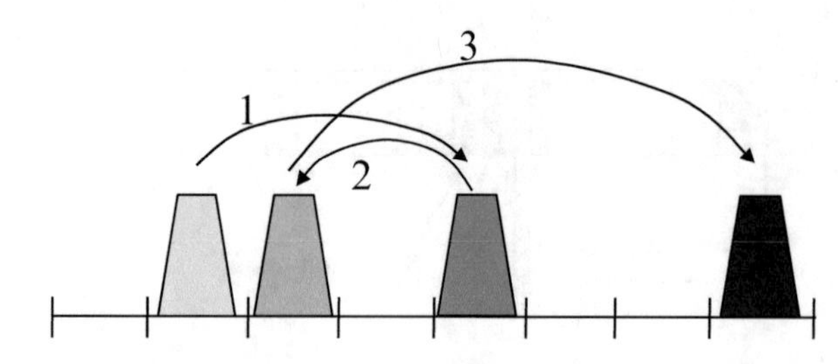

Fig. 6.39: Spreading process with FH-CDMA (scrambling process)

Even if an interfering signal is present in a portion of the bandwidth of the spread signal, the receiver de-spreads the useful signal and spreads on a larger bandwidth the interfering signal, which becomes similar to background noise.

The DS-CDMA technology is preferred to the FH-CDMA one since it is expensive to obtain frequency synthesizers able to switch the transmission frequency quickly.

In DS-CDMA-based cellular systems, the different DS-CDMA signals (users) of a cell can be perfectly separated from one another in the case of synchronous transmissions with orthogonal codes (null cross-correlation). These signals do not interfere with one another. However, if synchronism is lost, we have to consider partial cross-correlations among different codes so that orthogonality of users is lost and Multiple Access Interference (MAI) is experienced: the de-spreading process with the single-user receiver is unable to cancel the interference completely coming from simultaneous transmissions. Downlink transmissions (i.e., from the base station to its mobile users) are synchronous in a cell so that intra-cell MAI is negligible (there is some residual MAI due to multipath phenomena). Instead, there is intra-cell MAI in uplink (i.e., from mobile users to the base station of the cell). Inter-cell MAI is present for both uplink and downlink. Any technique able to reduce MAI increases capacity with DS-CDMA. For instance, we can consider the following:

- Squelching of transmissions during inactivity phases
- Directional antennas or smart antennas (multi-sectored cells) at the base stations
- Multi-user receivers, which reduce MAI coming from the users in the same cell (intra-cell interference).

With CDMA, it is possible to use a special receiver, named RAKE, which combines the signal contributions from different paths (micro-diversity). This receiver is particularly useful in the multipath environment of mobile communications to improve the bit error rate performance [10].

In CDMA-based mobile systems, a power control scheme must be adopted to avoid a user close to the base station being received with a much higher power level than users at cell borders (*near-far problem*) [25]. All transmissions in a cell should be received with the same power level for both uplink and downlink, unless complex multi-user receivers are adopted.

CDMA well supports powerful coding schemes, which can contribute in part to the spreading process. Accordingly, CDMA achieves greater robustness and a higher capacity than other multiple access schemes, like TDMA and FDMA. DS-CDMA has been adopted by third-generation (3G) mobile communication systems, such as the Universal Mobile Telecommunications System (UMTS).

6.4.4 Orthogonal Frequency Division Multiple Access (OFDMA)

Orthogonal Frequency Division Multiplexing (OFDM) is a digital modulation technique that is particularly effective in reducing the effects of frequency-selective fading [27]. This is possible because OFDM divides one high-speed signal into numerous slow signals (*multi-carrier signal*) that can be transmitted using orthogonal sub-carriers without being subject to the same multipath fading distortion of high-speed single-carrier transmissions. The data coming from numerous sub-carriers are then recombined at the receiver in order to form one high-speed transmission. The transmitter at the base station uses N available sub-carriers in the assigned bandwidth. These sub-carriers are evenly divided into C sub-channels, each consisting of $P = N/C$ sub-carriers. There are different techniques to select the sub-carriers in the spectrum to form a given sub-channel.

The generation of the multicarrier signal is based on an FFT process. Each transmitted OFDM symbol has a Cyclic Prefix (CP) that eliminates Inter-Symbol Interference (ISI) as long as the CP duration is longer than the channel delay spread. An OFDM symbol is made up of three different types of sub-carriers: data, pilot, and null. OFDM allows sub-carriers to be adaptively modulated depending on distance and noise level. Different modulation and coding combinations are possible depending on the technology considered. The basic OFDM parameters are the FFT size, the number of data sub-carriers in the available bandwidth, the total bandwidth, the oversampling factor, and the CP parameter.

Orthogonal Frequency Division Multiple Access (OFDMA) is a multi-user version of the OFDM digital modulation scheme. Multiple access is achieved in OFDMA by assigning different subsets of sub-carriers to individual users.

OFDM/OFDMA is adopted by many current wireless technologies, such as IEEE 802.11{a, g, n, ac}, IEEE 802.16 (WiMAX), Digital Video Broadcasting-Terrestrial (DVB-T), and Long Term Evolution (LTE), a fourth-generation cellular system. OFDM is also used in ADSL based on the ITU G.992.1 G.DMT (Discrete Multi-tone Modulation) standard. OFDM together with massive Multiple-Input/Multiple-Output (MIMO) antennas is adopted by the 5G standard defined by 3GPP to increase the efficiency of transmissions.

6.4.5 Resource Reuse in Cellular Systems

First-generation (1G), second-generation (2G), and also fourth-generation (4G) cellular systems adopt the reuse concept [28]. In particular, due to the limited number of radio resources, it is necessary to reuse the same carrier among sufficiently distant cells of radio coverage so that the inter-cell interference is negligible. In 4G and fifth-generation (5G) systems, the whole spectrum can be reused in every cell, with the possibility to differentiate the spectrum usage at the cell border to reduce the effects of interference in those areas where the user signal is weaker. This is the soft frequency reuse scheme.

The *reuse distance* D is the distance between two cells that can simultaneously use the same resources (carriers). Assuming a hexagonal regular cellular layout, the resource reuse implies dividing the total number of resources into K groups, distributed among the different cells as in a *mosaic*. Depending on the possible D values, the corresponding K values are [28] 1, 3, 4, 7, 9, ... Based on the reuse pattern K, if we have N_c system resources (i.e., carriers with FDMA, TDMA, and OFDMA), we may assign N_c/K resources per cell (*fixed channel allocation*). Hence, there can be at most $Q = m \times N_c/K$ simultaneous phone conversations per cell, where m denotes the capacity of phone calls per carrier. A call generated in a cell where all its Q resources are busy is blocked and cleared. If we assume that calls arrive in a cell according to a Poisson process with mean rate λ and that the channel holding time, X, is generally distributed with mean value $E[X]$, the blocking probability P_b experienced by calls is given by the well-known Erlang-B formula (see Sect. 4.9), according to an M/G/Q/Q model:

$$P_b(\rho,\ Q) = \frac{\rho^Q}{Q!\sum_{n=0}^{Q}\frac{\rho^n}{n!}} \tag{6.59}$$

where $\rho = \lambda E[X]$ is the traffic intensity offered to a cell in Erlangs.

6.4.6 Non-orthogonal Multiple Access (NOMA)

NOMA has recently attained strong attention from academia and industry for 5G and beyond systems. Compared to Orthogonal Multiple Access (OMA) techniques (i.e., FDMA, TDMA, and CDMA), NOMA achieves better spectral efficiency [29].

With OMA, users' access is orthogonal and, ideally, they do not interfere with one another, while they share the communication channel. In these schemes, orthogonal radio resources in time, frequency, and code domains or their combinations are assigned to multiple users. However, as the number of orthogonal resources is limited, the OMA systems cannot serve many users. The best solution for 5G systems is represented by NOMA that allows inter-user interference in the resource allocation (multiple users are served using the same resource block). Interference cancellation schemes such as Successive Interference Cancellation (SIC) are applied.

In general, the existing NOMA schemes can be classified into two categories: power-domain NOMA and code-domain NOMA. The former assigns a unique power level to a user, and multiple users transmit their signals, sharing the same time–frequency–code resources, each using its allocated power. The power level for a user is decided based on its channel gain: a user with higher channel gain is often assigned a lower power level. At the receiving ends, various users' signals can be separated by exploiting the users' power difference based on SIC: the decoding of a strong user allows subtracting its interference for the signal to decode the next user. Code-domain NOMA relies on codebooks, spreading sequences, interleaving patterns, or scrambling sequences to non-orthogonally allocate resources to users. Spreading sequences are limited to sparse sequences (alternatively called low-density sequences) or non-orthogonal low cross-correlation sequences.

NOMA can be used for both uplink and downlink transmissions and achieves a capacity increase with respect to classical FDMA, TDMA, and CDMA schemes. NOMA can increase spectral efficiency and fairness. Today, there is an increasing interest by 3GPP in NOMA for massive sensors (IoT) and ultra-low-latency reliable applications for 5G systems.

6.5 LAN/WLAN Technologies and MAC Protocols Implemented

6.5.1 IEEE 802.3 Standard

The Ethernet LAN dates back to 1976 when Xerox adopted the CSMA/CD protocol to implement a network at 1.94 Mbit/s to connect more than 100 terminals using cables. There was immediately a significant success for this technology, so that Digital, Intel, and Xerox (DIX) joined in a consortium, DIX, to define the Ethernet LAN specifications at 10 Mbit/s, using a thick copper coaxial cable as the physical medium. In the same period, the IEEE 802 committee started to develop a LAN standard based on CSMA/CD, similar to the Ethernet, and called IEEE 802.3. Ethernet and IEEE 802.3 have significant similarities, although there are some differences.

The IEEE 802.3 standard specifies both physical and MAC layers. The IEEE 802.3 standard essentially envisages a bus topology on a broadcast medium, but a star topology is also possible and nowadays very important. Correspondingly, two operation modes are allowed by the MAC layer [30]:

- *Half-duplex transmissions* for bus topology (broadcast medium): terminals contend for the use of the physical medium using CSMA/CD; this protocol is used in all the Ethernet configurations but is necessary only for those broadcast configurations (bus topology) where simultaneous transmission and reception are impossible without interference, such as 10Base-2 and 100Base-T4 that are shown later in this section.
- *Full-duplex transmissions* for star topology (point-to-point links): it has been introduced later (approved in 1987). It consists of a central switch with a dedicated connection for each terminal: different pairs of wires are used to support simultaneous transmission and reception for each terminal without interference. This is the so-called switched Ethernet. Examples are 10Base-T and 100Base-TX/FX, as described later in this section. The CSMA/CD protocol is not necessary in the full-duplex case since there are no collisions on the medium.

The following description refers to the IEEE 802.3 half-duplex operation mode that is characterized as follows [30]:

- 1-persistent CSMA/CD access protocol with truncated BEB algorithm.
- Baseband transmissions of bits with Manchester line encoding,[8] as shown in Fig. 6.40.

According to the 1-persistent CSMA/CD protocol, when a packet is ready for transmission if the station does not reveal a carrier, it waits for an Inter-Frame Gap (IFG) and then transmits (no further carrier sense verification is performed). Instead, if the medium is busy, the station defers the transmission.

If the receiver interface reveals a signal when a station is transmitting, a collision event is assumed. The Network Interface Card (NIC) can detect a collision by revealing an increase in the average voltage level on the line. Suppose

[8] With Manchester encoding, each bit contains a transition in the middle: a transition from low to high represents a "0 bit" and a transition from high to low represents a "1 bit" (also the opposite convention is possible). The bandwidth needed to transmit the signal practically doubles with respect to the case without this encoding. The advantage is that we have transitions on a predictable basis that are useful for synchronization purposes.

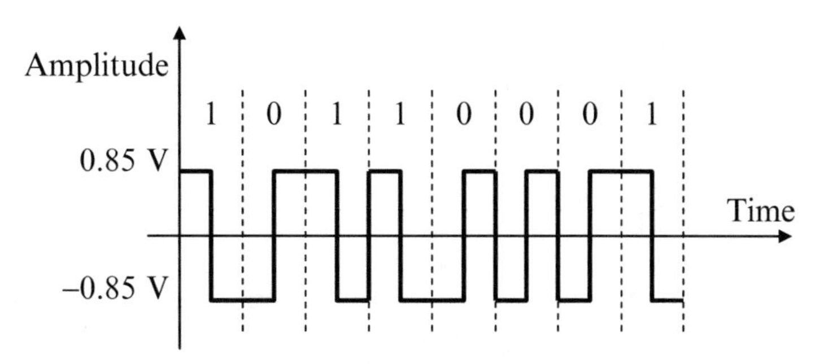

Fig. 6.40: Example of Manchester encoding

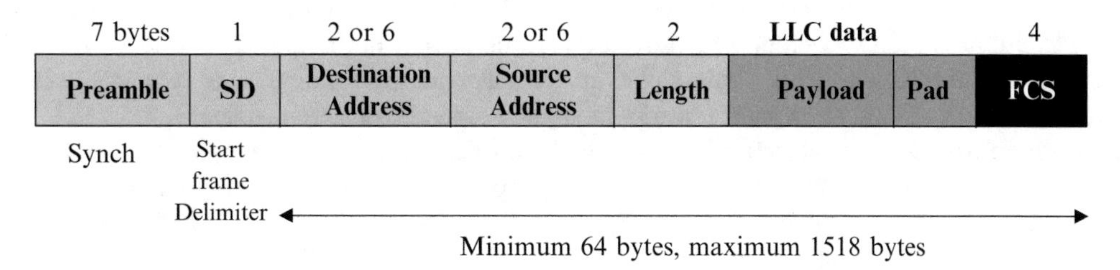

Fig. 6.41: IEEE 802.3 frame format (including the overhead of 18 bytes)

a collision is detected, according to the CSMA/CD protocol. In that case, the station revealing a collision sends a 32-bit jam message (also 48-bit jam messages are possible) to make sure that all the other stations involved abandon their transmissions (corrupted frames) [30]. Then, a retransmission procedure is performed based on a truncated BEB algorithm. Soon after the first collision, time is slotted. The *time slot* T_s corresponds to the time to transmit the minimum frame of 64 bytes, 512 bits, at the speed of R_b bit/s allowed by the medium: $T_s = 512/R_b$; this time also corresponds to the maximum possible round-trip propagation delay on the LAN, 2τ. Then, at the first reattempt with the truncated BEB algorithm, the network adapter chooses the retransmission time R equal to 0 slots with probability 1/2 and equal to 1 slot with probability 1/2. Let us assume that the adapter chooses $R = 0$. After transmitting the jam signal, it immediately senses the channel and transmits if the medium is free; if the medium is busy, it waits for a free medium and then transmits. After a second (subsequent) collision, R is chosen with equal probability within the values {0, 1, 2, 3} slots. After three collisions, R is selected with equal probability within the values of {0, 1, 2, 3, 4, 5, 6, 7} slots. After ten or more collisions, R is chosen with equal probability within the values of $\{0, 1, 2, \ldots, 1023\}$ slots. After 16 retransmission attempts, the packet is discarded: a congested LAN may drop packets.

All the stations connected to the same physical medium (through hubs or repeaters, both operating at the physical layer) share collisions and form a *collision domain*. The definition of a *segment* depends on the Ethernet type. However, a segment can be considered as a collision domain or a part of it.

Repeaters are adopted in Ethernet LANs to counteract the attenuation due to the transmission medium. A repeater operates at the physical layer and must receive the digital signal and has to regenerate or simply amplify it before retransmitting. A repeater is used to connect two network segments within a collision domain.

At the end of the successful transmission of a packet, a "silent" IFG period is inserted to allow the transmitting station to switch its circuitry from transmission to reception mode so that it does not miss a frame. Moreover, the IFG interval is also needed by the receiving station to recognize the end of a frame. If, for any reason, the IFG reduces or collapses, it is possible that two subsequent frames overlap so that they may be considered as a unique packet by a receiver. This can be caused by repeaters encountered in the LAN since they may alter the time separation between two subsequent packets. Such a problem can be avoided by selecting a sufficiently large IFG value and by imposing a maximum number of repeaters in a LAN. The selected IFG corresponds to 96-bit times. Then, the IFG duration depends on the speed of the transmission medium: IFG = 96 μs for a transmission speed of 1 Mbit/s, IFG = 9.6 μs for a transmission speed of 10 Mbit/s, IFG = 0.96 μs for a transmission speed of 100 Mbit/s, and, finally, IFG = 0.096 μs for a transmission speed of 1 Gbit/s.

The 802.3 MAC frame (packet) format is shown in Fig. 6.41.

The Ethernet frame is prefixed by a header (preamble plus SD field) of 8 bytes.

The MAC addresses of destination and source can have a length of 2 or 6 bytes (6 bytes is the most common case). If the most significant bit of the destination address is equal to "0" (equal to "1"), the address is a "normal address" ("multicast address"). If all the bits of the destination address are equal to 1, we have a broadcast transmission.

The length field in the frame format denotes the length of the payload from 0 to 1500 bytes; Ethernet-based LANs have an MTU of 1500 bytes (i.e., maximum IP packet size allowed). The minimum frame length from destination address to FCS (Frame Check Sequence) must be 64-byte long. Let us refer to the case of addresses of 6 bytes. The Pad field is used to guarantee that the payload plus Pad is not shorter than 46 bytes. The maximum frame length from destination address to FCS is 1518 bytes (also considering preamble and SD field, we have 1526 bytes in total). The FCS field contains a cyclic code to reveal the presence of errors in the frame.

A maximum frame length has been imposed to avoid a station occupying the transmission medium for too long. The minimum frame length is defined so that a collision can be revealed before the packet transmission ends: this is a prerequisite to apply the CD scheme. The minimum frame length imposes constraints on both the maximum distance d between two terminals in the same collision domain and the number n of repeaters along the path in the collision domain. Note that each repeater works at layer 1 and introduces a delay δ.

We refer to a minimum frame of 64 bytes (neglecting the additional header of 8 bytes for a conservative study). Let R_b denote the transmission bit-rate; let v denote the propagation speed of the signal on the physical medium (in general, we can consider v about equal to $2c/3$, where c is the speed of the light in the vacuum; however, $v = 2.3 \times 10^8$ m/s for a thick coaxial cable and $v = 1.77 \times 10^8$ m/s for a UTP cable). According to (6.20), the following condition can be used to bound the maximum distance d between two terminals belonging to a collision domain (i.e., *network diameter*) with CSMA/CD:

$$T_s = \frac{64 \times 8}{R_b} \geq 2 \times \left(\frac{d}{v} + n\delta\right) = 2\tau \Rightarrow d \leq \frac{512v}{2R_b} - n\delta v \tag{6.60}$$

For instance, let us consider a thick coaxial cable (i.e., 10Base-5) with $v = 2.3 \times 10^8$ m/s, $\delta = 3.5$ µs per repeater, $n = 4$ repeaters (i.e., a LAN composed of five segments), and $R_b = 10$ Mbit/s. Correspondingly, the maximum value of d is about 2.5 km. If the bit-rate R_b is increased by a factor of 10, the diameter is divided by a factor of 10 if we maintain the same protocol characteristics and frame format. For instance, the maximum distance d is limited to 223 m for $v = 1.77 \times 10^8$ m/s (UTP cable), $\delta = 1.3$ µs per repeater, $n = 1$ repeater, and $R_b = 100$ Mbit/s.

Stations connected via repeaters and hubs are in the same collision domain and share the same bandwidth. To increase the capacity of the LAN, a possibility is to divide the LAN into different collision domains (thus reducing collisions) within which users are assumed to communicate more frequently. To this end, we must use network elements operating above the physical layer, such as bridges (layer 2), switches (layer 2), and routers (layer 3) [1].

A bridge connects physically separated segments. A bridge transfers packets between different collision domains only when necessary. Another solution is to adopt switches that can forward frames to all ports at wire speed. This approach requires a star topology with dedicated links to connect the different stations to the different ports of a switch. The switch is a sort of "advanced hub" using the destination address of a data packet to direct it to the specific station on the LAN. The Ethernet full-duplex mode of operation is based on switches.

There are two typical topologies for IEEE 802.3 LANs, i.e., "hub Ethernet" (half-duplex mode) and "switched Ethernet" (full-duplex mode), both sharing the same packet format.

- *Hub Ethernet*:
  - The various stations are connected to a hub, acting as a broadcast repeater.
  - All stations are in the same collision domain so that the topology is equivalent to a bus.
  - CSMA/CD is adopted for the access to the shared medium.
- *Switched Ethernet*:
  - The different stations are connected to the switch (one switch has 20–40 ports) so that the topology is a star.
  - There is not a shared collision domain; only point-to-point connections are adopted between stations and switch. There is no need for CSMA/CD.
  - The stations transmit whenever they want and at full speed. The switch queues the packets and sends them to the destinations. There could be packet losses due to buffer overflows; this is more critical when the speed increases, as in the Gigabit Ethernet (Gbit/s).

- This is the best solution to increase the data rate up to Gbit/s and this is the current version implemented for Ethernet.

As already anticipated, the IEEE 802.3 standard has several variants, which are differentiated primarily based on the physical medium and, consequently, based on the available bit-rate. In all these variants, both MAC protocol and frame format are not modified for compatibility reasons. The different standards are denoted by a name, like "sTYPE-t or l", which can be described as

- *s*: speed in Mbit/s
- TYPE: BASE meaning baseband (the physical medium is dedicated to the Ethernet) or BROAD meaning broadband (the physical medium can simultaneously support Ethernet and other possibly non-Ethernet services)
- *l*: maximum segment length in multiples of 100 m in the case of coaxial medium
- *t*: media type used.

In particular, we can consider the LAN technologies described below on the basis of the bit-rate allowed by the physical medium [30].

Classical Ethernet technologies at 10 Mbit/s:

- 10Base-5: Original Ethernet with large thick coaxial cable
- 10Base-2: Thin coaxial cable version
- 10Base-T: Voice-grade UTP
- 10Base-F: Two optical fibers in a single cable.

Fast Ethernet technologies at 100 Mbit/s:

- 100Base-TX: Two pairs of STP cables or category 5 (or above) UTP cables
- 100Base-T4: Four pairs of UTP cables (category 3, 4, or 5)
- 100Base-T2: Two pairs of UTP cables (category 3, 4, or 5)
- 100Base-FX: Multimode fiber.

Main Gigabit Ethernet technologies at 1 Gbit/s:

- 1000Base-SX: Pair of multimode optical fibers using short-wavelength optical transmissions (range up to 500 m)
- 1000Base-LX: Pair of optical fibers using long-wavelength optical transmissions (range up to 2 km)
- 1000Base-CX: Two pairs of specialized copper cabling, called "twinaxial" cabling (they are similar to coaxial cables but with two inner conductors instead of one)
- 1000Base-T: Four pairs of UTP cables of categories 5, 6, or 7 (range of 100 m).

Main 10Gigabit Ethernet technologies at 10 Gbit/s:

- 10GBase-CX4: Twinaxial cables, 4 lanes (range of 15 m)
- 10GBase-ER: Single-mode fiber with a laser wavelength of 1550 nm to cover distances up to 40 km
- 10GBase-LR: Single-mode fiber with a laser wavelength of 1310 nm to provide communications up to 10 km
- 10GBase-LRM: Multimode fiber and a laser wavelength of 1310 nm with a range of 260 m
- 10GBase-LX4: Single-mode fiber with a laser wavelength of 1310 nm to cover distances up to 10 km
- 10GBase-SR Multimode 850 nm fiber with a maximum distance of 65 m.

Further details on Ethernet technologies are provided below:

10Base-5 details (1983)
Topology: bus
Transmission medium: thick shielded coaxial cable (RG8, yellow cable)
Interconnection with the cable: vampire taps
Transceiver in charge of carrier sensing and collision detection
Bit-rate: 10 Mbit/s
Maximum length of a segment: 500 m
Maximum number of segments: 5
Maximum number of stations per segment: 100
Maximum number of stations in the network: 1023
Maximum distance covered: 2.5 km

Stations can only be connected at distances multiple of 2.5 m to avoid that reflections from multiple taps are in phase (hence, the minimum distance between two adjacent stations is 2.5 m)
Maximum number of repeaters between two stations in the network: 2.

10Base-2 details (1985)
Topology: bus
Transmission medium: thin shielded coaxial cable (RG58)
Interconnection with the cable: BNC (Bayonet Neill–Concelman) connector
Transceiver integrated on the Ethernet board
Bit-rate: 10 Mbit/s
Maximum length of the cable connecting a station to the network: 50 m
Maximum length of a segment: 185 m
Maximum number of stations per populated segment: 30
Maximum distance covered: 925 m
Minimum distance between two adjacent stations: 0.5 m
The advantage of 10Base-2 is the reduced hardware cost; the disadvantages are signal reflections and attenuation caused by BNC connectors.

10Base-T details (1990)
Topology: star
Transmission medium: one UTP cable (a couple of twisted wires) is for transmitting and another one for receiving; category 3, 4, 5, or 6
The RJ45 connector is used.
Transceiver integrated on the Ethernet board
Bit-rate: 10 Mbit/s
Maximum length of a segment: 100 m
The advantage of this solution is that UTP cables are ubiquitous in buildings, so that this technology allows us to take advantage of the wiring already installed in the walls.

10Base-F details (1993)
The maximum length of a segment is 2 km. The transmission medium is the optical fiber. There are three different possibilities for cabling:

- 10Base-FB (Fiber Backbone): only for links between repeaters
- 10Base-FL (Fiber Link): an active star topology of point-to-point connections between repeaters
- 10Base-FP (Fiber Passive): a star topology where transceivers are the sole active elements.

Fast Ethernet details (1995)
In the middle of 1990s, the Fast Ethernet technology (IEEE 802.3u) emerged to increase the available bit-rate to 100 Mbit/s. Fast Ethernet obtained significant success since it was compatible with the previous 10 Mbit/s Ethernet version and maintained all the parameters of the CSMA/CD protocol and the frame format. On the basis of (6.60), the network diameter results one-tenth of that possible with previous technologies at 10 Mbit/s.

- 100Base-TX employs two couples of UTP cables of category 5 or above. One couple is used for transmission and the other for reception according to a full-duplex operation mode with start topology. The typical maximum length of a segment is 100 m.
- 100Base-T4 uses four couples of balanced UTP cables of category 3, 4, or 5. The topology is a star. The maximum distance is 100 m.
- 100Base-T2 uses two pairs of cables according to a full-duplex operation mode. This technology is not widely used.
- 100Base-FX uses a 1300 nm wavelength transmitted on two fibers (two directions). The maximum segment distance is 400 m for half-duplex connections or 2 km for full-duplex multimode optical fibers.
- 100Base-SX uses two multimode fibers, reaching a distance up to 550 m (it is a cheaper solution than 100Base-FX).
- 100Base-BX uses one optical fiber operating in a single mode. It uses a multiplexer, which divides the wavelength between transmitting and receiving directions (1310/1550 nm). Covered distances can be 10, 20, and 40 m.
- 100Base-LX10 uses two single-mode fibers to reach 10 km with a wavelength of 1310 nm.

Hubs and repeaters can be of class I or II, which are differentiated in terms of long or short latency. To realize a Fast Ethernet with greater distances covered than in the above constraints, a multi-segment structure needs to be used: the LAN is divided into separate collision domains using a switch, a bridge, or a router.

Gigabit Ethernet details (1998 and most recent developments)

Gigabit Ethernet at 1 Gbit/s is the evolution of the Fast Ethernet. First, the IEEE 802.3z standard (1998) and then the IEEE 802.3ab standard (1999) have defined this network technology. The possible physical media are STP, UTP of category 5, 5e, or 6, and optical fibers. Even if a half-duplex mode is supported, the most common case is full-duplex (star topology with switches). The *carrier extension* method is adopted to extend the maximum distance in the half-duplex case: the minimum frame size is still 64 bytes, but the corresponding slot size is extended to have a duration corresponding to equivalent 512 bytes. Some special symbols are transmitted after the frame FCS field. This approach reduces the Ethernet efficiency when short packets are transmitted. However, many short packets can be sent together (up to a maximum size of 1500 bytes) through the *packet bursting* scheme, so that carrier extension (if needed) is applied only to the first packet of this train. There are different distances covered depending on the various technologies. For instance, 1000BASE-CX reaches 25 m with twinaxial cabling; 1000BASE-T has a maximum range of 100 m; 1000BASE-LX has a range of 550 m with multimode fiber and 5 km with single-mode fiber; and finally, 1000BASE-ZX reaches 70 km with single-mode fiber at 1550 nm wavelength. Ethernet is evolving from pure LAN use to metropolitan and Wide Area Network (WAN) applications.

The 10 Gigabit Ethernet technology is now available commercially, as originally defined in the IEEE 802.3ae standard (2002). The different 10 Gigabit Ethernet versions are denoted by acronyms like 10GBASE. 10 Gigabit Ethernet operates only over point-to-point links in full-duplex mode: there is no need for the CSMA/CD protocol, but the Ethernet framing is maintained. The IEEE 802.3ae standard defines two different types of PHYs: LAN PHY and WAN PHY. The LAN PHY transmits Ethernet frames directly over a 10 Gbit/s serial interface and is suitable for an enterprise LAN. The WAN PHY encapsulates Ethernet packets in OC-192/STM-64 SONET/SDH frames without the need for any rate adaptation. Both optical fibers (10GBASE-SR, 10GBASE-LR, 10GBASE-LRM, 10GBASE-ER, 10GBASE-EW, 10GBASE-ZR, 10GBASE-LX4) and copper medium (10GBASE-CX4, 10GBASE-KX4, 10GBASE-KR, 10GBASE-T) are possible. Multimode fibers are used for shorter distances (300 m); instead, single-mode fibers (1550 nm) allow greater distances (40 km) using the 10GBASE-EW for WAN applications. As for copper cabling, categories 6 and 7 are used for a maximum range of 100 m.

100 Gigabit Ethernet and 40 Gigabit Ethernet are high-speed network standards. They support the transmission of Ethernet frames at 40 Gbit/s (LAN applications) and 100 Gbit/s (Internet backbone) over multiple 10 or 25 Gbit/s lanes, according to the IEEE 802.3ba standard (2010). For instance, 100GBASE-ER4 uses four lines at 25 Gbit/s, each one requiring a laser and a single-mode fiber, thus reaching a distance of 40 km (Extended Range). A standard variant, defined by IEEE 802.3bg (2011), concerns 40 Gbit/s serial transmissions over a single-mode (1550 nm) optical fiber (40GBASE-FR standard) for a maximum distance of 2 km. Finally, IEEE 802.3bm (2015) also concerns 40/100 Gbit/s transmissions.

Planning rules

There are specific planning rules for the realization of Ethernet networks. For instance, referring to the classical 10 Mbit/s Ethernet (10Base-5 and 10Base-2), we can consider the so-called 5-4-3 rule for the number of repeaters and segments in the LAN. This rule is characterized as follows, referring to one collision domain (shared access medium, bus topology):

- The maximum number of Ethernet segments between two stations in the same network cannot be greater than 5 (each segment should have a maximum length of 500 and 185 m, for 10Base-5 and 10Base-2, respectively; hence, the maximum distance covered by a collision domain is 2.5 km and 925 m for 10Base-5 and 10Base-2, respectively).
- The maximum number of repeaters between two stations in the network is 4.
- No more than three populated segments. The remaining two segments are not populated and are used just as inter-repeater links.

This rule guarantees that a signal sent out over the LAN reaches every point of the network within a specified maximum time.

In the Fast Ethernet case of the 100Base-TX type, a collision domain can have

- up to three segments and up to two class II repeaters/hubs and
- up to two segments and up to one class I repeater/hub.

There are, then, constraints on the maximum possible distance for each collision domain.

Finally, in the Gigabit Ethernet case, planning constraints are related to the distance covered by each segment.

6.5.2 IEEE 802.11 Standard

Wireless LANs (WLANs) are an emerging technology. They can be structured (i.e., with a central controller managing the access protocol) or unstructured (i.e., without a central controller). In the unstructured case, we have the so-called ad hoc networks, suitable for providing connectivity in dynamic WLAN scenarios and also used for sensor networks.

IEEE 802.11x is a family of standards for wireless access networks [31]. We have different technologies that are designated commercially under the name of WiFi (Wireless Fidelity).[9] The IEEE 802.11 standard defines only the lower layers of the ISO model, i.e., (1) the physical layer (PHY) and (2) the data link layer composed of two sublayers: the 802.2 Logical Link Control (LLC) and the Medium Access Control (MAC). MAC and LLC sublayers are common to all these systems under the WiFi umbrella.

The classical IEEE 802.11 system (1997) is characterized by a channel bit-rate of 1 or 2 Mbit/s in the Industrial, Scientific, and Medical (ISM) frequency band of 2.4–2.4835 GHz, with two different wireless transmission techniques: Direct Sequence Spread Spectrum (DSSS) and Frequency Hopping Spread Spectrum (FHSS). Note that these spread spectrum-techniques are needed to reduce the interference with other devices, using ISM frequencies (e.g., microwave ovens, cordless phones, Bluetooth, and other appliances). Moreover, there is also the possibility of infrared transmissions with a wavelength in the range from 850 to 950 nm.

The IEEE 802.11a standard (1999) operates in the ISM frequency bands at 5.15–5.35 GHs and 5.725–5.825 GHz with a physical layer based on Orthogonal Frequency Domain Multiplexing (OFDM) at a bit-rate of 54 Mbit/s.

The IEEE 802.11b standard (1999) is an enhancement of the DSSS physical layer, named High-Rate DSSS (HR-DSSS), operating in the 2.4 GHz ISM band and delivering up to 11 Mbit/s. Note that IEEE 802.11b supports both the DSSS mode at lower bit-rates of 1 and 2 Mbit/s and the HR-DSSS mode at 5.5 and 11 Mbit/s.

The IEEE 802.11g amendment (2003) is a standard for WLANs still in the 2.4 GHz band, which achieves high bit-rate transmissions (the maximum bit-rate is 54 Mbit/s) with an OFDM-based physical layer. IEEE 802.11g is fully interoperable with IEEE 802.11b.

IEEE 802.11n (2009) adopts an OFDM air interface and MIMO antennas. 802.11n operates in both 2.4 and 5 GHz ISM bands. The maximum data rate goes from 54 up to 600 Mbit/s, ten times faster than IEEE 802.11g.

The further evolution is represented by the IEEE 802.11ac standard frozen in 2013, providing high-throughput wireless LANs on the 5 GHz band. This standard has been retroactively labeled as WiFi 5 by Wi-Fi Alliance (being WiFi 1 the IEEE 802.11b standard). The specification has multi-station throughput of at least 1 Gbit/s and single-link throughput of at least 0.5 Gbit/s. This is achieved by extending the MIMO-OFDM air interface concepts of 802.11n.

The most advanced WiFi specification is represented by the IEEE 802.11ax protocol (2021), denoted as WiFi 6. This project aims at providing 4 times the throughput of 802.11ac at the user layer, having just 37% higher data rates at the PHY layer. OFDM is used together with higher-order modulations. Multiple User-MIMO (MU-MIMO) is adopted. This WiFi version allows a user bit-rate up to 9.6 Gbit/s in the 2.4 GHz band.

The actual bit-rate experienced by the users can be regarded roughly as half the air interface physical bit-rate due to overheads, collisions, and packet headers.

The ISM band is divided into sub-bands (e.g., at 2.4 GHz, 5 GHz, etc.). Each range is split into many channels with a spacing of 5 MHz and an amplitude of about 20 MHz. Channels can also be combined together (bonded) to form larger channels (i.e., 40 MHz, 80 MHz, 180 MHz). The 2.4 GHz ISM band is divided into 14 overlapping channels (only 11 are available in North America), each with a bandwidth of 20 or 22 MHz and a 5 MHz offset. The minimum channel separation for installations in close proximity is 3 (e.g., channels 1, 4, 7, 10, and 13, only in Europe), but the recommended separation is 5 (e.g., channels 1, 6, and 11). The channel reuse is possible for sufficiently spaced cells of radio coverage. The 5 GHz ISM band is divided into 23 non-overlapping channels.

Three different topologies are available for IEEE 802.11 according to the *service set* concepts (i.e., a logical grouping of wireless terminals, also called stations, STAs):

[9] The IEEE 802.11 family includes different technologies (i.e., legacy 802.11, 802.11a, 802.11b, 802.11g, 802.11n, 802.11ac, and 802.11ax) under the name of WiFi. Other standards in this family (from letter c to f and from letter h to j) are service enhancements and modifications of previous specifications.

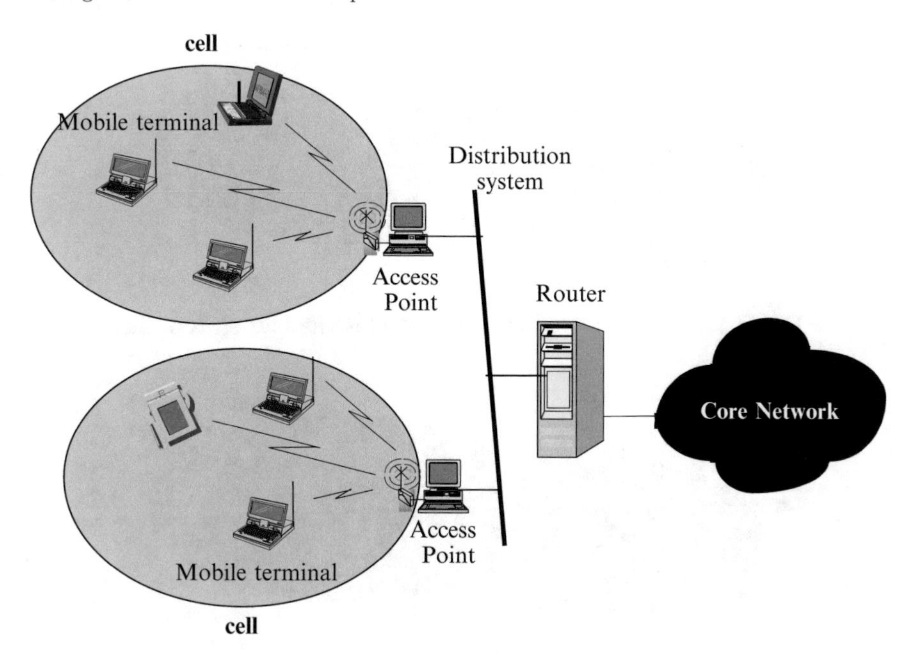

Fig. 6.42: ESS with infrastructure BSS

- Independent Basic Service Set (IBSS)
- infrastructure Basic Service Set (BSS)
- Extended Service Set (ESS).

An IBSS consists of a group of 802.11 STAs communicating directly with one another in an *ad hoc mode* (peer-to-peer operation mode). A BSS is a group of 802.11 STAs, which do not communicate directly with one another, but only through the Access Point (AP), a specialized STA forwarding the frames to the destination STA. Generally, the AP is connected to a wired network, and for this reason, a BSS is also referred to as *infrastructure mode*. Many APs with the related BSSs can be interconnected to a backbone system, named Distribution System (DS). The set of BSSs and DS forms the Extended Service Set (ESS), as shown in Fig. 6.42. The DS could be a wired Ethernet or another IEEE 802.11 wireless system.

WiFi is a worldwide success; it is also used to create wireless mesh networks according to the IEEE 802.11s amendment to the WiFi standard. In this case, there is not an AP with a centralized architecture, but the different WiFi nodes act as routers. Suitable routing protocols need to be adopted.

Typical Access Problems in Wireless Systems

Typical access problems in wireless systems are the *hidden terminal* and the *exposed terminal* [1]. Figure 6.43 describes the hidden terminal problem: while node A is transmitting to node B, node C verifies that there is no colliding transmission and decides to send a message to node D. However, the transmissions of A and C collide at terminal B, which, therefore, cannot correctly receive the message sent by A. The problem is that terminal C (i.e., hidden terminal) cannot "see" the simultaneous transmission of A since C is beyond the radius of coverage of A.

Figure 6.44 describes the exposed terminal problem on the same "topology" envisaged for illustrating the hidden terminal problem. While terminal B is transmitting to terminal A, terminal C must send data to terminal D, but C decides not to transmit, since C perceives an occupied channel due to the transmission of B (false carrier sensing). Hence, C does not transmit to D even if it could do so without generating any collision with the transmission of B to A.

The WiFi wireless systems are half-duplex: they cannot transmit and receive at the same time. WiFi stations send data or listen to the channel to receive data; they cannot do both functions simultaneously.

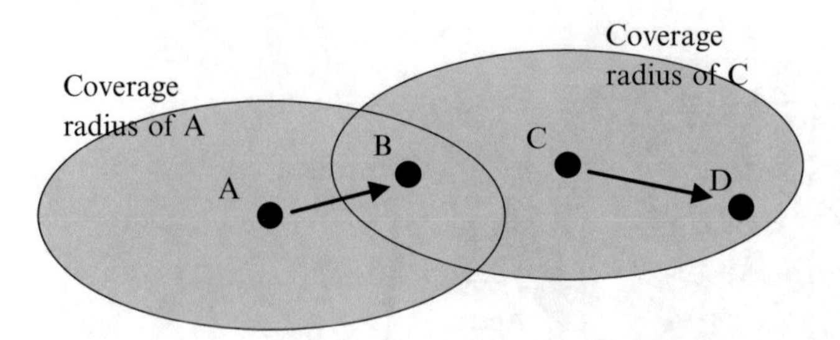

Fig. 6.43: Hidden terminal problem; C is the hidden terminal

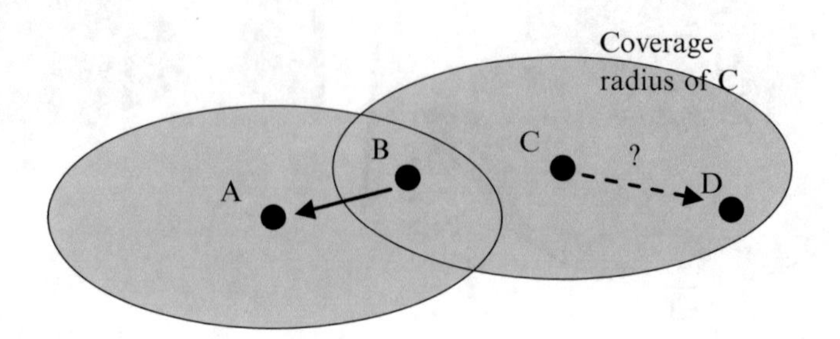

Fig. 6.44: Exposed terminal problem (note that B is in the coverage area of C; by reciprocity, C can hear messages sent by B); C is the exposed terminal

6.5.2.1 The IEEE 802.11 MAC Sublayer

The MAC sublayer is responsible for frame addressing and formatting, error checking, channel allocation procedures, fragmentation, and reassembly. The MAC standard envisages two access modes:

- Distributed Coordination Function (DCF), a *mandatory* contention-based access scheme for the asynchronous delivery of delay-insensitive data (e.g., e-mail and FTP).
- Point Coordination Function (PCF), an *optional* contention-free access scheme for time-bounded delay-sensitive transmissions (e.g., real-time audio and video), used in combination with DCF.

In the first mode, wireless terminals have to contend for the use of the channel for each data packet transmission. DCF is the most important access method; it is based on a Carrier Sense Multiple Access with Collision Avoidance (CSMA/CA) scheme (see the part below). PCF optional method is only usable in an infrastructure BSS (i.e., a wireless network with a coordinating AP). PCF is implemented on top of DCF and is based on a polling scheme: the AP polls the stations to access the medium, thus eliminating contentions. PCF is very rarely implemented in the devices. In the following description, numerical values are referred to the IEEE 802.11b standard.

The CSMA/CD protocol cannot be used directly in the wireless case since there can be situations where simultaneous transmitters cannot hear each other (hidden terminal problem). In addition to this, a collision detection scheme would require that the sender can simultaneously receive, thus resulting in costly solutions. These are the reasons why a modified CSMA scheme has been proposed for WLANs, i.e., CSMA with Collision Avoidance (CSMA/CA). Two CSMA/CA versions are available for DFC operations: the primary (and mandatory) CSMA/CA scheme and the optional CSMA/CA version with Request To Send (RTS) and Clear To Send (CTS) messages. Both versions are described below.

DCF mode with basic CSMA/CA

Let us refer to Fig. 6.45, which describes the essential characteristics of the CSMA/CA access protocol. A station needing to send data starts sensing the medium: if the medium is free for the duration of an Inter-Frame Space (IFS)[10] (and Network Allocation Vector, NAV = 0, as explained later), the station can start sending data (i.e., direct access); otherwise, if the medium is busy, the station has to wait for a free medium plus an IFS time plus a random backoff time within a *contention window*. This window is divided into slots of duration "SlotTime," which

[10] The duration of an IFS period depends on the packet type; in case of an information packet, IFS becomes a DCF Inter-Frame Space (DIFS).

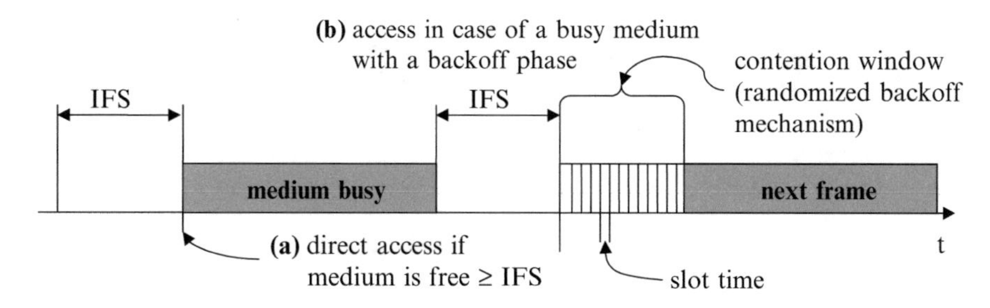

Fig. 6.45: Basic operations of CSMA/CA: (**a**) direct access and (**b**) access with backoff phase

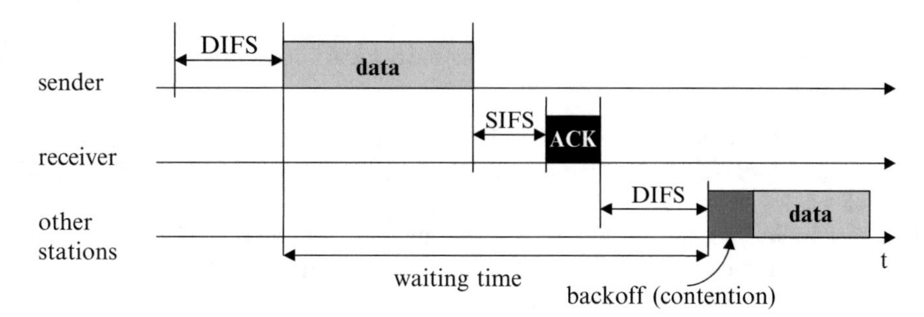

Fig. 6.46: CSMA/CA scheme of the DFC mode

depends on PHY characteristics. Thus, a station defines a random backoff value within the window and decrements its backoff time counter as long as the channel is sensed free. The counter is stopped when a transmission is detected on the channel and reactivated when the channel is sensed free again for more than DIFS. A station transmits its packet when its backoff counter reaches zero.

IFSs are mandatory idle time intervals between the transmissions of subsequent packets. Different IFS types have been defined:

- Short Inter-Frame Spacing (SIFS) has the shortest duration (thus entailing the highest access priority) and is used for sending ACKnowledgment (ACK), CTS, and polling response. The SIFS duration (SIFSTime) and SlotTime lengths depend on the PHY layer. SIFS is also the time needed for a station to switch from transmission to reception and vice versa.
- PCF IFS (PIFS) has a medium priority and is used for PCF transmissions: PIFSTime = SIFSTime + SlotTime.
- DCF IFS (DIFS) has the lowest priority and is used to transmit asynchronous data (i.e., messages): DIFSTime = SIFSTime + 2 × SlotTime.

Note that SIFS < PIFS < DIFS. For instance, we have the following IFS values in the IEEE 802.11b standard: SIFSTime = 10 μs, SlotTime = 20 μs, PIFSTime = 30 μs, and DIFSTime = 50 μs.

The basic CSMA/CA access scheme of DCF is described below, referring to Fig. 6.46.

- A station can send data if the medium is free for a DIFS time.
- The receiving station checks the correctness of the received packet (Cyclic Redundancy Check, CRC).
- If the packet has been received correctly, the receiver waits for a SIFSTime and then sends back an ACK packet.
- In case of a transmission error or collisions, no ACK is sent and the packet is retransmitted automatically after a timeout.

The MAC layer receives a MAC SDU (MSDU) from the higher layer that is fragmented into MAC PDUs (MPDUs). The maximum MSDU size is 4095 bytes for FHSS and HR-DSSS and 8191 bytes for DSSS. There are three different types of frames or packets (MPDUs): data, management (i.e., beacon and probe response), and control (i.e., ACK, RTS, CTS). Each data frame has a payload with a maximum length of 2312 bytes, a MAC header of 30 bytes, and a trailer with an FCS of 4 bytes. If an MSDU is larger than 2312 bytes, it must be fragmented into more MPDUs. The physical layer adds a preamble plus a header to the MPDU.

In IEEE 802.11, carrier sensing is needed to determine if the medium is available. There are two methods: physical carrier sense mechanism and virtual carrier sense mechanism:

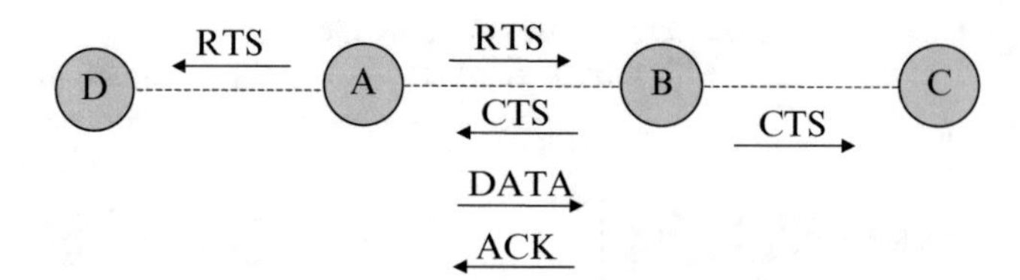

Fig. 6.47: RTS/CTS protocol

- A physical carrier sense function is provided by the PHY layer and depends on the medium and modulation used.
- A virtual carrier sense mechanism is obtained using the Network Allocation Vector (NAV), set in a specific field of the MAC frame header.

The channel is busy if one of the two mechanisms indicates it to be. In particular, physical carrier sensing detects the channel activity using the signal strength received from other sources. In contrast, virtual carrier sensing is performed by setting the NAV value in the MAC header of data frames (as well as of RTS/CTS messages). In particular, NAV is a timer indicating the amount of time the medium will be reserved until the current transmission is over. The ACK transmission time is included in the NAV period: after a packet is successfully received, there are a SIFS time[11] and the ACK transmission time. A transmitting station sets the NAV. Other stations count down from the current NAV value to 0. When NAV reaches 0, the virtual carrier sense mechanism indicates that the medium is free. NAV is used in both the basic access scheme and its improved version with RTS/CTS, as described below (see Fig. 6.48).

If a packet transmission is detected on the air (through one of the two above methods), the sending device must choose a random backoff time and wait for the backoff to expire before trying to send its packet (see Fig. 6.45). A sender must also select a random backoff time if its packet is not acknowledged by the receiver due to a collision or a packet corruption on the radio medium (CSMA/CA implicitly detects a collision or a packet error when a transmitter does not receive an expected ACK).

Suppose two stations (i.e., STA#1 and STA#2) need to transmit, while another STA is transmitting. In that case, they stop any attempt and define random backoff timers to avoid that they transmit simultaneously when the medium becomes free, thus causing a collision. Let us assume that STA#1 selects a backoff timer of four slots and STA#2 selects a backoff timer of two slots. Then, STA#2 begins to transmit without collisions. Moreover, STA#1 hears a new NAV duration from the STA#2 frame so that STA#1 sets its NAV value. STA#1 must wait for its NAV to reach 0 and for its PHY to report that the medium is free before resuming its backoff countdown.

The backoff time in slots of an STA is an integer value with a uniform distribution over the interval $[0, c_w]$, where c_w is called *contention window*. At the first attempt, the contention window has a minimum value: $c_w = c_{wmin}$ equal to 31 slots. The backoff time of an STA is decreased as long as the channel is free. If the channel becomes busy, the backoff time is frozen; the countdown restarts as soon as the channel becomes free. When the backoff time reaches zero, the STA transmits its packet. If a collision occurs, the STA has to compute a new random backoff time doubling the previous c_w value to reduce the probability of a new collision. The maximum c_w value is $c_{wmax} = 1023$ slots. Note that c_w is reset to c_{wmin} after a successful transmission or after reaching the maximum number of attempts (retry limit).

The WiFi DCF scheme has been analyzed in [11] using a discrete-time Markov chain in saturated conditions.

DCF mode with CSMA/CA and RTS/CTS

In conditions of heavy traffic with significant collisions, it is convenient to use an improved version of the CSMA/CA scheme, based on RTS and CTS messages. Figure 6.47 depicts a simple example to explain the operation of the RTS/CTS protocol.

STA A willing to transmit data to STA B sends an RTS message to B. STA B replies with a CTS message (if the medium is free). Upon receipt of CTS, STA A can start transmissions. B acknowledges (ACK) each data packet received from A. If A fails to receive the ACK, A retransmits its packet, assuming an error. Both C and D remain quiet until an ACK is delivered to avoid collisions with this important message; this is obtained through the NAV contained in RTS and CTS messages. In particular, RTS and CTS contain sender address, receiver address, and

[11] SIFS is shorter than DIFS to prioritize the transmission of the ACK by the receiving station over other possible transmissions by other stations.

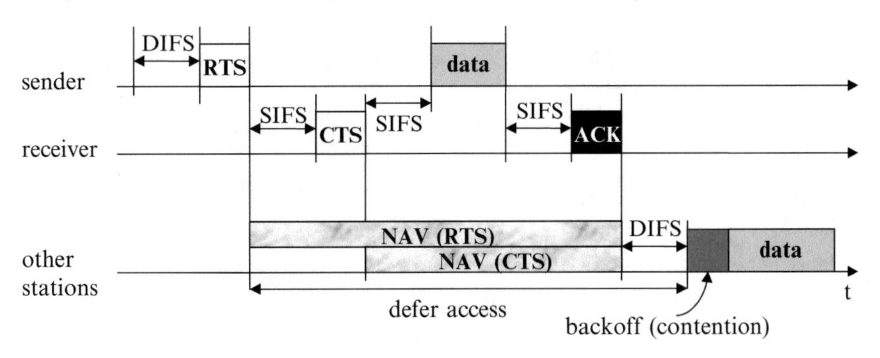

Fig. 6.48: CSMA/CA protocol with RTS/CTS

the NAV, specifying the expected transmission duration, including the ACK transmission time. All STAs receiving RTS and/or CTS will set their NAV for the given duration and will use this information together with the physical carrier sensing to determine when the medium is free.

Since RTS and CTS are short frames, they allow a more efficient access phase by reducing the time wasted due to collisions. Moreover, the RTS/CTS scheme allows overcoming the hidden terminal problem, as shown in Fig. 6.47, since STA C can hear the CTS message (with NAV) sent from B to A, and then STA C avoids transmissions.

The detailed procedure to send a packet with CSMA/CA and RTS/CTS is provided as follows, referring to Fig. 6.48:

- If the medium is free for DIFS, an STA can send an RTS with the NAV, which determines the amount of time the data packet needs the medium; the receiver STA replies with a CTS message (containing the NAV) after SIFS. The other STAs store the NAV distributed via RTS and CTS. The sending STA can now transmit data. The receiving STA sends an ACK after a SIFS interval.
- Otherwise, if the medium is not free, the collision avoidance scheme is adopted: once the channel becomes free, the sending STA waits for a DIFS time plus a randomly chosen backoff time before attempting to transmit (the backoff time is needed to avoid collisions among waiting transmissions). The backoff interval is doubled at each new attempt and suspended if the medium is sensed busy. The effect of this procedure is that when multiple STAs are deferring their transmissions, the STA selecting the smallest backoff time will win the contention.

Since RTS and CTS messages entail some overhead, this mechanism is not adequate to send short data packets.

PCF mode

PCF is a contention-free access that is adopted in Infrastructure BSS. PCF is an optional feature. PCF has a higher priority than DCF because the coordinating AP (also called Point Coordinator, PC) takes the control of the medium after a busy period and after an interval, called PIFS, which is shorter than DIFS. The PCF protocol is on top of the DCF one. PCF is based on a *superframe structure* that contains in order: a Beacon Frame (BF), a Contention-Free Period (CFP), and a Contention Period (CP). During CFP, PCF is adopted for accessing the medium, while DCF is used during CP. The superframe term is not used in the standard.

The BF is a management frame, which maintains the synchronization of the local timers in the STAs and delivers protocol-related parameters. Note that BF is also required in pure DCF, but in this case, it is used just for network admission and authentication. The beacon is like an advertisement message, which informs the STAs that an AP is alive. If an STA receives a BF and wants to join an existing cell, it will transmit a probe request frame to the AP.

Let us describe the PCF mode in detail. The coordinating AP schedules the transmission of BFs at regular intervals so that every STA knows (more or less) when it will receive a BF. The coordinating AP schedules the BF transmission based on a Target Beacon Transition Time (TBTT); the TBTT interval basically corresponds to the above superframe description. Whenever an STA receives the BF, it sets its NAV to the maximum possible (i.e., TBTT) duration to block any DCF traffic on the wireless medium. To make even surer that DCF traffic cannot be generated during the CFP period, all PCF transmissions are separated by PIFS or SIFS that are both shorter than DIFS so that no DCF STA can grab the medium.

Note that the coordinating AP schedules the transmission of the BF according to the TBTT time, but the BF is transmitted at this target time only if the medium has been free for at least a time PIFS; otherwise, the BF transmission is delayed.

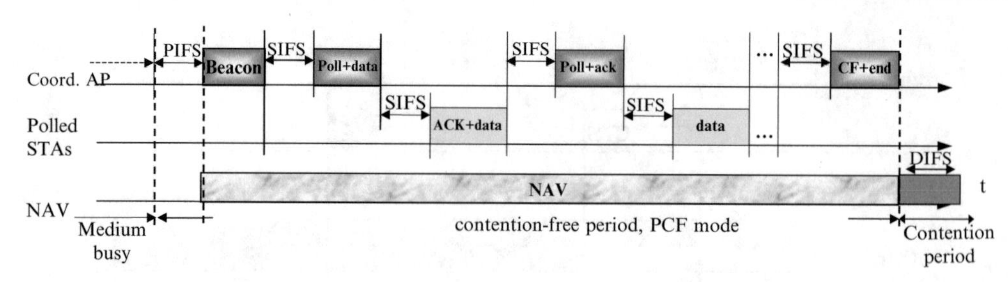

Fig. 6.49: PCF (contention-free) operation mode

The coordinating AP keeps a polling list (see Sect. 6.3.1). Once the AP has acquired control of the medium (transmitting the BF), it polls any associated STA for data transmissions. If an STA wants to transmit during a CFP, it can do so only if the coordinating AP invites the STA to send data with a polling frame (CF-Poll). In particular, soon after the beacon, the coordinating AP polls (CF-Poll) an STA asking for a pending frame (see Fig. 6.49). If the coordinating AP itself has pending data for this STA, it uses a combined frame by piggybacking the poll frame on the data one. The polled STA acknowledges the successful reception. Every CF-Poll sent by the coordinating AP allows the transmission of one data frame. If the coordinating AP does not receive a response from a polled STA, after a time PIFS, it polls the next STA or ends the CFP. Thus, no idle period longer than PIFS can occur during CFP. The coordinating AP continues to poll STAs until CFP expires. A specific control frame, called CF-End, is transmitted by the coordinating AP to notify the end of the CFP phase. The PCF scheme is rarely implemented.

Enhanced MAC of WiFi with QoS support: IEEE 802.11e

The main problem with DCF (used in IEEE 802.11a/b/g/n/ac) is that all traffic flows are managed as best effort. Hence, real-time traffic cannot be supported with adequate QoS since collisions delay transmissions. Moreover, even if PCF avoids time wasted in collisions, there are also some problems with PCF. Among many others, there are unpredictable beacon delays and unknown durations of the transmissions of the polled STAs. This may severely affect the QoS since time delays are unpredictable in each superframe.

These are the reasons why the IEEE 802.11e amendment (2005) has been proposed to support QoS in WiFi. IEEE 802.11e envisages the Hybrid Coordination Function (HFC) mechanism at the MAC layer. HCF has both a contention-based access phase, called Enhanced Distributed Channel Access (EDCA), and a contention-free (polling-based) phase, named HCF Controlled Channel Access (HCCA). EDCA and HCCA operate together according to the HCF superframe structure. HCF is a mandatory function in IEEE 802.11e. A new feature of HCF is the Transmission OPportunity (TXOP), denoting the maximum time interval for which an STA (now called QoS-enabled STA, QSTA) is authorized to grab the medium to send data frames. The aim of TXOP is to limit the time interval for which a QSTA is allowed to transmit frames.

- EDCA is an extension of the DCF mechanism, supporting eight priority levels and four Access Categories (ACs), typically for voice, video, best effort, and background traffic. Different priority levels can be used within an AC. Depending on the AC class, the following quantities are characterized: minimum and maximum contention window size, maximum TXOP value, Persistence Factor (PF), and the time interval between the transmissions of frames, now called Arbitration Inter Frame Space (AIFS); it substitutes the DIFS interval: AIFS $\geq$ DIFS. These parameters can be dynamically updated by the AP (now named Hybrid Coordinator, HC) for each AC through BFs or in probe and reassociation response frames. Shorter backoff intervals can be considered for high-priority traffic sources so that they contend successfully. The AIFS length gives another priority mechanism: if two QSTAs need to transmit at the same time, the QSTA with the shorter AIFS will grab the medium. After a collision, a new contention window value c_{w} is calculated with the help of PF. In the classical 802.11 standard, c_{w} is always doubled after an unsuccessful transmission. Instead, c_{w} is increased by a factor PF at each attempt with 802.11e, where the PF value depends on the AC class. A QSTA has four queues at the MAC EDCA level (one queue per AC). Each queue provides frames to an independent channel access function, implementing the EDCA contention algorithm. This allows a sort of scheduling function within each QSTA to decide each time the highest priority frame to be transmitted. Admission control is mandatory in EDCA.
- HCCA uses the HC to manage the access to the medium centrally, according to a polling-like scheme. However, there are many differences between HCCA and PCF. In particular, HCCA is more flexible than PCF, where

the interval between two beacon frames is divided between contention phase and contention-free phase. The HC can take control of the channel to start a contention-free phase any time during a contention phase using the PIFS interval (SIFS < PIFS < DIFS). Resources are managed by the HC in a QoS-aware way so that there are significant differences with respect to the PCF mechanism.

6.6 Exercises

The following exercises exploit the characteristics of arrival processes and queuing theory to study access protocols' behavior.

Ex. 6.1 Let us consider an access system where terminals spread over a particular area transmit packets (duration T s) on a radio channel to a remote central controller. Transmissions are at random but can only start at synchronization instants (i.e., slots). New packets arrive according to exponentially distributed interarrival times with a mean value of $1/\lambda$ s. When a terminal transmits a packet, we have that:

- With probability $1 - P_c$, this packet reaches the remote central controller with a significantly attenuated power level (due to the random attenuation phenomena of the radio channel; e.g., shadowing effects) so that (1) the packet cannot be decoded correctly and (2) the packet cannot collide with other packets received simultaneously.
- With probability P_c, the packet is received with an adequate power level and can also collide with other packets, which are received with a sufficient power level.
- If a packet is not received correctly (due to either radio channel effects or collisions), it is retransmitted after a random delay (backoff).

It is requested to model this system by determining the relation between the carried traffic load (throughput), S, and the total circulating traffic, G. Finally, we have to determine the maximum traffic load supported by this access system.

Ex. 6.2 We have N stations, each generating packets with a mean arrival rate λ of 10 pkts/s and mean packet transmission time $T = 1$ ms. Stations must exchange traffic with a master station using a suitable LAN technology.

It is requested to choose (providing adequate justifications) a random access scheme (among Aloha, S-Aloha, non-persistent CSMA, and 1-persistent CSMA) to manage the traffic generated by the different stations in each of the following cases:

1. $N = 20$, one-way propagation delay $\tau = 20$ ms;
2. $N = 95$, one-way propagation delay τ negligible with respect to T.

Referring to the access scheme selected in the second case ($\tau \ll T$), but assuming $N = 10$, we have to determine the mean number of packets in the system in the case where the mean packet transmission delay (from measurements) is equal to $T_p = 2$ ms.

Ex. 6.3 Let us consider an optical fiber ring LAN based on the *token ring* protocol. There are $N = 10$ stations in the LAN. Considering that the transmission on the optical fiber is at a rate $R = 100$ Mbit/s, that each station generates traffic of $\lambda = 100$ pkts/s, that each packet contains $m = 10{,}000$ bits, and that the time to send the token from one station to another is $\delta = 10$ µs, it is requested to determine the mean cycle length.

If a packet arrives at an empty buffer of a station, how long, on average, must this packet wait for the service (i.e., before starting its transmission in the ring)? May this ring network support $N = 100$ stations, all with the same traffic as the previous ones?

Ex. 6.4 We have remote stations using radio transmissions to send control packets to a remote controller. Packets are generated according to exponentially distributed intervals with a mean value T. When a station has a packet ready, it is sent immediately without any form of coordination and synchronization with the other stations. Let Δ denote the packet transmission time. Partly overlapping packets experience a destructive collision. However, a packet sent by a station without collisions can be received with errors (thus requiring retransmissions) according to the two following independent effects:

- Errors due to the radio channel, with probability p;

- Lack of synchronization at the receiver of the remote controller, with probability q.

We have to model this access protocol and to determine the relation between the offered traffic load, S, and the total circulating traffic, G. Finally, it is requested to evaluate the maximum traffic intensity in Erlangs that this system can support.

Ex. 6.5 Let us refer to a ring LAN with $M = 6$ stations where the *token ring* protocol of the exhaustive type is adopted. We know that the time to send the token from one station to another is $\delta = 0.5$ ms, equal for all stations. The rate according to which packets of fixed length are sent in the ring is $\mu = 20$ pkts/s. The arrival process of messages at a station is Poisson with a mean rate of $\lambda = 1$ msgs/s. Messages have a length l_p (≥ 1) in packets according to the following distribution:

$$\text{Prob}\{l_p = n \text{ pkts}\} = \frac{1}{1-(1-0.3)^5}\binom{5}{n}0.3^n(1-0.3)^{5-n}, \quad n \in \{1,2,3,4,5\}$$

It is requested to determine the following quantities:

- The mean cycle duration;
- The stability condition for the buffers of the stations on the ring;
- The mean transfer delay from the message arrival at the buffer of a station to the instant when the message is delivered to another station on the ring. In this case, we have to refer to an exhaustive service policy for the buffers of the stations.

Ex. 6.6 We have a random access scheme of the Slotted-Aloha type where stations are divided into two groups:

- *Group #1*: Stations generate messages composed of one packet (transmitted in a slot of length T); the total message arrival process (first generation and retransmissions after collisions) for group #1 stations is Poisson with a mean rate Λ_1.
- *Group #2*: Stations generate messages composed of two packets (transmitted in two slots); the total message arrival process (first generation and retransmissions after collisions) for group #2 stations is Poisson with a mean rate Λ_2.

Assuming that Λ_1 and Λ_2 are known quantities, it is requested to determine the probability P_{s1} that a transmission attempt of a type #1 station is successful and the probability P_{s2} that a transmission attempt of a type #2 station is successful.

Ex. 6.7 Different remote stations transmit packets to a central controller using a synchronous random access scheme on multiple carriers (=m carriers), as explained below:

- There are infinite users generating packets, according to a Poisson process with a mean rate λ.
- The transmissions on the different carriers are synchronous.
- Two packets collide destructively if they are transmitted on the same slot (slot length = T) and the same carrier.
- When a new packet (or a collided packet) has to be (re)sent, a carrier is selected randomly with equal probability among the m carriers.

Note that this access protocol is characterized by multiple Aloha channels according to an FDMA/Aloha format. Such a protocol has been studied by Abramson under the acronym of MAMA (Multiple Aloha Multiple Access) protocol [32].

We have to determine the relationship between the total offered traffic (on m carriers), S, and the total circulating traffic, G. What is the maximum traffic load carried out by this access protocol?

Ex. 6.8 We have a carrier shared by different users employing synchronous TDMA: the frame has a length T_f and contains N slots. Each user generates messages that are queued to be transmitted on the assigned slot resources of the TDMA frame. Messages are composed of a fixed number L of packets (one packet is transmitted in one time slot). Let us assume that each user has assigned one slot per frame. If the mean interarrival time of messages is equal to T slots, it is requested to determine the traffic intensity for a generic user's buffer. What is the maximum traffic intensity (stability limit) supported by the user queue?

Ex. 6.9 Let us consider a random access system with synchronous access. We have an infinite number of elementary stations, which generate new packets according to a Poisson process with mean rate λ. Let T denote the packet transmission time. The different stations perform uncoordinated transmission attempts as described below.

As soon as a station has a packet ready to be transmitted (either a new packet or a retransmission), the station sends the packet on a slot with probability p (*permission probability*) or repeats this procedure in the next slot with probability $1-p$. Two packets transmitted simultaneously collide and must be retransmitted.

It is requested to determine the relationship between the offered traffic S and the total circulating traffic G. What is the maximum throughput (in Erlangs) that this protocol can support with a stable behavior? Are there some differences with respect to the maximum throughput achievable by the classical Slotted-Aloha scheme?

Ex. 6.10 We have a LAN adopting the unslotted non-persistent CSMA protocol with $N=10$ stations. Each station generates new packets according to exponentially distributed interarrival times with a mean value $D=1$ s. The packets transmission time is $T=10$ ms. The maximum propagation delay is $\tau=0.6$ µs.

- Determine the approximate relation between the offered traffic, S, and the total circulating traffic, G.
- Determine the total traffic generated by the N stations in Erlangs.
- Study the stability of the non-persistent protocol in this particular case and in general.

Ex. 6.11 Let us consider a WLAN adopting an access protocol of the Slotted-Aloha type. The arrival process of new packets is Poisson with a mean arrival rate λ. The mean packet transmission time is $T=1$ ms. This protocol adopts a form of regulation according to which the central controller broadcasts a synchronization pulse and a probability value $1-p$ to be used by the remote stations to block (and discard) the transmissions of some packets in case of congestion. We neglect the propagation delays from the central controller to remote stations (i.e., remote stations instantly know the value of $1-p$ to use).

It is requested to determine an ideal strategy to select the value of p as a function of λ so that the maximum possible traffic load is admitted into the network under stability conditions. In particular, we have to determine the regulation law of p as a function of λ and the behavior of the carried traffic intensity, S, as a function of λ.

Ex. 6.12 Let us consider a Slotted-Aloha system, where packets arrive according to a Poisson process with mean rate λ and are transmitted in a time T. The packet transmission power is selected between two levels (namely P_1 and P_2, with $P_1 \gg P_2$) with the same probability. This mechanism allows a partial capture effect, as follows:

- Two simultaneously transmitted packets of the same power level class collide destructively (i.e., both packets are destroyed).
- A packet transmitted at power level P_1 is always received correctly if it collides with any number of simultaneous transmissions with power level P_2 (partial capture effect).

It is requested to determine the relation between the intensity of the offered traffic, S, and the intensity of the total circulating traffic, G. Can this access protocol support an input traffic intensity of 0.5 Erlangs? Finally, it is requested to derive the mean packet delay.

Ex. 6.13 Let us consider a Reservation-Aloha access protocol with m minislots per frame for the access phase. Let us assume to have k terminals, which attempt to transmit in the same access phase of a frame, randomly selecting one of the minislots. We consider two different case studies:

1. *Case #1, No capture effect*: Two transmissions on the same minislot collide destructively.
2. *Case #2, Ideal capture*: Among all transmissions on the same minislot, one is always received correctly.

It is requested to determine in both cases the mean number of successful attempts per access phase.

Ex. 6.14 Let us consider a Fast Ethernet LAN with UTP cabling (100Base-TX). We have to determine the maximum distance between two terminals to have CSMA/CD operating correctly in the half-duplex case. In the LAN, each repeater contributes a delay $\delta=1.3$ µs, and the propagation speed in the UTP cable is $v=1.77\times10^8$ m/s. It is requested to determine the maximum distance allowed by the CSMA/CD protocol with one repeater. Is it possible to have two repeaters?

Ex. 6.15 Referring to the IEEE 802.3 standard, it is requested to evaluate the minimum and the maximum MAC layer efficiency allowed by the 10Base-2 LAN technology, considering a continuous flow of frames spaced regularly by IFGs.

Ex. 6.16 Let us consider a random access scheme, which implements an evolved version of the Slotted-Aloha protocol with Successive Interference Cancellation (SIC), whereby it is possible to recover packets that have undergone a collision. We model SIC adoption by simply considering that this scheme can successfully recover all colliding packets up to three simultaneous transmissions. We consider that the whole packet arrival process from the stations is Poisson with a mean rate λ. Let T denote the packet transmission time. We have to determine the relationship between the total offered traffic, S, and the total circulating traffic, G. What is the maximum traffic load carried out by this access protocol?

References

1. Tanenbaum AS (2003) Computer networks, 4th edn. Pearson Education International, New Jersey
2. Hayes JF (1986) Modeling and analysis of computer communication networks. Plenum, New York
3. Schwartz M (1987) Telecommunication networks: modeling, protocols and analysis. Addison Wesley, Boston
4. IEEE 802 standard family official Web site with http://www.ieee802.org/
5. Abramson N (1970) The ALOHA system-another alternative for computer communications. In: Fall Joint computer conference
6. Mroue H, Nasser A, Hamrioui S, Parrein B, Motta-Cruz E, Rouyer G (2018) MAC layer-based evaluation of IoT technologies: LoRa, SigFox and NB-IoT. In: IEEE middle east and North Africa communications conference (MENACOMM) 2018
7. Casini E, De Gaudenzi R, Herrero O (2007) Contention resolution diversity slotted ALOHA (CRDSA): an enhanced random access scheme for satellite access packet networks. IEEE Trans Wireless Commun 6(4):1408–1419
8. Roberts L (1972) ARPANET Satellite System, Notes 8 (NIC Document 11290) and 9 (NIC Document 11291), available from the ARPA Network Information Center, Stanford Research Institute, Menlo Park, June 26, 1972
9. Kleinrock L, Lam SS (1975) Packet switching in a multiaccess broadcast channel: performance evaluation. IEEE Trans Commun 23(4):410–423
10. Proakis JG (1995) Digital communications. McGraw-Hill International Editions, Singapore
11. Bianchi G (2000) Performance analysis of the IEEE 802.11 distributed coordination function. IEEE J Sel Areas Commun 18(3):535–547
12. Nanda S, Goodman DJ, Timor U (1991) Performance of PRMA: a packet voice protocol for cellular systems. IEEE Trans Veh Technol 40(3):584–598
13. Andreadis A, Giambene G (2002) Protocols for high-efficiency wireless networks. Kluwer Academic Publishers, New York
14. Nanda S (1994) Stability evaluation and design of the PRMA joint voice data system. IEEE Trans Commun 42(3):2092–2104
15. Giambene G, Zoli E (2003) Stability analysis of an adaptive packet access scheme for mobile communication systems with high propagation delays. Int J Satell Commun Netw 21:199–225
16. Saunders PT (1980) An introduction to catastrophe theory. Cambridge University Press, New York
17. Kleinrock L, Tobagi F (1975) Packet switching in radio channels: part I—carrier sense multiple access and their throughput-delay characteristics. IEEE Trans Commun 23(12):1400–1416
18. Guérin R, Peris V (1999) Quality-of-service in packet networks: basic mechanisms and directions. Comp Netw 31:169–189
19. Levy H, Kleinrock L (1991) Polling systems with zero switch-over periods: a general method for analyzing the expected delay. Perform Eval 13(2):97–107
20. Hayes JF, Ganesh Babu TVJ (2004) Modeling and analysis of telecommunication networks. Wiley, Hoboken
21. Roberts LG (1973) Dynamic allocation of satellite capacity through packet reservation. In: Proceedings of the national computer conference, AFIPS NCC73 42, pp 711–716
22. Stallings W (2003) Data and computer communications. Prentice Hall, Upper Saddle River (see Chapter 14: "LAN Systems")
23. Lee WCY (1991) Overview of cellular CDMA. IEEE Trans Veh Technol 40(2):291–302
24. Pickholtz RL, Milstein LB, Schilling DL (1991) Spread spectrum for mobile communications. IEEE Trans Veh Technol 40(2):313–322
25. Prasad R, Ojanpera T (1998) An overview of CDMA evolution toward wideband CDMA. IEEE Commun Surv 1:2–29, Fourth quarter

26. Viterbi J (1993) Erlang capacity of a power controlled CDMA system. IEEE J Sel Areas Commun 11:892–900
27. Mouly M, Pautet M-B (1992) The GSM system for mobile communications. Cell & Sys., Palaiseau
28. Yaacoub E, Dawy Z (2012) A survey on uplink resource allocation in OFDMA wireless networks. IEEE Commun Surv Tutor 14(2):322–337
29. Islam SMR, Zeng M, Dobre OA, Kwak K-S (2019) Nonorthogonal multiple access (NOMA): How it meets 5G and beyond. Wiley Online Library, Online Pre-publication, December 2019
30. IEEE 802.3 standard, publicly available at the following http://standards.ieee.org/findstds/standard/802.3ba-2010.html
31. Roshan P, Leary J (2003) 802.11 wireless LAN fundamentals, 1st edn. Cisco Press, Indianapolis. ISBN:1-58705-077-3, December 2003
32. Abramson N (2000) Internet access using VSATs. IEEE Commun Mag 7:60–68

Chapter 7
Networks of Queues

Abstract In the previous chapters, we have investigated systems composed of a single queue that can be solved using Markov chains or Imbedded Markov chains. This chapter considers analytical methods that allow us to model the mean delay of networks of queues (considered as *store-and-forward networks*), where the output from a queue can become the input to the next one. The Burke Theorem has been introduced to deal with tandem queues. Instead, the Jackson Theorem has been adopted to study general network queues where feedback traffic (and traffic loops) is possible. It has been shown how this queuing network model can be applied to any network for which we know the traffic flow relations from input to output for the different destinations. Finally, an application example of queuing networks to multipath routing and traffic offloading has been proposed, showing a possible traffic engineering approach.

Key words: Network of queues, Burke theorem, Jackson theorem

7.1 Introduction

In Chaps. 4 and 5, we have focused on studying problems where the analytical models involve the use of a single queue. The interest is now in considering problems where queues exchange traffic as in a network. We can have both *open networks* of queues, where the traffic can be received and sent outside of the network, or *closed networks*, where traffic cannot be exchanged with external nodes [1]. Closed networks are more related to the modeling of computer systems. Therefore, our study deals with open networks, which can be well suited to model *store-and-forward networks*, where different nodes (modeled using queues) exchange data traffic in the form of variable-length messages. This is, for instance, the case of IP networks.

Figure 7.1 below describes an example of an open network with four nodes and the corresponding model in terms of a network of queues, where queues exchange data traffic. The output process of a queue becomes the input process of the next queue. In the model, the generic ith *node* receives input traffic with a mean rate λ_i from outside of the network and also receives internal traffic flows routed from other network nodes with the total mean input rate denoted by Λ_i [1–7]. Each arrival generally corresponds to a message with a random length. The total arrival process at the ith node is randomly split among the different outgoing *links* from the ith node. Each link entails a buffer and a transmission line of adequate capacity so that it can be modeled by one queue. We assume *stochastic routing* at the nodes of the network. Let q_{ij} denote the split probability for the total traffic of the ith node to be routed to the jth node of the network; note that $1 - \Sigma q_{ij}$ denotes the probability that the traffic leaves the network at the ith node. Of course, there is no stochastic routing in real networks. Therefore, our network model with stochastic routing represents a macroscopic description of how traffic flows are distributed inside the network. Routing probabilities q_{ij} are determined based on how traffic is routed at the ith node towards adjacent

---

Electronic Supplementary Material The online version contains supplementary material available at (https://doi.org/10.1007/978-3-030-75973-5_7).

G. Giambene, *Queuing Theory and Telecommunications*, Textbooks in Telecommunication Engineering,
https://doi.org/10.1007/978-3-030-75973-5_7

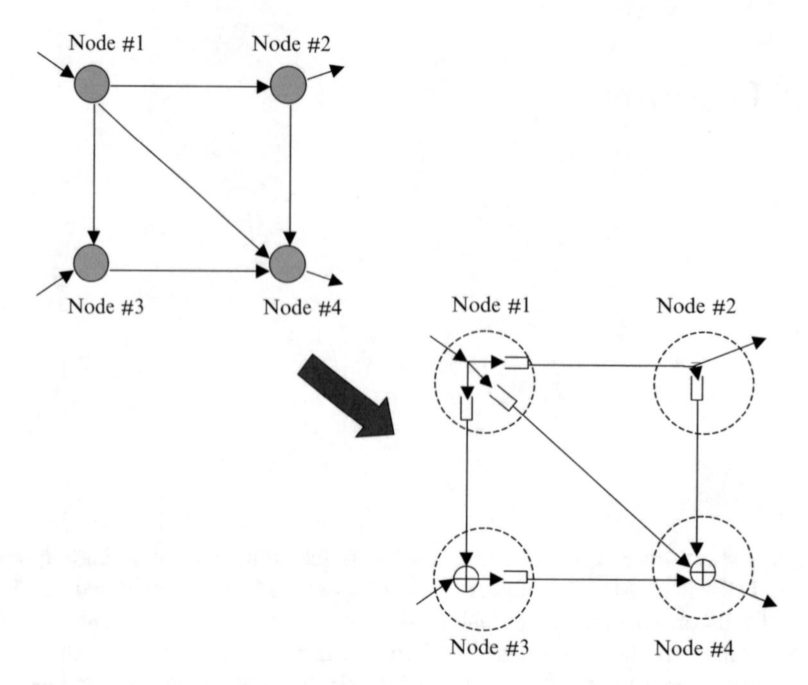

Fig. 7.1: Telecommunication network with nodes and links and related model in terms of network of queues

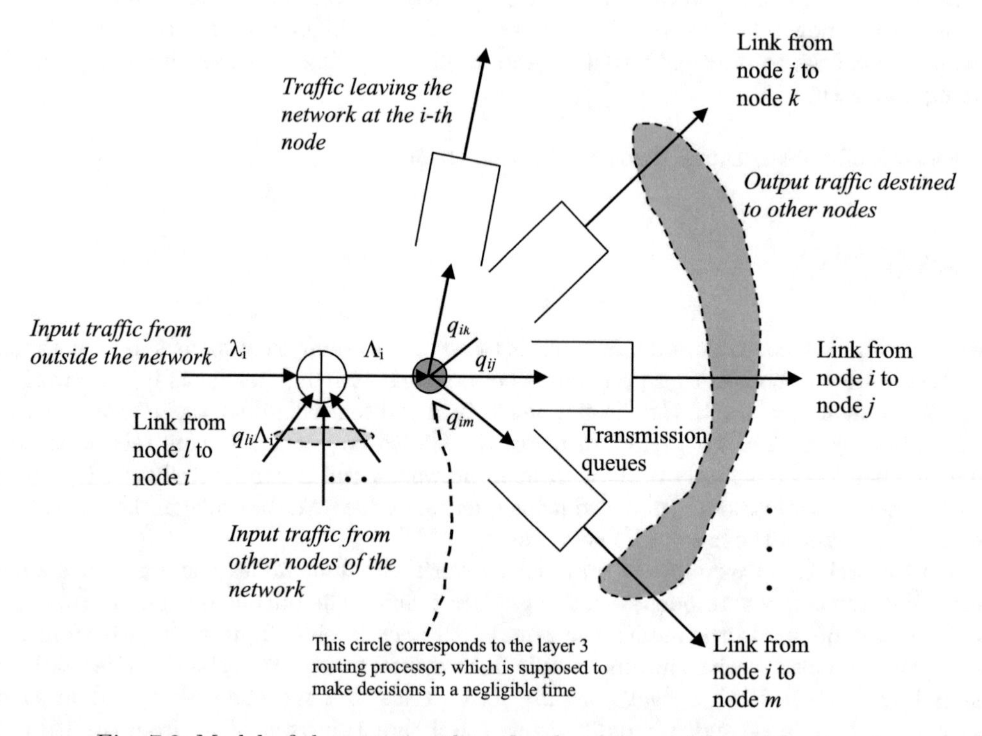

Fig. 7.2: Model of the generic ith node of a store-and-forward network

nodes and the traffic loads for the different destinations of the network. An empirical method to derive the q_{ij} values is shown in Sect. 7.4.

The details of the model for each node of an open network are provided in Fig. 7.2. In this model, we neglect both input queues at the node and layer 3 processing times for routing each incoming message. Moreover, we consider that output (link) transmission queues have infinite rooms and one sever. Under stability assumptions, the carried traffic of the generic link from node i to node j has mean rate $\Lambda_i q_{ij}$.

In the first part of this chapter, we will not use any Poisson assumption for the input arrival processes at the nodes from outside of the network. However, in the case of Poisson arrivals with mean rates λ_i (uncorrelated from node to node), the total arrival processes at the different nodes may lose the Poisson characteristic because of feedback loops, which cause a *peaked* arrival process, as described in Sect. 4.15. A network that allows (does not allow) feedback loops is called *cyclic* (*acyclic*). The Poisson characteristic of the input traffic can also be lost because of queues with finite rooms since full queues drop newly arriving packets. In this case, the traffic in the network is *smoothed*. In conclusion, if we avoid feedback loops and blocking phenomena inside the network, the Poisson characteristic of input processes is also maintained within the network because of the stochastic routing at each node. We will address this case in Sect. 7.2.

In a real network, there is a strong correlation in the behaviors of the different queues due to two main reasons: (1) the correlation of the arrival processes and the service processes because of feedback loops (cyclic network); (2) the correlation in the behaviors of the different nodes because the same message crosses several nodes in the network and has the same (or proportional) service time in all of them.

Let us consider a network of queues with the set of nodes $\{1, 2, \ldots, N\}$. For the generic ith node, we have

$$\sum_{j=1}^{N} q_{ij} \leq 1 \tag{7.1}$$

In general, we can have $q_{ii} > 0$ if there is traffic that is looped by the ith node onto itself. In (7.1), the equality holds (i.e., $\Sigma q_{ij} = 1$) if there is no traffic leaving the network at the ith node (otherwise, messages leave the network at the ith node with probability $1 - \Sigma q_{ij}$).

We can have different sub-cases for the study of networks of queues and, in particular:

- *Tandem queues*, where the whole output traffic from one queue is at the input of another queue; in this configuration, queues are chained, as considered in Sect. 7.2.
- *Feed-forward network of queues*, where there are no feedback loops (acyclic networks): A traffic flow crossing the network from input to output passes through a queue at most once. See the example in Fig. 7.1.
- *Generic network of queues*, where feedback loops are allowed, as described in Sect. 7.3.

7.1.1 Traffic Rate Equations

Let us consider a network of queues with the set of nodes $\{1, 2, \ldots, N\}$ and where each node is modeled as shown in Fig. 7.2. We can write the following balance (i.e., traffic rate equation) for the total input traffic of the ith node with a mean rate Λ_i:

$$\Lambda_i = \lambda_i + \sum_{j=1}^{N} \Lambda_j q_{ji} \tag{7.2}$$

We can write an equation like (7.1) for the N different nodes of the network. We obtain a system of N equations (i.e., the system of traffic rate equations) in the N unknown terms Λ_i since we assume that the input arrival rates from outside of the network, λ_i, and the split probabilities q_{ij} are known. The system of traffic rate equations studies the network on a node basis (not on a queue/link-basis).

Note that this system can be solved under general assumptions for the input traffic from outside of the network (i.e., in general, it is not requested that the input traffic is Poisson).

7.1.2 The Little Theorem Applied to the Whole Network

As already introduced in Chap. 4, the Little theorem [8] can be applied not only to a queue but also to a whole network of queues, as envisaged in this section. We refer here to the queues modeling the transmission on the different links of the network. Links (queues) are numbered according to the following set $\{1, 2, \ldots, L\}$. Let $\Im_k$ denote the mean number of messages in the kth queue. Let T denote the mean message delay from the input to the output of the network. The Little theorem applied to the whole network of queues can be expressed as

$$T = \frac{\Im_{\text{tot}}}{\lambda_{\text{tot}}} \tag{7.3}$$

where $\Im_{\text{tot}} = \sum_{k=1}^{L} \Im_k$ and $\lambda_{\text{tot}} = \sum_{k=1}^{N} \lambda_k$, i.e., the total mean arrival rate from outside of the network.

Thus, we need to express $\Im_k$ as a function of the total traffic intensity offered to the kth queue, ρ_k ($= \Lambda_i q_{ij}/\mu_k$, considering that the kth queue corresponds to the link from node i to node j and denoting with $1/\mu_k$ the mean service time of a message on the kth link). In the following sections, we will be able to do so through the Burke theorem and the Jackson theorem.

7.2 Tandem Queues and the Burke Theorem

We study two tandem queues or, in general, a network of tandem queues. The first queue admits an M/M/S model (Poisson arrivals/exponentially distributed service times/S servers, infinite rooms); its output process is the input process of the next queue (i.e., all messages completing the service in the first queue arrive at the second queue to be served). The system under consideration is depicted in Fig. 7.3.

According to the Burke theorem, the whole output process from the first M/M/S queue is Poisson with mean arrival rate[1] λ [1]. This output process is the process of service completions for the first queue. Then, it is possible to prove that the intervals between the service completion instants from the first M/M/S queue are exponentially distributed with mean rate λ. A related study (under the assumption of general service times but with one server) has already been carried out in Exercise 5.9 of Chap. 5; the interested reader should refer to this exercise in the solution manual for a proof of this theorem in the case $S = 1$.

Thus, based on the Burke theorem, the second queue has a Poisson input process. Hence, also this queue is of the M/M/S type if we assume again an exponential service time for messages (in a real system, the message length distribution does not change from one queue to another; only a change of scale is possible, considering different bit-rate capacities for the transmission lines associated with the different queues).

Actually, the Burke theorem would require a more complex proof because the above considerations concern only the Poisson characteristics of the output/input processes but do not prove that each queue in tandem behaves independently of the other(s). Indeed, the Burke theorem proves that a *product form expression* is valid for the joint state probability distribution of the system composed of the two queues: the tandem queues behave independently. In particular, let n_1 denote the number of messages in the first queue with related distribution P_{n_1}. Let n_2 denote the number of messages in the second queue with related distribution P_{n_2}. The joint state probability (n_1, n_2) is characterized by the distribution P_{n_1,n_2} as

$$P_{n_1,n_2} = P_{n_1} \times P_{n_2} \tag{7.4}$$

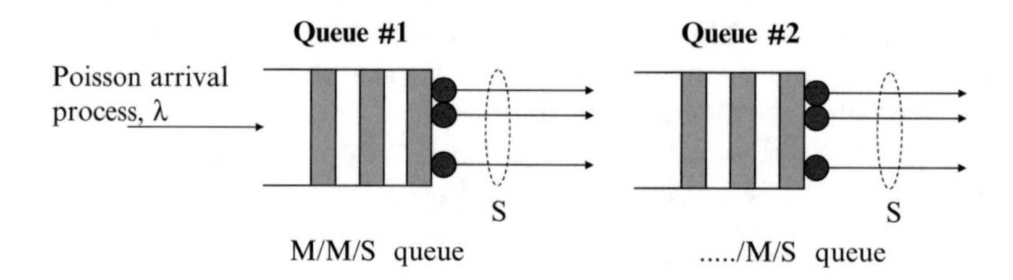

Fig. 7.3: Tandem queues

[1] Under stability conditions for the first queue, we can state that the mean output rate is λ, even without considering the Poisson nature of the input process and the statistics of the service time. The Burke theorem yields a more detailed characterization of the output process under special assumptions.

This result of the Burke theorem allows that there are Poisson arrivals at the different queues of the network if the nodes fulfill the model in Fig. 7.2 and under the following standard assumptions:

- Sum of independent Poisson processes at the input of the nodes
- Random splitting at the nodes (i.e., stochastic routing at the nodes)
- No losses
- No loops (i.e., acyclic network)
- Exponentially distributed service times

Hence, each queue in the chain of tandem queues admits an M/M/... model. The Burke theorem can also be applied to feed-forward networks (see Fig. 7.1), where, unlike the case of tandem queues, the output process of a queue is divided among different queues on the basis of a stochastic routing decision; there are no loops in feed-forward networks.

7.3 The Jackson Theorem

To study the behavior of an entire network we can refer to the Jackson theorem detailed below [1, 9]. We consider the following hypotheses for this theorem:

1. An open network with independent Poisson arrivals of messages at each node
2. Single-server queues[2] modeling the transmissions on L links with infinite rooms (no packet loss) and stable behavior
3. Exponential message service times at the nodes with FIFO discipline[3]
4. *Arrival process and service time process are independent*
5. Probabilistic routing: after service completion, the next node is independently selected from message to message

The thesis of this theorem is as follows:

- The joint probability distribution function of queue occupancies has a *product form*, where each factor corresponds to a queue and admits an M/M/1 characterization[2] (as if the input processes at the queues of the network were Poisson):

$$P\left(n_1, n_2, \ldots, n_L\right) = \left(1-\rho_1\right)\rho_1^{n_1}\left(1-\rho_2\right)\rho_2^{n_2}\ldots\left(1-\rho_L\right){\rho_L}^{n_L}$$

- The mean number of messages in each queue and the corresponding delay are according to the classical M/M/1 theory.

Traffic rate system (7.2) can be used to derive the total arrival rates of messages Λ_i at the different nodes. Then, we can determine the arrival rates $\Lambda_i q_{ij}$ and the traffic intensities offered to each queue.

The Jackson theorem–hypothesis #4 is based on an abstract concept of the network: service times are associated with the servers (and not properly with the messages, crossing several serves along their paths) and servers are independent [9]. This hypothesis cannot be true in "real" store-and-forward networks because the service time depends on the length of the message, which is the same from queue to queue. Feedback loops further increase this correlation since a server may receive the same message several times, but the service time of this message is always the same. Hence, there are dependencies between the arrival process and the service time in real networks.

To apply the Jackson theorem to store-and-forward networks, Kleinrock made the additional *independence assumption* in 1964 [10, 11]: *the service time of a message is selected independently each time it passes through a node, being it a new node or one already crossed due to a loop.* This assumption allows us to reapply hypothesis #4 of Jackson networks to real networks. This assumption is strong and can be more acceptable if there is a sufficiently good mix of different traffic sources in the network and the network has a high number of nodes, as

[2] In our model, each link in a node corresponds to a queue with one server. Hence, the Jackson theorem proves that each link can be modeled as an M/M/1 queue. Nevertheless, we could have that each link has S parallel servers so that it corresponds to an M/M/S queue according to this theorem.

[3] In some literature, this theorem is stated by adding the FIFO assumption for the service at the queues, even if this additional hypothesis is not strictly needed to prove the Jackson theorem on the product form solution. Nevertheless, the FIFO assumption is needed if we like to derive the distribution of the end-to-end delay on the basis of the M/M/1 queuing delay distribution shown in Sect. 4.14.1 [1].

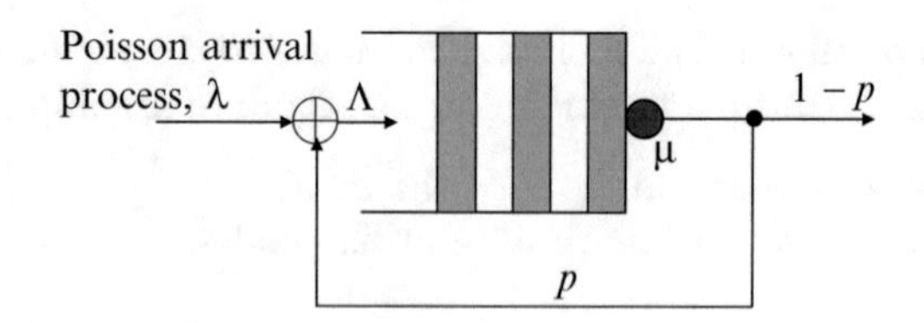

Fig. 7.4: Queue with feedback

verified through simulations in [10, 11]. Note that under the above hypotheses, network acyclicity is not requested so that we can have feedback loops in the network. Hence, in general, we have not Poisson processes into the network.

Based on Jackson theorem's outcomes, we can express the mean delay T experienced by a message to cross the network from input to output. Referring to the generic network model in Fig. 7.2, we recall that nodes are labeled with numbers from 1 to N; instead, links (modeled as queues with one server) are labeled with numbers from 1 to L. Let k be the index for the generic link. We use the following notations:

- μ_k the mean completion rate for the kth link
- α_k the mean arrival rate for the kth link (if this link connects, let us say, node i to node j, $\alpha_k = \Lambda_i q_{ij}$)
- d_k the mean delay for the queue of the kth link
- τ_k the mean propagation delay for the transmission line of the kth link

Based on the Little theorem applied to the kth link, the mean number of messages in this link results as $\Im_i = \alpha_k(d_k + \tau_k)$. Hence, we apply the Little theorem to the whole network according to (7.3) to derive the mean message delay T:

$$T = \frac{1}{\lambda_{\text{tot}}} \sum_{k=1}^{L} \alpha_k \left(d_k + \tau_k\right) = \sum_{k=1}^{L} \frac{\alpha_k}{\lambda_{\text{tot}}} \left(d_k + \tau_k\right) \tag{7.5}$$

where d_k can be expressed using an M/M/1 formula according to the Jackson theorem:

$$d_k = \frac{1}{\mu_k - \alpha_k} \tag{7.6}$$

Finally, it is important to note that the mean service time of a message on the kth link, $1/\mu_k$, is given by the mean message length $E[M]$ in bits divided by the link transmission capacity C_k in bit/s as

$$\frac{1}{\mu_k} = \frac{E\left[M\right]}{C_k} \tag{7.7}$$

7.3.1 Analysis of a Queue with Feedback

As an application of the Jackson theorem, we consider the particular case of a queue with one server where the requests completing the service have a form of stochastic routing according to which they may be fed back to the same queue with probability p. The arrival of messages from outside of the network is according to a Poisson process with a mean rate λ. The message duration (i.e., message service time) is exponentially distributed with a mean rate μ. This system is depicted in Fig. 7.4.

We can apply the traffic rate equation (7.2) to the system in Fig. 7.4 to derive the total mean arrival rate Λ. Under the stability assumption, Λ also represents the mean output rate from the queue. We have

$$\Lambda = \lambda + \Lambda p \Rightarrow \Lambda = \frac{1}{1-p}\lambda \tag{7.8}$$

According to the notations for the network of queues, the link traffic α corresponds here to Λ and λ_{tot} corresponds to λ. Then, we can note that $\alpha/\lambda_{\text{tot}} \equiv \Lambda/\lambda = 1/(1-p)$ represents the mean number of passes through the queue.

Under the Kleinrock assumption, *the message service time is independently regenerated each time the message is fed back to the queue.* Then, based on the Jackson theorem, the queue in Fig. 7.4 admits an M/M/1 model [2]. The Kleinrock assumption entails a significant approximation since the same message is considered to have different service times for each new pass through the queue; the Kleinrock assumption works better for larger networks with a high number of mixed traffic flows. Another approximation is because the feedback queue is studied as if its total input process were Poisson. However, the input traffic is not Poisson but peaked: a new message may pass several times through the queue depending on the p value. Hence, the total input process is characterized by bursts of arrivals that can be more or less evident depending on p, λ, and μ values.

However, applying the Jackson theorem to the system in Fig. 7.4, the mean delay d experienced by a message at each pass through the queue results as follows on the basis of the M/M/1 model:

$$d = \frac{1}{\mu - \Lambda} \tag{7.9}$$

The stability of the queue is assured under the ergodicity condition: $\Lambda/\mu < 1$ Erlang.

From (7.5) with zero propagation delays, we can express the mean delay T experienced by a message to cross the network from input to output as

$$T = \frac{1}{\lambda} \times \Lambda d \tag{7.10}$$

Combining (7.10) with (7.9) and (7.8) we obtain

$$\begin{aligned} T &\approx \frac{1}{\lambda} \times \frac{1}{1-p}\lambda \times \frac{1}{\mu - \dfrac{1}{1-p}\lambda} = \frac{1}{1-p} \times \frac{1}{\mu - \dfrac{1}{1-p}\lambda} \\ &= \frac{1}{1-p} \times \frac{1-p}{\mu(1-p)-\lambda} = \frac{1}{\mu(1-p)-\lambda} \end{aligned} \tag{7.11}$$

The result in (7.11) on the mean message delay T can be interpreted as follows. A message entering the system from outside passes through the same queue for a number of times n (first arrival from outside and subsequent further passes due to the stochastic feedback) according to a modified geometric distribution. The mean number of passes is therefore $E[n] = 1/(1-p)$. Each time a message goes through the queue, it experiences a mean delay d as in (7.9), i.e., $(1-\mathrm{p})/[\mu(1-p)-\lambda]$. The product $E[n] \times d$ yields the mean message delay T, as expressed in (7.11).

It is interesting to note that we could solve the system in Fig. 7.4 as an application of the M/G/1 theory with imbedding points at the instants when messages leave the system (see also Exercise 5.15 of Chap. 5): we consider a cumulative service time Y due to the sum of the service times of the different passes. Then, we apply the Pollaczek–Khinchin formula in (5.18) with mean and mean square value of the service time, respectively, given as follows:

$$\begin{aligned} E[Y] &= E[n] \times E[X] = \frac{1}{\mu(1-p)} \\ E\left[Y^2\right] &= E\left[n^2\right] \times E\left[X^2\right] = \frac{2(1+p)}{\mu^2(1-p)^2} \end{aligned}$$

We consider that a message has *the same service time at each pass through the queue* with this approach. We achieve the following (more correct) result for the mean message delay T:

$$T = \frac{1 + \dfrac{\lambda p}{\mu(1-p)}}{\mu(1-p)-\lambda} \tag{7.12}$$

We can note that (7.12) is different from (7.11) by a factor

$$1+\frac{\lambda p}{\mu\,(1-p)}.$$

Of course, the stability limits are the same in both cases. If we have $p=0$, (7.11) and (7.12) become equal to the M/M/1 mean delay term since there is no feedback in this case.

Finally, suppose we adopt the Kleinrock assumption that *the message service time is independently regenerated at each pass through the queue.* In this case, the M/G/1 approach gives the same approximate solution for T as in (7.11).

7.4 Traffic Matrices

For a real network with nodes labeled with numbers from 1 to N, we assume to know the traffic matrix $\{\lambda_{mn}\}$ from measurements, where λ_{mn} denotes the mean arrival rate of messages entering the network at node m and leaving the network at node n [1]. Hence, based on the notations introduced in Sect. 7.1, we have

$$\lambda_m=\sum_{n=1}^{N}\lambda_{mn},\quad \lambda_{\text{tot}}=\sum_{m=1}^{N}\sum_{n=1}^{N}\lambda_{mn} \tag{7.13}$$

Let us assume: (1) a given routing algorithm is adopted with consequently fixed routing tables in the network; (2) there are no routing loops (i.e., feed-forward network). Then, we can determine both the total mean input rate Λ_i for each node i and the mean arrival rate for the link from node i to adjacent node j (previously denoted as α_k) due to the traffic contributions λ_{mn} that are routed through this link. Let Φ_{ij} denote the set of couples of source m and destination n so that traffic λ_{mn} is routed through the link from node i to node j. Hence, routing probabilities q_{ij} (stochastic routing) can be obtained as

$$q_{ij}=\frac{\sum_{\Phi_{ij}}\lambda_{mn}}{\sum_j\sum_{\Phi_{ij}}\lambda_{mn}}=\frac{\alpha_k}{\Lambda_i} \tag{7.14}$$

Then, assuming that the arrival processes corresponding to the mean rates λ_{mn} are Poisson and independent so that the total input processes from outside at the nodes are Poisson, we can apply the Jackson theorem with the above q_{ij} and Λ_i terms to determine the mean message delay T according to (7.5). In this case, the generic term $\alpha_k/\lambda_{\text{tot}}$ in (7.5) represents the probability that one arrival from outside of the network is routed through the kth link. Hence, the mean message delay T according to (7.5) results as the weighted sum of the mean delays experienced in the different queues of the network (M/M/1 terms); weights are given by probabilities $\alpha_k/\lambda_{\text{tot}}$.

7.5 Network Planning Issues

Network planning and dimensioning with QoS support is a multistep process, which involves the consideration of the following aspects:

- Identification of node locations
- Definition of the network topology
- Definition of a routing strategy
- Capacity allocation to the links so that suitable QoS metrics (mean end-to-end delay, jitter, etc.) are met

These steps are connected among them. For instance, capacity allocation to links depends on traffic loads on the links and, then, on traffic routing. On the other hand, traffic routing can also be adapted to take account of traffic bottlenecks resulting from capacity shortage on some links. As it is evident from these considerations, network planning is quite a complex optimization process. The analysis carried out in this chapter can provide a useful tool to allocate capacity to the links of the network once nodes, input traffic, and routing characteristics have been defined. An optimization method in this respect is shown in [11].

7.6 Traffic Engineering and Network Optimization

As introduced in Chap. 1, traffic engineering encompasses the application of scientific principles and technology to the measurement, modeling, characterization, and management of multimedia multi-class traffic and the application of such knowledge and techniques to achieve specific performance objectives, including the planning of network capacity under QoS guarantee, and the efficient, reliable transfer of information as specified in RFC 3272 (then updated by RFC 5462) [12].

The primary objective of traffic engineering is to improve network performance while maintaining QoS requirements by optimizing the use of network resources. The need to allocate and balance resources among different traffic classes to achieve the best use of network resources is a crucial traffic engineering problem.

Traffic engineering entails different combined techniques such as traffic classes, connection admission control, prioritization, resource reservation, and routing optimization. Traffic engineering has a broad scope, including capacity allocation, traffic conditioning, queue management, and scheduling [12]. We are concerned here with the traffic engineering task to optimize the routing of traffic in the network according to some criterion. A class of routing schemes, called QoS routing, selects the path to be used by a flow based on its QoS requirements. These routing protocols can also take specified traffic attributes, network constraints, and policy constraints into account when making routing decisions. For path-oriented technologies such as MPLS, routing can be further controlled by the manipulation of relevant traffic engineering parameters. In MPLS, the path of an explicit Label Switched Path (LSP) can be computed using constraint-based routing implementing traffic engineering concepts. Adopting the new SDN concepts (see Chap. 2) can facilitate the implementation of teletraffic engineering policies on a wide scale.

A special routing case is investigated in the following sub-section, adopting a path diversity scheme (multipath scheme) to allow traffic offloading based on some QoS parameters, like the (mean) end-to-end delay that is an effective parameter to represent the path congestion.

7.6.1 Multipath Routing and Traffic Offloading

We consider that there are multiple paths between the source and destination nodes in the SDN, which allow implementing centralized schemes for path selection and path scheduling through the OpenFlow protocol. In SDN networks, routing is centralized at the SDN controller. In the SDN architecture, packet forwarding devices (switches) are manipulated by the logically centralized controller through a suitable interface. The paper in [13] provides a queue-based analytical model for SDN switches and the SDN controller.

Multipath routing is a scheme that adopts various alternative paths through a network. This approach yields many benefits, such as fault tolerance, increased bandwidth, and better security. Extensions of distance-vector routing and link-state routing are available to support the multipath schemes to maximize the throughput [14]. We consider traffic offloading that is a form of routing scheme able to distribute a traffic flow from a source to a destination using multiple paths that have different characteristics in terms of (bottleneck link) capacity, propagation delay, and congestion. The use of multiple paths in the network not only requests the adoption of suitable routing schemes but also calls for the use of multi-homing schemes (meaning the possibility for a node to belong simultaneously to different networks; this is quite common in the case of wireless networks where the same end-user device can have access to both WiFi and a 4G/5G cellular system). The multiple paths can be used alternatively or simultaneously by the routing scheme. The multiple paths might be overlapped, edge-disjointed, or node-disjointed with each other. The Linux operating system supports multipath routing by allowing the specification of multiple next hops to reach a certain destination.

Protocols to identify multiple paths are very common in Mobile Ad hoc NETworks (MANETs). We can consider for instance the Split Multipath Routing (SMR), which is a multipath version of Dynamic Source Routing (DSR), a self-organizing and self-configuring routing scheme [15]. Unlike many prior multipath routing protocols, which keep multiple paths as backups routes, SMR is designed to utilize multipath concurrently by splitting traffic onto two maximally disjoint routes. Two routes are said to be maximally disjoint if the number of shared links is minimum.

However, multipath routing would introduce extra overhead in the Internet. Exchanging the extra topology or path information required for multipath routing would consume additional bandwidth and processing resources. Moreover, there is storage overhead at each router because of the number of paths [16]. The Internet today is destination-based hop-by-hop forwarding. Then, not only the routing protocol has to determine multiple paths, but there is the need for the routing table to keep multiple next hop for the same destination. The novel SDN approach can facilitate multipath routing: as we have seen in Chap. 2, SDN enables the network to be programmable and application-aware. The SDN controller could calculate multiple shortest paths between source and destination nodes [17]. Once a multipath routing protocol is defined for the Internet, there is the need to distribute the traffic packets on the different concurrent paths that are determined for a given source–destination couple.

The enormous growth of mobile data traffic has surprised mobile operators earlier than expected. Novel technologies have emerged quickly to manage routing optimization of data flows, divert user packets around network bottlenecks and critical entities of the architecture, and redirect traffic. All these techniques are called *data offloading*. For end-users, the gain of applying offloading schemes relies on better data service cost control and the availability of higher bandwidth. From operators' perspective, offloading schemes are needed to avoid network congestion.

Traffic offloading using multiple paths can be adopted in wireless communication systems assuming that the mobile device can connect simultaneously via both WiFi and 4G/5G systems (multi-homing). In this case, the mobile device is receiving the traffic via both wireless systems. This technique is typically adopted to offload some traffic from the 4G/5G cellular system in favor of the WiFi local access. The simultaneous use of multiple networks together is opportunistic. Similarly, we can have to offload traffic from macro-cells to smaller cells, providing focused coverage. Some offloading strategies of this type are Selective IP Traffic Offload (SIPTO), Local IP Access (LIPA), and IP Flow Mobility (IFOM). Additional details are provided below [18]:

- The SIPTO function enables an operator to offload certain traffic types at a network node close to that User Equipment (UE) point of attachment to the access network. SIPTO is a network optimization technique involving traffic analysis being undertaken at the eNB (evolved Node B); each traffic flow is transferred across the most convenient path in the 4G/5G core network or bypasses the core network completely (for instance, being sent straight to the Internet).
- LIPA is a technology related to 4G/5G, which allows subscribers attached to femto- or pico-cells to access devices on their local network without their traffic flows entering the service provider's network. Practically, the traffic destined to another entity on the same local network arrives at the femto-cell and is bounced straight back into the local network itself for immediate delivery.
- IFOM is a technique used in WiFi offload to allow selective traffic offload, allowing some traffic streams to traverse the 4G/5G network whilst others traverse the WiFi network. It explicitly calls for the use of simultaneous connections to both 4G/5G cells and WiFi. The mobile terminal needs to be dual stack. This is the most advanced scheme of the three described here [19, 20].

In what follows, we propose a modelization approach for traffic offloading, referring generically to the availability of multiple paths with different characteristics to connect the traffic source with its destination. We consider that a Poisson traffic flow with a mean rate λ in pkts/s from a source to a destination has two alternative paths available in the network. The delay in crossing each path is modeled using its bottleneck link congestion and the propagation delay. We assume an M/M/1 model to describe the effect of congestion for each path and adopt the following settings (see Fig. 7.5):

- Path #1: with capacity of μ_1 pkts/s and propagation delay of δ_1 s
- Path #2: with capacity of μ_2 pkts/s and propagation delay of δ_2 s

In SDN context, as described in Chap. 2, we can expect that a system orchestrator knows from measurements μ_1, μ_2, δ_1, and δ_2. Then, it can apply an offloading scheme to this traffic so that only a portion α of the flow exploits path #1, and the rest uses path #2. In particular, we consider that each Poisson arrival of packets with rate λ is delivered via path #1 with probability α and via path #2 with probability $1 - \alpha$. Hence, the Poisson traffic flow with mean rate λ is split into two parts: the Poisson traffic flow with a rate $\alpha\lambda$ via path #1 and the Poisson traffic flow with a rate $(1 - \alpha)\lambda$ via path #2; α is the so-called *split probability*.

We consider that the mean delay experienced on path #1, D_1, can be expressed as follows using the classical M/M/1 theory that is very simple and can allow capturing the delay performance on a coarse scale:

$$D_1 = \frac{1}{\mu_1 - \alpha\lambda} + \delta_1 \quad [s] \tag{7.15}$$

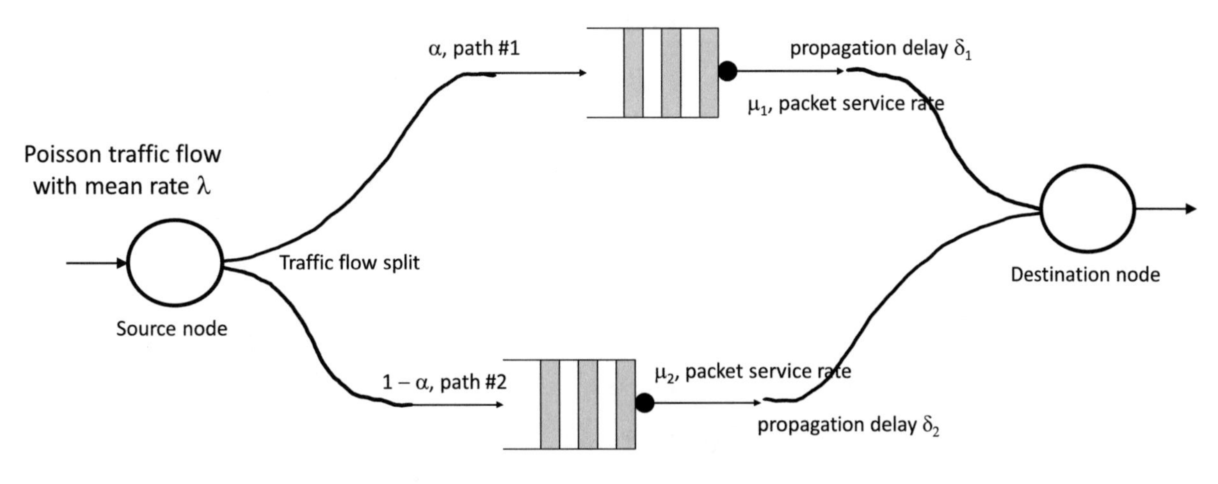

Fig. 7.5: Model for traffic offloading using two paths. The queues model the effects of the bottleneck links on each path

under the stability condition $\alpha\lambda < \mu_1$.

Similarly, the mean delay experienced on path #2, D_2, can be expressed as follows through the classical M/M/1 theory:

$$D_2 = \frac{1}{\mu_2 - (1-\alpha)\,\lambda} + \delta_2 \quad [s] \tag{7.16}$$

under the stability condition $(1-\alpha)\,\lambda < \mu_2$.

Then, combining the stability conditions of (7.15) and (7.16), we have

$$\lambda < \min\left\{\frac{\mu_1}{\alpha},\ \frac{\mu_2}{1-\alpha}\right\} \tag{7.17}$$

The first criterion to determine the value of α is to minimize the total mean end-to-end delay, D_{tot}, obtained as

$$D_{\text{tot}} = \alpha D_1 + (1-\alpha)\,D_2 \tag{7.18}$$

In this case, we have to consider the null-derivative condition with respect to α to perform the optimization. We have

$$\frac{\partial D_{\text{tot}}}{\partial \alpha} = D_1 + \alpha\frac{\partial D_1}{\partial \alpha} - D_2 + (1-\alpha)\frac{\partial D_2}{\partial \alpha} = 0 \tag{7.19}$$

Since this condition yields a 4th-degree equation in α, it is better to solve it graphically, searching for possible solutions in the range of α values that guarantee stability, as discussed later in this section.

Alternatively, since a traffic flow is split into two paths and assuming that the information cannot be managed until all data traveling on the two paths are received (for instance, this is the case of TCP-based traffic, where the data sent through the two paths with different delays have to be reordered at the receiver before they can be delivered to higher layers), we can consider a different definition of D_{tot} as follows:

$$D_{\text{tot}} = \max\{D_1,\ D_2\} \tag{7.20}$$

In this case, we can adopt an alternative method to select the α value, balancing the delays on the two paths so that $D_1 = D_2$.

$$\frac{1}{\mu_1 - \alpha\lambda} + \delta_1 = \frac{1}{\mu_2 - (1-\alpha)\,\lambda} + \delta_2 \tag{7.21}$$

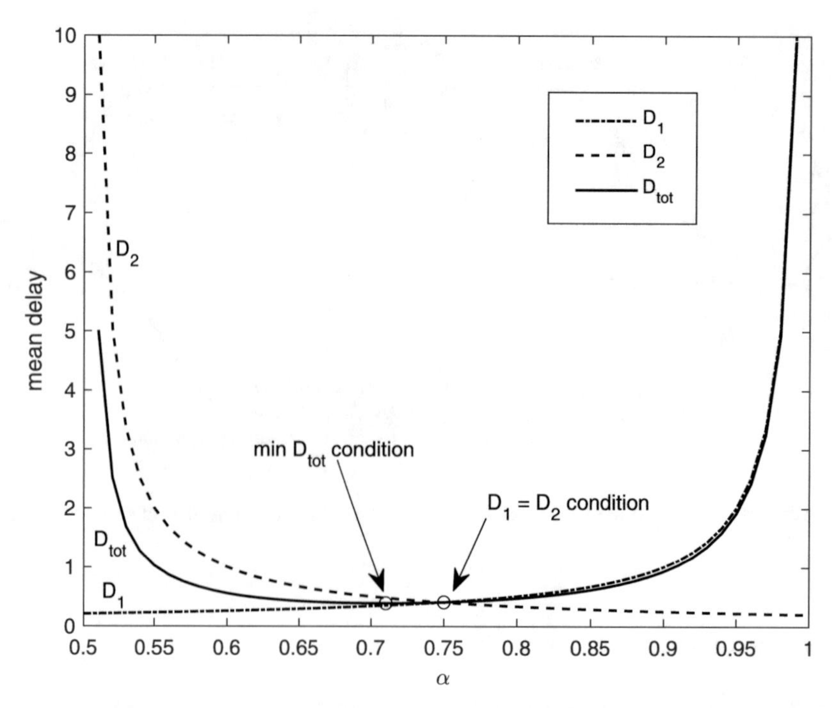

Fig. 7.6: Graphical solution of the two methods for the optimized choice of α

This condition yields a 2nd-order equation in α.

We can refer to the following numerical example where we determine the split probability α: $\lambda = 10$ pkts/s, $\mu_1 = 10$ pkts/s, $\mu_2 = 5$ pkts/s, $\delta_1 = 8$ ms (one-way propagation delay for path #1), and $\delta_2 = 5$ ms (one-way propagation delay for path #2). In this case, without the help of path #2, path #1 would be completely congested; hence, traffic offloading is needed.

If λ is given and α is variable, we have to rearrange the stability condition (7.17) in terms of α as

$$1 - \frac{\mu_2}{\lambda} < \alpha < \frac{\mu_1}{\lambda} \quad \Rightarrow \quad \frac{1}{2} < \alpha < 1 \tag{7.22}$$

Note that the determined range of α values is consistent with the fact that path #1 has a much larger capacity than path #2. At the extreme values of the range for α we have vertical asymptotes for the mean total delay D_{tot}.

Figure 7.6 shows the graphical solutions of the two optimization alternatives in (7.19) and (7.21). In this case, the two solutions are quite close. In particular, for the condition (7.19), the α value to be selected is about 0.71. Instead, for the condition (7.21), the α value to be selected is about 0.75. We can justify these results because if the system is well designed (optimal α value selection), we have no congestion for both paths #1 and #2. In these circumstances, D_1 and D_2 are quite flat around the optimal α value so that their derivatives practically vanish in (7.19) that then becomes almost equivalent to (7.21).

Note that if δ_1 and δ_2 have negligible values (so then their difference) in comparison to the other terms in the condition (7.21), we can linearize the problem and determine a simple solution of (7.21) as

$$\alpha_{opt} \approx \frac{\mu_1 - \mu_2}{2\lambda} + \frac{1}{2} \tag{7.23}$$

and according to our numerical setting, we have $\alpha_{opt} \approx 0.75$.

Based on (7.23), as expected, we can see that α_{opt} decreases as λ increases. Moreover, in the particular case with $\mu_1 = \mu_2$ and $\delta_1 = \delta_2$, we have $\alpha_{opt} = 0.5$; this occurs when the two paths are perfectly balanced in terms of service rate and propagation delay.

A refinement of the above study could also consider differentiated Packet Loss Rates (PLR) on the two paths, denoted, respectively, as PLR_1 and PLR_2. These PLR values can represent different network congestion conditions (due to buffer overflows and the RED policy) or are caused by noisy wireless links along the paths. Of course,

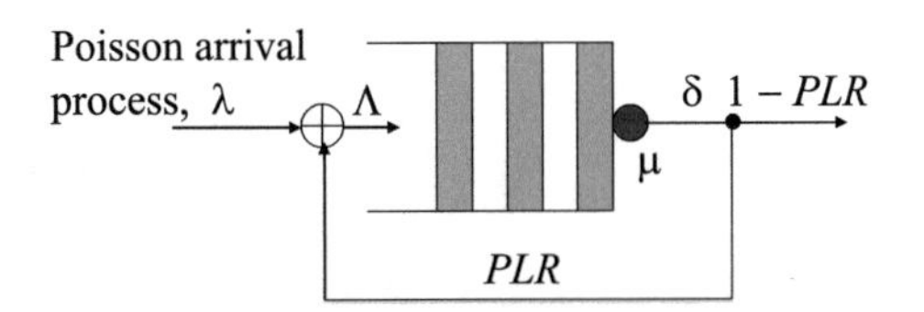

Fig. 7.7: Model of a path also considering the effects of PLR

a link with a large capacity and a high PLR value is not convenient. To study this case, the single-queue model of each path is substituted by a feedback queue to account for the retransmissions of the lost packets. A similar model has been adopted in Sect. 7.3.1, but now we also take the propagation delay δ into account. Therefore, we adopt the path model in Fig. 7.7, where, for reliability reasons, we retransmit a packet on a given path until it is successfully delivered.

To study this system, we can consider the M/G/1-based approach assuming that a new message entails multiple transmissions until the message is successfully received and that the message transmission time is the same for the different passes through the queue. However, we cannot use the mean delay in (7.12) because that analysis of the queue with feedback does not take the propagation delay into account. To do so, we have to assume a stop-and-wait message retransmission policy according to which no new packet is transmitted until the previous one has been received successfully. Then, we consider that the service time Y is the time of n rounds and that each round needs a time X with exponential distribution (rate μ) plus the propagation delay δ. Then, the service time can be expressed as

$$Y = n \times (X + \delta) \tag{7.24}$$

where n has a modified geometric distribution independent of X:

$$\text{Prob}\{n = k\} = (1 - PLR)\, PLR^{k-1}. \tag{7.25}$$

We determine the mean queue delay D that models a certain path using the Pollaczek–Khinchin formula (5.18), where

$$E[Y] = E[n]\,(E[X] + \delta) \tag{7.26}$$

$$E\left[Y^2\right] = E\left[n^2\right] E\left[(X+\delta)^2\right] = E\left[n^2\right]\left(E\left[X^2\right] + 2\delta E[X] + \delta^2\right) \tag{7.27}$$

and where

$$E[n] = \frac{1}{1 - PLR}$$

$$E\left[n^2\right] = \frac{1 + PLR}{(1 - PLR)^2}$$

$$E[X] = \frac{1}{\mu}$$

$$E\left[X^2\right] = \frac{2}{\mu^2}$$

Through some algebraic manipulations, we have

$$D = \frac{\frac{1}{\mu} + \delta}{1 - PLR} + \frac{\lambda \frac{1+PLR}{1-PLR}\left(2 + 2\mu\delta + \mu^2\delta^2\right)}{2\mu\left[\mu\,(1 - PLR) - \lambda\,(1 + \mu\delta)\right]} \tag{7.28}$$

Note that this model is not comparable with the previous M/M/1 one in the absence of PLR because of the effect of the propagation delays and retransmissions caused by packet losses. The two models become comparable only if $PLR_1 = PLR_2 = 0$ and $\delta_1 = \delta_2 = 0$.

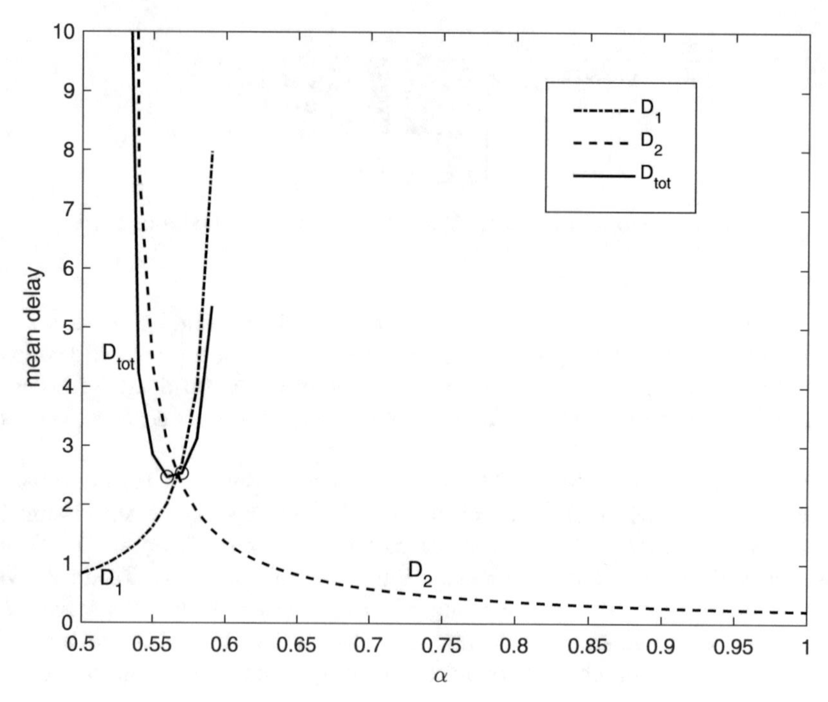

Fig. 7.8: Graphical solution of the two methods for the optimized choice of α with packet losses on the two paths

With this new model of the path delay performance, we can apply again the path selection probability α according to the conditions (7.19) and (7.21). We determine the optimized solutions using a graphical approach as done before.

Let us now consider the example of before where we add the following settings: $PLR_1 = 0.34$ and $PLR_2 = 0.03$. We get the results shown in Fig. 7.8 for what concerns the delays D_1, D_2, and D_{tot}. We can see that the effect of heavy packet losses on the first path is that we need to reduce the value of α with respect to the previous optimization. The new optimal value is $\alpha \approx 0.56$ for both the approaches that also in this case give similar results.

We have seen that the optimized α value depends on the knowledge of the bottleneck link capacity, the round-trip propagation delay, and the path packet loss rate. We assume that the SDN controller knows these parameters so that it can estimate (measure) the α on the fly. We quickly survey below some basic network measurement techniques that could be used to provide the necessary input parameters to determine the α value of the prospected offloading scheme:

- Round-Trip Time: It can be estimated using timestamps. This approach is, for instance, supported by the Internet Control Message Protocol (ICMP).
- Packet Loss Rate: Two dominant methods for measuring packet loss, based on ICMP, are ping and traceroute: these protocols send probe packets to a host and measure packet losses by observing whether or not response packets arrive within a specific time limit. These methods, however, cannot distinguish between losses on the forward or the reverse path. Moreover, several host operating systems limit the rate of ICMP responses, thus making this approach not wholly reliable. More effective methods use TCP that can signal the data received through ACKs.
- Bottleneck link bandwidth: The end-to-end network bandwidth between two hosts at the endpoints of a given path can be estimated through measurement of packet interarrival times of some probe packets sent of known size.

7.7 Exercises

This section contains some examples concerning the Little theorem applied to the networks, the Burke theorem, and the Jackson theorem.

Ex. 7.1 Let us refer to the acyclic network of queues shown in Fig. 7.9. Considering that the input arrival processes from outside of the network are independent and Poisson with mean rates shown in Fig. 7.9 and that the message transmission times are exponentially distributed with mean rates for the different queues shown in Fig. 7.9, it is requested to determine the mean delay experienced by a message from input to output of the network.

Ex. 7.2 We have to study the network of queues with feedback shown in Fig. 7.10. We know that:

- The message arrival process at queue #1 from outside of the network is Poisson with a mean rate λ.
- The message service times at queues #1 and #2 are independent and exponentially distributed with mean rates μ_1 and μ_2, respectively.
- Queues have infinite rooms.
- The routing is stochastic at the output of queue #2.

It is requested to determine:

- The stability conditions for the different queues
- The state probability distribution for each queue
- The mean number of messages in each queue
- The mean message delay from input to output of the network

Ex. 7.3 Concerning the network of queues with feedback in Fig. 7.11, we have to determine the stability conditions for the different queues and the mean delay experienced by a message from input to output, considering that:

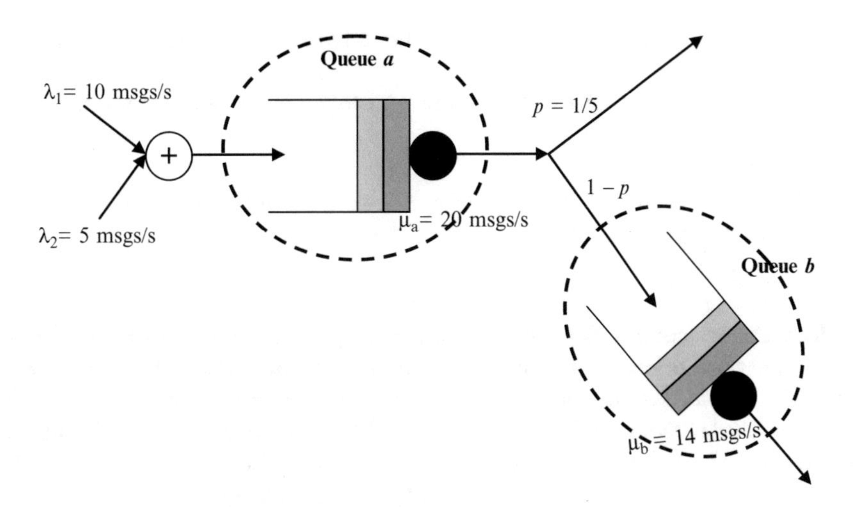

Fig. 7.9: A network of queues

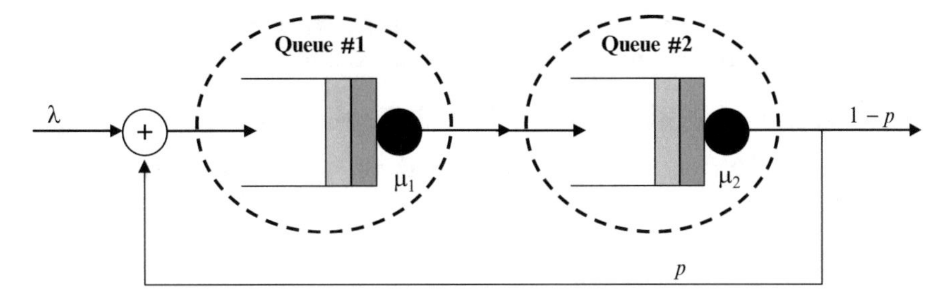

Fig. 7.10: A network of tandem queues with feedback

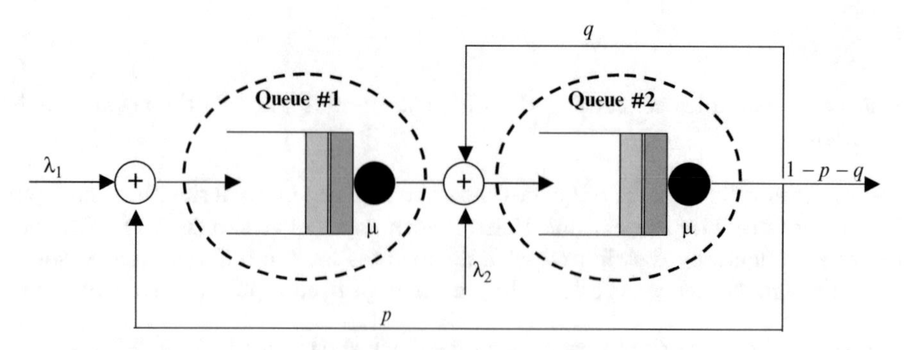

Fig. 7.11: A network of queues with feedback

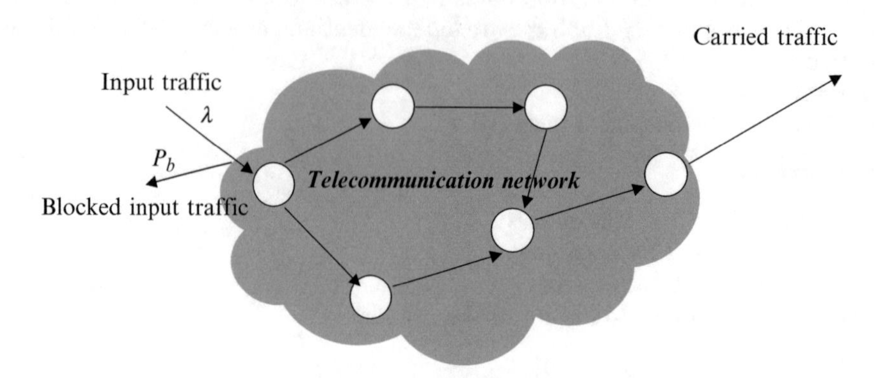

Fig. 7.12: Network with input traffic, blocked traffic, and carried traffic

- The input traffic flows to the queues from outside of the network are Poisson with mean rates λ_1 and λ_2 for queues #1 and #2, respectively.
- Message service times for both queues are independent and exponentially distributed with the same mean rate μ.
- Queues have infinite capacity.
- There is a random splitting at the output of queue #2: an arriving message is fed back to queue #1 with probability p and is provided back to queue #2 with probability q.
- $0 < p,\ q < 1$.

Ex. 7.4 Let us consider the telecommunication network in Fig. 7.12. This network is composed of nodes interconnected by links. The network operates a form of connection admission control on the arriving messages from outside to block the excess traffic. We model this control by considering that an arriving message is refused (i.e., not admitted, blocked) by the network with probability P_{b}. Knowing that the total input traffic to the network has a mean rate λ and that the total mean number of messages in the whole network is N, it is requested to evaluate the mean delay experienced by an accepted message to cross the network.

Ex. 7.5 We study the acyclic network of queues (a feed-forward network of queues) in Fig. 7.13. The input processes from outside of the system are Poisson and independent. Determine the mean number of messages in each queue of the network and apply the Little theorem to obtain the mean message delay from input to output.

Ex. 7.6 Let us consider the network of queues in Fig. 7.14. We know that:

- The message arrival processes from outside for queues #1 and #2 are Poisson and independent with mean rates λ_1 and λ_2, respectively.

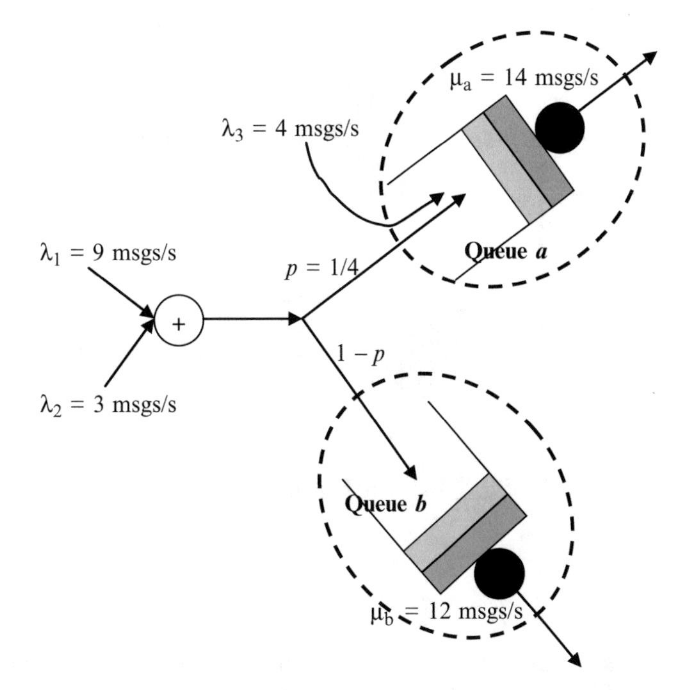

Fig. 7.13: Network of queues

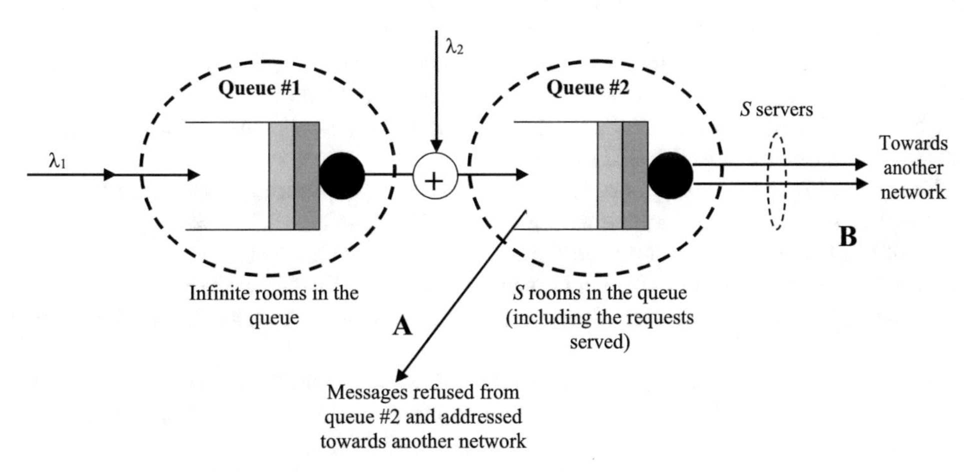

Fig. 7.14: Tandem queues

- Queue #1 has infinite rooms and exponentially distributed message service times with a mean rate μ.
- Queue #2 has S rooms, S servers, and generally distributed message service times with a mean value $E[\mathrm{X}]$.

It is requested to determine the mean number of messages in each queue and the mean message delay T from the message arrival at the network to the instant when this message leaves the network from A or B.

Ex. 7.7 Let us consider the cyclic network of queues in Fig. 7.15 where: (1) the input processes from outside of the system are Poisson and independent; (2) message transmission times at the nodes are exponentially distributed with the mean rates shown in Fig. 7.15. It is requested to determine the mean number of messages in the whole system and the mean message delay from input to output.

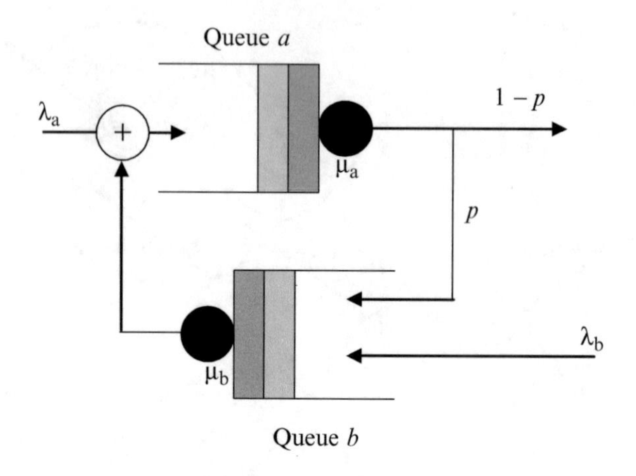

Fig. 7.15: A network of queues

References

1. Hayes JFH (1986) Modeling and analysis of computer communication networks. Plenum Press, New York
2. Buttò M, Colombo G, Tofoni T, Tonietti A (1991) Ingegneria del traffico nelle reti di telecomunicazioni [Traffic engineering in telecommunications networks]. Scuola Superiore G. Reiss Romoli (SSGRR), L'Aquila
3. Kleinrock L (1976) Queuing systems, vol I and II. Wiley, New York
4. Gross D, Harris CM (1985) Fundamentals of queueing theory, 2nd edn. Wiley, New York
5. Walrand J (1988) Queueing networks. Prentice-Hall, Englewood Cliffs
6. Baskett F, Chandy KM, Muntz RR, Palacios FG (1975) Open, closed and mixed networks of queues with different classes of customers. J ACM 22:248–260
7. Disney RL, Konig D (1985) Queueing networks: a survey of their random processes. SIAM Rev 27:335–403
8. Little JDC (1961) A proof of the queueing formula $L = \lambda W$. Oper Res 9:383–387
9. Jackson JR (1963) Jobshop-like queueing systems. Manag Sci 10(1):131–142
10. Kleinrock L (1964) Communication nets. Ph.D. thesis, McGraw-Hill, New York
11. Kleinrock L (2008) Communication nets: Stochastic message flow and delay. Dover Publications, New York, reprinted
12. Awduche D, Chiu A, Elwalid A, Widjaja I, Xiao X (2002) Overview and principles of internet traffic engineering
13. Xiong B, Yang K, Zhao J, Li W, Li K (2016) Performance evaluation of openflow-based software-defined networks based on queueing model. Computer Networks 102:172–185
14. Chen J (1999) New approaches to routing for large-scale data networks. Ph.D. thesis, Rice University
15. Parissidis G, Lenders V, May M, Plattner B (2006) Multi-path routing protocols in wireless mobile Ad Hoc networks: A quantitative comparison. In: Koucheryavy Y, Harju J, Iversen VB (eds) NEW2AN 2006, LNCS 4003. Springer, Berlin, Heidelberg, pp 313–326
16. He J, Rexford J (2008) Toward internet-wide multipath routing. IEEE Network 22(2):16–21
17. Jiawei W, Xiuquan Q, Guoshun N (2018) Dynamic and adaptive multi-path routing algorithm based on software-defined network. Int J Distrib Sensor Networks. https://doi.org/10.1177/1550147718805689
18. Zheng T, Gu D (2014) Traffic offloading improvements in mobile networks. In: The tenth international conference on networking and services, ICNS 2014
19. 3GPP (2018) IP flow mobility and seamless wireless local area network (WLAN) Offload; Stage 2, TS 23.261 V15.0.0
20. Soliman H (2009) Mobile IPv6 support for dual stack hosts and routers, RFC 5555

Index

G. Giambene, *Queuing Theory and Telecommunications*, Textbooks in Telecommunication Engineering,
https://doi.org/10.1007/978-3-030-75973-5

Printed in the United States
by Baker & Taylor Publisher Services